Handbook of
INDUSTRIAL BIOCATALYSIS

T0199632

Handbook of

INDUSTRIAL BIOCATALYSIS

Ching T. Hou

CRC Press
Taylor & Francis Group
Boca Raton London New York

CRC Press is an imprint of the
Taylor & Francis Group, an **informa** business
A TAYLOR & FRANCIS BOOK

CRC Press
Taylor & Francis Group
6000 Broken Sound Parkway NW, Suite 300
Boca Raton, FL 33487-2742

First issued in paperback 2019

ISBN-13: 978-0-8247-2423-8 (hbk)
ISBN-13: 978-0-367-39267-3 (pbk)
Library of Congress Card Number 2005051435

Library of Congress Cataloging-in-Publication Data

Hou, Ching T. (Ching-Tsang), 1935-
  Handbook of industrial biocatalysis / Ching T. Hou.
      p. cm.
  Includes bibliographical references and index.
  ISBN 0-8247-2423-2 (alk. paper)
  1. Enzymes--Biotechnology. I. Title.

TP248.65.E59H68 2005
660.6'34--dc22   2004064922

Visit the Taylor & Francis Web site at
http://www.taylorandfrancis.com

and the CRC Press Web site at
http://www.crcpress.com

Preface

Pasteur initiated the scientific study of fermentation. When it was appreciated that microbes catalyzed the chemical reactions used in the production of wine, cheese, and other foods, it became reasonable to expect that they could be put to work in the manufacture of chemicals for industry. Beginning in the late 1960s, both chemical industry and government agencies moved away from petroleum-based nonrenewable feedstocks for production of commodity and specialty chemicals to emphasize the use of renewable resources such as carbohydrates, oils, and fats. In addition, in recent years, the oil and fat industry has started to emphasize quality rather than quantity of oil and fat for human consumption. The definition of biocatalysis includes enzyme catalysis, biotransformation, bioconversion, fermentation, and biotechnology. It deals not only with one-step catalytic reaction, but also includes many sequential reaction steps to produce a product. Biocatalysis is a bioprocess including molecular manipulation of enzymes, the reaction itself, and product recovery.

This handbook was assembled with the intent of bringing together all types of industrial biocatalysis. It consists of 29 chapters whose authors are the world's most famous and most active researchers in this field. The basic information and theoretical considerations for a specific topic area or a specific biotechnological application are provided, and every effort has been made to include the current information. This is the most up-to-date handbook on industrial applications of biosciences and biotechnology.

The book is divided into three sections. The first describes the world's newest biotechnology, including bioprocesses on producing potential industrial products from hydrophobic substrates such as oils and fats. The products include healthy food, nutritional supplements, neutraceuticals, chiral synthons, specialty chemicals, surfactants, biopolymers, and antimicrobial and physiologically active agents. Metabolic pathways and function of polyunsaturated fatty acids (PUFAs) in mammals as well as transgenic production of long-chain PUFA-enriched oils are presented by Vic Huang of Abbott Labs. J. Ogawa and S. Shimizu of Kyoto University, Japan introduce examples of bioprocess development that started from process design stemming from the discovery of the unique metabolites hydrantoin, cyclic imide, microbial nucleosides, and conjugated fatty acids. Tsuneo Yamane of Nagoya University, Japan describes biocatalysis in microaqueous organic media including lipase, esterase, protease, and so on. Rolf Schmid of the University of Stuttgart, Germany contributes a chapter on biocatalysts for the epoxidation and hydroxylation of fatty acid alcohols including fermentation/bioreactor process using oxygenases. Kumar Mukherjee of Munster, Germany uses lipase specificities toward fatty acids and their derivatives—alcohols, alkanethiols, and sterols—to enrich n-3 and n-6 PUFAs, very long-chain monounsaturated fatty acids, other acids, and alcohols. K. Lee and J. Shaw of the Academia Sinica, Taiwan describe a successful story about recombinant *Candida rugosa* lipase with improved catalytic properties and stabilities. Ching Hou of NCAUR, USDA shares his discovery of novel oxygenated fatty acids and their potential industrial application from vegetable oils. Yuji Shimada of Osaka Municipal Technical Research Institute (OMTRI) describes many examples of the application of lipase to industrial-scale purification of oil- and fat-related compounds including production of PUFA-enriched oil, conversion of waste edible oil to biodiesel fuel, and purification of tocopherols, sterols, and steryl esters.

Casmir Akoh's group at University of Georgia give a thorough overview on lipase modification of lipids. Shuji Adachi of Kyoto University describes lipase-catalyzed condensation in an organic solvent including substrate selectivity of lipase for various carboxylic acids and continuous production of esters such as acyl ascobates. Naoto Yamada of Kao Corporation, Japan presents enzymatic production of diacyl glycerol and its beneficial physiological function. Diacyl glycerol functions like oil for all food preparation including frying, yet prevents the accumulation of body fat. Satoshi Negishi of Nisshin OilliO, Japan Ltd., describes the use of nonimmobilized lipase for industrial esterification of food oils. M. Hosokawa and K. Takahashi of Hokkaido University, Japan describe their design for industrial production of polyunsaturated phospholipids and their biological functions (health application). Dan Solaiman's group at ERRC, USDA presents the production of biosurfactants by fermentation of fats, oils, and their coproducts including microbial glycolipids, sophorolipids, and rhamnolipids. Tsunehiro Aki of Hiroshima University, Japan describes current metabolic engineering on development and industrialization of transgenic oils. Tom Foglia's group at ERRC, USDA presents lipase-catalyzed production of structured lipids as low-calorie fats. Toro Nakahara of National Institute of Advanced Industrial Science and Technology, Japan describes microbial polyunsaturated fatty acid production including lipids from bacteria, yeasts, and fungi. Gudmundur Haraldsson of University of Iceland describes lipase-catalyzed production of EPA or DHA containing triacylglycerols derived from fish oil. Rich Ashby et al. of ERRC, USDA describe biopolyesters derived from the fermentation of renewable resources including polylactic acid, polytrimethylene terephthalate, and polyhydroxyalkanoates.

The second section of the handbook deals with producing value-added products from carbohydrate substrates. The scope includes ethanol production, oligosaccharides and glycosides, utilization of hemicelluloses, and carbohydrate-active enzymes. Hajime Taniguchi of Chubu University, Nagoya, Japan presents carbohydrate-active enzymes for the production of oligosaccharides including enzymes for most of the oligosaccharides, such as isomalto-, nigero-, gentio-, fructo-, galacto-, chitosan-, and xylo-oligosaccharides, trehalose, palatinose, trehalulose, and lactosucrose. Peter Biely and Gregory Cote of NCAUR, USDA describe a special group of carboxylic acid esters that operate on highly hydrated substrates such as partially acylated polysaccharides. H. Nakano and S. Kitahata of OMTRI, Osaka, Japan describe industrial-scale production of various cyclodextrins from starch by cyclodextrin glucotransferase. Bruce Dien of NCAUR, USDA contributes a review on converting herbaceous energy crops to bioethanol with emphasis on pretreatment processes. Badal Saha of NCAUR, USDA describes enzymes as biocatalysts for conversion of lignocellulosic biomass to fermentable sugars. Its substates include various agricultural residues such as corn fiber, corn stover, wheat straw and rice straw.

The third section deals with other potential industrial bioprocesses. Gregory Zeikus of Michigan State University describes applications of bioelectrocatalysis for synthesis of chemicals, fuels, and drugs. Ramesh Patel of Bristol-Myers Squibb, New Jersey uses bioprocesses to synthesize chiral intermediates for drug development including anticancer drugs (paclitaxel, orally active taxane, deoxyspergualin and antileukemic agent), antiviral drugs (BMS-186318, HIV protease inhibitor, Atzanavir, crixivan), reverse transcriptase inhibitor (Abacavir, Lobucavir), antihypertensive drugs (angiotensin converting enzyme inhibitor, captopril, monopril), neutral endopeptidase inhibitors, squalene synthase inhibitors, thromboxane A2 antagonist, calcium channel blockers, potassium channel blockers, β-3-receptor agonists, melatonin receptor agonists, anti-Alzheimers drugs, anti-infective drugs, respiratory and allergic diseases, acyloin condensation, enantioselective and enzymatic deprotection. R. Sakaguchi and L. Junejia of Taiyo Kagaku Company, Japan present a novel nutrition delivery system that also preserves the stability of food components and flavor. Sima Sariaslani et al. of Dupont Central Research & Development, Experimental

Station, Wilmington, Delaware describe pathway engineering for production of trans-para-hydroxycinnamic acid from renewable resources. Finally, Chiara Schiraldi and Mario De Rosa of Second University of Naples, Italy describe industrial applications of extremophiles.

The *Handbook of Industrial Biocatalysis* is intended for teachers, postdoctorate and graduate students, and industrial scientists who conduct research in biosciences and biotechnology. It is therefore expected that it will serve as a valuable reference for researchers in the field and as a complementary text for graduate-level reading and teaching.

Ching T. Hou, Ph.D.
Peoria, Illinois

Contributors

Shuji Adachi
Division of Food Science and
  Biotechnology
Graduate School of Agriculture
Kyoto University
Japan

Tsunehiro Aki
Department of Molecular
  Biotechnology
Graduate School of Advanced
  Sciences of Matter
Hiroshima University
Japan

Casimir C. Akoh
Department of Food Science
  & Technology
University of Georgia
Athens, GA

Richard D. Ashby
Eastern Regional Research
  Center
ARS
USDA Wyndmoor, PA

Arie Ben-Bassat
Biochemical Sciences &
  Engineering
DuPont Central Research and
  Development
Experimental Station
Wilmington, DE

Peter Biely
Institute of Chemistry
Slovak Academy of Sciences
Slovakia

Gregory L. Côté
National Center for Agricultural
  Utilization Research
Agricultural Research Service
U.S. Department of Agriculture
Peoria, IL

Bruce S. Dien
National Center for Agricultural
  Utilization Research
USDA, Agricultural Research
  Service
Peoria, IL

Thomas A. Foglia
Eastern Regional Research
  Center
ARS
USDA Wyndmoor, PA

Anthony Gatenby
Biochemical Sciences &
  Engineering
DuPont Central Research and
  Development
Experimental Station
Wilmington, DE

**Gudmundur G.
  Haraldsson**
Science Institute
University of Iceland
Reykjavik, Iceland

Masashi Hosokawa
Graduate School of Fisheries
  Sciences
Hokkaido University
Japan

Ching T. Hou
Microbial Genomics and
  Bioprocessing Research Unit
National Center for Agricultural
  Utilization Research
Agricultural Research Service
USDA
Peoria, IL

Lisa Huang
Biochemical Sciences &
  Engineering
DuPont Central Research and
  Development
Experimental Station
Wilmington, DE

Yung-Sheng Huang
Ross Products Division
Abbott Laboratories
Columbus, OH

Loren B. Iten
National Center for Agricultural
  Utilization Research
USDA
Agricultural Research Service
Peoria, IL

L.R. Juneja
Nutritional Foods Division
Taiyo Kagaku Co. Ltd.
Mie, Japan

Seiji Kawamoto
Department of Molecular
  Biotechnology
Graduate School of Advanced
  Sciences of Matter
Hiroshima University
Japan

Sumio Kitahata
Department of Bioscience and
  Biotechnology
Shinshu University
Nagano, Japan

Guan-Chiun Lee
Institute of Botany
Academia Sinica
Nankang, Taipei, Taiwan

Ki-Teak Lee
Department of Food Science and
  Technology
Chungnam National University
Yusung-Gu, Taejon
Republic of Korea

Jeung-Hee Lee
Department of Food Science and
  Technology
Chungnam National University
Yusung-Gu, Taejon
Republic of Korea

Amanda E. Leonard
Ross Products Division
Abbott Laboratories
Columbus, OH

Noboru Matsuo
Biological Science Laboratories
Kao Corporation
Japan

Steffen C. Maurer
Institute for Technical
  Biochemistry
University of Stuttgart
Germany

Kumar D. Mukherjee
Institute for Lipid Research
Federal Centre for Cereal
Potato and Lipid Research
Münster, Germany

Toshihiro Nagao
Osaka Municipal Technical
  Research Institute
Osaka, Japan

Toro Nakahara
National Institute of Advanced
  Industrial Science and
  Technology (AIST)

Hirofumi Nakano
Osaka Municipal Technical
  Research Institute
Osaka, Japan

Satoshi Negishi
Research Laboratory of The
  Nisshin OilliO Group, Ltd.
Yokosuka, Kanagawa, Japan

Jun Ogawa
Division of Applied Life Sciences
Graduate School of Agriculture
Kyoto University
Japan

Kazuhisa Ono
Department of Molecular
  Biotechnology
Graduate School of Advanced
  Sciences of Matter
Hiroshima University
Japan

Ramesh N. Patel
Bristol-Myers Squibb
  Pharmaceutical Research
  Institute
New Brunswick, NJ

Suzette L. Pereira
Ross Products Division
Abbott Laboratories
Columbus, OH

TP Rao
Nutritional Foods Division
Taiyo Kagaku Co. Ltd.
Mie, Japan

Badal C. Saha
Fermentation Biotechnology
  Research Unit
National Center for Agricultural
  Utilization Research
Agricultural Research Service
USDA
Peoria, IL

N. Sakaguchi
Nutritional Foods Division
Taiyo Kagaku Co. Ltd.
Mie, Japan

Sima Sariaslani
Biochemical Sciences &
  Engineering
DuPont Central Research and
  Development
Experimental Station
Wilmington, DE

Rolf D. Schmid
Institute for Technical
    Biochemistry
University of Stuttgart
rolf.d.schmid@rus.uni-
    stuttgart.de

Subramani Sellappan
Department of Food Science
    & Technology
University of Georgia
Athens, GA

Jei-Fu Shaw
Institute of Botany
Academia Sinica
Nankang
Taipei, Taiwan

Chwen-Jen Shieh
Department of Bioindustrial
    Technology
Dayeh University
Taiwan

Yuji Shimada
Osaka Municipal Technical
    Research Institute
Japan

Sakayu Shimizu
Division of Applied Life Sciences
Graduate School of Agriculture
Kyoto University
Japan

Christopher D. Skory
National Center for Agricultural
    Utilization Research
USDA, Agricultural Research
    Service
Peoria, IL

Daniel K.Y. Solaiman
Eastern Regional Research
    Center
ARS, USDA
Wyndmoor, PA

Koretaro Takahashi
Graduate School of Fisheries
    Sciences
Hokkaido University
Japan

Hajime Taniguchi
Department of Biological
    Chemistry
Chubu University
Japan

Tina Van Dyk
Biochemical Sciences &
    Engineering
DuPont Central Research and
    Development
Experimental Station
Wilmington, DE

Takaaki Watanabe
Processing Development
    Research Laboratories
Kao Corporation
Japan

Yomi Watanabe
Osaka Municipal Technical
    Research Institute
Osaka, Japan

Naoto Yamada
Production Quality
    Management Division
Kao Corporation
Japan

Tsuneo Yamane
Laboratory of Molecular
    Biotechnology
Graduate School of Bio- &
    Agro-Sciences
Nagoya University
Japan

Teruyoshi Yanagita
Department of Applied
    Biological Science
Saga University
Japan

J. Gregory Zeikus
Department of Biochemistry &
    Molecular Biology
Michigan State University
East Lansing, MI

Contents

1

Enzymes for the Transgenic Production of Long-Chain Polyunsaturated Fatty Acid-Enriched Oils

Yung-Sheng Huang

Suzette L. Pereira

Amanda E. Leonard

1.1 Introduction

Polyunsaturated fatty acids (PUFAs) are fatty acids of 18 carbons or more in length with two or more methylene-interrupted double bonds in the *cis* position. Depending on the position of the first double bond proximate to the methyl end of fatty acids, PUFAs can be designated by the omega (ω-) or (n-) number, and classified into two major groups: ω6 (or n-6) and ω3 (or n-3) families. For example, linoleic acid (LA) in the n-6 family is designated as C18:2n-6 to indicate that this fatty acid contains 18 carbons and two double bonds, with the first double bond at the sixth carbon from the methyl end. Similarly, α-linolenic acid (C18:3n-3) in the n-3 family has 18 carbons and three double bonds, with the first double bond located at the third carbon from the methyl end (Figure 1.1).

1.1.1 Metabolism of Linoleic Acid and α-Linolenic Acid

Animals are incapable of synthesizing both linoleic acid (LA, C18:2n-6) and α-linolenic acid (ALA, C18:3n-3) due to lack of the Δ12 and Δ15-desaturases. However, animals can metabolize these two fatty acids obtained from the diet to form longer and more unsaturated PUFAs to meet the metabolic needs. Since these two fatty acids must be obtained from the diet, they are considered to be essential fatty acids.

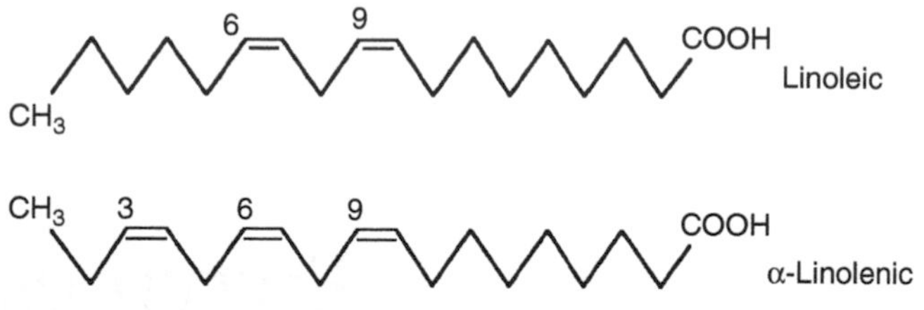

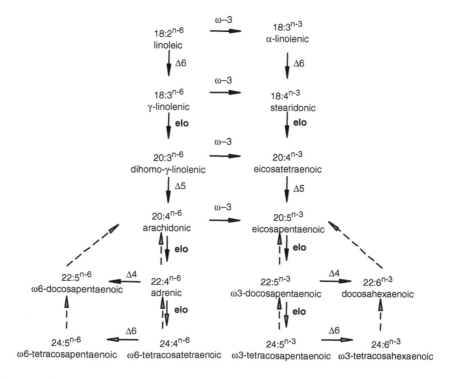

FIGURE 1.1 Nomenclature of polyunsaturated fatty acids

In most eukaryotes, the biosynthesis of PUFAs involves a complex series of desaturation and elongation steps (Figure 1.2).[1] For example, eicosapentaenoic acid (EPA, C20:5n-3) is synthesized from ALA by the addition of a double bond by a Δ6-desaturase to form stearidonic acid (SDA, C18:4n-3); the elongation of SDA to form ω3-eicosatetraenoic acid (ω3-ETA, C20:4n-3); and the addition of another double bond by a Δ5-desaturase to form EPA.[2] The formation of DHA from EPA occurs via different mechanisms in eukaryotes. In higher eukaryotes like mammals, EPA is elongated to ω3-docosapentaenoic acid (ω3-DPA, C22:5n-3), which is further elongated to ω3-tetracosapentaenoic acid (ω3-TPA, C24:5n-3). ω3-TPA is then desaturated by a Δ6-desaturase to generate ω3-tetracosahexaenoic acid (THA, C24:6n-3) in the microsomes. The THA is then transported to the peroxisomes, where it is β-oxidized to form DHA.[3,4] However, in lower eukaryotes like the *thraustochytrid* sp., EPA is elongated to ω3-DPA followed by the addition of a double bond directly to ω3-DPA, by Δ4-desaturase, to generate DHA.[5] The synthesis of long-chain n-6 PUFAs from LA occurs via similar alternating desaturation and elongation steps (Figure 1.2).

1.1.2 Function of PUFAs

In mammals, PUFAs are important structural components that modulate membrane fluidity and permeability.[6] For example, docosahexaenoic acid (DHA, C22:6n-3), a long-chain n-3 PUFA, and arachidonic

FIGURE 1.2 Metabolic pathway of linoleic and α–linolenic acids

acid (AA, C20:4n-6), a long-chain n-6 PUFA, are found in high proportions in neuronal tissues such as brain and retina, and testis.[7,8] PUFAs also serve as precursors for a number of biologically active molecules, such as eicosanoids, growth regulators, and hormones.[9] In mammals, eicosanoids such as prostaglandins, leukotrienes, and thromboxanes act locally on various signaling mechanisms that have effects on numerous cellular functions including chemotaxis, vascular permeability, inflammation, vasoconstriction, etc.[9] Thus, PUFAs have profound effects on various physiological processes, such as cognitive function, and immunosuppressive and anti-inflammatory actions.[10]

1.1.3 PUFA Production and Chronic Diseases

The availability of long-chain PUFAs depends on the diet providing the precursors, such as LA and ALA, and the activity of enzymes involved in the biosynthesis.[11] Generally, only a small proportion of dietary linoleate and α-linolenate (3.0% and 1.5%, respectively) can be converted to longer PUFAs, with the rest getting β-oxidized to provide energy.[12] The slow formation of PUFAs can be further compromised by various nutritional and hormonal factors.[13] For example, the activity of the Δ6-desaturase *in vivo* is regulated by certain dietary components, age, and hormones.[14] In addition, in chronic diseases like cancer and diabetes, altered expression levels of this enzyme in different tissues have been observed.[15–21] The activity of the Δ5-desaturase is also regulated by diet,[22] and altered expression levels of this enzyme have been associated with various disease conditions, including eye disorders, Alzheimer's disease, and diabetes.[23–25] Thus, low levels of long-chain PUFAs have been associated with disorders of the neurovisual development and other complications of premature birth[26–29] as well as implicated with incidence of chronic diseases, such as diabetes, hypercholesterolemia, rheumatoid arthritis, autoimmune disorders, Crohn's disease, and cancer.[13,30,31]

Clinical evidence has shown that dietary supplementation of PUFAs, such as γ-linolenic acid (GLA, C18:3n-6) and EPA/DHA, can exert the anti-inflammatory, antithrombotic, and antiarrhythmic activities, and provide beneficial effects on glucose and lipid metabolism.[32–38] These findings have received much attention from food manufacturers and pharmaceutical companies, as well as the general public. As a result, sales of these long-chain PUFAs as supplements and fortified foods, such as "DHA plus" eggs, and DHA- and AA-fortified infant formulas have drastically increased in the past few years.

1.2 Commercial Sources of PUFAs

1.2.1 Fish Oil

Currently, the richest sources of EPA and DHA are derived from fish oils obtained from mackerel, herring, salmon, and sardines. Fish obtain these long-chain PUFAs (LC-PUFAs) from the LC-PUFA-rich microalage and phytoplankton they consume. Commercially, fish oils are available in the form of gelatin capsules or oily preparations. Fish oils obtained from fish liver (e.g., cod liver oil) are rich in vitamin A and D, and contain lower amounts (13% to 22%) of EPA/DHA. In contrast, fish oils obtained from fish bodies (e.g., salmon oil) contain 20% to 30% EPA and DHA and are low in cholesterol and vitamin A and D(reviewed in reference 39). These fish oils are used to enrich food products, animal feeds, and aquaculture feeds, in addition to their use for direct human consumption. These oils are not very economical due to the high costs involved in processing, refining, and stabilizing the oils. In addition, the effects of overfishing and the vulnerability of global fisheries to environmental and climatic changes have resulted in decreased yields, which has further driven up the cost of fish oils.[40]

1.2.2 Plant Oils

Plants do not produce EPA or DHA. However, certain plants can produce oils rich in GLA or ALA, and serve as the current commercial sources of these PUFAs. Plant oils derived from borage, evening primrose, and black currant are found to be rich in GLA (reviewed in reference 41). Borage oil is derived from the seed of *Borago officinalis* and contains ~23% GLA. Evening primrose oil obtained from the seed of

Oenothera biennis contains ~9% GLA. Black currant oil derived from the seeds of *Ribes nigrum* is attractive in that it contains 12% ALA in addition to 16% GLA. However, these oils are expensive due to high costs of cultivation, seed harvesting, and oil extraction. Linseed oil (flax) is the richest source of ALA (57% of total fatty acids). Dietary ALA can also be obtained from oils of canola, soybean, wheat germ, and walnut.

1.2.3   Microbial Oils

LC-PUFAs can also be extracted from single cell organisms like microalgae and fungi that can be commercially cultivated in fermentors (heterotrophic producers), or in photoautotrophic cultivation systems.[42,43] Oleagenous fungi such as *Mortierella alpina* accumulate up to 40% (by wt.) oil, of which up to 40% represents AA.[44] Thus, this organism is commercially used in production of AA.[45] Diatoms such as *Nitzschia*, a good producer of EPA, and dinoflagellates such as *Crypthecodinium cohnii* that produce large amounts of DHA, are currently commercially utilized for the production of EPA- and DHA-enriched oils.[42] Currently, marine protists such as the *Thraustochytrids* that make large amounts of DHA are being explored for their potential to make DHA-rich oils for human consumption.[42,46,47] The costs of production of these oils are still considerably high, and further work involving strain-improvement and cultivation-optimization need to be carried out to make these oils more economical.

1.3   Transgenic Production of PUFAs

Although LC-PUFAs-enriched oils are commercially available as discussed in Section 1.2, the cost of production of these oils is generally very high, and the sources of supplies are often unreliable or nonrenewable. The increase in demand for these PUFAs has raised interest in obtaining these from alternate sources that are more economical and sustainable. One attractive option is the production of LC-PUFA-enriched vegetable oils in oilseed crops like soybean, canola, and others. Since these plants can only synthesize 18-carbon (C_{18}) PUFAs such as LA and ALA, it is necessary to genetically manipulate their lipid biosynthetic pathways in order to produce long-chain PUFAs such as GLA, EPA, and DHA. For this, genes encoding the enzymes involved in LC-PUFA biosynthesis need to be isolated from LC-PUFA-rich organisms, and transgenically expressed in oilseed crops. These include the various desaturases and elongases outlined in Figure 1.3.

Desaturases are enzymes that catalyze the addition of a double bond (unsaturation) in a fatty acyl chain (reviewed in reference 48). These enzymes are specific to the location, number, and stereochemistry of double bonds already present in fatty acids.[49] In addition, they have specificity for their substrate carriers, which can be CoA-linked substrates, acyl carrier protein (ACP)-linked substrates, or glycerolipid-linked substrates.

Elongases are enzymes that are responsible for the addition of two-carbon units to the carboxyl end of a fatty acid chain. In both plants and animals, the elongase system is composed of four enzymes: a condensing enzyme, β-ketoacyl CoA synthase (also referred to as elongase), β-ketoacyl CoA reductase, β-hydroxyacyl CoA dehydrase, and *trans*-2-enoyl CoA reductase. Fatty acid elongation is initiated by the condensation of malonyl-CoA with a long chain acyl-CoA, yielding a β-ketoacyl-CoA in which the acyl moiety has been elongated by two carbon atoms. This reaction is catalyzed by the condensing enzyme β-ketoacyl CoA synthase (also referred to as "elongase"). β-ketoacyl-CoA is then reduced, dehydrated, and further reduced by the remaining enzymes in the system to yield the elongated acyl-CoA.[50] The condensing enzyme (elongase) is known to be the rate-limiting enzyme,[50–52] which regulates the systems specificity for the fatty acid substrate in term of chain length and degree of unsaturation. The elongases involved in the elongation of LC-PUFAs are distinct from the plant and yeast elongases that are involved in the elongation of saturated or monounsaturated fatty acids (reviewed in reference 53).

The following sections will focus on the characterization and transgenic expression of nonmammalian sources of LC-PUFA biosynthetic genes, with applications for the production of PUFA-enriched transgenic oils. Enzymes known to exist in all plants, such as the Δ9- and the Δ12-desaturase that are involved in LA production (Figure 1.3), will not be described here since these enzymes are highly active in native

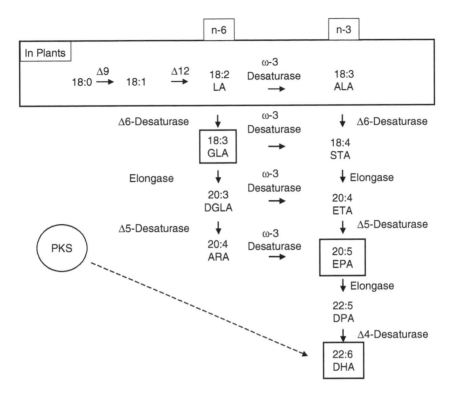

FIGURE 1.3 Pathway and transgenic production of LC–PUFAs

oil-seed crops. Although the discussion will focus on production of transgenic plant oils, it should be noted that these enzymes also have applications for the production of transgenic oils in other oleaginous organisms such *Yarrowia* and *Rhodotorula*.

1.3.1 Enzymes Required for Transgenic GLA Production

Commercially available sources of GLA-enriched oils from borage, evening primrose, and black currant[54] are not economical for large-scale production. Hence efforts are ongoing to transgenically produce GLA in an oil-seed crop. The key step in GLA production involves the insertion of a double bond between carbon #6 and #7 of LA to generate GLA (Figure 1.3). This reaction is mediated by the Δ6-desaturase, a membrane-bound enzyme located in the endoplasmic reticulum. This enzyme is classified as a "front-end" desaturase because it is capable of introducing a double bond between a preexisting double bond and the "front" (carboxyl end) of the fatty acid. It also contains a fused cytochrome b_5 domain at the N-terminus, which plays a role as an electron donor during desaturation, and this domain is essential for activity.[55]

Δ6-desaturases has been isolated from several fungal, plant, microbial, and mammalian sources that produce GLA.[48] Some of these include *M. alpina*,[56] *Mucor rouxii*,[57] *Phytium irregulare*,[58] *Physcomitrella patens*,[59] borage,[60] *Echium* plant sp.,[61] *Primula* sp.,[62] and *Synechcocystis*.[60,63] Most of these Δ6-desaturases are thought to act exclusively on the phospholipid-linked LA substrate.[44,64] In contrast, the Δ6-desaturases from mammalian sources are thought to recognize CoA-linked LA substrates.[65]

Most of the Δ6-desaturases have been functionally expressed in yeast, and many have also been tested in plants.[56,58,60,61,66] In addition, some Δ6-desaturases have also been expressed in oil-seed crops, resulting in production of GLA in seeds. In the early study conducted in our laboratory, expression of the *M. alpina* Δ6-desaturase in a low-linolenic acid variety of *Brassica napus* resulted in the generation of low amounts (~13%) of GLA in addition to the production of an uncommon fatty acid, Δ6,9-18:2.[67,68] This uncommon fatty acid was derived by the desaturation of oleic acid (OA, 18:1) by the Δ6-desaturase. This problem was resolved by coexpressing the *M. alpina* Δ6-desaturase with its Δ12-desaturase, an enzyme that converts oleic acid to LA (Figure 1.3). This resulted in the accumulation of >40% GLA in the transgenic canola oil, with no detectable Δ6,9-18:2.[67,68] Subsequent studies have been carried out using the *Phytium irregulare* Δ6-desaturase in *Brassica juncea*, and have also resulted in successfully generation of 25% to 40% GLA in the transgenic seed.[58] This oil, however, was also found to contain 2% to 10% stearidonic acid (SDA, 18:4n-3) in addition to the uncommon fatty acid Δ6,9-18:2.[58]

1.3.2 Enzymes Required for Transgenic ARA and EPA Production

The production of the C_{20}-PUFA arachidonic acid (ARA, 20:4n-6) from LA involves the desaturation of LA to GLA (see Section 1.3.1.), followed by the elongation of GLA to dihomo-γ-linolenic acid (DGLA, 20:3n-6), and a subsequent desaturation of DGLA to ARA (Figure 1.3). Thus three major enzymes are involved in this process: a Δ6-desaturase, a C_{18}-PUFA-specific elongase, and a Δ5-desaturase. These same enzymes also function on the n-3 pathway intermediates and are thus also involved in the biosynthesis of EPA (20:5n-3) (Figure 1.3). Since Δ6-desaturase has just been discussed in the previous section (Section 1.3.1), this section will focus only on the C_{18}-PUFA-specific elongase and the Δ5-desaturase.

The first C_{18}-PUFA-specific elongase to be isolated was identified in our laboratory from the ARA-rich fungus, *Mortierella alpina*.[67] This enzyme when tested in baker's yeast specifically recognized and elongated the n-6 and n-3 C_{18}-PUFA substrates, GLA and stearidonic acid (SDA, 18:4n-3), respectively, whereas it demonstrated no activity on monounsaturated or saturated fatty acid substrates.[69–72] Enzymes with similar elongating activity have been isolated from *Caenorhabditis elegans*,[73] *Physcomitrella patens*,[74] and from the marine protist, *Thraustochytrium* sp.[75] In addition, several PUFA-specific elongases have been isolated from mammalian sources.[53] All these PUFA-specific elongases contain five hydrophobic regions predicted to be membrane-spanning regions. In addition, they contain a highly conserved histidine-box motif composed of three histidine residues (HXXHH) embedded in the fourth membrane spanning region.[76–78] These features distinguish them from the plant elongases that are involved in elongation of very long chain saturated and monounsaturated fatty acids, but not PUFAs.[79,80] In addition, the C_{18}-PUFA-specific elongases are thought to recognize CoA-linked substrates[64] as opposed to the plant elongases that recognize acyl carrier protein (ACP)-linked substrates.

The Δ5-desaturase catalyzes the final step in the production of the C_{20}-PUFAs, ARA and EPA. This enzyme is so called because it introduces a double bond at the Δ5-position of the fatty acid. This desaturase is also considered a "front-end" desaturase and shares all the conserved structural characteristics displayed by other front-end desaturases such as the Δ6-desaturase. Δ5-desaturase genes have been identified from fungi and algae such as *Mortierella alpina*.[81,82] *Thraustochytrium* sp.,5 and *Phaeodactylum tricornutum*,[83] and these have been functionally characterized in yeast. Additional Δ5-desaturases have been identified from *C. elegans*,[84] human,[85] and rat.[86] All the Δ5-desaturases identified so far are capable of desaturating both the n-6 and n-3 PUFA substrates, DGLA and eicosatetraenoic acid (ETA, 20:4n-3), respectively. Coexpression of the *M. alpina* Δ5-desaturase along with its C_{18}-PUFA-specific elongase in yeast, in the presence of exogenously supplied free fatty acid substrate, resulted in the production of significant amount of ARA or EPA.[69] When introduced into a low linolenic variety of *Brassica napus*, the *M. alpina* Δ5-desaturase was capable of desaturating oleic acid (OA, 18:1n-9) to taxoleic acid (Δ5,9-18:2), and LA to pinolenic acid (Δ5,9,12-18:3).[81] This demonstrates its functionality in desaturating fatty acids at the Δ5-position in higher plants, even in the absence of its preferred PUFA substrates.

Since plant oils often contain LA and ALA, the transgenic expression of the Δ6-desaturase, C_{18} PUFA-specific elongase, and Δ5-desaturase would result in ARA and EPA. For production of an EPA-enriched transgenic oil that does not contain ARA, it is necessary to shunt the n-6 PUFA metabolites to their n-3 counterparts. This reaction is catalyzed by a group of enzymes designated the ω3-desaturases. These enzymes are absent from mammals, but can be found in some plants, lower eukaryotes, and cyanobacteria.[87] These enzymes are so called because they introduce a double bond at carbon atom #3 when counted from the methyl- (ω-) end of the fatty acyl chain. ω3-desaturases share all the conserved features present in other membrane-bound desaturases. These include the presence of two long stretches of hydrophobic residues that traverse the lipid bilayer, and three histidine-rich motifs proposed to be involved in the ligation of iron atoms within the active site domain of these enzymes.[88] This protein also contains the C-terminal motif, KAKSD, proposed to be a retention signal for many transmembrane proteins in the ER.[89] Unlike the front-end desaturases, the ω3-desaturases do not contain a fused cytochrome b_5 domain at their N-terminus, and are thus assumed to interact with a separate cytochrome b_5 for their activity.

All plant and cyanobacterial ω3-desaturases act exclusively on the C_{18}-PUFA substrate, LA, converting it to ALA (Figure 1.3). Although many oilseed crops do contain endogenous ω3-desaturases, these enzymes do not efficiently convert LA to ALA as evidenced by a high LA-to-ALA ratio in their total lipids.[54] Thus for transgenic EPA production, it might be necessary to transgenically express ω3-desaturases with high enzymatic activity, in order to increase the shunt through the n-3 PUFA pathway. A novel ω3-desaturase was identified from *C. elegans* that was capable of recognizing multiple n-6 PUFA substrates, which included the C_{18}-PUFAs, LA and GLA, as well as the C_{20}- PUFA, DGLA.[90,91] This enzyme was found to be functional in plants[90] and thus has potential applications for transgenic EPA production. In addition, a novel fungal ω3-desaturase was recently identified in our laboratory that could specifically convert ARA to EPA, and this enzyme was found to be functional in an oilseed crop.[92] This enzyme has applications for the removal of ARA from EPA- and DHA-enriched transgenic oils, by converting ARA to EPA. This is especially necessary if the transgenic oils are targeted for adult nutrition, since ARA is a precursor for synthesis of proinflammatory eicosanoids that are implicated in inflammatory and cardiovascular disease development.

Thus the transgenic production of EPA will be contingent on the success in coexpressing at least four different enzymes in a single system. Coexpression of three of the PUFA biosynthetic enzymes, the Δ6-desaturase, the PUFA-specific elongase, and the Δ5-desaturase, has been successfully demonstrated in reconstituted baker's yeast, resulting in ARA and EPA production when their respective substrate, LA or ALA, was supplied exogenously.[73,83] However, the yields of ARA or EPA obtained in these studies were poor. Similar results were reported by Domergue et al.[64] in their attempt to coexpress the Δ6- and Δ5-desaturase from *P. tricornutum* along with the C_{18}-PUFA elongase from *P. patens* in transgenic linseed. From these experiments, it appears that there is an accumulation of the Δ6-desaturated fatty acids in the membrane fractions and almost none in the acyl-CoA pool. It is thought that the Δ6-desaturase from most fungi and algae function on phospholipid-linked (mainly phosphatidylcholine-linked) LA or ALA substrates. However, the consecutive step is catalyzed by a PUFA-specific elongase that requires its substrates to be present in the acyl-CoA pool. Thus it appears that a bottleneck in the pathway is created due to the inefficient transfer of the Δ6-desaturated products from the phospholipids to the acyl-CoA pool.[64] To overcome this bottleneck, it might be necessary to identify and coexpress additional enzymes that are involved in the transfer of phospholipid-linked PUFAs to the acyl CoA pools.

1.3.3   Enzymes Required for Transgenic DHA Production

The pathway for the biosynthesis of DHA varies among different groups of organisms. In lower eukaryotes such as DHA-rich algae and fungi, it is thought that DHA is synthesized from EPA in a two-step process (Figure 1.3): a) An initial elongation step that is catalyzed by a C_{20}-PUFA recognizing elongase, wherein EPA is elongated to ω3-DPA; b) The desaturation of ω3-DPA, catalyzed by a Δ4-desaturase, resulting in the generation of DHA. Thus the production of transgenic DHA would involve coexpressing these two new genes in addition to the four previously described genes (Section 1.3.2.) needed for EPA production.

The only enzymes identified to date that can recognize and elongate C_{20}-PUFAs are from mammals.[53] Expression of these genes in baker's yeast revealed that some of them could elongate C_{18}-PUFAs in addition to C_{20}- and C_{22}-PUFAs, whereas others had a specificity for C_{20}- and C_{22}-PUFA substrates. None of these enzymes acted on monounsaturated fatty acids or saturated fatty acid substrates.[53] Attempts are currently under way to identify similar C_{20}-PUFA recognizing elongases from lower eukaryotes, which can then be used for the production of transgenic oils.

The first Δ4-desaturase to be described was identified from a marine protist, *Thraustochytrium*, which produces copious amounts of DHA.[5] Like the Δ5- and Δ6-desaturase, the Δ4-desaturase is also a front-end desaturating enzyme capable of introducing a double bond at carbon # 4 of ω3-DPA. In addition, this enzyme can also desaturate the n-6 substrate adrenic acid (ADA, C22:4n-6) to generate ω6-DPA (C22:5n-6) (Figure 1.3). Expression of the Δ4-desaturase gene in a oilseed crop, *Brassica juncea*, in the presence of exogenously supplied ω3-DPA substrate resulted in the production of 3–6% DHA in the leaves, stems, and roots of the transgenic *Brassica*.[5] A new Δ4-desaturase was recently described from *Euglena gracilis*, and this enzyme was found to desaturate C_{16}-fatty acids in addition to C_{22}-PUFAs.[93]

1.3.4 Alternate Enzymes for Transgenic EPA/DHA Production: The PKS System

LC-PUFA biosynthesis in bacteria such as *Shewanella* and *Vibrio* occurs via a novel polyketide synthase (PKS) pathway.[94–96] This pathway is thought to be analogous to the fatty acid synthase (FAS) pathway involved in the synthesis of short-chain fatty acids.[97] This system was also identified to be involved in DHA production in the marine eukaryote, *Schizochytrium*.[98] Here, PUFA production is thought to be initiated by the condensation between a short-chain starter unit like acetyl CoA, and an extender unit like malonyl CoA. The four-carbon acyl chain formed is covalently attached to an acyl carrier protein (ACP) domain of the PKS complex, and goes through successive rounds of reduction, dehydration, reduction, and condensation, with the acyl chain growing by two carbon units with each round. A novel dehydratase/isomerase has been proposed to exist in this PKS complex that can catalyze trans- to cis-conversion of the double bonds, thus generating double bonds in the correct position of EPA and DHA.[98] Genes involved in this PKS system exist sequentially on long (20–30 Kb) open reading frames (ORFs), and the identity of every region within these ORFs are still unknown.[95,96,98] Expression of the *Shewanella* PKS system in an *E coli* or *Synechococcus* expression system resulted in EPA production, although the levels of EPA produced were low.[94,98] Although none of these PUFA-PKS genes have been expressed in plants as yet, this system offers an attractive alternative to the desaturase/elongase system for the production of EPA/DHA-enriched transgenic oils.

1.4 Conclusion and Future Perspectives

Fatty acids are critical for the normal development and function of all organisms, and in particular, very long chain PUFAs are necessary for the health and maintenance of higher organism such as mammals. Although the biosynthetic pathway of long-chain PUFAs has been studied for a while, detailed biochemical analysis of the enzymatic machinery has been especially hard. This is because the extreme hydrophobicity of the desaturases and elongases creates difficulties during purification of large amounts of these enzymes that are required for biochemical characterization. However, much progress has been made over the last few years in the cloning and identification of genes encoding the PUFA biosynthetic enzymes from different organisms. These findings have important biotechnological applications. For example, these genes can be used in the production of PUFA-rich transgenic oils to meet the increasing demands of the chemical, pharmaceutical, and nutraceutical industry for therapeutic and prophylactic use. Advances in understanding gene regulation in PUFA biosynthesis will also impact the single-cell oil industry. This in turn will affect the marine fish-farming industry, which depends on PUFAs generated by microalgae and fungi for enhancing the levels of PUFAs in fish.

However, some challenges still need to be overcome with respect to transgenic production of PUFAs. Although the overall scheme of PUFA biosynthesis appears to be common for most organisms, enzymes from different organisms may not necessarily be compatible. This may result in unanticipated bottlenecks in the pathway leading to lower yields of LC-PUFAs. This had already been observed during preliminary studies on transgenic EPA production (Section 1.3.2). In addition, many unknowns still need to be addressed with respect to coexpressing several genes from multiple sources simultaneously. It is still early to predict if the transgenic oils thus generated will have a fatty acid profile reflective of natural fish oil, without the accumulation of unwanted fatty acid byproducts. In addition, it is not known if the transgenically produced LC-PUFAs will indeed accumulate in the triacylglycerol (TAG) fraction. Once these challenges have been overcome however, these transgenic PUFA-enriched oils will afford the public an economical source of desirable PUFAs that will greatly impact general health and nutrition in the future.

References

1. HE Bazan, MM Careaga, H Sprecher, NG Bazan. Chain elongation and desaturation of eicosapentaenoate to docosahexaenoate and phospholipid labeling in the rat retina *in vivo*. *Biochim Biophys Acta* 712:123–128, 1982.
2. H Sprecher. Biochemistry of essential fatty acids. *Prog Lipid Res* 20:13–22, 1982.
3. H Sprecher. Metabolism of highly unsaturated n-3 and n-6 fatty acids. *Biochim Biophys Acta* 1486:219–231, 2000.
4. S Ferdinandusse, S Denis, PA Mooijer, Z Zhang, JK Reddy, AA Spector, RJ Wanders. Identification of the peroxisomal β-oxidation enzymes involved in the biosynthesis of docosahexaenoic acid. *J Lipid Res* 42:1987–1995, 2001.
5. X Qiu, H Hong, SL Mackenzie. Identification of a Δ4 desaturase from *Thraustochytrium* sp. involved in the biosynthesis of docosahexaenoic acid by heterologous expression in *Saccharomyces cerevisiae* and *Brassica juncea*. *J Biol Chem* 276:31561–31566, 2001.
6. R Uauy, P Peirano, D Hoffman, P Mena, D Birch, E Birch. Role of essential fatty acids in the function of the developing nervous system. *Lipids* 31 (Suppl.):167S–176S, 1996.
7. JM Bourre, OS Dumont, MJ Piciotti, GA Pascal, GA Durand. Dietary alpha-linolenic acid deficiency in adult rats for 7 months does not alter brain docosahexaenoic acid content, in contrast to liver, heart and testes. *Biochim Biophys Acta* 1124:119–122, 1992.
8. K Retterstol, TB Haugen, BO Christophersen. The pathway from arachidonic to docosapentaenoic acid (20:4n-6 to 22:5n-6) and from eicosapentaenoic to docosahexaenoic acid (20:5n-3 to 22:6n-3) studied in testicular cells from immature rats. *Biochim Biophys Acta* 1483:119–131, 2000.
9. DB Jump. The biochemistry of n-3 polyunsaturated fatty acids. *J Biol Chem* 277:8755–8758, 2002.
10. AP Simopoulos. Omega-3 fatty acids in inflammation and autoimmune diseases. *J Am Coll Nutr* 21:495–505, 2002.
11. AA Spector. Essentiality of fatty acids. *Lipids* 34:1S–3S, 1999.
12. SC Cunnane, MJ Anderson. The majority of dietary linoleate in growing rats is b-oxidized or stored in visceral fat. *J Nutr* 127:146–152, 1997.
13. RR Brenner. Endocrine control of fatty acid desaturation. *Biochem Soc Trans* 18:773–775, 1990.
14. MT Nakamura, HP Cho, SD Clarke. Regulation of hepatic delta-6 desaturase expression and its role in the polyunsaturated fatty acid inhibition of fatty acid synthase gene expression in mice. *J Nutr* 130: 1561–1565, 2000.
15. JE Brown, RM Lindsay, RA Riemersma. Linoleic acid metabolism in the spontaneously diabetic rat: Delta6-desaturase activity vs. product/precursor ratios. *Lipids* 35:1319–1323, 2000.
16. S Abel, CM Smuts, C de Villiers, WC Gelderblom. Changes in essential fatty acid patterns associated with normal liver regeneration and the progression of hepatocytes nodules in rat hepatocarcinogenesis. *Carcinogenesis* 22:795–804, 2001.
17. G Agatha, R Hafer, F Zintl. Fatty acid composition of lymphocyte membrane phospholipids in children with acute leukemia. *Cancer Lett* 173:139–144, 2001.
18. E Demcakova, E Sebokova, J Ukropec, D Gasperikova, I Klimes. Delta-6 desaturase activity and gene expression, tissue fatty acid profile and glucose turnover rate in hereditary hypertriglyceridemic rats. *Endocr Regul* 35: 179–186, 2001.
19. OJ Rimoldi, GS Finarelli, RR Brenner. Effects of diabetes and insulin on hepatic delta-6 desaturase gene expression. *Biochem Biophys Res Commun* 283:323–326,2001.

20. M Tsimaratos, TC Coste, A Djemli-Shipkolye, P Vague, G Pieroni, D Raccah. Gamma-linolenic acid restores renal medullary thick ascending limb Na(+), K(+)-ATPase activity in diabetic rats. *J. Nutr* 131:3160–3165, 2001.

21. MB Hansen-Petrik, MF McEntee, BT Johnson, MG Obukowicz, J Masferrer, B Zweifel, CH Chiu, J Whelan. Selective inhibition of delta-6 desaturase impedes intestinal tumorigenesis. *Cancer Lett* 175:157–163, 2002.

22. HP Cho, M Nakamura, SD Clarke. Cloning, expression, and nutritional regulation of the mammalian Δ6-desaturase. *J Biol Chem* 274:471–477, 1999.

23. T Nakada, IL Kwee, WG Ellis. Membrane fatty acid composition shows Δ6-desaturae abnormalities in Alzheimer's disease. *Clin Neurosci Neuropath* 1:153–155, 1990.

24. DR Hoffman, JC DeMar, WC Heird, DG Birch, RE Anderson. Impaired synthesis of DHA in patients with X-linked retinitis pigmentosa. *J Lipid Res* 42:1395–1401, 2001.

25. S Nishida, T Segawa, I Murai, S Nakagawa. Long-term melatonin administration reduces hyperinsulinemia and improves the altered fatty-acid compositions in type 2 diabetic rats via the restoration of delta-5 desaturase activity. *J Pineal Res* 32:26–33, 2002.

26. SE Carlson, SH Werkman, JM Peeples, RJ Cooke, EA Tolley. Arachidonic acid status correlates with first year growth in preterm infants. *Proc Natl Acad Sci USA* 90: 1073–1077, 1993.

27. MA Crawford, K Costeloe, K Ghebremeskel, A Phylactos, L Skirvin, F Stacey. Are deficits of arachidonic and docosahexaenoic acids responsible for the neural and vascular complications of preterm babies? *Am J Clin Nutr* 66:1032S–1041S, 1997.

28. SM Innis, H Sprecher, D Hachey, J Edmond, RE Anderson. Neonatal polyunsaturated fatty acid metabolism. *Lipids* 34:139–149, 1999.

29. L Lauritzen, HS Hansen, MH Jorgensen, KF Michaelsen. The essentiality of long chain n-3 fatty acids in relation to development and function of the brain and retina. *Prog Lipid Res* 40:1–94, 2001.

30. AP Simopoulos. Essential fatty acids in health and chronic disease. *Am J Clin Nutr* 70:560S–569S, 1999.

31. WE Connor. Importance of n-3 fatty acids in health and disease. *Am. J. Clin. Nutr* 71:171S–175S, 2000.

32. T Babcock, WS Helton, NJ Espat. Eicosapentaenoic acid (EPA): an anti-inflammatory ω-3 fat with potential clinical applications. *Nutrition* 16:1116–1118, 2000.

33. JS Charnock, GL Crozier, J Woodhouse. Gamma-linolenic acid, black currant seed and evening primrose oil in the prevention of cardiac arrhythmia in aged rats. *Nutr Res* 14:1089–1099, 1994.

34. GA Jamal, HA Carmichael, AI Weir. Gamma-linolenic acid in diabetic neuropathy. *Lancet* i: 1098, 1986.

35. JM Kremer, DA Lawrence, W Jubiz, R DiGiocomo, R Rynes, LE Bartholomew, M Sherman. Dietary fish oil and olive oil supplementation in patients with rheumatoid arthritis. Clinical and immunologic effects. *Arthritis Rheum* 33:810–820, 1990.

36. JX Kang, A Leaf. Antiarrhythmic effects of polyunsaturated fatty acids. *Circulation* 94:1774–1780, 1996.

37. R Zurier, RG Rossetti, EW Jacobson, DM DeMarco, NY Liu, JE Temming, BM White, M Laposata. Gamma-linolenic acid treatment of rheumatoid arthritis. A randomized, placebo-controlled trial. *Arthritis Rheum* 39:1808–1817, 1996.

38. RJ Woodman, TA Mori, V Burke, IB Puddey, GF Wats, LJ Beilin. Effects of purified eicosapentaenoic acid and docosahexaenoic acid on glycemic control, blood pressure, and serum lipids in type 2 diabetic patients with treated hypertension. *Am J Clin Nutr* 76:1007–1015, 2002.

39. EA Trautwein. n-3 Fatty acids-physiological and technical aspects for their role in food. *Eur J Lipid Sci Technol* 103:45–55, 2001.

40. JR Sargent, AGJ Tacon. Development of farmed fish: A nutritionally necessary alternative to meat. *Proc Nutr Soc* 58:377–383, 1999

41. DE Barre. Potential of evening primrose, borage, black currant, and fungal oils in human health. *Ann Nutr Metab* 45:47–57, 2001.

42. WR Barclay, KM Meager, JR Abril. Heterotrophic production of long-chain omega-3 fatty acids utilizing algae and algae-like microorganisms. *J Appl Phycol* 6:123–129, 1994.

43. ZY Wen, F Chen. Heterotrophic production of eicosapentaenoic acid by microalgae. *Biotechnol Adv* 21:273–94, 2003.

44. A Kendrick, C Ratledge. Lipids of selected molds grown for production of n-3 and n-6 polyunsaturated fatty acids. *Lipids* 27:15–20, 1992.

45. H Streekstra. On the safety of *Mortierella alpina* for the production of food ingredients, such as arachidonic acid. *J Biotechnol* 56:153–65, 1997.

46. PK Bajpai, P Bajpai, OP Ward. Optimization of production of Docosahexaenoic acid (DHA) by *Thraustochytrium aureum* ATCC 34304. *J Am Oil Chem Soc* 68:509–514, 1991.

47. P Bajpai, PK Bajpai, OP Ward. Production of docosahexaenoic acid by *Thraustochytrium aureum*. *Appl Microbiol Biotech* 35:706–710, 1991.

48. SL Pereira, AE Leonard, P Mukerji. Recent advances in the study of fatty acid desaturases from animals and lower eukaryotes. *Prostaglandins Leukot Essent Fatty Acids* 68:97–106, 2003.

49. E Heinz. Biosynthesis of polyunsaturated fatty acids. In: Moore, T.S., ed. *Lipid Metabolism in Plants*. CRC Press, Boca Raton, FL, 1993, pp. 33–89.

50. DL Cinti, L Cook, MN Nagi, SK Suneja. The fatty acid chain elongation system of mammalian endoplasmic reticulum. *Prog Lipid Res* 31:1–51, 1992.

51. DH Nugteren. The enzymatic chain elongation of fatty acids by rat–liver microsomes. *Biochim Biophys Acta* 160:280–290, 1965.

52. JT Bernert, H Sprecher. Studies to determine the role rates of chain elongation and desaturation play in regulating the unsaturated fatty acid composition of rat liver lipids. *Biochim Biophys Acta* 398:354–363, 1975.

53. AE Leonard, SL Pereira, Y-S Huang. Elongation of long-chain fatty acids. *Prog. Lipid Res* 43:36–54, 2004.

54. FB Padley, FD Gunstone, JL Harwood. Occurrence and characteristics of oils and fats, In: FD Gunstone, JL Harwood, FB Padley, eds. *The Lipid Handbook*, 2nd ed., Chapman and Hall, London, 1994, pp. 47–223.

55. O Sayanova, PR Shewry, JA Napier. Histidine-41 of cytochrome b5 domain of the borage Δ6 fatty acid desaturase is essential for enzyme activity. *Plant Phys* 121:641–646, 1999.

56. Y-S Huang, S Chaudhary, JM Thurmond, EG Bobik, Jr., L Yuan, GM Chan, SJ Kirchner, P Mukerji, DS Knutzon. Cloning of Δ12- and Δ6-desaturases from Mortierella alpina and recombinant production of gamma-linolenic acid in Saccharomyces cerevisiae. *Lipids* 34:649–659, 1999.

57. K Laoten, R Mannontarat, M Tanticharoen, S Cheevadhanarak. Δ6-desaturase of *Mucor rouxii* with high similarity to plant D6-desaturase and its heterologous expression in *Saccharomyces cerevisiae*. *Biochem Biophys Res Comm* 279:17–22, 2000.

58. H Hong, N Datla, DW Reed, PS Covello, SL MacKenzie, X Qiu. High-level production of gamma-linolenic acid in Brassica juncea using a delta6 desaturase from Pythium irregulare. *Plant Physiol* 129:354–62, 2002.

59. T Girke, H Schmidt, U Zähringer, R Reski, E Heinz. Identification of a novel Δ6-acyl-group desaturase by targeted gene disruption in *Physcomitrella patens*. *Plant J* 15:39–48, 1998.

60. AS Reddy, ML Nuccio, LM Gross, TL Thomas. Isolation of a delta 6-desaturase from the cyanobacterium Synechocystis sp. strain PCC 6803 by gain-of-function expression in Anabaena sp. strain PCC7120. *Plant Mol Biol* 27:293–300, 1993.

61. F Garcia–Maroto, JA Garrido-Cardenas, J Rodriguez-Ruiz, M Vilches-Ferron, AC Adam, J Polaina, DL Alonso. Cloning and molecular characterization of the delta6-desaturase from two Echium plant species: production of GLA by heterologous expression in yeast and tobacco. *Lipids* 37:417–26, 2002.

62. OV Sayanova, F Beaudoin, LV Michaelson, PR Shewry, JA Napier. Identification of primula fatty acid delta 6-desaturases with n-3 substrate preferences. *FEBS Lett* 542:100–104, 2003.

63. O Sayanova, MA Smith, P Lapinskas, AK Stobart, G Dobson, WW Christie, PR Shewry, JA Napier. Expression of a borage desaturase cDNA containing an N-terminal cytochrome b5 domain results in the accumulation of high levels of delta 6-desaturated fatty acids in transgenic tobacco. *Proc Natl Acad Sci USA* 94:4211–4216, 1997.

64. F Domergue, A Abbadi, C Ott, TK Zank, U Zahringer, E Heinz. Acyl carriers used as substrates by the desaturases and elongases involved in very long-chain polyunsaturated fatty acids biosynthesis reconstituted in yeast. *J Biol Chem* 278:35115–26, 2003

65. T Okayasu, M Nagao, T Ishibashi, Y Imai. Purification and partial characterization of linoleoyl-CoA desaturase from rat liver microsomes. *Arch Biochem Biophys* 206:21–28, 1981

66. X Qiu, H Hong, N Datla, SL MacKinzie, DC Taylor, TL Thomas. Expression of borage Δ6 desaturase in Saccharomyces cerevisiae and oilseed crops. *Can J Bot* 80:42–49, 2002.

67. DS Knutzon, G Chan, P Mukerji, JM Thurmond, S Chaudhary, Y–S Huang. Genetic engineering of seed oil fatty acid composition. IX International Congress on Plant Tissue and Cell Culture, Jerusalem, Israel. 1998a.

68. J-W Liu, Y-S Huang, S DeMichele, M Bergana, E Bobik, C Hastilow, L-T Chuang, P Mukerji, D Knutzon. Evaluation of the seed oils from a canola plant genetically transformed to produce high levels of γ-linolenic acid. In: Y-S Huang, VA Ziboh, eds., *Gamma-Linolenic Acid: Recent Advances in Biotechnology and Clinical Applications*, AOCS Press, Champaign, IL. 2001, pp. 61– 71.

69. JM Parker-Barnes, T Das, E Bobik, AE Leonard, JM Thurmond, L-T Chuang, Y-S Huang, P Mukerji. Identification and characterization of an enzyme involved in the elongation of n-6 and n-3 poly-unsaturated fatty acids. *Proc Natl Acad Sci USA* 97:8284–8289, 2000.

70. T Das, Y-S Huang, P Mukerji. Delta 6-desaturase and GLA biosynthesis: A biotechnology perspective. In: Y-S Huang, VA Ziboh, eds. *Gamma-Linolenic Acid: Recent Advances in Biotechnology and Clinical Applications.* AOCS Press, Champaign, IL, 2000, pp. 6–23.

71. T Das, JM Thurmond, E Bobik, AE Leonard, JM Parker-Barnes, Y-S Huang, P Mukerji. Polyun-saturated fatty acid-specific elongation enzymes. *Biochem Soc Trans* 28:658–660, 2000.

72. T Das, JM Thurmond, AE Leonard, JM Parker-Barnes, E Bobik, L-T Chuang, Y-S Huang, P Mukerji. In: Y.-S. Huang and V.A. Ziboh, eds. γ-*Linolenic Acid: Recent Advances in Biotechnology and Clinical Applications*, AOCS Press, Champaign, IL, 2002.

73. F Beaudoin, LV Michaelson, SJ Hey, MJ Lewis, PR Shewry, O Sayanova, JA Napier. Heterologous reconstitution in yeast of the polyunsaturated fatty acid biosynthetic pathway. *Proc Natl Acad Sci USA* 97:6421–6426, 2000.

74. TK Zank, U Zahringer, J Lerchl, E Heinz. Cloning and functional expression of the first plant fatty acid elongase specific for Delta(6)-polyunsaturated fatty acids. *Biochem Soc Trans* 28:654–658, 2000.

75. E Heinz, T Zank, U Zaehringer, J Lerchl, A Renz. Patent: WO 0159128-A 16 Aug 2001.

76. AE Leonard, EG Bobik, J Dorado, PE Kroeger, L-T Chuang, JM Thurmond, JM Parker-Barnes, T Das, Y-S Huang, P Mukerji. Cloning of a human cDNA encoding a novel enzyme involved in the elongation of long-chain polyunsaturated fatty acids. *Biochem J* 350:765–770, 2000.

77. P Tvrdik, R Westerberg, S Silve, A Asadi, A Jakobsson, B Cannon, G Loison, A Jacobsson. Role of a new mammalian gene family in the biosynthesis of very long chain fatty acids and sphingolipids. *J Cell Biol* 149:707–718, 2000.

78. YA Moon, NA Shah, S Mohapatra, JA Warrington, JD Horton. Identification of a mammalian long chain fatty acyl elongase regulated by sterol regulatory element-binding proteins. *J Biol Chem* 276:45358–45366, 2001.

79. C Cassagne, R Lessire, JJ Bessoule, P Moreau, A Creach, F Schneider, B Sturbois. Biosynthesis of very long chain fatty acids in higher plants. *Prog Lipid Res* 33:55–69, 1994.

80. P Von Wettstein-Knowles, JG Olsen, K Arnvig, S Larsen. In: JL Harwood, PJ Quinn, eds., Recent Advances in the Biochemistry of Plant Lipids, Portland Press, 2001, pp. 601–607.

81. DS Knutzon, JM Thurmond, Y-S Huang, S Chaudhary, EG Bobik, Jr., GM Chan, SJ Kirchner, P Mukerji. Identification of Δ5-desaturase from *Mortierella alpina* by heterologous expression in bakers' yeast and canola. *J Biol Chem* 273:29360–29366, 1998.

82. LV Michaelson, CM Lauzarus, G Griffiths, JA Napier, AK Stobart. Isolation of a Δ5-fatty acid desaturase gene from Mortierella alpina. *J Biol Chem* 273:19055–19059, 1998.

83. F Domergue, J Lerchl, U Zahringer, E Heinz. Cloning and functional characterization of Phaeo-dactylum tricornutum front-end desaturases involved in eicosapentaenoic acid biosynthesis. *Eur J Biochem* 269:4105–4113, 2002.

84. JL Watts, J Browse. Isolation and characterization of a Δ5-fatty acid desaturase from Caenorhabditis elegans. *Arch Biochem Biophys* 362:175–182, 1999.

85. AE Leonard, B Kelder, EG Bobik, L-T Chuang, JM Parker-Barnes, JM Thurmond, PE Kroeger, JJ Kopchick, Y-S Huang, P Mukerji. cDNA cloning and characterization of human Δ5-desaturase involved in the biosynthesis of arachidonic acid. *Biochem J* 347:719–724, 2000.

86. R Zolfaghari, CJ Cifelli, MD Banta, AC Ross. Fatty acid delta(5)-desaturase mRNA is regulated by dietary vitamin A and exogenous retinoic acid in liver of adult rats. *Arch Biochem Biophys* 391:8–15, 2001.

87. DR Tocher, MJ Leaver, PA Hodgson. Recent advances in the biochemistry and molecular biology of fatty acyl desaturases. *Prog Lipid Res* 37:73–117, 1998.

88. J Shanklin, E Whittle, BG Fox. Eight histidine residues are catalytically essential in a membrane-associated iron enzyme, stearoyl-CoA desaturase, and are conserved in alkane hydroxylase and xylene monooxygenase. *Biochemistry* 33:2787–2794, 1994.

89. MR Jackson, T Nilsson, PA Peterson. Identification of a consensus motif for retention of trans-membrane proteins in the endoplasmic reticulum. *EMBO J* 9:3153–3162, 1990.

90. JP Spychalla, AJ Kinney, J Browse. Identification of an animal omega-3 fatty acid desaturase by heterologous expression in Arabidopsis. *Proc Natl Acad Sci USA* 94:1142–1147, 1997.

91. D Meesapyodsuk, DW Reed, CK Savile, PH Buist, SJ Ambrose, PS Covello. Characterization of the regiochemistry and cryptoregiochemistry of a *Caenorhabditis elegans* fatty acid desaturase (FAT–1) expressed in *Saccharomyces cerevisiae*. *Biochemistry* 39:11948–11954, 2000.

92. SL Pereira, Y-S Huang, EG Bobik, AJ Kinney, KL Stecca, JCL Packer, P Mukerji. A novel omega 3- (ω3-) fatty acid desaturase involved in the biosynthesis of eicosapentaenoic acid. *Biochem J* 2003 (in press).

93. A Meyer, P Cirpus, C Ott, R Schlecker, U Zahringer, E Heinz. Biosynthesis of docosahexaenoic acid in *Euglena gracilis*: biochemical and molecular evidence for the involvement of a delta 4-fatty acyl group desaturase. *Biochemistry* 42:9779–9788, 2003.

94. H Takeyama, D Takeda, K Yazawa, A Yamada, T Matsunaga. Expression of the eicosapentaenoic acid synthesis gene cluster from Shewanella sp. in a transgenic marine Cyanobacterium, Synecho-coccus sp. *Microbiology* 143:2725–2731, 1997.

95. K Yazawa. Production of Eicospentaenoic acid from marine bacteria. *Lipids* 31:S297–S300, 1996

96. N Morita, A Ueno, M Tanaka, S Ohgiya, T Hoshino, K Kawasaki, I Yumoto, K Ishizaki, H Okuyama. Cloning and sequencing of clustered genes involved in fatty acid biosynthesis from the docosa-hexaenoic acid-producing bacterium, *Vibrio marinus* strain MP-1. *Biotechnol Lett* 21:641–646, 1999.

97. DA Hopwood, DH Sherman. Molecular genetics of polyketides and its comparison to fatty acid biosynthesis. *Annu Rev Genet* 24:37–66, 1990.

98. JG Metz, P Roessler, D Facciotti, C Levering, F Dittrich, M Lassner, R Valentine, K Lardizabal, F Domergue, A Yamada, K Yazawa, V Knauf, J Browse. Production of polyunsaturated fatty acids by polyketide synthases in both prokaryotes and eukaryotes. *Science* 293:290–293, 2001.

2

Screening for Unique Microbial Reactions Useful for Industrial Applications

Jun Ogawa

Sakayu Shimizu

2.1 Introduction

In the coming postpetrochemical period, production processes will be required to save energy and to reduce environmental damage. In this sense, biological reactions are now widely recognized as practical alternatives to conventional chemical reactions.[1,2] On the other hand, novel catalytic procedures are necessary to produce the emerging classes of organic compounds that are becoming the targets of molecular and biomedical research. Therefore, screening for novel biocatalysts that are capable of catalyzing new reactions is constantly needed.

A bioprocess is sometimes designed from an organic chemistry standpoint, regardless of whether or not a suitable biocatalyst has already been found. This forces screening from a certain level to ascertain the existence of a desirable biocatalyst. Thus, it is also important to increase the catalog of biocatalysts to maintain the motivation for such a steady search. One of the most efficient and successful means of finding new biocatalysts is to screen large numbers of microorganisms, because of their characteristic diversity and versatility.[3–6] The keys for increasing the probability of discovery in the screening are to examine as many samples as possible and to maintain a thoughtful insight.

Usually, screening is simply focused on a one-step target reaction. In such screening, the cells of microorganisms are incubated with target substrates under various reaction conditions and their transformation is monitored. Successful examples of such screening are the finding of novel carbonyl reductases and lactonohydrolases. You can refer to the recent reviews presenting the details of these enzymes.[7,8] Also, information obtained on detailed analysis of a microbial metabolic process leads to unexpected new reactions and substrates. Examples of the latter, i.e., unique reactions found in the microbial metabolism of cyclic amides, nucleosides, and fatty acids, are described together with their applications in this chapter.

2.2 Analysis and Application of Microbial Cyclic Amide Metabolism

2.2.1 Overview of Microbial Cyclic Amide Metabolism

The chemical structure of cyclic amides can be found in many natural and unnatural compounds. The transformation of naturally occurring cyclic amides of pyrimidines and purines plays an important role in nucleobase metabolism. These metabolic activities comprise those of various cyclic amide hydrolases (EC 3.5.2.-), such as dihydropyrimidinase in reductive pyrimidine metabolism,[9] barbiturase in oxidative pyrimidine metabolism,[10] dihydroorotase in pyrimidine biosynthesis,[11] and allantoinase in purine metabolism.[12] In the 1970s, studies on rat liver dihydropyrimidinase showed that the enzyme hydrolyzed 5-monosubstituted hydantoin to N-carbamoyl amino acid and that the reaction proceeded D-stereospecifically.[13,14] Later on, Yamada and coworkers showed that microbial cells are good catalysts for this reaction.[15] Based on these observations, hydantoin metabolism in microorganisms was intensively studied and applied to optically active amino acid production.[16,17] *Blastobacter* sp. A17p-4 was screened from soil as a hydantoin-assimilating bacterium for the purpose of D-amino acid production from DL-5-monosubstituted hydantoins.[18] During the course of studies on hydantoin metabolism in this bacterium, it was found that it showed not only hydantoin- but also cyclic imide-metabolizing activity.[19] A recent study revealed that the strain has not only hydantoin-metabolizing enzymes but also enzymes specific to cyclic imide derivatives. Since then, cyclic imide metabolism and specific enzymes have been widely found and studied in bacteria, yeasts, and molds.

2.2.2 Analysis and Application of Microbial Hydantoin Metabolism

2.2.2.1 Diversity of Hydantoin-Metabolizing Enzymes

Hydantoin is metabolized to an amino acid through two-step hydrolysis via an N-carbamoyl amino acid. The enzyme catalyzing the first step, hydrolysis of hydantoin to N-carbamoyl amino acid, is called hydantoinase. Three typical hydantoinases with stereospecificity to D-, L-, and DL-5-monosubstituted hydantoin are named D-hydantoinase, L-hydantoinase, and DL-hydantoinase, respectively (Figure 2.1).[17,20–22] D-Hydantoinases have been found in various genera of bacteria such as *Pseudomonas*, *Bacillus*, *Blastobacter*, and *Arthrobacter*, and most of them show dihydropyrimidinase activity. The existence of L- and DL-hydantoinases might be rarer in nature than that of D-hydantoinase. These enzymes can be divided into two groups: one needing ATP for activity and the other not. The ATP-requiring enzyme from *Pseudomonas putida* 77, which functions in creatinine metabolism, showed L-hydantoinase activity.[23]

The second step of hydantoin metabolism, N-carbamoyl amino acid hydrolysis to amino acid, ammonia, and carbon dioxide, is catalyzed by carbamoylase. A variety of carbamoylases have been reported (Figure 2.2). Two typical carbamoylases with stereospecificity to N-carbamoyl D- and L-amino acids are named D-carbamoylase and L-carbamoylase, respectively.[17] D-Carbamoylase generally shows wide substrate specificity to both aromatic and aliphatic N-carbamoyl-D-amino acids.[24] L-Carbamoylase shows rather limited specificity to aromatic or aliphatic N-carbamoyl-L-amino acids. An L-carbamoylase with relatively broad substrate specificity has been found in *Alcaligenes xylosoxidans*.[25] β-Ureidopropionase from

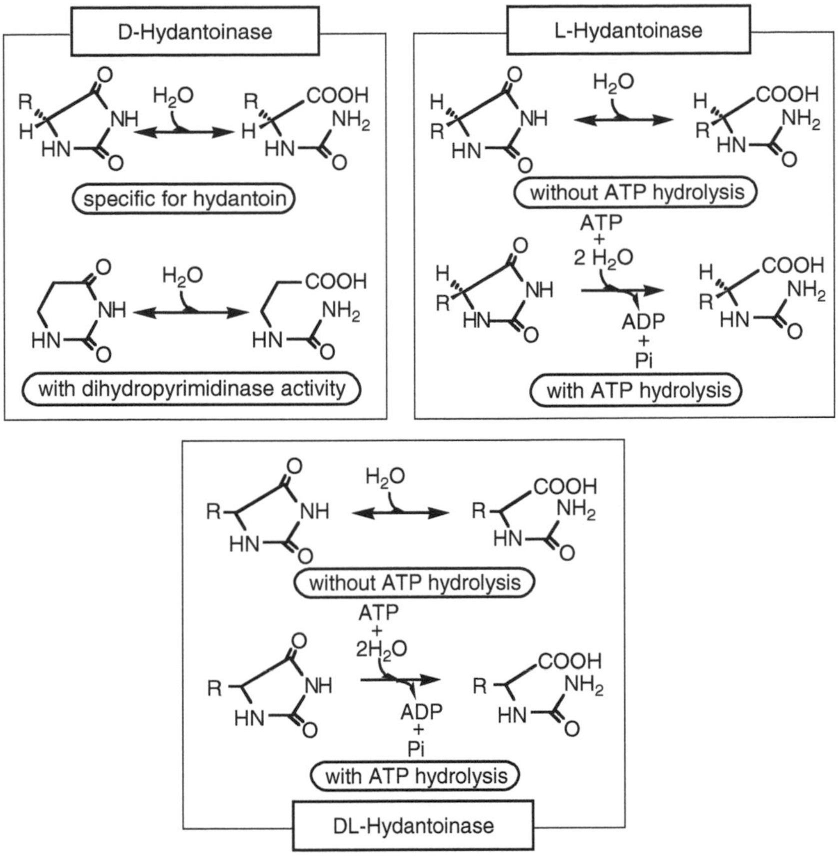

FIGURE 2.1 Reactions catalyzed by typical hydantoinases.

P. putida IFO 12996, which functions in the pyrimidine degradation during *N*-carbamoyl-β-alanine hydrolysis, showed broad substrate specificity not only toward *N*-carbamoyl-β-amino acids, but also toward *N*-carbamoyl-γ-amino acids and several *N*-carbamoyl-α-amino acids.[26] The hydrolysis of *N*-carbamoyl-α-amino acids is strictly L-stereospecific (Figure 2.2).

2.2.2.2 Optically Active Amino Acid Production by Hydantoin-Metabolizing Enzymes

Different combinations of hydantoin-metabolizing enzymes, i.e., hydantoinases and carbamoylases, provide a variety of processes for the production of optically pure α-amino acids (Figure 2.3).[17] The broad substrate range of the processes is valuable, especially for the production of D-amino acids and unnatural L-amino acids.[27,28] Some enzymes recognize multichiral centers other than the α-carbons of amino acids, so they enable simultaneous resolution of multichiral amino acids such as β-methylphenylalanine.[29] Recent research on hydantoin-metabolizing enzymes has been concentrated on newly isolated or improved enzymes, and has included directed evolution techniques, structure elucidation, studies on fusion proteins, and the use of specially designed whole cell biocatalysts.[30]

A practical representative is the production of D-*p*-hydroxyphenylglycine, a building block for semisynthetic penicillins and cephalosporins. The process involves one chemical step and two enzymatic steps (Figure 2.4). The substrate, DL-5-(*p*-hydroxyphenyl)hydantoin, is synthesized by an efficient chemical method involving the amidoalkylation reaction of phenol with glyoxylic acid and urea under acidic conditions. Then, the D-5-(*p*-hydroxyphenyl)hydantoin is hydrolyzed enzymatically to *N*-carbamoyl-D-*p*-hydroxyphenylglycine with a thermostable immobilized D-hydantoinase under alkaline conditions.

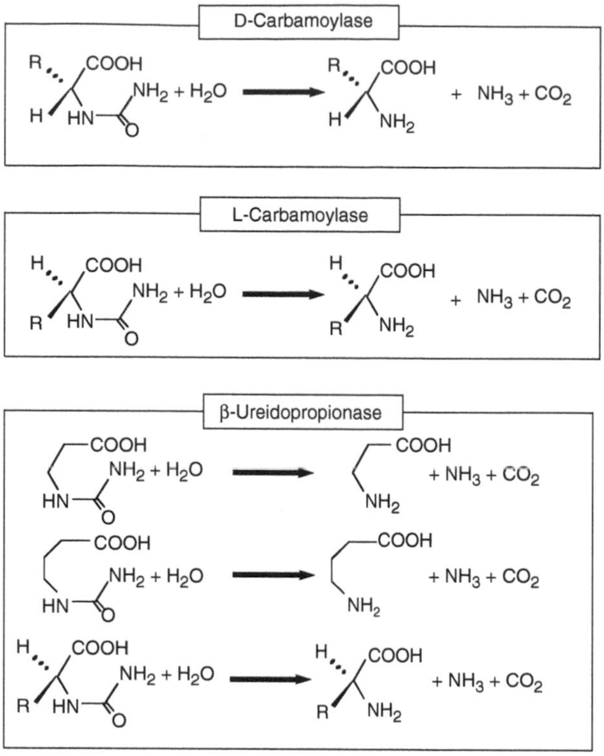

FIGURE 2.2 Reactions catalyzed by typical carbamoylases.

Under these conditions, the L-isomer of the remaining 5-(p-hydroxyphenyl)hydantoin is racemized through base catalysis. Therefore, racemic hydantoins can be converted quantitatively into N-carbamoyl-D-p-hydroxyphenylglycine through this step. Decarbamoylation to D-p-hydroxyphenylglycine is performed with an immobilized mutant D-carbamoylase. D-Carbamoylase from *Agrobacterium* sp. KNK712 has been improved into a mutant enzyme showing 20°C-higher thermal stability.[31,32] This process involving immobilized hydantoinase and carbamoylase has been used for the commercial production of D-p-hydroxyphenylglycine since 1995.

2.2.3   Analysis and Application of Microbial Cyclic Imide Metabolism

2.2.3.1   Analysis of Microbial Cyclic Imide Metabolism

Based on the finding of cyclic imide-hydrolyzing activity in *Blastobacter* sp.,[19] the metabolism of various cyclic imides by microorganisms was systematically investigated. The fact that *Blastobacter* sp. grows well in a synthetic minimum medium containing succinimide as the sole carbon source indicates that the bacterium has a metabolic system for the assimilation of cyclic imides as energy sources and nutrients.[33] The bacterium can metabolize various cyclic imides with structures similar to that of succinimide such as maleimide, 2-methylsuccinimide and glutarimide, and sulfur-containing cyclic imides such as 2,4-thiazolidinedione and rhodanine. Further investigation of the metabolic fate of these cyclic imides showed that they were metabolized through a novel metabolic pathway (Figure 2.5). This pathway comprises in turn the hydrolytic ring-opening of cyclic imides into half-amides, hydrolytic deamidation of the half-amides to dicarboxylates, and dicarboxylate transformation similar to that in the tricarboxylic acid (TCA) cycle. Two novel enzymes, imidase and half-amidase, and D-hydantoinase were found to function in this pathway.

The nature of imidase, which hydrolyzes a cyclic imide to a half-amide, was further investigated in detail. Three types of enzymes with different substrate specificities were found (Figure 2.6). An imidase

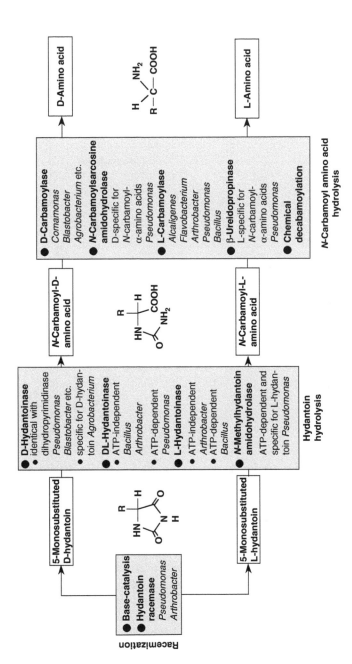

FIGURE 2.3 Processes for optically active α-amino acid production with combinations of various hydantoin–metabolizing enzymes.

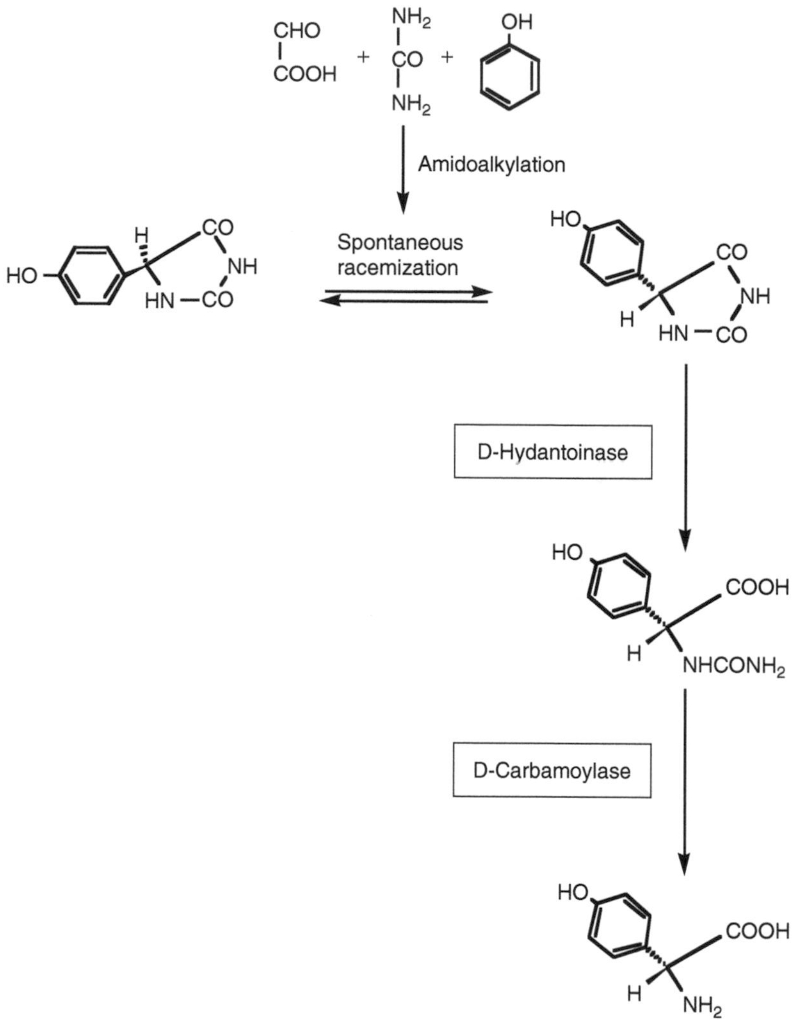

FIGURE 2.4 Industrial process for D-*p*-hydroxyphenylglycine production with hydantoin-metabolizing enzymes.

with specificity toward simple cyclic imides was purified from *Blastobacter* sp.[34] This enzyme is also active toward sulfur-containing cyclic imides such as 2,4-thiazolidinedione and rhodanine. However, bulky cyclic imides or monosubstituted cyclic ureides are not hydrolyzed. Bulky cyclic imides are hydrolyzed by the D-hydantoinase of *Blastobacter* sp. and mammalian dihydropyrimidinases.[35] Another imidase, phthalimidase, with specificity toward phthalimide derivatives was found in *Alcaligenes ureafaciens*.[36] Half-amides, the products of imidase, were further metabolized to dicarboxylates by half-imidase. The enzyme was purified from *Blastobacter* sp. and found to be specific toward half-amides.[37] These enzyme activities were widely distributed among bacteria, yeast, and molds.[38] Cyclic imide metabolism and the enzymes involved have practical potential for the production of high-value organic acids such as pyruvate from cyclic imides or their metabolites, and also the stereo- and regiospecific production of half-amides and dicarboxylates.

2.2.3.2   Application of Microbial Cyclic Imide Metabolism to Pyruvate Production

Cyclic imide metabolism has been applied to the production of a high-value organic acid, pyruvate.[39] The commercial demand for pyruvate has been increasing because of its use as an effective precursor in the synthesis of various drugs and agrochemicals in addition to as a component of mammalian-cell

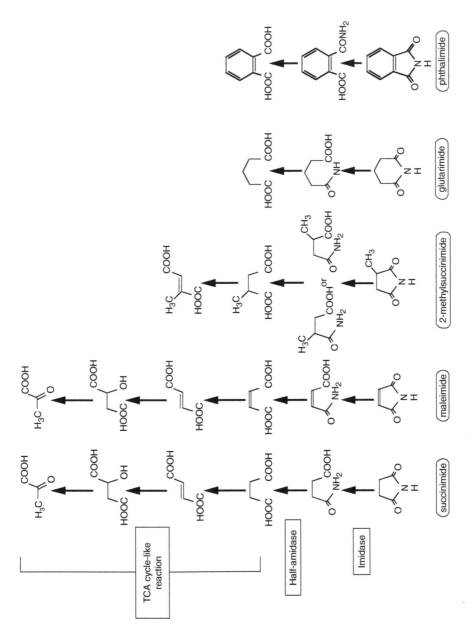

FIGURE 2.5 Pathways of microbial cyclic imide-metabolism.

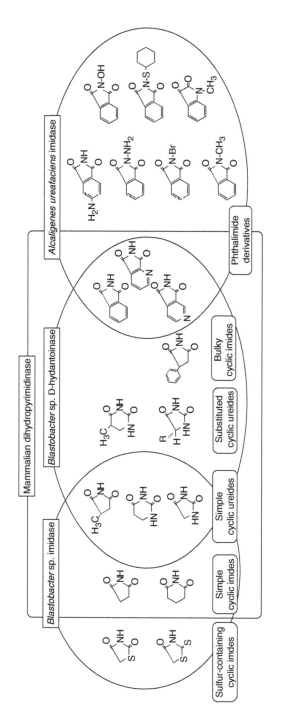

FIGURE 2.6 Substrate spectra of typical imidases.

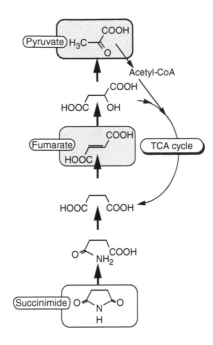

FIGURE 2.7 Pyruvate production from fumarate through microbial cyclic imide metabolism.

culture media. *Pseudomonas putida* s52 isolated with succinimide as the sole carbon source exhibits highly active cyclic imide metabolism. This activity has been used for pyruvate production from fumarate, a cheap cyclic imide metabolism intermediate (Figure 2.7). Using cells cultivated in medium containing 2% (w/v) fumarate as the catalyst, 286 mM pyruvate was produced from 500 mM fumarate in 27 h. Bromopyruvate, a malic enzyme inhibitor, inhibited the pyruvate production and also the growth of *Pseudomonas putida* s52 in the medium with fumarate as the sole carbon source. Bromopyruvate-resistant mutants were derived from *Pseudomonas putida* s52, and their pyruvate production was examined. One of the mutants showed much higher pyruvate production than the parent strain. Using the mutant cells cultivated in medium containing 2% (w/v) fumarate as the catalyst, 770 mM pyruvate was produced from 1000 mM fumarate in 96 h.[40]

2.2.3.3 Application of the Imidase-Catalyzing Reaction to Fine Organic Synthesis

In case of a half-amide, a useful building block for organic synthesis, there is synthetic difficulty in selective amidation at one of two equivalent carboxyl groups. 3-Carbamoyl-α-picolinic acid (α-3CP) is one of the regioisomeric half-amides of 2,3-pyridinedicarboxylic acid (PDC). α-3CP is a promising intermediate for modern insecticide synthesis. Chemical synthesis of α-3CP from PDC via the dimethylester involves troublesome regiospecific diester hydrolysis to the half-ester. Enzymatic regiospecific hydrolysis of 2,3-pyridinedicarboximide (PDI) (Figure 2.8) is one of the attractive methods for overcoming this disadvantage. Some imidases and D-hydantoinases (dihydropyrimidinases) have been found to hydro- lyze aryl-substituted cyclic imides such as PDI, 3,4-pyridinedicarboximide and phthalimide.[35,41] Based on these findings, potential imidases that are applicable to the regiospecific hydrolysis of PDI to α-3CP were screened for. Phthalimide-assimilating microorganisms have been isolated as possible catalysts for the regiospecific hydrolysis of PDI to α-3CP. *Arthrobacter ureafacience* O-86 was selected as the best strain and applied to the cyclohexanone-water two-phase reaction system, pH 5.5, where the spontaneous random hydrolysis of PDI was avoided and the enzyme maintained its activity. Under the optimized conditions, with the periodical addition of PDI (in total, 40 mM), 36.6 mM α-3CP accumulated in the water phase with a molar conversion yield of 91.5% and a regioisomeric purity of 94.5% in 2 h.[36]

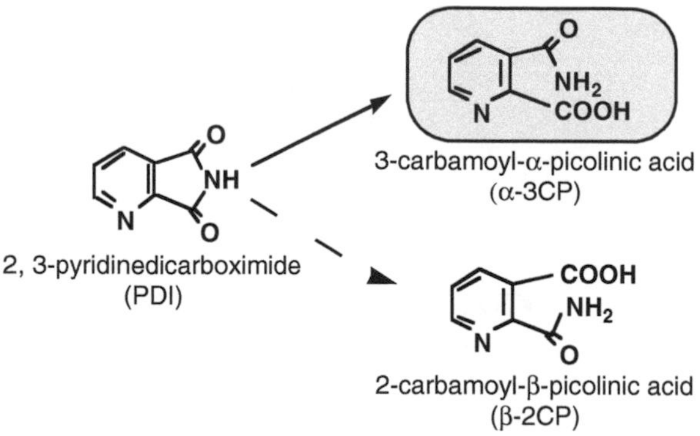

FIGURE 2.8 Imidase-catalyzing regioselective hydrolysis of 2,3-pyridinedicarboximide (PDI) to 3-carbamoyl-α-picolinic acid (α-3CP).

2.3   Analysis and Application of Microbial Nucleoside Metabolism

2.3.1   Overview of Microbial Nucleoside Metabolism

Recently, nucleosides and a variety of chemically synthesized nucleoside analogs have attracted a great deal of interest as they have antibiotic, antiviral, and antitumor effects.[42] In light of this trend, the microbial metabolism of nucleosides was reevaluated in detail, although it had been well studied as an assimilation or salvage pathway.[43] The first reaction in nucleoside metabolism is N-riboside cleavage. Two kinds of enzymes, nucleosidase (nucleoside hydrolase; EC 3.2.2.-) and nucleoside phosphorylase (EC 2.4.2.-), are known to catalyze this reaction (Figure 2.9). Nucleosidase catalyzes the irreversible hydrolysis of nucleosides and participates mainly in the assimilation pathway. A nucleosidase from *Ochrobactrum anthropi*, which specifically catalyzes the N-riboside cleavage of purine nucleosides, has been purified and characterized.[44] The enzyme was revealed to be useful for the decomposition of purine nucleosides in foodstuffs, with these nucleosides causing hyperuricemia, an increasingly common disease in adults.[45] On the other hand, nucleoside phosphorylase catalyzes the phosphorolytic cleavage of nucleosides and shows ribosyl transferase activity.[46] This enzyme functions mainly in the salvage pathway. Nucleoside phosphorylase has been well studied and used as a catalyst for the synthesis of nucleoside analogs through base exchange reactions.[46,47]

In the case of 2'-deoxyribonucleoside degradation, the product of the nucleoside phosphorylase reaction is 2-deoxyribose 1-phosphate. This is further transformed into D-glyceraldehyde 3-phosphate and acetaldehyde via 2-deoxyribose 5-phosphate. These reactions are reversible and successively catalyzed

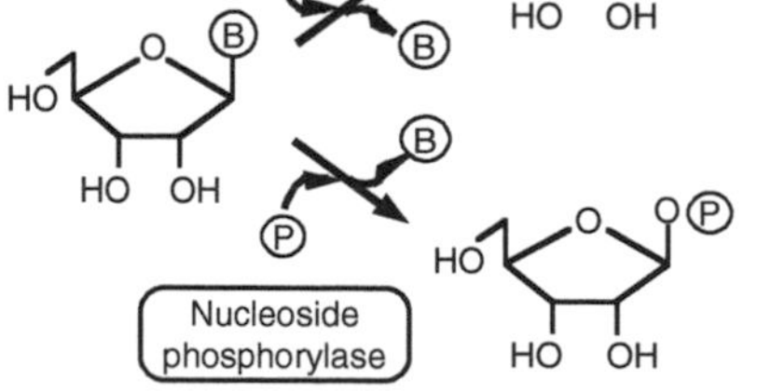

FIGURE 2.9   Reactions catalyzed by nucleosidase and nucleoside phosphorylase.

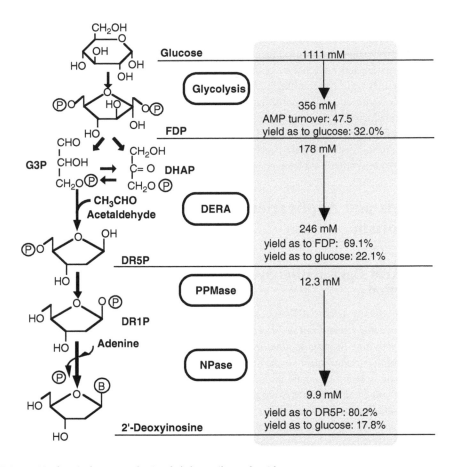

FIGURE 2.10 Biochemical retrosynthesis of 2'-deoxyribonucleoside.

by phosphopentomutase and deoxyriboaldolase, and the products of these reactions, D-glyceraldehyde 3-phosphate and acetaldehyde, finally flow into a central metabolic process such as the glycolytic pathway. Recently, a unique application of these reversible reactions to nucleoside synthesis was investigated, as described below.

2.3.2 Biochemical Retrosynthesis of 2'-Deoxyribonucleoside Through Microbial Nucleoside Metabolism

With the spread of PCR techniques and new antiviral nucleosides, and the advent of antisense drugs for cancer therapy, there will be an urgent need for a DNA building block, 2'-deoxyribonucleoside, on a large scale in the near future. Classical 2'-deoxyribonucleoside sources are hydrolyzed herring and salmon sperm DNA. These sources, however, will not allow us to meet future demands for a stable and economical supply of 2'-deoxyribonucleoside. A possible microbial method for 2'-deoxyribonucleoside production from easily available materials, glucose, acetaldehyde, and a nucleobase, has been examined, that is, the use of reversible reactions involved in nucleoside degradation. In this process, microorganisms possessing glycolytic enzymes, deoxyriboaldolase, phosphopentomutase, and nucleoside phosphorylase, were used as catalysts. The glycolytic enzymes produce D-glyceraldehyde 3-phosphate from glucose. Subsequently, deoxyriboaldolase, phosphopentomutase, and nucleoside phosphorylase cooperatively produce 2'-deoxyribonucleosides from D-glyceraldehyde 3-phosphate, acetaldehyde, and a nucleobase via 2-deoxyribose 5-phosphate (Figure 2.10). A deoxyriboaldolase suitable for 2-deoxyribose 5-phosphate synthesis with tolerance to acetaldehyde has been found on screening.[48] A potential enzyme was found in *Klebsiella pnuemoniae* and transformed into

E coli.[49] Using the *E coli* transformant as a source of glycolytic enzymes and deoxyriboaldolase, 2-deoxyribose 5-phosphate was produced from glucose and acetaldehyde in the presence of ATP, which is required for D-glyceraldehyde 3-phosphate generation from glucose by glycolytic enzymes. The energy derived from ATP was replaced by the energy derived from alcohol fermentation by baker's yeast in the form of fructose 1,6-diphosphate, which was further utilized as a D-glyceraldehyde 3-phosphate precursor by the *E coli* transformant.[50] The 2-deoxyribose 5-phosphate produced was further transformed to 2'-deoxyribonu-cleoside by *E coli* transformants expressing phosphopentomutase and nucleoside phosphorylase.[51] Typical results of such synthesis are also presented in Figure 2.10. It is noteworthy that the glycolytic pathway supplies the important substrates, fructose 1,6-diphosphate and D-glyceraldehyde 3-phosphate, for 2'-deoxyribonucleoside synthesis via 2-deoxyribose 5-phosphate.

2.4 Analysis and Application of Microbial Fatty Acid Metabolism

2.4.1 Fatty Acid Desaturation Systems for Polyunsaturated Fatty Acid Production

C20 polyunsaturated fatty acids (PUFAs), such as 5,8,11-*cis*-eicosatrienoic acid (mead acid, MA), dihomo-γ-linolenic acid, arachidonic acid, and 5,8,11,14,17-*cis*-eicosapentaenoic acid (EPA), exhibit unique biological activities. Because food sources rich in these PUFAs are limited to a few seed oils and fish oils, the screening for alternative sources of these PUFAs in microorganisms has been conducted, which resulted in the isolation of an arachidonic acid-producing fungus, *Mortierella alpina* 1S-4.[52] This fungus produces 30–60 g/l of mycelia (dry weight) containing about 60% lipids. The lipids mainly consist of triacylglycerol, which contains arachidonic acid. The amount of arachidonic acid is 40–70% of the lipids—approximately 13 kg/kl on large-scale fermentation. The fungus operates unique fatty acid desaturation systems involving at least five desaturases with different specificities. EPA has been produced through the ω3 route from α-linolenic acid via successive Δ6 desaturation, elongation and Δ5 desaturation using the same fungus.[53] Further investigation led to the isolation of desaturase-defective mutants and opened routes for the production of other PUFAs (Figure 2.11).[54,55] For example, dihomo-γ-linolenic acid and MA can be produced through the ω6 and ω9 routes using Δ5 and Δ12-desaturase-defective mutants, respectively. A recent review presented the details of this research.[56]

2.4.2 Fatty Acid Metabolism Useful for Conjugated Fatty Acid Production

Conjugated fatty acids have attracted much attention as a novel type of biologically beneficial functional lipids. For example, dietary conjugated linoleic acid (CLA) reduces carcinogenesis, atherosclerosis, and body fat.[57,58] Today, CLA is produced through chemical isomerization of linoleic acid, which results in the by-production of unexpected isomers. Considering the use of CLA for medicinal and nutraceutical purposes, an isomer-selective and safe process is required. A bioprocess is a potential alternative for this purpose.

Dairy products are among the major natural sources of CLA, of which *cis*-9,*trans*-11-octadecadienoic acid is the main isomer.[59] CLA has been shown to be produced from polyunsaturated fatty acids by certain rumen microorganisms such as *Butyrivibrio* species. *cis*-9,*trans*-11-Octadecadienoic acid has been suggested as an intermediate in the biohydrogenation of linoleic acid to octadecaenoic acid by the anaerobic rumen bacterium *Butyrivibrio fibrisolvens*.[60] It has also been reported that *Propionibacterium freudenreichii*, which is commonly used as a dairy starter culture, can produce CLA from free linoleic acid.[61] Recently, the ability to produce CLA from linoleic acid was extensively screened for in lactic acid bacteria.[62] Many strains were found to produce CLA from linoleic acid, and the mechanism of CLA production was investigated with *Lactobacillus acidophilus* AKU 1137 as a representative strain.[63] The

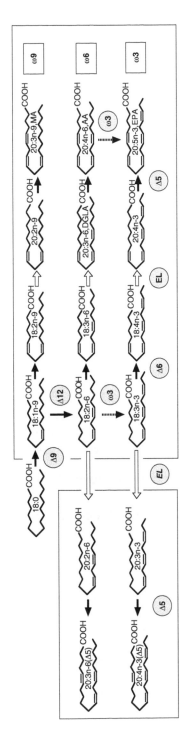

FIGURE 2.11 Polyunsaturated fatty acid–synthesizing systems in *M. alpina* 1S–4 and mutants of it. EL, elongase; AA, arachidonic acid; MA, Mead acid; ALA, α-linolenic acid; EPA, 5,8,11,14,17-*cis*-eicosapentaenoic acid; DGLA, dihomo-γ-linolenic acid.

CLAs produced by *L. acidophilus* were identified as *cis*-9,*trans*-11-octadecadienoic acid (CLA1) and *trans*-9,*trans*-11-octadecadienoic acid (CLA2).[64] Preceding the production of CLA, hydroxy fatty acids identified as 10-hydroxy-*cis*-12-octadecaenoic acid and 10-hydroxy-*trans*-12-octadecaenoic acid were accumulated. The isolated 10-hydroxy-*cis*-12-octadecaenoic acid was transformed to CLA on incubation with washed cells of *L. acidophilus*, suggesting that this hydroxy fatty acid is one of the intermediates of CLA production from linoleic acid (Figure 2.12). Based on these results, the transformation of hydroxy fatty acids by lactic acid bacteria was investigated. Lactic acid bacteria transformed ricinoleic acid (12-hydroxy-*cis*-9-octadecaenoic acid) into CLA (a mixture of CLA1 and CLA2).[65] There are two possible pathways for CLA synthesis from ricinoleic acid by lactic acid bacteria: 1) direct transformation of ricinoleic acid into CLA through dehydration at the $\Delta 11$ position, and 2) dehydration of ricinoleic acid at the $\Delta 12$ position to linoleic acid, which is a potential substrate for CLA production by lactic acid bacteria (Figure 2.12). In a similar manner on linoleic acid transformation to CLA, lactic acid bacteria transformed α- and γ-linolenic acid into the corresponding conjugated trienoic acids.[66,67] Those produced from α-linolenic acid were identified as *cis*-9,*trans*-11,*cis*-15-octadecatrienoic acid (18:3) and *trans*-9,*trans*-11,*cis*-15-18:3, and those from γ-linolenic acid as *cis*-6,*cis*-9,*trans*-11-18:3 and *cis*-6,*trans*-9,*trans*-11-18:3 (Figure 2.13). Washed cells of lactic acid bacteria exhibiting high productivity of conjugated fatty acids were obtained by cultivation in medium supplemented with polyunsaturated fatty acids such as linoleic acid and α-linolenic acid, indicating that these enzyme systems are induced by polyunsaturated fatty acids, maybe for their detoxication.[60]

2.4.2.1 Preparative CLA Production from Linoleic Acid by Lactic Acid Bacteria

After screening 14 genera of lactic acid bacteria, *L. plantarum* AKU 1009a was selected as a potential strain for CLA production from linoleic acid.[62] Washed cells of *L. plantarum* exhibiting a high level of CLA production were obtained by cultivation in a nutrient medium containing linoleic acid. Under the optimum reaction conditions with the free form of linoleic acid as the substrate, washed cells of *L. plantarum* produced 40 mg/ml CLA (33% molar yield) from 12% (w/v) linoleic acid in 108 h. The resulting CLA comprised a mixture of CLA1 (38% of total CLA) and CLA2 (62% of total CLA), and accounted for 50% of the total fatty acids obtained. A higher yield (80% molar yield as to linoleic acid) was attained with 2.6% (w/v) linoleic acid as the substrate in 96 h, resulting in CLA production of 20 mg/ml [consisting of CLA1 (2%) and CLA2 (98%)] and accounting for 80% of the total fatty acids obtained. Most of the CLA produced was associated with the washed cells, and mainly as a free form.[62]

2.4.2.2 Preparative CLA Production from Ricinoleic Acid and Castor Oil by Lactic Acid Bacteria

The ability to produce CLA from ricinoleic acid is widely distributed in lactic acid bacteria. Washed cells of *L. plantarum* JCM 1551 were selected as a potential catalyst for CLA production from ricinoleic acid.[68] Cells cultivated in a medium supplemented with a mixture of α-linolenic acid and linoleic acid showed enhanced CLA-productivity. Under the optimum reaction conditions, with the free acid form of ricinoleic acid as the substrate and washed cells of *L. plantarum* as the catalyst, 2.4 mg/ml CLA was produced from 3.4 mg/ml ricinoleic acid in 90 h, the molar yield as to ricinoleic acid being 71%. The CLA produced, which was obtained in the free fatty acid form, consisted of CLA1 (21% of total CLA) and CLA2 (79% of total CLA), and accounted for 72% of the total fatty acids obtained.[68]

Ricinoleic acid is abundant in a plant oil, castor oil. Castor oil is an economical source of ricinoleic acid. About 88% of the total fatty acids in castor oil is ricinoleic acid. Unfortunately, CLA cannot be directly produced from castor oil by lactic acid bacteria. Lactic acid bacteria only use the free form of ricinoleic acid for CLA production, i.e., not its triacylglycerol form, which is mainly found in castor oil. However, in the presence of lipase, castor oil became an effective substrate for CLA production by lactic acid bacteria.[65] The addition of a polyhydroxy-type detergent enhanced the CLA production from castor oil. Under the optimum conditions with castor oil as the substrate and washed cells of *L. plantarum* JCM 1551 as the catalyst, 2.7 mg/ml CLA was produced from 5.0 mg/ml castor oil in 99 h. The CLA produced accounted for 46% of the total fatty acids obtained, and consisted of CLA1 (26%) and CLA2 (74%).[69]

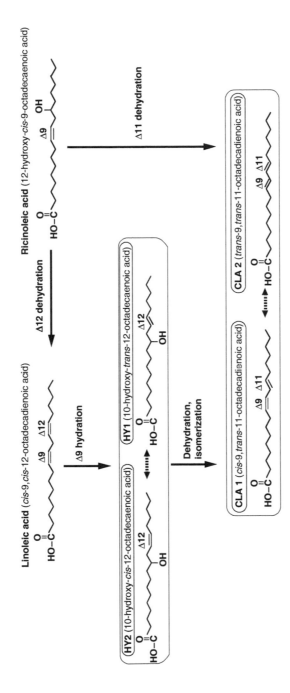

FIGURE 2.12 Conjugated linoleic acid (CLA)–producing systems in lactic acid bacteria.

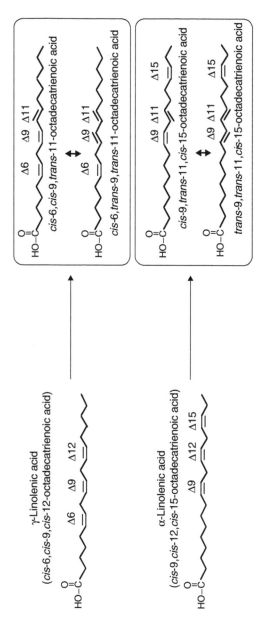

FIGURE 2.13 α-and γ-Linolenic acid transformation into conjugated fatty acids by lactic acid bacteria.

2.5 Conclusions: For Expansion of Biocatalysts for Practical Purposes

In this chapter, the authors introduced several examples of unique microbial reactions useful for practical purposes. The reactions resulted from detailed observation of microbial cyclic amide, nucleoside, and fatty acid metabolisms, and have paved the way to new bioprocesses for the production of optically active amino acids, organic acids, half-amides, 2'-deoxyribonucleosides, polyunsaturated fatty acids, conjugated fatty acids, etc.

Modern society requests the development of processes exhibiting environmental harmonization, economic efficiency, and specificity. This trend is causing the application of biological reactions to a greater variety of industries. Future bioprocesses will generally not be limited by the available technology or the nature of the substrates and products. Instead, the feasibility of new bioprocesses will often be determined by the availability of biocatalysts, the search for which needs patience for steady research but has a deep impression when a new biocatalyst is encountered.

Recently, some rational methods creating new biocatalysts have been rapidly developed. Modern gene technology, crystal structure analysis, and bioinformatics enable the modulation of enzyme function through site-directed mutation, DNA shuffling, etc. One example is modification of the substrate specificity of a monooxygenase, P450 BM-3, from *Bacillus megaterium*. P450 BM-3 is a fatty acid monooxygenase. Its substrate specificity was expanded to aromatic hydrocarbons, and phenolic and arylalkyl compounds by means of crystal structure-based directed mutation, which resulted in the creation of novel catalysts for regiospecific and stereospecific alcohol synthesis.[70–72] However, "rationality" is not the only answer for developing new biocatalysts. Classical screening of microbial diversity and versatility is still important. Such screening is something like a midnight walk without moonlight; however, detailed observation and deep insight with a well-considered strategy will lead to a new biocatalyst. This philosophy has now been succeeded by in vitro random evolution technology.[73]

The industrial success of biocatalysts unfortunately depends on the economics of the specific processes. However, once successful, it provides enormous opportunities. With the introduction of each new process accumulating experience and confidence, it becomes easier to develop and justify the next new bioprocess. Thus, it is important to increase the catalog of biocatalysts waiting to be examined for practical purposes.

References

1. KM Koeller, CH Wong. Enzymes for chemical synthesis. *Nature* 409:232–240, 2001.
2. A. Schmid, JS Dordick, B Hauer, A Kiener, M Wubbolt, B Witholt. Industrial biocatalysis today and tomorrow. *Nature* 409:258–268, 2001.
3. H Yamada, S Shimizu. Microbial and enzymatic processes for the production of biologically and chemically useful compounds. *Angew Chem Int Ed Engl* 27:622–642, 1988.
4. S Shimizu, J Ogawa, M Kataoka, M Kobayashi. Screening of novel microbial enzymes for the production of biologically and chemically useful compounds. *Adv Biochem Eng Biotechnol* 58:45–87, 1997.
5. J Ogawa, S Shimizu. Microbial enzymes: new industrial applications from traditional screening methods. *Trends Biotechnol* 17:13–21, 1999.
6. J Ogawa, S Shimizu. Industrial microbial enzymes: their discovery by screening and use in large-scale production of useful chemicals in Japan. *Curr Opin Biotechnol* 13:367–375, 2002.
7. S Shimizu, M Kataoka, K Honda, K Sakamoto. Lactone-ring-cleaving enzymes of microorganisms: their diversity and applications. *J Biotechnol* 92:187–194, 2001.
8. M Kataoka, K Kita, M Wada, Y Yasohara, J Hasagawa, S Shimizu. Novel bioreduction system for the production of chiral alcohols. *Appl Microbiol Biotechnol,* in press, 2003.
9. GD Vogels, C Van der Drift. Degradation of purines and pyrimidines by microorganisms. *Bacteriol Rev* 40: 403–468, 1976.

10. CL Soong, J Ogawa, E Sakuradani, S Shimizu. Barbiturase, a novel zinc-containing amidohydrolase involved in oxidative pyrimidine metabolism. *J Biol Chem* 277:7051–7058, 2002.

11. J Ogawa, S Shimizu. Purification and characterization of dihydroorotase from *Pseudomonas putida*. *Arch Microbiol* 164:353–357, 1995.

12. GPA Bongaerts, GD Vogels. Uric acid degradation by *Bacillus fastidiosus* strains. *J Bacteriol* 125:689–697, 1976.

13. KH Dudley, DL Bius, TC Butler. Metabolic fates of 3-ethyl-5-phenylhydantoin (ethotoin, peganone), 3-methyl-5-phenylhydantoin and 5-phenylhydantoin. *Pharmacol Exp Ther* 175:27–37, 1970.

14. F Cecere, G Galli, F Morisi. Substrate and steric specificity of hydropyrimidine hydrase. *FEBS Lett* 57:192–194, 1975.

15. H Yamada, S Takahashi, Y Kii, H Kumagai. Distribution of hydantoin hydrolyzing activity in microorganisms. *J Ferment Technol* 56:484–491, 1978.

16. J Ogawa, S Shimizu. Diversity and versatility of microbial hydantoin-transforming enzymes. *J Molec Catal B: Enzymatic* 2:163–176, 1997.

17. J Ogawa, S Shimizu. Stereoselective synthesis using hydantoinases and carbamoylases. In: RN Partel, ed. *Stereoselective Biocatalysis*. New York: Marcel Dekker, 2000, pp. 1–21.

18. J Ogawa, MCM Chung, S Hida, H Yamada, S Shimizu. Thermostable *N*-carbamoyl-D-amino acid amidohydrolase; screening, purification and characterization. *J Biotechnol* 38:11–19, 1994.

19. J Ogawa, M Honda, CL Soong, S Shimizu. Diversity of cyclic ureide compound-, dihydropyrimidine-, and hydantoin-hydrolyzing enzymes in *Blastobacter* sp. A17p-4. *Biosci Biotechnol Biochem* 59:1960–1962, 1995.

20. GJ Kim, HS Kim. Identification of the structural similarity in the functionally related amidohydrolases acting on the cyclic amide ring. *Biochem J* 15: 295–302, 1998.

21. O May, A Habenicht, R Mattes, C Syldatk, M Siemann. Molecular evolution of hydantoinases. *Biol Chem* 379:743–747, 1998.

22. C Syldatk, O May, J Altenbucher, R Mattes, M Siemann. Microbial hydantoinases—industrial enzymes from the origin of life? *Appl Microbiol Biotechnol* 51:293–309, 1999.

23. J Ogawa, JM Kim, W Nirdnoy, Y Amano, H Yamada, S Shimizu. Purification and characterization of an ATP-dependent amidohydrolase, *N*-methylhydantoin amidohydrolase, from *Pseudomonas putida* 77. *Eur J Biochem* 229: 284–290, 1995.

24. J Ogawa, S Shimizu, H Yamada. *N*-Carbamoyl-D-amino acid amidohydrolase from *Comamonas* sp. E222c: Purification and characterization. *Eur J Biochem* 212:685–691, 1993.

25. J Ogawa, H Miyake, S Shimizu. Purification and characterization of *N*-carbamoyl-L-amino acid amidohydrolase with broad substrate specificity from *Alcaligens xylosoxidans*. *Appl Microbiol Biotechnol* 43:1039–1043, 1995.

26. J Ogawa, S Shimizu. β-Ureidopropionase with *N*-carbamoyl-α-L-amino acid amidohydrolase acitivity from an aerobic bacterium, *Pseudomonas putida* IFO 12996. *Eur J Biochem* 223:625–630, 1994.

27. C Syldatk, R Müller, M Pietzsch, F Wagner. Microbial and enzymatic production of D-amino acids from D,L-monosubstituted hydantoins. In: D Rozzell, F Wagner, eds. *Biocatalytic Production of Amino Acids and Derivatives*. München: Hanser Publishers, 1992, pp. 75–128.

28. C Syldatk, R Müller, M Siemann, K Krohm, F Wagner. Microbial and enzymatic production of L-amino acids from D,L-monosubstituted hydantoins. In: D Rozzell, F Wagner, eds. *Biocatalytic Production of Amino Acids and Derivatives*.München: Hanser Publishers, 1992, pp. 129–176.

29. J Ogawa, A Ryono, SX Xie, RM Vohra, R Indrati, H Miyakawa, T Ueno, S Shimizu. Separative preparation of the four stereoisomers of β-methylphenylalanine with *N*-carbamoyl amino acid amidohydrolases. *J Molec Catal B: Enzymatic* 12:71–75, 2001.

30. J Altenbuchner, M Siemann-Herzberg, C. Syldatk. Hydantoinases and related enzymes as biocatalysts for the synthesis of unnatural chiral amino acids. *Curr Opin Biotechnol* 12:559–563, 2001.

31. H Nanba, Y Ikenaka, Y Yamada, K Yajima, M Takano, K Ohkubo, Y Hiraishi, K Yamada, S Takahashi. Immobilization of *N*-carbamyl-D-amino acid amidohydrolase. *Biosci Biotechnol Biochem* 62:1839–1844, 1998.

32. Y Ikenaka, H Nanba, K Yajima, Y Yamada, M Takano, S Takahashi. Thermostability reinforcement through a combination of thermostability-related mutations of *N*-carbamyl-D-amino acid amidohydrolase. *Biosci Biotechnol Biochem* 63:91–95, 1999.

33. J Ogawa, CL Soong, M Honda, S Shimizu. Novel metabolic transformation pathway for cyclic imides in *Blastobacter* sp. A17p-4. Appl *Environ Microbiol* 62:3814–3817, 1996.

34. J Ogawa, CL Soong, M Honda, S Shimizu. Imidase, a dihydropyrimidinase-like enzyme involved in the metabolism of cyclic imides. *Eur J Biochem* 243:322–327, 1997.

35. CL Soong, J Ogawa, M Honda, S Shimizu. Cyclic-imide-hydrolyzing activity of D-hydantoinase from *Blastobacter* sp. strain A17p-4. *Appl Environ Microbiol* 65:1459–1462, 1999.

36. J Ogawa, CL Soong, M Ito, T Segawa, T Prana, MS Prana, S Shimizu. 3-Carbamoyl-a-picolinic acid production by imidase-catalyzed regioselsective hydrolysis of 2,3-pyridinedicarboximide in a water-organic solvent, two-phase system. *Appl Microbiol Biotechnol* 54:331–334, 2000.

37. CL Soong, J Ogawa, S Shimizu. A novel amidase (half-amidase) for half-amide hydrolysis involved in the bacterial metabolism of cyclic imides. *Appl Environ Microbiol* 66:1947–1953, 2000.

38. CL Soong, J Ogawa, H Sukiman, T Prana, MS Prana, S Shimizu. Distribution of cyclic imide-transforming activity in microorganisms. *FEMS Microbiol Lett* 158:51–55, 1998.

39. J Ogawa, CL Soong, M Ito, S Shimizu. Enzymatic production of pyruvate from fumarate–an application of microbial cyclic-imide-transforming pathway. *J Molec Catal B: Enzymatic* 11:355–359, 2001.

40. J Ogawa, W Tu, CL Soong, M Ito, T Segawa, S Shimizu. Pyruvate production through microbial cyclic imide transformation. Proceedings of 4th International Symposium on Green Chemistry in China, Jinan, 2001, p. 291.

41. YS Yang, S Ramaswamy, WB Jakoby. Rat liver imidase. *J Biol Chem* 268:10870–10875, 1993.

42. H Mitsuya, S Border. Inhibition of the *in vivo* infectivity and cytopathic effect of human T-lymphotrophic virus type III/lymphadenopathy-associated virus (HTLV-III/LAV) by 2',3'-dideoxynucleosides. *Proc Natl Acad Sci USA* 83:1911–1915, 1986.

43. S Koga, J Ogawa, LY Chen, YM Choi, H Yamada, S Shimizu. Nucleoside oxidase, a hydrogen peroxide–forming oxidase, from *Flavobacterium meningosepticum*. *Appl Environ Microbiol* 63:4282–4286, 1997.

44. J Ogawa, S Takeda, SX Xie, H Hatanaka, T Ashikari, T Amachi, S Shimizu. Purification, characterization, and gene cloning of purine nucleosidase from *Ochrobactrum anthropi*. *Appl Environ Microbiol* 67:1783–1787, 2001.

45. NL Edwards, IH Fox. Disorders associated with purine and pyrimidine metabolism. *Spec Top Endocrinol Metab* 6:95–140, 1984.

46. TA Krenitsky, GW Koszalka, JV Tuttle. Purine nucleoside synthesis, an efficient method employing nucleoside phosphorylases. *Biochemistry* 20:3615–3621, 1981.

47. K Yokozeki, H Shirae, K Kubota. Enzymatic production of antivial nucleosides by the application of nucleoside phosphorylase. *Ann N Y Acad Sci* 613:757–759, 1990.

48. J Ogawa, K Saito, T Sakai, N Horinouchi, T Kawano, S Matsumoto, M Sasaki, Y Mikami, S Shimizu. Microbial production of 2-deoxyribose 5-phosphate from acetaldehyde and triosephosphate for the synthesis of 2'-deoxyribonucleosides. *Biosci Biotechnol Biochem* 67:933–936, 2003.

49. N Horinouchi, J Ogawa, T Sakai, T Kawano, S Matsumoto, M Sasaki, Y Mikami, S Shimizu. Construction of deoxyriboaldolase-overexpressing *Escherichia coli* and its application to 2-deoxyribose 5-phosphate synthesis from glucose and acetaldehyde for 2'-deoxyribonucleoside production. *Appl Environ Microbiol* 69:3791–3797, 2003.

50. N Horinouchi, S Sakai, T Kawano, J Ogawa, S Shimizu. Efficient production of 2-deoxyribose 5-phosphate from glucose using glycolysis of baker's yeast. Proceedings of Annu Meet 2002 Soc Biotechnol Japan, Osaka, 2002, p. 125.

51. J Ogawa, T Kawano, N Horinouchi, S Sakai, S Shimizu. Enzymatic synthesis of 2'-deoxyribonucleoside from microbially prepared 2-deoxyriobse 5-phosphate. Proceedings of Annu Meet Soc Biosci Biotechnol Agrochem Japan, Tokyo, 2003, p. 182.

52. Y Shinmen, S Shimizu, K Akimoto, H Kawashima, H Yamada. Production of arachidonic acid by Mortierella fungi: selection of a potent producer and optimization of culture conditions for large-scale production. *Appl Microbiol Biotechnol* 31:11–16, 1989.

53. S Shimizu, H Kawashima, K Akimoto, Y Shinmen, H Yamada. Conversion of linseed oil to an eicosapentaenoic-acid containing oil by *Mortierella alpina* 1S-4 at low temperature. *Appl Microbiol Biotechnol* 21:1–4, 1989.

54. M Certik, E Sakuradani, S Shimizu. Desaturase defective fungal mutants: useful tools for the regulation and overproduction of polyunsaturated fatty acids. *Trends Biotechnol* 16:500–505, 1998.

55. M Certik, S Shimizu. Biosynthesis and regulation of microbial polyunsaturated fatty acid production. *J Biosci Bioeng* 87:1–14, 1999.

56. J Ogawa, E Sakuradani, S Shimizu. Production of C20 polyunsaturated fatty acids by an arachidonic acid-producing fungus *Mortierella alpina* 1S-4 and related strains. In: TM Kuo, HW Gardner, eds. *Lipid Biotechnology*. New York: Mercel Dekker, 2002, pp. 563–574.

57. MW Pariza, Y Park, ME Cook. The biologically active isomers of conjugated linoleic acid. *Prog Lipid Res* 40:283–298, 2001.

58. JM Ntambi, Y Choi, Y Park Y, JM Peters, MW Pariza. Effects of conjugated linoleic acid (CLA) on immune responses, body composition and stearoyl-CoA desaturase. *Can J Appl Physiol* 27:617–628, 2002.

59. SF Chin, W Liu, JM Storkson, YL Ha, MW Pariza. Dietary sources of conjugated dienoic isomers of linoleic acid, a newly recognized class of anticarcinogens. *J Food Compos Anal* 5:185–197, 1992.

60. CR Kepler, KP Hirons, JJ Mcneill, SB Tove. Intermediates and products of the biohydrogenation of linoleic acid by *Butyrivibrio fibrisolvens*. *J Biol Chem* 241:1350–1354, 1966.

61. J Jiang, L Bjorck, R Fonden. Production of conjugated linoleic acid by dairy starter cultures. *J Appl Microbiol* 85:95–102, 1998.

62. S Kishino, J Ogawa, Y Omura, K Matsumura, S Shimizu. Conjugated linoleic acid production from linoleic acid by lactic acid bacteria. *J Am Oil Chem Soc* 79:159–163, 2002.

63. J Ogawa, K Matsumura, S Kishino, Y Omura, S Shimizu. Conjugated linoleic acid accumulation via 10-hydroxy-12-octadecaenoic acid during microaerobic transformation of linoleic acid by *Lactobacillus acidophilus*. *Appl Environ Microbiol* 67:1246–1252, 2001.

64. S Kishino, J Ogawa, A Ando, T Iwashita, T Fujita, H Kawashima, S Shimizu. Structural analysis of conjugated linoleic acid produced by *Lactobacillus plantarum*, and factors affecting isomer producton. *Biosci Biotechnol Biochem* 67:179–182, 2003.

65. S Kishino, J Ogawa, A Ando, Y Omura, S Shimizu. Ricinoleic acid and castor oil as substrates for conjugated linoleic acid production by washed cells of *Lactobacillus plantarum*. *Biosci Biotechnol Biochem* 66:2283–2286, 2002.

66. S Kishino, J Ogawa, S Shimizu. α- and γ-Linolenic acid transformation to conjugated fatty acids by *Lactbacillus plantarum* AKU1009a. Proceedings of Annu Meet Soc Biosci Biotechnol Agrochem Japan, Sendai, 2002, p. 296.

67. S Kishino, J Ogawa, A Ando, S Shimizu. Conjugated α-linolenic acid production from α-linolenic acid by *Lactobacillus plantarum* AKU1009a. *Eur J Lipid Sci Technol* 105:572–577, 2003.

68. A Ando, J Ogawa, S Kishino, S Shimizu. Conjugated linoleic acid production from ricinoleic acid by lactic acid bacteria. *J Am Oil Chem Soc* 80:889–894, 2003.

69. A Ando, S Kishino, S Sugimoto, J Ogawa, S Shimizu. Conjugated linoleic acid production from castor oil by *Lactobacillus plantarum* JCM 1551. *Enzyme Microb Technol* 35:40–45.

70. QS Li, J Ogawa, S Shimizu. Critical role of the residue size at position 87 in H_2O_2-dependent substrate hydroxylation activity and H_2O_2 inactivation of cytochrome P450BM-3. *Biochem Biophys Res Commun* 280:1258–1261, 2001.

71. QS Li, J Ogawa, RD Schmid, S Shimizu. Engineering cytochrome P450 BM-3 for oxidation of polycyclic aromatic hydrocarbons. *Appl Environ Microbiol* 67:5735–5739, 2001.

72. QS Li, J Ogawa, RD Schmid, S Shimizu. Residue size at position 87 of cytchrome P450 BM-3 determines its stereoselectivity in propylbenzene and 3-chlorostyrene oxidation. *FEBS Lett* 508:249–252, 2001.
73. FH Arnold. Combinatorial and computational challenges for biocatalyst design. *Nature* 409:253–257, 2001.

3

Biocatalyses in Microaqueous Organic Media

Tsuneo Yamane

3.1   Introduction

Enzymatic or microbial reactions are generally performed by enzymes or microorganisms that exist in the large excess of water. However, there are cases where yield or rate or productivity increases significantly by reducing water content in the reaction system. Most biocatalytic reactions using organic solvents are involved in this category.

Performing enzymatic or microbial reactions in organic media has several advantages as opposed to using aqueous medium, including the following:

- Shifting of thermodynamic equilibrium to favor synthesis over hydrolysis.
- Reduction in water-dependent side reaction (such as hydrolysis reaction in transfer reactions).
- Immobilization of the enzyme is often unnecessary (even if it is desired, merely physical deposition onto solid surfaces is enough).
- Elimination of microbial contamination (quite critical in view of industrial scale process).
- Increasing solubility of hydrophobic substrates.
- Recovery of product from low boiling-point solvents is easy and also the insoluble biocatalyst is easily separated.

Here organic media as the reaction system are classified into two categories:[1]

1. Solvent systems
2. Solvent-free systems

In the former system one or more substrates are dissolved in an inert organic solvent that does not participate in the reaction in any respect, but to provide an environment for the biocatalyst to exert its action on the dissolved substrate(s).

There are a number of cases where the former is the system of choice: for example, when the substrate is solid at the temperature of the reaction, when high concentration of the substrate is inhibitory for the reaction, when the solvent used yields a better environment (accelerating effect) for the enzyme, and so forth.

In the latter system, no other compounds but substrate(s) and enzyme are present in a reactor. In principle, one substrate can be used in a large excess over another, and if so, it may also act as a solvent for the other reactant. This may be alternatively named "neat" biotransformation.

One of the big attractions of enzymatic solvent-free synthesis is potentially very high volumetric productivity. This, however, does not apply to all reactions, and in many instances it may actually take a longer time to achieve the desired degree of conversion in the absence of added solvent. In this case, volumetric productivity in the reactor [(mass of product formed) (reactor volume)$^{-1}$ (time)$^{-1}$] should be calculated for both solvent-based and solvent-free systems using the same volume of the reaction mixture and the same amount of the enzyme in order to make an economically justified choice between the two. Similarly, no risk of solvent-induced inactivation of the biocatalyst in solvent-free system is another advantage, but the overall loss of the enzyme activity can still be significant if the reaction time is too long. It should be added that avoidance of organic solvents is particularly advantageous to the food industry where stringent legal regulations related to the use of organic solvents are in force. Also, no fireproof and explosion-proof equipment/procedures are necessary for the solvent-free processing and the environment in the factory is less hazardous to the health of workers.

Biocatalytic reactions in organic media are currently being studied very actively as interdisciplinary fields between organic chemistry and enzyme engineering, and between oleochemistry and enzyme engineering to synthesize or convert lipids, saccharides, peptide, chiral compounds, polymers, etc. A comprehensive monograph was published in 1996 reviewing the progress,[2] and methods and protocols related to this topic were summarized in monographs.[3–5]

Types of reactions involve: 1) bond formations of ester, amide, and glycoside by synthetic reaction as reverse ones of hydrolyses or by transfer reactions (transesterification, transpeptidylation, transglycosylation, etc), 2) oxidation and reduction, 3) C-O and C-N bond formation by addition/substitution reactions, 4) C-C bond formation, 5) polymerizations, etc. The use of biocatalysts for these reactions results in the best advantages in their selectivities involving enantio-selectivity, regio-selectivity, functional group selectivity, etc. The process may be made simpler because of these selectivities without having to introduce a protecting group and its deprotection afterward.

3.2 Enzymes

Many enzymes have been used for studying their actions in organic media, including lipases whose origins are bacteria, molds, yeasts, and mammals and plants to lesser extents, esterases, proteases such as thermolysin, -chymotrypsin, and subtilisin, peroxidase, phenol oxidase, alcohol dehydrogenase (yeast), and so on. As a microorganism in organic media, baker's yeast has been most often used.

3.2.1 Techniques Enabling Biocatalyst Highly Active in Organic Media

A number of techniques have been developed that enable biocatalysts active in organic media. These include:

1. Enzyme molecules are dissolved freely (some enzymes are soluble in glycerol or dimethylsulfoxide).
2. Enzyme molecules are derivatized with polyethylene glycol (PEG), or are formed as a complex with surfactant (they are called lipid-coated enzyme or surfactant-modified enzymes, or surfactant-enzyme complexes). All these modified enzymes are soluble in some organic solvents.
3. Enzymes are suspended or dispersed in organic media. If their dispersion is not uniform, it is deposited on the surface of fine solid particles beforehand.
4. The enzymes are confined in a reverse micelle.

5. The enzymes exist in aqueous phase inside micropores of porous solid particles. Enzymes are either free or immobilized.
6. The enzyme or microbial cells are entrapped in hydrophobic gel.
7. Microbial cells (wet or dry) having a particular enzyme are suspended in organic media, or they are first immobilized in macroporous supports followed by suspension or packed in a bioreactor.

Since a number of techniques have been developed as mentioned above, it can be said confidently that any enzyme can make its activity high in organic media. Techniques #2 and #4 exert high activity of the enzyme preparation, but its recovery or continuous usage will be difficult. Technique #3 will give us biocatalysts suitable for industrial usage because of its ease in recovery or of ability of continuous operation, although one must invent a method of pretreatment or addition of activity enhancer to get higher activity.

3.2.2 Reactivation

Powders of some commercial enzyme preparations exhibit low activities when they are suspended in organic media, but they are drastically reactivated when they have been re-lyophilized from their aqueous solutions containing proper amount of sugar alcohols,[6-8] surfactants,[9,10] fatty acids,[11] hydrocarbons,[12] etc. This phenomenon is related to the second technique mentioned above.

3.2.3 Purity

A factor that many researchers have paid little attention to, but that is very important, is the purity of the enzyme preparations they use. As will be mentioned later, a profile of the effect of a trace amount of water depends on the purity of the enzyme preparation, and the enzyme molecules are surrounded by large amounts of impurities remaining around them and influence directly the enzyme catalysis when the crude enzyme preparations are suspended in organic media (Figure 3.1a).[13] Therefore their catalytic activities are affected strongly by the nature and amount of the impurities. Crude enzyme powder exhibit enough activity when it is uniformly dispersed in organic media, but pure enzyme powder is not uniformly dispersed, and shows no activity unless three factors are optimized: water content, proper support material, and the addition of a reactivator (or activity enhancer) (Figure 3.1b).[6] Figure 3.2 indicates that

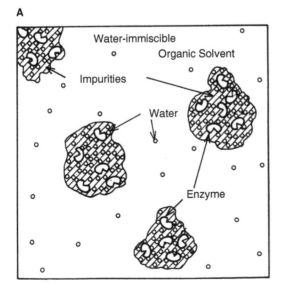
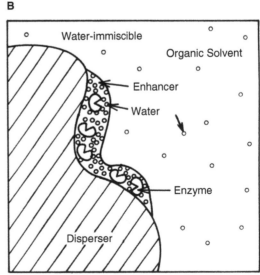

FIGURE 3.1 Schematic pictures of crude enzyme powder particles suspended in microaqueous medium (a), and the system of (pure enzyme + activity enhancer) deposited on a disperser particle which is suspended in microaqueous medium (b).[6]

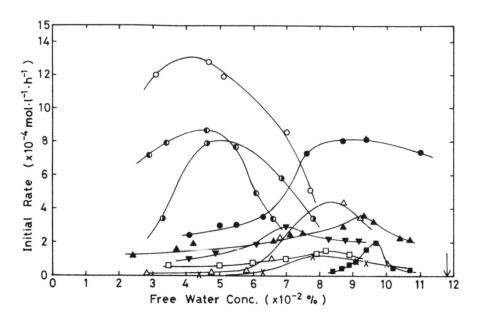

FIGURE 3.2 Effect of additive on the lactonization activity of the pure enzyme at various concentrations of free water. The vertical arrow indicates the solubility of water in benzene at 40°C (0.118%). —△—, no addition, i.e., enzyme plus celite; —O—, enzyme plus arabitol plus celite; —◑—, enzyme plus sorbitol plus celite; —◐—; enzyme plus erythritol plus celite; —●—, enzyme plus phosphotidylcholine plus celite; —▲—, enzyme plus lactose plus celite; —×—, enzyme plus BSA plus celite; —■—, enzyme plus casein plus celite; —▽—, enzyme plus PVA plus celite; —□—, enzyme plus dextran plus celite.

the rate of lactonization by *Pseudomonas fluorescens* lipase is increased significantly when a sugar alcohol such as erythritol, arabitol, or sorbitol is added before freeze-drying together with celite powder.[6] In this case, the pure lipase is not dispersed at all without celite powder, and even if the enzyme is deposited on the celite powder, the catalytic activity is very low.

3.2.4 Properties of Enzymes

Properties of an enzyme change more or less in organic media, including thermal stability and various specificities (selectivities):

1. Thermal stability (half life), $t_{1/2}$—Inactivation of an enzyme is regarded as change of its three-dimensional structure from an active form to an inactive form, and often involves water molecules during this structural change. Such structural changes do not occur in a complete anhydrous state so that its thermostability is enhanced. Therefore, the value of its half life, $t_{1/2}$, becomes much longer in very anhydrous conditions than in an aqueous solution, and approaches to $t_{1/2}$ value of its aqueous solution when the water content increases.[14] Bear in mind, however, that its catalytic activity in nearly complete anhydrous state drops down to a low level, as will be mentioned later. In any cases, $t_{1/2}$ should be evaluated under a strict microaqueous condition, also taking into consideration the nature of the organic solvent.

2. Substrate specificity, k_{cat}/K_m—Substrate specificity of a given enzyme is quantitatively evaluated by the value of k_{cat}/K_m. In nonaqueous enzymology its value changes in various organic solvents.[15] This implies that the specificity of an enzyme can be varied by selecting the kind of organic solvent without applying protein engineering and/or screening another new enzyme from nature.

3. Enantioselectivity, E—Enantioselectivity of a given enzyme is quantitatively evaluated by the value of the so-called enatiomeric ratio, or E value, which is defined as the relative ratio of k_{ca}/K_m values

of *R*-and *S*-isomers.[16] Optical purity of the product, the *ee* value, increases with any increase in the *E* value. It is said that the *E* value should be more than 100 to produce optically pure product industrially. The *E* value of an enzyme changes in various organic solvents. Attempts have been made to correlate the *E* value of enzymes with physicochemical properties of organic solvents, such as dielectric constant, dipole moment, and molar volume.[17,18]

A number of factors affect enzymatic reactions in organic media, among which a trace amount of water and the nature of organic solvents will be discussed briefly.

3.3 Water

Special attention should be paid to the water content when one carries out biocatalytic reactions in organic media because a trace amount of water strongly affects, among other things: 1) reaction rate, 2) yield and selectivity, and 3) operational stability. Numerous studies in many laboratories have shown that a complete depletion of water from the reaction system results in the nonoccurrence of the biochemical reaction. Water seems essential for a biocatalyst to display its full catalytic activity. Enzyme as a protein needs a fluctuation or perturbation to exert its catalytic activity, and water bound to the protein allows its fluctuation. Completely dried enzymes cannot fluctuate.

3.3.1 The "Microaqueous" Concept

In order to emphasize the importance of a trace amount of water for biocatalyst utilization in organic media, the author proposed a novel technical term "microaqueous" to depict the reaction system more accurately.[19–21] This technical term implies that the system is neither aqueous nor nonaqueous/anhydrous. In its strictly scientific sense, nonaqueous/anhydrous implies the complete absence of water. In between these extremities (aqueous and nonaqueous), there is a state where the system has little water. This state could be called "microaqueous." Thus, strictly speaking, an organic solvent in which the biocatalyst works is a microaqueous organic solvent, or more simply a microaqueous solvent. Putting the adjective "microaqueous" in front of (organic) solvent suggests properly that the (organic) solvent contains a trace amount of water.

The effects of a trace amount of water on lipase-catalyzed esterification and transesterification are shown in Figure 3.3a[13] and Figure 3.3b, respectively. In the range of very low water content, the reaction rates are limited by hydration of the enzyme proteins (hydration-limited), whereas in the range of excess water, they are limited by the reverse reaction (reverse reaction-limited). In the transesterification reaction (Figure 3.3b), the hydrolytic side reaction appears at a higher water content, resulting in a decrease of the product yield.

3.3.2 Existing Places of Water and Equilibrium

In the system of enzyme powder suspended in organic media (state #3 in Section 3.2.1) as the simplest case, a general moisture balance is given by

$$C_{w, total} = Y_{ap} C_{ap} + Y_{im} C_{im} + C_{w, free} \qquad (3.1)$$

where

$C_{w, total}$ = total water concentration [g H$_2$O/mL]
Y_{ap} = amount of water bound to the active protein [g H$_2$O/g dry protein]
C_{ap} = concentration of the active protein (enzyme)[g dry protein/mL]
Y_{im} = amount of water bound to the inert material [g H$_2$O/g dry inert material]
C_{im} = concentration of the inert material [g dry material/mL]
$C_{w, free}$ = concentration of the free water dissolved in organic medium [g H$_2$O/mL]

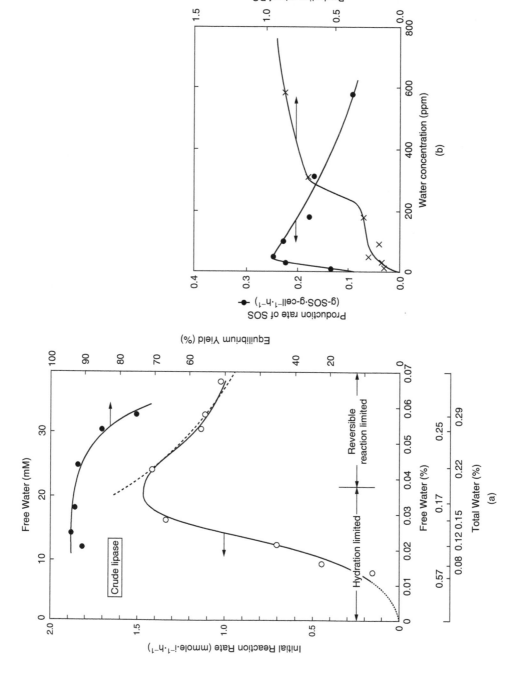

FIGURE 3.3 Effect of trance amount of water on lipase-catalyzed esterification (a)[13] and transesterification (b).[30]

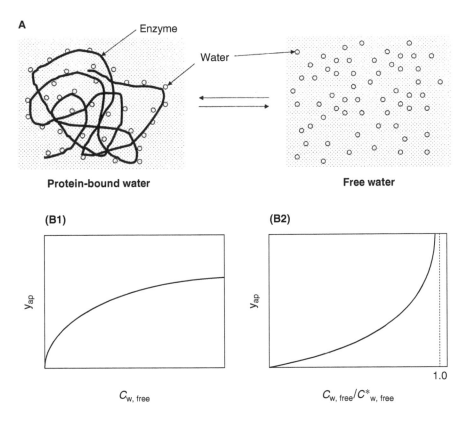

FIGURE 3.4 Dynamic equilibrium between enzyme-bound water and free water in an organic medium (A), and general profiles of the adsorption isotherms of water between the enzyme and water-miscible organic solvent (B1) and between enzyme and water-immiscible organic solvent (B2).[20] $C^*_{w,free}$ is water solubility of the organic solvent.

By using Equation 3.1, it is possible to explain the characteristic difference in dependency of the reactivity of variously graded enzyme on the total moisture content. The optimal moisture level will decrease with an increase in the purity of the enzyme preparation, and the optimal total moisture content will increase when the crude enzyme preparation is used. In any case, the optimal total moisture content will increase as the amount of the enzyme added to the reaction mixture increases. Some researchers discuss the effect of water as a function of added water or total water, but they are not scientific variables. It is important to realize that the water only bound to the enzyme molecule, y_{ap}, affects directly its catalytic activity, whereas the water dissolved freely in an organic medium, $C_{w,free}$, only participates in the reaction as a substrate or as a product so that the conversion or yield is determined mostly by the free water concentration $C_{w,free}$.

There is a thermodynamic equilibrium between Y_{ap} and $C_{w,free}$ (Figure 3.4A).[20] Water-miscible organic solvents exhibit Langmuir-type adsorption isotherms having saturation phenomena (Figure 3.4B1):

$$y_{ap} = \frac{y_{ap,max} C_{w,free}}{K + C_{w,free}} \tag{3.2}$$

On the other hand, water-immiscible organic solvents generally show simply rising curves having no saturation (Figure 3.4B2).

They are expressed by BET-type multiple-layer adsorption isotherm:

$$Y_{ap} = \frac{y_{ap,max} C_{w,free}/C^*_{w,free}}{(1-C_{w,free}/C^*_{w,free})(1-C_{w,free}/C^*_{w,free}+KC_{w,free}/C^*_{w,free})} \tag{3.3}$$

The curves shown in Figure 3.4 suggest that water hydrated on the enzyme, Y_{ap}, can be controlled by changing $C_{w,free}$.

3.3.3   Water Activity, a_w

Effects of water content on the activities of an enzyme in various organic solvents are partly summarized by applying a parameter named water activity, a_w.[22] a_w is a thermodynamic parameter, defined as in an equilibrium the water vapor pressure, of gas phase, p_w, divided by the saturated water vapor pressure of

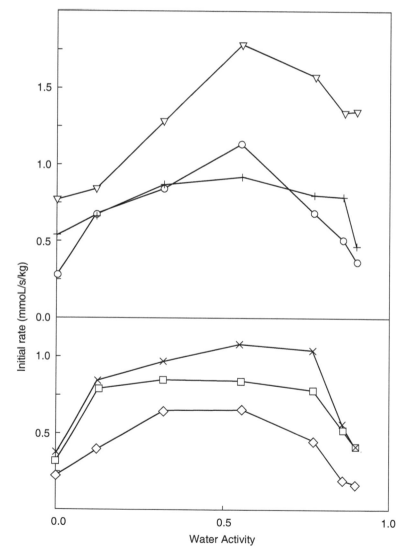

FIGURE 3.5 Activity of Lipozyme catalyst as a function of water activity in a range of solvents.[24] Hexane(∇), toluene($\times$), trichloroethylene(O), isopropyl ether( ), pentan-3-one($\Diamond$), and none, i.e., liquid reactant mixture(+).

the gas phase p_w^*, in a closed vessel of a constantly controlled temperature, in which substances are put for a long time to reach an equilibrium.

$$a_w \, p_w/p_w^*, \, 0 < a_w < 1 \tag{3.4}$$

The substances, powdered enzyme, organic solvent-containing substrates, and an aqueous saturated solution of a mineral salt are kept separate for the study of biocatalysis in organic solvent. After reaching an equilibrium at which all the components have some value of a_w, the powdered enzyme and the reaction solvent are mixed to measure the initial reaction rate. By changing the kind of the mineral salts, one can change the value of a_w between 0 and 1. A curve is obtained between a_w and $C_{w,free}$ (or more precisely molar fraction of water), which is characteristic for each of various organic solvents.[23]

The reaction rate with suspended enzyme catalyst shows similar dependence on a_w in different organic solvents, as seen in Figure 3.5.[24] However, as also seen in Figure 3.4, not all the data are on a single curve, but the absolute rates are scattering depending on the nature of the organic solvents, indicating that a_w is not almighty. There is a case where a_w fails to predict critical hydration level for enzyme activity in polar organic solvents.[23]

3.4 Organic Solvents

When wanting to apply a solvent system, one must choose a suitable solvent from the vast kinds of organic solvents prior to carrying out an enzymatic reaction in an organic solvent. From active basic research having been carried out in the past two decades, there has been progress in our understanding of properties of enzymes in organic media (mentioned in Section 3.2), and in how organic solvents affect them. Some researchers call the achievement "medium engineering."

Among numerous kinds of organic solvents, those often used for enzymatic reactions are not many, and may be classified into three categories (Table 3.1),[25] in view of the importance of water solubility of the organic solvents concerned:

1. Water-miscible organic solvents—Any cosolvent system having 0–100% ratio of the solvent/water can be prepared from this kind of solvent. Note that some organic solvents having limited solubility

TABLE 3.1 Classification of Solvents Commonly Used for Biocatalytic Reactions in Organic Media

1. Water-miscible organic solvents
   - Methanol, ethanol, ethylene glycol, glycerol, N,N'-dimethylformamide, dimethylsulfoxide, acetone,
   - formaldehyde, dioxane, etc.
2. Water-immiscible organic solvents (water solubility·g/l at the temperature indicated)
   - alcohols
   - (*n*-, *iso*-) proppyl alcohol, (*n*-, *s*-, *t*-) butyl alcohol, (*n*-, *s*-, *t*-) amyl alcohol, *n*-octanol, etc.
   - esters
   - methyl acetate, ethyl acetate (37.8, 40°C), *n*-butyl acetate, hexyl acetate, etc.
   - alkyl halides
   - methylene chloride (2, 30°C), chloroform, carbon tetrachloride, trichloroethane (0.4, 40°C), chlorobenzene, (*o*-, *m*-, *p*-) dichlorobenzene, etc.
   - ethers
   - diethyl ether (12, 20°C; 14.7, 25°C), dipropyl ether, diisopropyl ether, dibutyl ether, dipentyl ether, tetrahydrofuran, etc.
3. Water-insoluble organic solvents (water solubility·ppm at the temperature indicated)
   - aliphatic hydrocarbons
   - *n*-hexane (320, 40°C), *n*-heptane (310, 30°C), *n*-octane, isooctane (180, 30°C), etc.
   - aromatic hydrocarbons
   - benzene (600, 25°C; 1200, 40°C), toluene (300, 25°C; 880, 30°C), etc.
   - alicyclic hydrocarbons
   - cyclohexane (160, 30°C), etc.

at an ambient temperature, and hence not regarded as water-miscible, become miscible at an elevated temperature.

2. Water-immiscible solvents—These organic solvents have noticeable but limited solubility of water, ranging roughly 0.1–10% of their solubilities. The water solubility is of course increased as the temperature is raised.

3. Water-insoluble organic solvents—These solvents are also water-immiscible and have very low water solubility, so that they are regarded as water-insoluble, i.e., water is practically insoluble in these organic solvents. Most aliphatic and aromatic hydrocarbons belong to this category.

In Table 3.1 organic solvents often used for enzymatic reactions are listed together with their water solubilities (although not for all of them).

As physico-chemical properties that may affect enzyme activity, hydrophobicity parameter and dielectric constant have been studied.

3.4.1 Hydrophobicity, Log*P*

A hydrophobicity parameter, log*P*, was first proposed for microbial epoxidation of propene and 1-butene.[26,27] Log*P* is the logarithm of *P*, where *P* is defined as the partition coefficient of a given compound in the standard *n*-octanol/water two phase:

$$P \equiv \frac{\text{Solubility of a given compound in } n\text{-octanol phase}}{\text{Solubility of a given compound in water phase}} \tag{3.5}$$

As a general rule, biocatalysis in organic solvents is low in polar solvents having a log*P* < 2, moderate in solvents having a log*P* between 2 and 4, and high in apolar solvents having a log*P* > 4.27. The three divisions correspond roughly to the three classifications in Table 3.1. The correlation between polarity and activity parallels the ability of organic solvents to distort the essential water layer bound to the enzyme that stabilizes the enzyme. Since log*P* can easily be determined experimentally, or can be estimated from hydrophobic fragmental constants, many biotechnologists have tried to correlate the effects of organic solvents on biocatalysts they have studied with the log*P* approach. Their results have been successful, not completely but only partially. A number of exceptions to the "log*P* rule" have been reported. In discussing the enzyme activity by the log*P* approach, water content should be strictly controlled.

3.4.2 Dielectric Constant (or Dipole Moment), *ε* (or *D*)

Interactions between an enzyme and a solvent in which the enzyme is suspended are mostly noncovalent ones as opposed to interactions in water. These strong noncovalent interactions are essentially of electrostatic origin, and thus, according to Coulomb's law, their strength is imposed dependently on the dielectric constant, ε (which is higher for water than for almost all organic solvents). It is likely that enzymes are more rigid in anhydrous solvents of low ε than in those of high ε. Thus, ε of a solvent can be used as a criterion of rigidity of the enzyme molecule. For the enzyme to exhibit its activity, it must be dynamically flexible during its whole catalytic action so that its activity in a solvent of low ε should be less than in a solvent of higher ε. On the other hand, its selectivity (or specificity) becomes higher when its flexibility decreases so that the enzyme selectivity in a solvent of lower ε should be higher than in a solvent of higher ε.

3.5 Bioreactor System of Microaqueous Organic Media

When finding out a highly active and long-lived biocatalyst for a useful biochemical reaction in an organic medium on laboratory scale, one will construct a bioreactor system with an aim of its commercialization on large scale. In this instance, what is different from an ordinary aqueous bioreactor system is an optimal control of a trace amount of water involved in the system. Among vast varieties of biochemical reactions, esterifications and transesterifications both catalyzed by lipases have been most extensively studied.

3.5.1 Esterification

There are many industrially important esters, including wax esters (composed of aliphatic mono alcohols and fatty acids), monoglycerides, fatty acid esters of polyglycerol, sugar esters, steroid esters such as cholesterol palmitate, alcohol ester of terpenoids such as glanyl butylate, and so forth. All these esters can be synthesized by lipases. Since lipase's action is reversible, ester can be synthesized by acid and alcohol. The anhydrous state is thermodynamically preferable in terms of the ester yield, but lipases require a trace amount of water to exhibit their catalytic activity so that the reaction system should be microaqueous. Water liberated during the ester synthesis must be removed to achieve the high ester yield. During the early stage of the ester reaction, relatively higher water content is allowable to attain a higher reaction rate, but at the later and final stages it must be minimized to raise the product yield even with sacrifice of the reduced reaction rate. Thus, time-dependent optimal water content profile is inferred.

To eliminate water from the reaction system, a number of techniques are known, including reduced pressure at room temperature, distillation at reduced pressure, azeotropic distillation, flash evaporation at ambient pressure or at reduced pressure, purging dry gas (dry gas bubbling), pervaporation, and so forth. The reduced pressure at an ambient temperature (vacuum) is the most feasible, but an invention must be involved when the reaction mixture becomes viscous with the progress of the reaction. Such a case can be seen in a lipase-catalyzed estolides synthesis by oligo-condensation or licinoleic acid.[28] The bioreactor system consisted of two units: the enzymatic reaction unit and the water control unit. The water content control unit was an aspirator with which a thin film flash evaporator was connected (Figure 3.6). Highly qualified estolides without coloration could be produced at a high rate by repeated batch operation using the same immobilized lipase.

Pervaporation is a kind of new membrane separation technique that has rapidly been developed recently. Its major advantage is the ability to remove trace water selectively while confining organic solvent without loss in the reactor even if it is volatile.[29]

Figure 3.7 illustrates conceptually a bioreactor system of bioesterification incorporating a pervaporation unit.

3.5.2 Transesterification

Transesterification is another lipase-catalyzed reaction intensively studied and is subclassified into alco-holysis, acidolysis, interesterification, and aminolysis. Products by some of them have been implemented industrially, including coco-butter substitutes, 1,3-dibehenoyl-2-oleoyl glycerol, and 1,3-dioleoyl-2-palmitoylglycerol.

Transesterification reaction has various options with respect to 1) substrate (free fatty acid or fatty acid ester which reacts with fats and oils or ester); 2) state of biocatalyst (free or immobilized lipase, or free or immobilized dry microbial mass having lipase); 3) reaction system (solvent-based or solvent-free); 4) bioreactor configuration (stirred tank, or packed bed, or fluidized bed); and 5) bioreactor operation (batch, or repeated batch, or fed-batch, or continuous). Whichever option is adopted, an optimal control of the trace amount of water is critical to achieve both high yield and high productivity.

The merits of using immobilized dry fungal mass having lipase are 1) an omission of both separation and purification of the lipase enzyme from the culture and 2) a longer life of the cell-bound lipase than the separated free or immobilized one, both of which are expected to result in the reduction of the production cost. *Rhizopus chinensis* was cultivated in the presence of biomass support particles (BSPs) to get dry fungal mass having lipase, and a sophisticated bioreactor system shown in Figure 3.8 was constructed to investigate quantitatively the effect of water concentration.[30] Later operational stability of the cell-bound lipase was studied by keeping microaqueous level at various values, indicating that its half life was ca. 50 days at the optimal free water concentration of 100 ppm.[31]

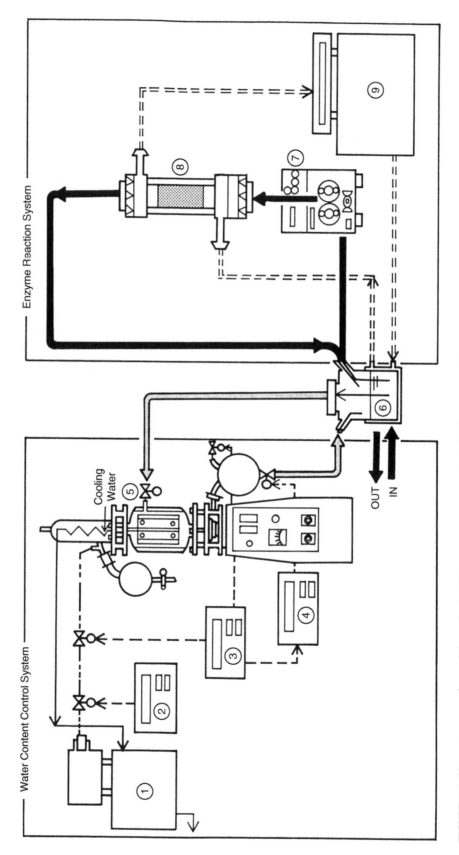

FIGURE 3.6 Bioreactor system for estolide synthesis with water content control unit.[28] (1) Aspirator, (2) vacuum controller, (3) and (4) timers, (5) flash evaporator, (6) substrate tank, (7) feed pump, (8) reaction column, (9) water bath.

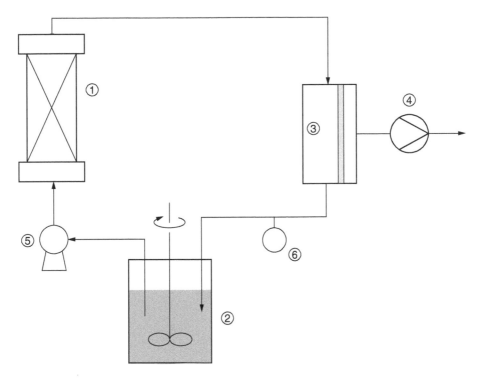

FIGURE 3.7 A schematic bioreactor system of bioesterification incorporating a pervaporation unit. (1) packed-bed bioreactor containing immobilized lipase, (2) storage tank of the reaction mixture, (3) pervaporation membrane module, (4) vacuum pump, (5) liquid circulation pump, (6) pressure gauge.

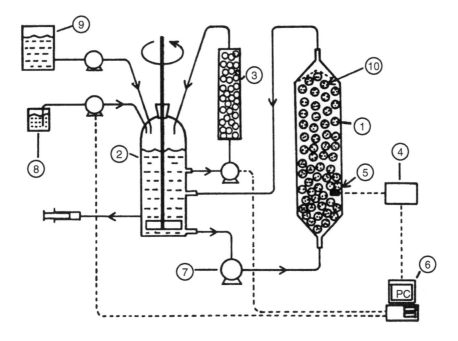

FIGURE 3.8 A laboratory–scale bioreactor system for the interesterification of fats and oils by immobilized fungus at constant water concentration.[30] (1) Fluidized-bed type bioreactor, (2) agitator, (3) silica gel column, (4) moisture analyzer, (5) sensor, (6) personal computer, (7) pump, (8) storage tank of water, (9) reactants and diluents, (10) immobilized fungi.

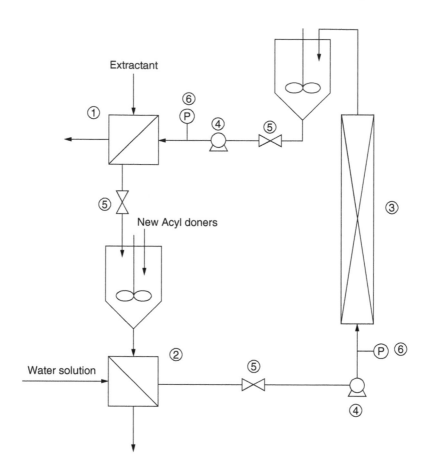

FIGURE 3.9 Production process of structured lipids, including a packed-bed reactor for reaction and two membrane modules for separation and water control.[32] (1) membrane module for removing the released medium-chain fatty acids, (2) membrane module for regulating the water control of the substrates, (3) packed-bed column of the immobilized lipase, (4) pumps, (5) valves, (6) pressure gauges.

Membrane bioreactors are also one of promising bioreactor configurations. A bioreactor system using two hollow-fiber membrane modules together with a packed-bed column in the system has been proposed for the production of structured lipids form fats and fatty acid (Figure 3.9).[32] One membrane module is used for separating undesired fatty acids, and the other is used to regulate water content of the substrate. To control the microaqueous level of the substrate feed, contacting with a saturated salt solution having a definite water activity through the membrane is possible.

References

1. T Yamane. Solvent-free biotransformations of lipids. In: EN Vulfson, PJ Halling, HC Holland, ed. *Enzymes in Nonaqueous Solvents, Methods and Protocols*. New Jersey: Humana Press, 2001, pp. 509–516.
2. AMP Koshinen, AM Klibanov, ed. *Enzymatic Reactions in Organic Media*. London: Academic & Professional (An Imprint of Chapman & Hall), 1996.
3. MN Gupta, ed. *Methods in Non-Aqueous Enzymology*. Basel-Boston-Berlin: Birkhäuser Verlag, 2000.
4. EN Vulfson, PJ Halling, HC Holland, ed. *Enzymes in Nonaqueous Solvents, Methods and Protocols*. New Jersey: Humana Press Inc., 2001.

5. UT Bornscheuer, ed. *Enzymes in Lipid Modification*. Weinheim: Wiley-VCH, 2000.

6. T Yamane, T Ichiryu, M Nagata, A Ueno, S Shimizu. Intermolecular esterification by lipase powder in microaqueous benzene: Factors affecting activity of pure enzyme. *Biotechnol Bioeng* 36: 1063–1069, 1990.

7. K Dabulis, AM Klibanov. Dramatic enhancement of enzymatic activity in organic solvents by lyoprotechtants. *Biotechnol Bioeng* 41: 566–571, 1993.

8. Aö Triantafyllou, E Wehtje, P Aldercreutz, B Mattiasson. Effects of sorbitol addition on the action of free and immobilized hydrolytic enzymes in organic media. *Biotechnol Bioeng* 45: 406–414, 1995.

9. M Goto, H Kameyama, M Goto, M Miyata, F Nakashio. Design of surfactants suitable for surfactant-coated enzymes as catalysts in organic media. *J Chem Eng Japan* 26: 109–111, 1993.

10. S Basheer, K Mogi, M Nakajima. Surfactant-modified lipase for the catalysis of the interesterification of triglycerides and fatty acids. *Biotechnol Bioeng* 45: 187–195, 1995.

11. DG Kenneth, M Nakajima. Evaluation of immobilized modified lipase: Aqueous preparation and reaction studies in *n*-hexane. *J Am Oil Chem Soc* 75: 1519–1526, 1998.

12. T Maruyama, M Nakajima. Effect of hydrocarbon-water interfaces on synthetic and hydrolytic activities of lipases. *J Biosci Bioeng* 92: 242–247, 2001.

13. T Yamane, Y Kojima, T Ichiryu, M Nagata, S Shimizu. Intermolecular esterification by lipase powder in microaqueous benzene: Effect of moisture content. *Biotechnol Bioeng* 34: 838–843, 1989.

14. A Zaks. New enzymatic properties in organic media. In: AMP Koshinen, AM Klibanov, ed. *Enzymatic Reactions in Organic Media*. London: Blackie Academic & Professional (An Imprint of Chapman & Hall), 1996, pp. 70–93.

15. A. Zaks, AM Klibanov. Substrate specificity of enzymes in organic solvents vs. water is reversed. *J Am Chem Soc* 108: 2767–2768, 1986.

16. CS Chen, Y Fujimoto, G Girdaukas, CJ Sih. Quantitative analysis of biochemical kinetic resolutions of enantiomers. *J Am Chem Soc* 104: 7294–7299, 1982.

17. T Sakurai, AL Margolin, AJ Russell, AM Klibanov. Control of enzyme enantioselectivity by the reaction medium. *J Am Chem Soc* 110: 7236–7237, 1988.

18. K Nakamura, M Kinoshita, A Ohno. *Tetrahedron* 50: 4681–4690, 1994.

19. T Yamane. Enzyme technology for the lipids industry: An engineering overview. *J Am Oil Chem Soc* 64: 1657–1662, 1987.

20. T Yamane, Y Kozima, T Ichiryu, S Shimizu. Biocatalysis in a microaqueous organic solvent. *Ann New York Acad Sci* 542: 282–293, 1988.

21. T Yamane. Importance of moisture control for enzymatic reactions in organic solvents: A novel concept of 'microaqueous'. *Biocat* 2: 1–9, 1988.

22. P Halling. High-affinity binding of water by proteins is similar in air and in organic solvents. *Biochim Biophys Acta* 1040: 225–228, 1990.

23. G Bell, AEM Janssen, PJ Halling. Water activity fails to predict critical hydration level for enzyme activity in polar organic solvents: Interconversion of water concentrations and activities. *Enz Microb Technol* 20: 471–477, 1997.

24. RH Valivety, P Halling, AR Macrea. Reaction rate with suspended lipase catalyst shows similar dependence on water activity in different organic solvents. *Biochim Biophys Acta* 1118: 218–222, 1992.

25. T Yamane. Factors affecting activity of enzyme in organic solvent (in Japanese). *Nippon Nogeikagaku Kaishi,* 65: 1103–1106, 1991.

26. C Laane, S Baeren, K Vos. On optimizing organic solvents in multi-liquid-phase biocatalysis. *Trends Biotechnol* 3: 251–252, 1985.

27. C Laane, S Boeren, K Vos, C Feeger. Rules for optimization of biocatalysis in organic solvents. *Biotechnol Bioeng* 30: 81–87, 1987.

28. Y Yoshida, M Kawase, C Yamaguchi, T Yamane. Enzymatic synthesis of estolides by a bioreactor. *J Am Oil Chem Soc* 74: 261–267, 1997.

29. SJ Kwon, KM Song, WH Hong, JS Rhee. Removal of water produced from lipase-catalyzed esterification in organic solvent by pervaporation. *Biotechnol Bioeng* 46: 393–395, 1995.
30. S Kyotani, H Fukuda, Y Nojima, T Yamane. Interesterification of fats and oils by immobilized fungus at constant water concentration. *J Ferment Technol* 66: 567–575, 1988.
31. S Kyotani, T Nakashima, E Izumoto, H Fukuda. Continuous interesterification of oils and fats using dried fungus immobilized in biomass support particles. *J Ferment Bioeng* 71: 286–288, 1991.
32. X Xu. Enzyme bioreactors for lipid modifications. *Inform* 11: 1004–1012, 2000.

4

Biocatalysts for the Epoxidation and Hydroxylation of Fatty Acids and Fatty Alcohols

Steffen C. Maurer

Rolf D. Schmid

4.1 Introduction

4.1.1 Attractiveness of Enzymatic Oxyfunctionalization of Fatty Acids and Their Derivatives

Whereas most products of the chemical industry are based on petrochemical feedstocks, considerable efforts have been made during the past few decades to use renewable resources as industrial raw materials. Besides polysaccharides and sugars, plant oils and animal fats play an important role in such programs because of their ready availability (present production is >100 million t/a, which could be increased on demand) of which the lion's share is used for nutrition (~85 million t/a), whereas ~15–20 million t/a are used for the synthesis of polymers, surfactants, emollients, lubricants, bio-diesel, emulsifiers, etc.[1]

From a chemical point of view, most natural triglycerides offer just two reactive sites: the ester group and the double bonds of unsaturated fatty acids. In fact, the chemistry of fats and oils is largely focused on the ester group, which can be hydrolyzed or catalytically reduced, leading to glycerol and fatty acids or fatty alcohols, respectively. Reactions involving the alkyl chain or double bonds of triglycerides, fatty acids, fatty alcohols or their derivatives represent far less than 10% of today's oleochemistry, with the production of sulfonated fatty alcohols and their derivatives being a major process of this kind. Oxidation

FIGURE 4.1 Prileshajev epoxidation: generation of short-chain peroxy acids.

reactions at the alkyl or alkenyl chains would be highly desirable as they would lead to oleochemicals with new properties, but the methods available today lack selectivity and require harsh conditions. Notable exceptions are the epoxidation of unsaturated plant oils and the synthesis and use of a few hydroxy fatty acids.

State of the art: Plant oil epoxidation—The Prileshajev epoxidation (Figure 4.1 and Figure 4.2) of unsaturated plant oils (predominantly soybean oil) is used worldwide for the production of more than 200,000 t/a of epoxidized soybean oil.

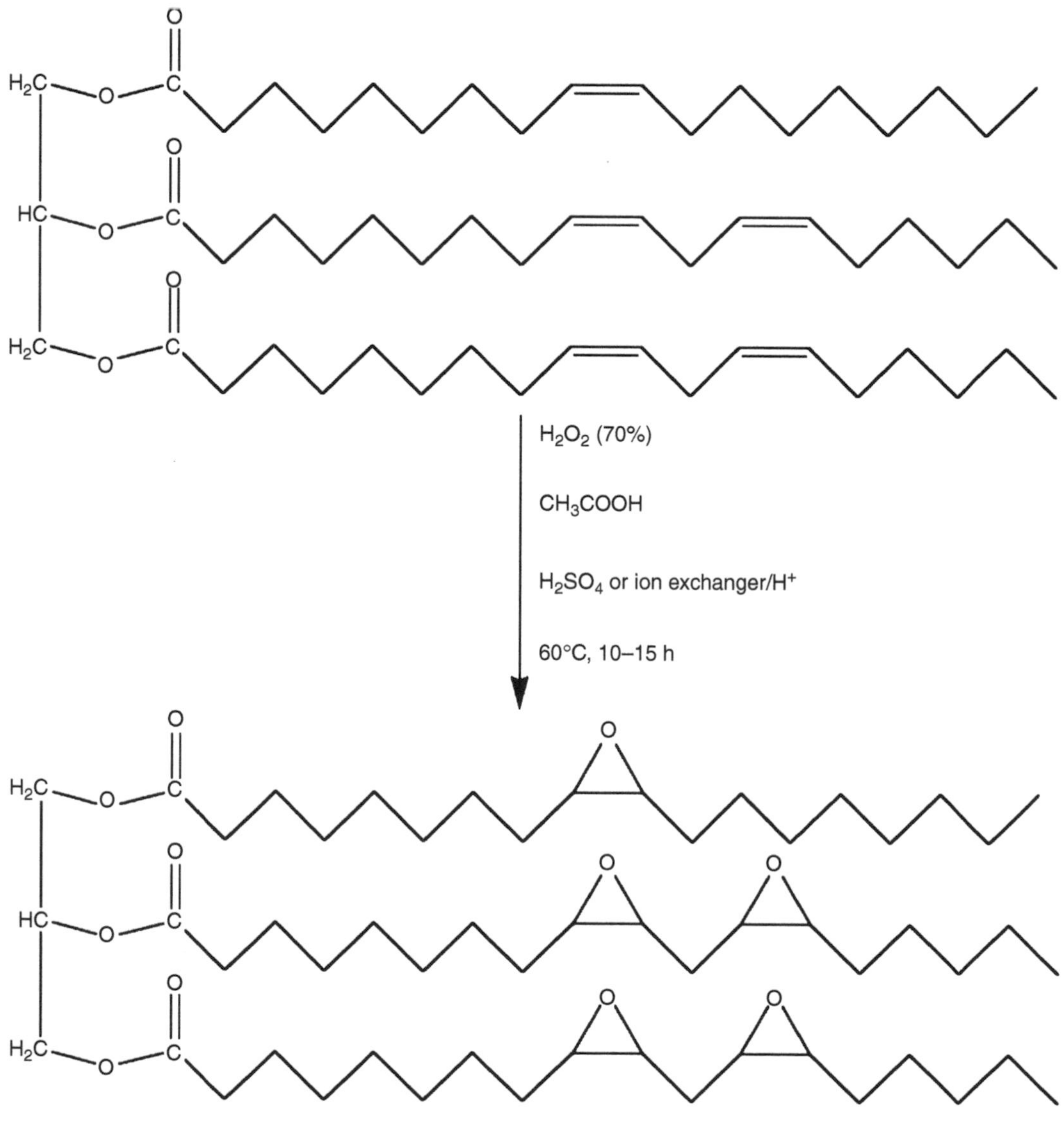

FIGURE 4.2 Prileshajev epoxidation: use of short-chain peroxy acids for epoxidation of unsaturated plant oils.

In this reaction, peroxy acids such as peracetic acid are generated from the corresponding acid and hydrogen peroxide in the presence of a strong mineral acid (Figure 4.1). Due to the potential danger of handling peroxy acids, the intermediate peroxy acids are not isolated during large-scale epoxidation of unsaturated triglycerides (Figure 4.2). There are considerable side reactions via oxirane ring opening, leading to diols, hydroxyesters, estolides, and other dimers, which are believed to be catalyzed by the presence of a strong mineral acid. As a result, the selectivity of this process never exceeds 80%.[2] Further- more, the presence of a strong acid in an oxidative environment causes corrosion of the equipment. Currently fatty acid epoxides are predominantly used as PVC-plasticizers and -stabilizers, because of their ability to scavenge free HCl, thus slowing down degradation. In addition, epoxidized derivatives of fatty acids are used as reactive dilutants for paints and as intermediates for polyurethane-polyol production. Fatty epoxides also represent valuable raw materials for the production of glues and other surface coatings.

State of the art: Hydroxy fatty acids and diacids—At present, the only commercial source of a hydroxy fatty acid is the castor bean (*Ricinus communis*), a naturally occurring oil crop extensively cultivated in countries like India, Brazil, and China. Castor oil contains up to 90% ricinoleic acid (12R-hydroxy-9Z-octadecenoic acid), which can easily be obtained by hydrolysis of the corresponding triglycerides and is used in a variety of applications, such as the manufacture of speciality lubricants, paints, and cosmetics.[3] Some other polyhydroxylated fatty acids might be valuable pharmaceuticals or antimicrobial agents. For example 7S,10S-dihydroxy-8E-octadecenoic acid isolated from *Pseudomonas aeruginosa* stops growth of the pathogenic yeast *Candida albicans*, whereas 7S,10S,12R-trihydroxy-8E-octadecenoic acid, also isolated from *Pseudomonas aeruginosa*, exhibits antimicrobial activity and curtails the rice blast fungus (*Magnaporthe grisea*).[4]

Diacids are important chemical building blocks that are used for the preparation of polyesters, polya- mides, adhesives, etc.[5] On a commercial scale, adipic acid is produced from petrochemical feedstock (through Bayer-Villiger rearrangement of cyclohexanone) and from sebacic acid, which is obtained by ozonolysis of the Δ9-double bond of erucic acid (13Z-docosenoic acid). ω-hydroxy fatty acids are produced on a small scale as raw materials or intermediates in the synthesis of α,ω-dicarboxylic acids.

4.1.2 Oxyfunctionalization of Oleochemicals: Biotechnological Routes

Selective biological hydroxylation and epoxidation reactions of fatty acids and fatty alcohols have been shown to be feasible in principle and, as a result, are now a field of active research both in academic and industrial laboratories.

At present, three major routes are being explored (Figure 4.3):

1. *Transgenic oil crops.* Many plants produce oxygenated lipids as a component of their seed oils. An obvious way to prepare selectively oxygenated fatty acids is thus the isolation of plant oil that contains the desired compounds, the optimization of yields, or the genetic engineering of standard oil plants to accumulate oxyfunctionalized seed oils. Whereas the identification of oxygenated fatty acids in plant oils dates back by several decades, genetic engineering of oil crops producing oxygenated fatty acids is a very young discipline; most reports originate from the year 2000 or

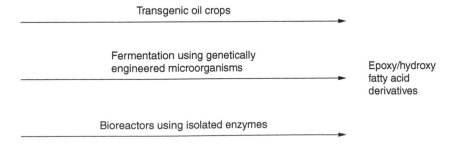

FIGURE 4.3 Biological/biotechnological routes for production of epoxy and hydroxy fatty acids.

later, investigating the use of the model plant *Arabidopsis thaliana* (whose genome has been completely sequenced)[6] as a host system for seed-specific expression of hydroxylases or epoxidases.[3,7,8] More recently, the soybean (*Brassica napis*) was successfully engineered for the production of fatty acid epoxides.[9] Common to these reported processes is the relatively low content (<60%) of the desired oxyfunctionalized oils. Moreover, the fatty acid spectrum found in the seed oil was observed to contain oxygenated fatty acids different from those which were expected. Intensive research is still needed to obtain "chemical factories on a field," for valuable specialty products as much as for oxyfunctionalized commodities required at a large scale.

2. *Fermentation processes.* The use of whole microbial cells for biotransformation requires the transport of reactants and products across the cell walls. Yields may be reduced by side reactions within the cell and by the expensive recovery process from dilute fermentation broths. Obvious advantages are that natural cofactors required for oxidation are available inside the cells, and that the oxidative enzymes are continuously expressed, even if they are bound to the cell's membranes. Thus, oxidation processes may ensue as long as the cell is alive.

   Candida yeasts have been widely used for alkane or fatty acid oxygenation e.g., for the production of α,ω-dicarboxylic acids. First reports on the use of *Candida* strains in biotransformations focused on alkane functionalization, but it was shown that the same strains efficiently produced α,ω-dicarboxylic acids from saturated or unsaturated fatty acids. Mutant strains engineered for higher productivity were reported to produce up to ~300 g of dioic acids per liter of fermentation broth.[10,11] Apart from yeasts, several bacteria such as Pseudomonades, Bacilli, or Rhodococci are able to hydroxylate fatty acids in a terminal or subterminal manner.[4,12,13] While these processes have been studied to some extent, they do not yet allow to prepare oxyfunctionalized fatty acids at a sufficiently low price for commercial exploitation. For the synthesis of hydroxy fatty acids, *E coli* was transformed with a suitable hydroxylase (P450 BM-3) and a fatty acid uptake system.[14,15]

3. *Bioreactors using isolated enzymes.* Biocatalytic epoxidation and hydroxylation of fatty acids using isolated enzymes in a bioreactor up to now is limited to either lipoxygenases (LOX, see Chapter 4.1.6) or lipase-catalyzed perhydrolysis (see Section 4.2.3) with subsequent self-epoxidation.

As cytochrome P450 monooxygenases and diiron cluster-containing monooxygenases are able to hydroxylate nonactivated carbon-hydrogen bonds, they offer a particularily interesting possibility for selective fatty acid or fatty alcohol oxidation. However, up to now this is an issue of academic research only, as both enzyme classes require stochiometric amounts of the costly nicotinamide cofactors NAD(P)H for reductive activation of dioxygen. Various strategies for replacement of the natural cofactors by cheaper sources of reduction equivalents have been proposed (see Section 4.2.2).

In conclusion it must be stated that biocatalytic hydroxylation and epoxidation of fatty acids is still in its infancy. Large-scale applications seem to be possible in the near future only in very few cases (such as dicarboxylic acids by fermentation or fatty acid epoxides by lipase-catalyzed perhydrolysis), whereas most methods are to date predominantly an issue of academic interest. In the following sections a more detailed analysis is provided. Due to space limitations and the authors' interest, special emphasis given on P450 monooxygenases. For this review, the patent literature has been covered only as to those references available from Chemical Abstract Services online.

4.1.3 Natural Functions of Oxygenated Fatty Acids

This section briefly discusses the occurrence and the possible functions of oxygenated fatty acids, recently termed oxylipins,[16] in diverse forms of life. Readers who would like to learn more about the physiological role and metabolism of oxygenated fatty acids are referred to the literature cited. The natural sources and functions of oxygenated fatty acids are interesting as they show: (1) which organisms may be particularly useful for the prospecting of oxygenating enzymes and (2) which natural function oxygenated fatty acids have, suggesting possible applications in a technical context.

While at least one hydroxy fatty acid (ricinoleic acid) is available from castor oil, there is no natural source of fatty epoxides worth exploitation. Nevertheless some plants, notably *Vernonia galamensis* and

Euphorbia lagascae, produce up to 60% vernolic acid (12S, 13R-epoxy-9-*cis*-octadecenoic acid) in their seed oils. Because these plants are poor oil producers, applications of this functionalized fatty acid have not been studied to a significant extent, and the manufacture of this oil is not yet commercialized.

On the other hand, the occurrence of a wide variety of oxygenated fatty acids has been reported in mammals, plants, fungi and bacteria.

In mammals, arachidonate (C20:4) ω-hydroxylation is the first step in the arachidonic acid cascade. Oxygenated metabolites of arachidonic acid play a major role in blood pressure regulation and in the inflammatory process. They are signaling molecules in stress response to infection, allergy, exposure to food, drug and environmental harmful substances.[17] Prominent signaling molecules such as leucotrienes or prostaglandines are derived from oxidized arachidonic acid.[18] α-Hydroxy fatty acids are also found in the sphingolipids of a wide variety of organisms, where they decrease membrane fluidity, in particular as a reaction to increasing temperature.[19] In nervous tissues of vertebrates, α-hydroxy fatty acids play an indispensable role as part of hydroxycerebrosides.[20]

In plants, arachidonic acid is quite rare, but phytooxylipins derived from linole(n)ic acid via the so-called lipoxygenase or oxylipin pathway are usually found. Thus, a whole series of compounds are derived from highly reactive 9- or 13-hydroperoxyoctadecadi(tri)enoic acid.[16] Among these are natural pesticides, termed phytoalexins, which exhibit antibacterial and antifungal effects, as well as volatile aldehydes influencing the flavor or fragrance of many vegetables (reported for olive oil, tomatoes, cucumbers, etc.).[21] Most of these products seem linked to a plant's response toward attack: apart from the phytoalexins, also the cutin monomers (the structural unit of the cuticle protecting the surface of all aerial parts of plants) and the wound hormone traumatin originate from the lipoxygenase pathway.[16,22]

Gram-negative bacteria contain β-hydroxy fatty acids as acyl moieties of cell wall lipids such as lipid A and ornithine lipid,[23] whereas gram-positive bacteria such as *Bacillus* strains may produce acylpeptides containing β-hydroxy fatty acid as antibiotic compounds (e.g., surfactin).[24,26] Hydroxy fatty acids occur as corynomycolic acids in cell walls of pathogenic Mycobacteria (responsible for tuberculosis and leprosy),[25] and in sugar esters produced as biosurfactants by a number of alkane-degrading bacteria.[26,27]

4.1.4 Some Enzymes Involved in Oxyfunctionalization

The CH-bond of a fatty acid methylene group is one of the most stable chemical bonds whose breakage requires approximately 98 kcal mol^{-1}. This energy is beyond the range of simple enzyme reactions and either requires a series of enzymatic reactions (as in the well-known fatty acid-degrading β–oxidation pathway)[18] or a metal cofactor that harnesses the oxidative power of dioxygen to break this bond.[28] In view of fatty acid oxygenation, there are three main reaction types involved: (1) oxidation by P450 monooxygenases, (2) oxidation by diiron center oxygenases, and (3) oxidation by lipoxygenases.

1. *P450 monooxygenases (EC 1.14.x.y)* are an enzyme superfamily reported in all kingdoms of life. Cytochromes P450 contain a heme-thiolate prosthetic group and incorporate one atom of oxygen from dioxygen into their substrate while reducing the other oxygen atom to water. The reducing equivalents are delivered by the cofactor NAD(P)H. P450 enzymes are related to catabolic as well as anabolic metabolism, catalyzing oxidative transformations of exogenous and endogenous compounds. With respect to fatty acids, P450s can act both as hydroxylases or as epoxidases of unsaturated fatty acids.[29]

2. *Diiron-center oxygenases* (mainly in plants and bacteria) use a diiron-center for reductive oxygen activation, abstract a hydrogen atom from a CH-bond and insert oxygen following a radical rebound mechanism, yielding epoxy as well as hydroxy compounds.[28] They share high sequence homology with desaturases.

3. *Lipoxygenases (linoleate: oxygen oxidoreductases; EC 1.13.11.12; LOXs)* are limited to eukaryotes for fatty acid oxyfunctionalization.[30] They contain one non-heme iron per protein molecule and catalyze the regio- and stereoselective dioxygenation of polyenoic fatty acids forming S-configurated hydroperoxy derivatives. These highly reactive and cytotoxic hydroperoxides are immediately used for the biosynthesis of a vast range of compounds including epoxides, mono-, di-, and trihydroxyderivatives, allene oxides, epoxyalcohols, aldehydes, oxoacids, etc.

In the following, all three groups of enzymes will be further discussed. However, as many reviews can be found on LOX,[31,32] the focus will be on the P450 monooxygenases and the diiron cluster enzymes.

4.1.5 P450 Monooxygenases

4.1.5.1 Introduction

Cytochrome P450 enzymes belong to the class of monooxygenases (EC 1.14.x.y). They are widely distributed in nature[33] and play a key role in primary and secondary metabolism as well as in the detoxification of xenobiotic compounds. Common to all enzymes of the P450 superfamily is a heme group in the catalytic center which contains—in contrast to other hemoproteins—a fifth cysteine ligand coordinated to the iron atom. This feature is responsible for the characteristic spectral properties, which gave P450 systems their name:[34] an absorption maximum at 450 nm in the presence of carbon monoxide.

From a functional point of view, all P450 enzymes catalyze the transfer of molecular oxygen to nonactivated aliphatic, to allylic or aromatic XH-bonds (X: -C, -N, -S). Moreover, a remarkable number of P450 enzymes are capable of epoxidizing C=C-double bonds.[33] In certain cases the oxygenated compounds are not stable (e.g., hemi-acetals) and undergo subsequent reactions such as demethylation.

P450 enzymes play a pivotal role in several metabolic pathways such as the metabolism of arachidonic acid to prostaglandines, leucotrienes, and thromboxanes;[35] the formation of cortisone by 11ß-hydroxylation of progesterone;[36,37] the biosynthesis of insect and plant hormones;[38,39] the formation of colors and odors of plants;[40] and ergosterol biosynthesis in yeast.[41] In the mammalian liver, P450s act as phase I enzymes,[42] activating water-insoluble or barely water-soluble compounds for conjugation and elimination. Subsequent reactions are performed by phase II enzymes such as glutathione transferases, N-acetyltransferases, or sulfotransferases, which add further polar groups, rendering these metabolites water-soluble.[42]

P450s recently came into the focus of biotechnologists, as they are able to carry out regio- and stereospecific oxidations at nonactivated C-H- and C=C-double bonds—reactions that are possible only with molecules bearing certain functional groups (e.g., allylic alcohols) in synthetic organic chemistry. They might also be used in bioremediation. Thus, genetically engineered microorganisms containing suitable P450 monooxygenases have been proposed to, e.g., detoxify soil contaminated by polycyclic aromatic hydrocarbons.[43,44]

4.1.5.2 Nomenclature and Classification of Cytochromes P450

The P450 superfamily is one of the largest and oldest gene families.[45] By the end of 2002, more than 2500 putative P450 sequences had been reported (http://drnelson.utmem.edu/CytochromeP450.html). The number of P450 genes grows rapidly: In 2000, about 1000 sequences had been published,[46] but recently the rice genome project alone has led to the identification of 481 new putative P450 genes. Despite a low sequence homology, all P450s adopt a characteristic three-dimensional structure, as revealed by the 12 crystal structures known to date.[47–58]

The nomenclature of P450 genes is based on primary sequence homologies. All P450 genes with a protein sequence homology >40 % belong to the same gene family, those with a sequence homology >55% constitute a subfamily. To describe a P450 gene, it is recommended that one use the italicized abbreviation *"CYP"* for all P450 genes except for mouse and *Drosophila*, which are represented by italicized *"Cyp"* letters. The CYP abbreviation (nonitalicized referring to the protein) is followed by an arabic number denoting the family, a letter designating the subfamily, and a second arabic number representing the individual gene within the subfamily. For example, *CYP*102A2 represents the second gene identified within the P450 subfamily A of the P450 family 102.

Cytochromes P450 show a complex protein architecture, usually involving several cofactors and auxiliary proteins. Depending on the architecture of the overall protein complex, in particular the electron transfer (reductase) system, which transfers reduction equivalents from the cofactors NAD(P)H, they have been divided into four classes:[59,60]

- Class I: Electrons are supplied from a flavoprotein reductase via an iron-sulfur protein to the P450. This type occurs mainly in mitochondrial systems and in most bacteria.

- Class II: Electrons are supplied by a single FAD/FMN-reductase. This type of P450s is often located in the endoplasmatic reticulum.
- Class III: They do not require reduction equivalents, as they use peroxygenated substrates which have already incorporated "activated" oxygen.
- Class IV receives its electrons directly from NADH. This class is represented by a single member, nitric oxide synthase.

4.1.5.3 P450 Reaction Cycle and Its Implications for Synthetic Applications

The postulated reaction cycle of P450 monooxygenases[33,61] is shown in Figure 4.4. Current knowledge about this cycle is mainly based on investigations using P450cam from *Pseudomonas putida*, which catalyzes the regio- and stereospecific hydroxylation of camphor to 5-exo-hydroxycamphor.

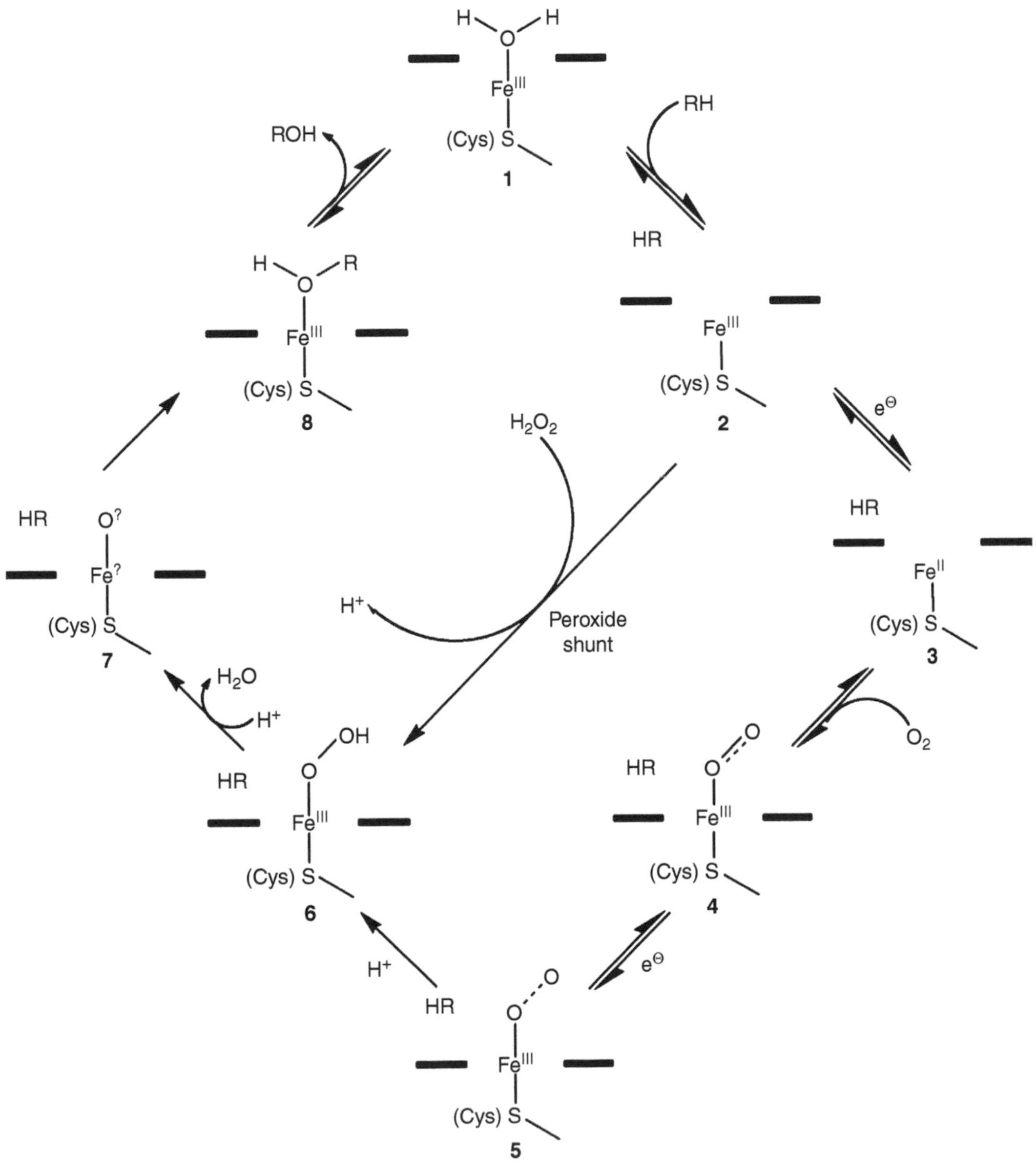

FIGURE 4.4 Postulated reaction cycle of P450 monooxygenases.

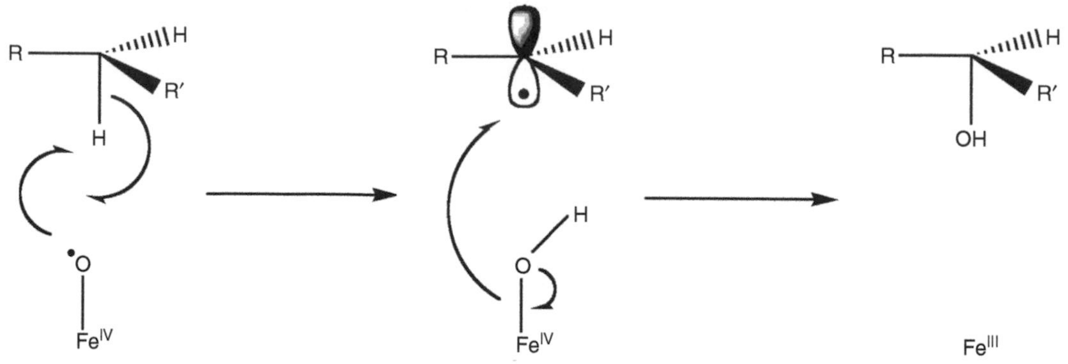

FIGURE 4.5 Rebound mechanism.

In the inactive and substrate-free forms of P450s, the low-spin Fe^{III}-center (d^5, $S = 0.5$) is sixfold coordinated via a protoporphyrin IX system, a fifth cysteinate ligand and a water molecule (**1** in Figure 4.4). On binding of a substrate molecule near to the heme the water ligand is displaced (**2**). The substrate binding induces a spin-state shift: the initial low-spin Fe^{III} is converted to a high-spin Fe^{III}-complex, with the iron center far away from the porphyrin system plane, called out-of-plane structure. The redox potential is thereby increased from ~ −300 mV to ~ −170 mV, facilitating the one-electron reduction of the iron center. An artificial "shortcut" called the shunt-pathway, mediated by strong oxidants such as peroxides, periodate, or peracids, is leading to direct oxygenation of compound **2** to form the hydroperoxy-iron species **6**. Under physiological conditions, one-electron reduction results in the formation of the high-spin Fe^{II} center ($S = 2$) **3**. This configuration with four unpaired electrons is well suited to bind triplet oxygen. The low-spin dioxy-iron(III) complex **4** was isolated and characterized by cryocrystallography, revealing the iron atom in-plane to the porphyrin system. The next step, generating the peroxo-iron complex **5**, requires another one-electron reduction. The protons required for cleavage of the iron-bound dioxygen are delivered by a protein-water hydrogen-bonding network. The oxy-ferryl species **7**, produced with simultaneous formation of water, has a short iron-to-oxygen distance, suggesting an Fe=O bond. As knowledge about this complex originates mainly from crystal structures, the oxidation state of the iron and the electronic state of the heme could not be determined. The last step of the reaction cycle is the oxygenation, usually hydroxylation of the substrate.

The postulated rebound mechanism (Figure 4.5) is currently under debate due to mechanistic studies with ultrafast radical clocks to probe the presence of free radicals, suggesting a concerted oxene-insertion mechanism. This is especially true for epoxidation reactions, which are now suspected to be mediated by three different iron-oxygen species. The versatility in oxidative reactions may, in part, be attributed to the ability of P450 systems to use the peroxo-, hydroperoxo-, or oxenoid-iron species as the active oxidant depending on the substrate and the type of reaction effected.[61,62]

In a living organism, the two electrons required at distinct steps in this cycle are ultimately derived from the nicotinamide cofactors NAD(P)H. One major challenge for all attempts to construct a bioreactor with isolated P450s is to engineer an artificial electron supply system, as NAD(P)H is far too expensive for industrial applications.

4.1.5.4 P450 Monooxygenases in Fatty Acid–Alcohol Modification

Among the ~2500 known P450s are many fatty acid hydroxylases and epoxidases. Most enzymes showing hydroxylation activity with saturated fatty acids can also mediate epoxidation when unsaturated fatty acids are used as substrates.

High regioselectivity generally implies high steric demands of the enzyme with respect to substrate fixation, orientation, and control of the hydroxylation cycle. In the case of P450 monooxygenases, rational analysis is hampered by the lack of crystallographic data of enzyme-substrate complexes. From the few data available, e.g., a complex between CYP102A1 and palmitoleic acid[51] (which does not necessarily describe the productive state of enzyme–substrate interaction), it can be implied that the acid group binds to positively charged or hydrogen-bonding residues near the entrance of the hydrophobic substrate acess channel, while the alkyl chain penetrates the channel.[63] Astonishingly, the distance between the fatty acid and the catalytic center is too long for the hydroxylation reaction to take place. Many fatty acid-metabolizing P450 enzymes additonally accept fatty alcohols as substrates, though often with lower affinity due to weaker interactions of their functional group with the carboxylate binding site.

Fatty acid-hydroxylating P450 enzymes can be subdivided into terminal and subterminal fatty acid hydroxylases. With some enzymes the regiochemical outcome of the reaction depends on the chain length of the fatty acid used. This may be due to a carboxylate binding site in a fixed distance from the catalytic iron, allowing terminal hydroxylation of the fatty acid spanning exactly this distance. The more a fatty acid exceeds this critical chain length, the further in-chain the reaction will occur. The hydroxylation mechanism (rebound) most likely involves a carbon-centered radical, which is lower in energy when located at a secondary $-CH_2$-group. Thus it is surprising that ω-hydroxylases generally show higher regiospecificity than their subterminally hydroxylating counterparts. On the other hand, the possibility to perform the less favored reaction specifically implies that for each position in the fatty acid carbon chain a specific P450 hydroxylase might be accessible either from natural sources or from genetic engineering techniques.

A large number of P450 enzymes, predominantly members of the gene families CYP2, CYP4, CYP52, CYP505, and CYP102, use fatty acids and their derivatives as substrates. Only well-characterized, readily available, and biotechnologically interesting hydroxylases and epoxidases are presented in the following survey.

4.1.5.5 The CYP102 Family as a Model for Natural Monooxygenase/Reductase Fusions

CYP102A1, called P450 BM-3, is one of the most intensely studied P450 monooxygenases. This 118 kDa enzyme, originally cloned from *Bacillus megaterium*, represents a natural fusion protein, incorporating an FAD- and FMN-containing reductase (class II P450) and the P450-domain on a single peptide chain.[64,65] The P450 BM-3 heme domain was one of the first P450s whose structure was determined using x-ray crystallography.[66] By now, structures of various P450 BM-3 mutants as well as of the FMN-domain have been published.[67] CYP102A2 and CYP102A3 from *Bacillus subtilis*[68] show about 60% amino acid sequence identity when compared to P450 BM-3. These fatty acid monooxygenases are currently being investigated with respect to their biotechnological potential.[69,70] CYP116, a novel class of self-sufficient P450 from Rhodococci, displays properties closely paralleling those of the CYP102 family.[71] An eukaryotic counterpart of the CYP102 fusion proteins has lately been cloned, overexpressed, and characterized: P450foxy (CYP505) from the fungus *Fusarium oxysporum* strongly resembles P450 BM-3 in terms of sequence homology and catalytic activities.[72,73]

As the CYP102 monooxygenases are fusion proteins, the experimental setup for their application in organic syntheses is a lot easier compared to other P450s, which require one or two additional electron transport proteins for activity. Advantages of bacterial P450s are their solubility and higher stability.[74] In addition, all self-sufficient P450 monooxygenases (fusion proteins not requiring further proteinaceous electron transport components) characterized to date exhibit rather high turnover numbers (>1000 s^{-1}) with their preferred substrates.

P450 BM-3 has been subjected to numerous mutational studies,[75–78] leading to enzyme variants hydroxylating shorter-chain fatty acids than the wildtype enzyme. Oliver et al. showed that the amino acid at position 87 controls the regioselectivity of the fatty acid hydroxylation.[79] A single mutation (F87A) shifted the regioselectivity from subterminal to nearly exclusively terminal myristic and lauric acid hydroxylation. An amino acid exchange F87V converted P450 BM-3 into a stereo- and regioselective arachidonic acid (14S,15R)-epoxygenase.[80]

Generally, it was found that the regioselectivity of P450 BM-3 depends strongly on the chain length of the fatty acid substrate.

Truan and coworkers determined the absolute configuration of the three hydroxylation products obtained from palmitic acid. They found a high enantiomeric excess for both 15R- and 14R-hydroxypalmitic acid (98% ee) and a somewhat lower selectivity for ω-3 hydroxylation yielding 13R-hydroxypalmitic acid (72% ee).[81] Arachidonic acid is oxidized by P450 BM-3 to nearly enantiomerically pure (R)-18-hydroxyeicosatetraenoic acid (80% of total products) and 14S,15R-epoxyeicosatrienoic acid.[82] The arachidonic acid analogues eicosapentaenoic and eicosatrienoic acid were quantitatively converted to 17S,18R-epoxytetraenoic acid or a mixture of 17-, 18-, and 19-hydroxyeicosatrienoic acid.

Recently P450 BM-3 has been the subject of several studies using laboratory evolution techniques aimed at improving stability against and activity with hydrogen peroxide, stability against organic solvents, pH-stability as well as thermostability.[78,83–85] Screening for improved mutants was frequently based on the pNCA activity assay,[86] which allows high-throughput colorimetric determination of subterminal fatty acid hydroxylation. In all cases considerable improvements of the desired properties could be achieved.

4.1.5.6   The *CYP*52 Family

Members of the *CYP*52 family have been isolated from *Candida* species such as *C. maltosa*,[87,88] *C. apicola*,[89] and *C. tropicalis*.[90] They catalyze the conversion of *n*-alkanes to α,ω-dicarboxylic acids, where terminal hydroxylation of *n*-alkanes is rate-limiting, and ω-hydroxylation of fatty acids ensues.[91] *Candida* species have long been known to produce α,ω-diacids. The potential of these conversions is stressed by the fact that most research in the field was done by industrial research laboratories, e.g., at Cognis (formerly Henkel) or Nippon Mining.[5,10,92,93,94] *CYP*52 is the only P450 family that exclusively shows terminal hydroxylation, even if alkanes, alcohols, or fatty acids of various chain lengths are used as substrates. The reactions leading to dicarboxylic acids are performed by a large class of P450 enzymes catalyzing all oxidation steps from terminal methyl groups to carboxylates. The further oxidation of terminal alcohols is catalyzed partially by P450s, but predominantly by fatty alcohol oxidases and dehydrogenases.[95] In the yeasts *Candida maltosa* and *Candida tropicalis*, eight[88] respectively ten[5] structurally related CYP52A genes and the corresponding reductase systems have been identified. The characterization of the *Candida maltosa* multigene CYP52 family led to a phylogenetic tree that describes the evolutionary distance among the members.[96] This is also reflected by the differences in their substrate specificity. For instance, *CYP*52A3 isoenzymes (P450 Alk1A, P450 Cm1) prefer alkanes, *CYP*52A4 (P450 Alk3A, P450 Cm2) and *CYP*52A5 (P450 Alk2A) hydroxylate alkanes and fatty acids to a similar extent, whereas *CYP*52A9 (P450 Alk5A), *CYP*52A10, and *CYP*52A11 prefer fatty acids as substrates. Scheller and co-workers found that a single P450 enzyme, *CYP*52A3 from *C. maltosa*, catalyzes the complete oxygenation cascade starting from *n*-alkanes to α,ω-dicarboxylic acids.[91]

Compared to the fusion proteins regarded in the last chapter, the catalytic efficiency of the CYP52 family is at least a factor ten lower, reaching 80 s^{-1} with the best substrate.[46] Thus their application is more promising in genetically engineered yeasts than in bioreactor applications, as a low activity *in vivo* can be compensated by a high level of gene expression and high cell densities.

4.1.5.7   The *CYP*4 Family

In order to present this large P450 family adequately, general characteristics of the *CYP*4 family and of a mainly investigated member with respect to biotechnological applications, CYP4A1, are discussed in the following paragraphs.

*CYP*4 is one of the evolutionary oldest P450 families and contains 22 subfamilies.[97] *CYP*4 enzymes are primarily involved in hydroxylation of fatty acids, prostaglandins, leukotrienes, and other eicosanoids in mammalian species.[98] The major fatty acid hydroxylating *CYP* enzymes are all part of the *CYP*4A subfamily. These enzymes all show strong preference for hydroxylation of the thermodynamically disfavored ω-position of arachidonic acid, suggesting a role in the mammalian arachidonic acid cascade. Another common feature of many *CYP*4 enzymes is their inducibility by hypolipidemic agents.

For P450s of the *CYP4A* subfamily, the preferred substrates are C_{12} to C_{20} fatty acids. Shorter (C_7-C_{10}) ones are not hydroxylated, or only at low turnover. These C_7-C_{10} fatty acids are converted by *CYP4B1* isozymes with turnover numbers between 0.8 and 11 equiv min^{-1}.[99] Reaction products of *CYP4B1* are ω-1 hydroxylated fatty acids or 2-hydroxy-alkanes.

The *CYP4A1* enzyme, which was first isolated from rat liver,[100] is one of the most investigated and active fatty acid hydroxylases of the *CYP4* family. Inspired by the natural fusion protein P450 BM-3, Fisher and coworkers fused the rat liver NADPH reductase to the rat liver P450 4A1.[101] Subsequent expression of the fused enzymes in *E. coli* resulted in fatty acid hydroxylation activity ranging from 10 to 30 equiv min^{-1}. The activity of this fusion protein was enhanced tenfold.[102]

By the use of a reconstituted system with the human P450 reductase, Hoch and co-workers[103] reported high turnover numbers for rat P450 4A1 with values up to 649 equiv min^{-1} for the preferred substrate lauric acid, thus showing about tenfold higher ω-hydroxylation activity than the enzymes from the *CYP52* family. Hoch and coworkers identified the amino acid residues that enable *CYP4A* enzymes to bind fatty acids tightly enough to hydroxylate them nearly exclusively at the less reactive ω-position.[104]

Remarkably, *CYP4A1* hydroxylates not only lauric acid but also lauroyl alcohol, whereas dodecane is not a substrate.[105]

4.1.6   Diiron-Cluster Containing Proteins

4.1.6.1   Introduction

The topic of catalysis by proteins containing binuclear nonheme iron clusters is currently under intense study, as these enzymes catalyze a diverse set of reactions including hydroxylation, desaturation, and epoxidation.[106] Diiron enzymes can be divided into two classes: soluble enzymes and a class of integral membrane enzymes, which show only very low sequence similarity to each other.[28] Because of the difficulties in obtaining large quantities of purified membrane proteins, progress in understanding the membrane-bound class has lagged behind that of the soluble class. In the soluble enzymes, the two iron atoms are coordinated via four acidic and two histidine residues, whereas the membrane-bound enzymes use six histidines, as revealed by site-directed mutagenesis studies. The crystal structures of some soluble diiron enzymes have been resolved, providing precise active site geometries.[107–109] This has led to models of the catalytic cycle of this enzyme class.[110] The reactions mediated by the integral membrane diiron enzymes are believed to proceed via a very similar mechanism.

Most, if not all, proteins containing a binuclear diiron cluster react with dioxygen as part of their functional processes.[111] Thus they are often referred to as "diiron-oxo" proteins.

The most intensively studied soluble hydroxylating diiron-oxo protein is methane monooxygenase (MMO), which produces methanol from methane and oxygen.[110,112] As this process is beyond standard chemistry, the mechanism of this enzyme has attracted great attention in the last few years.

A rather well-characterized member of the integral membrane enzymes is the alkane ω-hydroxylase from *Pseudomonas oleovorans*.[113] This enzyme is responsible for the oxygen- and rubredoxin-dependent oxidation of the methyl group of an alkane to produce the corresponding alcohol in a reaction that closely parallels that of MMO. In addition, this enzyme produces epoxides from alkenes, suggesting that epoxidation of double bonds can also be mediated by diiron enzymes.[110]

With respect to fatty acid modification, diiron-oxo proteins were for a long time only recognized as desaturases.[28] This may be explained by the fact that all higher plants contain at least one membrane-bound oleate desaturase that catalyzes the oxygen-dependent insertion of a double bond between carbons 12 and 13 of lipid-linked oleic acid to produce linoleic acid. In contrast, only 14 species in 10 plant families have been found to accumulate the structurally related hydroxy fatty acid, ricinoleic acid, which is synthesized by an oleate hydroxylase that exhibits a high degree of sequence similarity to oleate desaturases.[114] Similarity between desaturases and epoxidases is emphasized by the fact that as few as four amino acid substitutions can convert an oleate 12-desaturase to a hydroxylase.[114] Recently histidine-motif-containing genes encoding fatty acid epoxygenases have been isolated.[115–117] The corresponding enzymes regioselectively introduce epoxy groups into the 12-position of linoleic acid.

From a biotechnological point of view, the diiron-oxo monooxygenases share a considerable drawback with P450 enzymes: both classes use electrons originating from the costly cofactor NAD(P)H for reductive oxygen activation. Moreover, just as P450s do, diiron monooxygenases require electron transport chains delivering these reduction equivalents. In these electron transport systems pairs of electrons arising from NADH or NADPH are simultaneously transferred to a flavoprotein (ferredoxin-NADP$^+$ oxidoreductase) that releases them one at a time to a carrier protein (ferredoxin) capable of carrying only a single electron. In photosynthetic tissues electrons arise from photosystem I and are directly transferred to ferredoxin, which in turn supplies the monooxygenase independently of ferredoxin-NADP$^+$ oxidoreductase.[28] For overcoming the biotechnological disadvantage of a complicated electron supply system with an extremely expensive electron source, strategies paralleling those developed for P450 systems can be suggested.

4.1.6.2 Mechanism of Monooxygenation

Details of the reaction mechanism of the soluble MMO are currently better understood than those of the membrane-bound fatty acid hydroxylases, epoxygenases, and desaturases.[110] The proposal has been made that all of these enzymes share a common activated diiron-oxygen intermediate.[28] Therefore, the same potential mechanisms can be envisaged for the integral membrane hydroxylases as for their soluble counterparts. As the exact reaction mechanism of diiron monooxygenases remains under debate, the following mechanistic consideration will only describe intermediates, which are commonly accepted.

In the resting form, the diiron center is in the oxidized (diferric or Fe^{III}-Fe^{III}) form (**1** in Figure 4.6). Activation is initiated by 2-electron reduction via the corresponding reductase to produce the reduced (diferrous or Fe^{II}-Fe^{II}) form **2**. After reduction, molecular oxygen binds to the iron center resulting in the peroxo complex **3**. Scission of the O-O bond gives rise to compound **4**, the key oxidizing intermediate responsible for subsequent hydrogen abstraction. At some point during oxygen activation, a molecule of water is lost, but the precise timing of this step remains to be defined; consequently, it is

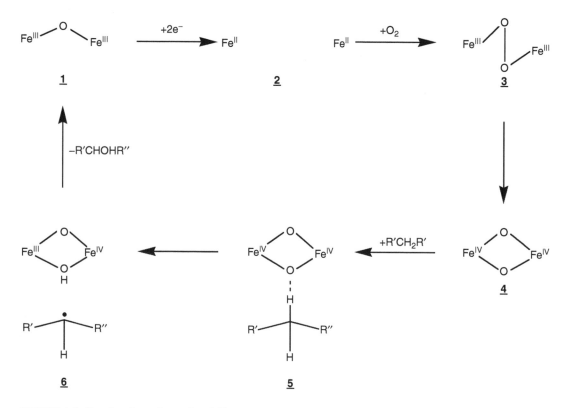

FIGURE 4.6 Postulated reaction cycle of diiron monooxygenases.

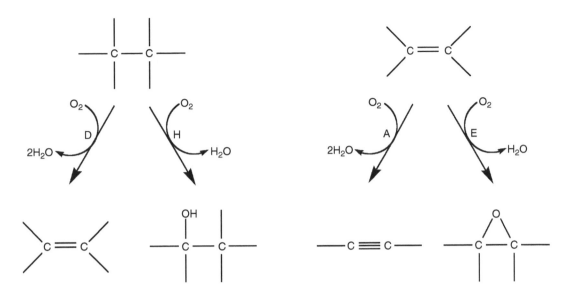

FIGURE 4.7 General scheme for fatty acid modification by diiron-oxo enzymes. Hydroxylation and dehydration are catalyzed by closely related proteins.

not shown in Figure 4.6. According to current models, this oxidizing species **4** then abstracts a hydrogen atom forming a caged hydroxyl intermediate **6** and a carbon-centered radical. This in turn undergoes oxygen rebound in the fashion described for cytochrome P450 hydroxylases (see Chapter 4.1.4.3) to yield the hydroxylated product.

Compound **6** is also believed to be an intermediate of the desaturation mechanism.[28] Instead of rebound, the abstraction of a second hydrogen would result in formation of a transient diradical that would spontaneously recombine to form the olefinic double bond. This close mechanistic similarity is reflected in the high sequence homology between hydroxylases and desaturases. Whether the reaction outcome is hydroxylation or dehydration is a matter of subtle differences in substrate positioning relative to the active center. Thus, it can be rationalized why desaturases can be converted to hydroxylases and vice versa. Even bifunctional enzymes are known: the *Lesquerella fendleri* 12-hydroxylase also exhibits desaturase activity *in vitro*.

The same relationship was proposed for epoxidases and acetylenases, which are both acting on double bonds. This leads to the proposed general scheme for fatty acid modification by diiron-oxo enzymes (Figure 4.7).[28]

4.1.6.3   Applications in Fatty Acid Oxygenation

Despite their occurrence in most plant species producing oils enriched in epoxy acids (only Euphorbiacea species use a P450 monooxygenase for fatty acid epoxidation), attempts to use diiron cluster-containing fatty acid monooxygenases in biotechnology are still limited.[117] As most available sequences originate from plants and are coding for integral membrane proteins, the primary focus was on attempts to engineer plants containing unusual fatty acids in their seed oils (see 4.2.1).

4.1.7   Lipoxygenases

Lipoxygenases (EC 1.13.11.12; LOXs) are nonheme iron dioxygenases that catalyze the stereospecific incorporation of dioxygen into the 1*Z*,4*Z* pentadienyl system of polyunsaturated fatty acids to generate optically active (*S*)-dienic hydroperoxides (Figure 4.8).[31]

These compounds are potentially cytotoxic, because they can induce radical chain reactions. Thus, in living organisms they are immediately converted to a vast array of secondary metabolites (see Section 4.1.2). Beside

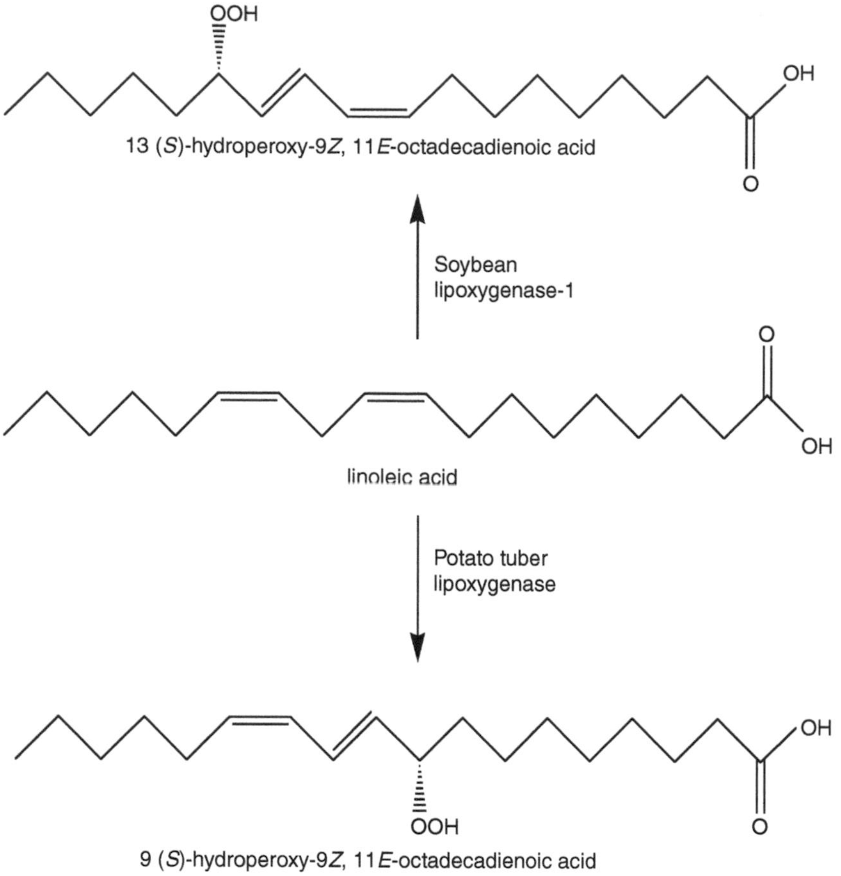

FIGURE 4.8 Lipoxygenase reactions and their regiospecificities.

the dioxygenase reaction, LOXs catalyze the secondary conversion of hydroperoxy lipids (hydroperoxidase reaction) and the formation of epoxy leukotrienes (leukotriene synthase reaction).[32]

As LOXs are not directly hydroxylating or epoxidizing fatty acids, sequential reactions must be performed to yield hydroxy or epoxy fatty acids. Thus they are not strictly subject to this review. Contrary to most monooxygenases discussed above however, they are quite stable and active, and they posses a high degree of regio- and stereospecificity, leading to compounds that are difficult to obtain by chemical synthesis. Because of this potential in the production of oxylipins, a brief overview on biotechnological applications of LOXs will be given in the following paragraphs. Interested readers may find more information in reviews by Iacazio and Feussner.[31,32]

LOXs are classified according to their regiospecificity using either linoleic acid with vegetal LOXs or arachidonic acid with other LOXs. For example, soybean LOX isoenzyme-1 is classified as a 13-LOX, whereas potato tuber LOX is classified as a 9-LOX when—according to the conventions—linoleic acid is used as substrate (Figure 4.8). It should be noted that despite an opposite regiochemistry, the two enzymes both yield a hydroperoxide of S absolute configuration, bearing a $2E,4Z$ conjugated dienic system.

The LOX reaction may lead to various regioisomers. For example, a fatty acid such as arachidonic acid, which contains three allylic methylene groups, can be oxygenated by a LOX to 6 regio-isomeric hydroperoxy derivatives, namely the 15- and 11- hydroperoxy derivative (originating from C_{13} hydrogen removal), the 12- and 8-hydroperoxy derivative (C_{10} hydrogen removal) and the 9- and 5-dioxygenated derivatives (C_7 hydrogen removal). LOXs displaying all these positional specificities are accessible, either from natural sources or by enzyme variants obtained by site-directed mutagenesis.[118]

4.1.7.1 Applications of LOXs in Organic Synthesis

It should be noted that the lipid hydroperoxides formed by LOXs can either be reduced to hydroxy compounds or processed further by enzymes of the lipoxygenase pathway. The need for reduction is not necessarily a disadvantage compared to other oxygenating enzymes like heme- and diiron monooxygenases. While these monooxygenases require reduction equivalents from extremely expensive cofactors or artificial reduction systems, the reductive step in transformations using LOX can be performed by relatively inexpensive chemical reducing agents such as $NaBH_4$, $SnCl_2$ or—after extraction of the reaction mixture—with triphenylphosphine (TPP).

4.2 Application of Enzymes in the Oxygenation of Fatty Acid Derivatives

We will now discuss the use of enzymes in lipid oxyfunctionalization during the preparation of transgenic oil crops, during microbial fermentations and in bioreactors in more detail.

4.2.1 Transgenic Oil Crops

In an attempt to design new oilcrops yielding epoxy fatty acids, CYP726A1 from the euphorbiacea *Euphorbia lagasca* was expressed in the seeds of various plant hosts. Some euphorbiaceae species have long been known to contain 12-epoxy-cis-9-octadecenoic acid (vernolic acid) in their seed oils. Expressing CYP726A1 under seed-specific promotors in transgenic tobacco or somatic soybean embryos resulted in formation of up to 15% (w/w) fatty acid epoxides in the seed oils.[9,117] Despite the low yields obtained yet, using a highly productive oilcrop as the soybean is a highly promising step for future developments.

A Δ12-epoxygenase isolated from *Crepis palaestina*[115,119] has also been expressed in the model plant *Arabidopsis thaliana* under a seed-specific promotor, resulting in production of up to 8% of epoxy fatty acids, mainly 12,13-epoxy-cis-9-octadecenoic acid.[8] Developing seeds of *Arabidopsis* were also used as host system for the expression of castor bean (*Ricinus communis*) oleate Δ12-hydroxylase and the corresponding enzyme from *Lesquerella fendleri*.[120] Arabidopsis lines lacking the FAD3 ER Δ-15-desaturase accumulated up to 50% hydroxy fatty acids (34% 18:1-OH; 16% 20:1 OH) in their seed oils, when the castor hydroxylase was expressed.[7]

Interest in using plants expressing P450 genes for production of oxygenated fatty acids is generally growing and the coming years will possibly see significant improvements in this field.[121,122]

4.2.2 Microbial Fermentations

By blocking the β-oxidation pathway in *Candida tropicalis* and enhancing CYP52 expression, this species efficiently produces both saturated or unsaturated terminal diacids (C_{12} to C_{22}) from alkanes or fatty acids.[5,10] This process is highly productive: 300 g of diacids per liter of fermentation broth can be obtained.[94] *Candida tropicalis* M25 has been shown to produce 3-hydroxy dienedioic acids from linoleic acid.[123]

Pseudomonas aeruginosa strain PR3 was used for hydroxylation of oleic and ricinoleic acid, yielding 7S,10S-dihydroxy-8E-octadecenoic acid, and 7S,10S,12R-trihydroxy-8E-octadecenoic acid, respectively. 7S,10S-dihydroxy-8E-octadecenoic, which is of pharmacological interest (see Section 4.1.1), was obtained by this method in concentrations of nearly 10 g/l.[4]

Bacillus sphaericus strains were shown to produce 10-ketostearic acid from oleic acid.[13] Under optimized conditions, conversion of oleic acid reaches up to 60%. Alternatively, up to 10 g/l of 10-ketostearic acid can be obtained by fermentation of oleic acid with *Sphingobacterium thalpophilum*.[12]

P450 BM-3, heterologously expressed in *E coli* in combination with a fatty acid uptake system from *Pseudomonas oleovorans* has been used *in vivo* to produce mixtures of chiral 12-, 13-, and 14-hydroxy-pentadecanoic acid in preparative scale at high optical purities.[14,15]

4.2.3 Bioreactors

Oxidation reactions using isolated enzymes may be most readily performed using LOXs, as they do not require reductive cofactors. Some applications are listed in the following paragraphs.

In 1989, Corey and coworkers[124] exploited the capacity of soybean lipoxygenase-1 to carry out double dioxygenation of arachidonic acid to realize a simple synthesis of lipoxin A$_4$, an important physiologically active eicosanoid of the arachidonic acid cascade. After reduction with NaBH$_4$, two of the three asymmetric carbon centers of lipoxin A$_4$ in correct configuration had been introduced.

An analogous two-step chemo-enzymatic procedure using TPP for reduction was applied to the synthesis of the natural products coriolic acid[125] and dimorphecolic acid and to other hydroxy-polyunsaturated fatty acids.[126,127]

Viewed from the perspective of potential industrial applications, it is important to understand that in the absence of oxygen, LOXs can catalyze an anaerobic reaction between their products and their substrates (hydroperoxidase reaction). This reaction is thought to generate radicals that are deleterious to the enzyme. Thus it is vital to maintain a sufficient concentration of dissolved oxygen in the reaction medium either by oxygen bubbling or by pressurization. Optimal reaction conditions were investigated by Martini et al.[125] using a SOTELEM (Rueil Malmaison, France) stainless steel chemical reactor (MU 4004), which proved to be particularly suited to carry out lipoxygenation reactions. Thus, linoleic acid could be converted almost quantitatively to 13S-hydroperoxy-9Z,11E-octadecadienoic acid (95.5%, 98% ee) at 5°C and 250 kPa oxygen pressure, pH11 (0.1 M borate buffer), 0.1 M substrate, and 4 mg ml^{-1} soybean lipoxygenase. Especially remarkable is that quite high substrate concentrations of 0.1 M were used, which largely exceeded the solubility of unsaturated fatty acids in the aqueous buffer system. Still, the reaction was complete within 30 min.

In 1999 Hsu et al.[128] developed a packed-bed bioreactor for continuous oxygenation of linoleic acid. LOX immobilized in sol-gel matrices was used for these experiments.

Very recently, an industrial research group at Cognis (formerly Heukel), a company that for decades has been highly interested in oleochemistry, devised a new and efficient method to produce fatty acid hydroperoxides based on LOX. This process is conducted in an oil (or fatty acid)/water two-phase system containing LOX and catalase. Molecular oxygen as primary oxidant was substituted by hydrogen peroxide, resulting in higher turnover numbers.[129]

Hydrogen peroxide is also used as oxidant in the chemoenzymatic synthesis of fatty epoxides by lipase-catalyzed perhydrolysis. A group at Novo Nordisk A/S first discovered that unbranched saturated fatty acids with 4 to 22 carbon atoms can be converted to peroxy fatty acids using hydrogen peroxide, and Novozyme 435, an immobilized lipase B from *Candida antarctica*.[130] If an unsaturated fatty acid or its ester is treated with H$_2$O$_2$ in the presence of Novozyme 435, "self"-epoxidation occurs.[131] First, the unsaturated fatty acid is converted to an unsaturated peroxy fatty acid (only this step is catalyzed by the lipase). Subsequently, the unsaturated peroxy acid epoxidizes "itself" in an uncatalyzed Prileshajev (Figure 4.1 and Figure 4.2) epoxidation. Depending on the chain length, peroxy fatty acid yields from 70% to 95% were obtained.[131,132] In addition to the production of partially or completely epoxidized fatty acids, this method is capable of producing epoxidized plant oils[132] and fatty alcohol epoxides.[133] The epoxidation processes described by Warwel and Rüsch have already been carried out on the kilogram scale. If further successful, they may allow one to substitute the problematic Prileshajev-epoxidation with lipase-catalyzed perhydrolysis.

Several years ago an enzymatic activity termed peroxygenase was found and characterized in plant extracts by Blee[16] and Hamberg[134] while investigating the fate of lipid hydroperoxides originating from the lipoxygenase reaction. Peroxygenase, a hemoprotein, catalyzes the inter- and intramolecular transfer of oxygen from a fatty acid hydroperoxide to form epoxides from unsaturated fatty acids. It was found that an external oxidant such as hydrogen peroxide or cumene hydroperoxide could be used as an alternative source of oxygen. Hamberg demonstrated that oat (*Avena sativa*) seeds are a good source of peroxygenase. Starting from this point, Piazza et al. devised a method for rapid isolation and immobilization of this enzyme on membranes and conducting epoxidation reactions in organic solvents. Using *t*-butyl hydroperoxide as oxidant, oleic acid, linoleic acid, and arachidonic acid could be converted nearly quantitatively to their corresponding epoxides.[135]

The enzymatic reactions described so far are performed easily and without involvement of cofactors. This situation changes, if reactions mediated by cytochromes P450 are considered. Thus, the following considerations to date are predominantly an issue of academic interest.

One possibility for applications of isolated P450 enzymes is to use the "shunt pathway." Addition of cheap and readily available hydrogen peroxide (or of an organic peroxide) to the initial enzyme-substrate complex can supply both the electrons and the oxygen atom required to form the reactive iron-oxygen intermediate, the species that inserts the oxygen atom into the substrate (Figure 4.4). Unfortunately, the stoichiometric use of peroxides is quite inefficient, and may quickly lead to inactivation of the enzyme.[85]

In this context the recent identification of two unusual peroxygenase P450s is interesting. They lack a threonine residue that is highly conserved among P450 monooxygenases and thought to be essential for dioxygen activation. Thus these enzymes are not cofactor-dependent and are well suited for *in vitro* applications. Fatty acid hydroxylase from *Sphingomonas paucimobilis* (*CYP*152B1) efficiently produces (*S*)-α-hydroxy fatty acids,[136,137] whereas its close relative *CYP*152A1 (alternatively called *ybdT* gene) from *Bacillus subtilis* attacks the α-carbon as well as the β-carbon of myristic acid.[138] As both enzymes possess very high affinity to hydrogen peroxide, these biocatalysts require only very low H_2O_2-concentrations, thus tempering the deleterious effects of H_2O_2 on the enzyme. Enhanced stability against and activity with H_2O_2 has also been achieved by laboratory evolution of P450 BM-3.[139]

Direct electron supply from electrodes[140,141] is another option for delivering reduction equivalents to the catalytic iron of P450s. It was found to result in very low productivity, but could be enhanced by the use of soluble electron mediators such as cobalt(III)-sepulchrate[142] in combination with a *CYP*4A1/ reductase fusion protein.[102] Recently the electrochemical reduction of the prosthetic flavin of a reductase mediated by an organometallic rhodium-complex was investigated and applied to the synthesis of styrene oxide.[143] This method has the potential to be applied to all flavin-containing reductases and thus can possibly be used for the synthesis of oxygenated fatty acid derivatives.

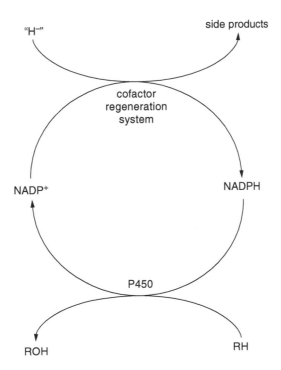

FIGURE 4.9 Cofactor recycling systems for cytochromes P450; "H⁻"may be derived from enzymatic oxidation (dehydrogenases), from oxidation by organometallic complexes or from electrochemical sources.

Schwaneberg and coworkers used the electron mediator cobalt(III)-sepulchrate[142] and zinc dust as electron source to drive the catalytic cycle of P450 BM-3, resulting in turnover numbers reaching 50% of those with the natural cofactor.[144,145]

Finally, another method to circumvent the stochiometric need of reduced cofactors is their regeneration (Figure 4.9), which must be highly efficient to reduce the cofactor costs to an economically acceptable level.

In this context enzymatic[146] and organometallic[147] approaches have been suggested for hydride transfer to NAD(P)$^+$. The required electrons may be derived from formate or glucose oxidation using the respective dehydrogenases(formate or glucose dehydrogenase), or from electrochemical sources in case of organometallic rhodium-complexes.

Sol-gel encapsulation of P450 BM-3 mutants in combination with genetically engineered NADP$^+$-dependent formate dehydrogenase (FDH)[148] resulted in a heterogeneous, stable, and self-sufficient hydroxylation biocatalyst capable of recycling NADPH from NADP$^+$ (Figure 4.9).[149] The only side products of this hydroxylation cycle are carbon dioxide and water.

4.3 Outlook

The information provided in this chapter, and especially the examples provided, demonstrate that fatty acid-oxygenating enzymes are indeed a versatile and useful class of biocatalysts. On the other hand, industrial applications of this class of enzymes are still scarce. This is not surprising if the drawbacks— need of expensive cofactors, low stability under reaction conditions, or low turnover numbers—associated with the biocatalysts under investigation are considered. On the other hand, *in vivo* processes are, with the remarkable exception of dioic acid production, not developed far enough to be established at industrial scale. However, in academic as well as industrial research laboratories, many efforts are being made to overcome the problems in biocatalytic oxidation. Especially the tremendous development of directed evolution techniques, eventually in combination with rational enzyme engineering based on the growing insights into protein structures and functions, could help to overcome the problems yet unresolved. Cytochromes P450 and diiron cluster-containing monooxygenases could be used in preparative synthesis as soon as an efficient method for substitution of their cofactors NAD(P)H is found.

As organic chemistry to date does not present efficient tools for (stereo) selective oxygenation of fatty acid carbon chains, biocatalytic processes have a great potential to be the first technique providing access to compounds yet not obtainable at commercial scale. This would undoubtedly open up the possibility to develop new products based on oleochemistry. It is the consideration of the authors that at least some of the enzymes presented here will find applications in the future.

Acknowledgment

The authors are grateful to Dr. Vlada Urlacher for helpful discussions.

References

1. K Hill. Fats and oils as oleochemical raw materials. *J Oleo Sci* 50:433–444, 2001.
2. M Rüsch gen Klaas, S Warwel. Complete and partial epoxidation of plant oils by lipase-catalyzed perhydrolysis. *Ind Crops and Products* 9:125–132, 1999.
3. M Smith, H Moon, L Kunst. Production of hydroxy fatty acids in the seeds of arabidopsis thaliana. *Biochem Soc Trans* 28:947–950, 2000.
4. TM Kuo, AC Lanser. Factors influencing the production of a novel compound, 7,10-dihydroxy-8(e)-octadecenoic acid, by pseudomonas aeruginosa pr3 (nrrl b-18602) in batch cultures. *Curr Microbiol* 47:186–191, 2003.
5. CR Wilson, DL Craft, LD Eirich, M Eshoo, KM Madduri, CA Cornett, AA Brenner, M Tang, JC Loper, M Gleeson. Cytochrome p 450 monooxygenase and nadph cytochrome p 450 oxidoreductase genes and proteins related to the w-hydroxylase complex of *candida tropicalis*. PCT Int. Appl. (Henkel Corporation, USA; et al.). WO. 200 pp., 2000.

6. Arabidopsis Genome Initiative. Analysis of the genome sequence of the flowering plant arabidopsis thaliana. *Nature* 408:796–815, 2000.

7. MA Smith, H Moon, G Chowrira, L Kunst. Heterologous expression of a fatty acid hydroxylase gene in developing seeds of *arabidopsis thaliana*. *Planta* 217:507–516, 2003.

8. S Singh, S Thomaeus, M Lee, A Green, S Stymne. Inhibition of polyunsaturated fatty acid accumulation in plants expressing a fatty acid epoxygenase. *Biochem Soc Trans* 28:940–942, 2000.

9. B Cahoon Edgar, G Ripp Kevin, E Hall Sarah, B McGonigle. Transgenic production of epoxy fatty acids by expression of a cytochrome p450 enzyme from euphorbia lagascae seed. *Plant Phys* 128:615–624, 2002.

10. S Picataggio, T Rohrer, LD Eirich. Method for increasing the w-hydroxylase activity of *candida tropical* is in the manufacture of a,w-dicarboxylic acids. *PCT Int. Appl.* (Henkel Research Corp., USA). WO. 52 pp. 1991.

11. KD Green, MK Turner, JM Woodley. Candida cloacae oxidation of long-chain fatty acids to dioic acids. *Enzyme Microb Technol* 27:205–211, 2000.

12. TM Kuo, AC Lanser, LK Nakamura, CT Hou. Production of 10-ketostearic acid and 10-hydroxystearic acid by strains of sphingobacterium thalpophilum isolated from composted manure. *Curr Microbiol* 40:105–109, 2000.

13. TM Kuo, LK Nakamura, AC Lanser. Conversion of fatty acids by bacillus sphaericus-like organisms. *Curr Microbiol* 45:265–271, 2002.

14. S Schneider, MG Wubbolts, G Oesterhelt, D Sanglard, B Witholt. Controlled regioselectivity of fatty acid oxidation by whole cells producing cytochrome p450bm-3 monooxygenase under varied dissolved oxygen concentrations. *Biotechnol Bioeng* 64:333–341, 1999.

15. S Schneider, MG Wubbolts, D Sanglard, B Witholt. Biocatalyst engineering by assembly of fatty acid transport and oxidation activities for *in vivo* application of cytochrome p-450bm-3 monooxygenase. *Appl Environ Microbiol* 64:3784–3790, 1998.

16. E Blee. Biosynthesis of phytooxylipins. The peroxygenase pathway. *Fett/Lipid* 100:121–127, 1998.

17. KC Nicolaou, JY Ramphal, NA Petasis, MH Serhan. Lipoxins and related eicosanoids: Biosynthesis, biological properties and chemical synthesis. *Angew Chem Int Ed Engl* 30:1100–1116, 1991.

18. D Voet, JG Voet. Biochemie. *Biochemie*. Weinheim: VCH 2000, pp. 620–631.

19. K Kaya, CS Ramesha, GA Thompson, Jr. On the formation of alpha-hydroxy fatty acids. Evidence for a direct hydroxylation of nonhydroxy fatty acid-containing sphingolipids. *J Biol Chem* 259:3548–3553, 1984.

20. PF Ki, Y Kishimoto, EE Lattman, EF Stanley, JW Griffin. Structure and function of urodele myelin lacking alpha-hydroxy fatty acid-containing galactosphingolipids: Slow nerve conduction and unusual myelin thickness. *Brain Res* 345:19–24, 1985.

21. JJ Salas, J Sanchez. Hydroperoxide lyase from olive (olea europaea) fruits. *Plant Sci* (Shannon, Ireland) 143:19–26, 1999.

22. E Blee, F Schuber. Biosynthesis of cutin monomers: Involvement of a lipoxygenase/peroxygenase pathway. *Plant Journal* 4:113–123, 1993.

23. Y Nakagawa, K Kishida, Y Kodani, T Matsuyama. Optical configuration analysis of hydroxy fatty acids in bacterial lipids by chiral column high-performance liquid chromatography. *Microb Immunol* 41:27–32, 1997.

24. K Hosono, H Suzuki. Acylpeptides, the inhibitors of cyclic adenosine 3'.5'-monophosphate phosphodiesterase. *J Antibiot* 36:667–673, 1983.

25. E Janczura, C Abou-Zeid, C Gailly, C Cocito. Chemical identification of some cell-wall components of microorganisms isolated from human leprosy lesions. *Zentralbl Bakteriol Mikrobiol Hyg* [A] 251:114–125, 1981.

26. AK Koch, O Kappeli, A Fiechter, J Reiser. Hydrocarbon assimilation and biosurfactant production in pseudomonas aeruginosa mutants. *J Bacteriol* 173:4212–4219, 1991.

27. JC Philp, MS Kuyukina, IB Ivshina, SA Dunbar, N Christofi, S Lang, V Wray. Alkanotrophic rhodococcus ruber as a biosurfactant producer. *Appl Microbiol Biotechnol* 59:318–324, 2002.

28. J Shanklin, EB Cahoon. Desaturation and related modifications of fatty acids. *Ann Rev Plant Phys and Plant Mol Biol* 49:611–641, 1998.

29. D Werck-Reichhart, R Feyereisen. Cytochromes p450: A success story. *Genome Biol.* 1:3001–3012, 2000.

30. AR Brash. Lipoxygenases: Occurrence, functions, catalysis and acquisition of substrate. *J Biol Chem* 274:23679–23682, 1999.

31. I Feussner, H Kühn, Application of lipoxygenases and related enzymes for the preparation of oxygenated lipids. In *Enzymes in Lipid Modification*, UT Bornscheuer, Editor. 2000, Wiley-VCH: Weinheim, New York. pp. 309–336.

32. G Iacazio, D Martini-Iacazio, Properties and applications of lipoxygenases, In *Enzymes in Lipid Modification*, UT Bornscheuer, Editor. 2000, Wiley-VCH: Weinheim, New York. pp. 337–359.

33. DF Lewis. Cytochromes p450: *Structure, Function and Mechanism*, vol 1. London: Taylor & Francis 1996.

34. T Omura, RJ Sato. The carbon monoxide-binding pigment of liver microsomes. I. Evidence for its hemoprotein nature. *J Biol Chem* 239:2370–2378, 1964.

35. D Kupfer, KA Holm. Prostaglandin metabolism by hepatic cytochrome p450. *Drug Metab Rev* 20:753–764, 1989.

36. A el-Monem, H el-Refai, AR Sallam, H Geith. Microbial 11-hydroxylation of progesterone. *Acta Microbiol Pol B* 4:31–36, 1972.

37. L Sallam, N Naim, A Zeinel-Abdin Badr, A El-Refai. Bioconversion of progesterone with cell preparations with aspergillus niger 171. *Rev Latinoam Microbiol* 19:151–153, 1977.

38. R Feyereisen. Insect p450 enzymes. *Annu Rev Entomol* 44:507–533, 1999.

39. F Durst, DP O'Keefe. Plant cytochromes p450: An overview. *Drug Metab Drug Interact* 12:171–187, 1995.

40. TA Holton, F Brugliera, DR Lester, Y Tanaka, CD Hyland, JG Menting, CY Lu, E Farcy, TW Stevenson, EC Cornish. Cloning and expression of cytochrome p450 genes controlling flower colour. *Nature* 366:276–279, 1993.

41. D Berg, M Plempel, K Buchel, G Holmwood, K Stroech. Sterol biosynthesis inhibitors. Secondary effects and enhanced *in vivo* efficacy. *Ann N Y Acad Sci* 544:338–347, 1988.

42. JA Goldstein, MB Faletto. Advances in mechanisms of activation and deactivation of environmental chemicals. *Environ Health Perspect* 100:169–176, 1993.

43. S Harayama. Polycyclic aromatic hydrocarbon bioremediation design. *Curr Opin Biotechnol* 8:268–273, 1997.

44. DG Kellner, SA Maves, SG Sligar. Engineering cytochrome p450s for bioremediation. *Curr Opin Biotechnol* 8:274–278, 1997.

45. DR Nelson, HW Strobel. Evolution of cytochrome p450 proteins. *Mol Biol Evol* 5:199–220, 1988.

46. U Schwaneberg, UT Bornscheuer. Fatty acid hydroxylations using p450 monooxygenases. In *Enzymes in Lipid Modification*, UT Bornscheuer, ed. Wiley-VCH: Weinheim, New York, 2000. pp. 394–414.

47. TL Poulos, BC Finzel, AJ Howard. High-resolution crystal structure of cytochrome p450cam. *J Mol Biol* 195:687–700, 1987.

48. CA Hasemann, KG Ravichandran, JA Peterson, J Deisenhofer. Crystal structure and refinement of cytochrome p450terp at 2.3 a resolution. *J Mol Biol* 236:1169–1185, 1994.

49. K Nakahara, H Shoun, S Adachi, T Iizuka, Y Shiro. Crystallization and preliminary x-ray diffraction studies of nitric oxide reductase cytochrome p450nor from fusarium oxysporum. *J Mol Biol* 239:158–159, 1994.

50. JR Cupp-Vickery, TL Poulos. Structure of cytochrome p450eryf involved in erythromycin biosynthesis. *Nat Struct Biol* 2:144–153, 1995.

51. H Li, TL Poulos. The structure of the cytochrome p450bm-3 haem domain complexed with the fatty acid substrate, palmitoleic acid. *Nat Struct Biol* 4:140–146, 1997.

52. SY Park, K Yamane, S Adachi, Y Shiro, KE Weiss, SG Sligar. Crystallization and preliminary x-ray diffraction analysis of a cytochrome p450 (cyp119) from sulfolobus solfataricus. *Acta Crystallogr D Biol Crystallogr* 56:1173–1175, 2000.

53. PA Williams, J Cosme, A Ward, HC Angove, D Matak Vinkovic, H Jhoti. Crystal structure of human cytochrome p450 2c9 with bound warfarin. *Nature* 424:464–468, 2003.

54. LM Podust, TL Poulos, MR Waterman. Crystal structure of cytochrome p450 14alpha -sterol demethylase (cyp51) from mycobacterium tuberculosis in complex with azole inhibitors. *Proc Natl Acad Sci USA* 98:3068–3073, 2001.

55. LM Podust, Y Kim, M Arase, BA Neely, BJ Beck, H Bach, DH Sherman, DC Lamb, SL Kelly, MR Waterman. The 1.92-a structure of streptomyces coelicolor a3(2) cyp154c1. A new monooxygenase that functionalizes macrolide ring systems. *J Biol Chem* 278:12214–12221, 2003.

56. JK Yano, F Blasco, H Li, RD Schmid, A Henne, TL Poulos. Preliminary characterization and crystal structure of a thermostable cytochrome p450 from thermus thermophilus. *J Biol Chem* 278:608–616, 2003.

57. EE Scott, YA He, MR Wester, MA White, CC Chin, JR Halpert, EF Johnson, CD Stout. An open conformation of mammalian cytochrome p450 2b4 at 1.6-a resolution. *Proc Natl Acad Sci USA* 100:13196–13201, 2003.

58. S Nagano, H Li, H Shimizu, C Nishida, H Ogura, PR Ortiz De Montellano, TL Poulos. Crystal structures of epothilone d-bound, epothilone b-bound, and substrate-free forms of cytochrome p450epok. *J Biol Chem.* [Epub ahead of print], 2003.

59. KW Degtyarenko. Structural domains of p450-containing monooxygenase systems. *Protein Eng* 8:737–747, 1995.

60. JA Peterson, SE Graham. A close family resemblance: The importance of structure in understanding cytochromes p450. *Structure* 6:1079–1085, 1998.

61. PR Ortiz de Montellano, JJ De Voss. Oxidizing species in the mechanism of cytochrome p450. *Nat. Prod. Rep.* 19:477–493, 2002.

62. ADN Vaz, DN McGinnity, MJ Coon. Epoxidation of olefins by cytochrome p450: Evidence from site-specific mutagenesis for hydroperoxo-iron as an electrophilic oxidant. *Proc Natl Acad Sci USA* 95:3555–3560, 1998.

63. PR Ortiz de Montellano, WK Chan, SF Tuck, RM Kaikus, NM Bass, JA Peterson. Mechanism-based probes of the topology and function of fatty acid hydroxylases. *FASEB J.* 6:695–699, 1992.

64. LO Narhi, AJ Fulco. Characterization of a catalytically self-sufficient 119,000-dalton cytochrome p-450 monooxygenase induced by barbiturates in bacillus megaterium. *J Biol Chem* 261:7160–7169, 1986.

65. LP Wen, AJ Fulco. Cloning of the gene encoding a catalytically self-sufficient cytochrome p-450 fatty acid monooxygenase induced by barbiturates in bacillus megaterium and its functional expression and regulation in heterologous (escherichia coli) and homologous (bacillus megaterium) hosts. *J Biol Chem* 262:6676–6682, 1987.

66. KG Ravichandran, SS Boddupalli, CA Hasermann, JA Peterson, J Deisenhofer. Crystal structure of hemoprotein domain of p450bm-3, a prototype for microsomal p450's. *Science* 261:731–736, 1993.

67. IF Sevrioukova, H Li, H Zhang, JA Peterson, TL Poulos. Structure of a cytochrome p450-redox partner electron-transfer complex. *Proc Natl Acad Sci USA* 96:1863–1868, 1999.

68. MC Gustafsson, CN Palmer, CR Wolf, C von Wachenfeldt. Fatty-acid-displaced transcriptional repressor, a conserved regulator of cytochrome p450 102 transcription in bacillus species. *Arch Microbiol* 176:459–464, 2001.

69. O Lentz, V Urlacher, RD Schmid. Substrate specificity of native and mutated cytochrome P450 (*CYP* 102A3) from *Bacillus subtilis. Biotechnology* 108:41–49, 2004.

70. M Budde, SC Maurer, RD Schmid, VB Urlacher. Cloning, expression and characterization of *CYP* 102A2, a self-sufficient P450 in monooxygenase from *Bacillus subtilis. Appl Microbiol Biotechnol* 66:180–186, 2004.

71. R De Mot, AH Parret. A novel class of self-sufficient cytochrome p450 monooxygenases in prokaryotes. *Trends Microbiol* 10:502–508, 2002.

72. T Kitazume, A Tanaka, N Takaya, A Nakamura, S Matsuyama, T Suzuki, H Shoun. Kinetic analysis of hydroxylation of saturated fatty acids by recombinant p450foxy produced by an escherichia coli expression system. *Eur J Biochem* 269:2075–2082, 2002.

73. T Kitazume, N Takaya, N Nakayama, H Shoun. Fusarium oxysporum fatty-acid subterminal hydroxylase (cyp505) is a membrane-bound eukaryotic counterpart of bacillus megaterium cytochrome p450bm3. *J Biol Chem* 275:39734–39740, 2000.

74. V Urlacher, RD Schmid. Biotransformations using prokaryotic p450 monooxygenases. *Curr Opin Biotechnol* 13:557–564, 2002.

75. O Lentz, QS Li, U Schwaneberg, S Lutz-Wahl, P Fischer, RD Schmid. Modification of the fatty acid specificity of cytochrome p450 bm-3 from bacillus megaterium by directed evolution: A validated assay. *J Mol Cat* B 15:123–133, 2001.

76. QS Li, U Schwaneberg, M Fischer, J Schmitt, J Pleiss, S Lutz-Wahl, RD Schmid. Rational evolution of a medium chain-specific cytochrome p-450 bm-3 variant. *Biochim Biophys Acta* 1545:114–121, 2001.

77. TW Ost, CS Miles, J Murdoch, Y Cheung, GA Reid, SK Chapman, AW Munro. Rational re-design of the substrate binding site of flavocytochrome p450 bm3. *FEBS Lett* 486:173–177, 2000.

78. ET Farinas, U Schwaneberg, A Glieder, FH Arnold. Directed evolution of a cytochrome p450 monooxygenase for alkane oxidation. *Adv Synth Catal* 343:601–606, 2001.

79. CF Oliver, S Modi, MJ Sutcliffe, WU Pimrose, LY Lian, GC Roberts. A single mutation in cytochrome p450 bm-3 changes substrate orientation in a catalytic intermediate and the regioselectivity of hydroxylation. *Biochemistry* 36:1567–1572, 1997.

80. SE Graham-Lorence, G Truan, JA Peterson, JR Falck, S Wei, C Helvig, JH Capdevila. An active site substitution f87v, converts cytochrome p450 bm-3 into a regio- and stereoselective (14s, 15r)-arachidonic acid epoxygenase. *J Biol Chem* 262:801–810, 1997.

81. G Truan, MR Komandla, JR Falck, JA Peterson. P450bm-3: Absolute configuration of the primary metabolites of palmitic acid. *Arch Biochem Biophys* 366:192–198, 1999.

82. JH Capdevila, S Wei, C Helvig, JR Falck, Y Belosludtsev, G Truan, SE Graham-Lorence, JA Peterson. The highly stereoselective oxidation of polyunsaturated fatty acids by cytochrome p450 bm-3. *J Biol Chem* 271:22663–22671, 1996.

83. PC Cirino, Y Tang, K Takahashi, DA Tirrell, FH Arnold. Global incorporation of norleucine in place of methionine in cytochrome p450 bm-3 heme domain increases peroxygenase activity. *Biotechnol Bioeng* 83:729–734, 2003.

84. A Glieder, ET Farinas, FH Arnold. Laboratory evolution of a soluble, self-sufficient, highly active alkane hydroxylase. *Nat Biotechnol* 20:1135–1139, 2002.

85. GCK Roberts. The power of evolution: Accessing the synthetic potential of p450s. *Chemistry & Biology* 6:269–272, 1999.

86. U Schwaneberg, C Schmidt-Dannert, J Schmitt, RD Schmid. A continuous spectrophotometric assay for p450 bm–3, a fatty acid hydroxylating enzyme, and its mutant f87a. *Anal Biochem* 269:359–366, 1999.

87. WH Schunck, E Kargel, B Gross, B Wiedmann, S Mauersberger, K Kopke, U Kiessling, M Strauss, M Gaestel, HG Muller. Molecular cloning and characterization of the primary structure of the alkane hydroxylating cytochrome p450 from the yeast candida maltosa. *Biochem Biophys Res Commun* 161:843–850, 1989.

88. M Ohkuma, S Muraoka, T Tanimoto, M Fujii, A Ohta, M Takagi. Cyp52 (cytochrome p450alk) multigene family in candida maltosa: Identification and characterization of eight members. *DNA Cell Biol* 14:163–173, 1995.

89. K Lottermoser, WH Schunck, O Asperger. Cytochromes p450 of the sophorose lipid-producing yeast candida apicola: Heterogeneity and polymerase chain reaction-mediated cloning of two genes. *Yeast* 12:565–575, 1996.

90. W Sghezzi, C Meili, R Ruffiner, R Kuenzi, D Sanglard, A Fiechter. Identification and characterization of additional members of the cytochrome p450 multigene family cyp52 of candida tropicalis. *DNA Cell Biol* 11:767–780, 1992.

91. U Scheller, T Zimmer, D Becher, F Schauer, WH Schunck. Oxygenation cascade in conversion of n-alkanes to α,ω-dioic acids catalyzed by cytochrome p450 52a3. *J Biol Chem* 273:32528–32534, 1998.

92. TL Rohrer, SK Picataggio. Targeted integrative transformation of candida tropicalis by electroporation. *Appl Microbiol Biotechnol* 36:650–654, 1992.

93. S Picataggio, T Rohrer, K Deanda, D Lanning, R Reynolds, J Mielenz, LD Eirich. Metabolic engineering of candida tropicalis for the production of long-chain dicarboxylic acids. *Biotechnology* (NY) 10:894–898, 1992.

94. DL Craft, KM Madduri, M Eshoo, CR Wilson. Identification and characterization of the *CYP52* family of candida tropicalis atcc 20336, important for the conversion of fatty acids and alkanes to alpha,omega-dicarboxylic acids. *Appl Environ Microbiol* 69:5983–5991, 2003.

95. FM Dickinson, C Wadforth. Purification and some properties of alcohol oxidase from alkane grown candida tropicalis. *Biochem J* 282:325–331, 1992.

96. T Zimmer, T Iida, WH Schunck, Y Yoshida, A Ohta, M Takagi. Relation between evolutionary distance and enzymatic properties among the members of the *CYP52a* subfamily of candida maltosa. *Biochem Biophys Res Commun* 251:244–247, 1998.

97. AE Simpson. The cytochrome p450 4 (*CYP4*) family. *Gen Pharmacol* 28:351–359, 1997.

98. S Rendic, FJ Di Carlo. Human cytochrome p450 enzymes: A status report summarizing their reactions, substrates, inducers and inhibitors. *Drug Metab Rev* 29:413–580, 1997.

99. MB Fisher, YM Zheng, AE Rettie. Positional specificity of rabbit *CYP4b1* for ω-1 hydroxylation of short-medium chain and hydrocarbons. *Biochem Biophys Res Commun* 248:352–355, 1998.

100. PP Tamburini, HA Masson, SK Bains, RJ Makowski, B Morris, GG Gibson. Multiple forms of hepatic cytochrome p450. Purification, characterisation and comparison of a novel clofibrate-induced isozyme. *Eur J Biochem* 139:235–246, 1984.

101. CW Fisher, MS Shet, DL Caudle, CA Martin-Wixtrom, RW Estabrook. High-level expression in escherichia coli of enzymatically active fusion proteins containing the domains of mammalian cytochromes p450. *Proc Natl Acad Sci* USA 89:10817–10821, 1992.

102. CS Chaurasia, MA Altermann, P Lu, RP Hanzlik. Biochemical characterization of lauric acid w-hydroxylationby a *CYP4a1*/nadph-cytochrome p450 reductase fusion protein. *Arch Biochem Biophys* 317:161–169, 1995.

103. U Hoch, Z Zhang, DL Kroetz, PR Ortiz de Montellano. Structural determination of the substrate specificities and regioselectivities of the rat and human fatty acid w-hydroxylases. *Arch Biochem Biophys* 373:63–71, 2000.

104. U Hoch, JR Falck, PRO Ortiz de Montellano. Molecular basis for the ω-regiospecificity of the cyp4a2 and cyp4a3 fatty acid hydroxylases. *J Biol Chem* 275:26952–26958, 2000.

105. MA Altermann, CS Chaurasia, P Lu, JP Hardwick, RP Hanzlik. Fatty acid discrimination and w-hydroxylation by cytochrome p450 4a1 and a cytochrome p4504a1/nadph-p450 reductase fusion protein. *Arch Biochem Biophys* 320:289–296, 1995.

106. BG Fox, Catalysis by non–heme iron, in *Comprehensive Biological Catalysis*, M Sinott, ed. 1997, Academic: London. pp. 261–348.

107. Y Lindqvist, WJ Huang, G Schneider, J Shanklin. Crystal structure of a δ^9 stearoyl-acyl carrier protein desaturase from castor seed and its relationship to other diiron proteins. *EMBO J* 15:4081–4092, 1996.

108. N Elango, R Radhakrishnan, WA Froland, BJ Wallar, CA Earhart, JD Lipscomb, DH Ohlendorf. Crystal structure of the hydroxylase component of methane monooxygenase from methylosinus trichosporium ob3b. *Protein Sci* 6:556–568, 1997.

109. P Nordlund, H Eklund. Structure and function of the escherichia coli ribonucleotide reductase protein r2. *J Mol Biol* 232:123–164, 1993.

110. BJ Wallar, JD Lipscomb. Dioxygen activation by enzymes containing binuclear non-heme iron clusters. *Chem Rev* 96:2625–2657, 1996.

111. DJJ Kurtz. Structural similarity and functional diversity in diiron-oxo proteins. *JBIC* 2:159–167, 1997.

112. LJ Shu, JC Nesheim, K Kauffmann, E Münck, JD Lipscomb, L Que. An $fe_2^{iv}o_2$ diamond core structure for the key intermediate q of methane monooxygenase. *Science* 275:515–518, 1997.

113. EJ McKenna, MJ Coon. Enzymatic ω-oxidation. Iv. Purification and properties of the ω-hydroxylase from *pseudomonas oleovorans*. *J Biol Chem* 245:3882–3889, 1970.

114. P Broun, J Shanklin, E Whittle, C Somerville. Catalytic plasticity of fatty acid modification enzymes underlying chemical diversity of plant lipids. *Science* 282:1315–1317, 1998.

115. S Stymne, A Green, S Singh, M Lenman. Genes for fatty acid d12-epoxygenase of fatty acid epoxide-containing plants and the development of useful producers of fatty acid epoxides. *PCT Int App.*:150 pp., 1998.

116. WD Hitz. *Fatty acid modifying enzymes from developing seeds of vernonia galamenensis and fatty acid desaturase gene DNA sequences, U.S.* (E. I. Du Pont De Nemours and Company, USA). US. 21 pp., 1998.

117. EB Cahoon. *Cloning and sequence of euphorbia lagascae cytochrome p 450 associated with the synthesis of d12-epoxy fatty acids in plants and construction of chimeric gene for production of d12-epoxy fatty acids, U.S. Pat. Appl. Publ.* (USA). US. 30 pp., Cont. of U.S. Ser. No. 219,833, 2003.

118. E Hornung, S Rosahl, H Kühn, I Feussner. Creating lipoxygenases with new positional specificities by site-directed mutagenesis. *Biochem Soc Trans* 28:825–826, 2000.

119. M Lee, M Lenman, A Banas, M Bafor, S Singh, M Schweizer, R Nilsson, C Liljenberg, A Dahlqvist, PO Gummeson, S Sjodahl, A Green, S Stymne. Identification of non-heme diiron proteins that catalyze triple bond and epoxy group formation. *Science* 280:915–918, 1998.

120. M Smith, H Moon, L Kunst. Production of hydroxy fatty acids in the seeds of *arabidopsis thaliana*. *Biochem Soc Trans* 28:947–950, 2000.

121. N Tijet, F Pinot, I Benveniste, BR Le, C Helvig, Y Batard, F Cabello-Huartado, D Werck-Reichhart, J-P Salaun, F Durst. *Plant cytochrome p450-dependent fatty acid hydroxylase genes and their expression in yeast for manufacture of hydroxylated fatty acids, PCT Int. Appl.* (The Centre National de Recherche Scientifique, Fr.). WO. 157 pp., 1999.

122. JB Ohlrogge. Design of new plant products: Engineering of fatty acid metabolism. *Plant Physiol* 104:821–826, 1994.

123. D Fabritius, HJ Schafer, A Steinbuchel. Biotransformation of linoleic acid with the candida tropicalis m25 mutant. *Appl Microbiol Biotechnol* 48:83–87, 1997.

124. EJ Corey, W-G Su, MB Cleaver. A simple and efficient synthesis of (7e,9e,11z,13e)-(5s,6r,15s)-trihydroxyeicosatetraenoic acid (6r-lipoxin a). *Tetrahedron Lett* 30:4181–4184, 1989.

125. D Martini, G Iacazio, D Ferrand, G Buono, C Triantaphylides. Optimization of large scale preparation of 13(s)-hydroperoxy-9z,11e-octadecadienoic acid using soybean lipoxygenase. Application to the chemoenzymatic synthesis of (+)-coriolic acid. *Biocatalysis* 11:47–63, 1994.

126. D Martini, G Buono, G Iacazio. Regiocontrol of soybean lipoxygenase oxygenation. Application to the chemoenzymatic synthesis of methyl 15(s)-hete and methyl 5(s),15(s)-dihete. *J Org Chem* 61:9062–9064, 1996.

127. D Martini, G Buono, J-L Montillet, G Iacazio. Chemo-enzymatic synthesis of methyl-9(s)-hode (dimorphecolic acid methyl ester) and methyl-9(s)-hote catalysed by barley seed lipoxygenase. Tetrahedron: *Asymmetry* 7:1489–1492, 1996.

128. AF Hsu, E Wu, S Shen, TA Foglia, K Jones. Immobilized lipoxygenase in a packed-bed column bioreactor: Continuous oxygenation of linoleic acid. *Biotechnol Appl Biochem* 30 (Pt 3):245–250, 1999.

129. A Weiss, U Schoerken, V Candar, Y Eryasa, M Wunderlich, N Buyukuslu, C Beverungen. *Process for the enzymic generation and recovery of fatty acid hydroperoxides, Eur. Pat. Appl.* (Cognis Deutschland GmbH & Co. K.-G., Germany). EP. 22 pp., 2003.

130. F Björkling, SE Godtfredsen, O Kirk. Lipase-mediated formation of peroxy carboxylic acids used in catalytic epoxidation of alkenes. *J Chem Soc, Chem Commun*:1301–1303, 1990.

131. S Warwel, MR Klaas. Chemoenzymic epoxidation of unsaturated carboxylic acids. *J Mol Cat B: Enzymatic* 1:29–35, 1995.

132. M Rüsch gen. Klaas, S Warwel. Enzymic preparation of peroxy-, epoxy- and peroxy-epoxy fatty acids. Oils-Fats-Lipids 1995, *Proceedings of the World Congress of the International Society for Fat Research, 21st*, The Hague, Oct. 1–6, 1995 3:469–471, 1996.

133. M Rüsch gen Klaas, S Warwel. Lipase-catalyzed peroxy fatty acid generation and lipid oxidation, In: Enzymes in Lipid Modification, UT Bornscheuer, Editor. 2000, VCH: Weinheim, New York. pp 116–127.

134. M Hamberg, G Hamberg. *Arch Biochem Biophys* 283:409–, 1990.

135. GJ Piazza, A Nunez, TA Foglia. Epoxidation of fatty acids, fatty methyl esters, and alkenes by immobilized oat seed peroxygenase. *J Mol Cat* B: Enzymatic 21:143–151, 2003.

136. I Matsunaga, T Sumimoto, E Kusunose, K Ichihara. Phytanic acid a-hydroxylation by bacterial cytochrome p450. *Lipids* 33:1213–1216, 1998.

137. I Matsunaga, T Sumimoto, A Ueda, E Kusunose, K Ichihara. Fatty acid-specific, regiospecific, and stereospecific hydroxylation by cytochrome p450 (cyp152b1) from sphingomonas paucimobilis: Substrate structure required for a–hydroxylation. *Lipids* 35:365–371, 2000.

138. I Matsunaga, A Ueda, N Fujiwara, T Sumimoto, K Ichihara. Characterization of the ybdt gene product of bacillus subtilis: Novel fatty acid b-hydroxylating cytochrome p450. *Lipids* 34:841–846, 1999.

139. PC Cirino, FH Arnold. A self-sufficient peroxide-driven hydroxylation biocatalyst. *Angew Chem Int Ed Engl* 42:3299–3301, 2003.

140. BD Fleming, Y Tian, SG Bell, LL Wong, V Urlacher, HA Hill. Redox properties of cytochrome p450bm3 measured by direct methods. *Eur J Biochem* 270:4082–4088, 2003.

141. J Kazlauskaite, ACG Westlake, L-L Wong, HAO Hill. Direct electrochemistry of of cytochrome p450cam. *Chem Commun* 18:2189–2190, 1996.

142. KM Faulkner, MS Shet, CW Fisher, RW Estabrook. Electrocatalytically driven ω-hydroxylation of fatty acids using cytochrome p450 4a1. *Proc Natl Acad Sci* USA 92:7705–7709, 1995.

143. F Hollmann, PC Lin, B Witholt, A Schmid. Stereospecific biocatalytic epoxidation: The first example of direct regeneration of a fat-dependent monooxygenase for catalysis. *J Am Chem Soc* 125:8209–8217, 2003.

144. U Schwaneberg, D Appel, J Schmitt, RD Schmid. P450 in biotechnology: Zinc driven omega-hydroxylation of p-nitrophenoxydodecanoic acid using p450 bm-3 f87a as a catalyst. *J Biotechnol* 84:249–257, 2000.

145. B Hauer, RD Schmid, U Schwaneberg. *Electron donor system for redox enzymes and its use for preparation of hydroxylated fatty acids*, PCT Int. Appl. (BASF Aktiengesellschaft, Germany). WO. 94 pp., 2001.

146. P Fernandez–Salguero, C Gutierrez–Merino, AW Bunch. Effect of immobilization on the activity of rat hepatic microsomal cytochrome p450 enzymes. *Enzyme Microb Technol* 15:100–104, 1993.

147. F Hollmann, B Witholt, A Schmid. [cp*rh(bpy)(h2o)]2+: A versatile tool for efficient and non-enzymatic regeneration of nicotinamide and flavin coenzymes. *J Mol Cat* B 791:1–10, 2002.

148. VI Tishkov, AG Galkin, VV Fedorchuk, PA Savitsky, AM Rojkova, H Gieren, MR Kula. Pilot scale production and isolation of recombinant nad+− and nadp+−specific formate dehydrogenases. *Biotechnol Bioeng* 64:187–193, 1999.

149. S Maurer, V Urlacher, H Schulze, RD Schmid. Immobilisation of p450 bm-3 and an nadp+ cofactor recycling system: Towards a technical application of heme-containing monooxygenases in fine chemical synthesis. *Adv Synth Catal* 345:802–810, 2003.

5

Lipase-Catalyzed Kinetic Resolution for the Fractionation of Fatty Acids and Other Lipids

Kumar D. Mukherjee

5.1   Introduction

Lipases (triacylglycerol acylhydrolases, EC 3.1.1.3) from various organisms have been used successfully as biocatalysts in bioorganic synthesis[1–6] and modification of fats and other lipids via hydrolysis, esterification, and interesterification.[7–10] Since many of the commercially available lipase preparations are rather expensive, their industrial use is justified mainly for the manufacture of products of good commercial value,[10,11] such as purified fatty acids and lipids from inexpensive starting materials. This chapter covers the specificity of lipases in view of their application as biocatalyst in the fractionation of fatty acids and other lipids via kinetic resolution,[12–15] which utilizes the difference in reactivities of individual constituents of a mixture in lipase-catalyzed reactions that enables the enrichment of the desired component in either the product or the unreacted substrate.

5.2   Fractionation of Fatty Acids and Other Lipids by Lipase-Catalyzed Reactions

The ability of certain lipases to prefer or discriminate against particular fatty acids or acyl moieties of lipids has been utilized for the enrichment of such fatty acids or their esters from natural fats and oils via selective hydrolysis, esterification, and interesterification, i.e., via kinetic resolution, as shown schematically

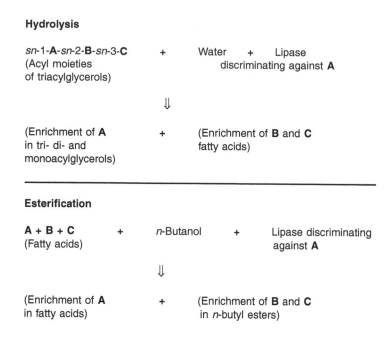

FIGURE 5.1   Scheme of kinetic resolution for the enrichment of fatty acids/acyl moieties via lipase-catalyzed hydrolysis of triacylglycerols or esterification of fatty acids with *n*-butanol.

in Figure 5.1. Thus, in the hydrolysis of triacylglycerols, catalyzed by a lipase that discriminates against the acyl moieties A over B and C, the fatty acids cleaved are enriched with the fatty acids B and C, whereas A is enriched in the unhydrolyzed triacylglycerols and di- as well as monoacylglycerols. If the lipase is regioselective for the *sn*-1,3 positions of the triacylglycerols, the acyl moieties A and C are preferentially hydrolyzed to fatty acids and B is enriched in the unhydrolyzed triacylglycerols and *sn*-1,2(2,3)-diacylglycerols as well as *sn*-2-monoacylglycerols. The substrate specificities of triacylglycerol lipases are also utilized for the enrichment of a specific fatty acid or group of fatty acids from their mixtures with other fatty acids via kinetic resolution through selective esterification or transesterification. For example, as shown in Figure 5.1, in the lipase-catalyzed esterification of a mixture of fatty acids A, B, and C with *n*-butanol, the fatty acid A which is discriminated against by the lipase, is enriched in the unesterified fatty acids while the fatty acids B and C are enriched in the *n*-butyl esters formed.

5.2.1   Lipase Specificities

Regioselectivity, stereoselectivity (enantioselectivity), and fatty acid specificity of triacylglycerol lipases are well documented in the literature.[5,12–15]

5.2.1.1   Regioselectivity and Stereoselectivity

Regioselectivity of a lipase, e.g., selectivity toward acyl moieties located at the primary (*sn*-1-(and *sn*-3) positions) of the glycerol backbone *versus* those at the secondary (*sn*-2 position), can be determined from the regiospecific analysis of triacylglycerols formed by lipase-catalyzed esterification or interesterification. For regiospecific analysis, the triacylglycerols are generally subjected to partial hydrolysis, catalyzed by a regiospecific (*sn*-1,3-specific) lipase, e.g., from porcine pancreas and the composition of fatty acids liberated from the *sn*-1,3-positions of the triacylglycerols and those retained in the *sn*-2-monoacylglycerols are separated by thin-layer chromatography on silicagel, impregnated with boric acid, converted to methyl esters, and analyzed by gas chromatography.[16] Alternatively, the triacylglycerols are subjected to Grignard degradation and the *sn*-1,3-diacylglycerols and the *sn*-2-monoacylglycerols formed are separated by thin-layer chromatography on silicagel, impregnated with boric acid.[16] Each of the *sn*-1,3-diacylglycerols and the *sn*-2-monoacylglycerols are converted to methyl esters and analyzed by gas chromatography to determine the regioselective distribution of the acyl moieties on the triacylglycerols.

Using the same methodology as above, the stereoselectivity of a lipase for *sn*-1 and *sn*-3 positions of the glycerol backbone can be determined via Grignard degradation by separation and quantitation of *sn*-1- and *sn*-3-monoacylglycerol enantiomers as their 3,5-dinitrophenylurethane derivatives on a chiral HPLC column.[17]

For the determination of stereoselectivity of a lipase for *sn*-1 and *sn*-3 positions of the glycerol backbone, racemic mixtures of alkylglycerols, e.g., *rac*-1(3)-*O*-octadecylglycerols have been esterified with oleic acid using various lipases as biocatalyst.[18] Subsequently, the unesterified mixture of enantiomeric substrates were converted to their 3,5-dinitrophenylurethane derivatives, which were resolved into *sn*-1 and *sn*-3 enantiomers by HPLC on a chiral stationary phase.[18] The relative proportion of *sn*-1 and *sn*-3 enantiomers in the unreacted alkylglycerols revealed the enantioselectivity of the lipase. Lipases from porcine pancreas, *Rhizopus* sp., *Pseudomonas* sp., *Candida rugosa*, *Chromobacterium viscosum*, and *Penicillium cyclopium* had a distinct preference for 1-*O*-octadecyl-*sn*-glycerol over its enantiomer, indicating stereoselectivity for the *sn*-3 position, whereas lipase from *Rhizomucor miehei* had a slight stereoselectivity for the *sn*-1 position. The stereoselectivity of most of the above lipases for the *sn*-3 position in esterification parallels the *sn*-3 stereoselectivity found in the hydrolysis of triacylglycerols by rabbit and human lingual lipases,[19,20] rabbit and human gastric lipases,[21] and lipase from American cockroach.[22] Moreover, *sn*-3 stereoselectivity has also been observed in the hydrolysis of *rac*-1(3)-acyl-2-acyl-amino-2-deoxyglycerols by porcine pancreatic lipase and that of *rac*-1(3)-*O*-alkyl-2,3-diacylglycerols by rabbit and human gastric lipase[23] and rat lingual lipase.[24] In analogy, the *sn*-1 stereoselectivity observed in the esterification using *Rhizomucor miehei* lipase[18] has also been found in the hydrolysis of triacylglycerols by lipoprotein lipase[25] and lipases from bovine[25] and human[26] milk. Moreover, *sn*-3 stereoselectivity has also been observed in the hydrolysis of *rac*-1(3)-*O*-alkyl-2,3-diacylglycerols by lipases from bovine milk and adipose tissue.[24]

Partial hydrolysis of both chiral and racemic mixtures of triacylglycerols and analysis of the fatty acids released have been carried out in order to determine whether a lipase from *Carica papaya* is truly stereoselective.[27]

In a rapid method for the determination of regiospecific distribution of acyl moieties in triacylglycerols, they are subjected to transesterification with 1-butanol or 2-butanol, catalyzed by an *sn*-1,3-specific lipase, e.g., from *R. miehei*, and the resulting butyl esters of fatty acids from the *sn*-1- and *sn*-3-positions analyzed by gas chromatography.[28] Similarly, the regiospecific distribution of acyl moieties in enzymatically structured triacylglycerols has been determined by their transesterification with ethyl acetate, catalyzed by a *sn*-1,3-specific lipase, followed by direct gas chromatography of the reaction products.[29] Acyl moieties located at the *sn*-2 position are determined as triacylglycerols containing two acetate groups in the *sn*-1,3-positions, whereas the acyl moieties located at the *sn*-1,3-positions are simultaneously determined as their ethyl esters.

Using the above methods, regioselectivity and stereoselectivity of lipases have been determined as summarized in Table 5.1.

5.2.1.2 Lipase Specificities Toward Fatty Acids and Their Derivatives

Fatty acid selectivity and activity of several commercially important lipase preparations have been evaluated[35–37] with a view to assess their potential as biocatalysts for the enrichment of definite fatty acids that are of considerable interest due to their biomedical[38] and technical properties.[10,11] Similarly, specificity of lipases toward alcohols is of interest in view of enrichment of a particular alcohol from mixtures via kinetic resolution. Specificity of lipases toward individual fatty acids is determined from the overall reaction rates in hydrolysis, esterification, and interesterification. Using mixtures of triacylglycerols or fatty acids as substrates, preferences for or discrimination against definite acyl moieties/fatty acids are determined by comparing their relative rates of conversion. Table 5.1 also summarizes the fatty acid chain-length specificities of lipases.

TABLE 5.1   Chain-Length Specificity and Regioselectivity of Triacylglycerol
Lipases from Different Sources

Source of Lipase [Reference]	Fatty Acid Chain-Length Specificity[a]	Positional Specificity Regioselectivity
Microorganisms		
Aspergillus niger [30]	S, M, L	sn-1, 3 >> sn-2
Candida antarctica[30]	S > M, L	sn-3
Candida rugosa (syn.*Candida cylindracea*) [30]	S, L > M	sn-1, 2, 3
Chromobacterium viscosum [30]	S, M, L	sn-1, 2, 3
Rhizomucor miehei [30]	S > M, L	sn-1, 3 >> sn-2
Rhizomucor miehei [31,32]	S < M < L	
Penicillium roqueforti[30]	S, M >> L	sn-1, 3
Pseudomonas aeruginosa [27]	S, M, L	sn-1
Pseudomonas flourescens[30]	S, L > M	sn-1, 2, 3
Rhizopus delemar[30]	S, M, L	sn-1, 2, 3
Rhizopus oryzae[30]	M, L > S	sn-1, 3 >> sn-2
Plants		
Rapeseed [*Brassica napus*] [33]	S > M, L	*sn*-1, 3 > *sn*-2
Papaya [*Carica papaya*] latex [27]	*sn*-3	*sn*-3
Papaya [*Carica papaya*] latex [34]	S < M < L	
Animal Tissues		
Porcine pancreatic	S > M, L	*sn*-1, 3
Rabbit gastric [27]	S, M, L	*sn*-3

[a] S, short-chain; M, medium-chain; L, long-chain

The effect of the chemical structure of the acid substrate on its relative reactivity in esterification with
1-octanol, catalyzed by Lipozyme, has been systematically studied by Miller et al.[39] The rate of esterification of medium-chain saturated acids has been shown to increase with their chain length from C4 to
C8, whereas the presence of a methyl, ethyl, cyclohexyl, or phenyl substituent on the ß-position of the
carbon chain reduces the reaction rate to a great extent. The presence of an olefinic bond in hexanoic
acid results in greatly reduced reactivity as compared to the saturated analog, i.e., hexanoic acid.

In the esterification of straight-chain saturated fatty acids of different chain lengths with 1-hexanol,
catalyzed by papaya (*Carica papaya*) lipase, the initial reaction rate has been found to increase with
increasing chain length of the acid from C4:0 to C18:0, followed by a slight decrease with C20:0.[34]

Kinetic analysis under competitive conditions can be carried out according to Rangheard et al.[37] For
example, kinetic analysis of lipase-catalyzed esterification of fatty acids with *n*-butanol has been carried
out by reacting 250 mM of each of the fatty acids, individually, together with 250 mM of myristic or
oleic acid, the reference standard, and 500 mM *n*-butanol in 1 ml hexane in the presence of 5–10% w/w
of substrate of the lipase preparation.[35] The reaction products, consisting of butyl esters and unesterified
fatty acids, are treated with diazomethane to convert the unreacted fatty acids to their methyl esters. The
resulting mixtures of butyl esters and methyl esters are analyzed by gas chromatography. Overall reaction
rates are calculated from the composition of the reaction products as µmol butyl esters formed per g
lipase per min.

Specificity constants are calculated according to reference 37 as follows. For two substrates competing
for the lipase, the ratio of the reaction rates of each substrate ($v1$ and $v2$) is derived from

$$v1/v2 = \alpha.(Ac1X)/(Ac2X)$$

where (Ac1X) and (Ac2X) are concentrations of the two substrates at time X and α is the competitive
factor that is defined by the following equation:

$$\alpha = (VAc1X/KAc1X)/(VAc2X/KAc2X)$$

TABLE 5.2 Specificity Constants of Lipases from Various Sources in the Esterification of Fatty Acids with *n*-Butanol Under Competitive Conditions

Fatty Acid	Specificity Constant of Lipase From				
	Candida rugosa	*Rhizopus arrhizus*	*Rhizomucor miehei*	*Porcine pancreas*	*Brassica napus*
Petroselinic(*cis*-6 18:1)	0.04	0.02	0.09	0.11	0.08
gamma-Linolenic (all *cis*-6,9,12-18:3, n-6 18:3)	0.02	0.02	0.06	0.32	0.09
Dihomogamma-linolenic (all *cis*-6,9,12-20:3, n-6 20:3)	0.32	0.06	0.15	0.89	0.16
Stearidonic (all *cis*-6,9,12,15-18:4, n-3 18:4)	0.13	0.07	0.15	0.38	0.07
Docosahexaenoic (all *cis*-4,7,10,13,16,19-22:6, n-3 22:6)	0.13	0.05	0.02	0.49	0.09
Hydnocarpic (11-[cyclopent-2-en-1-yl]undecanoic)	1.02	1.24	1.47	1.34	2.23
Chaulmoogric (13-[cyclopent-2-en-1-yl]tridecanoic	0.50	0.73	0.96	0.98	1.43
Ricinoleic(12-hydroxy-*cis*-9-octadecenoic)	1.90	2.63	2.03	6.10	1.60
12-Hydroxy-octadecanoic		0.69	0.97	3.50	0.86
cis-9,10-Epoxy octadecanoic		1.46	1.70	0.80	
trans-9,10-Epoxy-octadecanoic	1.77	2.63	1.59	2.30	1.46
Oleic (*cis*-9-octadecenoic)	1.00	1.00	1.00	1.00	1.00

Source: I Jachmanián, E Schulte, KD Mukherjee. Substrate selectivity in esterification of less common fatty acids catalysed by lipases from different sources. *Appl Microbiol Biotechnol* 44: 563–567, 1996.

where V is the maximal velocity and *K* is Michaelis constant. The competitive factor is calculated from the substrate concentrations Ac1X0 and Ac2X0 at time zero as follows:

$$\alpha = Log[Ac1X0/Ac1X]/Log[Ac2X0/Ac2X].$$

The specificity constant is calculated from the competitive factor as $1/\alpha$ with reference to the specificity constant of the reference standard, e.g., myristic or oleic acid, taken as 1.00. The higher the specificity constant for a particular fatty acid, the higher the preference of the lipase for that particular fatty acid as substrate.

Table 5.2 summarizes the fatty acid selectivity of triacylglycerol lipases from various sources, expressed as specificity constant in the lipase-catalyzed esterification of these fatty acids with *n*-butanol. Lipases from *C. rugosa*, *R. miehei*, *R. arrhizus* (Table 5.2 and reference 35), *Penicillium cyclopium* and *Penicillium* sp. (Lipase G).[35] and those from germinating rapeseed (Table 5.2 and reference 33), papaya (*Carica papaya*) latex[40] and bromelain from pineapple as well as *Rhizopus* sp.[41] have closely resembling substrate specificities in esterification with *n*-butanol. All these lipases strongly discriminate against unsaturated fatty acids having the first double bond from the carboxyl end at an even number carbon, i.e., *cis*-4, e.g. all-*cis*-4,7,10,13,16,19-docosahexaenoic (n-3 22:6), *cis*-6, e.g., γ-linolenic (all-*cis*-6,9,12-octadecatrienoic, n-6 18:3), petroselinic (*cis*-6-octadecenoic, n-12 18:1) and stearidonic (all-*cis*-6,9,12,15-octadecatetraenoic, n-3 18:4) or *cis*-8, e.g. dihomo-γ-linolenic (all-*cis*-8,11,14-eicosatrienoic, n-6 20:3) acid. Recently, potato tuber lipid acyl hydrolase has also been found to strongly discriminate against n-3 22:6 and all-*cis*-5,8,11,14,17-eicosapentaenoic (n-3 20:5) acid in the esterification with glycerol and 1,3-propanediol.[42]

The selective discrimination of the lipase from porcine pancreas against fatty acids having a *cis*-4, *cis*-6, and *cis*-8 double bond is, however, not as pronounced as the corresponding selectivities of the above lipases (Table 5.2 and reference 35).

The lipase from *Chromobacterium viscosum* exhibits different substrate specificity compared to the above lipase preparations. This lipase utilizes fatty acids with a *cis*-6 double bond, e.g., n-12 18:1, n-6 18:3, and n-3 18:4, as substrates equally well or even better than myristic acid or α-linolenic acid (n-3 18:3).[35]

In interesterification reactions, lipase from *R. miehei* discriminates against *cis*-6-18:1,[43] n-6 18:3,[37,43] and n-3 22:6.[44] Lipases from *C. rugosa*, porcine pancreas, and *Geotrichum candidum* also discriminate against n-6 18:3 in interesterification reactions.[37]

In hydrolytic reactions catalyzed by lipases from *R. arrhizus* and porcine pancreas, discrimination against *cis*-6-18:1[45] and n-3 22:6[46] moieties has been observed. Similarly, lipase from rape hydrolyzes tripetroselinin and tri-γ-linolenin at much lower rate than triolein.[33]

It appears from the above data that discrimination against fatty acids having the first double bond from the carboxyl end as a *cis*-4, *cis*-6 or a *cis*-8 is a common feature of many lipases. It has been suggested that the lipase from rape discriminates against *anti*-oriented *cis*-4 and *cis*-6 unsaturated fatty acids due to the direction of twist of the carbon chain after the first double bond, which might hinder binding of the reactive group to the lipase.[33] It is conceivable that the same argument is valid for the selective discrimination of many lipases against fatty acids having a *cis*-4, *cis*-6, or a *cis*-8 double bond as the first olefinic bond at the carboxyl end of the fatty acid.

In the esterification of common and unusual fatty acids with *n*-butanol, lipases from rape, porcine pancreas, *C. rugosa*, *R. miehei* and *R. arrhizus* (Table 5.2), papaya[40] and pineapple as well as *Rhizopus* sp.[41] have been shown to have strong preference for fatty acids having hydroxy groups, e.g., ricinoleic (12-hydroxy-*cis*-9-octadecenoic) and 12-hydroxystearic acid, epoxy groups, e.g., *trans*-9,10-epoxystearic acid, and cyclopentenyl fatty acids having saturated alkyl chains, e.g., hydnocarpic [11-(cyclopent-2-en-1-yl)undecanoic] and chaulmoogric [13-(cyclopent-2-en-1-yl)tridecanoic] acid, whereas a cyclopentenyl fatty acid having a *cis*-6 olefinic bond, i.e., gorlic [13-(cyclopent-2-en-1-yl)tridec-6-enoic] acid is strongly discriminated against by several lipases.[36,47]

Various lipases have been found to exhibit different selectivity toward geometrical and positional isomers of linoleic acid. Thus, in the lipase-catalyzed esterification of the individual isomers of conjugated linoleic acids with *n*-butanol, lipase from *C. rugosa* exhibits a distinct preference for *cis*-9, *trans*-11-octadecadienoic acid, and the lipase B from *C. antarctica* a strong selectivity toward 9-*trans*-,11-*trans*-octadecadienoic acid, whereas *C. antarctica* lipase A and *Mucor miehei* lipase do not exhibit strong selectivity for individual isomers.[48] Lipase from *Geotrichum candidum*[49,50] as well as that from *C. rugosa*[51] have a strong selectivity toward *cis*-9, trans-11-octadecadienoic acid over *trans*-10, *cis*-11-octadecadienoic acid.

A partially purified lipase from *Vernonia galamensis* seeds has been shown to catalyze the hydrolysis of trivernolin (tri-*cis*-12,13-epoxy-*cis*-9-octadecenoin), the predominant constituent of the seed oil of *V. galamensis*, much faster than triolein or other triacylglycerols.[52] Similarly, in the transesterification of tricaprylin with fatty acids, catalyzed by purified *V. galamensis* lipase, a strong preference for vernolic (*cis*-12,13-epoxy-*cis*-9-octadecenoic) acid has been observed.[52]

Lipase contained in the seeds of *Cuphea procumbens* has over 20-fold selectivity for capric acid,[53] which can be utilized for the enrichment of this acid via selective hydrolysis of an oil-containing capric acid. In the esterification of short-chain, medium-chain, and long-chain acids with glycerol, catalyzed by potato tuber lipid acyl hydrolase in isooctane, the selectivities are of the order caprylic > myristic > linoleic > butyric.[54]

The above substrate specificities of various lipases have been utilized for the enrichment of specific unsaturated fatty acids or derivatives via kinetic resolution from their mixtures, obtained from naturally occurring fats and other lipids, as summarized in Table 5.3.

5.2.1.3 Lipase Specificities Toward Alcohols

Specificity of lipases toward alcohols can be determined in a similar manner as described above for fatty acids from relative reactivity of individual alcohols from their equimolar mixtures with others during lipase-catalyzed esterification and interesterification with acids and esters, respectively, under competitive conditions.

TABLE 5.3 Specificity of Lipases in View of Kinetic Resolution of Fatty Acids and Other Lipids

Lipase From	Reaction	Specificity	Reference
Porcine pancreas	Hydrolysis of seed oil triacylglycerols	Discrimination against acyl moieties having a *trans*-3 olefinic bond	[55]
G. candidum	Hydrolysis of triacylglycerols containing isomeric *cis*-octadecenoyl moieties	Strong preference for *cis*-9-18:1	[56]
Porcine pancreas	Hydrolysis of triacylglycerols containing isomeric *cis*-octadecenoyl moieties	Discrimination against *cis*-2-18:1 to *cis*-7-18:1	[57]
G. candidum	Hydrolysis of triacylglycerols	Distinct preference for C_{18} acyl moieties having *cis*-9- or *cis*-9, *cis*-12-bonds	[58]
G. candidum	Hydrolysis of triacylglycerols	Distinct preference for C_{18} acyl moieties having *cis*-9- or *cis*-9, *cis*-12-bonds and discrimination against *cis*-13-22:1	[59]
G. candidum (lipase B)	Hydrolysis of fatty acid methyl esters	Highly selective for *cis*-9-18:1 as compared to *cis*-11-18:1 or *trans*-9-18:1	[60]
Oat (*Avena sativa*) seed lipase	Hydrolysis of triacylglycerols	Discrimination against tripetroselinin as compared to triolein, trilinolein and tri-α-linolenin	[61]
G. candidum	Esterification of fatty acids with *n*-butanol	Distinct preference for C_{18} acyl moieties having *cis*-9- or *cis*-9, *cis*-12-bonds and discrimination against *cis*-13-22:1, *trans*-9-18:1, n-6 18:3, ricinoleic and 9-docosynoic acids	[62]
G. candidum [lipase B]	Hydrolysis of triacylglycerols and esterification of fatty acids with *n*-octanol	Highly selective for triolein and for *cis*-9-18:1 as compared to *cis*-6-18:1, *cis*-11-18:1 or *trans*-9-18:1	[63]
Pythium ultimum	Hydrolysis of triacylglycerols	Preference for tri-α-linolenin > trilinolein > triolein	[64]
Oat [*Avena sativa*] caryopses	Hydrolysis of milk fat triacylglycerols	Preferential hydrolysis of C_6-C_{10} acids	[65]
R. miehei	Alcoholysis of fatty acid ethyl esters with *n*-propanol	Preference for n-9 20:1 and n-9 22:1, strong discrimination against n-3 22:6, and moderate discrimination against n-3 20:5	[66]
Candida parapsilosis	Hydrolysis of fatty acid methyl esters	Substrate preference: *cis*-9-16:1 ≈ *cis*-9-18:1 >> n-6 18:3 > *cis*-11-18:1 >> *cis*-6-18:1	[67]
C. rugosa	Hydrolysis of borage oil	Discrimination against n-6 18:3	[68]
R. miehei	Hydrolysis of fatty acid ethyl esters	Discrimination against n-3 22:6 as compared to n-6 18:2 and n-3 18:3	[69]
Flavobacterium odoratum	Hydrolysis of *p*-nitro-phenyl esters of fatty acids	Substrate preference: n-6 18:2 > *cis*-9-18:1 ≈ n-6 18:3 >> *cis*-6-18:1 > *trans*-9-18:1	[70]

(continued)

TABLE 5.3   Specificity of Lipases in View of Kinetic Resolution of Fatty Acids and Other Lipids (Continued)

Lipase From	Reaction	Specificity	Reference
R. delemar	Interesterification of randomized triacylglycerols with caprylic acid	Strong discrimination against n-6 18:3 and n-3 22:6; moderate discrimination against n-6 20:4 and n-3 20:5	[71]
C. rugosa, Pseudomonas cepacia, porcine pancreas	Esterification of acetylenic fatty acids with *n*-butanol	Preference for 10-undecynoic acid and strong discrimination against 6-octadecynoic acid	[72]
C. rugosa, R. miehei, porcine pancreas	Esterification of medium- and long-chain olefinic and acetylenic alcohols with pentanoic acid and stearic acid	Discrimination against medium- and long-chain olefinic alcohols in the esterification with pentanoic acid; olefinic and acetylenic alcohols well accepted in the esterification with stearic acid catalyzed by lipase from porcine pancreas	[73]
Burkholderia sp.	Hydrolysis of *t*-butyl esters	Strong preference for *t*-butyl octanoate as compared to *t*-butyl palmitate and stearate	[74]
C. rugosa, R. miehei, C. antarctica lipase A	Esterification of *cis*- and *trans*-octadecenoic acids with *n*-butanol	Preference for oleic acid over elaidic acid by *C. cylindracea* and *R. miehei*, whereas strong preference for elaidic acid by *C. antarctica* lipase A	[75]
C. rugosa, Mucor miehei, C. antarctica lipase A	Transesterification of *cis-trans*-isomers of linoleic acid with *n*-butanol	Varying selectivity of the lipases toward isomeric octadecadienoic acids	[76]
C. rugosa	Hydrolysis of menhaden oil	Preference for palmitoleic (n-7 16:1) acid	[77]
Pseudomonas aeruginosa	Hydrolysis of a single-cell oil from a *Mortierella alpina* mutant	Preference for linoleic acid (n-9 18:2) over Mead's acid (n-9 20:3)	[78]

The effect of the chemical structure of the alcohol substrate on its relative reactivity in esterification with myristic acid, catalyzed by immobilized *Rhizomucor miehei* lipase (Lipozyme RM IM), has been studied systematically by Miller et al.[39] In general, primary alcohols react better than secondary alcohols, and most of the alcohols containing cyclic saturated or aromatic substituents are well accepted as substrate by this enzyme, although to a different degree, e.g., cyclohexylmethanol reacts much faster than its aromatic analog benzyl alcohol. Numerous other substances containing a hydroxy group, such as methoxyethanol, ethylene cyanohydrin, glycolic acid, methoxypropanediol, 3-phenylpropanol, cinnamyl alcohol, geraniol, and allyl alcohol, are good substrates for esterification. In the Lipozyme RM-catalyzed esterification of a mixture of *cis*- and *trans*-4-methyl cyclohexanol with myristic acid, the *trans*-isomer reacts more than twice as fast as the *cis*-isomer, which should form the basis of kinetic resolution of these two substances.

Esterification of various short-chain acids with C1-C6 alcohols and terpene alcohols using *R. arrhizus* and *R. miehei* lipases has been reported.[31]

An optimal chain length of C4 has been found in the esterification of C2 to C16 primary alcohols with oleic (*cis*-9-octadecenoic) acid, catalyzed by lipase from rape (*Brassica napus*).[33] The same lipase discriminates against secondary and tertiary alcohols.

Also in the geranyl ester synthesis catalyzed by *Rhizopus oryzae* lipase, C2, C3, and C4 acids are converted to the extent of 31%, 86%, and 95%, respectively.[79]

In the esterification of dodecanoic acid with C2-C10 alcohols using a lipase from *Thermomyces lanuginosus* (*syn*. Humicola *lanuginosa*), an increase in conversion with increasing chain length of alcohols from C2 to C8 is observed, which is followed by a marginal decrease in the case of 1-decanol.[80] A linear decrease of initial rate of esterification with increase in alcohol chain length has been observed in the synthesis of alkyl oleates (*cis*-9-octadecenoates) of C3-C16 alcohols using the immobilized *R. miehei* lipase.[81] Similarly, in the synthesis of alkyl oleates, gondoates (*cis*-11-eicosenoates) and erucates (*cis*-13-docosenoates), the above lipase gives an increase in synthetic yields with increasing alcohol chain length from C10 to C18.[32] However, with C6, C8 or C10 acids, the esterification appears to be independent of the chain length of the alcohol.[32]

In the esterification of oleic acid with long-chain saturated alcohols, catalyzed by rat pancreatic lipase, porcine pancreatic carboxylester lipase, and *Pseudomonas fluorescence* lipase, the extent of formation of wax esters steeply increases with increasing chain length of the alcohol from C4 and C6 to C10 to C18.[82] On the other hand, in the esterification of arachidonic acid with alcohols, catalyzed by *C. antarctica* lipase, the rate of esterification increases with decreasing chain length of alcohol from 1-heptanol to methanol.[83]

In the esterification of saturated primary alcohols (C2-C16) with 1-octanoic acid, catalyzed by papaya (*Carica papaya*) lipase, maximum conversion has been obtained with 1-octanol, whereas short-chain alcohols and secondary as well as tertiary alcohols show low reactivity.[34]

In the esterification of 1-octanoic acid with terpene alcohols, catalyzed by papaya lipase, β-citronellol [3,7-dimethyl-6-octen-1-ol] and geraniol [(2E)-3,7-dimethylocta-2,6-dien-1-ol] are more reactive than nerol [(2Z)-3,7-dimethylocta-2,6-dien-1-ol].[34] The highest reaction rate is found for the aromatic benzyl alcohol (phenylmethanol).

In the alcoholysis of triacylglycerols of crambe or camelina oil, catalyzed by Lipozyme RM IM or Novozym 435 (lipase B from *C. Antarctica*), long-chain alcohols, such as *cis*-9-octadecenyl (oleyl) alcohol and those derived from crambe (*Crambe abyssinica*) oil and camelina (*Camelina sativa*) oil, yield higher proportions of esters as compared to medium-chain (e.g., *n*-octanol) and short-chain alcohols (e.g., isopropanol).[84]

Straight-chain saturated C4 to C18 alcohols and unsaturated C18 alcohols, such as *cis*-9-octadecenyl (oleyl), *cis*-6-octadecenyl (petroselinyl), *cis*-9,*cis*-12-octadecadienyl (linoleyl), all-*cis*-9,12,15-octadecatrienyl (α-linolenyl) and all-*cis*-6,9,12,-octadecatrienyl (γ-linolenyl) alcohols, have been esterified with caprylic acid using papaya (*Carica papaya*) latex lipase (CPL) and immobilized lipases from *C. antarctica* (lipase B, Novozym) and *R. miehei* (Lipozyme RM) as biocatalysts.[85] With papaya lipase, the highest activity is found for octyl and decyl caprylate syntheses, whereas both Novozym and Lipozyme RM show a broad chain length specificity toward the alcohol substrates. Papaya lipase strongly discriminates against all C18-alcohols studied, relative to *n*-hexanol, whereas the microbial lipases accept the C18 alcohols as substrates nearly as good as *n*-hexanol. Both petroselinyl and γ-linolenyl alcohol are very well accepted as substrates by Novozym as well as Lipozyme RM,[85] although the corresponding fatty acids, i.e., petroselinic and γ-linolenic acid, are strongly discriminated against by several microbial and plant lipases, including Lipozyme RM[35] and papaya lipase.[40]

Among the interesterification reactions of tripalmitin with various medium-chain substrates, catalyzed by papaya lipase, the alcoholysis with *n*-octanol has been found to be the fastest reaction, followed by transesterification with *n*-butyl and *n*-propyl, ethyl and methyl caprylates, whereas acidolysis of tripalmitin with caprylic acid is the slowest of all the reactions.[86]

In the esterification of fatty acids from borage oil with alcohols, catalyzed by *R. miehei* lipase, the extent of enrichment of γ-linolenic acid in the unesterified fatty acid fraction has been found to be higher with 1-butanol than with its higher homologues.[87]

Although lipase B from *Candida antarctica* has been reported to be nonregiospecific,[88] primary alcohols react at distinctly higher rates than the secondary alcohols in their esterification, catalyzed by this lipase.[89,90]

Specificity of short-, medium-, and long-chain acids in multicompetitive esterifications with glycerol, 1,2-propanediol or 1,3-propanediol has been studied using *Pseudomonas cepacia*, *Rhizomucor miehei*, or

Candida antarctica B lipases as biocatalysts.[91] With *Pseudomonas cepacia* lipase each alcohol substrate (with the exception of glycerol) shows maximum reactivity at acid chain length of C8 and C16, whereas with both *R. miehei* and *C. antarctica* B lipases each of the three alcohol substrates shows maximum reactivity at an acid chain length of C8.[91]

5.2.1.4 Lipase Specificities Toward Miscellaneous Substrates

5.2.1.4.1 *Alkanethiols*

Long-chain saturated acyl thioesters (thio wax esters) have been prepared by lipase-catalyzed thioesterification of fatty acids and transthioesterification of fatty acid alkylesters with alkanethiols.[92,93] Similarly, *cis*- and *trans*-unsaturated acyl thioesters have been prepared by lipase-catalyzed thioesterification and transthioesterification of the corresponding fatty acids and their alkyl esters, respectively, in the presence of antioxidants that prevent stereomutation (*cis*-/*trans*-isomerization) and thioether formation.[94] In the esterification of long-chain saturated fatty acids with dodecanethiol, catalyzed by lipase B from *C.antarctica* (Novozym 435) in the presence of *t*-butanol and molecular sieve 4Å, the extent of thioesterification increases with increasing chain length of the saturated fatty acid from C14 to C18.[92] In the esterification of palmitic acid with long-chain thiols, catalyzed by Novozym 435 in the presence of molecular sieve 4Å with[92] or without[93] *t*-butanol as solvent, the extent of thioesterification decreases with increasing chain length of the alkanethiol from C10 to C16. Using the same lipase preparation under vacuum, but without *t*-butanol and molecular sieve 4Å, the rate of thioesterification is not affected substantially by the chain length of the alkanethiol.[95]

5.2.1.4.2 *Sterols*

It has long been known that the ingestion of plant sterols (phytosterols, phytostanols) and their fatty acid esters can reduce total as well as LDL-cholesterol concentrations of blood by inhibiting the intestinal absorption of both dietary and endogenously formed cholesterol and thus reduce the risk of coronary heart diseases.[96,97] A recent meta-analysis of randomized, placebo-controlled, double-blind intervention studies has revealed that the daily ingestion of 2 g of phytosterols, phytostanols, or their esters reduces the LDL-cholesterol concentration of blood by 9%–14% without affecting the concentrations of HDL-cholesterol and triacylglycerols.[98]

Fatty acid esters of phytosterols and phytostanols, rather than the unesterified phytosterols and phytostanols, are generally used as cholesterol-reducing supplements in fat-containing functional foods, such as margarine, mayonnaise, and salad dressing, due to higher solubility of the phytosteryl and phytostanyl esters in fat.[99] Fatty acid esters of sterols, stanols, and steroids are usually prepared from the corresponding sterols by chemical esterification with fatty acids or interesterification with fatty acid alkyl esters as well as by their reaction with fatty acid halogenides or anhydrides.[100,101]

Sterols (sitosterol, cholesterol, stigmasterol, ergosterol, and 7-dehydrocholesterol) and sitostanol have been converted in high- to near-quantitative yields to the corresponding long-chain acyl esters via esterification with fatty acids or transesterification with methyl esters of fatty acids or triacylglycerols using lipase from *Candida rugosa* as biocatalyst in vacuo (20–40 hPa) at 40°C; neither an organic solvent nor water or a drying agent, such as molecular sieve, is added in these reactions.[102]

The time courses of formation of sitostanyl oleate and cholesteryl oleate during esterification of the corresponding sterols with oleic acid and their transesterification with methyl oleate or triolein, each catalyzed by *C. rugosa* lipase in vacuo, has shown that with both sterols the esterification with oleic acid is almost complete within 2 h reaction, whereas the rates of transesterification of both sitostanol and cholesterol with either methyl oleate or triolein are distinctly lower.[102]

The time course of formation of steryl esters by esterification of an equimolar mixture of cholesterol and sitostanol with oleic acid, catalyzed by *C. rugosa* lipase in vacuo, has shown that, under competitive conditions, initially (30-min conversion) cholesteryl oleate is formed at a higher rate than sitostanyl oleate; however, after prolonged reaction (>24 h) both sterols are almost quantitatively esterified.[102] Esterification of an equimolar mixture of oleic and linoleic acids with sitostanol, catalyzed by *C. rugosa* lipase *in vacuo*, shows that both fatty acids react at the same rate and both oleic and linoleic acids are almost quantitatively esterified in about 4 h.[102]

Conversion of two steroids, i.e., 5α-pregnan-3β-ol-20-one and 5-pregnen-3β-ol-20-one, to their pro-pionic acid esters via transesterification with tripropionin, catalyzed by *C. rugosa* lipase in vacuo, is moderate (about 35–55%) as compared to the conversion of cholesterol to cholesteryl butyrate by transesterification with tributyrin, catalyzed by the same lipase.[102]

Studies using immobilized lipases from *R. miehei* (formerly Lipozyme IM, *syn.* Lipozyme RM) and *C. antarctica* (lipase B, Novozym 435) as biocatalysts have shown that sitostanol is converted in high- to near-quantitative yields to the corresponding long-chain acyl esters via esterification with oleic acid or transesterification with methyl oleate or trioleoylglycerol in vacuo (20–40 hPa) at 80°C.[103] Corresponding rates of conversions observed with papaya (*Carica papaya*) latex lipase are generally lower.

Transesterification of methyl oleate with sterols, catalyzed by Lipozyme RM under competitive con-ditions, shows that saturated sterols such as sitostanol and 5α-cholestan-3β-ol are the preferred substrates as compared to Δ⁵-unsaturated cholesterol.[103]

Transesterification of methyl oleate with sitostanol, 5α-cholestan-3β-ol, 5α-cholestan-3β-ol, and cho-lesterol catalyzed by Lipozyme TL (immobilized lipase from *Thermomyces lanuginosus*) reveals that, under competitive conditions, saturated sterols such as sitostanol and 5α-cholestan-3β-ol as well as Δ⁵-unsat-urated cholesterol react with methyl oleate at similar rates with slight preference for the two stanols.[104] After prolonged reaction time (≥ 24 h) all three sterols are almost quantitatively esterified. It is worth noting that the enzyme activity of Lipozyme TL is around threefold higher than that of Lipozyme RM.

5.3 Applications of Kinetic Resolution

Figures 5.2–5.7 show some examples of applications of lipases for the enrichment of fatty acids and other lipids via kinetic resolution.

5.3.1 Enrichment of n-6 Polyunsaturated Fatty Acids

Applications of lipase-catalyzed kinetic resolution for the enrichment of n-6 polyunsaturated fatty acids are summarized in Table 5.4 and shown schematically in Figure 5.2 and 5.3.

γ-Linolenic acid is of considerable commercial interest due to its beneficial biomedical properties.[126] Seed oils of evening primrose, *Oenothera biennis*,[127,128] borage, *Borago officinalis*,[129] and *Ribes* spp.,[130] are

Fatty acids from oils containing **GLA, DHA or PET**
+   *n*-Butanol +   Lipase (discriminating against
GLA, DHA or **PET**)

⇓ *Esterification*

Fatty acids   +   *n*-Butyl esters
(enriched with   (enriched with fatty acids
GLA, DHA or PET)   other than **GLA, DHA or PET**)

⇓ *Downstream processing*

Fatty Acid Concentrate
enriched with **GLA, DHA** or **PET**

FIGURE 5.2 Scheme of kinetic resolution for the enrichment of γ-linolenic (GLA), n-3 docosahexaenoic (DHA) or petroselinic (PET) acid via lipase-catalyzed selective esterification of fatty acid mixtures with n-butanol.

Oil containing **GLA, DHA, ERUC** or **PET**
+ Water + Lipase (discriminating against **GLA, DHA,
            ERUC** or **PET**)

⇓ *Hydrolysis*

Tri- + Di + Monoacylglycerols   +   Fatty acids
(enriched with **GLA, DHA,**   (other than **GLA,**
ERUC or **PET**)   **DHA, ERUC** or **PET**)

⇓ *Downstream processing*

Acylglycerol-Concentrate
enriched with **GLA, DHA,**
ERUC or **PET**

FIGURE 5.3 Scheme of kinetic resolution for the enrichment of γ-linolenic (GLA), n-3 docosahexaenoic (DHA), erucic (ERUC) or petroselinic (PET) acid via lipase-catalyzed selective hydrolysis of triacylglycerols.

Oil containing **ERUC** in *sn*-1,3-positions
+ Water + Lipase (specific for *sn*-1,3-positions)

⇓ *Hydrolysis*

Tri- + Di + Monoacylglycerols   +   Fatty acids
(enriched with Fatty acids   (enriched with **ERUC**)
other than **ERUC**)

⇓ *Downstream processing*

Fatty Acid Concentrate enriched with **ERUC**

FIGURE 5.4 Scheme of kinetic resolution for the enrichment of erucic (ERUC) acid via lipase-catalyzed regioselective hydrolysis of triacylglycerols.

Oil containing **ERUC** in *sn*-1,3-positions
+ *n*-Butanol + Lipase (specific for *sn*-1,3-positions)

⇓ *Transesterification*

Di + Monoacylglycerols   +   *n*-Butyl esters of fatty acids
(enriched with fatty acids   (enriched with **ERUC**)
other than **ERUC**)

⇓ *Downstream processing*

Butyl Ester Concentrate
enriched with **ERUC**

FIGURE 5.5 Scheme of kinetic resolution for the enrichment of erucic (ERUC) acid via lipase-catalyzed regioselective transesterification (alcoholysis) of triacylglycerols with n-butanol.

Hydrolysis

Oil containing **OL** and **LINOL**
+ Water + Lipase (specific for fatty acids with
a *cis* -9 double bond)

⇓ *Hydrolysis*

Tri- + Di + Monoacylglycerols + Fatty acids
(enriched with fatty acids other (enriched with
than **OI** and **LINOL**) **OL** and **LINOL**)

⇓ *Downstream processing*

Fatty Acid Concentrate
enriched with **OL + LINOL**

FIGURE 5.6 Scheme of kinetic resolution for the enrichment oleic (OL) and linoleic (LINOL) acids via selective hydrolysis of triacylglycerols, catalyzed by a lipase specific for fatty acids/acyl moieties having a cis-9 olefinic bond.

some common sources of γ-linolenic acid. Moreover, fungal oils containing γ-linolenic acid, such as those from *Mortierella* spp.[131] and *Mucor ambiguus*,[132] are now commercially available. γ-Linolenic acid has been enriched from natural resources via urea adduct formation,[133] separation on Y-Zeolite,[134] and solvent winterization.[135]

The ability of the lipases from rape (*Brassica napus*) seedlings, *R. miehei*, *C. rugosa*, and *G. candidum* to discriminate against γ-linolenic acid over other fatty acids has been utilized for the enrichment of this acid from the mixture of fatty acids derived from evening primrose (*Oenothera biennis*) seed oil or borage (*Borago officinalis*) seed oil via lipase-catalyzed selective esterification with *n*-butanol (Figure 5.2 and Table 5.4). Most of the fatty acids from evening primrose oil or borage oil are converted to butyl esters,

Mixture of Fatty acids containing **CLA-isomers**
(*cis*-9,*trans*-11-18:2 and *trans*-10,*cis*-12-18:2)

+ Lauryl alcohol + Lipase (selective toward
or octanol ***cis*-9,*trans*-11-18:2**)

⇓ *Esterification*

Lauryl or octyl esters + **Fatty acids** (enriched with
(enriched with ***trans*-10,*cis*-12-18:2**)
***cis*-9, *trans*11-18:2**)

⇓ *Downstream processing*

Fatty Acid Concentrate enriched with
***trans*-10,*cis*-12-18:2**

FIGURE 5.7 Scheme of kinetic resolution for the separation of cis-9,trans-11-octadecadienoic acid from trans-10,cis-11-octadecadienoic acid from a commercial CLA mixture via selective esterification of the cis-9,trans-11-isomer with lauryl alcohol or octanol, catalyzed by a lipase having preference for cis-9,trans-11-octadecadienoic acid.

TABLE 5.4   Enrichment of n-6 Unsaturated Fatty Acids and Their Derivatives via Kinetic
Resolution Catalyzed by Lipases

Lipase From	Reaction	Fatty Acid Enriched	Reference
Rapeseed (*Brassica napus*), *R. miehei*, *G. candidum*, *C. rugosa*,	Selective esterification of fatty acids of evening primrose oil or borage oil with *n*-butanol	γ-Linolenic acid in unesterified fatty acids	[47,87,105–111]
Rapeseed (*Brassica napus*), *R. miehei*, *C. rugosa*, black cumin (*Nigella sativa*)	Selective hydrolysis of evening primrose oil or borage oil	γ-Linolenoyl moieties in tri-, di- and monoacylglycerols	[105,112–118]
C. rugosa	Selective hydrolysis of triacylglycerols of fungal oil from *Mortierella* sp.	Arachidonoyl moieties in tri-, di- and mono-acylglycerols	[119]
R. delemar	Selective esterification of fatty acids of borage oil with lauryl alcohol	γ-Linolenic acid in unesterified fatty acids	[120,121]
R. miehei	Selective esterification of fatty acids of fungal oil from *Mortierella* sp. with *n*-butanol	γ-Linolenic acid in unesterified fatty acids	[122,123]
R. miehei	Selective esterification of fatty acids of blackcurrant oil with *n*-butanol	γ-Linolenic acid in unesterified fatty acids	[124]
C. rugosa	Selective esterification of fatty acids of fungal oil from *Mortierella* sp. with lauryl alcohol	Arachidonic acid in unesterified fatty acids	[125]

with the exception of γ-linolenic acid, which is enriched as a concentrate in the unesterified fatty acids.[33,105,107] Selective hydrolysis of the triacylglycerols of evening primrose oil or borage oil, catalyzed by lipases from rapeseed,[105] *R. miehei*, and *C. rugosa*,[113] also leads to enrichment of γ-linolenoyl moieties in the unhydrolyzed acylglycerols, i.e., tri-, di-, and monoacylglycerols, whereas the other acyl moieties are cleaved to yield fatty acids (Figure 5.3 and Table 5.4).

Fatty acid concentrates containing about 75% γ-linolenic acid are prepared via selective esterification of fatty acids of evening primrose oil and borage oil with *n*-butanol, catalyzed by lipase from *R. miehei*.[107]

Selective hydrolysis of triacylglycerols of evening primrose oil using lipase from *C. rugosa* leads to enrichment of γ-linolenic acid from about 10% in the starting material to about 47% in the unhydrolyzed acylglycerols.[113] Similarly, hydrolysis of borage oil triacylglycerols using lipase from *C. cylindracea* results in enrichment of γ-linolenic acid from about 20% in the starting material to about 48% in the unhydrolyzed acylglycerols.[113]

The selectivity of the lipases from *R. miehei* and *C. rugosa* toward γ-linolenic acid has been utilized to concentrate this fatty acid from fatty acids of fungal oil *via* selective esterification with *n*-butanol.[122] Using the lipase from *C. rugosa*, after 1 h reaction as much as 92% of the fungal oil fatty acids are esterified, and concomitantly the level of γ-linolenic acid in the unesterified fatty acids is raised from 10% to about 47%; virtually none of the γ-linolenic acid is converted to butyl esters.[122] With the lipase from *R. miehei* 91% esterification of the fungal oil fatty acids occurs after 4 h of reaction that results in an increase in the level of γ-linolenic acid in the unesterified fatty acids to 69% (yield of γ-linolenic acid 59%).[122] With both lipases the enrichment of γ-linolenic acid is paralleled by a decrease in the levels of palmitic, oleic, and linoleic acid in the fatty acid fraction and some increase in the proportion of these acids in the butyl esters.

It has been shown using several lipases that selective esterification of fungal oil fatty acids with *n*-butanol, rather than selective hydrolysis of fungal oil triacylglycerols, is the method of choice for enrichment of γ-linolenic acid.[122]

5.3.2   Enrichment of n-3 Polyunsaturated Fatty Acids

n-3 polyunsaturated fatty acids are extensively used in nutraceutical preparations and functional foods due to their interesting biomedical properties.[38]

The ability of the lipase from rape to discriminate against n-3 docosahexaenoic acid (n-3 22:6) has been utilized for the enrichment of this acid from a mixture of fatty acids derived from cod liver oil via lipase-catalyzed selective esterification with *n*-butanol,[106] as shown schematically in Figure 5.2. Moreover, the ability of the lipase from *R. miehei* to discriminate against n-3 22:6 has been utilized for the enrichment of this fatty acid from fish oil fatty acids via selective esterification with methanol.[44] Selective esterification of fish oil fatty acids with methanol, catalyzed by *R. miehei* lipase, raises the level of n-3 22:6 from about 8% in the initial fatty acid mixture to about 48% in the unesterified fatty acids.[44] Selective hydrolysis of fish oil triacylglycerols, catalyzed by lipases from *Aspergillus niger* and *C. rugosa*, has been employed to concentrate n-3 polyunsaturated fatty acids, i.e., n-3 docosapentaenoic acid (n-3 20:5) and n-3 22:6, that are enriched in the tri- + di- + monoacylglycerols (Figure 5.3).[136]

Since the above studies, numerous publications have appeared (Table 5.5) on selective hydrolysis of marine oil triacylglycerols for the enrichment of n-3 long-chain polyunsaturated fatty acids (n-3 LCP-UFA) in the unhydrolyzed acylglycerols (Figure 5.3) and selective esterification of marine oil fatty acids for the enrichment of the n-3 LCPUFA in the unesterified fatty acids (Figure 5.2).

Two successive hydrolyses of tuna oil triacylglycerols, catalyzed by lipase from *G. candidum*, result in enrichment of n-3 fatty acids, i.e., n-3 20:5 and n-3 22:6, from about 32% in the untreated oil to about 49% in the fraction consisting of tri-, di-, and monoacylglycerols.[159] In a commercial process,[140] fish oil is partially hydrolyzed by *C. rugosa* lipase to yield an acylglycerol fraction enriched in n-3 20:5, and especially in n-3 22:6; the acylglycerol fraction is subsequently isolated by evaporation and converted to triacylglycerols via hydrolysis and reesterification, both catalyzed by *R. miehei* lipase.

Selective esterification of fatty acids from tuna oil with lauryl alcohol, catalyzed by *R. delemar* lipase, yields an unesterified fatty acid fraction containing 73% n-3 22:6 as compared to 23% n-3 22:6 in the starting material.[162] Using the same lipase, selective esterification of tuna oil fatty acids with lauryl alcohol, extraction of the unreacted fatty acids, and their repeated reesterification with lauryl alcohol result in an unesterified fatty acid fraction containing as much as 91% n-3 22:6.[172]

Using lipase from *Pseudomonas* sp. marine oil triacylglycerols have been interesterified with marine oil fatty acid fractions, which have been enriched with n-3 polyunsaturated fatty acids, raising the level of total n-3 LCPUFA in the triacylglycerols to 65%.[152]

Lipase-catalyzed selective alcoholysis of marine oil triacylglycerols or alkyl esters of marine oil fatty acids with short- or long-chain alcohols results in the enrichment of n-3 LCPUFA in the acylglycerol fraction or alkyl ester fraction, respectively. Thus, selective alcoholysis of ethyl esters of tuna oil fatty acids with ethanol using *R. delemar* lipase and *R. miehei* lipase as biocatalyst leads to enrichment of n-3 22:6 in the ethyl ester fraction from 23 mol% to about 50 mol%[163] and from 60 mol% to 93 mol%.[167] However, with *R. miehei* lipase as biocatalyst selective interesterification of tuna oil triacylglycerols with ethanol yields an acylglycerol fraction containing 49% n-3 22:6, whereas selective esterification of tuna oil fatty acids with ethanol yields an unesterified fatty acid fraction containing 74% n-3 22:6.[168]

5.3.3 Enrichment of Very Long-Chain Monounsaturated Fatty Acids (VLCMFA)

Very long-chain (>C18) monounsaturated fatty acids (VLCMFA), such as gondoic (*cis*-11-eicosenoic, 20:1) acid and erucic (*cis*-13-docosenoic, 22:1) acid,[173,174] nervonic (*cis*-15-tetracosenoic, 24:1) acid,[175] and *cis*-5-eicosenoic acid,[176] are abundant constituents of seed oils, especially from Cruciferae. Erucic acid and its derivatives are extensively used for the manufacture of a wide variety of oleochemical and technical products[177] and other VLCMFA might also find novel applications in hitherto less explored areas.

Erucic acid is currently isolated from high-erucic oils by fat splitting followed by fractional distillation, which are rather energy-consuming processes. Lipase-catalyzed selective hydrolysis of high-erucic triacylg-lycerols has been suggested as an alternative process for the isolation of erucic acid.[59,112,178] Selective hydrolysis of high-erucic oils using lipases from *G. candidum*[59] and *C. rugosa*[112,178] leads to enrichment of erucic acid in the tri-, di-, and monoacylglycerols (Figure 5.3). The level of erucic acid is thus raised from 43% in the high-erucic rapeseed oil to 66% in the tri-, di-, and monoacylglycerols.[59] In the esterification of fatty acids

TABLE 5.5   Enrichment of n-3 Unsaturated Fatty Acids and Their Derivatives via Kinetic Resolution Catalyzed by Lipases

Lipase from	Reaction	Fatty Acid Enriched	Reference
Porcine pancreas	Hydrolysis of whale oil triacylglycerols	Discrimination against n-3 22:6 as compared to C_{16} and C_{18} acyl moieties	[137]
R. miehei	Selective esterification of fatty acids of marine oil with methanol	n-3 Docosahexaenoic acid in unesterified fatty acids	[43]
Rapeseed [*Brassica napus*], *R. miehei*	Selective esterification of fatty acids of fish oil with *n*-butanol	n-3 Docosahexaenoic acid in unesterified fatty acids	[47,106]
C. rugosa	Selective hydrolysis of fish oil triacylglycerols	n-3 Docosahexaenoyl moieties in tri-, di-, and monoacyl-glycerols	[136,138–143]
Rhizopus niveus	Selective hydrolysis of fish oil triacylglycerols	n-3 Docosahexaenoyl moieties in monoacylglycerols	[144]
R. miehei	Interesterification of cod liver oil with a fatty acid mixture or its ethyl esters enriched in n-3 20:5 and n-3 22:6	n-3 Eicosapentaenoyl and n-3 docosahexaenoyl moieties in triacylglycerols	[145]
R. miehei	Interesterification of cod liver oil with a fatty acid mixture enriched in n-3 20:5 and n-3 22:6	n-3 Docosahexaenoyl moieties in acylglycerols	[146,147]
R. miehei	Interesterification of vegetable oils with n-3 20:5 and n-3 22:6 fatty acids or their methyl or ethyl esters	n-3 Eicosapentaenoyl and n-3 docosahexaenoyl moieties in triacylglycerols	[148–151]
Pseudomonas sp.	Interesterification of sardine oil with a fatty acid mixture enriched in n-3 20:5 and n-3 22:6	n-3 Eicosapentaenoyl and n-3 docosahexaenoyl moieties in acylglycerols	[152]
Pseudomonas sp.	Interesterification of fish oil with alcohols	n-3 Eicosapentaenoyl and n-3 docosahexaenoyl moieties in monoacylglycerols	[153,154]
C. antarctica	Esterification of glycerol or alkylglycerols with n-3 20:5 and n-3 22:6 fatty acids	n-3 Eicosapentaenoyl and n-3 docosahexaenoyl moieties in triacylglycerols and alkylacylglycerols, respectively	[155,156]
C. viscosum	Esterification of acylglycerols, enriched in n-3 22:6, with a fatty acid fraction enriched in n-3 22:6	n-3 Docosahexaenoyl moieties in triacylglycerols	[157]
Pseudomonas sp.	Selective hydrolysis of fish oil triacylglycerols	n-3 Docosahexaenoyl moieties in tri-, di- and monoacylglycerols	[158]
G. candidum	Selective hydrolysis of fish oil triacylglycerols	n-3 Docosahexaenoyl moieties in tri-, di- and monoacylglycerols	[159,160]
C. antarctica	Selective interesterification of cod liver oil with ethanol in supercritical carbon dioxide	n-3 Docosahexaenoyl moieties in monoacylglycerols	[161]
Rapeseed (*Brassica napus*)	Selective hydrolysis of fish oil triacylglycerols	n-3 Docosahexaenoyl moieties in tri-, di- and monoacylglycerols	[114]
R. delemar	Selective esterification of fatty acids of fish oil with lauryl alcohol	n-3 Docosahexaenoic acid in unesterified fatty acids	[162]
R. delemar	Selective interesterification of fatty acid ethyl esters of fish oil with lauryl alcohol	n-3 Docosahexaenoyl moieties in fatty acid ethyl esters	[163]
Pseudomonas sp.	Interesterification of fish oil triacylglycerols with ethanol	n-3 Eicosapentaenoyl and n-3 docosahexaenoyl moieties in acylglycerols	[164,165]

(continued)

TABLE 5.5 Enrichment of n-3 Unsaturated Fatty Acids and Their Derivatives via Kinetic Resolution Catalyzed by Lipases (Continued)

Lipase from	Reaction	Fatty Acid Enriched	Reference
R. miehei	Selective hydrolysis of phospholipids	n-3 Docosahexaenoyl moieties in unhydrolyzed phospholipids	[166]
R. miehei	Interesterification of ethyl esters of tuna oil fatty acids with lauryl alcohol	n-3 Docosahexaenoyl moieties in fatty acid ethyl esters	[167]
R. miehei	Interesterification of fish oil triacylglycerols with ethanol	n-3 Docosahexaenoyl moieties in acylglycerols	[168]
R. miehei	Selective esterification of fatty acids from sardine cannery effluents with *n*-butanol	n-3 Docosahexaenoic and n-3 eicosapentaenoic acid in unesterified fatty acids	[169]
C. rugosa	Selective esterification of fatty acids from sardine cannery effluents with *n*-butanol	n-3 Eicosapentaenoic acid in unesterified fatty acids	[170]
R. miehei	Selective esterification of fatty acids from marine algae (*Undaria pinnatifida* and *Uva pertusa*) with lauryl alcohol	n-3 Octadecatetraenoic acid in unesterified fatty acids	[171]
R. miehei	Selective esterification of fatty acids from marine algae (*Uva pertusa*) with lauryl alcohol	n-3 Hexadecatetraenoic acid in unesterified fatty acids	[171]

of high-erucic rapeseed oil with *n*-butanol, catalyzed by *G. candidum* lipase, strong discrimination against erucic acid is observed, which leads to enrichment of erucic acid from 47% in the fatty acid mixture to 83% in the unesterified fatty acids.[62] Partial hydrolysis of high-erucic rapeseed oil, catalyzed by *C. rugosa* lipase, yields dierucin in 73% purity, which is used for lipase-catalyzed synthesis of trierucin.[179]

Selective hydrolysis of triacylglycerols of meadowfoam (*Limnanthes alba*) oil containing over 60% *cis*-5-eicosenoic acid (*cis*-5-20:1) results in enrichment of *cis*-5-20:1 in tri-, di-, and monoacylglycerols when lipase from *Chromobacterium viscosum* is used as biocatalyst.[180] Selective esterification of meadowfoam oil fatty acids with *n*-butanol, catalyzed by *C. viscosum* lipase, leads to enrichment of *cis*-5-20:1 in the unesterified fatty acids in high yield (>95%) and purity (>99%).[180]

Partial hydrolysis and transesterification of high-erucic oils catalyzed by several lipases have been examined for the enrichment of VLCMFA utilizing the selectivity of these enzymes.[181] Lipase-catalyzed selective hydrolysis of high-erucic oils from white mustard (*Sinapis alba*) and oriental mustard (*Brassica juncea*) seeds has shown that the lipases studied can be broadly classified into three groups according to their substrate selectivity.

The first group includes lipases from *C. cylindracea* and *G. candidum* that selectively cleave the C18 acyl moieties from the triacylglycerols, which leads to enrichment of the C18 fatty acids in the fatty acid fraction. Concomitantly, the level of erucic acid and the other VLCMFA is raised in the mono- + di- + triacylglycerol fraction from 51% in the starting oil to about 80% and 72%, respectively, when lipases from *C. rugosa* and *G. candidum* are used as biocatalyst (Figure 5.3 and reference 181).

The second group includes lipases from porcine pancreas, *C. viscosum*, *R. arrhizus*, and *R. miehei*, with their known regioselectivity toward acyl moieties at the *sn*-1,3 positions of triacylglycerols.[8] With these lipases the VLCMFA, esterified almost exclusively at the *sn*-1,3 positions of the triacylglycerols of the high-erucic oils, are regioselectively cleaved and enriched in the fatty acid fraction to levels as high as 65–75% (Figure 5.4); concomitantly, the level of the C18 acyl moieties is increased in the acylglycerol fraction.[181]

The third group of lipases includes those from *Penicillium* sp. (Lipase G) and *Candida antarctica* (lipase B), which do not seem to exhibit any pronounced specificity toward either C18 fatty acids or VLCMFA.[181]

The lipases with strong fatty acid selectivity or regioselectivity have been tested for the selective hydrolysis of the triacylglycerols of honesty (*Lunaria annua*) seed oil, which is rich in VLCMFA, including nervonic acid, that are esterified almost exclusively at the *sn*-1,3-positions of the triacylglycerols.[175] The

C. rugosa lipase cleaves preferentially the C18 fatty acids which are enriched from 36% in the starting oil of honesty to 79% in the fatty acid fraction, while the VLCMFA are enriched in the di- and triacylglycerols (Figure 5.3). Especially the diacylglycerols are almost exclusively (>99%) composed of VLCMFA, whereas only traces of monoacylglycerols are formed.[181]

The lipases from porcine pancreas and *R. miehei* regioselectively cleave the VLCMFA, esterified almost exclusively at the *sn*-1,3-positions of honesty seed oil triacylglycerols, and consequently, the VLCMFA are extensively enriched in the fatty acid fraction, whereas the C18-fatty acids, esterified predominantly at the *sn*-2 position, are enriched in the mono- and diacylglycerols (Figure 5.4).[181]

The *sn*-1,3-regioselectivity of the *R. miehei* lipase has also been utilized for the enrichment of VLCMFA from high-erucic oils via regioselective transesterification of the triacylglycerols with alkyl acetates or regioselective alcoholysis of the triacylglycerols with *n*-butanol (Figure 5.5).

Transesterification of triacylglycerols of high-erucic mustard seed oil with ethyl, propyl, and butyl acetate, catalyzed by the lipase from *R. miehei*, yields alkyl (ethyl, propyl, and butyl, respectively) esters of fatty acids and a mixture of acetylacylglycerols (e.g., monoacetyldiacyl- + diacetylmonoacylglycerols) as well as acylglycerols (tri- + di + monoacylglycerols).[181] Regioselective transesterification of the triacylglycerols at the *sn*-1,3 positions leads to enrichment of the VLCMFA in the alkyl esters and decreases in their level of acetylacylglycerols and acylglycerols; concomitantly, the level of the C18 fatty acids is increased in the acetylacylglycerols and acylglycerols, and decreased in the alkyl esters.[181]

Similarly, alcoholysis of high-erucic mustard seed oil triacylglycerols with *n*-butanol, catalyzed by *R. miehei* lipase, yields butyl esters and mixtures of acylglycerols (mono- +di- +triacylglycerols) together with minor proportions of fatty acids,[181]] as shown in Figure 5.5. The VLCMFA, esterified at the *sn*-1,3 positions of the triacylglycerols, are regioselectively transesterified, resulting in their enrichment in the butyl ester fraction and decrease of their level in the acylglycerols. Concomitantly, the C18 fatty acids are enriched in the acylglycerols.

Regioselective transesterification of triacylglycerols with alkyl acetates or *n*-butanol is nearly as effective for the enrichment of VLCMFA in the alkyl ester fraction as regioselective hydrolysis of triacylglycerols for the enrichment of VLCMFA in the fatty acids or acylglycerols.[181] Ultimate choice of a process for the enrichment of VLCMFA depends on selectivity, cost, and reusability of the lipase preparation, as well as efficiency of downstream processing for the separation of the desired products from the reaction mixture, i.e., fatty acids from acylglycerols in the case of hydrolysis and alkyl esters from acylglycerols in the case of transesterification with alkyl esters or *n*-butanol. Acylglycerols and acetylacylglycerols could be useful by-products of such processes.

Selective hydrolysis of high-erucic rapeseed oil, catalyzed by lipases from *G. candidum*, results in the enrichment of erucic acid in diacylglycerols to an extent of 85%.[182] When hydrolysis, catalyzed by *C. rugosa* lipase, is carried out below 20°C, the reaction mixture solidifies and the diacylglycerols formed contain as much as 95% erucic acid.[182]

5.3.4   Enrichment of Other Acids and Alcohols

Lipase-catalyzed reactions have been employed for the enrichment of several common and unusual fatty acids via kinetic resolution. For example, oleic acid has been incorporated into a few selected plant oils by interesterification of the triacylglycerols with methyl oleate, catalyzed by *R. miehei* lipase.[183] By this process, oleoyl moieties replace the saturated acyl moieties and linoleoyl moieties of the triacylglycerols, yielding an oil with improved stability and nutritional properties.

Interesterification (glycerolysis) of fats such as beef tallow and lard with glycerol, catalyzed by *Pseudomonas fluorescens* lipase, results in enrichment of saturated acyl moieties in monoacylglycerols in high (45%–70%) yields, when the reaction is carried out at or below a "critical" temperature of 40°C.[184] Similarly, lipase-catalyzed glycerolysis of low-erucic rapeseed oil or soybean oil at 5°C, results in enrichment of palmitoyl and stearoyl moieties in monoacylglycerols.[184]

Short-chain fatty acids with desirable flavor have been produced by selective hydrolysis of butter fat fraction, catalyzed by *Penicillium roqueforti* lipase.[185]

A process aimed at the production of very low saturate oil from sunflower oil containing 12% saturated fatty acids is based on selective hydrolysis, catalyzed by *G. candidum* lipase B, to yield a fatty acid fraction containing >99% unsaturated fatty acids (Figure 5.6).[186] The unsaturated fatty acids are then recovered by evaporation and subsequently esterified to glycerol using *R. miehei* lipase with constant removal of the water formed by sparging with nitrogen under vacuum. This produces unsaturated triacylglycerols containing <1% saturated fatty acids in a yield of >95%.

Selective hydrolysis of fennel (*Foeniculum vulgare*) oil, catalyzed by a lipase from *R. arrhizus*, leads to enrichment of petroselinic acid in the acylglycerols (Figure 5.3).[45] Removal of the fatty acids using an ion exchange resin followed by hydrolysis of the resulting acylglycerols by the lipase from *C. cylindracea* yields a fatty acid concentrate containing 96% petroselinic acid.

Selective esterification of fatty acids of coriander (*Coriandrum sativum*) oil with *n*-butanol, catalyzed by rape lipase, leads to enrichment of petroselinic acid from about 80% in the starting material to > 95% in unesterified fatty acids (Figure 5.2).[47] Using the same lipase, selective esterification with *n*-butanol of fatty acids of *Hydnocarpus wightiana* seed oil containing about 10% gorlic acid yields an unesterified fatty acid fraction containing almost 50% gorlic acid.[47]

Hydroxy fatty acids, e.g., lesquerolic (14-hydroxy-*cis*-11-eicosenoic) and auricolic [14-hydroxy-*cis*-11-*cis*-17-eicosadienoic) acid, have been selectively cleaved from lesquerella (*Lesquerella fendleri*) seed oil triacylglycerols by hydrolysis catalyzed by lipases from *R. arrhizus*[187] or *R. miehei*,[188] yielding fatty acid concentrates containing 85% hydroxy acids.

Two unusual polyunsaturated fatty acids containing a *cis*-5-olefinic bond, e.g., all-*cis*-5,11,14-octadecatrienoic and all-*cis*-5,11,14,17-octadecatetraenoic acid, have been enriched from fatty acids of *Biota orientalis* seed oil via selective esterification with *n*-butanol, catalyzed by lipase from *C. rugosa*; thereby the content of total *cis*-5-polyunsaturated fatty acids is raised from 15% in the starting material to 73% in the unesterified fatty acids.[189] Similarly, selective hydrolysis of the *Biota orientalis* seed oil, catalyzed by *C. rugosa* lipase, leads to enrichment of total *cis*-5-polyunsaturated fatty acids in the acylglycerols to 41%.[189] It appears from these results and those reported by Shimada et al.[125] and Hayes and Kleiman[187,188] that fatty acids or acyl moieties having a *cis*-5 double bond are also discriminated against by some lipases.

Symmetrical triacylglycerols containing acetylenic acyl moieties of different chain lengths having the triple bond at various positions of the acyl chain are hydrolyzed at different rates by lipases from *C. rugosa*, *C. antarctica* (B), (*Thermomyces lanuginosus* (syn. *Humicola lanuginosa*), and *Pseudomonas cepacia*.[190]

Butter oil has been enriched with conjugated linoleic acids to an extent of 15% by interesterification of the triacylglycerols with conjugated linoleic acids, catalyzed by lipase from *C. antarctica*.[191]

cis-9, *trans*-11-Octadecadienoic acid and *trans*-10,*cis*-11-octadecadienoic acid, the major constituents of a synthetic mixture of conjugated linoleic acids, have been separated by kinetic resolution via selective esterification of the *cis*-9,*trans*-11-isomer with lauryl alcohol[49] or octanol,[50] both catalyzed by lipase from *Geotrichum candidum* (Figure 5.7). The *cis*-9,*trans*-11-isomer is enriched in the lauryl esters and octyl esters, respectively, whereas the *trans*-10, *cis*-11-octadecadienoic acid is enriched in the unesterified fatty acid fraction. Similarly, from a commercial preparation of methyl esters of conjugated linoleic acids, *cis*-9,*trans*-11-octadecadienoic acid has been separated via selective hydrolysis, catalyzed by *Geotrichum candidum* lipase, whereas the other isomers remains as methyl esters.[50]

A commercial mixture of conjugated linoleic acids containing almost equal amounts of *cis*-9,*trans*-11- and *trans*-10,*cis*-11-octadecadienoic acid has been subjected to esterification with lauryl alcohol, catalyzed by lipase from *Candida rugosa*, which results in selective conversion of the *cis*-9,*trans*-11-isomer to lauryl esters and enrichment of the *trans*-10,*cis*-11-octadecadienoic acid in the unesterified fatty acids.[51] Further enrichment of each isomer in lauryl esters and fatty acids has been achieved via repeated hydrolysis and selective enzymatic esterification.

Numerous racemic mixtures of a wide variety of organic compounds, including lipids, have been resolved by enantioselective hydrolysis, esterification, and transesterification reactions, catalyzed by lipases.[3,5] Earlier publications report enantioselective esterification with lauric acid, catalyzed by lipase from *Candida rugosa*, for the kinetic resolution of racemic alcohols, such as DL-*trans*-2-ethylcyclohexanol or DL-menthol into their enantiomers with preference for the esters having L-configuration.[192] In a similar manner, transesterification of racemic menthyl laurate with isobutanol or trilaurin has been carried out for resolution of the enantiomers of menthol, but with limited success.[193]

Stereoselective esterification of racemic menthol using acid anhydrides, such as butyric anhydride, as acylating agent and lipase AY-30 from *C. rugosa* as biocatalyst leads to selective conversion of L-menthol and its enrichment in the butyl esters, as determined by capillary gas chromatography on a chiral column.[194] Similarly, stereoselective transesterification of racemic menthol with vinyl propinate using lipase from *Pseudomonas cepacia* as biocatalyst leads to selective conversion of L-menthol and its enrichment in the butyl esters.[195] Furthermore, stereoselective esterification of DL-menthol with oleic acid, catalyzed by *C. rugosa* lipase, leads to selective esterification of L-menthol and its enrichment in the oleoyl esters.[196] L-menthol, having a characteristic aroma as compared to its enantiomer, is a commercially important flavoring agent.

Stereoselectivity has been investigated in the lipase-catalyzed reactions of racemic 2-octanol with octanoic acid[197,198] and its esters,[198] which is followed by conversion of the unreacted 2-octanol to its diastereomeric carbamates by reaction with (*S*)-α-methylbenzylisocyanate and subsequent analysis of the carbamates by gas chromatography. The above studies have shown good stereopreference for the *R*(-)-enantiomer of 2-octanol, especially by lipase from *R. miehei*.

Lipase OF 360 from *C. rugosa* has been used to stereoselectively esterify racemic mixtures of 2-(4-chlorophenoxy) propionic acids with various alcohols, in order to isolate the (*R*)-enantiomer, which is an important herbicide.[199] Long-chain alcohols, such as tetradecanol, have been found to be excellent substrates for such optical resolution.

5.4 Perspectives

Numerous applications of lipase-catalyzed reactions for the enrichment of particular fatty acids or lipids via kinetic resolution utilizing the fatty acid specificity and regioselectivity as well as stereoselectivity (enantioselectivity) of lipases have been seen in recent years. Fatty acid concentrates containing well over 70–80% of one particular fatty acid or its derivatives can be easily prepared in the laboratory and pilot plant scale by low energy-consuming processes. Further enrichment of such fatty acid concentrates can be carried out by repeated lipase-catalyzed kinetic resolutions. Several commercial applications of such processes, e.g., production of n-3 fatty acid concentrates from marine oils, have become known. Further commercial applications will probably be governed by the cost and reusability of lipases, economy of downstream processing, and the market value of the product. One potential market for fatty acid concentrates, prepared enzymatically, is the area of nutraceuticals and cosmetics. Especially the fatty acid concentrates can be used for the preparation of structured lipids using lipase-catalyzed esterification and interesterification.

References

1. C-S Chen, CJ Sih. General aspects and optimization of enantioselective biocatalysis in organic solvents: The use of lipases. *Angew Chem Int Ed Engl* 28: 695–707, 1989.
2. AM Klibanov. Asymmetric transformations catalyzed by enzymes in organic solvents. *Acc Chem Res* 23: 114–120, 1990.
3. GG Haraldsson. The application of lipases in organic synthesis. In: S Patai, ed. *The Chemistry of Acid Derivatives*, Vol 2. New York: John Wiley & Sons, 1992, pp. 1395–1473.
4. K-E Jaeger, MT Reetz. Microbial lipases form versatile tools for biotechnology. *Trends Biotechnol* 16: 396–403, 1998.
5. RJ Kazlauskas, UT Bornscheuer. Biotransformations with lipases. In: D.R Kelly, ed. *Biotechnology*, Vol 8a: *Biotransformations I*, Weinheim, Germany: Wiley-VCH, 1998, pp. 37–191.
6. J Kötting, H Eibl. Lipases and phospholipases in organic synthesis. In: P Woolley, SB Petersen, eds. *Lipases*, Cambridge, UK: Cambridge University Press, 1994, pp. 289–313.

7. AR Macrae. Lipase-catalyzed interesterification of oils and fats. *J Am Oil Chem Soc* 60: 243A–246A, 1983.

8. KD Mukherjee. Lipase-catalyzed reactions for modification of fats and other lipids. *Biocatalysis* 3: 277–293, 1990.

9. P Adlercreutz. Enzyme-catalysed lipid modification. *Biotechnol Genet Eng Revs* 12: 232–254, 1994.

10. NN Gandhi. Applications of lipase. *J Am Oil Chem Soc* 74: 621–634, 1997.

11. EN Vulfson. Industrial applications of lipases. In: P Woolley, SB Petersen, eds. *Lipases*, Cambridge, UK: Cambridge University Press, 1994, pp. 271–288.

12. KD Mukherjee. Fractionation of fatty acids and other lipids via lipase-catalyzed reactions. *J Franc Oleagineux Corps Gras Lipides* 5: 365–368, 1995.

13. KD Mukherjee. Lipase-catalyzed reactions for the fractionation of fatty acids. In: FX Malcata, ed. *Engineering of / with Lipases*, Dordrecht, Netherlands: Kluwer, 1996, pp. 51–64.

14. KD Mukherjee. Fractionation of fatty acids and other lipids using lipases. In: UT Bornscheuer, ed. *Enzymes in Lipid Modification*, Weinheim, Germany: Wiley-VCH, 2000, pp. 23–45.

15. KD Mukherjee. Lipid biotechnology. In: CC Akoh, DB Min, eds. *Food Lipids, Chemistry, Nutrition, and Biotechnology*, 2nd Edition, Revised and Expanded, New York: Marcel Dekker, 2002, pp. 751–812.

16. WW Christie. Lipid Analysis. *Isolation, Separation, Identification, and Structural Analysis of Lipids*, 3rd Edition. Bridgwater, UK: Oily Press, 2003, pp. 373–384.

17. T Takagi, Y Ando. Sereospecific analysis of triacyl-*sn*-glycerols by chiral high-performance liquid chromatography. *Lipids* 26: 542–547, 1991.

18. D Meusel, N Weber, KD Mukherjee. Stereoselectivity of lipases: esterification reactions of octade-cylglycerol. *Chem Phys Lipids* 61: 193–198, 1992.

19. RG Jensen, FA deJong, RM Clark, L Palmgren, TH Liao, M Hamosh. Stereospecificity of premature human infant lingual lipase. *Lipids* 17: 570–572, 1982.

20. RG Jensen, FA deJong, RM Clark. Determination of lipase specificity. *Lipids* 18: 239–252, 1983.

21. E Rogalska, S Ransac, R Verger. Stereoselectivity of lipases. II. Stereoselective hydrolysis of triglyc-erides by gastric and pancreatic lipases. *J Biol Chem* 265: 20271–20276, 1990.

22. AGD Hoffman, RGH Downer. End productivity of triacylglycerol lipases from intestine, fat body, muscle and haemolymph of the American cockroach, *Periplaneta americana* L. *Lipids* 14: 893–899, 1979.

23. S Ransac, E Rogalska, Y Gargouri, AM Deveer, F Paltauf, GH de Haas, R Verger. Stereoselectivity of lipases. I. Hydrolysis of enantiomeric glyceride analogues by gastric and pancreatic lipases, a kinetic study using the monomolecular film technique. *J Biol Chem* 266: 20263–20270, 1990.

24. F Paltauf, F Esfandi, A Holasek. Stereospecificity of lipases. Enzymic hydrolysis of enantiomeric alkyl diacylglycerols by lipoprotein lipase, lingual lipase and pancreatic lipase. *FEBS Lett* 40:119–123, 1974.

25. NH Morley, A Kuksis, D Buchnea, JJ Myher. Hydrolysis of diacylglycerols by lipoprotein lipase. *J Biol Chem* 250: 3414–3418, 1975.

26. CS Wang, A Kuksis, F Manganaro. Studies on the substrate specificity of purified human milk lipoprotein lipase. *Lipids*. 17: 278–284, 1982.

27. P Villeneuve, M Pina, D Montet, J Graille. *Carica papaya* latex lipase: sn-3 stereoselectivity or short-chain selectivity? Model chiral triglycerides are removing the ambiguity. *J Am Oil Chem Soc* 72: 753–755, 1995.

28. T Dourtoglou, E Stefanou, S Lalas, V Dourtoglou, C Poulos. Quick regiospecific analysis of fatty acids in triacylglycerols with GC using 1,3-specific lipase in butanol. *Analyst* 126: 1032–1036, 2001.

29. S Negishi, Y Arai, S Shirasawa, S Arimoto, T Nagasawa, H Kouzui, K Tsuchiya. Analysis of regiospe-cific distribution of FA of TAG using the lipase-catalyzed ester exchange. *J Am Oil Chem Soc* 80: 353–356, 2003.

30. T Godfrey. Lipases for industrial use. *Lipid Technol* 7: 58 (1995).

31. G Langrand, N Rondot, C Triantaphylides, J Baratti. Short-chain flavor esters synthesis by microbial lipase. *Biotechnol Lett* 12: 581–586, 1990.

32. E Ucciani, M Schmitt-Rozieres, A Debal, LC Comeau. Enzymatic synthesis of some wax-esters. *Fett/Lipid* 98: 206–210, 1996.

33. MJ Hills, I Kiewitt, KD Mukherjee. Lipase from *Brassica napus* L discriminates against *cis*-4 and *cis*-6 unsaturated fatty acids and secondary and tertiary alcohols. *Biochim Biophys Acta* 1042: 237–240, 1990.

34. NN Gandhi, KD Mukherjee. Specificity of papaya lipase in esterification with respect to the chemical structure of substrates. *J Agric Food Chem* 48: 566–570, 2000.

35. KD Mukherjee, I Kiewitt, MJ Hills. Substrate specificities of lipases in view of kinetic resolution of unsaturated fatty acids. *Appl Microbiol Biotechnol* 40: 489–493, 1993.

36. I Jachmanián, E Schulte, KD Mukherjee. Substrate selectivity in esterification of less common fatty acids catalysed by lipases from different sources. *Appl Microbiol Biotechnol* 44: 563–567, 1996.

37. M-S Rangheard, G Langrand, C Triantaphylides, J Baratti. Multi-competitive enzymatic reactions in organic media: a simple test for the determination of lipase fatty acid specificity. *Biochim Biophys Acta* 1004: 20–28, 1989.

38. SM Innis. Essential fatty acids in growth and development. *Prog Lipid Res* 30: 39–109, 1991.

39. C Miller, H Austin, L Posorske, J Gonzlez. Characteristics of an immobilized lipase for the commercial synthesis of esters. *J Am Oil Chem Soc* 65: 927–931, 1988.

40. KD Mukherjee, I Kiewitt. Specificity of *Carica Papaya* latex as biocatalyst in the esterification of fatty acids with 1-butanol. *J Agric Food Chem* 44: 1947–1952, 1996.

41. KD Mukherjee, I Kiewitt. Substrate specificity of lipases in protease preparations. *J Agric Food Chem* 46: 2427–2432, 1998.

42. P Pinsirodom, KL Parkin. Selectivity of potato tuber lipid acyl hydrolase toward long-chain unsaturated FA in esterification reactions with glycerol analogs in organic media. *J Am Oil Chem Soc* 80: 335–340, 2003.

43. E Osterberg, A-C Blomstrom, K Holmberg. Lipase catalyzed transesterification of unsaturated lipids in a microemulsion. *J Am. Oil Chem Soc* 66: 1330–1333, 1989.

44. P Langholz, P Andersen, T Forskov, W Schmidtsdorff. Application of a specificity of *Mucor miehei* lipase to concentrate docosahexaenoic acid. *J Am Oil Chem Soc* 66: 1120–1123, 1989.

45. K Mbayhoudel, L-C Comeau. Obtention sélective de l'acide pétrosélinique à partir de l'huile de fenouil par hydrolyse enzymatique. *Rev Franc Corps Gras* 36: 427–431, 1989.

46. L-Y Yang, A Kuksis, JJ Myher. Lipolysis of menhaden oil triacylglycerols and the corresponding fatty acid alkyl esters by pancreatic lipase *in vitro*: a reexamination *J Lipid Res* 31: 137–147, 1990.

47. I Jachmanián, KD Mukherjee. Esterification and Interesterification reactions catalyzed by acetone powder from germinating rapeseed. *J Am Oil Chem Soc* 73: 1527–1532, 1996.

48. S Warwel, R Borgdorf. Substrate selectivity of lipases in the esterification of *cis/trans*-isomers and positional isomers of conjugated linoleic acid (CLA). *Biotechnol Lett* 22: 1151–1155, 2000.

49. GP McNeill, C Rawlins, AC Peilow. Enzymatic enrichment of conjugated linoleic acid isomers and incorporation into triglycerides. *J Am Oil Chem Soc* 76: 1265–1268, 1999.

50. MJ Haas, JKG Kramer, G McNeill, K Scott, TA Foglia, N Sehat, J Fritsche, MM Mossoba, MP Yurawecz. Lipase-catalyzed fractionation of conjugated linoleic acid isomers. *Lipids* 34: 979–987, 1999.

51. T Nagao, Y Shimada, Y Yamauchi-Sato, T Yamamoto, M Kasai, K Tsutsumi, A Sugihara, Y Tominaga. Fractionation and enrichment of CLA isomers by selective esterification with *Candida rugosa* lipase. *J Am Oil Chem Soc* 79: 303–308, 2002.

52. I Ncube, T Gitlesen, P Adlercreutz, JS Read, B Mattiasson. Fatty acid selectivity of a lipase purified from *Vernonia galamensis* seed. *Biochim Biophys Acta* 1257: 149–156, 1995.

53. SA Hellyer, IC Chandler, JA Bosley. Can the fatty acid selectivity of plant lipases be predicted from the composition of the seed triglyceride. *Biochim Biophys Acta* 1440: 215–224, 1999.

54. P Pinsirodom, KL Parkin. Fatty acid product selectivities of potato tuber lipid acyl hydrolase in esterification reactions with glycerol in organic media. *J Am Oil Chem Soc* 76: 1119–1125, 1999.

55. R Kleimann, FR Earle, WH Tallent, IA Wolff. Retarded hydrolysis by pancreatic lipase of seed oils with *trans*-3 unsaturation. *Lipids* 5: 513–518, 1970.

56. RG Jensen, DT Gordon, WH Heimermann, RT Holman. Specificity of *Geotrichum candidum* lipase with respect to double bond position in triglycerides containing *cis*-octadecenoic acids. *Lipids* 7: 738–741, 1972.

57. WH Heimermann, RT Holman, DT Gordon, DE Kowalyshyn, RG Jensen. Effect of double bond position in octadecenoates upon hydrolysis by pancreatic lipase. *Lipids* 8: 45–46, 1973.

58. RG Jensen. Characteristics of the lipase from the mold, *Geotrichum candidum*: A review. *Lipids* 9, 149–157, 1974.

59. MW Baillargeon, PF Sonnet. Selective lipid hydrolysis by *Geotrichum candidum* NRRL Y-553 lipase. *Biotechnol Lett* 13: 871–874, 1991.

60. CM Sidebottom, E Charton, PP Dunn, G Mycock, C Davies, JL Sutton, AR Macrae, AR Slabas. *Geotrichum candidum* produces several lipases with markedly different substrate selectivities. *Eur J Biochem* 202: 485–491, 1991.

61. GJ Piazza, A Bilyk, DP Brower, MJ Haas. The positional and fatty acid selectivity of oat seed lipase in aqueous emulsions. *J Am Oil Chem Soc* 69: 978–981, 1992.

62. PE Sonnet, TA Foglia, MA Baillargeon. Fatty acid selectivity of lipases of *Geotrichum candidum*. *J Am Oil Chem Soc* 70: 1043–1045, 1993.

63. E Charton, AR Macrae. Specificities of immobilized *Geotrichum candidum* CMICC 335426 lipases A and B in hydrolysis and ester synthesis reactions in organic solvents. *Enzyme Microb Technol* 15: 489–493, 1993.

64. Z Mozaffar, JD Weete. Purification and properties of an extracellular lipase from *Pythium ultimum*. *Lipids* 28: 377–382, 1993.

65. S Parmar, EG Hammond. Hydrolysis of fats and oils with moist oat caryopses. *J Am Oil Chem Soc* 71: 881–886, 1994.

66. S Bech Pedersen, G Holmer. Studies of the fatty acid specificity of the lipase from *Rhizomucor miehei* toward 20:1n-9, 20:5n-3, 22:1n-9 and 22:6n-3. *J Am Oil Chem Soc* 72: 239–243, 1995.

67. D Briand, E Dubreucq, J Grimaud, P Galzy. Substrate specificity of the lipase from *Candida papapsilosis*. *Lipids* 30: 747–754, 1995.

68. F Ergan. Lipase specificities towards fatty acids. In: FX Malcata, ed. *Engineering of / with Lipases*, Dordrecht: Kluwer, 1996, pp. 65–72.

69. Y Kosugi, Q-l Chang, K Kanazawa, H Nakanishi. Changes in hydrolysis specificities of lipase from *Rhizomucor miehei* to produce polyunsaturated fatty acyl ethyl esters in different aggregation states. *J Am Oil Chem Soc* 74: 1395–1399, 1997.

70. RB Labuschagne, A van Tonder, D Litthauer. *Flavobacterium odoratum* lipase: Isolation and characterization. *Enzyme Microb Technol* 21: 52–58, 1997.

71. Y Shimada, A Sugihara, H Nakano, T Nagao, M Suenaga, S Nakai, Y Tominaga. Fatty acid specificity of *Rhizopus delemar* lipase in acidolysis. *J Ferment Bioeng* 83: 321–327, 1997.

72. MSF Lie Ken Jie, F Xun. Studies of lipase-catalyzed esterification reactions of some acetylenic fatty acids. *Lipids* 33: 71–75, 1998.

73. MSF Lie Ken Jie, F Xun. Lipase specificity toward some acetylenic and olefinic alcohols in the esterification of pentanoic and stearic acids. *Lipids* 33: 861–867, 1998.

74. S-H Yeo, T Nihira, Y Yamada. Screening and identification of a novel lipase from *Burkholderia* sp. YY62 which hydrolyzes *t*-butyl esters effectively. *J Gen Appl Microbiol* 44: 147–152, 1998.

75. R Borgdorf, S Warwel. Substrate selectivity of various lipases in the esterification of *cis*- and *trans*-octadecenoic acid. *Appl Microbiol Biotechnol* 51: 480–485, 1999.

76. S Warwel, R Borgdorf, L Brühl. Substrate selectivity of lipases in the esterification of oleic acid, linoleic acid, linolenoic acid and their all-*trans*-isomers and in the transesterification of *cis/trans*-isomers of linoleic acid. *Biotechnol Lett* 21: 431–436, 1999.

77. FW Cain, S Bouwer, MHW van den Hoek, A Menzel. Palmitoleic acid and its use in foods. U.S. Patent 6,461,662 B2, 2002.

78. Y Shimada, Y Watanabe, A Kawashima, K Akimoto, S Fujikawa, Y Tominaga, A Sugihara. Enzymatic fractionation and enrichment of n-9 PUFA. *J Am Oil Chem Soc* 60: 37–42, 2003.

79. F Molinari, G Marianelli, F Aragozzini. Production of flavor esters by *Rhizopus oryzae* lipase. *Appl Microbiol Biotechnol* 43: 967–973, 1995.

80. GP McNeill; R Berger. Lipase catalyzed synthesis of esters by reverse hydrolysis. *J Franc Oleagineux Corps Gras Lipides* 2: 359–363, 1995.

81. M Habulin, V Krmelj, Z Knez. Synthesis of oleic acid esters catalyzed by immobilized lipase. *J Agric Food Chem* 44: 338–342, 1996.

82. T Tsujita, M Sumiyoshi, H Okuda. Wax ester-synthesizing activity of lipases. *Lipids* 34: 1159–1166, 1999.

83. L Poisson, S Jan, F Ergan. Study on lipase-catalysed esterification of arachidonic acid in view of further PUFA enrichment of microalgae lipid extracts. In: G Baudimant, J Guézennec, P Roy, JF Samain, coordinators. *Marine Lipids*, Brest, France: Ifremer, 1998, pp. 204–211.

84. G Steinke, R Kirchhoff, KD Mukherjee. Lipase-catalyzed alcoholysis of crambe oil and camelina oil for the preparation of long-chain esters. *J Am Oil Chem Soc* 77: 361–366, 2000.

85. NN Gandhi, KD Mukherjee. Specificity of papaya lipase in esterification of aliphatic alcohols – A comparison with microbial lipases. *J Am Oil Chem Soc* 78: 161–165, 2001.

86. NN Gandhi, KD Mukherjee. Reactivity of medium-chain substrates in the interesterification of tripalmitin catalyzed by papaya lipase. *J Am Oil Chem Soc* 78: 965–968, 2001.

87. Y-H Ju, T-C Chen. High-purity γ-linolenic acid from borage oil fatty acids. *J Am Oil Chem Soc* 79: 29–32, 2002.

88. EM Anderson, KM Larsson, O Kirk. One biocatalyst — many applications: the use of *Candida antarctica* B lipase in organic synthesis. *Biocatal Biotransform* 16: 161–204, 1998.

89. J Arsan, KL Parkin. Selectivity of *Candida antarctica* B lipase toward fatty acid and (iso)propanol substrates in esterification reactions in organic media. *J Agric Food Chem* 48: 3738–3743, 2000.

90. M From, P Adlercreutz, B Mattiasson. Lipase catalyzed esterification of lactic acid. *Biotechnol Lett* 19: 315–317, 1997.

91. C-H Lee, KL Parkin. Comparative fatty acid selectivity of lipases in esterification reactions with glycerol and diol analogues in organic media. *Biotechnol Prog* 16: 372–377, 2000.

92. N Weber, E Klein, K Vosmann, KD Mukherjee. Preparation of long-chain acyl thioesters — thio wax esters — by the use of lipases. *Biotechnol Lett* 20: 687–691, 1998.

93. N Weber, E Klein, KD Mukherjee. Long-chain acyl thioesters prepared by solvent-free thioesterification and transthioesterification catalysed by microbial lipases. *Appl Microbiol Biotechnol* 51: 401–404, 1999.

94. N Weber, E Klein, KD Mukherjee. Antioxidants eliminate stereomutation and thioether formation during lipase-catalyzed thioesterification and transthioesterification for the preparation of uniform *cis*- and *trans*-unsaturated thioesters. *Chem Phys Lipids* 105: 215–223, 2000.

95. N Weber, E Klein, KD Mukherjee. Solvent-free lipase-catalyzed thioesterification and transthioesterification of fatty acids and fatty acid esters with alkanethiols *in vacuo*. *J Am Oil Chem Soc* 76: 1297–1300, 1999.

96. OJ Pollak. Reduction of blood cholesterol in man. *Circulation* 7: 702–706, 1953.

97. OJ Pollak, D Kritchevsky. Sitosterol. In: TB Clarkson, D Kritchevsky, OJ Pollak, eds. *Monographs on Atherosclerosis*, Vol. 10, Basel, Switzerland: Karger, 1981, pp. 1–219.

98. M Law. Plant sterol and stanol margarines and health. *Br Med J* 320: 861–864, 2000.

99. S Kochhar. Influence of processing of sterols of edible vegetable oils. *Prog Lipid Res* 22, 161–188, 1983.

100. HK Mangold, T Muramatsu. Preparation of reference compounds. In: HK Mangold, ed. *CRC Handbook of Chromatography*, Vol. II, Boca Raton, FL: CRC Press, 1984, pp. 319–329.

101. F Spener. Preparation of common and unusual waxes. *Chem Phys Lipids* 24: 431–448, 1979.

102. N Weber, P Weitkamp, KD Mukherjee. Fatty acid steryl, stanyl and steroid esters by esterification and transesterification in vacuo using *Candida rugosa* lipase as catalyst. *J Agric Food Chem* 49: 67–71, 2001.

103. N Weber, P Weitkamp, KD Mukherjee. Steryl and stanyl esters of fatty acids by solvent-free esterification and transesterification in vacuo using lipases from *Rhizomucor miehei*, *Candida antarctica* and *Carica papaya*. *J Agric Food Chem* 49: 5210–5216, 2001.

104. N Weber, P Weitkamp, KD Mukherjee. Steryl esters by transesterification reactions catalyzed by lipase from *Thermomyces lanuginosus*. *Eur J Lipid Sci Technol*, 105: 624–626, 2003.

105. MJ Hills, I Kiewitt, KD Mukherjee. Enzymatic fractionation of evening primrose oil by rape lipase: Enrichment of γ-linolenic acid. *Biotechnol Lett* 11: 629–632, 1989.

106. MJ Hills, I Kiewitt, KD Mukherjee. Enzymatic fractionation of fatty acids: Enrichment of γ-linolenic acid and docosahexaenoic acid by selective esterification catalyzed by lipases. *J Am Oil Chem Soc* 67: 561–564, 1990.

107. MSK Syed Rahmatullah, VKS Shukla, KD Mukherjee. γ-Linolenic acid concentrates from borage and evening primrose oil fatty acids via lipase-catalyzed esterification. *J Am Oil Chem Soc* 71: 563–567, 1994.

108. T Foglia, PE Sonnet. Fatty acid selectivity of lipases: γ-Linolenic acid from borage oil. *J Am Oil Chem Soc* 72: 417–420, 1995.

109. F-C Huang, Y-H Ju, C-W Huang. Enrichment of γ-linolenic acid from borage oil *via* lipase-catalyzed reactions. *J Am Oil Chem Soc* 74: 977–981, 1997.

110. M Schmitt-Rozieres, G Vanot, V Deyris, L-C Comeau. *Borago officinalis* oil: Fatty acid fractionation by immobilized *Candida rugosa* lipase. *J Am Oil Chem Soc* 76: 557–562, 1999.

111. E Van Heerden, D Litthauer. The comparative discriminating abilities of lipases in different media and their application in fatty acid enrichment. *Biocatal Biotransf* 16: 461–474, 1999.

112. F Ergan, S Lamare, M Trani. Lipase specificity against some fatty acids. *Ann N Y Acad Sci* 672: 37–44, 1992.

113. MSK Syed Rahmatullah, VKS Shukla, KD Mukherjee. Enrichment of γ-linolenic acid from evening primrose oil and borage oil via lipase-catalyzed hydrolysis. *J Am Oil Chem Soc* 71: 569–573, 1994.

114. I Jachmanián, KD Mukherjee. Germinating rapeseed as biocatalyst: Hydrolysis of oils containing common and unusual fatty acids. *J Agric Food Chem* 43: 2997–3000, 1995.

115. F-C Huang, Y-H Ju, C-W Huang. Enrichment in γ-linolenic acid of acylglycerols by the selective hydrolysis of borage oil. *Appl Biochem Biotechnol* 67: 227–236, 1997.

116. F-C Huang, Y-H Ju, J-C Chiang. γ-Linolenic acid-rich triacylglycerols derived from borage oil *via* lipase-catalyzed reactions. *J Am Oil Chem Soc* 76: 833–837, 1999.

117. Y Shimada, N Fukushima, H Fujita, Y Honda, A Sugihara, Y Tominaga. Selective hydrolysis of borage oil with *Candida rugosa* lipase: Two factors affecting the reaction. *J Am Oil Chem Soc* 75: 1581–1586, 1998.

118. M Tuter, HA Aksoy, G Ustun, S Riva, F Secundo, S Ipekler. Partial purification of *Nigella sativa* L. seed lipase and its application in hydrolytic reactions. Enrichment of γ-linolenic acid from borage oil. *J Am Oil Chem Soc* 80: 237–241, 2003.

119. Y Shimada, A Sugihara, K Maruyama, T Nagao, S Nakayama, H Nakano, Y Tominaga. Enrichment of arachidonic acid: Selective hydrolysis of a single-cell oil from *Mortierella* with *Candida cylindracea* lipase. *J Am Oil Chem Soc* 72: 1323–1327, 1995.

120. Y Shimada, A Sugihara, M Shibahiraki, H Fujita, H Nakano, T Nagao, T Terai, Y Tominaga. Purification of γ-linolenic acid from borage oil by a two-step enzymatic method. *J Am Oil Chem Soc* 74: 1465–1470, 1997.

121. Y Shimada, N Sakai, A Sugihara, H Fujita, Y Honda, Y Tominaga. Large-scale purification of γ-linolenic acid by selective esterification using *Rhizopus delemar* lipase. *J Am Oil Chem Soc* 75: 1539–1543, 1998.

122. KD Mukherjee, I Kiewitt. Enrichment of γ-linolenic acid from fungal oil by lipase-catalysed reactions. *Appl Microbiol Biotechnol* 35: 579–584, 1991.

123. PO Carvalho, GM Pastore. Enrichment of γ-linolenic acid from fungal oil by lipases. *Food Biotechnol* 12: 57–71, 1998.

124. M Vacek, M Zarevúcka, Z Wimmer, K Stránský, K Demnerová, M-D Legoy. Selective enzymatic esterification of free fatty acids with *n*-butanol under microwave irradiation and under classical heating. *Biotechnol Lett* 22: 1565–1570, 2000.

125. Y Shimada, A Sugihara, Y Minamigawa, K Higashiyama, K Akimoto, S Fujikawa, S Komemushi, Y Tominaga. Enzymatic enrichment of arachidonic acid from *Mortierella* single cell oil. *J Am Oil Chem Soc* 75: 1213–1217, 1998.

126. DF Horrobin. Nutritional and medical importance of γ-linolenic acid. *Prog Lipid Res* 31: 163–194, 1992.

127. BJF Hudson. Evening primrose (*Oenothera* spp.) oil and seed. *J Am Oil Chem Soc* 61: 540–543, 1984.

128. KD Mukherjee, I Kiewitt. Formation of γ-linolenic acid in the higher plant evening primrose (*Oenothera biennis* L.). *J Agric Food Chem* 35: 1009–1012, 1987.

129. A Whipkey, JE Simon, J Janick. *In vivo* and *in vitro* lipid accumulation in *Borago officinalis* L. *J Am Oil Chem Soc* 65: 979–984, 1988.

130. H Traitler, H Winter, U Richli, Y Ingenbleek. Characterization of γ-linolenic acid in *Ribes* seeds. *Lipids* 19: 923–928, 1984.

131. L Hansson, M Dostálek. Effect of culture conditions on mycelial growth and production of γ-linolenic acid by the fungus *Mortierella ramanniana*. *Appl Microbiol Biotechnol* 28: 240–246, 1988.

132. H Fukuda, H Morikawa. Enhancement of γ-linolenic acid production by *Mucor ambiguus* with nonionic surfactants. *Appl Microbiol Biotechnol* 27: 15–20, 1987.

133. H Traitler, HJ Wille, A Studer. Fractionation of black current seed oil. *J Am Oil Chem Soc* 65: 755–760, 1988.

134. T Yokochi, MT Usita, Y Kamisaka, T Nakahara, O Suzuki. Increase in the γ-linolenic acid content by solvent winterization of fungal oil extracted from *Mortierella* genus. *J Am Oil Chem Soc* 67: 846–851, 1990.

135. M Arai, H Fukuda, H. Morikawa. Selective separation of γ-linolenic acid ethyl ester using Y-zeolite. *J Ferment Technol* 65: 569–574, 1987.

136. T Hoshino, T Yamane, S Shimizu. Selective hydrolysis of fish oil by lipase to concentrate n-3 polyunsaturated fatty acids. *Agric Biol Chem* 54: 1459–1467, 1990.

137. NR Bottino, GA Vandenburg, R Reiser. Resistance of certain long-chain polyunsaturated fatty acids of marine oils to pancreatic lipase hydrolysis. *Lipids* 2: 489–493, 1967.

138. Y Tanaka, J Hirano, T Funada. Concentration of docosahexaenoic acid in glyceride by hydrolysis of fish oil with *Candida cylindracea* lipase. *J Am Oil Chem Soc* 69: 1210–1214, 1992.

139. Y Tanaka, T Funada, J Hirano, R Hashizume. Triglyceride specificity of *Candida cylindracea* lipase: Effect of docosahexaenoic acid on resistance of triglyceride to lipase. *J Am Oil Chem Soc* 70: 1031–1034, 1993.

140. SR Moore, GP McNeill. Production of triglycerides enriched in long-chain n-3 polyunsaturated fatty acids from fish oil. *J Am Oil Chem Soc* 73: 1409–1414, 1996.

141. GP McNeill, RG Ackman, SR Moore. Lipase-catalyzed enrichment of long-chain polyunsaturated fatty acids. *J Am Oil Chem Soc* 73: 1403–1407, 1996.

142. UN Wanasundara, F Shahidi. Concentration of ω-3 polyunsaturated fatty acids of marine oils using *Candida cylindracea* lipase: Optimization of reaction conditions. *J Am Oil Chem Soc* 75: 1767–1774, 1998.

143. KE Rice, J Watkins, CG Hill, Jr. Hydrolysis of menhaden oil by a *Candida cylindracea* lipase immobilized in a hollow-fiber reactor. *Biotechnol Bioeng* 63: 33–45, 1999.

144. VB Yadwad, OP Ward, LC Noronha. Application of lipase to concentrate the docosahexaenoic acid (DHA) fraction of fish oil. *Biotechnol Bioeng* 38: 956–959, 1991.

145. GG Haraldsson, Ö Almarsson. Studies on the positional specificity of lipase from *Mucor miehei* during interesterification reactions of cod liver oil with *n-3* polyunsaturated fatty acids. *Acta Chem Scand* 45: 723–730, 1991.

146. T Yamane, T Suzuki, Y Sahashi, L Vikersveen, T Hoshino. Production of n-3 polyunsaturated fatty acid enriched fish oil by lipase-catalyzed acidolysis without solvent. *J Am Oil Chem Soc* 69: 1104–1107, 1992.

147. T Yamane, T Suzuki, T Hoshino. Increasing the n-3 polyunsaturated fatty acid content of fish oil by temperature control of lipase-catalyzed acidolysis. *J Am Oil Chem Soc* 70: 1285–1287, 1993.

148. R Sridhar, G Lakshminarayana. Incorporation of eicosapentaenoic and docosahexaenoic acids into groundnut oil by lipase-catalyzed ester exchange. *J Am Oil Chem Soc* 69: 1041–1042, 1992.

149. L Zu-yi, OP Ward. Enzyme catalysed production of vegetable oils containing omega-3 polyunsaturated fatty acid. *Biotechnol Lett* 15: 185–188, 1993.

150. K-s Huang, CC Akoh. Lipase-catalyzed incorporation of n-3 polyunsaturated fatty acids into vegetable oils. *J Am Oil Chem Soc* 71: 1277–1280, 1994.

151. K-s Huang, CC Akoh, M Erickson. Enzymatic modification of melon seed oil: Incorporation of eicosapentaenoic acid. *J Agric Food Chem* 42: 2646–2648, 1994.

152. S Adachi, K Okumura, Y Ota, M Mankura. Acidolysis of sardine oil by lipase to concentrate eicosapentaenoic and docosahexaenoic acids in glycerides. *J Ferment Bioeng* 75: 259–264, 1993.

153. L Zuyi, OP Ward. Lipase-catalyzed alcoholysis to concentrate the n-3 polyunsaturated fatty acid of cod liver oil. *Enzyme Microb Technol* 15: 601–606, 1993.

154. L Zuyi, OP Ward. Stability of microbial lipase in alcoholysis of fish oil during repeated enzyme use. *Biotechnol Lett* 15: 393–398, 1993.

155. GG Haraldsson, A Thorarensen. The generation of glyceryl ether lipids highly enriched with eicosapentaenoic acid and docosahexaenoic acid by lipase. *Tetrahedron Lett* 35: 7681–7684, 1994.

156. GG Haraldsson, BÖ Gudmundsson, Ö Almarsson. The preparation of homogeneous triglycerides of eicosapentaenoic acid and docosahexaenoic acid by lipase. *Tetrahedron Lett* 34: 5791–5794, 1993.

157. Y Tanaka, J Hirano, T Funada. Synthesis of docosahexaenoic acid-rich triglyceride with immobilized *Chromobacterium viscosum* lipase. *J Am Oil Chem Soc* 71: 331–334, 1994.

158. H Maehr, G Zenchoff, DL Coffen. Enzymic enhancement of n-3 fatty acid content in fish oils. *J Am Oil Chem Soc* 71: 463–467, 1994.

159. Y Shimada, K Maruyama, S Okazaki, M Nakamura, A Sugihara, Y Tominaga. Enrichment of polyunsaturated fatty acids with *Geotrichum candidum* lipase. *J Am Oil Chem Soc* 71: 951–954, 1994.

160. Y Shimada, K Maruyama, M Nakamura, S Nakayama, A Sugihara, Y Tominaga. Selective hydrolysis of polyunsaturated fatty acid-containing oil with *Geotrichum candidum* lipase. *J Am Oil Chem Soc* 72: 1577–1581, 1995.

161. H Gunnlaugsdottir, B Sivik. Lipase-catalyzed alcoholysis of cod liver oil in supercritical carbon dioxide. *J Am Oil Chem Soc* 72: 399–405, 1995.

162. Y Shimada, A Sugihara, H Nakano, T Kuramoto, T Nagao, M Gemba, Y Tominaga. Purification of docosahexaenoic acid by selective esterification of fatty acids from tuna oil with *Rhizopus delemar* lipase. *J Am Oil Chem Soc* 74: 97–101, 1997.

163. Y Shimada, A Sugihara, S Yodono, T Nagao, K Maruyama, H Nakano, S Komemushi, Y Tominaga. Enrichment of ethyl docosahexaenoate by selective alcoholysis with immobilized *Rhizopus delemar* lipase. *J Ferment Bioeng* 84: 138–143, 1997.

164. H Breivik, GG Haraldsson, B Kristinsson. Preparation of highly purified concentrates of eicosapentaenoic acid and docosahexaenoic acid. *J Am Oil Chem Soc* 74: 1425–1429, 1997.

165. GG Haraldsson, B Kristinsson, R Sigurdardottir, GG Gudmundsson, H Breivik. The preparation of concentrates of eicosapentaenoic acid and docosahexaenoic acid by lipase-catalyzed transesterification of fish oil with ethanol. *J Am Oil Chem Soc* 74: 1419–1424, 1997.

166. M Ono, M Hosokawa, Y Inoue, K Takahashi. Water activity-adjusted enzymatic partial hydrolysis of phospholipids to concentrate polyunsaturated fatty acids. *J Am Oil Chem Soc* 74: 1415–1417, 1997.

167. Y Shimada, K Maruyama, A Sugihara, T Baba, S Komemushi, S Moriyama, Y Tominaga. Purification of ethyl docosahexaenoate by selective alcoholysis of fatty acid ethyl esters with immobilized *Rhizomucor miehei* lipase. *J Am Oil Chem Soc* 75: 1565–1571, 1998.

168. GG Haraldsson, B Kristinsson. Separation of eicosapentaenoic acid and docosahexaenoic acid in fish oil by kinetic resolution using lipase. *J Am Oil Chem Soc* 75: 1551–1556, 1998.

169. M Schmitt-Rozieres, M-C Guilhem, R Phan Tan Loo, LC Comeau. Enrichment of -3 polyunsaturated fatty acids of sardine cannery effluents using Lipozyme™. *Biocatal Biotransf* 18: 355–371, 2000.

170. M Schmitt-Rozieres, V Deyris, LC Comeau. Enrichment of polyunsaturated fatty acids from sardine cannery effluents by enzymatic selective esterification. *J Am Oil Chem Soc* 77: 329–332, 2000.

171. K Ishihara, M Murata, M Kaneniwa, H Saito, W Komatsu, K Shinohara. Purification of stearidonic acid (18:4(n-3)) and hexadecatetraenoic acid (16:4(n-3)) from algal fatty acid with lipase and medium pressure liquid chromatography. *Biosci Biotechnol Biochem* 64: 2454–2457, 2000.

172. Y Shimada, K Maruyama, AS Sugihara, Y Tominaga. Purification of docosahexaenoic acid from tuna oil by a two-step enzymatic method: Hydrolysis and selective esterification. *J Am Oil Chem Soc* 74: 1441–1446, 1997.

173. KL Mikolajczak, TK Miwa, FR Earle, IA Wolff, Q Jones. Search for new industrial oils. V. Oils of cruciferae. *J Am Oil Chem Soc* 38: 678–681, 1961.

174. PF Knowles, TE Kearney, DB Cohen. Species of rapeseed and mustard as oil crops in California. In: EH Pryde, LH Princen, KD Mukherjee, eds. *New Sources of Fats and Oils*, Champaign, IL: American Oil Chemists' Society, 1981, pp. 255–268.

175. KD Mukherjee, I Kiewitt. Lipids containing very long chain monounsaturated acyl moieties in seeds of *Lunaria annua*. *Phytochemistry* 25: 401–404, 1986.

176. GD Jolliff. Development and production of meadowfoam (*Limnanthes alba*). In: EH Pryde, LH Princen, KD Mukherjee, eds. *New Sources of Fats and Oils*, Champaign, IL: American Oil Chemists' Society, 1981, pp. 269–285.

177. NOV Sonntag. Erucic, behenic: Feedstocks of the 21st century. *Int News Fats Oils Relat Mater* 2: 449–463, 1991.

178. TNB Kaimal, RBN Prasad, T Chandrasekhara Rao. A novel lipase hydrolysis method to concentrate erucic acid glycerides in cruciferae oils. *Biotechnol Lett* 15: 353–356, 1993.

179. M Trani, R Lortie, F Ergan. Enzymatic synthesis of trierucin from high-erucic rapeseed oil. *J Am Oil Chem Soc* 70: 961–964, 1993.

180. DG Hayes, R Kleiman. The isolation and recovery of fatty acids with Δ5 unsaturation from meadowfoam oil by lipase-catalyzed hydrolysis and esterification. *J Am Oil Chem Soc* 70: 555–560, 1993.

181. KD Mukherjee, I Kiewitt. Enrichment of very long-chain monounsaturated fatty acids by lipase-catalyzed hydrolysis and transesterification. *Appl Microbiol Biotechnol* 44: 557–562, 1996.

182. GP McNeill, PE Sonnet. Isolation of erucic acid from rapeseed oil by lipase-catalyzed hydrolysis. *J Am Oil Chem Soc* 72: 213–218, 1995.

183. R Sridhar, G Lakshminarayana, TNB Kaimal. Modification of seleced edible vegetable oils to high oleic oils by lipase-catalyzed ester interchange. *J Agric Food Chem* 39: 2069–2071, 1991.

184. GP McNeill, D Borowitz, RG Berger. Selective distribution of saturated fatty acids into the monoglyceride fraction during enzymatic glycerolysis. *J Am Oil Chem Soc* 69: 1098–1103, 1992.

185. RW Lencki, N Smink, H Snelting, J Arul. Increasing short-chain fatty acid yield during lipase hydrolysis of a butterfat fraction with periodic aqueous extraction. *J Am Oil Chem Soc* 75: 1195–1200, 1998.

186. RMM Diks, MJ Lee. Production of a very low saturate oil based on the specificity of *Geotrichum candidum* lipase. *J Am Oil Chem Soc* 76: 455–462, 1999.

187. DG Hayes, R Kleiman. Recovery of hydroxy fatty acids from lesquerella oil with lipases. *J Am Oil Chem Soc* 69: 982–985, 1992.

188. DG Hayes, R Kleiman. 1,3-Specific lipolysis of *Lesquerella fendleri* oil by immobilized and reverse-micellar encapsulated enzymes. *J Am Oil Chem Soc* 70: 1121–1127, 1992.

189. MSF Lie Ken Jie, MSK Syed Rahmatullah. Enzymatic enrichment of C20 *cis*-5 polyunsaturated fatty acids from *Biota orientalis* seed oil. *J Am Oil Chem Soc* 72: 245–249, 1995.

190. MSF Lie Ken Jie, F Xun, MML Lau, ML Chye. Lipase-catalyzed hydrolysis of TG containing acetylenic FA. *Lipids* 37: 997–1006, 2002.

191. HS Garcia, JM Storkson, MW Pariza, CG Hill, Jr. Enrichment of butteroil with linoleic acid via enzymatic interesterification (acidolysis) reactions. *Biotechnol Lett* 20: 393–395, 1998.

192. G Langrand, M. Secchi, J Baratti, C Triantaphylides. Lipase-catalyzed ester formation in organic solvents. An easy preparative resolution of of α-substituted cyclohexanols. *Tetrahedron Lett* 26: 1857–1860, 1985.

193. G Langrand, J Baratti, G Buono, C Triantaphylides. Lipase-catalyzed reactions and strategy for alcohol resolution. *Tetrahedron Lett* 27: 29–32, 1986.

194. W-H Wu, CC Akoh, RS Phillips. Lipase-catalyzed stereoselective esterification of DL-menthol in organic solvents using acid anhydrides as acylating agents. *Enzyme Microb Technol* 18: 536, 1996.

195. W-H Wu, CC Akoh, RS Phillips. Stereoselective acylation of DL-menthol in organic solvents by an immobilized lipase from *Pseudomonas cepacia* with vinyl propionate. *J Am Oil Chem Soc* 74: 435–439, 1997.

196. Y Shimada, Y Hirota, T Baba, S Kato, A Sugihara, S Moriyama, Y Tominaga, T Terai. Enzymatic synthesis of L-menthyl esters in organic solvent-free system. *J Am Oil Chem Soc* 76: 1139–1142, 1999.

197. PE Sonnet. Lipase selectivities. *J Am Oil Chem Soc* 65: 900–904, 1988.

198. PE Sonnet, GG Moore. Esterifications of 1- and *rac*-2-octanols with selected acids and acid derivatives using lipases. *Lipids* 23: 955–960, 1988.

199. S-H Pan, T Kawamoto, T Fukui, K Sonomoto, A Tanaka. Stereoselective esterification of halogen-containing carboxylic acids by lipase in organic solvent: effects of alcohol chain length. *Appl Microbiol Biotechnol* 34: 47–51, 1990.

6

Protein Engineering of Recombinant *Candida rugosa* Lipases

Guan-Chiun Lee

Chwen-Jen Shieh

Jei-Fu Shaw

6.1 Introduction

Candida rugosa (formerly *Candida cylindracea*) lipase (CRL) is a very important industrial enzyme widely used in biotechnological applications such as the production of fatty acid, the synthesis of various esters, and the resolution of racemic mixtures.[1–7] However, crude enzyme preparations obtained from the various commercial suppliers exhibit remarkable variation in their catalytic efficiency and stereospecificity.[8] Seven lipase genes, namely *LIP1* to *LIP7*, have been described in *C. rugosa*. However, only three lipases (LIP1, LIP2, and LIP3) have been identified[9–13] in commercial crude enzyme preparations. The purified isoenzymes displayed different substrate specificities and thermal stabilities.[10,12,14–16] Native CRLs produced by a conventional fermentation process contain variable mixtures of isoforms due to the differential responses of different genes of CRL isoforms in wild type and mutant strains to fermentation conditions.[17,18] This results in serious problems in biotechnological applications due to the irreproducibility brought about by using enzymes from various suppliers, or even different batches, and the complications in interpretations.

Separation of CRL isozymes is highly desirable as it allows their use under well-defined conditions. However, a high identity in their protein sequences causes similarities in the physical properties of the lipases that make isolation difficult. In addition, the purification procedure may affect the properties of different isoforms,[19] and the differential expression level of the five lipase genes makes it difficult to purify each isoenzyme directly from cultures of *C. rugosa* on an industrial scale. Therefore, cloning and expression can be suggested as the most suitable approach for the production, characterization, and optimization of the biocatalytic properties of pure isoforms.

Unfortunately, despite the general availability of the cloned genes, the non-spore-forming yeast *C. rugosa* utilizes the nonuniversal codon; namely, the triplet CTG, a universal codon for leucine, is read as serine.[20] CTG triplets encode most of the serine residues in five *C. rugosa* lipases, including the catalytic Ser-209 (Figure 6.1). Therefore the heterologous expression of such genes may result in the production

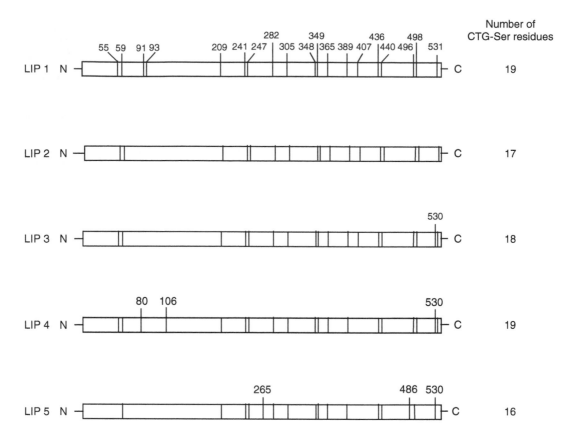

FIGURE 6.1 Distribution of CTG triplets encoding the serine residues in five *C. rugosa* lipases.

of inactive lipases. Thus the conversion of several or even all of the CTG codons into universal serine triplets is required for the expression of a functional lipase protein in heterologous hosts.

Recently, amended *LIP4* and *LIP2* gene by site-directed mutagenesis have been successfully expressed in *Escherichia coli* and *Pichia pastoris*.[21-23] Alternatively, the *LIP1* gene by complete synthesis with optimized codons has also been successfully expressed in *P. pastoris*.[24] These breakthroughs open the way to the large-scale industrial production of individual recombinant *C. rugosa* LIP isoforms and to the further engineering of catalytic properties desired for applications of the enzymes. This chapter describes some of the recent applications of CRL lipase and the genetic engineering of their isoforms.

6.2   Differential Expression of CRL Isoforms

C. rugosa lipase preparations from different suppliers were reported to show different catalytic efficiency and stereospecificity in various applications such as the resolution of racemic 2-(4-hydroxyphenoxy) propionic acid.[25] We speculated that this might be due to the difference in lipase isozyme expression resulting from different cultural conditions at different lipase production companies. Indeed, we discovered the existence of multiple enzyme forms in a commercial *C. rugosa* lipase preparation.[26] The purified multiple enzyme forms had a diversity of values for properties such as substrate specificity and thermal stability. Therefore, different compositions of lipase isozymes in different preparations would lead to different enzyme stabilities and specificities in applications.

Different multiple form patterns of lipolytic enzymes were identified in three commercial *C. rugosa* lipase preparations.[17] This accounted for the differences in catalytic efficiency and specificity of *C. rugosa* lipase reported in the literature. Tween 80 and Tween 20 not only promote lipase productivity, but also change the expression of multiple forms in cultured *C. rugosa*. These lipase preparations show quite different substrate specificities and thermal stabilities. This suggests that the specificity and stability of lipase preparations used in biotechnological applications can be engineered by cultural conditions that change the multiple form compositions.

To date, five lipase-encoding genomic sequences from *C. rugosa* have been characterized.[27,28] The five lipase-encoding genes (*LIP1, 2, 3, 4,* and *5*) have been isolated from a *Sac*I genomic library of the yeast *C. rugosa* by colony hybridization. The five genes encode for mature proteins of 534 residues with putative signal peptides of 15 (in *LIP1, 3, 4,* and *5*) and 14 (in LIP 2) amino acids in length, respectively. The five deduced amino acid sequences share an overall identity of 66% and similarity of 84%. Due to a high-sequence homology among the five deduced amino acid sequences and the differential expression level of the five lipase genes, it is difficult to purify each isozyme directly from the cultures of *C. rugosa* on an industrial scale.

For highly related genes, the conventional methods in mRNA analysis are not specific and sensitive enough to distinguish and quantitate individual mRNAs. It is also difficult to distinguish the transcription pattern of genes with a high degree of identity by Northern blot analysis. Although the nuclease protection assay has the ability to discriminate among closely related genes, this method, like Northern blot, is not sensitive enough to detect low amounts of mRNA and permits only crude quantitation. The competitive reverse transcription-PCR (RT-PCR) technique was developed to obtain quantitative information on highly related *C. rugosa LIP* genes at the transcriptional level owing to its high sensitivity and specificity.[18] As shown in Table 6.1, the relative abundance of *LIP* mRNAs was found to be (in decreasing order) *LIP1, LIP3, LIP2, LIP5,* and *LIP4. LIP1* and *LIP3* achieve higher expression, while expression of *LIP2, LIP4,* and *LIP5* is only 0.1–0.5% of the expression of *LIP1* transcript under YM culture conditions. These expression profiles are consistent with the findings that LIP1 and LIP3 are the major lipase proteins obtained by purification methods.[10,12]

TABLE 6.1   Comparison of Specific mRNA Expression Under Various Culture Conditions by Competitive RT-PCR

Target mRNA	Culture Conditions	Fold Induction[a]	Amount Relative to *LIP1*[b]
LIP1	YM	1	1
	YM-olive oil	1.07	1
	YM-oleic acid	0.56	1
	YM-Tween 20	1.08	1
LIP2	YM	1	0.005
	YM-olive oil	2.69	0.013
	YM-oleic acid	1.24	0.009
	YM-Tween 20	1.17	0.005
LIP3	YM	1	0.68
	YM-olive oil	2.07	1.32
	YM-oleic acid	1.53	1.83
	YM-Tween 20	0.88	0.56
LIP4	YM	1	0.001
	YM-olive oil	2.13	0.002
	YM-oleic acid	4.05	0.007
	YM-Tween 20	7.89	0.008
LIP5	YM	1	0.004
	YM-olive oil	2.09	0.007
	YM-oleic acid	2.03	0.011
	YM-Tween 20	1.95	0.006

[a] The fold-induction of individual *LIP* mRNA in YM was normalized to 1.
[b] Relative amount is defined as the amount of individual mRNA versus *LIP1* under each culture condition.
Modified from Lee et al., *Appl Environ Microbiol* 69: 3888–3895 (1999)

All five *LIP* genes are transcriptionally active. Different inducers may change the expression profile of individual genes. The constitutively expressed *LIP1* and *LIP3* showed few changes at the transcriptional level in various culture media while suppression of *LIP1* by oleic acid and induction of *LIP3* by olive oil was demonstrated. Olive oil and oleic acid also promoted the expression of inducible *LIP2, 4*, and *5* even in the presence of glucose, previously reported to be a repressing carbon source.[29] Obviously, Tween 20 had a significant inducing effect on *LIP4* only. This experiment unequivocally demonstrates that the expression profiles of *C. rugosa LIP* genes can be altered by different culture conditions and even by batch-to-batch culture differences. Different lipase isoforms from *C. rugosa* displayed quite different substrate specificities and thermal stabilities. The production of different lipase isoforms in response to different growth conditions is physiologically important for *C. rugosa*, enabling it to grow on various substrates and in different environments. Traditionally, the culture conditions in fermentation are optimized for the maximal production of enzyme activity units. Our results indicate that quality is as important as quantity in enzyme preparation since different culture conditions might result in heterogeneous compositions of the isozymes displaying different catalytic activities and specificities. By engineering the culture conditions, we can obtain enzyme preparations enriched in selected isozymes for particular biotechnological applications.

Recently, we cloned and sequenced the five CRL gene promoters and showed that their sequences are quite different and contain different nutrient-related transcriptional controlling elements (unpublished results). By assaying β-galactosidase activities of promoter-*lacZ* fusions in *Saccharomyces cerevisiae*, our unpublished data showed that the promoter activities of *LIP3* under various culture conditions were much higher than those of *LIP4* promoter. These findings suggest that the expression profile of *LIP* genes could be accompanied by different regulation of the *LIP* promoter activities. Studies of the transcriptional controlling elements of *C. rugosa LIP* genes are needed to further elaborate the mechanism of differential regulation of *LIP* genes by various inducers.

6.3   Structures of Purified CRL Isoforms

The crystal structures of three native CRL isoforms (LIP1, LIP2, and LIP3) purified from crude enzyme mixtures have been determined (Table 6.2).[9,30–32,35,49–50] The catalytic triad residues (Ser209, Glu341, and His449) and the distinct hydrogen bonding patterns stabilizing the triads are well conserved in all three isoforms. The oxyanion hole predicted to be formed by the backbone NH groups of Gly124 and Ala210 are also conserved among the three isoforms. Therefore, the basic catalytic machinery appears to be well conserved for all CRL isoforms.

All the CRL isoforms have an amphipathic α-helix which serves as a lid (or flap) covering the active site. The flap regions are quite different among the CRL isoforms. For example, the LIP2 has 11 and 9 different amino acid residues compared with LIP1 and LIP3, respectively, in the 30 amino acid-flap region.[32] The differences in the hinge point presumably affect the opening of the flap of different isoforms and the hydrophobicity differences of these residues in the internal side were suggested to affect the substrate binding. The hinge point amino acid residues are Glu66, Met 66, and Glu66 for LIP1, LIP2, and LIP3, respectively. The flap of LIP2 is more hydrophobic than LIP1 or LIP3. The flap interaction with the substrate-binding pocket has been proposed to be important in the different catalytic properties among various CRL isoforms.[32,33]

Structural comparison of the substrate binding sites among three isoforms reveals important different amino acid variations that might affect substrate specificity and catalytic properties.[32] It was proposed that the Phe content at the mouth of the hydrophobic tunnel (acyl binding site) greatly affected the substrate-recognition properties of the CRL isoforms. The Phe content of this region for LIP1, LIP2, and LIP3 are 5, 3, 4, respectively. The residues 296 and 344 for LIP1, LIP2, and LIP3 are Phe-Phe, Val-Leu, and Phe-Ile, respectively. Coincidently, the lipase/esterase activity ratio toward triacetin for LIP1, LIP2, and LIP3 are 43.7, 7.3, and 7.6, respectively. In contrast, the cholesterol esterase activity decreases from LIP2 >>LIP3>LIP1. It appears that the Phe content in the mouth of the hydrophobic tunnel significantly

TABLE 6.2   A Tabular Report of 10 Structures of *C. rugosa* Lipases in the Current PDB Release

PDB ID	Experimental Technique	Title	Chain Identifier	Chain Length	pH Value	Resolution [Å]	Ligand name	Ligand formula	Ligand ID	Authors	Reference
1CLE	X-ray diffraction	Structure of uncomplexed and linoleate-bound Candida cylindracea cholesterol esterase.	A B	534 534	n/a	2.000	Phosphate ion Cholesteryl linoleate N-acetyl-d-glucosamine	2(O4 P1 3-) 2(C45 H76 O2) 6(C8 H15 N1 O6)	PO4 CLL NAG	D. Ghosh, Z. Wawrzak, V.Z. Pletnev, N. Li, R. Kaiser, W. Pangborn, H. Jornvall, M. Erman, W.L. Duax	31
1CLE	X-ray diffraction	Insights into interfacial activation from an open structure of Candida rugosa lipase.	n/a	534	n/a	2.060	N-acetyl-d-glucosamine	3(C8 H15 N1 O6)	NAG	P. Grochulski, Y. Li, J.D. Schrag, F. Bouthillier, P. Smith, D. Harrison, B. Rubin, M. Cygler	9
1GZ7	X-ray diffraction	Structural insights into the lipase/esterase behavior in the candida rugosa lipases family: crystal structure of the lipase 2 isoenzyme at 1.97a resolution	A B C D	534 534 534 534	4.500	1.970	Glycerol N-acetyl-d-glucosamine	8(C3 H8 O3) 8(C8 H15 N1 O6)	GOL NAG	J.M. Mancheno, M.A. Pernas, M.J. Martinez, B. Ochoa, M.L. Rua, J.A. Hermoso	32
1LLF	X-ray diffraction	Three-Dimensional Structure of Homodimeric Cholesterol Esterase-Ligand Complex at 1.4Å Resolution	A B	534 534	7.300	1.400	Tricosanoic acid N-acetyl-d-glucosamine	2(C23 H46 O2) 8(C8 H15 N1 O6)	F23 NAG	V. Pletnev, A. Addlagatta, Z. Wawrzak, W. Duax	49
1LPM	X-ray diffraction	A structural basis for the chiral preferences of lipases.	n/a	549	n/a	2.200	Calcium ion N-acetyl-d-glucosamine (1r)-menthyl hexyl phosphonate	2(CA1 2+) 3(C8 H15 N1 O6) C16 H32 O2 P1	CA NAG MPA	M. Cygler, P. Grochulski, R.J. Kazlauskas, J.D. Schrag, F. Bouthillier, B. Rubin, A.N. Serreqi, A.K. Gupta	50

TABLE 6.2 (*Continued*)

PDB ID	Experimental Technique	Title	Chain Identifier	Chain Length	pH Value	Resolution [Å]	Ligand name	Ligand formula	Ligand ID	Authors	Reference
1LPN	X-ray diffraction	Analogs of reaction intermediates identify a unique substrate binding site in Candida rugosa lipase.	n/a	549	n/a	2.200	Calcium ion Dodecanesulfonate N-acetyl-d-glucosamine	2(CA1 2+) 2(C12 H25 O2 S1) 3(C8 H15 N1 O6)	CA DSC NAG	P. Grochulski, F. Bouthillier, R.J. Kazlauskas, A.N. Serreqi, J.D. Schrag, E. Ziomek, M. Cygler	35
1LPO	X-ray diffraction	Analogs of reaction intermediates identify a unique substrate binding site in Candida rugosa lipase.	n/a	549	n/a	2.200	Calcium ion N-acetyl-d-glucosamine 1-hexadecanosulfonic acid	2(CA1 2+) 3(C8 H15 N1 O6) C16 H34 O3 S1	CA NAG HDS	P. Grochulski, F. Bouthillier, R.J. Kazlauskas, A.N. Serreqi, J.D. Schrag, E. Ziomek, M. Cygler	35
1LPP	X-ray diffraction	Analogs of reaction intermediates identify a unique substrate binding site in Candida rugosa lipase.	n/a	549	n/a	2.050	Calcium ion N-acetyl-d-glucosamine 1-hexadecanosulfonic acid	2(CA1 2+) 3(C8 H15 N1 O6) 2(C16 H34 O3 S1)	CA NAG HDS	P. Grochulski, F. Bouthillier, R.J. Kazlauskas, A.N. Serreqi, J.D. Schrag, E. Ziomek, M. Cygler	35
1LPS	X-ray diffraction	A structural basis for the chiral preferences of lipases.	n/a	549	n/a	2.200	Calcium ion N-acetyl-d-glucosamine (1s)-menthyl hexyl phosphonate	2(CA1 2+) 3(C8 H15 N1 O6) C16 H32 O2 P1	CA NAG MPC	M. Cygler, P. Grochulski, R.J. Kazlauskas, J.D. Schrag, F. Bouthillier, B. Rubin, A.N. Serreqi, A.K. Gupta	50
1TRH	X-ray diffraction	Two conformational states of Candida rugosa lipase.	n/a	534	n/a	2.100	N-acetyl-d-glucosamine	2(C8 H15 N1 O6)	NAG	P. Grochulski, Y. Li, J.D. Schrag, M. Cygler	30

affects the environment of the catalytic triad and the lipase/esterase catalytic properties and substrate specificities. This should be further confirmed by site-directed mutagenesis.

In order to understand the molecular basis of substrate specificity, we compared geometry and properties of the CRL isoforms. Three-dimensional (3-D) structures of the LIP4 and LIP5 isoenzymes were obtained from the Web-based SWISS-MODEL server (Version 3.5; Glaxo Smith Kline S.A., Geneva, Switzerland; www.expasy.org/swissmod/SWISS-MODEL.html), which utilizes the comparative modeling

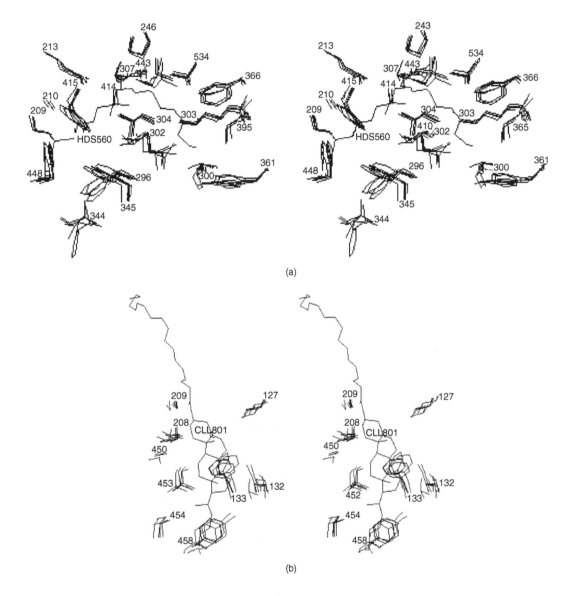

FIGURE 6.2 The substrate-binding regions of CRL isoforms and structural differences in these regions. Amino acid changes in these regions among different lipases are shown in Table 6.3. (a) Superposition of the fatty acid-binding tunnels of the *C. rugosa* lipases 1, 2, 3, 4, and 5. The same residue numbers among the five isoforms are indicated. The inhibitor hexadecane sulfonate is labeled in HDS560. (b) Superposition of the alcohol-binding sites of CRL isoforms. The cholesteryl linoleate molecule is labeled in CLL801. (c) Superposition of the flaps (residue 66-92) of CRL isoforms. Amino acid residues facing the substrate pocket are indicated in residue number and most of them are nonpolar (black). The opposite sides of the flap facing outside are mostly basic, acid, and polar (gray). The catalytic triad residues (Ser209, Glu341, and His449) are also shown. The images were produced using Swiss-Pdb Viewer (Glaxo Smith Kline S.A.).

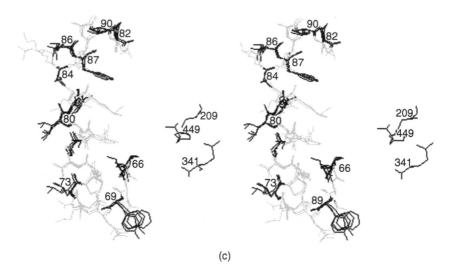

(c)

FIGURE 6.2 (*Continued*)

approach.[34] The structures of molecular complexes of LIP1 with hexadecanesulphonyl chloride,[35] LIP2,[32] and LIP3 with cholesteryl linoleate[31] were obtained from the RCSB protein data bank[36] under the entry numbers 1LPO, 1GZ7, and 1CLE, respectively. These structures were used as the corresponding templates for comparative protein modeling. Display and analysis of modeled structures was performed using a Swiss-Pdb Viewer (Glaxo Smith Kline S.A.). Figure 6.2 shows the comparative 3-D topography of substrate-binding sites for CRL isoforms. The results allowed analysis of the cholesterol linoleate-binding site and the acyl-binding tunnel. Amino acid changes in these regions may determine changes in the interactions affecting the stability and specificity of the CRL isoforms. Most of the residues located in close contact with the cholesterol linoleate molecule and in place of the acyl chain (within the 4 Å cut-off distance) were identical and shared the same conformations as those of the template residues. Some of the substrate-binding residues showed differences between LIP isoforms (Table 6.3). Four definite differences were found at the 20 amino acids of acyl-binding sites. The alcohol-binding sites show more variations, and five out of eight amino acids are not conserved. The hydrophobic face of the flap, which also plays an important role in contact with substrate, has five variations out of eleven residues.

6.4 Expression of Active Recombinant CRL in Heterologous Hosts

The difficulty in the heterologous expression of lipase genes stems from *C. rugosa*'s nonuniversal codon usage, in which the triplet CTG, the universal codon for leucine, is read as serine. The recombinant LIP1 isoenzyme has been functionally expressed in *S. cerevisiae* and *P. pastoris* by completely synthesizing the *LIP1* gene with an optimized nucleotide sequence.[24] We replaced the 19 CTG codons in *LIP4* to universal serine codons by site-directed mutagenesis and functionally expressed the mutated gene of LIP4.[21,22] A more efficient overlap extension PCR-based multiple site-directed mutagenesis method was further developed to convert the 17 non-universal serine codons (CTG) of LIP2 gene into universal serine codons (TCT) (Figure 6.3).[23] An active recombinant LIP2 was overexpressed in *P. pastoris* and secreted into the culture medium. The same method was successfully applied to produce highly active LIP1, LIP3, and two other newly identified CRL isoforms (LIP8 and LIP9) (patents pending, unpublished results).

The characterization of the recombinant CRL isoforms (LIP1, LIP2, and LIP4) reveals interesting findings. LIP1 and LIP3 have been purified from commercial preparation or cultures of *C. rugosa* by complex procedures.[10–12,37] The substrate specificity of LIP1 toward fatty acid methyl esters differed

TABLE 6.3 Important Amino Acid Changes Producing Structural
Differences Among *C. rugosa* Lipases

Residue	LIP1	LIP2	LIP3	LIP4	LIP5
Fatty acid binding site					
210	Ala	Ala	Ala	Ala	Ala
213	Met	Met	Met	Met	Met
246	Pro	Pro	Pro	Pro	Pro
296	Phe	Val	Phe	Ala	Phe
300	Ser	Pro	Ser	Pro	Thr
302	Leu	Leu	Leu	Leu	Leu
303	Arg	Arg	Arg	Arg	Arg
304	Leu	Leu	Leu	Leu	Leu
307	Leu	Leu	Leu	Leu	Leu
344	Phe	Leu	Ile	Val	Leu
345	Phe	Phe	Phe	Phe	Phe
361	Tyr	Tyr	Tyr	Tyr	Tyr
365	Ser	Ser	Ser	Ser	Ser
366	Phe	Phe	Phe	Phe	Phe
410	Leu	Leu	Leu	Leu	Leu
413	Leu	Leu	Leu	Leu	Leu
414	Gly	Ala	Ala	Ala	Thr
415	Phe	Phe	Phe	Phe	Phe
449	His	His	His	His	His
534	Val	Val	Val	Val	Val
Alcohol binding site					
127	Val	Leu	Ile	Val	Ile
132	Thr	Leu	Ile	Leu	Ile
133	Phe	Phe	Phe	Phe	Phe
208	Glu	Glu	Glu	Glu	Glu
450	Ser	Gly	Ala	Ala	Ala
453	Ile	Ile	Ile	Ile	Ile
454	Val	Ile	Val	Val	Val
458	Tyr	Tyr	Tyr	Phe	Phe
Hydrophobic face of the flap					
66	Glu	Met	Glu	Leu	Glu
69	Tyr	Phe	Phe	Trp	Tyr
73	Leu	Leu	Leu	Leu	Leu
77	Ala	Ala	Ala	Ala	Ala
80	Leu	Leu	Leu	Ser	Leu
81	Val	Val	Val	Leu	Val
84	Ser	Ser	Ser	Ser	Ser
86	Val	Ile	Val	Leu	Val
87	Phe	Phe	Phe	Phe	Phe
90	Val	Val	Val	Val	Val
92	Pro	Pro	Pro	Pro	Pro

*The nonconserved positions are shown in bold.

significantly from that of commercial CRLs (CL), which have specificity toward short-chain fatty acids.[24,38] However, recombinant LIP1 has catalytic activities similar to CL, suggesting that the major component of CL should be LIP1.[24] In contrast, the recombinant LIP4 favors substrates of longer (C16 and C18) acyl chains while LIP1 favors shorter ones (C8 and C10). This confirms individual CRL lipase isoforms possess distinctive enzyme activities, despite high degrees of sequence similarity (81% identity in LIP1 and LIP4). Moreover, variations of substrate specificity in commercial preparations of lipases from *C. rugosa* may be due to the different combinations of catalytic activities of individual LIPs.

The commercial lipase and recombinant glycosylated LIP4 from *P. pastoris* has a higher molecular mass than recombinant LIP4 from *E coli*.[22] This phenomenon is due to glycosylation of these CRLs. By

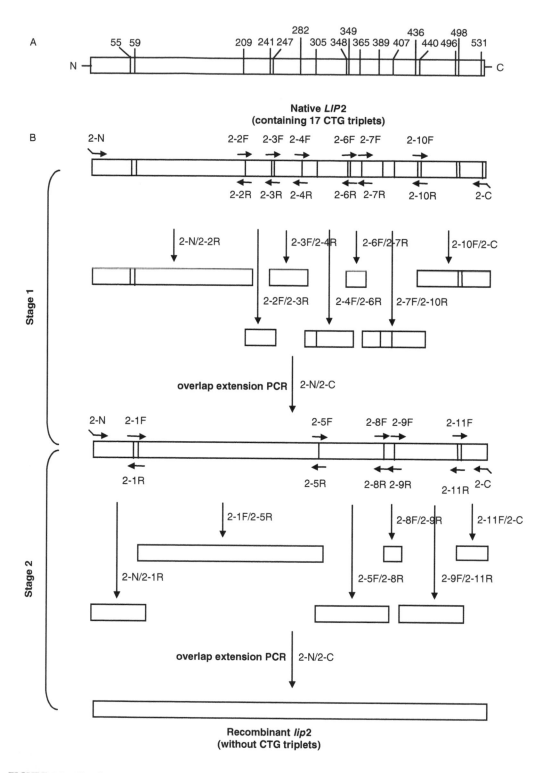

FIGURE 6.3 Simultaneous multiple mutagenesis introduced by overlap extension PCR for the replacement of the nonuniversal serine codons in the *C. rugosa LIP2* gene. (A) The 17 CTG serine residues are indicated with their residue numbers along the polypeptide chain of LIP2. (B) The arrows indicate the mutagenic primers that were used to alter the 17 CTG triplets in the *LIP2* gene. (From Lee et al., *Biochem J* 366:603–611, 2002).

deducing from these DNA sequences, LIP1, LIP3, and LIP5 have been proposed to have three glycosylation sites while LIP2 and LIP4 have only one.[39] After analysis the DNA sequence using a computer program, we found that LIP4 has only a potential N-glycosylation motif (Asn-Val-Thr) and the Asn residue is located at position 351. Based on these findings, Asn-351 may be the major glycosylation site in CRLs. Although Asn-291 and Asn-314 have also been considered potential sites, they may not undergo glycosylation in LIP1, LIP3, and LIP5.

According to the analysis of crystal structure, carbohydrate seems to play an active role in the activation of CRL by providing an additional stabilization for the open conformation of the flap.[30] Although LIP4 expressed in *E coli* is not glycosylated, its enzymatic activity is similar to that of the glycosylated lipase from *P. pastoris*. This finding indicates that glycosylation may not significantly affect the enzyme activities of LIP4. However, glycosylated LIP4 has higher thermal stability than unglycosylated LIP4. It is possible that the glycosylation level of CRLs may affect their structure, stability, and movement through the secretory pathways. Replacing the three N-linked glycosylation sites Asn-291, 314, and 351 of recombinant LIP1 with Gln by site-directed mutagenesis demonstrates the importance of glycosidic chains in the activation of LIP1.[40] Comparison of the activity of mutants Asn314Gln and Asn351Gln with that of the wild-type lipase indicates that both mutants influence the enzyme activity both in hydrolysis and esterification reactions, but they do not alter the enzyme water activity profiles in organic solvents or temperature stability. Replacing Asn351 by Gln probably disrupts the stabilizing interaction between the sugar chain and residues of the inner side of the lid in the enzyme active conformation. The effect of deglycosylation at position 314 suggests a more general role of the sugar moiety for the structural stability of lipase 1. Conversely, Asn291Gln substitution does not affect the lipolytic or the esterase activity of the mutant that behaves essentially as a wild-type enzyme.

Recombinant LIP2 overexpressed in *P. pastoris* has a molecular mass similar to the native LIP2 isoenzyme, which had been purified from a *C. rugosa* fermentation.[41] These two enzymes showed similar substrate specificities for the hydrolysis of long-chain *p*-nitrophenyl esters. However, the recombinant enzyme displays higher specific activity than the native enzyme towards all *p*-nitrophenyl esters tested. In particular, the chain-length preference for triacylglycerols was quite different between these two enzymes. Recombinant LIP2 preferred to hydrolyze the short-chain triacylglycerol (tributyrin) while the native LIP2 was more active on the long-chain triacylglycerol (triolein). These variations may be due to differences in glycosylation, the additional N-terminal peptide or amino acid substitutions between the recombinant and native LIP2.[23]

CRLs have been widely used in biotechnological applications, and almost all the applications use crude commercial enzymes. A comparison of pure recombinant LIP2 and LIP4 lipases with a crude preparation will provide an evaluation of the potential uses of pure recombinant enzymes in biotechnological applications. The recombinant LIP2 possesses unique and remarkable catalytic properties that essentially differ from those of LIP4 and CL. For example, the high specific activity of LIP2 towards cholesteryl esters might be very useful for the determination of cholesterol in clinical and food analyses. Since 70–80% of serum cholesterol is esterified by various fatty acids,[42] if coupled to cholesterol oxidase and peroxidase, LIP2 could be useful as an enzymatic sensor of serum cholesterol. LIP2 showed high activity toward long-chain alcohols in esterification with myristic acid. This enzyme could be used in the industrial production of wax esters, such as hexadecyl myristate and octadecyl myristate, used in lubricants and cosmetics. In the esterification of propanol with various fatty acids, LIP2 preferred short-chain fatty acids such as butyric acid and, therefore, might be useful in the industrial production of flavor esters.

Three-dimensional structures of the substrate-binding sites of *C. rugosa* LIP2 and LIP4 lipases were obtained through comparative modeling. Six out of ten residues at the alcohol-binding site and two out of 15 residues at the acyl-binding site were not identical among the four *C. rugosa* isoenzymes considered in the previous study.[23] It is suggested that the alcohol-binding site shows lower conservation than the acyl-binding tunnel and, therefore, might play an important role in substrate specificity. For example, some replacements obviously changed hydrophobicity in the alcohol-binding site, reflected in the higher activity of LIP2 than LIP4 in the hydrolysis of cholesterol esters and in the synthesis of myristate esters.

Our preliminary results also suggest that the flap domain could greatly affect enzyme specificity and other biochemical properties (GC Lee, SJ Tang, and JF Shaw, unpublished work).

6.5 Protein Engineering of Recombinant CRL Isoforms

The availability of individual recombinant CRL isoforms and the 3-D structures of three isoforms (LIP1, LIP3, and LIP2) open the way for further improving the CRL catalytic activity, substrate specificity, and enzyme stability for industrial applications by gene shuffling (directed evaluation) or rationally designed protein engineering.

Manetti et al.[43] modified the molecular recognition of 2-arylpropionic esters by recombinant LIP1 using site-directed mutagenesis of Phe344 and Phe345. The enzyme activity and selectivity were significantly decreased by Phe345Val mutation and Phe344, 345Val double mutation. This suggests that Phe345 plays important role in the S-enantiomer preference of (±)-2-(3-benzoylphenyl) propionic acid (Ketoprofen) and (±)-2-(6-methoxy-2-naphthyl) propionic acid (Naproxen).

Schmitt et al.[44] modified the chain-length specificity of recombinant LIP1 by engineering the amino acids at the substrate binding site. Different chain length specificity of the enzyme can be achieved by mutating different amino acid residues of CRL inside the tunnel. Computer modeling revealed P246 and L413 are located near atom C6/C8 of the bound fatty acid. P246F and L413F block chains longer than C6/C8, and the mutant enzymes strongly favor the hydrolysis of short chain triglycerides. P246F mutant enzyme has a 67- and 14-fold relative activity toward C4 and C6 chains, respectively, compared to the activity toward C8 chains. In contrast, the corresponding activity of wide-type CRL is 0.78 and 0.16, respectively.

Interestingly, a mutant L304F, which blocks the entrance of the tunnel, did not accept C4 and C6 short-chain triglycerides, but hydrolyzed longer-chain triglycerides ≥ 8. An alternative binding site outside the tunnel for medium- and long-chain fatty acids was proposed. The short-chain fatty acids preferably bind to the tunnel, while medium- and long-chain fatty acids can bind to either the tunnel or the alternative site, which has no specificity.

Lee et al.[45,46] reported the effect of the lid region on the recombinant CRL LIP4 lipase activity and specificity by exchanging the lid regions from the other four *C. rugosa* isoforms (LIP1, 2, 3, and 5; and corresponding lids 1, 2, 3, and 5) with that of LIP4 and expressed as chimeric proteins Trx-LIP4/lid1, Trx-LIP4/lid2, Trx-LIP4/lid3, and Trx-LIP4/lid5, respectively. The lipase hydrolysis activities of Trx-LIP4/lid2 and Trx-LIP4/lid3 increased 14% and 32%, respectively, while Trx-LIP4/lid1 and Trx-LIP4/lid5 decreased 85% and 20%, respectively, compared with native LIP4 with tributyrin as a substrate. The best substrate for Trx-LIP4, Trx-LIP4/lid2, and Trx-LIP4/lid3 is cholesterol caprate, but the best for Trx-LIP4/lid1 and Trx-LIP4/lid5 is cholesterol stearate. In contrast, when *p*-nitrophenyl esters were used as substrates, both *p*-nitrophenyl caprate and stearate were the best substrates for Trx-LIP4 and Trx-LIP2, while only *p*-nitrophenyl caprate was the best substrate for Trx-LIP4/lid1, Trx-LIP4/lid3, and Trx-LIP4/lid5. The lid change also affected the substrate specificity of enzymes on the selectivity of cholesterol esters of various desaturated fatty acids. The cholesteryl oleate (18:1) was the best substrate for Trx-LIP4, followed by cholesteryl linoleate (18:2, relative activity 68%) while cholesteryl stearate (18:0) was a poor substrate (relative activity 7%). Trx-LIP4/lid2 and Trx-LIP4 had a similar substrate preference pattern. The lid domain also affected the enantioselectivity of lipase. The *C. rugosa* lipase favored the hydrolysis of *l*-menthyl acetate over *d*-menthyl acetate. The recombinant Trx-LIP4 and all the chimeric LIP4s showed much better enantioselectivities than a commercial *C. rugosa* lipase (Lipase Type VII, Sigma). Only the enantioselectivity of Trx-LIP4/lid3 was similar to Trx-LIP4. Other chimeric proteins (Trx-LIP4/lid1, Trx-LIP4/lid2, and Trx-LIP4/lid5) showed substantial decreases in enantioselectivity with methyl acetate as a substrate. The enantioselectivity preference order might possibly change if other chiral substrates are used.

Brocca et al.[47] also report the sequence of the lid affects the activity and specificity of recombinant CRL LIP1 isoform by a similar lid swapping method. They have shown that swapping the LIP3 lid to LIP1 isoform confers the cholesterol esterase activity to LIP1, which has no cholesterol esterase in its native form.

6.6 Concluding Remarks

The *Candida rugosa* lipase has been widely used in various applications for nearly 40 years since its discovery in 1962.[48] However, a serious discrepancy concerning the biochemical properties and substrate specificities among various reports using commercial crude enzyme preparations from various suppliers has long been a disputed issue. Purifications and characterizations of several isoforms from crude enzyme preparations clearly showed CRL are mixtures of multiple isoforms with significantly different biochemical properties. The compositions of isoforms in crude enzymes varied greatly due to the differential expression of isoforms in response to culture conditions. The cloning and sequence analysis of five genes revealed the high sequence identity (77–88%), the same 534 amino acid residues, and predicted molecular weight ranging from 56,957 to 57,744. The 3-D structures of purified native LIP1, LIP2, and LIP3 provide insights into the structure-function relationship of various CRL isoforms.

The recent exciting breakthroughs in overcoming unusual serine codon usage problems by either total synthesis of gene or site-directed mutagenesis and expressing various functional recombinant isoforms in conventional heterologous hosts such as *E coli* and *P. pastoris* pave the way to high-level industrial production of individual isoforms for various biotechnological applications. Recombinant CRL enzymes with improved catalytic properties and stabilities will become possible by DNA shuffling or rationally designed site-directed mutagenesis.

References

1. YJ Wang, JY Sheu, FF Wang, JF Shaw. The lipase catalyzed oil hydrolysis in the absence of added emulsifier. *Biotechnol Bioeng* 31:628–633, 1988.
2. AM Klibanov. Asymmetric transformations catalyzed in organic solvents. *Acc Chem Res* 23:114–120, 1990.
3. JF Shaw, RC Chang, FF Wang, YJ Wang. Lipolytic activities of a lipase immobilized on six selected supporting materials. *Biotechnol Bioeng* 35:132–137, 1990.
4. RH Valivety, PJ Halling, AD Peilow, AR Macrae. Relationship between water activity and catalytic activity of lipases in organic media. Effects of supports, loading and enzyme preparation. *Eur J Biochem* 222:461–466, 1994.
5. EN Vulfson. Industrial applications of lipases. In: P Woolley, and SB Petersen, eds. *Lipase—Their Structure, Biochemistry and Application*. Cambridge: Cambridge University Press, 1994, pp. 271–288.
6. NN Gandhi. Applications of lipase. *J Am Oil Chem Soc* 74:621–634, 1997.
7. S Benjamin, A Pandey. Candida rugosa lipases: molecular biology and versatility in biotechnology. *Yeast* 14:1069–1087, 1998.
8. MJ Barton, JP Hamman, KC Fichter, GJ Calton. Enzymatic resolution of (R,S)-2-(4-hydroxyphenoxy) propionic acid. *Enzyme Microb Technol* 12:577–583, 1990.
9. P Grochulski, Y Li, JD Schrag, F Bouthillier, P Smith, D Harrison, B Rubin, M Cygler. Insights into interfacial activation from an open structure of Candida rugosa lipase. *J Biol Chem* 268:12843–12847, 1993.
10. ML Rúa, T Diaz Maurino, VM Fernandez, C Otero, A Ballesteros. Purification and characterization of two distinct lipases from Candida cylindracea. *Biochim Biophys Acta* 1156:181–189, 1993.
11. R Kaiser, M Erman, WL Duax, D Ghosh, H Jornvall. Monomeric and dimeric forms of cholesterol esterase from Candida cylindracea. Primary structure, identity in peptide patterns, and additional microheterogeneity. *FEBS Lett* 337:123–127, 1994.
12. MA Diczfalusy, U Hellman, SE Alexson. Isolation of carboxylester lipase (CEL) isoenzymes from Candida rugosa and identification of the corresponding genes. *Arch Biochem Biophys* 348:1–8, 1997.
13. P Ferrer, JL Montesinos, F Valero, C Sola. Production of native and recombinant lipases by Candida rugosa: a review. *Appl Biochem Biotechnol* 95:221–255, 2001.
14. YY Linko, XY Wu. Biocatalytic production of useful esters by two forms of lipase from Candida rugosa. *J Chem Technol Biotechnol* 65:163–170, 1996.

15. MJ Hernaiz, JM Sanchez Montero, JV Sinisterra. New differences between isoenzymes A and B from Candida rugosa lipase. *Biotechnol Lett* 19:303–306, 1997.

16. C López, NP Guerra, ML Rua. Purification and characterization of two isoforms from Candida rugosa lipase B. *Biotechnol Lett* 22:1291–1294, 2000.

17. RC Chang, SJ Chou, JF Shaw. Multiple forms and functions of Candida rugosa lipase. *Biotechnol Appl Biochem* 19:93–97, 1994.

18. GC Lee, SJ Tang, KH Sun, JF Shaw. Analysis of the gene family encoding lipases in Candida rugosa by competitive reverse transcription-PCR. *Appl Environ Microbiol* 65:3888–3895, 1999.

19. MJ Hernaiz, M Rua, B Celda, P Medina, JV Sinisterra, JM Sanchez-Montero. Contribution to the study of the alteration of lipase activity of Candida rugosa by ions and buffers. *Appl Biochem Biotechnol* 44:213–229, 1994.

20. Y Kawaguchi, H Honda, J Taniguchi-Morimura, S Iwasaki. The codon CUG is read as serine in an asporogenic yeast Candida cylindracea. *Nature* 341:164–166, 1989.

21. SJ Tang, KH Sun, GH Sun, TY Chang, GC Lee. Recombinant expression of the Candida rugosa lip4 lipase in Escherichia coli. *Protein Expr Purif* 20:308–313, 2000.

22. SJ Tang, JF Shaw, KH Sun, GH Sun, TY Chang, CK Lin, YC Lo, GC Lee. Recombinant expression and characterization of the Candida rugosa lip4 lipase in Pichia pastoris: Comparison of glycosylation, activity, and stability. *Arch Biochem Biophys* 387:93–98, 2001.

23. GC Lee, LC Lee, V Sava, JF Shaw. Multiple mutagenesis of non-universal serine codons of the Candida rugosa LIP2 gene and biochemical characterization of purified recombinant LIP2 lipase overexpressed in Pichia pastoris. *Biochem J* 366:603–611, 2002.

24. S Brocca, C Schmidt-Dannert, M Lotti, L Alberghina, RD Schmid. Design, total synthesis, and functional overexpression of the Candida rugosa lip1 gene coding for a major industrial lipase. *Protein Sci* 7:1415–1422, 1998.

25. MJ Barton, JP Hamman, KC Fichter, GJ Calton. Enzymatic resolution of (R,S)-2-(4-hydroxyphenoxy) propionic acid. *Enzyme Microb Technol* 12:577–583, 1990.

26. JF Shaw, CH Chang, YJ Wang. Characterization of three distinct forms of lipolytic enzyme in a commercial Candida lipase preparation. *Biotechnol Lett* 11:779–784, 1989.

27. S Longhi, F Fusetti, R Grandori, M Lotti, M Vanoni, L Alberghina. Cloning and nucleotide sequences of two lipase genes from Candida cylindracea. *Biochim Biophys Acta* 1131:227–232, 1992.

28. M Lotti, R Grandori, F Fusetti, S Longhi, S Brocca, A Tramontano, L Alberghina. Cloning and analysis of Candida cylindracea lipase sequences. *Gene* 124:45–55, 1993.

29. M Lotti, S Monticelli, JL Montesinos, S Brocca, F Valero, J Lafuente. Physiological control on the expression and secretion of Candida rugosa lipase. *Chem Phys Lipids* 93:143–148, 1998.

30. P Grochulski, Y Li, JD Schrag, M Cygler. Two conformational states of Candida rugosa lipase. *Protein Sci* 3:82–91, 1994.

31. D Ghosh, Z Wawrzak, VZ Pletnev, N Li, R Kaiser, W Pangborn, H Jornvall, M Erman, WL Duax. Structure of uncomplexed and linoleate-bound Candida cylindracea cholesterol esterase. *Structure* 3:279–288, 1995.

32. JM Mancheno, MA Pernas, MJ Martinez, B Ochoa, ML Rua, JA Hermoso. Structural insights into the lipase/esterase behavior in the Candida rugosa lipases family: crystal structure of the lipase 2 isoenzyme at 1.97A resolution. *J Mol Biol* 332:1059–1069, 2003.

33. MA Pernas, C Lopez, ML Rua, J Hermoso. Influence of the conformational flexibility on the kinetics and dimerisation process of two Candida rugosa lipase isoenzymes. *FEBS Lett* 501:87–91, 2001.

34. T Schwede, J Kopp, N Guex, MC Peitsch. SWISS-MODEL: An automated protein homology-modeling server. *Nucleic Acids Res* 31:3381–3385, 2003.

35. P Grochulski, F Bouthillier, RJ Kazlauskas, AN Serreqi, JD Schrag, E Ziomek, M Cygler. Analogs of reaction intermediates identify a unique substrate binding site in Candida rugosa lipase. *Biochemistry* 33:3494–3500, 1994.

36. HM Berman, J Westbrook, Z Feng, G Gilliland, TN Bhat, H Weissig, IN Shindyalov, PE Bourne. The Protein Data Bank. *Nucleic Acids Res* 28:235–242, 2000.

37. MC Brahimi-Horn, ML Guglielmino, L Elling, LG Sparrow. The esterase profile of a lipase from Candida cylindracea. *Biochim Biophys Acta* 1042:51–54, 1990.

38. C Schmidt-Dannert, J Pleiss, RD Schmid. A toolbox of recombinant lipases for industrial applications. *Ann N Y Acad Sci* 864:14–22, 1998.

39. M Lotti, A Tramontano, S Longhi, F Fusetti, S Brocca, E Pizzi, L Alberghina. Variability within the Candida rugosa lipases family. *Protein Eng* 7:531–535, 1994.

40. S Brocca, M Persson, E Wehtje, P Adlercreutz, L Alberghina, M Lotti. Mutants provide evidence of the importance of glycosydic chains in the activation of lipase 1 from Candida rugosa. *Protein Sci* 9:985–990, 2000.

41. MA Pernas, C Lopez, L Pastrana, ML Rua. Purification and characterization of Lip2 and Lip3 isoenzymes from a Candida rugosa pilot-plant scale fed-batch fermentation. *J Biotechnol* 84:163–174, 2000.

42. P Röschlau, E Bernt, W Gruber. Enzymatische Bestimmung des Gesamt-Cholesterins im Serum. *Z Klin Chem Klin Biochem* 12:403–407, 1974.

43. F Manetti, D Mileto, F Corelli, S Soro, C Palocci, E Cernia, I D'Acquarica, M Lotti, L Alberghina, M Botta. Design and realization of a tailor-made enzyme to modify the molecular recognition of 2-arylpropionic esters by Candida rugosa lipase. *Biochim Biophys Acta* 1543:146–158, 2000.

44. J Schmitt, S Brocca, RD Schmid, J Pleiss. Blocking the tunnel: engineering of Candida rugosa lipase mutants with short chain length specificity. *Protein Eng* 15:595–601, 2002.

45. JF Shaw, GC Lee, SJ Tang. Recombinant *Candida rugosa* lipases. EPC European Patent Application 02009616.0, 2002.

46. GC Lee, SJ Tang, JF Shaw. Structure and function of *Candida rugosa* lipase. 90th Annual Meeting of American Oil Chemists' Society. May 9–13, Orlando, FL, 1999.

47. S Brocca, F Secundo, M Ossola, L Alberghina, G Carrea, M Lotti. Sequence of the lid affects activity and specificity of Candida rugosa lipase isoenzymes. *Protein Sci* 12:2312–2319, 2003.

48. K Yamada, H Machida. Studies on the production of lipase by microorganisms. I. The selection and identification of a new strain [Candida cylindrica sp. n.]. *Agric Biol Chem* 26:A69, 1962.

49. V Pletnev, A Addlagatta, Z Wawrzak, W Duax. Three-dimensional structure of homodimeric cholesterol esterase-ligand complex at 1.4 A resolution. *Acta Crystallogr D Biol Crystallogr* 59:50–56, 2003.

50. M Cygler, P Grochulski, RJ Kazlauskas, JD Schrag, F Bouthillier, B Rubin, AN Serreqi, AK Gupta. A structural basis for the chiral preferences of lipases. *J Am Chem Soc* 116:3180–3186, 1994.

7

Production of Value-Added Industrial Products from Vegetable Oils: Oxygenated Fatty Acids

Ching T. Hou

Masashi Hosokawa

7.1 Introduction

The United States produces more than 18 billion pounds of soybean oil (SBO) annually with a yearly carryover of more than 300 million pounds. How to utilize this surplus oil effectively becomes a large economic issue in the agricultural community. SBO is a relatively cheap raw material at 22 to 25 cents per pound and an attractive candidate for bioindustries. The content of unsaturated fatty acids such as oleic and linoleic acids are 22% and 55% for soybean oil and 26% and 60% for corn oil, respectively. Currently, the major use of vegetable oils is for food products such as shortenings, salad and cooking oils, and margarines, with small quantities serving industrial applications. Total nonfood uses of vegetable oils have grown little during the past 30 years. Although markets for epoxidized oils have increased, other markets have lost to competitive petroleum products. It is only through continued development of new industrial products or commercial processes that vegetable oils will be able to maintain their market share. For applications such as cosmetics, lubricants, and chemical additives, soybean oil and other vegetable oils are too viscous and too reactive toward atmospheric oxygen to establish significant markets. For other uses, including coatings, detergents, polymers, flavors, agricultural chemicals, and pharmaceuticals, reactivity of vegetable oils needs to be enhanced by introducing additional functionalities or cleaving the fatty acid molecules.

Oxygenated fatty acids are important industrial materials. The hydroxy, keto, or epoxy group gives a fatty acid special properties, such as higher viscosity and reactivity compared to other fatty acids. Because of their special chemical attributes, oxygenated fatty acids are used in a wide range of products, including resins, waxes, nylons, plastics, corrosion inhibitors, cosmetics, coatings, and recently in biomedical.

7.2   α-, β-, and ω- Hydroxy Fatty Acids

7.2.1   α-Hydroxy Fatty Acids

Hydroxy fatty acids are common in nature. In mammals, oxidation of long-chain fatty acids to α-hydroxy fatty acids has been demonstrated in microsomes of brain and other tissues. α-Hydroxy long-chain fatty acids are constituents of brain lipids. α-Hydroxy fatty acids are major components of fatty acids found in *Arthrobacter simplex,* where the cell-free extracts contain enzyme activity to convert palamitic acid to α-hydroxypalmitic acid.[1] The sphingolipids of a wide variety of organisms are rich in α-hydroxy fatty acids, and in *Sphingobacterium paucimobilis,* α-hydroxymyristic acid is the major fatty acid component.[2] The α-hydroxylase of *S. paucimobilis* responsible for converting myristate to α-hydroxymyristate is a member of the P450 superfamily,[3] but the enzyme uses hydrogen peroxide instead of NADH and molecular oxygen.[4] Moreover, the enzyme exhibits high substrate specificity for α-hydroxylation reactions.[5] A mixture of 2(R)-hydroxy fatty acids (C_{14}–C_{18}) possesses antimicrobial activity against *Vibrio tyrogenuses.*[6] Short-chain branched 2-hydroxyalkanoic acids are part of the ring structure of several peptide antibiotics such as sporidesmolides, amidomycin, and valinomycin.[7–9] 2(R)-Hydroxyhexadecanoic acid is a metabolite with unknown function in yeast.[10] 2-Hydroxy saturated C_{20}–C_{25} fatty acids are constituents found in the phospholipids of marine sponges.[11] 2-Hydroxy saturated and monounsaturated C_{22}–C_{24} fatty acids are constituents of phospholipids in a Caribbean sea urchin.[12]

7.2.2   β- Hydroxy Fatty Acids

β-Oxidation is the major fate of fatty acid metabolism in mammals. During β-oxidation, fatty acids are oxidized in mitochondria by a sequence of reactions in which the fatty acyl chain is shortened two carbon atoms at a time to produce β-hydroxy fatty acids. Recently, β-hydroxy fatty acids became popular health supplements.

3-Hydroxydodecanoic acid is a constituent of the peptide antibiotics isariin[13] and serratamolide.[14] 3-Hydroxy C_{10}–C_{12} fatty acids are components of the peptide antibiotic esperin.[15] 3 (R)-Hydroxypalmitic acid is an extracelluar metabolite of the yeast *Saccharomycopsis malanga* NRRL Y-6954.[16] In *Mucor* sp. A-73, formation of 3-hydroxyalkanoic acid may serve as a physiological mechanism to prevent intracellular accumulation of undesirable metabolites.[17] A direct hydroxylation of oleic acid at the C-3 position occurs in *Alcaligenes* sp. 5-18, producing 3-hydroxyoleic acid and 3-hydroxyhexadecenoic acid.[18] The yeast *Dipodascopsis uninucleata* UOFS-Y128 can transfer exogeneous arachidonic acid to a stable 3-hydroxy-5,8,11,14-eicosatetraenoic acid (3-HETE),[19] a compound possessing signal trasduction activity in human neutrophils.[20] Formation of 3-HETE is inhibited by aspirin and, apparently, differs from cyclooxygenases and lipoxygenase enzyme systems that initiate the formation of other arachidonic acid metabolites. Chiral-phase high-performance liquid chromatography analysis of 3-HETE and other metabolites reveals that they are 3(R)-hydroxy fatty acids. The regiospecificity of 3-HETE hydroxylation rules out the biosynthesis via a normal β-oxidation pathway but perhaps suggests a direct monooxygenase reaction at C-3 or a 2-enoyl-CoA hydratase with opposite steric specificity.[21] *Dipodascopsis* also converts exogenous linoleic acid and 11(Z), 14(Z), 17(Z)-eicosatrienoic acid to the 3(R)-hydroxylated metabolites of shorter chain length, but hydroxylates neither oleic acid, linolelaidic acid, γ-linolenic acid, nor eicosanoic acid.[21]

7.2.3 ω-Hydroxy Fatty Acids

Long-chain fatty acids may also undergo ω-oxidation to produce ω-hydroxy fatty acids that are subsequently converted to dicarboxylic acids. This series of reactions has been observed with enzymes in liver microsomes and with soluble enzyme preparations from bacteria. The cell-free enzyme system of *Pseudomonas oleovolans* catalyzes ω-oxidation of saturated, even-numbered fatty acids ranging from C8 to C18. Medium-chain fatty acids (octanoate, decanoate, laurate, and myristate) are the most active substrates, whereas long-chain fatty acids (palmitate and stearate) are oxidized at lower rates. Alkanoates less than C6 are not oxidized.[22] The same enzyme system can also catalyze the epoxidation of alkenes and hydroxylation of alkanes, but does not react on carbon chain lengths of less than 6.[23,24] The hydroxylation and epoxidation of alkanes and alkenes with carbon chain lengths of less than 6, such as methane through hexane and ethylene through hexene, are oxidized by methane monooxygenase and propane monooxygenase systems.[25,26] The ω-hydroxylation system of *P. oleovolans* (ω-hydroxylase) requires Fe^{2+} ions, NADH, molecular oxygen and two other coenzymes, rubredoxin and an NADH-rubredoxin reductase.[27] The ω-hydroxylase is purified and shown to be a nonheme iron protein requiring phospholipids and ferrous irons for full activity.[28–30] The enzyme complex transfers electrons in the following sequence: NADH → reductase → rubredoxin → ω-hydroxylase → O_2.[31] The enzyme was also reported to have a di-iron cluster for catalytic activity.[32] The ω-hydroxylase belongs to a large family of such functionally diverse enzymes as desaturase, hydroxylase, acetylenase, and epoxidase that may share a generalized mechanism for modification of fatty acids.[33]

7.2.4 ω-1, ω-2 and ω-3 Hydroxy Fatty Acids

Oxygen-dependent hydroxylations also occur on carbon atoms adjacent to the terminus of long-chain fatty acids. ω- and (ω-1)-Hydroxy fatty acids are components of yeast sorphorose glycolipids belonging to a group of surface-active compounds.[34] In the production of extracellular hydroxy fatty acid sophorosides, strains of *Torulopsis* spp. hydroxylate palmitic, stearic, and oleic acids to their corresponding 15-hydroxy- or 17-hydroxy-fatty acids.[34,35] The ω-1 hydroxylation is stereospecific in the introduction of an oxygen atom from molecular oxygen.[35,36] The involvement of cytochrome P-450 in the hydroxylation is also well characterized in *Candida apicola*.[37,38] Under aerobic conditions, strains of *Bacillus pumilus* hydroxylate the ω-1, ω-2, and ω-3 carbon atoms of oleic acid to produce 15-, 16-, and 17-hydroxy-9-octadecenoic acid, having 17-hydroxyoctadecenoic acid as the most abundant product.[39] A strain of *Bacillus* sp. U88 converts 12-hydroxyoctadecanoic acid to 12,15-, 12,16-, and 12,17-dihydroxyoctadecanoic acids.[40]

A soluble cell-free system from *Bacillus megaterium* ATCC 4581 hydroxylates saturated and monounsaturated fatty acids on the ω-1, ω-2 and ω-3 carbon atoms, primarily on the ω-2 position in the presence of NADH and oxygen.[41,42] It does not hydroxylate the terminal methyl (ω) group of either fatty acids, alcohols, or amides, and hydroxylation of methylene carbons beyond ω-3 position are insignificant.[42] Formation of ω-hydroxylate isomers involves a single species of ω-2 hydroxylase that requires a ferredoxin-type component and is dependent on P-450-type cytochrome.[43,44] The ω-2 hydroxylase is a unique, self-sufficient biocatalyst (designated as cytochrome P-450$_{BM-3}$), having both functions of an active cytochrome-c reductase and a monooxygenase residing in a single polypeptide with a molecular mass of 119 kDa.[45] The soluble recombinant P-450$_{BM-3}$, which is purified from *E coli* strain DH5α transformed with the P-450$_{BM-3}$ gene,[46] shows fatty acid chain-length specificity.[47] The soluble P-450$_{BM-3}$ converts lauric and myristic acids to their corresponding ω-2 hydroxy fatty acids, but fails to react with capric acid. When palmitic acid is the substrate, the resulting products are dependent on substrate concentrations,[47] and the fatty acid is converted to a mixture of ω-1, ω-2, and ω-3 hydroxy analogs at concentrations greater than 250 μM. When the concentration of palmitic acid is less than 250 μM, a mixture of 14-ketohexadecanoic acid, 15-ketohexadecanoic acid, 13-hydroxy-14-ketodecanoic acid, 14-hydroxy-15-ketohexadecanoic acid, and 13,14-dihydroxyhexadecanoic acid are produced, indicating that the soluble P-450$_{BM-3}$ can also function as dehydrogenase.[47] Cytochrome P-450$_{BM-3}$ also converts polyunsaturated arachidonic acid to 18 (R)-hydroxyeicosatetraenoic acid and 14(S), 15(R)-epoxyeicosatrienoic acid.[48]

7.3 Dicarboxylic Acids

Dicarboxylic acids (DCs) are useful raw materials for the preparation of fragrances, polyamides, adhesives, lubricants, marvrolide antibiotics, and polyesters.[49–51] The DCs can also be used by microorganisms to produce polyhydroxy alkanoates (PHAs), which can, in turn, be used to manufacture biodegradable plastics.[49] Tridecane-1,13-dicarboxylic acid (DC-15) is useful as a pharmaceutical stock chemical for synthesis of muscone, a cyclopentadecanone drug used in treating heart diseases and inflammation of joints.[52] Aliphatic dicarboxylic acids are formed by a reaction mechanism first described as diterminal oxidation by Kester and Foster[53] in the production of dicarboxylic acids from the C_{10}–C_{14} alkanes by a *Corynebacterium* sp. strain 7E1C. A mutant strain of *Candida cloacae* 310 unable to assimilate DC as a sole carbon source produces large amounts of DCs, predominantly with the same number of carbon atoms as those of n-alkanes (C_9–C_{10}) used in the culture.[54] Among the reaction products, n-tetradecane ω, ω-dicarboxylic acid (DC-16) from n-hexadecane, and DC-15 from n-heptadecane are the most abundant. Conversion of n-heptadecane to DC-15 is stimulated by organic solvent- and detergent-treated *Cryptococcus neoformans* and *P. aeruginosa* but is also inhibited by elevated levels of DC-15.[55] To avoid product inhibition, use of a continuous process with immobilized *Cryptococcus* cells can lead to a fivefold increase in yield as compared with the batch type of DE-15 production.[52]

Besides long-chain alkanes, diterminal oxidation also occurs in long-chain saturated and unsaturated fatty acids. The ω-hydroxy fatty acids formed in the cell-free enzyme preparation of *P. aeruginosa* can easily be oxidized further to the corresponding dicarboxylic acids by an NAD-dependent ω-hydroxy fatty acid dehydrogenase.[22] A mutant strain S_{76} of *Candida tropicalis* produces long-chain dioic acid, not only from alkanes but also from their alcohols, monoic acids, α, ω-diol, and ω-hydroxy acids of corresponding chain length by the reaction mechanisms of diterminal oxidation and/or ω-oxidation of the terminal methyl group.[56] The formation of α, ω-dioic acids from n-alkanes is carried out by a metabolic pathway that involves α, ω-diols as plausible intermediates.[56] The same microorganism also converts oleic acid and its alcohol and ester derivatives to form 9(Z)-1,18-octadecenoic acid by ω-oxidation of the terminal methyl group, which can be metabolized further to form shorter, even-numbered carbon atoms of saturated or unsaturated dioic acids.[57,58]

Conversion of n-alkanes to α,ω-dioic acids and ω- and (ω-1)- hydroxylation of fatty acids is well known for strains of *Candida maltosa*. Isoforms of cytochrome P-450 encoded by P450alk genes[59] are inducible by various long-chain n-alkanes and fatty acids and exhibit chain-length preferences in the terminal hydroxylation.[59–61] An active alkane monooxygenase system can be reconstituted from purified recombinant proteins of P-450 52A3 and the corresponding NADH-dependent reductase.[62] The enzyme system converts hexadecane to produce 1-hexadecanal, 1-hexadecanol, hexadecanoic acid, 1,16-hexadecanediol, 16-hydroxyhexadecanoic acid, 1,16-hexadecandioic acid, thus demonstrating the complete catalytic activity for sequentially converting n-alkanes to α,ω-dioic acids.

A genetically engineered *C. tropicalis,* in which the ω-oxidation pathway has been sequentially disrupted to fully redirect alkane and fatty acid substrates to the ω-oxidation pathway, greatly improves the production of long-chain dioic acids.[51] Amplification of genes encoding the rate-limiting ω-hydroxylase of the ω-oxidation-blocked strain further enhances productivity.[51] The addition of pristine to the culture to act as an inert carrier for the sparingly water-soluble fatty substrates improves the production of dioic acids from long-chain fatty acids in a bioreactor by *C. cloacae* FERM-P736, a selected mutant strain with an impaired ω-oxidation pathway.[63]

A mutant yeast strain M25 derived from *C. tropicalis* DSM 3152 was mutagenized with *N*-methyl-*N*-nitro-*N*′-nitrosoguanidine and selected with oleic acid as the sole carbon source.[64] Strain M25 converts oleic acid to produce 3-hydroxy-9(Z)-1,18-octadecendioic acid.[64] The mutant strain also transforms linoleic acid to (Z), (Z)-octadeca-6,9-dienedioic acid, (Z),(Z)-3-hydroxyoctadeca-9,12-dienedioic acid, and (Z),(Z)-3-hydroxy-tetradeca-5,8-dienedioic acid,[65] and ricinoleic acid to *R*-(Z)-7-hydroxy-9-octadecenedioic acid and (Z)-1,12-dihydroxy-9-octadecenedioic acid. In addition, sunflower oil and rapeseed oil are utilized to produce *R*-(Z)-3-hydroxy-9-octadecenedioic acid in the fermentation culture.[66] During bioconversion, therefore, the configuration of the double bond is not changed and the hydroxylation is site-specific and regiospecific.

7.4 Inner Carbon-Oxygenated Fatty Acids

Plant systems are also known to produce hydroxy fatty acids. In these cases, the hydroxy groups are usually located in the middle of the fatty acyl chain. These types of hydroxy fatty acids are important industrial materials. Hydroxy fatty acids are used in a wide range of products, including resins, waxes, nylons, plastics, corrosion inhibitors, cosmetics, and coatings. Furthermore, they are used in grease formulations for high-performance military and industrial equipment. Castor and lesquerella oils are the two major sources. Ricinoleic and sebacic acids, two of the castor oil derivatives, are classified by the U.S. Department of Defense as strategic and critical materials. The gene encoding oleoyl-12-hydroxylase for producing ricinoleic acid in castor beans recently has been cloned and expressed in transgenic plants with the yields of hydroxy oil relatively low at present. It is conceivable that the gene can be cloned into a bacterium for the commercial production of ricinoleic acid. Like ricinoleic acid, lesquerella's hydroxy fatty acids also have double bonds and a carboxyl group that provides sites where chemical reactions can occur. 12-Hydroxystearates (esters with C10-12 alcohols) are used in leather coatings requiring oil resistance and water imperviousness, and in roll leaf foils because of their alcohol solubility and excellent wetting and adhesion to metallic particle.

Microorganisms oxidize fatty acids either at the terminal carbon or inside of the acyl chain. Microbial conversion of oleic acid to 10-hydroxy and 10-keto fatty acids are known. Hydroxy- and keto-fatty acids are useful industrial chemicals applied in plasticizer, surfactant, lubricant, and detergent formulations. Saturated hydroxy fatty acids such as 14-hydroxyeicosanoic acid and 10-hydroxystearic acid (10-HAS) are used as commercial grease thickeners. Keto fatty acids or their derivatives are ingredients of multi-purpose greases. Newly discovered hydroxy unsaturated fatty acids were reported to have interesting physiological activity.[67–71] Recently, several microorganisms and biocatalytic processes have been identified that introduce new functional groups to the inner carbon atoms of fatty acids. These hydroxy fatty acids are classified into three types, namely: monohydroxy, dihydroxy, and trihydoxy fatty acids.

7.4.1 Monohydroxy Fatty Acids

7.4.1.1 10-Hydroxystearic Acid

In 1962, Wallen et al.[72] found that a Pseudomonad isolated from fatty material hydrated oleic acid at the *cis* 9 double bond to produce 10-hydroxystearic acid (10-HAS) with 14% yield. The 10-HSA is optically active and has the D-configuration. So far, microbial hydration of oleic acid has been found in *Pseudomonas, Nocardia (Rhodococcus), Corynebacterium, Sphingobacterium, Micrococcus, Absidia, Aspergillus, Candida, Mycobacterium,* and *Schizosaccharomyces.*[73] With a resting cell suspension of *Nocardia*, the production yield was greater than 95%.[74] Oleic acid and its esters were also converted to hydroxy fatty acid by *Sarcina lutea.*[75] A novel microbe, NRRL B-14797, isolated from compost also produces 10R-hydroxystearic acid.[76]

The stereospecificity of 10-HSA varied depending on the microorganism. *R. rhodochrous* ATCC 12674-mediated hydration of oleic acid gave mixtures of enantiomers 10(*R*)-HSA and 10(*S*)-HSA. *Pseudomonas* sp. NRRL B-3266 produced optically pure 10(*R*)-HSA. *Sphingobacterium* produced 10(*R*)-HAS containing 2–18% 10(*S*)-HAS, and *Flavobacterium* sp. DS5 produced 66% enantiomeric excess of 10(*R*)-HSA.

7.4.1.2 10-Ketostearic Acid (10-KSA)

Production of 10-KSA from oleic acid were reported from Staphylococcus sp. with greater than 90% yield,[77] from Flavobacterium sp. DS5 with 85% yield[78] and from Sphingobacterium sp. strain 022.[79] Data indicated that oleic acid is converted to 10-KSA via 10-HSA. Production of 10-KSA was also reported in Mycobacterium, Nocardia, Aspergillus terreus, Staphylococcus warneri, Sphingobacterium thalpophilum, and Bacillus sphaericus.

Nocardia cholesterolicum converts linoleic acid to 10-hydroxy-12(Z)-octadecanoic acid.[80] The *Flavobacterium* DS5 enzyme system also converts linoleic acid to 10-hydroxy-12(Z)-octadecenoic acid with 55% yield.[81] Two minor products of the reaction are 10-methoxy-12-octadecenoic acid and 10-keto-12-octadecenoic acid. Strain DS5 oxidizes unsaturated but not saturated fatty acids. The relative activities are in the following order: oleic > palmitoleic > arachidonic > linoleic > linolenic > gamma-linolenic > myristoleic acids.

7.4.1.3 Positional Specificity of Strain DS5 Hydratase

Strain DS5 converts α-linolenic acid to 10-hydroxy-12,15-octadecadienoic acid (Figure 7.1) and a minor product 10-keto-12,15-octadecadienoic acid.[80] It also converts γ-linolenic acid to 10-hydroxy-6(Z),12(Z)-octadecadienoic acid. The enzyme hydrates 9-unsaturation but does not alter the original 6,12-unsaturations. Strain DS5 converts myristoleic acid to 10-keto myristic and 10-hydroxymyristic acids, and similarly, palmitoleic acid to 10-ketopalmitic and 10-hydroxypalmitic acids. It is interesting to find that all unsaturated fatty acids tested are hydrated at the 9,10 positions with the oxygen functionality at C-10 despite their varying degree of unsaturations. Strain DS5 hydratase is not active on saturated fatty acids and other non-9(Z)-unsaturated fatty acids such as elaidic [9(E)-octadecenoic], arachidonic [5(E),8(E),11(E),14(E)-eicosatetraenoic], and erucic [13(E)-docosenoic] acids. It was found that all microbial hydratases hydrate oleic and linoleic acids at the C-10 position.

From the point of view of chemical structure similarity, 10-HSA and 10-KSA might be used to replace ricinoleic acid, the much higher valued fatty acid. The industrial application of other monohydroxy fatty acid products needs to be explored. For example, 10-hydroxy-12(Z)-octadecenoic acid decreased muscular tension in rat cardiac muscle.[82]

7.4.2 Dihydroxy Unsaturated Fatty Acids

A microbial culture, *Pseudomonas aeruginosa* PR3, isolated from a water sample at a pig farm in Morton, Illinois, was found to convert oleic acid to a new compound, 7,10-dihydroxy-8(E)-octadecenoic acid.[68] The yield of DOD was originally at 60% and is now improved to over 90%. The absolute configuration of DOD is 7(S),10(S)-dihydroxy-8(E)-octadecenoic acid.[83] The production of DOD from oleic acid is

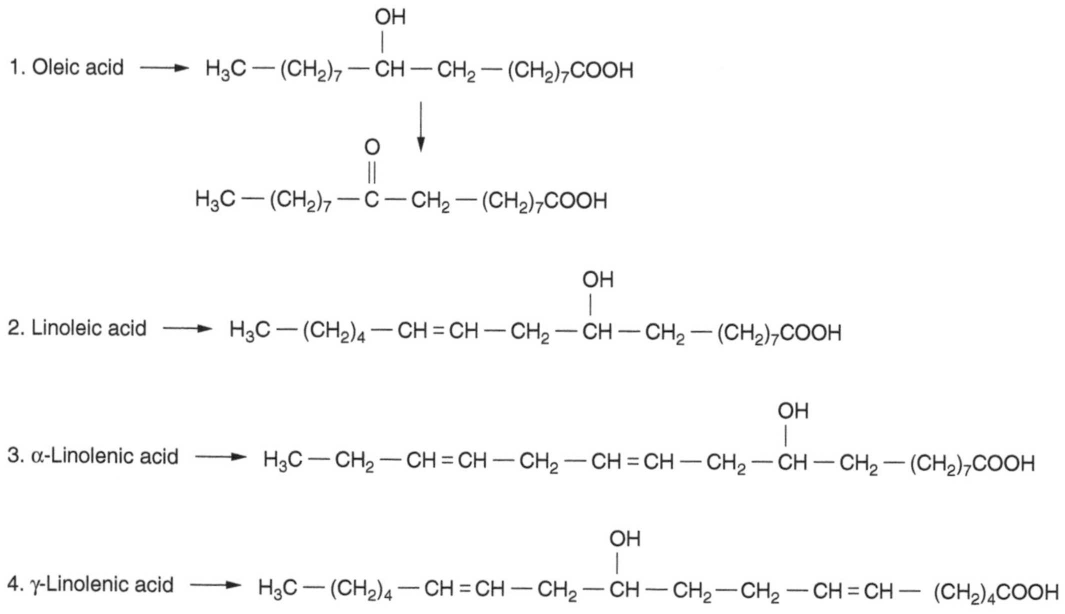

FIGURE 7.1 Bioconversion products by strain *Flavobacterium* sp. DS5.

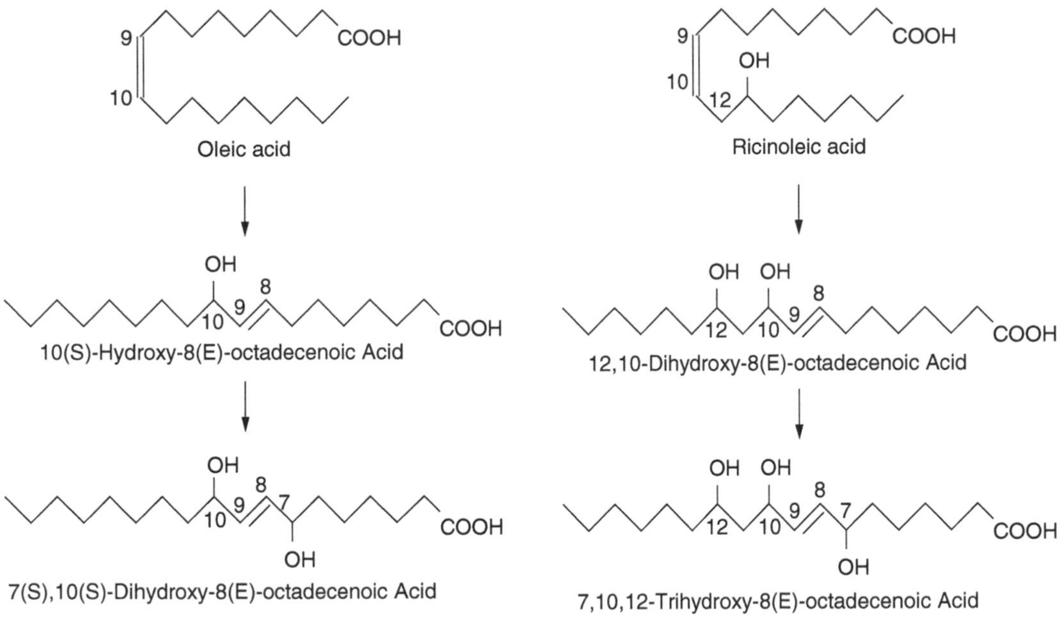

FIGURE 7.2 Bioconversion of oleic and recinoleic acids by *Pseudomonas aeruginosa* PR3.

unique in that it involves both hydroxylation and possibly isomerization, an addition of two hydroxy groups at two positions and a rearrangement of the double bond of the substrate molecule (Figure 7.2). Subsequent investigation of reactions led to the isolation of 10-hydroxy-8(*E*)-octadecenoic acid.[84] The absolute configuration of the hydroxy group at carbon 10 of HOD is in *S* form.[85] The overall bioconversion pathway of oleic acid to DOD by strain PR3 is postulated as follows: oleic acid is first converted to HOD during which one hydroxyl group is introduced at C-10(*S*) and the double bond is shifted from C-9 *cis* to C-8 *trans*, suggesting that there are possibly at least two or more enzymes involved in this first step of reaction. Further hydroxylation of HOD introduces a hydroxyl group at C-7(*S*). It is unlikely that a hydratase is involved in the PR3 reaction in that the double bond at C-9 of the substrate retained as a shifted *trans*-configured form during the hydroxylation until the formation of DOD.

Pseudomonas sp. 42A2 produces a dihydroxy unsaturated fatty acid using olive oil as the sole carbon source.[86] The compound was later identified as 7,10-dihydroxy-8(E)-octadecenoic acid.[87] Strain 42A2 produces the dihydroxy acid during the logarithmic phase and ceases production at the beginning of the stationary phase.[88] 10-Hydroperoxy-8(E)-octadecenoic acid and 10-hydroxy-8(E)-octadecenoic acid were also found in the reaction mixture.[89]

Pseudomonas sp. PR3 also converts ricinoleic acid to 7,10,12-trihydroxy-8(*E*)-octadecenoic acid (TOD) at 35% yield.[90] We isolated a new compound, 10,12-dihydroxy-8(*E*)-octadecenoic acid (DHOD) from the ricinoleic acid-PR3 system.[91] DHOD is an intermediate in the bioconversion reaction. The reaction mechanism is the same as that for the conversion of oleic acid to DOD (Figure 7.2).

With multiple functionality groups on the fatty acid molecule, DOD might be useful in plastics, coatings, cosmetics, and the lubricant industries. DOD has surface-active properties, and along with other chemically synthesized allylic mono- and dihydroxyfatty acids, these compounds may be suitable for usage in microemulsions or as additives to various commercial products. Physiological activity tests of DOD revealed that DOD has activity against *Bacillus subtilis* and a common yeast pathogen, *Candida albicans*. TOD showed antiplant pathogenic fungal activities, especially on the rice blast fungus.[89] Strain PR3 also converts linoleic acid to many isomeric 9,10,13 (9,12,13)-trihydroxy-11E (10E)-octadecenoic acids.[92] One kg of DOD was produced using a bioreactor with special control of the dissolved oxygen level during the reaction.[93] DOD was tested for its application in the production of polyurethane rigid foam,[94] and in skin-care products.

7.4.3 Trihydroxy Unsaturated Fatty Acids

We also discovered the production of a new compound, 12,13,17-trihydroxy-9(Z)-octadecenoic acid (12,13,17-THOA) from linoleic acid by a new microbial culture, which was isolated from a dry soil sample collected from McCalla, Alabama.[69] The strain was identified as *Clavibacter* sp. ALA2.[95] The yield of 12,13,17-THOD from linoleic acid was 35%.

Production of trihydroxy unsaturated fatty acids in nature is rare. These compounds are all produced in trace amounts by plants. Kato et al.[96] reported that hydroxy and epoxy unsaturated fatty acids present in some rice cultivars acted as antifungal substances and were active against the rice blast fungus. It was postulated that these fatty acids were derivatives of linoleic and linolenic acid hydroperoxides. Mixed hydroxy fatty acids were isolated from the *Sasanishiki* variety of the rice plant, which suffered from the rice blast disease and were shown to be active against the fungus. Their structures were identified as 9S,12S,13S-trihydroxy-10-octadecenoic acid and 9S,12S,13S-trihydroxy-10,15-octadecadienoic acid. 9,12,13-Trihydroxy-10(E)-octadecenoic acid was also isolated from potato, *Colocasia antiquorum* inoculated with black rot fungus, *Ceratocystis fimbriata*, and showed antiblack rot fungal activity.[97]

Other than extraction from plant materials, our discovery is the first report on production of trihydroxy unsaturated fatty acids by microbial transformation. The biological activity of 12,13,17-THOA was found[98] to inhibit the growth of the following plant pathogenic fungi: *Erisyphe graminis f. sp. tritici* (wheat powdery mildew); *Puccinia recondita* (wheat leaf rust); *Phytophthora infestans* (potato late blight); and *Botrytis cinerea* (cucumber botrytis). It seems that the position of the hydroxy groups on the fatty acids molecule plays an important role in the activity against certain specific plant pathogenic fungi.

7.4.4 Other Oxygenated Fatty Acids

In addition to the main product 12,13,17-THOA, the strain ALA2 system produced the following products from linoleic acid:[99,100] two tetrahydrofuranyl fatty acids (THFAs), 12-hydroxy-13,16-epoxy-9(Z)-octadecenoic acid and 7,12-dihydroxy-13,16-epoxy-9(Z)-octadecenoic acid; two diepoxy bicyclic fatty acids, 12,17;13,17-diepoxy-9(Z)-octadecenoic acid (DEOA) and 12,17;13,17-diepoxy-7-hydroxy-9(Z)-octadecenoic acid (hDEOA). Tetrahydrofuranyl compounds are known to have anticancer activity. The diepoxy bicyclic fatty acids are completely new chemical entities. With many functionality groups on the molecules, their application in biomedical as well as specialty chemical industries is expected. A small amount of 12,13,16-trihydroxy-9(Z)-octadecenoic acid (12,13,16-THOA) and 12,13-dihydroxy-9(Z)-octadecinoic acid (12,13-DHOA) was also detected in the bioconversion system.

These novel oxygenated fatty acids are of interest because they have unique chemical structures, and because they are expected to possess some biological functions. In general, metabolism of polyunsaturated fatty acids results in the biosynthesis of potent mediators with far-ranging physiological effects. These metabolites include prostaglandins, prostacyclines, thromboxanes, leukotrienes, lipoxins, and hydroxy, hydroperoxy, and epoxy fatty acids. To clarify the biological function and to develop an industrial production process for these fatty acids, it is important to understand their biosynthetic pathways.

7.4.5 Biosynthetic Pathways

By using pure 12,13,17-THOA and DEOA, the biosynthetic pathway of bicyclic fatty acids (DEOA and hDEOA) by *Clavibacter* sp. ALA2 was unraveled.[101] It was found that 12,13,17-THOA was a precursor of the biosynthesis of these bicyclic fatty acids, and that hDEOA was generated from DEOA. Therefore, the biosynthetic pathway depicts that linoleic acid is first converted to 12,13-DHOA and then through 12,13,17-THOA, to DEOA and then to hDEOA (Figure 7.3). A notable reaction step in the postulated pathway would be the formation of the bicyclic ring. Since the bicyclic moiety of DEOA is an intramolecular ketal structure, 12,13,17-THOA itself is not likely the direct precursor for DEOA. In other words,

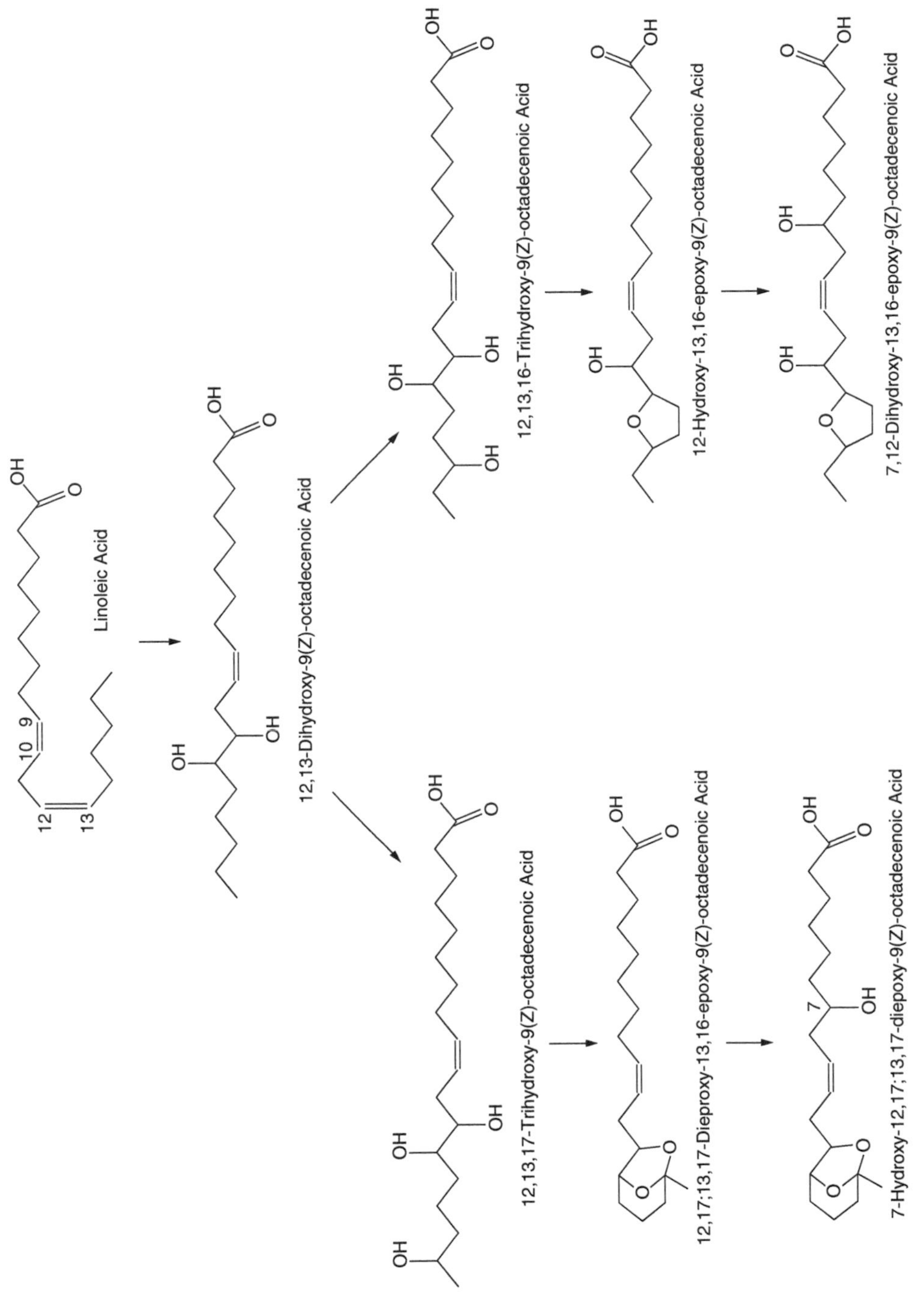

FIGURE 7.3 Bioconversion pathway of linoleic acid by *Clavibacter* sp. ALA2.

12,13,17-THOA might be first converted into another intermediate, and then into DEOA. A reasonable candidate for such an intermediate for DEOA might be a ketone-diol, possibly 12,13-dihydroxy-17-keto-9-octadecenoic acid, of which the keto group at C-17 reacts with the vicinal diol group at C-12 and C-13 to form the bicyclic ring.

For the biosynthetic pathway of linoleic acid to THFAs, it appears that 12,13-DHOA is the branch point. 12,13-DHOA is converted not only to 12,13,17-THOA, but is also converted to 12,13,16-THOA to a lesser extent.[102] We separated and purified 12,13,16-THOA from its isomer 12,13,17-THOA by silica gel column chromatography and by preparative thin layer chromatography. Purified 12,13,16-THOA was used as substrate to study the biosynthesis of THFAs. Within a 24-h incubation, cells of strain ALA2 converted 12,13,16-THOA to both 12-hydroxy-13,16-epoxy-9(Z)-octadecenoic acid (12-hydroxy-THFA) and 7,12-dihydroxy-13,16-epoxy-9(Z)-octadecenoic acid (7,12-dihydroxy-THFA). The relative abundance of 7,12-dihydroxy-THFA increased with incubation time, while those of 12,13,16-THOA and 12-dihydroxy-THFA decreased (Figure 7.3).[103] It also indicated that the conversion of 12-hydroxy-THFA to 7,12-dihydroxy-THFA was very fast and was not a rate-limiting step. Therefore, the biosynthetic pathway of THFAs from linoleic acid by strain ALA2 is as follows: linoleic acid ➭ 12,13-dihydroxy-9(Z)-octade-cenoic acid ➭ 12,13,16-THOA ➭ 12-hydroxy-THFA ➭ 7,12-dihydroxy-THFA. This is an analogous mechanism of DEOA biosynthesis that 12,13,16-THOA is cyclized to form THFAs (Figure 7.3).[101] In general, it is clear that *Clavibacter* sp. ALA2 is especially efficient at oxidizing C-12, -13 and -17 with hydroxyl groups, and to a lesser extent, hydroxyls also occurred at C-7 and -16.

7.4.6   Bioconversion of α-Linolenic Acid by *Clavibacter* sp. ALA2

Strain ALA2 is a very interesting microorganism. Its enzymes also converted other ω-3 and ω-6 polyun-saturated fatty acids (PUFAs) into a variety of oxylipins. For examples, it converted α-linolenic acid to novel THFAs: 13,16-dihydroxy-12, 15-epoxy-9(Z)-octadecenoic acid (13,16-dihydroxy-THFA) and 7,13,16-trihydroxy-12, 15-epoxy-9(Z)-octadecenoic acid (7,13,16-trihydroxy-THFA) (Figure 7.4).[104]

The crude extract obtained from incubation of α-linolenic acid and *Clavibacter* sp. ALA2 in culture media for 7 days was methylated with diazomethane. GC analyses of the methyl esters of the crude extract showed several product peaks. The GC retention time of the main product was at 17.8 min and a minor product at 25.6 min. Since retention times of α-linolenic acid, ricinoleic acid (12-hydroxy-9(Z)-octade-cenoic acid) and 9, 10-dihydroxy stearic acid were 8.1 min, 12.2 min and 16.5 min, respectively, we predicted that these new products have two or more hydroxy groups in their fatty acid molecules.

7.4.6.1   Isolation and Identification of GC Rt at 17.8 Min Product

The crude extract was treated with diazomethane and was then subjected to a preparative silica gel TLC plate to isolate the methyl ester of a product with GC Rt at 17.8 min. Ethyl acetate was used as development solvent. A band on TLC (R_f = 0.69) detected by UV was scraped with a razor blade and the material was eluted with ethyl acetate. The product band on TLC was also confirmed by cutting off a small side of the TLC plate and the corresponding product band was developed in iodine vapor. The recovered product was further purified by a second TLC with a different development solvent; n-hexane: ethyl acetate (1:4, vol/vol) (R_f = 0.64). The purity of the purified product was 85.5% by GC analysis.

7.4.6.2   Isolation and Identification of Product with GC Rt at 25.6 Min

To isolate the product with GC Rt at 25.6 min, the crude extract was applied onto a preparative silica gel TLC plate and was developed with methylene chloride: MeOH (9:1, vol/vol). The product band on TLC plate (R_f = 0.50) was recovered with the same method described above. The material was eluted from silica gel with MeOH. It was then methylated with diazomethane and was purified again by silica gel TLC with methylene chloride: MeOH (95:5, vol/vol) as development solvent (R_f = 0.35). The methyl ester of the product with GC Rt at 25.6 min was eluted with ethyl acetate from silica gel scrapped from the TLC plate. The purity of the purified product was 89.0% by GC analysis.

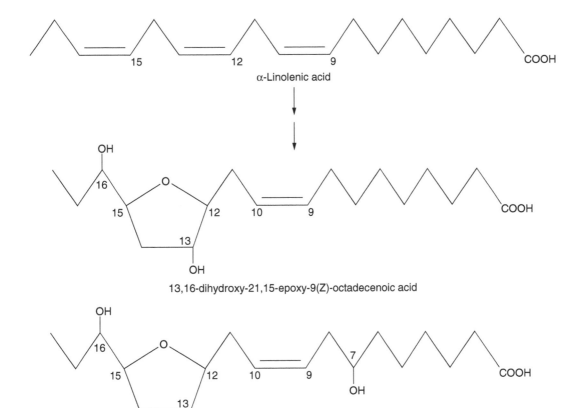

13,16-dihydroxy-21,15-epoxy-9(Z)-octadecenoic acid

7,13,16-trihydroxy-21,15-epoxy-9(Z)-octadecenoic acid

FIGURE 7.4 Bioconversion of α-linolenic acid by Clavibacter sp. ALA2.

7.4.6.3 Structure Analyses of Product with GC Rt at 17.8 Min

The structure was identified by GC-MS and NMR analyses. The electron impact mass spectrum (EI-MS) obtained from the methyl ester/O-trimethyl silyl ether (OTMSi) of the purified product gave a molecular ion at m/z 486 [M]$^+$ (relative intensity, 0.2). Fragment ions were interpreted as follows: 396 [M-TMSiOH]$^+$ (2), 381 [M- CH$_3$-TMSiOH]$^+$ (1), 355 [M-CH$_3$CH$_2$CHOTMSi]$^+$ (2), 289 [M-CH$_2$CH=CH(CH$_2$)7COOCH$_3$]$^+$ (10), 265 [355-TMSiOH]$^+$ (8), 199 [289-TMSiOH]$^+$ (50), 131 [CH2CH2CHOTMSi]$^+$ (80), 73 [TMSi]$^+$ (100) as well as many other ions (e.g. m/z 429, 337, 233, 181, 157, 145). The methyl ester was further hydrogenated by oxidation with pyridinium chlorochromate and applied onto GC-MS. The mass spectrum of the hydrogenated product showed that there was a gain of two hydrogens, which corresponded to one double bond as follows; m/z 473 [M-CH$_3$]$^+$(0.1), 398 [M-TMSiOH]$^+$(0.2), 383 [473-TMSiOH]$^+$(0.7), 357 [M-CH$_3$CH$_2$CHOTMSi]$^+$ (0.9), 267 [357-TMSiOH]$^+$ (40) (base ion m/z 73 (100)). As expected, a fragment ion with m/z 131 [CH$_2$CH$_2$CHOTMSi]$^+$(100), which corresponds to a hydroxy at the C-16 position, was observed even after hydrogenation and the relative intensity enhanced, while both fragment ions at m/z 289, 199 observed by a cleavage between tetrahydrofuranyl and the double bond became small after hydrogenation.

The structure of product with GC Rt at 17.8 min was further confirmed by ^{13}C and ^{1}H NMR analyses (Table 1). The ^{13}C NMR signals at 71.0 ppm (C-13) and 74.6 ppm (C-16) corresponded to C-13 and C-16 position hydroxy groups in the molecule. The positions of the hydroxy groups were also supported by ^{1}H NMR signals at 4.00 ppm (H-13) and 3.38 ppm (H-16). Resonance signals of the olefinic proton were observed at 5.47 ppm. ^{13}C NMR signals of 131.8 ppm (C-9) and 125.1 ppm (C-10) also indicated the presence of a double bond between C-9 and C-10. Since signals of methylene carbon next to the olefinic carbon were C-8 of 27.1 ppm and C-11 of 26.8 ppm; the double bond is in cis configuration.[7] ^{13}C NMR signals of C-12 and C-15 were observed at 83.5 ppm and 78.9 ppm corresponded to -CH, respectively. From these data, the structure of GC Rt at 17.8 min product is 13, 16-dihydroxy-12, 15-epoxy-9-(Z)-octadecenoic acid (13, 16-dihydroxy-THFA).

7.4.6.4   Structure Analyses of Product with GC Rt at 25.6 Min

The chemical structure of the GC Rt at 25.6 min product was also identified by GC-MS and NMR analyses. The mass spectrum of the methyl ester/OTMSi ether of the purified product was interpreted as follows: *m/z* 484 [M-TMSiOH]$^+$ (0.1), 469 [M-CH$_3$-TMSiOH]$^+$(0.1), 394 [484-TMSiOH]$^+$ (0.2), 379 [469-TMSiOH] + (0.3), 353 [M-CH$_3$CH$_2$CHOTMS-TMSiOH]$^+$ (0.4), 289 [M-CH$_2$CH=CH(CH$_2$)$_7$COOCH$_3$]+ (2), 263 [353-TMSiOH]$^+$ (0.9), 231 [CHOTMSi(CH$_2$)$_5$COOCH$_3$]$^+$ (98), 199 [289-TMSiOH]$^+$ (18), 131 [CH$_3$CH$_2$CHOTMSi]$^+$ (32), 73 [TMSi]$^+$ (100) as well as many other ions (e.g., *m/z* 310, 157, 145). A fragment ion at *m/z* 231 corresponded to a hydroxy residue at the C-7 position and was characteristic for the product with GC Rt at 25.6 min.

The hydroxy residue at the C-7 position was confirmed by ^{13}C NMR chemical shift at 70.9 ppm and ^{1}H NMR signals at 3.60 ppm. The ^{13}C NMR chemical shifts of C-6 and C-8 position were also shifted to higher field of 36.9 ppm (C-6) and 34.8 ppm (C-8) compared to those of 13,16-dihydroxy-THFA. Other ^{13}C NMR chemical shifts and ^{1}H NMR resonance patterns were very similar to that of 13,16-dihydroxy-THFA. From these data, the structure for the product with GC Rt at 25.6min is 7,13,16-trihydroxy-12, 15-epoxy-9-(Z)-octadecenoic acid (7,13,16-trihydroxy-THFA).

The optimum incubation temperature was 30°C for production of both hydroxy-THFAs. 13,16-Dihydroxy-THFA was detected after 2 days of incubation and reached 45 mg/50ml after 7 days of incubation. Whereas 7,13,16-trihydroxy-THFA wasn't detected after 2 days of incubation but reached 9 mg/50ml after 7 days of incubation. Total yield of both 13,16-dihydroxy-THFA and 7,13,16-trihydroxy-THFA reached 67% (w/w) after 7days of incubation at 30°C, 200rpm.

In our previous studies, we reported that *Clavibacter* sp. ALA2 oxidized C-7, C-12, C-13, C-16, and C-17 positions of linoleic acid (n-6) into hydroxy groups. In this case, the bond between C-16 and C-17 carbons is saturated. With α-linolenic acid (n-3), however, the bond between C-16 and C-17 carbons is unsaturated. It seems that enzymes of strain ALA2 oxidized the C-12/C-13 and C-16/C-17 double bonds into dihydroxy groups first, and then converted them to hydroxy-THFAs.

It is known that some microorganisms convert α-linolenic acid to hydroxy fatty acids. Brodowsky et. al.[105] reported the production of 8-hydroxy-octadecatrienoic acid, 17-hydroxy-octadecatrienoic acid and 7,8-dihydroxy-octadecatrienoic acid from α-linolenic acid by the fungus *Gaeumannomyces graminis*. Bioconversion of 10-hydroxy-12(Z)-octadecanoic acid by *Nocardia cholesterolicum* and *Flavobacterium* sp. DS5 were also reported.[80,81] However, these hydroxy fatty acids were monohydroxy and dihydroxy fatty acids with a straight carbon chain structure. In this study, *Clavibacter* sp. ALA2 oxidized α-linolenic acid. Both products, 13,16-dihydroxy-THFA and 7,13,16-trihydroxy-THFA, had a tetrahydrofuranyl ring in their molecule. 7,13,16-Trihydroxy-THFA is a new chemical entity with trihydroxy fatty acid containing a cyclic structure. *Clavibacter* sp. ALA2 can oxidize C-7, C-12, C-13, C-16 and C-17 positions of linoleic acid.[69,95,99,100] *Clavibacter* sp. ALA2 also placed hydroxyl groups at the C-7, C-13, C-16 positions of α-linolenic acid, an omega-3 polyunsaturated fatty acid. Therefore, it appears that the C-16 hydroxylation activity of *Clavibacter* sp. ALA2 is effective not only for double bond but also for single bonds.

When linoleic acid was used as substrate for bioconversion by *Clavibacter* sp. ALA2, both 12-hydroxy-THFA and 7,12-dihydroxy-THFA were produced.[100] These structures are different from 13,16-dihydroxy-THFA and 7,13,16-trihydroxy-THFA as to the position of their tetrahydrofuranyl ring and hydroxy groups. In addition, diepoxy bicyclic fatty acid products such as DEOA, 7-hydroxy-DEOA, and 16-hydroxy-DEOA weren't observed with α-linolenic acid as the substrate. These results indicate that the chemical structure of products produced by *Clabvibacter* sp. ALA2 depended remarkably on the type of polyunsaturated fatty acid substrate.

Moghaddam et. al.[106] reported that linoleic acid and arachidonic acid can be metabolized to their dihydroxy-THFA (tetrahydrofuran-diols) *in vitro* by microsomal cytochrome *P*-450 epoxidations followed by the reaction of microsomal epoxide hydrolase. In their metabolic pathways, saturated dihydroxy-THFAs are produced because 9,10(12,13)- dihydroxy-12,13(9,10)-epoxy-octadecanoate converted from linoleic acid methyl ester are cyclized.[106] These saturated dihydroxy-THFAs exhibit cytotoxic activity and mitogenic activity for breast cancer and prostate cancer cells.[107,108] Strain ALA2 produces unsaturated hydroxy-THFAs from linoleic acid through a biosynthetic pathway, which is different from the metabolic pathway by mouse liver microsomes. The biological functions of unsaturated hydroxy-THFAs are currently under investigation.

Other polyunsaturated fatty acids are metabolized to several unique products with various biological functions. These metabolites include prostaglandins, prostacyclines, thromboxanes, leukotrienes, lipoxins, and hydroxy, hydroperoxy, and epoxy fatty acids. Recently, it was reported that low doses of 9,(12)-oxy-10,13-dihydroxy-stearic acid and 10,(13)-oxy-9,12-dihydroxy-stearic acid isolated from corn stimulated breast cancer cell proliferation *in vitro* and disrupted the estrous cycle in female rats.[107] These dihydroxy-THFA inhibited breast cancer and prostate cancer cell proliferation at high doses.[108] However, little is known about biological functions and chemical properties of hydroxy-THFA. Therefore, it will be interesting to find out the physiological functions and industrial application of these novel hydroxy-THFAs.

7.4.7 Conversion of Other ω-3 and ω-6 Polyunsaturated Fatty Acids (PUFAs) by *Clavibacter* sp. ALA2

Clavibacter sp. ALA2 also bioconverts other ω-3 PUFAs, such as EPA and DHA, and ω-6 PUFAs, such as γ-linolenic acid and arachidonic acid, into new products.[108]

The crude lipid extracts were obtained from incubation of n-3 or n-6 PUFAs and *Clavibacter* sp. ALA2 in culture media for 7 days or longer. From EPA, one major product (Figure 7.5, product 1, GC retention time 24 min) and from DHA, one product (product 2, GC/RT 37 min) was obtained. Strain ALA2 converts γ-linolenic acid into three products (Figure 7.6), two major products (product 3, GC/RT 11

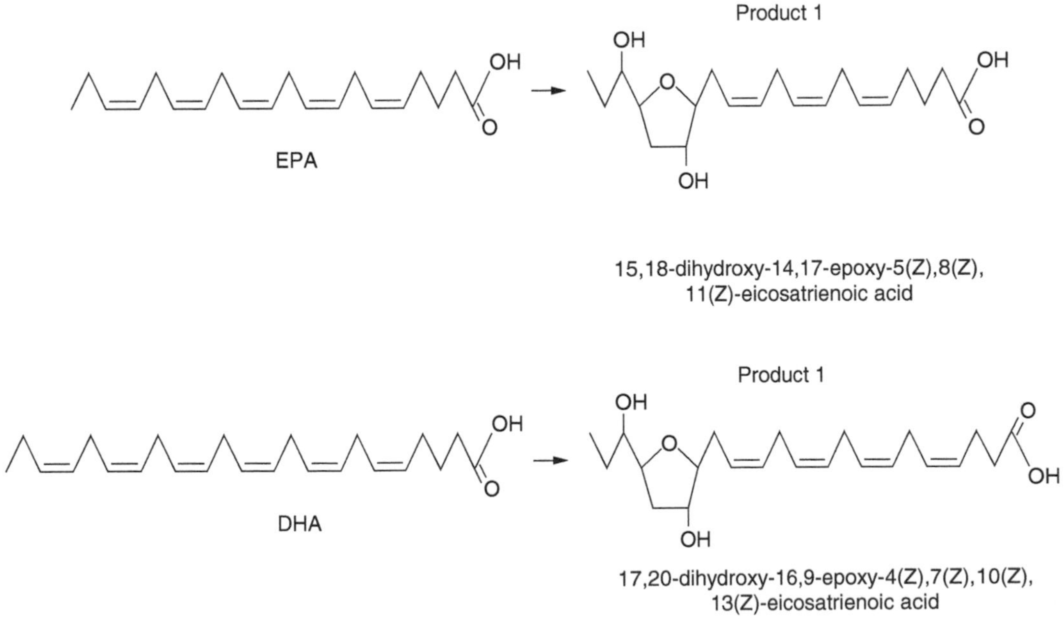

15,18-dihydroxy-14,17-epoxy-5(Z),8(Z),
11(Z)-eicosatrienoic acid

17,20-dihydroxy-16,9-epoxy-4(Z),7(Z),10(Z),
13(Z)-eicosatrienoic acid

FIGURE 7.5 Bioconversion of EPA and DHA by Clavibacter sp. ALA2.

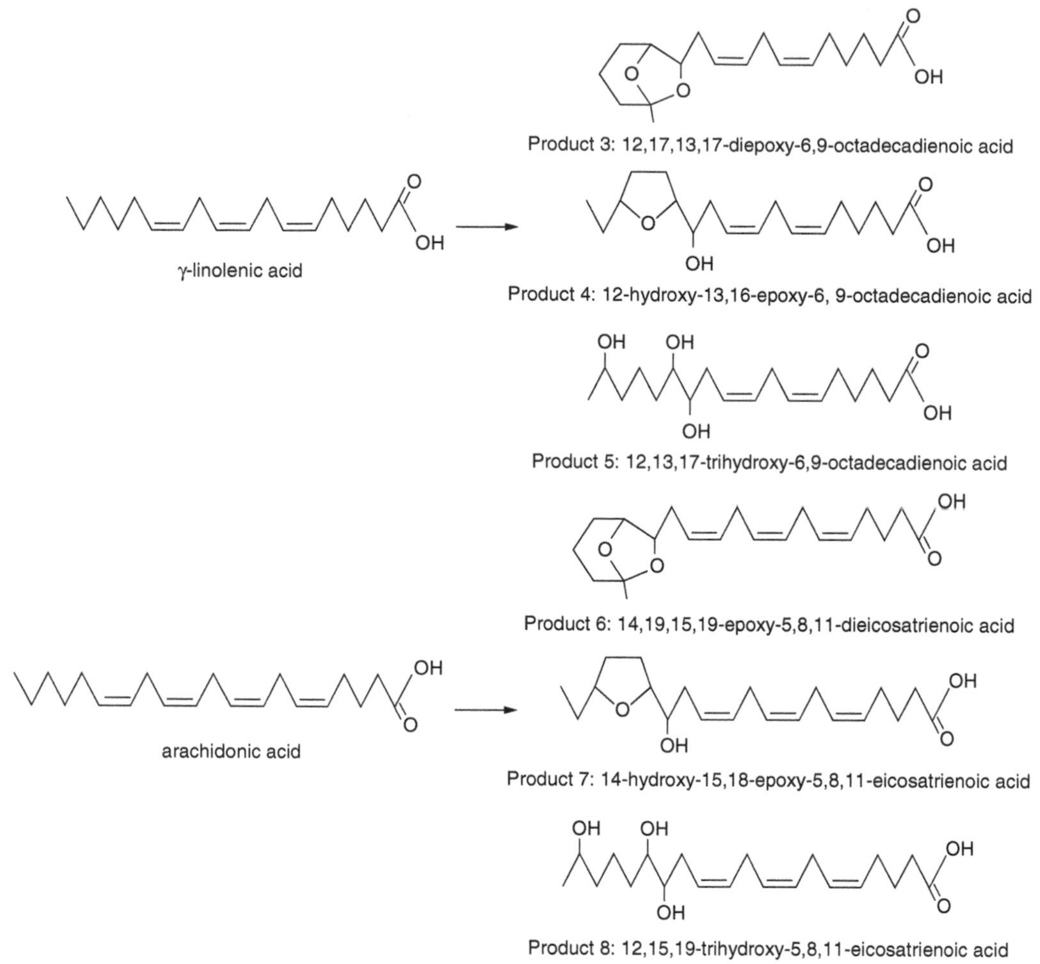

FIGURE 7.6 Bioconversion of γ-linolenic acid and arachidonic acid by Clavibacter sp. ALA2.

min; and product 5, GC/RT 22 min) and one minor product (product 4, GC/RT 13 min). Arachidonic acid was also converted into three products: product 6 GC/RT 15 min, product 7 GC/RT 18 min, and product 8 GC-MS/RT 18 min.

The crude lipid extracts were fractionated by column chromatography (2.5 cm i.d. x 35 cm length) packed with silica gel 60. Elutions were carried out using 500 ml of methylene chloride, 500–1000 ml methylene chloride:methanol (97:3, v/v), and 500–1000 ml methylene chloride:methanol (95:5, v/v). Products 1-4, product 6, and product 7 were obtained from the methylene chloride:methanol (97:3, v/v) fraction. Product 5 and product 8 were eluted by methylene chloride:methanol (95:5, v/v). Fractionated products were methylated with diazomethane and further purified on preparative silica gel 60 F_{254} TLC plates. Product 1 (converted from EPA): (developing solvent) n-hexane:ethyl ether (1:9, v/v), R_f = 0.44, purity (by GC analysis) 96%; Product 2 (converted from DHA): n-hexane:ethyl ether (1:9, v/v), R_f = 0.43, purity 92%; Product 3 (converted from γ-linolenic acid): methylene chloride, R_f = 0.22, purity 95%; Product 4 (converted from γ-linolenic acid): n-hexane:acetone (7:3, v/v), R_f = 0.52 purity 87%; Product 5 (converted from γ-linolenic acid): ethyl ether:acetone (4:1, v/v), R_f = 0.52, purity 97%; Product 6 (converted from arachidonic acid): methylene chloride:methanol (99:1, v/v), R_f = 0.70, purity 99%; Product 7 (converted from arachidonic acid): n-hexane:ethyl acetate (7:3, v/v), Rf = 0.53, purity 96%; Product 8 (converted from arachidonic acid): methylene chloride:methanol (9:1, v/v), R_f = 0.43, purity by GC-MS 95%. Structure analyses were conducted by GC-MS and NMR.

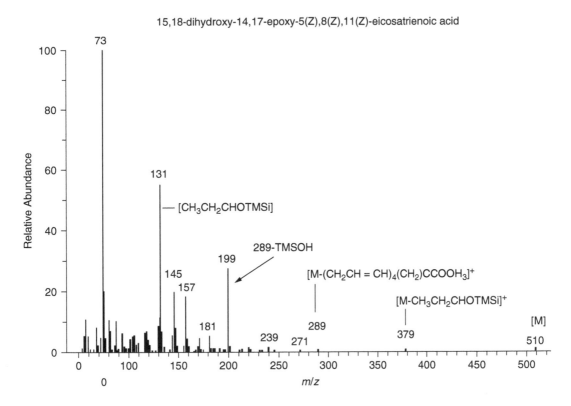

15,18-dihydroxy-14,17-epoxy-5(Z),8(Z),11(Z)-eicosatrienoic acid

FIGURE 7.7 GC/MS analysis of product 1 obtained from EPA biocnversion by Clavibacter sp. ALA2.

Product 1: The electron impact mass spectrum (EI-MS) of the methyl ester/trimethylsilyloxy ethers (OTMSi) of the isolated product 1 (Figure 7.7) gave a molecular ion at (relative intensity) m/z 510[M]⁺ (1). Typical fragment ions were interpreted as follows: 379 [M-CH₃CH₂CHOTMSi]⁺ (0.7), 289 [M-(CH₂CH=CH)₃(CH2)₃COOCH₃]+ (4), 271 (2), 239 (2), 199 [289-TMSiOH]⁺ (29), 181 (6), 157 (20), 145 (21), 131 [CH₃CH₂CHOTMSi]⁺ (59), 73 [TMSi]⁺ (100). The structure of product 1 (methyl ester derivative) was further confirmed by ¹³C and ¹H NMR analyses (Table 7.1). The ¹³C NMR signals at 71.2 ppm and 74.2 ppm corresponded to C-15 and C-18 positions hydroxy group in the molecule, respectively. ¹³C NMR signals of 129.8 ppm (C-5), 128.9 ppm (C-6), 128.2 ppm (C-8), 128.1ppm (C-9), 128.8 ppm (C-11), and 125.9 ppm (C-12) indicated the presence of three double bonds. The carbon peaks for C-7 (25.7 ppm), C-10 (25.6 ppm) and C-13 (27.1 ppm) indicated the three double bonds remained in cis configuration. Epoxy structure was found at C-14 (83.6 ppm) and C-17 (79.1 ppm). The positions of hydroxy group and double bonds were also supported by ¹H NMR signals (Table 7.1). From these data, the structure of product 1 converted from EPA was identified to be 15, 18-dihydroxy-14,17-epoxy-5(Z),8(Z),11(Z)-eicosatrienoic acid. Strain ALA2 hydroxylated the two double bonds between C-14 and C-15, C-17 and C-18, but did not alter the double bonds at C-5, C-8 and C-11 position.

Product 2: The EI-MS of the methyl ester/OTMSi of the isolated product 2 (Figure 7.8, from DHA) was interpreted as follows; EI-MS m/z (relative intensity): 536 [M]⁺ (0.8), 405 [M-CH₃CH₂CHOTMSi]⁺(0.3), 289 [M-(CH₂CH=CH) ₄(CH₂)₂COOCH₃]⁺ (2.), 271 (1), 199 [289-TMSiOH]⁺ (24), 181 (6), 157 (17), 145 (17), 131 [CH₃CH₂CHOTMSi]⁺ (52), 73 [TMSi]⁺ (100). Therefore, the chemical structure of product 2 was estimated to be 17, 19-dihydroxy-16,18-epoxy-4,7,10,13-docosatetraenoic acid because typical fragmentations such as m/z 289, 199, 131 were similar to those of product 1. Proton and ¹³C-NMR analyses (Table 7.1) confirmed this structure for the product 2. ¹³C-NMR signals of 129.1 ppm (C-4), 129.8 ppm (C-5), 128.0 ppm (C-7), 127.8 ppm (C-8), 128.2 ppm (C-10), 128.2 ppm (C-11), 129.3 ppm (C-13), and

TABLE 7.1   Proton and ^{13}C NMR Signals of Products Converted from EPA and DHA
by *Clavibacter* sp. ALA2

Carbon Number	Product 1 Converted from EPA Resonance Signals (ppm)			Product 2 Converted from DHA Resonance Signals (ppm)		
	Proton		^{13}C	Proton		^{13}C
1	—		174.3	—		174.0
2	2.34	t (J = 7.5 Hz)	33.4	2.30	m	33.9
3	1.70	m	24.8	2.30	m	22.7
4	2.12	m	26.5	5.40	m	128.1
5	5.40	m	129.8	5.40	m	129.8
6	5.40	m	128.9	2.85	m	26.7
7	2.87	m	25.7	5.40	m	128.0
8	5.40	m	128.2	5.40	m	127.8
9	5.40	m	128.1	2.85	m	25.7
10	2.87	m	25.6	5.40	m	128.2
11	5.40	m	128.8	5.40	m	128.2
12	5.40	m	125.9	2.85	m	25.4
13	2.45	m	27.1	5.40	m	129.3
14	3.64	m	83.6	5.40	m	125.9
15	4.03	m	71.2	2.30	m	27.0
16	1.84, 2.15	m	38.1	3.63	m	83.6
17	4.01	m	79.1	4.03	m	71.1
18	3.39	m	74.2	1.83, 2.30	m	37.9
19	1.60	m	26.9	4.00	m	79.2
20	0.99	t (J = 7.5 Hz)	10.2	3.37	m	74.7
21	—		—	1.59	m	25.6
22	—		—	0.98	t (J = 7.5 Hz)	10.1
OCH$_3$*	3.67	s	51.4	3.66	s	51.5

* Products were analyzed as methyl ester derivatives by ^1H and ^{13}C NMR.

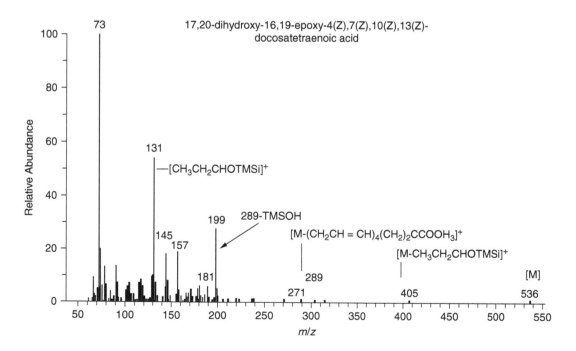

FIGURE 7.8 GC/MS analysis of product 2 obtained from DHA bioconversion by *Clavibacter* sp. ALA2.

125.9 ppm (C-14) indicated the presence of four double bonds. The ^{13}C-NMR signals of 26.7 ppm (C-6), 25.7 ppm (C-9), 25.4 ppm (C-12), 27.0 ppm (C-15) indicated the four double bonds remained in cis configuration. Therefore, the chemical structure of product 2 is 17, 19-dihydroxy-16,18-epoxy-4(Z),7(Z),10(Z),13(Z)-docosatetraenoic acid.

The identified products 1 and 2 had two hydroxy groups and a tetrahydrofuranyl (THF) ring in their molecules. We reported earlier that strain ALA2 converted α-linolenic acid to 13,16-dihydroxy-12,15-epoxy-9(Z)-octadecenoic acid, and 7,13,16-trihydroxy-12,15-epoxy-9(Z)-octadecenoic acid.[104] From product structures obtained from these n-3 PUFAs, it seems that strain ALA2 places hydroxyl groups at the same positions from the omega (ω)-terminal and cyclizes them to THF ring despite their varying degrees in carbon chain numbers and double bonds.

Product 3: The EI-MS of the methyl ester/ OTMSi derivative of product 3 gave a molecular ion at m/z (relative intensity) 322 [M] $^+$ (0.2). Typical fragment ions were interpreted as follows: 291 [M-CH$_3$O]$^+$ (1), 280 [M-CH$_2$CO]$^+$ (0.6), 262 (0.6), 194 (3), 180 (2), 127 [M-(CH$_2$CH=CH)$_2$(CH$_2$)$_4$COOCH$_3$]$^+$ (100), 99 (11), 81 (19), 67 (14), 55 (10). A fragmentation of m/z 127 was a characteristic ion of diepoxy bicyclic structure as reported previously.[98] Therefore, the chemical structure of product 3 is possibly 12,17;13,17-diepoxy-6,9-octadecadienoic acid. This structure was confirmed by proton and ^{13}C NMR analyses. The NMR data were similar to those of 12,17;13,17-diepoxy-9(Z)-octadecenoic acid obtained from linoleic acid by strain ALA2.[12] The singlet multiplicity of terminal methyl showed that there was no coupling with C-17 (Table 7.2), and the chemical shift of C-17 by ^{13}C NMR was consistent with diethyl functionality. The dihedral angle for H-13 and either of H-14 protons is about 60 degrees, resulting in small-unresolved splitting of the H-13 signal. ^{13}C NMR chemical shifts in a low field corresponding to C-6, C-7, C-9 and C-10 indicated to be two double bonds in the molecule.

Product 4: The mass spectrum of the methyl ester/OTMSi derivative of this product was as follows; EI-MS m/z (relative intensity): 396 [M]$^+$ (4), 306 [M-TMSiOH]+ (1), 297 [M-ethyltetrahydrofuranyl]$^+$ (6), 268 [rearrangement TMSi+(CH$_2$CH=CH)$_2$(CH$_2$)$_4$ COOCH$_3$]$^+$ (18), 201 [M-(CH$_2$CH=CH)$_2$(CH$_2$)$_4$COOCH$_3$]$^+$ (52), 175 (17), 129 (21), 99 [ethyltetrahydrofuranyl]$^+$ (28), 81 (39), 73 [TMSi] $^+$ (100), 55 (31). The m/z 99 indicated a terminal ethyl tetrahydrofuranyl group in its molecule as reported previously.[98] All these data indicated a possible structure of 12-hydroxy-13,16-epoxy-6,9-octadecadienoic acid. This structure was confirmed by ^{1}H and ^{13}C NMR (Table 7.2). Two double bonds were seen at C6-C7 and C9-C10. One hydroxyl group was seen at C12. And the epoxy structure was seen at C13 and C16. ^{1}H NMR of the terminal methyl group showed a triplet multiplicity indicating the structure to have a terminal THF ring. Other features of NMR data were also in agreement with the structure of 12-hydroxy-13,16-epoxy-6(Z),9(Z)-octadecadienoic acid.

Product 5: The GC-MS analysis of methyl ester/OTMSi ether derivative of the isolated product 5 gave the characteristic signal fragments as following: EI-MS, m/z (relative intensity) 468 [M-TMSiOH]$^+$ (0.1), 453 [M-CH$_3$-TMSiOH]$^+$(0.1), 437 [M-OCH$_3$- TMSiOH]$^+$ (0.2), 363 [M-CH$_3$-2TMSiOH or M-(CH$_2$CH=CH)$_2$(CH2)$_4$COOCH$_3$]$^+$ (3), 297 [CHOTMSi(CH$_2$CH=CH)$_2$(CH$_2$)$_4$COOCH$_3$]$^+$ (3), 268 (9), 261 [M-CHOTMSi (CH$_2$CH=CH)$_2$(CH$_2$)$_4$COOCH$_3$]$^+$ (13), 191 (12), 171 (14), 147 (25), 129 (45), 117 [CH$_3$CHOTMSi]$^+$(17), 73 [TMSi]$^+$ (100), 55 (7). All these data indicated that product 5 is 12,13,17-trihydroxy-6,9-octadienoic acid. Proton and ^{13}C NMR data confirmed this structure (Table 7.2). Two double bonds were seen at C6-C7 and C9-C10. Three hydroxyl groups were seen at C12, C13, and C17. Therefore, product 5 is identified as 12,13,17-trihydroxy-6(Z),9(Z)-octadienoic acid.

Products 6–8: GC/MS of product 6, product 7, and product 8 as well as their ^{1}H and ^{13}C NMR analyses (Table 7.3 and Table 7.4) confirmed their chemical structures as: 14,19;15,19-diepoxy-5(Z),8(Z),11(Z)-eicosatrienoic acid for product 6; 14-hydroxy-15,18-epoxy-5(Z),8(Z),11(Z)-eicosatrienoic acid for product 7; and 14,15,19-trihydroxy-5(Z),8(Z),11(Z)-eicosatrienoic acid for product 8. Thus, the products obtained from γ-linolenic acid and arachidonic acid by strain ALA2 consists of multiple compounds with diepoxy bicyclic structure, tetrahydrofuranyl ring or trihydroxy group in their molecules. These products had a very similar structure to those products obtained from linoleic acid by strain ALA2.[69,100]

It is interesting to note that the structures of the products are remarkably different between those obtained from n-3 and from n-6 PUFAs. It seems that substrate PUFAs with a double bond at the ω-3

TABLE 7.2 ¹H NMR Data for Bioconversion Products from γ-Linolenic Acid and Arachidonic Acid by *Clavibacter* sp. ALA2

Carbon no.	Product 3 d	mult	Hz	Product 4 d	mult	Hz	Product 5 d	mult	Hz	Product 6 d	mult	Hz	Product 7 d	mult	Hz	Product 8 d	mult	Hz
2	2.33	t	7.5	2.33	t	7.5	2.33	t	7.5	2.34	t	7.6	2.34	t	7.5	2.34	t	7.5
3	?	?		1.67	m		1.50	m		1.70	m		1.70	m		1.70	m	
4	1.39	m		1.42	m		1.50	m		2.12	m		2.12	m		2.11	m	
5	2.09	m		2.12	m		2.10	m		5.40	m		5.39	m		5.40	?	
6	5.40	m		5.40	m		5.40	m		5.40	m		5.39	m		5.40	?	
7	5.40	m		5.40	m		5.40	m		2.85	m		2.84	m		2.85	?	
8	2.82	m		2.82	m		2.84	m		5.40	m		5.39	m		5.40	?	
9	5.40	m		5.40	m		5.40	m		5.40	m		5.39	m		5.40	?	
10	5.40	m		5.40	m		5.50	m		2.85	m		2.84	m		2.85	m	
11	2.28	m		2.25	m		2.30	m		5.40	m		5.39	m		5.40	m	
12	4.04	t	6.7	3.44	m		3.00	m		5.40	m		5.39	m		5.40	m	
13	4.14	b,s		3.83	m		3.40	b,s		2.28	m		2.25	m		2.30	m	
14	1.40,1.79	?		1.60	m		1.50	m		4.04	t	6.7	3.45	m		3.45	m	
15	1.61	m		1.60	m		1.50	m		4.14	b,s		3.85	m		3.45	m	
16	1.65,1.79	?		3.83	m		1.50	m		1.50,1.80	m		1.54	m		1.50	m	
17	—	—		1.60	m		3.75	m		1.60,1.80	m		2.05	m		1.60	m	
18	1.41	s		0.94	t	7.5	1.18	?	6.2	1.60	m		3.85	m		1.45	m	
19	—			—			—			—	—		1.47,1.63	m		3.79	m	
20	—			—			—			1.41	s		0.93	t	7.4	1.17	?	6.2
OCH₃*	3.66	s		3.67	s		3.66	s		3.67	s		3.66	s		3.67	s	

* Products were analyzed as methyl ester derivatives by ¹H NMR.

TABLE 7.3 ¹³C NMR Chemical Shifts (ppm) for Bioconversion Products from γ-Linolenic Acid and Arachidonic Acid by *Clavibacter* sp. ALA2

Carbon no.	Product 3	Product 4	Product 5	Product 6	Product 7	Product 8
1	174.6	174.2	174.4	174.3	174.4	174.5
2	33.9	33.9	33.9	33.6	33.3	33.4
3	24.5	24.6	25.8	24.8	24.8	24.8
4	29.0	29.1	29.0	26.5	26.5	26.5
5	26.8	26.9	26.9	130.2	128.9	128.9
6	130.5	129.7	129.7	129.0	128.8	128.8
7	127.9	128.0	128.0	25.8	25.7	25.7
8	25.7	25.7	24.6	128.3	128.2	128.3
9	129.7	130.0	130.3	128.0	128.0	128.0
10	125.0	125.8	125.8	25.6	25.6	25.6
11	33.6	31.4	31.5	128.8	129.8	130.0
12	79.4	73.8	73.6	125.3	125.8	125.9
13	78.3	81.5	73.8	33.4	31.3	31.4
14	27.7	28.4	33.4	78.3	73.8	73.5
15	17.0	31.9	21.8	79.4	81.5	73.8
16	34.8	80.8	39.0	27.8	28.4	33.3
17	108.1	28.5	67.5	17.1	31.8	21.8
18	24.7	10.1	23.1	34.9	80.9	40.0
19	—	—	—	108.1	28.5	67.5
20	—	—	—	24.8	10.0	22.9
OCH₃*	51.3	54.0	51.4	51.4	51.4	51.4

* Products were analyzed as methyl ester derivatives by ¹³C NMR.

position affects the outcome of the bioconversion products by ALA2 enzymes. In addition, both the hydroxyl group and cyclic structure are placed at the same position from the ω-carbon terminal, despite differences in the number of carbons and double bonds of the substrate PUFAs. Strain ALA2 enzymes may be able to recognize the ω-carbon terminal of PUFAs and proceed with hydroxylation and cyclization accordingly. It is known that hydroxylation at the ω-1, ω-2, and ω-3 positions are catalyzed by cytochrome

TABLE 7.4 Relative Activity for Bioconversion of n-3 and n-6 Polyunsaturated Fatty Acids by *Clavibacter* sp. ALA2

Substrate	Product	Relative Activity (%)
n-3 PUFA[*1]		
α-Linolenic acid [*2]	7,13,16-Trihydroxy-12,15-epoxy-9(Z)-octadecadienoic acid	100
	13,16-Dihydroxy-12,15-epoxy-10(Z)-octadecadienoic acid	19
Eicosapentaenoic acid [*2]	15,18-Dihydroxy-14,17-epoxy-5,8,12-eicosatrienoic acid (1)	16
Docosahexaenoic acid [*2]	17,20-Dihydroxy-16,19-epoxy-4,7,10,13-docosatetraenoic acid (2)	14
n-6 PUFA[*1]		
Linoleic acid [*3]	Trihydroxy-9(Z)-octadecenoic acid	130
	12,17;13,17-Diepoxy-9(Z)-octadecenoic acid	19
	12-Hydroxy-13,16-epoxy-9(Z)-octadecenoic acid	2
g-Linolenic acid [*2]	Trihydroxy-6,9-octadecadienoic acid (5)	76
	12,17;12,17-Diepoxy-6,9-octadecadienoic acid (3)	27
	12-Hydroxy-13,16-epoxy-6,9-octadecadienoic acid (4)	9
Arachidonic acid [*2]	Trihydroxy-5,8,11-eicosatrienoic acid (8)	—
	14,19;15,19-Diepoxy-5,8,11-eicosatrienoic acid (6)	11
	14-Hydroxy-15,18-epoxy-5,8,11-eicosatrienoic acid (7)	7

[*1] PUFA : polyunsaturated fatty acid
[*2] PUFA 125 ml, 30°C, 7 days incubation
[*3] Linoleic acid 125 ml, 30°C, 2 days incubation

P-450. Their enzymes have been characterized in bacteria and fungi as well as mammals. Furthermore, Moghaddam et al. reported that P450 isolated from mouse catalyzed epoxydation at C-12 and C-13 position of linoleic acid, or at C-14 and C-15 position of arachidonic acid, corresponded to ω-5 and ω-6 position of their PUFAs, followed by conversion to diols.[106] Therefore, the bioconversion of strain ALA2 investigated in the present study may be related to P450 enzymes.

7.4.8 Substrate Specificity

The relative activities were measured by the formation of products detectable by GC. Strain ALA2 showed the highest relative activity for the bioconversion to trihydroxy-9(Z)-octadecenoic acid from linoleic acid. The relative activities for α- and γ-linolenic acids were lower than that of linoleic acid.[108] Increase of carbon chain number and double bonds made relative activity reduce remarkably (Table 7.4).

Some hydroxyl fatty acids are known to have biological activities such as antifungal activity and cytotoxic activity.[96-98,107-109] However, little is known about biological functions and chemical properties of tetrahydrofuranyl fatty acids and diepoxy bicyclic fatty acids.[100,101,103,104,110] Therefore, it will be interesting to find the biological functions of cyclic fatty acids identified in the present study.

7.5 Conclusion

Vegetable oils are a relatively cheap raw material at 22 to 25 cents per pound and are an attractive candidate for bioindustries. The content of unsaturated fatty acids such as oleic and linoleic acids are 22% and 55% for soybean oil and 26% and 60% for corn oil, respectively. The value-added products (oxygenated fatty acids) from these low-cost materials can be used as biomedical and specialty chemicals. Research and development of products in these areas have a great future. Recently, strain ALA2 was reclassification as *Bacillus megaterium* ALA2 (Hou et al., 2005).

References

1. I Yano, Y Furukawa, M Kusunose. α-Oxidation of long-chain fatty acids in cell-free extracts of *Arthrobacter simplex*. *Biochem Biophys Acta* 239: 513–516. 1971.
2. I. Matsunaga, E. Kusunose, I Yano, K Ichihara. Separation and partial characterization of soluble fatty acid α-hydroxylase from *Sphingomonas paucimobilis*. *Biochem Biophys Res Commun* 201: 1554–1460. 1994.
3. I Matsunaga, N Yokotani, O Gotoh, E Kusunose, M Yamada, K Ichihara. Molecular cloning and expression of fatty acid -hydroxylase from *Sphingomonas paucimobilis*. *J Biol Chem* 272: 23, 592–523, 596. 1997.
4. I Matsunaga, M Yamada, E Kusunose, T Miki, K Ichihara. Further characterization of hydrogen peroxide-dependent fatty acid α-hydroxylase from *Sphingomonas paucimobilis*. *J Biochem* 124: 105–110. 1998.
5. I Matsunaga, T Sumimoto, A ueda, E Kusunose, K Ichihara. Fatty acid-specific, regiospecific, and stereospecific hydroxylation by cytochrome P450 (CYP152B1) from *Sphingomonas paucimobilis*: Substrate structure required for α-hydroxylation. *Lipids* 35: 365–371. 2000.
6. CP Kurtzman, RF Vesonder, MJ Smiley. Formation of extracellular $C_{14} - C_{18}$ 2-D-hydroxy fatty acids by species of Saccharomycysis. *Appl Microbiol* 26: 650–652. 1973.
7. DW Russell. Depsipeptides of *Pithomyces chartarum*: The structure of *Sporidesmolide* I. *J Chem Soc* 1962: 753–761. 1962.
8. MM Shemyakin, YA Ovchinnikov, AA Kiryushkin,VT Ivanov. Concerning the structure of enniatin B. *Tetrahedron Lett* 28: 1927–1932. 1962.
9. MM Shemyakin, EL Vinogradova, MY Feigina, NA Aldanova. Structure of amidomycin and valinomycin. *Tetrahedron Lett* 6: 351–356. 1963.
10. RF Vesonder, FH Stodola, WK Rohwedder, DB Scott. 2-D-Hydroxyhexadecanoic acid: A metabolic product of the yeast *Hansenula sydowiorum*. *Can J Chem* 48: 1985–1986. 1970.

11. EM Carballeira, F Shalabi, V Negron. 2-Hydroxy fatty acids from marine sponges. 2. The phospholipid fatty acids of the Caribbean sponges *Verongula gigantean* and *Aplysina archeri*. *Lipids* 24: 229–232. 1989.

12. EM Carballeira, F Shalabi, M Reyes. New 2-hydroxy fatty acids in the Carribbean urchin *Trysneustes esculentus*. *J Nat Products* 57: 614–619. 1994.

13. LC Vining, WA Taber. Isariin, a new depsipeptide from *Isaria cretacea*. *Can J. Chem* 40: 1579–1584. 1962.

14. HH Wasserman, JJ Keggi, JE McKeon. The structure of Serratamolide. *J Am Chem Soc* 84: 2978–2982. 1962.

15. DW Thomas. The revised structure of the peptide antibiotic Esperin, established by mass spectrometry. *Tetrahedron* 25: 1985–1990. 1969.

16. RF Vesonder, LJ Wickerham, WK Rohweddeer. 3-D-Hydroxypalmitic acid: A metabolic product of the yeast NRRL Y-6954. *Can J Chem* 46: 2628–2629. 1968.

17. S Tahara, Y Suzuki, J Mizutani. Fungal metabolism of trans-2-octenoic acid using one of mucor species. *Agric Biol Chem* 41: 1643–1650. 1977.

18. N Esaki, S Ito, W Blank, K Soda. Biotransformation of oleic acid by *Alcaligenes* sp. 5-18, a bacterium tolerant to high concentrations of oleic acid. *J Ferment Bioeng* 77: 148–151. 1994.

19. MS van Dyk, JLF Kock, DJ Coetzee, OPH Augustyn, S Nigam. Isolation of a novel arachidonic acid metabolite 3-hydroxy-5,8,11,14-eicosatetraenoic acid (3-HETE) from the yeast *Dipodascopsis uninucleata* UOFS-Y128. *FEBS Lett* 283: 195–198. 1991.

20. JLF Kock, DJ van Vuuren, A Botha, MS van Dyk, DJ Coetzee, PJ Botes, N Shaw, J Friend, C Ratledge, AD Roberts, S. Nigam. Production of biologically active 3-hydroxy-5,8,11,14-eicosatetraenoic acid (3-HETE) and linoleic acid metabolites by Dipodascopsis. *System Appl Microbiol* 20: 39–49. 1997.

21. P Venter, JLF Kock, GS Kumar, A Botha, DJ Coetzee, PJ Botes, RK Bhatt, JR Falck, T Schewe, S Nigam. Production of 3R-hydroxy-polyenoic fatty acids by the yeast *Dipodascopsis uninucleata*. *Lipids* 32: 1277–1283. 1997.

22. M Kusunose, E Kusunose, MJ Coon. Enzymatic -oxidation of fatty acids. I. Products of octanoate, decanoate, and laurate oxidation. *J Biol. Chem* 239: 1374–1380. 1964.

23. BJ Abbott, CT Hou. Oxidation of 1-alkenes to 1,2-epoxyalkenes by *Pseudomonas oleovorans*. *Appl Microbiol* 26: 86–91. 1973.

24. SW May, BJ Abbott. Enzymatic epoxidation. II. Comparison between the epoxidation and hydroxylation reactions catalyzed by the -hydroxylation system of *Pseudomonas oleovorans*. *J Biol Chem* 248: 1725–1730. 1973.

25. CT Hou. Microbiology and biochemistry of methylotrophic bacteria. In C. T. Hou (ed.) *Methylotrophes: Microbiology, Biochemistry, and Genetics*, 1984. pp. 1–54, CRC Press, Boca Raton, FL.

26. CT Hou, RN Patel, AI Laskin, N Barnabe, I Barist. Purification and properties of a NAD-linked 1,2-propanediol dehydrogenase from propane-grown *Pseudomonas fluorescens* NRRL B-1244. *Arch Biochem. Biophys* 223: 297–308. 1983.

27. JA Peterson, D Basu, MJ Coon. Enzymatic ω-oxidation. I. Electron carriers in fatty acid and hydrocarbon hydroxylation. *J Biol Chem* 241: 5162–5164. 1966.

28. EJ McKenna, MJ Coon. Enzymatic ω-oxidation. IV. Purification and properties of the ω-hydroxylase of *Pseudomonas oleovorans*. *J Biol Chem* 245: 3882–3889. 1970.

29. RT Ruettinger, ST Olson, RF Boyer, MJ Coon. Identification of the ω-hydroxylase of *Pseudomonas oleovorans* as a nonheme iron protein requiring phospholipid for catalytic activity. *Biochem Biophys Res Commun* 57: 1011–1017. 1974.

30. RT Ruettinger, GR Griffith, MJ Coon. Characterization of the ω-hydroxylase of *Pseudomonas oleovorans* as a nonheme iron protein. *Arch Biochem Biophys* 183: 528–537. 1977.

31. GR Griffith, RT Ruettinger, EJ McKenna, MJ Coon. Fatty acid ω-hydroxylase (alkane hydroxylase) from *Pseudomonas oleovorans*. *Methods Enzymol* 53D: 356–360. 1978.

32. J. Shanklin, C Achim, H Schmidt, BG Fox, E Munck. Mossbauer studies of alkane ω-hydroxylase: Evidence for a diiron cluster in an integral-membrane enzyme. *Proc Natl Acad Sci USA*. 94: 2981–2986. 1997.

33. J Shanklin, EB Cahoon. Desaturation and related modifications of fatty acids. *Annu Rev Plant Physiol Plant Mol Biol* 49: 611–641. 1998.

34. AP Tulloch, JFT Spencer, PAJ Gorin. The fermentation of long-chain compounds by *Torulopsis magnoliae*. *Can J Chem* 40: 1326–1338. 1962.

35. E Heinz, AP Tulloch, JFT Spencer. Stereospecific hydroxylation of long-chain compounds by a species of *Torulopsis*. *J Biol Chem* 244: 882–888. 1969.

36. E Heinz, AP Tulloch, JFT Spencer. Hydroxylation of oleic acid by cell-free extract of a species of Torulopsis. *Biochem Biophys Acta* 202: 49–55. 1970.

37. RK Hommel, S Stegner, K Huse, HP Kleber. Cytochrome P-450 in the sophorose-lipid producing yeast *Candida (Torulopsis) apicola*. *Appl Microbiol Biotechnol* 40: 724–728. 1994.

38. K Lottermoser, WH Schunck, O Asperger. Cytochrome P-450 of the sophorose lipid-producing yeast *Candida apicola*: Heterogeneity and polymerase chain reaction-mediated cloning of two genes. *Yeast* 12: 565–575. 1996.

39. AC Lanser, RD Plattner, MO Bagby. Production of 15-,16-,and 17-hydroxy-9-octadecenoic acids by conversion of oleic acid with *Bacillus pumilus*. *J Am Oil Chem Soc* 69: 363–366. 1992.

40. JK Huang, KC Keudell, SJ Seong, WE Klopfenstein, L Wen, MO Bagby, RA Norton, RF Vesonder. Conversion of 12-hydroxyoctadecanoic acid to 12,15-; 12,16-; and 12,17-dihydroxyoctadecanoic acids with *Bacillus* sp. U88. *Biotechnol Lett* 18: 193–198. 1996.

41. Y Miura, AJ Fulco. ω-2 Hydroxylation of fatty acids by a soluble system from *Bacillus megaterium*. *J Biol Chem* 249: 1880–1888. 1974.

42. Y Miura, AJ Fulco. ω-1, ω-2, ω-3 Hydroxylation of long-chain fatty acids, amides, and alcohols by a soluble enzyme ststem from *Bacillus megaterium*. *Biochem Biophys Acta* 388: 305–317. 1975.

43. RS Hare, AJ Fulco. Carbon monoxide and hydroxymercuribenzoate sensitivity of a fatty acid ω-2 hydroxylase from *Bacillus megaterium*. *Biochem Biophys Res Commun* 65: 665–672. 1975.

44. PP Ho, AJ Fulco. Involvement of a single hydroxylase species in the hydroxylation of palmitate at the ω-1, ω-2, and ω-3 positions by a preparation from *Bacillus megaterium*. *Biochem Biophys Acta* 431: 249–256. 1976.

45. LO Narhi, AJ Fulco. Characterization of a catalytically self-sufficient 119,000-Dalton cytochrome P-450 monooxygenase induced by barbiturates in *Bacillus megaterium*. *J Biol Chem* 261: 7160–7169. 1986.

46. LP Wen, AJ Fulco. Cloning of the gene encoding a catalytically self-sufficient cytochrome P-450 fatty acid monooxygenase induced by barbiturates in *Bacillus megaterium* and its functional expression and regulation in heterologous (*Escherichia coli*) and homologous (*Bacillus megaterium*) hosts. *J Biol Chem* 262: 6676–6682. 1987.

47. SS Boddupalli, BC Pramanik, CA Slaughter, RW Estabrook, JA Peterson. Fatty acid monooxygenation by P-450$_{BM-3}$: Product identification and proposed mechanisms for the sequential hydroxylation reactions. *Arch Biochem Biophys* 292: 20–28. 1992.

48. JH Capdevila, S Wei, C Helvig, JR Falck, Y Belosludtsev, G Truan, SE Graham-Lorence, JA Peterson. The highly steroselective oxidation of polyunsaturated fatty acids by cytochrome P450$_{BM-3}$. *J Biol Chem* 271: 22, 663–622, 671. 1996.

49. M Akiyama, Y Doi. Production of poly (3-hydroxyalkanoates) from α, ω-alkanedioic acids and hydroxylated fatty acids by *Alcaligenes* sp. *Biotechol Lett* 15: 163–168. 1993.

50. I Shiio, R Uchio. Microbial production of long-chain dicarboxylic acids from n-alkanes. Part I. Screening and properties of microorganisms producing dicarboxylic acids. *Agric Biol Chem* 35: 2033–2042. 1971.

51. S. Picataggio, T Rohrer, K Decanda, D Lanning, R Reynolds, J Mielenz, LD Eirich. Metabolic engineering of *Candida tropicalis* for the production of long-chain dicarboxylic acids. *Biotechnology* 10: 894–898. 1992.

52. EC Chan, J Kuo. Biotransformation of dicarboxylic acid by immobilized *Cryptococcus* cells. *Enzyme Microb Technol* 20: 585–589. 1997.

53. AS Kester, JW Foster. Diterminal oxidation of long-chain alkanes by bacteria. *J Bacteriol* 85: 859–869. 1963.

54. R Uchio, I Shiio. Microbial production of long-chain dicarboxylic acids from n-alkanes. Part II. Production by *Candida cloacae* mutant unable to assimilate dicarboxylic acid. *Agric Biol Chem* 36: 426–433. 1972.

55. EC Chan, J Kuo, HP Lin, DG Mou. Stimulation of n-alkane conversion to dicarboxylic acid by organic-solvent- and detergent-treated microbes. *Appl Microbiol Biotechnol* 34: 772–777. 1991.

56. ZH Yi, HJ Rehm. A new metabolic pathway from n-dodecane to α, ω-dodecandioic acid in a mutant of *Candida tropicalis*. *Eur J Appl Microbiol Biotechnol* 15: 175–179. 1982.

57. ZH Yi, HJ Rehm. Formation and degradation of Δ^9 –1,18-octadecendioic acid from oleic acid by *Candida tropicalis*. *Appl Microbiol Biotechnol* 28: 520–526. 1988.

58. ZH Yi, HJ Rehm. Identification and production of Δ^9 –1,18-octadecendioic acid from oleic acid by *Candida tropicalis*. *Appl Microbiol Biotechnol* 30: 327–331. 1989.

59. M Ohkuma, T Zimmer, T Iida, WH Schunck, A Ohta, M Takagi. Isozyme function of n-alkane-inducible cytochromes P450 in *Candida maltosa* revealed by sequential gene disruption. *J Biol Chem* 273: 3948–3953. 1998.

60. U Scheller, T Zimmer, E Kargel, WH Schunck. Characterization of the n-alkane and fatty acid hydroxylating cytochrome P450 forms 52A2 and 52A4. *Arch Biochem Biophys* 328: 245–254. 1996.

61. T Zimmer, M Ohkuma, A Ohta, M Takagi, WH Schunck. The CYP52 multigene family of *Candida maltosa* encodes functionally diverse n-alkane-inducible cytochrome P450. *Bioochem Biophys Res Commun* 224: 784–789. 1996.

62. U Scheller, T Zimmer, D Becher, F Schauer, WH Schunck. Oxygenation cascade in conversion of n-alkanes to α, ω-dioic acids catalyzed by cytochrome P450 52A3. *J Biol Chem* 273: 32, 528–532, 534. 1998.

63. KD Green, MK Turner, JM Woodley. *Candida cloacae* oxidation of long-chain fatty acids to dioic acids. *Enzyme Microb Technol* 27: 205–211. 2000.

64. D Fabritius, HJ Schafer, A Steinbuchel. Identification and production of 3-hydroxy-Δ^9 –cis1,18-octadecenedioic acid by mutant of *Candida tropicalis*. *Appl Microbiol Biotechnol* 45: 342–348. 1996.

65. D Fabritius, HJ Schafer, A Steinbuchel. Biotransformation of linoleic aicd with the *Candida tropicalis* M25 mutant. *Appl Microbiol Biotechnol* 48: 83–87. 1997.

66. D Fabritius, HJ Schafer, A Steinbuchel. Bioconversion of sunflower oil, rapeseed oil and ricinoleic acid by *Candida tropicalis* M25. *Appl Microbiol Biotechnol* 50: 573–578. 1998.

67. T Kato, Y Yamaguchi, T Uyehara, T Yokoyama, T Namai, S Yamanaka. Self defensive substances in rice plant against rice blast disease. *Tetrahedron Lett* 24: 4715–4718. 1983.

68. CT Hou, MO Bagby, RD Plattner, S Koritala. A novel compound, 7,10-dihydroxy-8(E)-octadecenoic acid from oleic acid by bioconversion. *J Am Oil Chem So.* 68: 99–101. 1991.

69. CT Hou. A novel compound, 12,13,17-trihydroxy-9(Z)-octadecenoic acid, from linoleic acid by a new microbial isolate *Clavibacter* sp. ALA2. *J Am Oil Chem Soc* 73:1359–1362. 1996.

70. CT Hou. Biotransformation of unsaturated fatty acids to industrial products. *Advances in Applied Microbiology* 47: 201–220. 2000.

71. CT Hou. New uses of vegetable oils: Novel oxygenated fatty acids by biotransformation. *SIM News* 53: 56–61. 2003.

72. LL Wallen, RG. Benedict, RW Jackson. The microbial production of 10-hydroxystearic acid. *Arch Biochem Biophys* 99: 249–253. 1962.

73. SH El-Sharkawy, W Yang, L Dostal, JPN. Rosazza. Microbial oxidation of oleic acid. *Appl Environ Microbil* 58: 2116–2122. 1992.

74. S. Koritala, L Hosie, CT Hou, CW Hesseltine, MO Bagby. Microbial conversion of oleic acid to 10-hydroxystearic acid. *Appl Microbiol Biotechnol* 32: 299–304. 1989.

75. WH Blank, H Takayanagi, T Kido, F Meussdoerffer, N Esaki, K Soda. Transformation of oleic acid and its esters by *Sarcia lutea*, *Agric Biol Chem* 55: 2651–2652. 1991.

76. T Kaneshiro, J-K Huang, D Weisleder, MO Bagby. 10R-Hydroxystearic acid production by a novel microbe, NRRL B-14797, isolated from compost. *J Ind Microbiol* 13: 351–355. 1994.

77. AC Lanser. Conversion of oleic acid to 10-ketostearic acid by *Staphylococcus* sp. *J Am Oil Chem Soc* 70: 543–545. 1993.

78. CT Hou. Production of 10-ketostearic acid from oleic acid by a new microbial isolate, *Flavobacterium* sp. DS5, NRRL B-14859. *Appl Environ Microbil* 60: 3760–3763. 1994.

79. TM Kuo, AC Lanser, T Kaneshiro, CT Hou. Conversion of oleic acid to 10-ketostearic acid by *Sphingobacterium* sp. Strain 022. *J Am Oil Chem Soc* 76: 709–712. 1999.

80. S Koritala, MO Bagby. Microbial conversion of linoleic and linolenic acids to unsaturated hydroxy fatty acids. *J Am Oil Chem Soc* 69: 575–578. 1992.

81. CT Hou. Conversion of linoleic acid to 10-hydroxy-12(Z)-octadecenoic acid by *Flavobacterium* sp. DS5. *J Am Oil Chem Soc* 71: 975–978. 1994.

82. Y Yamada, H Uemura, H Nakaya, K Sakata, T Takatori, M Nagao, H Iwase, K Iwadate. Production of hydroxy fatty acid (10-hydroxy-12(Z)-octadecenoic acid) by *Lactobacillus plantarum* from linoleic acid and its cardiac effects to guinea pig papillary muscles. *Biochem Biophys Res Comm* 226: 391–395. 1996.

83. H Gardner, CT Hou. All (S) stereoconfiguration of 7,10-dihydroxy-8(E)-octadecenoic acid from bioconversion of oleic acid by *Pseudomonas aeruginosa*. *J Am Oil Chem Soc* 76: 1151–1156. 1999.

84. CT Hou, MO Bagby. 10-Hydroxy-8-(Z)-octadecenoic acid, an intermediate in the formation of 7,10-dihydroxy-8-(E)-octadecenoic acid from oleic acid by *Pseudomonas* sp. PR3. *J Ind Microbiol* 9:103: 107. 1992.

85. H Kim, HW Gardner, CT Hou. 10(S)-Hydroxy-8(E)-octadecenoic acid, an intermediate in the conversion of oleic acid to 7,10-dihydroxy-8(E)-octadecenoic acid. *J Am Oil Chem Soc* 77: 95–99. 2000.

86. E Mercade, M Robert, MJ Espuny, MP Bosch, A Manresa, JL Parra, J Guinea. New surfactant isolated from *Pseudomonas* sp. 42A2. *J Am Oil Chem Soc* 65: 1915–1916. 1988.

87. JL Parra, J Pastor, F Comelles, A Manresa, MP Bosch. Studies of biosurfactants obtained from olive oil. *Tenside Surfact Deterg* 27: 302–306. 1990.

88. C de Andres, E Mercade, J Guinea, A Manresa. 7,10-Dihydroxy-8E-octadecenoic acid produced by *Pseudomonas sp.*42A2: Evaluation of different cultural parameters of the fermentation. *World J Microbiol Biotechnol* 10: 106–109. 1994.

89. A Guerrero, I Casals, M Busquets, Y Leon, A Manresa. Oxydation of oleic acid to (E)-10-hydroperoxy-8-octadecenoic and (E)-10-hydroxy-8-octadecenoic acids by *Pseudomonas sp.* 42A2. *Biochem Biophys Acta* 1347: 75–81. 1997.

90. TM Kuo, H Kim, CT Hou. Production of a novel compound, 7,10,12-trihydroxy-8(E)-octadecenoic acid from ricinoleic acid by *Pseudomonas aeruginosa* PR3. *Current Microbiology* 43: 198–203. 2001.

91. H Kim, TM Kuo, CT Hou. Production of 10,12-dihydroxy-8(E)-octadecenoic acid, an intermediate in the conversion of ricinoleic acid to 7,10,12-trihydroxy-8(E)-octadecenoic acid by *Pseudomonas aeruginosa* PR3. *J Industrial Microbiol & Biotechnol* 24: 167–172. 2000.

92. H Kim, HW Gardner, CT Hou. Production of isomeric 9,10,13 (9,12,13)-trihydroxy-11E (10E)-octadecenoic acid from linoleic acid by *Pseudomonas aeruginosa* PR3. *J Industrial Microbiol & Biotechnol* 25: 109–115. 2000.

93. TM Kuo, KJ Ray and LK Manthey. A facile reactor process for producing 7,10-dihydroxy-8(E)-octadecenoic acid from oleic acid conversion by *Pseudomonas aeruginosa*. *Biotechnol Letts* 25: 29–33, 2003.

94. A Guo, TM Kuo, CT Hou, Z Petrovic. Rigid urethan foams from a soy polyol-DOD hybrid. The 94th AOCS annual meeting abstract IOP2, p 76. 2003.

95. CT Hou, W Brown, DP Labeda, TP Abbott, D Weisleder. Microbial production of a novel trihydroxy unsaturated fatty acid from linoleic acid. *J Ind Microbiol Biotechnol* 19: 34–38. 1997.

96. T Kato, Y Yamaguchi, N Abe, T Uyehara, T Nakai, S Yamanaka, N Harada. Unsaturated hydroxy fatty acids, the self-defensive substances in rice plant against blast disease. *Chem Lett* 25:409–412. 1984.

97. H Masui, T Kondo, M Kojima. An antifungal compound, 9,12,13-trihydroxy-(E)-10-octadecenoic acid, from *Colocasia antiquorum* inoculated with *Ceratocystis fimbriata. Phytochemistry* 28: 2613–2615. 1989.

98. CT Hou, RJ Forman III. Growth inhibition of plant pathogenic fungi by hydroxy fatty acids. *J Indust Microbiol Biotechnol* 24: 275–276. 2000.

99. CT Hou, H Gardner, W Brown. Production of multihydroxy fatty acids from linoleic acid by *Clavibacter* sp. ALA2. *J Am Oil Chem Soc* 75: 1483–1487. 1998.

100. HW Gardner, CT Hou, D Weisleder, W Brown. Biotransformation of linoleic acid by *Clavibacter* sp. ALA2: Cyclic and bicyclic fatty acids. *Lipids* 35: 1055–1060. 2000.

101. Y Iwasaki, W Brown, CT Hou. Biosynthetic pathway of diepoxy bicyclic fatty acids from linoleic acid by *Clavibacter sp.* ALA2. *J Am Oil Chem Soc* 79: 369–372. 2002.

102. CT Hou, HW Gardner, W Brown. 12,13,16-Trihydroxy-9(Z)-octadecenoic acid, a possible intermediate in the biosynthesis of linoleic acid to tetrahydrofuranyl fatty acids by *Clavibacter sp.* ALA2. *J Am Oil Chem Soc* 78: 1167–1169. 2001.

103. M Hosokawa, CT Hou. Biosynthesis of tetrahydrofuranyl fatty acids from linoleic acid by *Clavibacter* sp. ALA2. *J Am Oil Chem Soc* 80: 145–149. 2003.

104. M Hosokawa, CT Hou, D Weisleder, W Brown. Production of novel tetrahydrofuranyl fatty acids from α-linolenic acid by *Clavibacter* sp. ALA2. *Appl Environ Microbiol* 69: 3868–3873. 2003.

105. ID Brodowsky, EH Oliw. Metabolism of 18:2(n-6), 18:3(n-3), 20:4(n-6) and 20:5(n-3) by the fungus *Gaeumannomyces graminis*: identification of metabolites formed by 8-hydroxylation and by ω2 and ω3 oxygenation. *Biochem Biophys Acta* 1124: 59–65. 1992.

106. MF Moghaddam, K Matoba, B Borhan, F Pinot and BD Hammock. Novel metabolic pathways for linoleic and arachidonic acid metabolism. *Biochem Biophys Acta* 1290: 327–339. 1996.

107. BM Markaverich, MA Alejandro, D Markaverich, L Zitzow, N Casajuna, N Camarao, J Hill, K Bhirdo, R Faith, J Turk and JR Crowley. Identification of an endocrine disrupting agent from corn with mitogenic activity. *Biochem Biophys Res Commun* 291: 692–700. 2002.

108. B Markaverich, S Mani, MA Alejandro, A Mitchell, D Markaverich, T Brown, C Velez-Trippe, C Murchison, B O'Malley and R Faith. A novel endocrine-disrupting agent in corn with mitogenic activity in human breast and prostatic cancer cells. *Environ Health Perspect* 110: 169–177. 2002.

109. Y Miura, AJ Fulco. ω-1, ω–2 and ω-3 hydroxylation of long chain fatty acids, amides and alcohols by a soluble enzyme system from *Bacillus megaterium. Biochem Biophys Acta* 388: 305–317. 1975.

110. M Hosokawa, CT. Hou, D. Weisleder. Bioconversion of n-3 and n-6 polyunsaturated fatty acids by *Clavibacter* sp. ALA2. *J Am Oil Chem Soc* 80: 1085–1091. 2003.

111. CT Hou, DP Labeda, A Rooney. Evaluation of microbial strains for linoleic acid hydroxylation and reclassification of strain ALA2. *Antoxie van Leeuwenhoek.* 2005 (in press).

8

Application of Lipases to Industrial-Scale Purification of Oil- and Fat-Related Compounds

Yuji Shimada

Toshihiro Nagao

Yomi Watanabe

8.1 Introduction

Lipases are defined as enzymes that catalyze hydrolysis of triacylglycerols (TAGs). It is, however, well known that they also catalyze esterification and transesterification (acidolysis, alcoholysis, and interesterification) in the presence of only a small amount of water. In addition, lipases possess specificities for fatty acids (FAs), alcohols, and the position of ester bonds in TAG. Combination of these reactions and specificities can precisely modify natural oils and fats. Studies on application of lipases, which were begun in the 1980s, can be classified into two fields: production of functional lipids, and optical resolution of chiral compounds.

The application of lipase to lipid modification started from production of cocoa fat substitute (1,3-stearyl-2-oleoyl glycerol) in the middle of the 1980s. The reaction was performed by transesterification of palm olein with stearic acid using an immobilized 1,3-positional specific lipase as a catalyst.[1] This new technology drew considerable attention, and human milk substitute (1,3-oleoyl-2-palmitoyl glycerol),[2,3] docosahexaenoic acid (DHA; 22:6n-3)-rich oil,[4] diacylglycerol,[5] and medium-chain FA-containing oil[6] were commercialized as functional lipids thereafter. In addition, a great deal of attention is being focused on the application of lipases to production of structured lipids.[7–11]

The other application study is the use of lipases in chemical reactions. Studies in this field were started by the report of Zaks and Klivanov in 1984.[12] They found that a lipase acts even in nearly anhydrous water-immiscible organic solvents. This study has contributed to optical resolution of many chiral compounds, such as intermediates of pharmaceuticals and agrochemicals.[13–15] The optical resolution can be said in other words: purification of useful materials. As lipases are originally enzymes that act on oils and fats, they can be used for purification of useful oil- and fat-related compounds. Early in the 1990s, a feature of lipases that acts poorly on polyunsaturated fatty acids (PUFAs) attracted attention, and purification of PUFAs with lipases was started by several groups, including our laboratory.[16–18] In this chapter, we describe effectiveness of lipase reactions in purification of functional FAs from several FA mixtures and of useful compounds from unutilized biomass.

8.2 Purification of Desired Compounds by a Process Comprising Enzyme Reaction and Distillation

It is well known that enzyme-catalyzed reactions efficiently proceed under mild conditions. The reactions are therefore promised as procedures for conversion of unstable substances. Meanwhile, distillation is widely used for fractionation of oil- and fat-related compounds that have different molecular weights (boiling points). Molecular (short-path) distillation is particularly effective for purifying unstable compounds, because the operation is conducted under high vacuum and because the material is heated for only short time. We thus attempted to develop a new technology comprising enzyme reaction and distillation. If a desired component or contaminants are converted to different molecular forms by selective reaction with a lipase, the purification becomes relatively easy. The principle is schematically depicted in Figure 8.1. As decomposition of contaminants makes a difference in the molecular weights

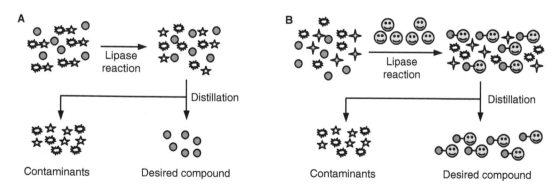

FIGURE 8.1 Outline of a process comprising enzyme reaction and distillation. **A,** Purification process by decomposing contaminants with a lipase. **B,** Purification process by converting a desired compound to its different molecular form with a lipase.

of a desired component and contaminants, the desired one can easily be purified by distillation of the reaction mixture (Figure 1a). The other example is modification of a desired component. If lipase treatment after addition of a substrate to the raw material induces a change in molecular form of a desired one, the component can easily be purified by distillation (Figure 1b). Purification of useful materials according to this principle is described hereafter.

8.3 Production of PUFA-Rich Oil by Selective Hydrolysis

8.3.1 Physiological Activities of PUFAs

PUFAs have various physiological functions and are used in various areas. For example, the ethyl ester of eicosapentaenoic acid (EPA; 20:5n-3) has been used for treatment of atherosclerosis and hyperlipemia since 1990 in Japan.[19] DHA plays a role in the prevention of a number of human diseases, including cardiovascular diseases,[20–22] inflammation,[23] and cancer.[24,25] DHA has also been reported to exhibit an important function in the brain[26] and retina,[27] and to accelerate the growth of preterm infants.[28,29] For these reasons, tuna oil containing DHA has been used as a food material, an ingredient in infant formulas, and a nutraceutical lipid.[30] Arachidonic acid (AA; 20:4n-6) is a precursor of hormones (prostaglandins, thromboxanes, and leukotrienes) involved in the AA cascade,[31,32] and accelerates the growth of preterm infants, as does DHA.[28,29] The use of an AA-rich oil produced by a microorganism[33,34] has begun in a formula for infants and in a nutraceutical food. γ-Linolenic acid (GLA; 18:3n-6) is a precursor of AA, and is effective for treating atopic eczema[35,36] and rheumatoid arthritis.[37,38] Thus, borage oil rich in GLA is used as an ingredient in infant formulas and a nutraceutical lipid.

8.3.2 Production of PUFA-Rich Oil

Oils containing high concentrations of PUFA can be expected to show greater physiological activities from even a small amount of intake. PUFA-rich oils have traditionally been produced by winterization. For example, the maximum DHA content is 30 wt% at most according to this method in industrial production of DHA-rich oil from tuna oil. Meanwhile, application of lipase can enrich PUFA more efficiently.[39–45] Lipases from *Candida rugosa*,[40,41,43,45] *Geotrichum candidum*,[42,44] and *Penicillium abeanum*[46] have strong hydrolysis activity and act very weakly on PUFAs. Hence, when PUFA-containing oils are hydrolyzed with these lipases, ester bonds of FAs except PUFA are hydrolyzed, resulting in enrichment of PUFA in undigested acylglycerols. The resulting reaction mixture consists of FFAs and acylglycerols, of which molecular weights are significantly different. The acylglycerols (PUFA-rich oil) are therefore easily purified from the reaction mixture by short-path distillation.[45] Figure 8.2 shows an industrially available process of producing PUFA-rich oil.

When tuna oil was hydrolyzed with *C. rugosa* lipase, the content of DHA in acylglycerols depends only on the degree of hydrolysis (Figure 8.3a).[47] The content of DHA increased from 23 wt% to 45 wt% in a 78% yield at 60% hydrolysis, and to 50 wt% in a 70% yield at 70% hydrolysis. The lipase acts on GLA as weakly as on DHA, and GLA-rich oil can be produced by selective hydrolysis of borage oil. However, the content of GLA was not raised over 46 wt% at above 65% hydrolysis (Figure 8.3b).[41,45,47]

Accumulation of FFAs in the reaction mixture did not raise the content of PUFA over a limit value by a single reaction. Hence, repeated hydrolysis was attempted. Single hydrolysis of borage oil increased the content of GLA in acylglycerols to 45 wt%. FFAs were removed from the reaction mixture by short-path distillation, and the resulting acylglycerols were hydrolyzed again. This repeated hydrolysis succeeded in production of oil containing 55.7 wt% GLA (Table 8.1). Similarly, an oil containing *ca.* 70 wt% DHA was produced by repeating hydrolysis of tuna oil three to four times.

Selective hydrolysis can also be applied to production of other PUFA-rich oils. *Mortierella alpina* produces a single-cell oil containing 40 wt% AA.[48] *C. rugosa* lipase acted weakly on AA, and hydrolysis of the oil with the lipase increased the content of AA to 57 wt% at 36% hydrolysis (recovery of AA, 91%).[49]

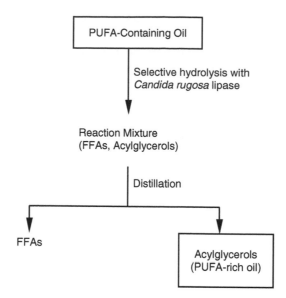

FIGURE 8.2   A process of producing PUFA-rich oil by selective hydrolysis of PUFA-containing oil. *C. rugosa* lipase is widely used in the oil and fat industry, which acts poorly on PUFAs. This lipase is useful as a catalyst for the selective hydrolysis. After the hydrolysis, acylglycerols in the reaction mixture are purified by short-path distillation.

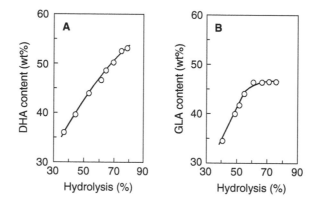

FIGURE 8.3   Correlation between the degree of hydrolysis and the content of PUFA in acylglycerols. **A**, Hydrolysis of tuna oil: a mixture of tuna oil, 50% water, and 20-2000 U/g-mixture of *C. rugosa* lipase was agitated at 35°C for 24 h. **B**, Hydrolysis of borage oil: a mixture of borage oil, 50% water, and 100 U/g-mixture of *C. rugosa* lipase was agitated at 35°C for 0.25–24 h.

8.4   Purification of PUFA by a Process Involving Selective Esterification

EPA ethyl ester (EPAEE) is purified from sardine oil by a combination of chemical ethanolysis, rectification under high vacuum, and urea adduct fractionation.[50] Its use as a medicine has called a great deal of attention to purification of functional PUFAs, and several procedures have been reported, such as distillation, urea adduct fractionation, high-performance liquid chromatography, and silver ion chromatography.[51,52] In addition to these methods, enzymatic methods using selective hydrolysis, esterification, and ethanolysis have been reported. As described in Section 8.3, the content of PUFA could not be raised to >70% by selective hydrolysis, owing to the moderate activity of the lipase toward the hydrolysis of PUFA ester

TABLE 8.1 Large-Scale Production of Oil Containing High Concentration of GLA by Repeated Hydrolysis

Procedure	Weight (kg)	Acid Value (mg KOH/g)	Acylglycerol (kg)	GLA Content (%)	GLA Recovery (%)
Borage oil	7.00	ND[1]	7.00	22.1	100
Hydrolysis 1[2]	6.29	122	2.44	45.2	71.2
Distillation[3]	2.28	10	2.17	45.4	63.6
Hydrolysis 2[4]	2.12	49	1.60	55.3	57.1
Distillation[5]	1.54	4	1.50	55.7	54.2

[1]Not detected.
[2]Borage oil was hydrolyzed at 35°C for 15 h in a mixture containing 50% water and 20 U/g-mixture of *C. rugosa* lipase.
[3]Oil layer recovered from the reaction mixture was applied to two-step short-path distillation: at 150°C and 0.05 mm Hg; at 160°C and 0.05 mm Hg. Acylglycerols were recovered in the residue.
[4]The residue (acylglycerol fraction) was hydrolyzed under the same conditions as those in hydrolysis 1.
[5]Distillation was conducted at 155°C and 0.04 mm Hg, and acylglycerols were recovered in the residue.

bonds. It was also reported that ethanolysis of sardine oil with *Pseudomonas* lipase enriched EPA and DHA in acylglycerol fraction. But the total content of EPA and DHA increased from 27 wt% to only 47.3 wt%.[53]

On the contrary, the FA specificity of a lipase is stricter in esterification than in hydrolysis and alcoholysis. Hills et al.[54] first applied lipase-catalyzed esterification to purification of PUFA. They esterified FFAs originating from evening primrose oil with butanol in *n*-hexane using immobilized *Rhizomucor miehei* lipase and succeeded in enriching GLA to 85% with a recovery of 64%. Similarly, DHA of cod liver oil was enriched from 9.4 wt% to 45 wt%. Other groups also reported that selective esterification was effective for increasing the PUFA content to near 90%, but the reaction systems required a large amount of organic solvents.[55,56] We thus attempted to develop an industrially available purification process for PUFA without organic solvent.

8.4.1 Strategy for Purification of PUFA

If selective esterification is involved in a process of industrial purification of PUFA, the reaction, of course, requires high FA specificity and high degree of esterification. In addition, FFAs have to be recovered from the reaction mixture composed of alcohol, FFAs, and FA esters by short-path distillation, which is adopted as a facile purification procedure. Lauryl alcohol (LauOH) is suitable as a substrate, which satisfies the two requirements for the following reasons: (1) Lipases generally recognize FFA and alcohol as substrates, but not the ester synthesized. Hence, the use of fatty alcohol as a substrate represses the reverse reaction and leads to a high degree of esterification. (2) LauOH is liquid state at the reaction temperature (30°C): lipases act strongly on liquid-state substrates but weakly on solid-state substrates. (3) Molecular weights (boiling points) of LauOH, FFA, and FA lauryl ester (FALE) are significantly different, indicating that short-path distillation can be adopted for separation of FFA from the reaction mixture. In addition, (4) LauOH is the cheapest among long-chain fatty alcohols. Fortunately, lipases showed the highest FA specificity in the esterification with LauOH among those with various fatty alcohols tested.[57]

We decided to use LauOH as a substrate, and planned a strategy for purification of PUFA (Figure 8.4). The first step is hydrolysis of PUFA-containing oil. The oil can be hydrolyzed by heating with a large amount of ethanol (EtOH) under alkaline conditions. However, this procedure requires a large-scale reactor and carries the risk of isomerization of PUFA. Also, the wastewater in the chemical process contains EtOH and has a high chemical oxygen demand (COD). Furthermore, pH of the reaction mixture has to be readjusted to acidic conditions to recover the FFAs. On the other hand, when industrial-scale hydrolysis is conducted using a lipase, FFAs can be recovered easily from the reaction mixture by short-path distillation. We thus adopt enzymatic hydrolysis. The second step is enrichment of PUFA in the FFA fraction by selective esterification of the resulting FFAs with LauOH using a lipase that acts weakly on PUFA. Because molecular weights of the components (LauOH, FFAs, and FALEs) in the reaction

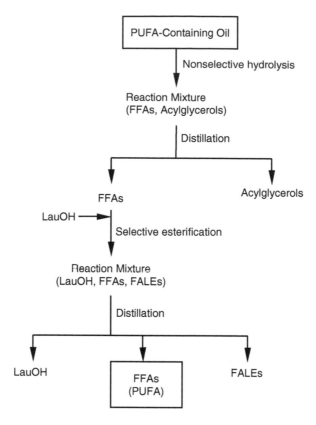

FIGURE 8.4 Strategy for purification of PUFA by a process comprising nonselective hydrolysis, selective esterification, and distillation. PUFA-containing oil is first hydrolyzed with a lipase that acts on PUFA as strongly as on the other FAs. FFAs recovered by short-path distillation are esterified with LauOH using a lipase that acts weakly on PUFA. FFAs rich in PUFA are finally recovered by distillation.

mixture are different, they can be easily separated by short-path distillation. GLA was purified from borage oil according to this strategy.

8.4.2 Purification of GLA from Borage Oil

An oil containing 45 wt% GLA (referred to as GLA45 oil) can be produced by selective hydrolysis of borage oil as described in Section 8.3.1. Purification of GLA was performed using 10 kg GLA45 oil as a starting material according to the strategy shown in Figure 8.4. The results are summarized in Table 8.2. As lipases from *Pseudomonas* and *Burkholderia* acted on GLA as strongly as on other C18 FAs, GLA45 oil was first hydrolyzed nonselectively with *Burkholderia cepacia* lipase. FFAs were recovered by short-path distillation, and were used for their esterification with LauOH.[58] Lipases from *C. rugosa* and *Rhizopus oryzae* acted very weakly on GLA but *R. oryzae* achieved higher degree of esterification than *C. rugosa* one.[57–59] We therefore selected *R. oryzae* lipase as a catalyst. Selective esterification of the FFAs with LauOH using the lipase increased the content of GLA from 46.3 wt% to 89.5 wt%. To further increase the purity, FFAs were recovered by distillation, and were esterified again with LauOH. The repeated esterification raised GLA purity in the FFA fraction to 97.3%. The FFA fraction recovered from the reaction mixture by distillation contained 13.5 wt% FALEs and 1.1 wt% LauOH. The FALEs were completely removed by urea adduct fractionation, but 0.8 wt% LauOH remained. The FFAs with 98.6 wt% GLA were prepared with a recovery of 49.4% of the initial content of GLA45 oil by a series of the purification procedures.[59] This result shows that the process comprising enzymatic method, short-path distillation, and urea adduct fractionation is effective for a large-scale purification of PUFA.

TABLE 8.2   Purification of GLA from GLA45 Oil

Step	Weight (kg)	Amount of FFA (kg)[1]	GLA in FFA Fraction		
			Content (wt%)	Amount(kg)	Recovery (%)
GLA45 oil	10.00	9.15[1]	45.1[2]	4.13[3]	100
Hydrolysis[4]	8.69	7.95	46.3	3.68	89.1
Distillation[5]	7.55	7.51	46.3	3.48	84.2
Esterification[6]	16.54	3.41	89.5	3.05	73.9
Distillation[7]	3.87	3.15	89.4	2.82	68.2
Esterification[8]	7.61	2.53	97.3	2.46	59.6
Distillation[7]	2.67	2.28	98.1	2.24	54.2
Urea adduct[9]	2.09	2.07	98.6	2.04	49.4

[1]The amount of FFA was calculated from its acid value, 200.
[2]The amount of FA in GLA45 oil.
[3]The content and amount of GLA in GLA45 oil.
[4]Reaction conditions: GLA45 oil/water, 2:1 (wt/wt); *B. cepacia* lipase, 250 U/g-mixture; 35°C; 24 h. Degree of hydrolysis was 91.5%.
[5]Reaction mixture was distilled at 180°C and 0.2 mm Hg, and the residue was then distilled at 200°C and 0.2 mm Hg. The two distillates were combined
[6]Reaction conditions: FFAs/LauOH, 1:2 (mol/mol); water, 20%; *R. oryzae* lipase, 50 U/g-mixture; 30°C; 16 h. Degree of esterification was 52.0%.
[7]LauOH was removed by distillation at 120°C and 0.2 mm Hg, and FFAs were then recovered in distillate by 185°C and 0.2 mm Hg.
[8]Reaction conditions: FFAs/LauOH, 1:2 (mol/mol); water, 20%; *R. oryzae* lipase, 70 U/g-mixture; 30°C; 16 h. Degree of esterification was 15.2%.
[9]FFA fraction (400 g) obtained by distillation was completely dissolved at 50°C in a solution of 2 L MeOH, 50 mL water, and 400 g urea. The solution was then cooled gradually to 5°C with agitating over *ca*. 10 h. After removing the precipitate, the volume of the filtrate was reduced to ca. 700 mL, and 300 mL of 0.2 N HCl was then added. The oil layer (FFAs) was washed with 300 mL water three times.

8.4.3   Purification of Other PUFAs

To confirm that lipase-catalyzed esterification with LauOH is effective for purification of other PUFAs, DHA and AA were purified from tuna oil and *M. alpina* single-cell oil, respectively. While distillation is suitable for industrial-scale purification of FFAs, *n*-hexane extraction is easy in the laboratory-scale purification. Hence, *n*-hexane extraction was adopted in this section for recovering FFAs from the reaction mixture.

Tuna oil containing 22.9 wt% DHA was hydrolyzed with a *Pseudomonas* lipase (hydrolysis, 79%). The lipase hydrolyzed a little strongly on DHA ester compared with EPA ester, resulting in increase of the DHA content in the FFA fraction to 24.2 wt%.[60] FFAs were recovered from the reaction mixture, and were then esterified with LauOH using *R. oryzae* lipase (esterification, 72.1%). Repeated esterification of the FFAs increased the content of DHA to 90.6 wt% (esterification, 30.4%). Recovery of DHA based on the content in tuna oil was 60.3%, when recovery of FFAs by *n*-hexane extraction was assumed to be 100%.[60]

AA was similarly purified from *M. alpina* single-cell oil containing 40 wt% AA. The oil was first hydrolyzed with *B. cepacia* lipase, and FFAs were recovered by *n*-hexane extraction. The FFA mixture included C22 and C24 saturated FAs, and was solid state at reaction temperature (30°C). These FAs were removed by urea adduct fractionation because lipase did not efficiently act in solid-state mixture, and were then subjected to selective esterification with LauOH. In the esterification, *R. oryzae* lipase acted moderately on AA. We thus selected *C. rugosa* lipase, which acted very weakly on AA.[61] The selective esterification twice raised the content of AA to 81.3 wt%. The single-cell oil contained 3.4 wt% GLA and 4.8 wt% di-homo-GLA (20:3n-6), and *C. rugosa* lipase acted on these FAs as weakly as on AA. Not only AA but also these FAs were therefore enriched in the FFA fraction, and total content of n-6 PUFAs reached 95.9%. Recoveries of AA and n-6 PUFAs were 52.9 and 51.9% of their initial contents in the single-cell oil, respectively.[49]

A mutant of *M. alpina* produces a single-cell oil containing n-9 FAs: 17.1 wt% Mead acid (MA; 20:3n-9), 14.3 wt% n-9 linoleic acid (n-9 LnA; 18:2n-9), and 35.4 wt% oleic acid (OA; 18:1n-9).[62,63] Lipase screening indicated that *Pseudomonas aeruginosa* lipase acted strongly on OA and n-9 LnA but weakly on MA, and that *C. rugosa* lipase acts weakly on MA and n-9 LnA but strongly on OA. Using the two enzymes, fractionation and enrichment of n-9 LnA and MA were conducted according to Figure 8.5. The single-cell oil was first hydrolyzed with *P. aeruginosa* lipase, and the reaction mixture was fractionated into FFAs containing 20.4 wt% n-9 LnA and 6.3 wt% MA, and acylglycerols containing 10.7 wt% n-9 LnA and 23.7 wt% MA. The FFA fraction was used for preparation of n-9 LnA-rich FFAs. After saturated FAs were removed by urea adduct fractionation, the FFAs were esterified with LauOH using *C. rugosa* lipase. Selective esterification in twice increased n-9 LnA content to 54.0 wt% (MA content, 19.0 wt%) with 38.2% recovery of the initial content of the single-cell oil.[64] The acylglycerol fraction obtained in hydrolysis with *P. aeruginosa* lipase was used for preparation of MA-rich FFAs. The acylglycerols were hydrolyzed under alkaline conditions in the presence of ethanol (EtOH), and saturated FAs were eliminated by urea adduct fractionation. Selective esterification of the FFAs with LauOH increased the MA content to 60.2 wt% with 53.5% recovery (n-9 LnA content, 20.1 wt%).[64] The two n-9 PUFAs were not highly purified, but the purities could significantly be increased if a lipase that distinguishes between n-9 LnA and MA was found.

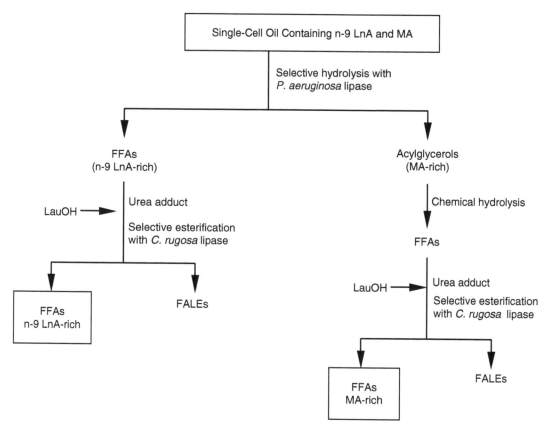

FIGURE 8.5 Strategy for enzymatic fractionation and enrichment of n-9 PUFAs. A single-cell oil containing n-9 PUFAs is first hydrolyzed with *P. aeruginosa* lipase. The reaction mixture is fractionated into FFAs rich in n-9 LnA and acylglycerols rich in MA. Saturated FAs in the FFA fraction are removed by urea adduct fractionation, and the resulting FFAs are estrified with LauOH using *C. rugosa* lipase to enrich n-9 LnA in the FFA fraction. On the other hand, acylglycerols rich in MA undergo chemical hydrolysis under alkaline conditions. After FFAs are recovered from the reaction mixture, saturated FAs are removed by urea adduct fractionation. The resulting FFAs are finally esterified with LauOH using *C. rugosa* lipase to enrich MA in the FFA fraction.

8.5 Purification of Conjugated Linoleic Acid (CLA) Isomers by a Process Involving Selective Esterification

CLA is a group of C18 FAs containing a pair of conjugated double bonds in either the *cis* or *trans* configuration. A commercialized product contains almost equal amounts of *cis*-9, *trans*-11 (*c9,t11*)-CLA, and *trans*-10, *cis*-12 (*t10,c12*)-CLA. The mixture of CLA isomers has various physiological activities, such as reduction of the incidence of cancer,[65–67] decrease in body fat content,[68–70] beneficial effects on atherosclerosis,[71,72] and improvement of immune function.[73] Also, it was recently reported that *c9,t11*-CLA has anticancer activity,[74] and that *t10,c12*-CLA decreases body fat content[75–77] and suppresses the development of hypertension.[78] These studies called a great deal of attention to the fractionation of CLA isomers.

Recently, an enzymatic method was found to be very effective for the fractionation of CLA isomers. Haas et al.[79] reported that *G. candidum* lipase recognized *c9,t11*-CLA more readily than *t10,c12*-CLA. They successfully enriched *c9,t11*-CLA in methyl esters from the products in the early stage of the reaction in which a mixture of CLA isomers was esterified with MeOH in an organic solvent system. In addition, *c9,t11*-CLA was enriched in FFAs by hydrolyzing methyl esters of CLA isomers. The purity of CLA isomers can be increased by their procedure, but the recovery is not good. At almost the same time, McNeill et al.[80] also showed that *c9,t11*-CLA and *t10,c12*-CLA were fractionated with *G. candidum* lipase. They esterified a mixture of CLA isomers with LauOH and separated the reaction mixture into the FFA fraction (*t10,c12*-CLA rich) and FALE fraction (*c9,t11*-CLA rich) by short-path distillation. The procedure allowed them to fractionate CLA isomers with a good recovery, but the purity was not high.

Our screening test showed that *C. rugosa* lipase acts strongly on *c9,t11*-CLA and poorly on *t10,c12*-CLA, as does *G. candidum* lipase.[81] The two CLA isomers were highly purified by selective esterification with *C. rugosa* lipase. The purification process is shown in Figure 8.6.[82] A mixture of CLA isomers was prepared by alkali conjugation of high purity (97.1%) of linoleic acid (LnA; 18:2n-6), and was used as a starting material, referred to as FFA-CLA. FFA-CLA contained 45.1 wt% *c9,t11*-CLA, 46.8 wt% *t10,c12*-CLA, and 5.3 wt% other CLA isomers. FFA-CLA was first esterified with LauOH using *C. rugosa* lipase, and FFA fraction containing 78.1 wt% *t10,c12*-CLA and FALE fraction containing 85.1 wt% *c9,t11*-CLA were recovered by short-path distillation. The FFA and FALE fractions were used for further purification of *t10,c12*- and *c9,t11*-CLAs, respectively.

To further increase in the purity of *t10,c12*-CLA, the FFA fraction was esterified again with an equimolar amount of LauOH. The reaction mixture was then subjected to short-path distillation. The FFA fraction recovered consisted of 0.3 wt% LauOH, 91.6 wt% FFAs, and 8.1 wt% FALEs, and FA composition in the fraction was 3.4 wt% *c9,t11*-CLA, 86.3 wt% *t10,c12*-CLA, and 9.7 wt% other CLA isomers. Finally, FALEs and CLAs except *c9,t11*- and *t10,c12*-CLAs were removed by urea adduct fractionation. This fractionation could completely remove FALEs, and decreased the content of CLAs except *c9,t11*- and *t10,c12*-isomers from 9.7 wt% to 1.3 wt%. Consequently, the purity of *t10,c12*-CLA reached 95.3% (the content of *t10,c12*-CLA based on the total content of *c9,t11*- and *t10,c12*-isomers, 96.9 wt%) (Table 8.3). Recovery of *t10,c12*-CLA by a series of purification procedures was 31% of the initial content.[82]

Another isomer, *c9,t11*-CLA, enriched in the FALE fraction was next purified according to Figure 8.6. The FALEs were chemically hydrolyzed under alkaline conditions, and FFAs were recovered by *n*-hexane extraction. The FFA fraction was esterified again with an equimolar amount of LauOH. After the reaction, FALEs were recovered by short-path distillation. The FALEs were hydrolyzed, and a mixture of LauOH and FFAs was recovered by *n*-hexane extraction. To the mixture was added safflower oil, which is a nonvolatile substance for the achievement of efficient distillation. FFAs purified by distillation consisted of 0.3 wt% LauOH, 99.4 wt% FFAs, and 0.3 wt% FALEs, and FA composition in the preparation was 93.1 wt% *c9,t11*-CLA, 3.5 wt% *t10,c12*-CLA, and 0.4 wt% other CLA isomers (the content of *c9,t11*-CLA based on the total content of *c9,t11*- and *t10,c12*-isomers, 96.4 wt%) (Table 8.3). Recovery of *c9,t11*-CLA by this process was 34% of the initial content.[82] The highly purified CLA isomers may be valuable for studies of their physicochemical and physiological properties.

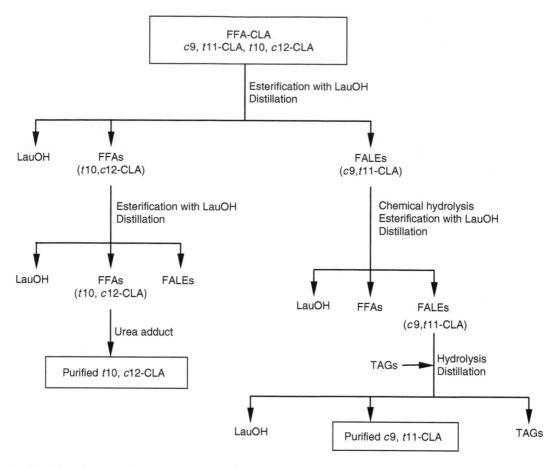

FIGURE 8.6 A process of purifying *c*9,*t*11- and *t*10,*c*12-CLA isomers. A 2.0-kg mixture of FFA-CLA/LauOH (1:1, mol/mol), 20% water, and 20 U/g-mixture of *C. rugosa* lipase was agitated at 30°C for 16 h (esterification, 47.1%). Short-path distillation of the reaction mixture fractionated into FFAs rich in *t*10,*c*12-CLA (516 g) and FALEs rich in *c*9,*t*11-CLA (751 g). The FFA fraction was esterified again at 30°C for 16 h with an equimolar amount of LauOH in a mixture containing 20% water and 30 U/g-mixture of the lipase (esterification, 20.1%). The FFA fraction recovered by distillation were finally subjected to urea adduct fractionation, resulting in recovery of 154 g of purified *t*10,*c*12-CLA. Another isomer, *c*9,*t*11-CLA, enriched in FALEs was next purified. The FALEs rich in *c*9,*t*11-CLA were hydrolyzed with NaOH in the presence of EtOH, and FFAs were recovered by *n*-hexane extraction. The FFAs was esterified at 30°C for 16 h with an equimolar amount of LauOH in a mixture containing 20% water and 10 U/g-mixture of the lipase (esterification, 53.0%), and FALEs were recovered by molecular distillation. After hydrolysis of the FALEs under alkaline conditions, safflower oil was added to the mixture and distillation was then performed. Consequently, 166 g of purified *c*9,*t*11-CLA was recovered.

TABLE 8.3   Composition in Purified Preparations of *t*10,*c*12-, and *c*9,*t*11-CLAs

| | | FFA (wt%) | | | | | |
| | LauOH | | | CLA | | | FALE |
Preparation	(wt%)	18:1	18:2	*c*9,*t*11	*t*10,*c*12	Others	(wt%)
FFA-CLA[1]	0	2.0	0.8	45.1	46.8	5.3	0
*t*10,*c*12-CLA	0.4	ND[2]	ND	3.0	95.3	1.3	ND
*c*9,*t*11-CLA	0.3	1.5	0.9	93.1	3.5	0.4	0.3

[1]Starting material that prepared by alkali conjugation of LnA (purity, 97.1%) in the presence of propylene glycol.
[2]ND, not detected.

8.6 Purification of PUFA Ethyl Ester (PUFAEE) by a Process Involving Selective Alcoholysis

8.6.1 Strategy for Purification of PUFAEE

Development of EPAEE as a pharmaceutical has necessitate an effective purification process of functional PUFAEEs. Free PUFAs, which were purified by a process involving selective esterification as described in Section 8.4, undergo 96% ethyl esterification using immobilized *Candida antarctica* lipase.[83] However, adoption of this process demands three steps: nonselective hydrolysis of PUFA-containing oil, selective esterification of the FFA mixture with LauOH, and ethyl esterification of the recovered PUFA. If selective alcoholysis of FA ethyl esters (FAEEs) with LauOH is possible, PUFAEE can be enriched through two steps: nonselective ethanolysis of PUFA-containing oil and selective alcoholysis of the resulting FAEEs. We thus drew out a strategy as shown in Figure 8.7.

The first step is the conversion of PUFA-containing oil to the corresponding ethyl esters. Of course, ethanolysis with an alkaline catalysis is possible, and sardine oil indeed undergoes chemical ethanolysis in the industrial purification of EPAEE. However, enzyme reactions under mild conditions are preferable, because heating under alkaline conditions often results in the isomerization of PUFA. FAEEs can roughly be fractionated into different carbon-chain length of FAs by short-path distillation. The second step is selective alcoholysis of the FAEE mixture, resulting in enrichment of PUFAEE in the FAEE fraction. If LauOH is used as a substrate, the reaction mixture will contain LauOH (molecular weight, 186), FAEEs (DHAEE, 328), and FALEs (>422). These compounds are easily separated by short-path distillation. Employing this strategy, we attempted to purify DHAEE from tuna oil.

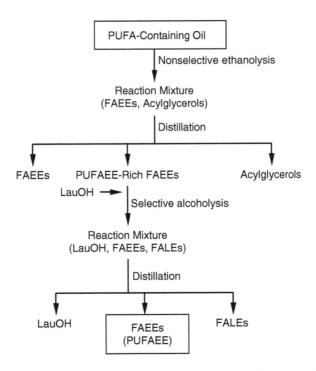

FIGURE 8.7 Strategy for purification of PUFAEE by a process involving selective alcoholysis. PUFA-containing oil undergoes nonselective ethanolysis, and the reaction mixture is subjected to short-path distillation to prepare FAEEs rich in PUFAEE. The PUFAEE fraction then undergoes selective alcoholysis with LauOH. Because FAEEs, except PUFAEEs, preferentially undergo alcoholysis, PUFAEEs can be enriched in the FAEE fraction. The fraction is finally purified by short-path distillation.

8.6.2   Purification of DHAEE from Tuna Oil

As described in detail in the following section (8.7), we have succeeded in an efficient conversion of waste edible oil to its methyl esters (biodiesel fuel) by stepwise addition of MeOH. We thus attempted to apply this reaction system to ethanolysis of tuna oil.[84] The first-step reaction was conducted at 40°C with a mixture of tuna oil and 1/3 molar equivalent of EtOH against total FAs in tuna oil using 4% immobilized *C. antarctica* lipase by weight of the reaction mixture. After 10-h reaction, EtOH was almost completely consumed and the content of FAEEs reached 33 wt%. To the reaction mixture was then added the second 1/3 molar equivalent of EtOH, and the reaction was continued for 14 h (total 24 h). The third 1/3 molar equivalent of EtOH was finally added to the reaction mixture, and the reaction was continued over 24 h (total 48 h). This three-step ethanolysis converted 96% of the oil to FAEEs. The stepwise reaction was recycled by transferring the immobilized lipase to a fresh substrate mixture, showing that the lipase preparation was used for more than 50 cycles (100 days) without significant decrease in the conversion (half-life of the lipase preparation, 65 days).[84]

The reaction mixture was subjected to short-path distillation. The fractionation enriched DHAEE from 23.7 wt% to 57.2 wt% without significant loss of DHAEE. The preparation is named DHAEE57. Selective alcoholysis of DHAEE57 with LauOH proceeded efficiently when either immobilized *R. oryzae* or *R. miehei* lipase was used in a nonaqueous system. *R. oryzae* lipase gave high recovery of DHAEE, and *R. miehei* lipase was suitable if we gave greater importance to the purity than to the recovery.[85,86] We selected immobilized *R. miehei* lipase, and continued to enrich DHAEE by the selective alcoholysis with LauOH.[87] Increasing quantities of LauOH enhanced alcoholysis over a long reaction time and improved FA specificity. Because the presence of a large amount of LauOH decreases the FAEE content in the reaction mixture, we fixed the amount of LauOH at 7 molar equivalents against FAEE. A substrate mixture of DHAEE57/LauOH was introduced to a column packed with 8.0 g of immobilized lipase at 30°C and a flow rate of 10 mL/h. The content of DHAEE in the FAEE fraction increased to 90 wt% with 58% alcoholysis. Even after 150 days, DHAEE content maintained 87 wt%, although alcoholysis decrease to 48% (half-life of the lipase preparation, 150 days). The reaction mixture flowing from the column was applied to short-path distillation, and FAEEs were recovered in a 82% yield. The FAEE fraction was contaminated with 2.4 wt% LauOH and 6.3 wt% FALEs, and FALEs could be completely removed by urea adduct fractionation. Through a series of purifications, DHAEE content was raised to 88 wt% in a 52% yield of the initial content in DHAEE57.[87]

8.7   Conversion of Waste Edible Oil to Biodiesel Fuel

A great deal of attention is being focused on the utilization of biomass with increasing interest in environment and renewable resources. In the oil and fat industry, conversion of waste edible oil and soapstock (a by-product generated in alkali refining of vegetable oils) to biodiesel fuel (FA methyl esters; FAMEs) has attracted a great deal of attention.[88,89] Our study on the production of biodiesel from waste edible oil has made the basis of ethyl esterification of DHA,[83] ethanolysis of tuna oil (Section 8.6), two-step *in situ* reaction for purification of tocopherols and sterols (Section 8.8.1), conversion of steryl ester to free sterol (Section 8.8.3), and conversion astaxanthin FA esters to free astaxanthin (Section 8.8.9). Hence, we hereafter describe enzymatic production of biodiesel fuel with immobilized *C. antarctica* lipase.

In Japan, 400,000 t of waste edible oils are discharged yearly. Half of this amount is estimated to be recycled as animal feed or raw materials for lubricants and paints. The remainder, however, is discharged into the environment. Hence, production of biodiesel fuel from waste edible oil is considered an important step in reducing and recycling waste oil. In this regard, several local governments in Japan have started collecting used frying oils from households and have converted them to biodiesel fuel for public transportation.

Industrial production of biodiesel fuel is presently performed by methanolysis of waste oil using alkaline catalysts. A by-product, glycerol, thus contains the alkali, and must be treated as a waste material. In addition, because waste oils contain a small amount of water and FFAs, the reaction generates FA alkaline salts (soaps). The soaps are removed by washing with water, which also removes glycerol, MeOH, and catalyst. Hence, disposal of the resulting alkaline water creates other environmental concerns. On the other hand, since enzymatic methanolysis of waste oil does not generate any waste materials, production of biodiesel fuel with a lipase is strongly desired.

8.7.1 Enzymatic Alcoholysis with Short-Chain Alcohol

Mittelbach[90] first conducted alcoholysis of sunflower oil with MeOH and EtOH using lipases from *Pseudomonas, Candida,* and *Mucor* in petroleum ether. Several groups have reported that enzymatic alcoholysis with longer than C3 of alcohols efficiently proceeds even in an organic solvent-free system,[91,92] but that methanolysis and ethanolysis of TAGs requires addition of *n*-hexane.[92,93] As stated in Section 8.6, we have also confirmed that alcoholysis of FAEEs with LauOH does not require organic solvent.

Alcoholysis of TAGs with short-chain alcohols in an organic solvent-free system was first attempted by Nelson et al.,[92] and ethanolysis of beef tallow with *R. miehei* lipase reached 65.5%. Using the same lipase, sunflower oil was also converted to their ethyl esters at >80% conversion.[94] In addition, the use of immobilized *C. antarctica* lipase led to success in efficient ethanolysis of sardine oil, although the reuse of the lipase was not studied.[53,95] However, because methanolysis system without organic solvent had not been reported, we attempted to develop an organic solvent-free methanolysis system.

8.7.2 Inactivation of Lipase by Insoluble MeOH

In general, lipases efficiently catalyze the reactions when the substrates dissolve each other. Preliminary investigation showed that the solubility of MeOH and EtOH in vegetable oil is only 1/2 and 2/3 of the stoichiometric amount, respectively. Disregarding the low solubility, all alcoholyses of TAGs so far reported were conducted with more than the stoichiometric amount of MeOH and EtOH. Proteins generally are unstable in short-chain alcohols, such as MeOH and EtOH. We thus hypothesized that low methanolysis (ethanolysis) is due to the inactivation of lipases by contact with insoluble MeOH (EtOH), which exists as drops in the oil. Actually, MeOH was completely consumed in methanolysis of vegetable oil with <1/3 molar equivalent of MeOH for the stoichiometric amount using immobilized *C. antarctica* lipase, but the methanolysis was decreased significantly by adding >1/2 molar equivalent of MeOH (Figure 8.8). In addition, the decreased activity did not restore in subsequent reaction with 1/3 molar equivalent of MeOH, showing that the immobilized lipase was irreversibly inactivated by contact with insoluble MeOH in the oil.[96]

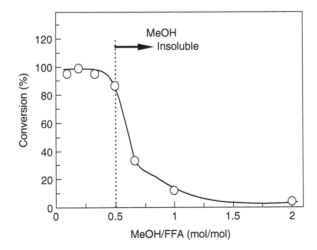

FIGURE 8.8 Methanolysis of vegetable oil with different amounts of MeOH using immobilized *C. antarctica* lipase. A mixture of 10 g vegetable oil/MeOH, 0.4 g immobilized lipase was shaken at 30°C for 24 h. The conversion is expressed as the amount of MeOH consumed for the ester conversion of the oil (when the molar ratio of MeOH/oil is <1.0), and as the ratio of FAMEs to oil (more than 1.0).

Incidentally, it was recently reported that the same immobilized lipase was efficiently catalyzed ethanolysis of TAGs with *ca.* 60 moles of EtOH, and that FAs at the 1,3-positions of TAGs were preferentially converted to FAMEs.[97,98] This result is inconsistent with our result that the immobilized lipase was inactivated by >2/3 moles of EtOH. Although we recycled the ethanolysis by transferring the enzyme to a fresh substrate mixture (TAGs/EtOH = 1:3, wt/wt), the lipase was confirmed to be stable. These results show that the lipase is stable in the presence of very large amounts of EtOH (MeOH).

8.7.3 Stepwise Batch Methanolysis of Vegetable Oil

At least a stoichiometric amount is required for the complete conversion of TAGs to their corresponding FAMEs. Immobilized *C. antarctica* lipase, however, was inactivated by adding >1/2 molar equivalent of MeOH for the stoichiometric amount, and the methanolysis ceased. Hence, we attempted the methanolysis of vegetable oil by three successive addition of 1/3 molar equivalent of MeOH (Figure 8.9). The reaction was started in a mixture containing 1/3 moles of MeOH, and the second and third 1/3 molar equivalents of MeOH were added after reaching steady state; 10 and 24 h, respectively. After 48 h in total, the conversion reached 97.3%, showing that the three-step methanolysis was effective for nearly complete conversion of the oil.[96]

Solubility of MeOH in TAG is low, but that in FAME is high. The first-step reaction product was composed of acylglycerols and 33% FAMEs, and 2/3 molar equivalent of MeOH against total FAs was completely soluble in the reaction mixture. Immobilized *C. antarctica* lipase was found not to inactivate even in the mixture of acylglycerols/FAMEs and the amount of MeOH. This finding led to success in a two-step methanolysis of TAG. Figure 8.9 shows a typical time course. The first-step methanolysis was started in a mixture containing 1/3 molar equivalent of MeOH. After the reaction reached steady state, the second 2/3 molar amount of MeOH was added. The conversion consequently reached 97% after 34 h in total.[99] These results showed that two-step methanolysis is effective for saving the reaction period. In addition, these reactions achieved a high degree of methanolysis using only an equimolar amount of MeOH, showing that this lipase recognizes the products, FAMEs, very weakly. Poor activity of a lipase toward FAME (FAEE) is the base for reaction systems described in Sections 8.8 and 8.9.

To study the stability of the lipase preparation, the two- and three-step methanolyses were repeated. More than 95% conversion was maintained during 52 cycles in three-step methanolysis (104 days), and

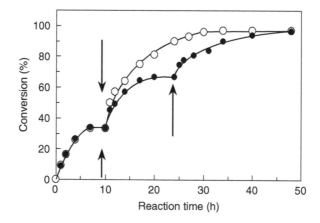

FIGURE 8.9 Time courses of stepwise methanolysis of vegetable oil. Three-step reaction (●): a mixture of 28.95 g vegetable oil, 1.05 g MeOH (1/3 molar equivalent for the stoichiometric amount), and 4 wt% immobilized *C. antarctica* lipase was incubated at 30°C with shaking, and 1.05 g MeOH was added at 10 h and 24 h. Two-step reaction (○): the first-step reaction was conducted under the same conditions as those of three-step reaction. After 10 h reaction, 2.10 g MeOH (2/3 molar equivalent) was added to the reaction mixture. Upward and downward arrows indicate the addition of 1/3 and 2/3 molar equivalents of MeOH, respectively.

during 70 cycles in two-step methanolysis (105 days).[99] These results indicated that *C. antarctica* lipase is very stable in these stepwise methanolysis systems.

8.7.4 Production of Biodiesel Fuel from Waste Edible Oil

Our aim is enzymatic conversion of waste edible oil to biodiesel fuel. A fresh vegetable oil and its waste differ significantly in water and FFA contents. A waste edible oil contained *ca.* 0.2 wt% water, 2.5 wt% FFAs, and 4.6 wt% partial acylglycerols. We reported earlier that >0.05 wt% water decreased the velocity of methanolysis of vegetable oil with immobilized *C. antarctica* lipase but did not affect the equilibrium of the reaction.[96] When the waste oil underwent methanolysis with 1/3 molar amount of MeOH, the conversion velocity was slow as expected. But the reuse of the immobilized enzyme increased the reaction velocity. This increase can be explained as follows. Water initially present in the waste oil transfers into the glycerol layer generated by methanolysis. Because the water goes out of the methanolysis system, the reaction velocity gradually increased. Actually, the water content in FAMEs/acylglycerols (oil layer) decreased to 0.05 wt%, and that of the glycerol layer was 4.1 wt% after five cycles.[100]

Three-step methanolysis of the waste oil was conducted. Time course of the first- and second-step methanolyses of the waste oil was completely the same as those of fresh vegetable oil. However, the conversion of waste oil reached 90.4% after the three-step methanolysis, although that of vegetable oil was 95.9% under the same conditions. The difference may be attributed to the oxidized FA compounds in waste oil. In general, when a vegetable oil is used for frying, some FAs are converted to peroxides, aldehydes, polymers, etc., by oxidation or thermal polymerization.[101,102] Because the lipase did not recognize these oxidized compounds, the conversion of waste oil decreased a little. In addition, the FFA content in the product after the three-step reaction (0.3 wt%) was lower than that in the waste oil (2.5 wt%), indicating that methylation of FFA occurred along with methanolysis of the oil.

To investigate the lipase stability, the three-step batch methanolysis was repeated by transferring the lipase to a fresh substrate (one cycle, 48 h). The conversion did not significantly decrease even after 50 cycles (100 days), showing that contaminants in waste oil did not affect the stability of the lipase preparation.[100]

8.8 Purification of Tocopherols, Sterols, and Steryl Esters

Natural tocopherols present in oilseeds are useful antioxidants. The α-isomer is used as a pharmaceutical substance, an ingredient in cosmetics, and a nutraceutical food. A mixture of α-, δ-, and γ-tocopherols (total content, *ca.* 60 wt%) is widely used as an additive to various kinds of oil and fat foods. Sterols and their FA esters (referred to as steryl esters) are known to reduce blood cholesterol level[102–104] and are added to salad oil and margarine. Tocopherols, sterols, and steryl esters are major components in deodorizer distillate by-produced in the deodorization step of vegetable oil refining. At present, tocopherols are purified from vegetable oil deodorizer distillate (VODD) by a combination of chemical methyl esterification of FAs, molecular (short-path) distillation, MeOH (EtOH) fractionation, ion exchange chromatography, and so on. Sterols are also purified from by-product in purification of tocopherols by fractionation with organic solvents, but the yield is not high. Meanwhile, a procedure for purifying steryl esters has not been developed, and all steryl esters are wasted.

A great deal of attention is focused on cholesterol-lowering effect of sterols and steryl esters, and salad oil including sterols and margarine with steryl esters has been commercialized. Application of these compounds strongly desires an efficient purification of sterols and steryl esters from VODD, resulting in development of a new process involving lipase-catalyzed reactions.

8.8.1 Purification of Tocopherols and Sterols

8.8.1.1 Strategy

A process comprising short-path distillation and EtOH fractionation is industrially adopted for purification of tocopherols from VODD. However, contamination of MAGs, DAGs, and sterols disturbs the purification of tocopherols, and the recovery is not satisfactory. If contaminants could be converted to other forms of which molecular weights are different, tocopherols would highly purify with higher recovery. We thus attempted conversion of sterols to steryl esters, and simultaneously of FFAs and acylglycerols to FAMEs. It should be favorable to convert FFA to FAME because boiling point of FAME is lower than that of FFA. Consequently, tocopherols can be efficiently purified by short-path distillation. Outline of purification of tocopherols according to this strategy is shown in Figure 8.10.

Our aim is to develop a process of efficiently purifying not only tocopherols, but also sterols. If components of which the boiling points are higher than tocopherols are only steryl esters, they are easily enriched to high purity in high yield. Therefore, the first step is fractionation of VODD into low and high boiling point substances. The low boiling point fraction includes FFAs, tocopherols, sterols, MAGs, and a part of DAGs, but not steryl esters. On the other hand, the high boiling point fraction includes steryl esters, DAGs, and TAGs.

Efficient removal of sterols is most strongly required for improvement of a conventional process of purifying tocopherols; thus, conversion of sterols to steryl esters should take precedence over conversion of FFAs to FAMEs. Preliminary experiment showed that a lipase treatment of low boiling point fraction in the presence of water and MeOH led to strong hydrolysis of acylglycerols and conversion of FFAs to FAMEs, but to poor esterification of sterols with FFAs. Hence, two-step *in situ* reaction is attempted: after esterification of sterols with FFAs is conducted, FFAs are converted to FAMEs by the addition of MeOH.

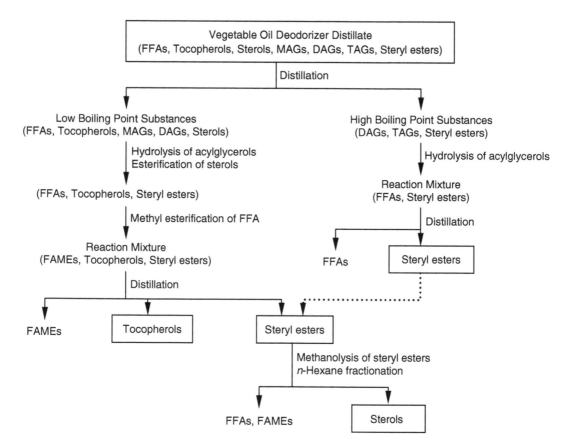

FIGURE 8.10 Strategy for purification of tocopherols, sterols, and steryl esters from VODD. VODD is fractionated into low boiling point substances (FFAs, MAGs, DAGs, tocopherols, and sterols) and high boiling point substances (DAGs, TAGs, and steryl esters). FFAs and acylglycerols in low boiling point fraction are converted to FAMEs by a lipase treatment and sterols are converted to steryl esters. The treatment does not, however, convert tocopherols to other molecular forms. After the treatment, FAMEs, tocopherols, and steryl esters are fractionated by distillation. On the other hand, high boiling point substances are used as materials for purification of steryl esters. Only acylglycerols in the fraction are hydrolyzed with a lipase, and the resulting FFAs and steryl esters are fractionated by distillation. Free sterols can be purified by *n*-hexane fractionation after conversion of steryl esters to free sterols through a lipase-catalyzed methanolysis.

The esterification of sterols and hydrolysis of acylglycerols proceed efficiently in the presence of water. The addition of MeOH after reaching a steady state in the esterification converts FFAs to FAMEs, and steryl esters synthesized does not convert to free sterols. The resulting reaction mixture involves FAMEs, tocopherols, and steryl esters. The three components can be separated by short-path distillation.

8.8.1.2 Purification of Tocopherols and Sterols from Soybean Oil Deodorizer Distillate (SODD)

Tocopherols and sterols were purified from SODD according to the strategy shown in Figure 8.10. SODD was first distilled at 240°C and 0.02 mm Hg, and was fractionated into low boiling point fraction (soybean oil deodorizer distillate tocopherol/sterol concentrate; SODDTSC) and high boiling point fraction (soybean oil deodorizer distillate steryl ester concentrate; SODDSEC). SODDTSC was used as a material for purification of tocopherols and sterols. Meanwhile, SODDSEC was used for the purification of steryl esters that is described in Section 8.8.2.

Purification of tocopherols and sterols from SODDTSC is summarized in Table 8.4. Esterification of sterols with *C. rugosa* lipase most efficiently proceeded in the presence of 20% water,[105,106] showing that the physical conditions of the reaction mixture might be the most suitable for the reaction. The esterification reached steady state after 16 h, *ca.* 80% of sterols were converted steryl esters. Acylglycerols (content in SODDTSC, 4.1 wt%) were simultaneously hydrolyzed in this treatment, and completely disappeared after 5 h. After 16-h reaction, 2 molar equivalents of MeOH against FFAs were added to the reaction mixture, and the treatment was further continued for 24 h (total, 40 h). The two-step *in situ* reaction decreased the contents of sterols from 12.0 wt% to 2.5 wt%, and the content of FFAs from 47.4 wt% to 11.6 wt%. Conversion of sterols to steryl esters can further be increased by repeating the reaction after removal of steryl esters. Hence, the reaction mixture was fractionated into FAME fraction (distillate 1-1), tocopherol fraction (distillate 1-2), and steryl ester fraction (residue 1-2) by short-path distillation.

Distillate 1-2 was treated under similar conditions to those of the first two-step *in situ* reaction. In this reaction, 56% of sterols were converted to steryl esters, showing that 91% sterols in SODDTSC were

TABLE 8.4 Purification of Tocopherols and Sterols by a Process Comprising Two-Step *in Situ* Enzymatic Reaction and Molecular Distillation

Step	Weight (g)	FFA (wt%)	FAME (wt%)	Tocopherol (wt%)	Tocopherol (g)	Sterol (wt%)	Sterol (g)	Steryl ester (wt%)	Steryl ester (g)
SODDTSC	3560	47.4	ND[1]	17.2	612	12.0	427	<0.1	—
Lipase treatment 1[2]	3520	11.6	35.8	17.0	598	2.5	88	15.3	539
Distillation 1									
Distillate 1–1[3]	1320	7.7	68.2	<0.1	—	<0.1	—	<0.1	—
Distillate 1–2[4]	1620	15.2	24.2	36.1	585	5.9	96	0.5	8
Residue 1–2[4]	522	0.3	<0.1	0.7	3	<0.1	–	97.6	509
Lipase treatment 2[5]	1570	6.4	33.1	35.7	561	2.6	41	6.2	97
Distillation 2									
Distillate 2–1[3]	684	11.1	75.3	<0.1	—	<0.1	—	<0.1	—
Distillate 2–2[6]	95	11.3	26.2	34.8	33	1.1	1	<0.1	—
Distillate 2–3[4]	675	1.8	<0.1	76.4	516	5.4	37	0.2	1
Residue 2–3[4]	99	0.2	<0.1	2.8	2.8	1.0	1	95.8	95

[1]ND, not detected

[2]A mixture of 3560 g SODDTSC, 890 g water, and 200 U/g-mixture of *C. rugosa* lipase was agitated at 30°C for 16 h. To the reaction mixture was added 2 molar equivalents of MeOH against FFAs, and the reaction was then continued for 24 h (total, 30 h).

[3]Distilled at 160°C and 0.2 mm Hg.

[4]Distilled at 240°C and 0.02 mm Hg.

[5]A mixture of 1620 g distillate 1–2, 405 g water, and 200 U/g-mixture of *C. rugosa* lipase was agitated at 30°C for 16 h. To the reaction mixture was added 2 molar equivalents of MeOH against FFAs, and the reaction was then continued for 24 h (total, 30 h).

[6]Distilled at 175°C and 0.02 mm Hg.

converted to steryl esters by the treatment twice. In addition, the ratio of FFAs/FAMEs in the repeated reaction reached 1:5.7 (wt/wt). To purify tocopherols, the reaction mixture was subjected to short-path distillation. FFAs and FAMEs were first removed in distillate 2-1 without a loss of tocopherols. The remaining FFAs and FAMEs were then removed almost completely in distillate 2-2. The residue contained mainly tocopherols and steryl esters, and they were fractionated into tocopherol fraction (distillate 2-3) and steryl ester fraction (residue 2-3). A series of procedures purified tocopherols to 76.4% in a 84% yield.[106] Distillate 2-2 contained 34.8 wt% tocopherols. If this fraction was recycled to next time distillation, the recovery of tocopherols would increase to nearly 90%. Meanwhile, steryl esters were recovered in residues 1-2 and 2-3. The purity of steryl esters in the mixture of two fractions was 97.2%, and the recovery was 86.3% for the sterol content in SODDTSC.[106] These results show that the process comprising two-step *in situ* reaction and molecular distillation is very effective for purification of not only tocopherols but also sterols.

8.8.2 Purification of Steryl Esters

Steryl esters in high boiling point fraction (SODDSEC; Figure 8.10) can easily be purified by a process involving a lipase reaction. Lipases hydrolyzed acylglycerols strongly, but steryl esters poorly in the presence of excess amounts of FFAs. SODDSEC was mainly composed of DAGs, TAGs, and steryl esters. Hence, steryl ester should be purified by hydrolysis of acylglycerols with a lipase, followed by distillation of the reaction mixture. Steryl esters were indeed purified according to this strategy shown in Figure 8.10. The purification is summarized in Table 8.5. A mixture of equal weights of SODDSEC and water was treated with *C. rugosa* lipase. Acylglycerols were almost completely hydrolyzed, although the content of steryl esters did not change. Distillation of the reaction mixture successfully purified steryl esters to 97.3% with 87.7% recovery.[107]

Hydrolysis of steryl esters with *C. rugosa* lipase in the presence of 50% water reached a steady state at 50% hydrolysis. But the ratio of sterols to steryl esters did not change before and after hydrolysis: 1:12 (wt/wt). This phenomenon is due to the fact that the ratio of sterols to steryl esters reached 1:12 (wt/wt) at the equilibrium state in the reaction with the equal weights of FFAs and steryl esters. It is also clarified that approximately 25% of constituent FAs in steryl esters are exchanged with FAs originating from acylglycerols. However, because FA composition in steryl esters is almost the same as that in acylglycerols, the FA composition in steryl esters is almost the same before and after the reaction.[107]

8.8.3 Conversion of Steryl Esters to Free Sterols

All sterols and steryl esters in VODD were recovered as steryl esters through the process described in Sections 8.8.1 and 8.8.2. As the oil and fat industry demands both of sterols and steryl esters, conversion of steryl esters to sterols was attempted.

We first conducted hydrolysis of steryl esters with various lipases. *C. rugosa, G. candidum, P. aeruginosa, P. stutzeri, Burkholderia glumae,* and *B. cepacia* hydrolyzed steryl esters, but their reactions reached steady

TABLE 8.5   Purification of Steryl Esters from SODDSEC

Step	Weight (g)					
	Total	FFA	DAG	TAG	Sterol	Steryl ester
SODDSEC[1]	1000	14	112	321	40	452
Hydrolysis[2]	910	410	ND[3]	ND	36	421
Distillation[4]	409	1.9	0.5	1.8	2.6	396

[1]Prepared from SODD as described in section 8.8.1.2.
[2]A mixture of 1.0 kg SODDSEC, 1.0 kg water, and 15 U/g-mixture of *C. rugosa* lipase was agitated at 30°C for 20 h.
[3]Not detected.
[4]Residue after removing FFAs by two-step distillation: at 180°C and 0.2 mm Hg and at 250°C and 0.02 mm Hg.

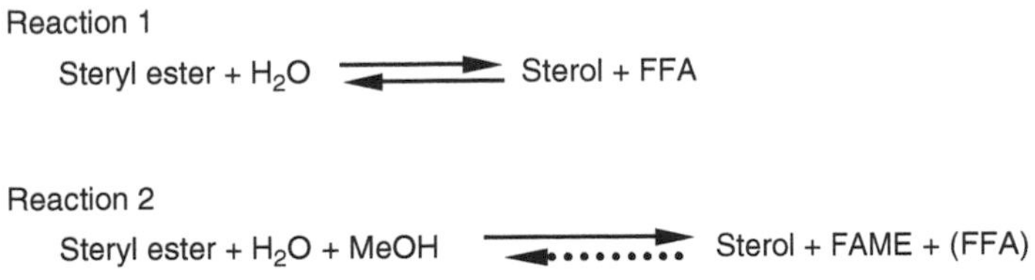

Reaction 1

Steryl ester + H$_2$O ⇌ Sterol + FFA

Reaction 2

Steryl ester + H$_2$O + MeOH ⇌ Sterol + FAME + (FFA)

FIGURE 8.11 Strategy for increase in conversion of steryl ester to sterol. Reaction 1 (without MeOH addition): Because lipase acts on sterol and FFA as strongly as on steryl ester, this reaction is reversible. Reaction 2 (with MeOH addition): Reaction products are sterol, FAME, and small amounts of FFA. Because FAME does not participate in synthesis of steryl ester, the equilibrium of the reaction shifts to accumulation of free sterols.

state at 50% hydrolysis, owing to the reversible reaction. Meanwhile, we have experienced from Sections 8.4, 8.5, 8.7, and 8.8.1.2 that lipases act more weakly on FAME and FALE than FFA. If reaction products, FFAs, in hydrolysis of steryl esters are converted to FAMEs, the reverse reaction (esterification of sterols) will scarcely occur because FAMEs does not participate in the reaction (Figure 8.11). Consequently, the equilibrium of the reaction shifts to accumulation of free sterols.

According to this strategy, conversion of OA steryl esters (OASEs) to free sterols was studied. The screening tests showed that *P. aeruginosa* lipase was suitable for this reaction, and that MeOH, butanol, octanol, and decanol are effective substrates. In addition, the velocity reached a constant value in the presence of >10% water, and the conversion was maximum at 10% water. Based on these results, the reaction conditions were determined as follows: the ratio of OASE/MeOH, 1:2 (mol/mol); water, 10%; amount of the lipase, 20 U/g-mixture; temperature, 30°C; and reaction period, 72 h.[108]

The next step is purification of free sterols from the reaction mixture, which mainly contains sterols and FAMEs. The melting point of sterols is >125°C, showing that purification of sterols by short-path distillation is difficult. Meanwhile, sterols are poorly soluble in organic solvents (*n*-hexane, MeOH, EtOH etc.) but FAMEs and FFAs are readily soluble in these solvents. Hence, sterols were purified by *n*-hexane fractionation. Purification of sterols from OASEs is summarized in Table 8.6. The recovery of sterols through a series of procedures was 87.2% of the initial content in OASEs (purity, 98.9%), showing that

TABLE 8.6 Purification of Sterols from OASEs by a Process Comprising Enzyme Reaction and *n*-Hexane Fractionation

Step	Composition (g)				
	Total	FFA	FAME	Sterol	OASE
Enzyme reaction[1]					
Before	90.4	0	0	3.5	83.7
After (oil layer)	86.9	1.7	31.3	51.1	1.6
Hexane fractionation[2]					
Supernatant	36.6	1.7	30.1	2.1	1.6
Precipitate	47.6	ND[3]	ND	47.1	ND

[1]Purity of OASEs was 92.6%. The composition of constituent sterols of OASEs was 4.0 wt% brassicasterol, 28.3 wt% campesterol, 17.0 wt% stigmasterol, and 50.7 wt% β-sitosterol; the FA composition of OASEs was 0.9 wt% palmitic acid, 3.1 wt% stearic acid, 90.0 wt% OA, and 4.4 wt% LnA. A 110-g mixture of OASEs/MeOH (1:2, mol/mol), 10% water, and 20 U/g-mixture of *P. aeruginosa* lipase was stirred at 30°C for 72 h (conversion, 98.1%)

[2]After separation of the reaction mixture into oil and water layers, *n*-hexane (450 mL) was added to the oil layer (86.9 g), and the suspension was kept at −20°C for 2 h with occasional agitation. The supernatant and precipitate were separated by centrifugation at 0°C. *n*-Hexane was removed under reduced pressure.

[3]Not detected.

n-hexane fractionation was effective for purification of sterols from the reaction mixture obtained by enzymatic conversion of steryl esters to sterols.[108] These procedures are also added to the overall process for purifying useful materials in VODD, which is shown in Figure 8.10.

8.9   Purification of Astaxanthin (Ax)

Ax (3,3'-dihydroxy-β,β-carotene-4,4'-dione) is a red pigment that is contained in marine products, such as salmon, shrimp, and crab. Ax is a precursor of vitamin A,[109] and has been reported to have various physiological activities: a quencher against free radicals and active oxygen species,[110,111] an anticancer agent,[112,113] an enhancer of the immune response,[114] and inhibitor of *Helicobacter pylori* infection.[115] These activities attracted a great deal of attention, and Ax has been used as a nutraceutical food, an ingredient in cosmetics, and a supplement for pigmentation of cultured fishes and shellfishes.

A commercially available Ax is prepared from *Haematococcus pluvialis* cells by acetone extraction. However, the content of Ax in the acetone extracts is not high (*ca.* 15 wt%), and the intrinsic odor also reduces the commercial value. Hence, enrichment of Ax and deodorization was attempted.

The composition in a commercially available cell extracts form *H. pluvialis* is shown in Table 8.7. Ax existed as free form, and FA mono and diesters, and the total content was 14.6 wt%. Meanwhile, most of contaminants were FFAs and acylglycerols, of which the total content reached 66.5 wt%. In addition, constituent FAs in acylglycerols were mainly C16 and C18. These facts shows that a process composed of a lipase-catalyzed hydrolysis and molecular distillation should be effective for enrichment of Ax (Figure 8.12). Because the process contains distillation, the odor should also be removed. Treatment of *H. pluvialis* cell extracts with *C. rugosa* lipase in the presence of 50% water hydrolyzed 97% of acylglycerols, but not Ax FA esters at all. The oil layer separated from the reaction mixture was applied to short-path distillation, and FFAs were removed in the distillate. Ax and its esters were recovered in the residue with a recovery of 92.5%, and the total content of Ax increased to 40.8 wt% (Table 8.8).[116]

The residue contained *ca.* 50% of unknown contaminants. Further purification was attempted by taking advantage of properties that free Ax is not soluble in *n*-hexane but Ax esters are soluble. Ax esters in the concentrate were first hydrolyzed with several lipases, but the content of free Ax did not exceed 40 mol% of total Ax content. Conversion of Ax esters to its free form by alcoholysis was then attempted (Figure 8.12). *P. aeruginosa* lipase was the most effective among lipases tested, and addition of C1-10 fatty alcohols successfully converted Ax esters to its free form. Based on studies of several factors affecting the reaction, the reaction conditions were decided as follows: Ax concentrate/water, 1:1 (wt/wt); water content, 50%; EtOH amount, 5 molar equivalents for FAs in Ax concentrate; amount of *P. aeruginosa* lipase, 1500 U/g-mixture; temperature, 30°C; reaction time, 68 h. Free Ax amount reached 89.3% of total Ax amount by the lipase treatment under these conditions. To purify Ax from the reaction mixture, ten volume parts of *n*-hexane were added to the oil layer, and insoluble free Ax was recovered by centrifugation. A series of procedures purified Ax to 70% with 64% recovery of the initial content of Ax in *H. pluvialis* cell extracts.[116] The purification is summarized in Table 8.8.

TABLE 8.7   Composition of Acetone Extracts from *H. pluvialis*

Component	Molecular Weight	Content (wt%)
FFA	256–282	24.9
MAG	330–356	4.3
DAG	568–620	8.0
TAG	806–884	29.3
Astaxanthin[1]		
Free form	597	0.5
Monoesters	853–861	11.3
Diesters	1073–1125	2.8

[1]Astaxanthin is a mixture of 5 mol% free form, 80 mol% monoesters,
   and 15 mol% diesters.

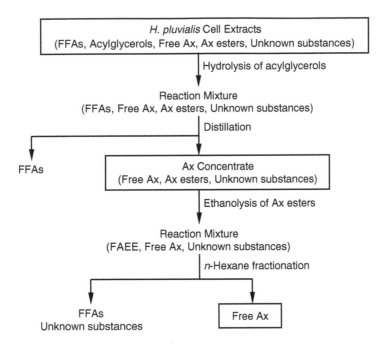

FIGURE 8.12 Strategy for purification of Ax from *H. pluvialis* cell extracts. Acylglycerols in the cell extract is hydrolyzed with *C. rugosa* lipase, and the reaction mixture is then subjected to short-path distillation. Free Ax and Ax esters are recovered in the residue. Ax esters in the residue undergoes ethanolysis with *P. aeruginosa* lipase and is converted to free Ax. The resulting free Ax is purified by *n*-hexane fractionation.

TABLE 8.8 Enrichment of Ax from *H. pluvialis* Cell Extracts by a Process including Lipase Reactions

Step	Total (g)	FFA (g)	FAEE (g)	Acylglycerol (g)	Ax Total[1] (wt%)	Free (g)	Monoester (g)	Diester (g)	Yield (%)
Cell extracts	1000	249	0	416	14.6	5.0	113.1	27.9	100
Hydrolysis[2]	871	637	0	19	15.8	4.7	106.6	26.3	94.2
Distillation[3]	331	10	0	14	40.8	4.6	104.6	25.8	92.5
Ethanolysis[4]	328	21	67	ND[5]	29.3	80.1	14.6	1.3	90.8
Hexane extraction[6]	94	ND	ND	ND	69.7	60.2	4.6	0.7	64.4

[1]Total content of free Ax and Ax mono and diesters.

[2]A mixture of 1.0 kg *H. pluvialis* cell extracts, 1.0 kg water, and 50 U/g-mixture of *C. rugosa* lipase was agitated at 30°C for 48 h.

[3]Reaction mixture (oil layer) was applied to short-path distillation. FFAs were removed in the distillates by two-step distillation: at 180°C and 0.2 mm Hg; and at 200°C and 0.2 mm Hg. Ax was recovered in the residue.

[4]A mixture of the distillation residue/water (1:1, wt/wt), 5 moles of EtOH for FAs in Ax esters, and 1500 U/g-mixture of *P. aeruginosa* lipase was agitated at 30°C for 48 h.

[5]Not detected.

[6]To the reaction mixture (oil layer) was added 10 volume parts of *n*-hexane. The resulting precipitate was recovered by centrifugation, and was then dried under reduced pressure.

8.10 Conclusion

It is only natural that repression of reverse reaction becomes a key factor for efficient production of a desired compound. To achieve this purpose, the following lipase-catalyzed reaction systems have been proposed: (1) repeated reaction after removal of a product from a reaction mixture; (2) removal of product using a membrane bioreactor; (3) removal of products (water and short-chain alcohols) under reduced pressure;

(4) solidification of product at low temperature: MAGs solidified at low temperature do not participate in the reaction, resulting in efficient accumulation in the reaction mixture; (5) synthesis of a product that the enzyme poorly recognizes: a high degree of esterification of FAs can be obtained using fatty alcohols and sterols as substrates; (6) conversion of a product (FA) to its alcohol ester (wax ester). In this chapter, we have described several examples regarding the first, fifth, and sixth systems. These reaction systems are based on the strategy that a product participating in the reaction is eliminated from the reaction system.

In addition, relatively easy purification can be obtained if contaminants or a desired compound is converted to different molecular forms by selective reaction with a lipase. When the difference in molecular sizes between a desired material and contaminants becomes large, distillation is useful for the purification; and when their solubility in an organic solvent is different, the solvent fractionation is preferable. In the application studies, the downstream purification procedure is very important. An enzyme reaction should therefore be developed by considering the purification process.

We have described purifications of useful oil- and fat-related compounds by processes including lipase-catalyzed reactions. Though we did not describe them in this chapter, lipase reactions are also useful for production of value-added lipids, such as structured TAGs and functional MAGs. In the future, we hope that lipase reactions will further be applied to the purification of useful materials as well as the improvement of oils and fats.

References

1. K Yokozeki, S Yamanaka, K Takinami, Y Hirose, A Tanaka, K Sonomoto, S Fukui. Application of immobilized lipase to regio-specific interesterification of triglyceride in organic solvent. *Eur J Appl Microbiol Biotechnol* 14:1–5, 1982.
2. CC Akoh, X Xu. Enzymatic production of Betapol and other specialty fats. In: TM Kuo, HW Gardner, Eds. *Lipid Biotechnology,* New York: Marcel Dekker, 2002, pp. 461–478.
3. T Nagao, Y Shimada, A Sugihara, A Murata, S Komemushi, Y Tominaga. Use of thermostable *Fusarium heterosporum* lipase for production of structured lipid containing oleic and palmitic acids in organic solvent-free system. *J Am Oil Chem Soc* 78:167–172, 2001.
4. K Maruyama, M Nishikawa. Function of fish oil and its application to foods. *Food Chemicals* 1995(4):31–37, 1995.
5. T Nagao, H Watanabe, N Goto, K Onizawa, H Taguchi, N Matsuo, T Yasukawa, R Tsushima, H Shimasaki, H Itakura. Dietary diacylglycerol suppresses accumulation of body fat compared to triacylglycerol in men in a double-blind controlled trial. *J Nutr* 130:792–797, 2000.
6. S Negishi, S Shirasawa, Y Arai, J Suzuki, S Mukataka. Activation of powdered lipase by cluster water and the use of lipase powders for commercial esterification of food oils. *Enz Microb Technol* 32:66–70, 2003.
7. BF Haumann. Structured lipids allow fat tailoring. *INFORM* 8:1004–1011, 1997.
8. CC Akoh, S Sellappan, LB Fomuso, VV Yankah. Enzymatic synthesis of structure lipids. In: TM Kuo, HW Gardner Eds. *Lipid Biotechnology,* New York: Marcel Dekker, 2002, pp. 433–460.
9. Y Shimada, A Sugihara, Y Tominaga. Production of functional lipids containing polyunsaturated fatty acids with lipase. In: UT Bornscheuer, Ed. *Enzymes in Lipid Modification,* Weinheim: Wiley-VCH, 2000, pp. 128–147.
10. T Yamane. Lipase-catalyzed synthesis of structure triacylglycerols containing polyunsaturated fatty acids – Monitoring of the reaction and increasing the yield. In: UT Bornscheuer, Ed. *Enzymes in Lipid Modification,* Weinheim: Wiley-VCH, 2000, pp. 148–169.
11. X Xu. Enzymatic production of structured lipids: process reactions and acyl migration. *INFORM* 11:1121–1131, 2000.
12. A Zaks, AM Klibanov. Enzymatic catalysis in organic media at 100 degrees C. *Science* 224:1249–1251, 1984.
13. RN Patel. Enzymatic preparation of chiral pharmaceutical intermediates by lipases. In: TM Kuo, HW Gardner, Eds. *Lipid Biotechnology,* New York: Marcel Dekker, 2002, pp. 527–562.

14. AM Klibanov. Asymmetric transformations catalyzed by enzymes in organic solvents. *Acc Chem Res* 23:114–120, 1990.

15. H Matsumae, M Furui, T Shibatani, T Tosa. Production of optically active 3-phenylglycidic ester by the lipase from *Serratia marcescens* on a hollow-fiber membrane reactor. *J Ferment Bioeng* 78:59–63, 1994.

16. Y Shimada, A Sugihara, Y Tominaga. Enzymatic purification of polyunsaturated fatty acids. *J Ferment Bioeng* 91:529–538, 2001.

17. KD Mukherjee. Fractionation of fatty acids and other lipids using Lipase. In: UT Bornscheuer, Ed. *Enzymes in Lipid Modification*, Weinheim: Wiley-VCH, 2000, pp. 23–45.

18. GG Haraldosson. Enrichment of lipids with EPA and DHA by lipase. In: UT Bornscheuer, Ed. *Enzymes in Lipid Modification*, Weinheim: Wiley-VCH, 2000, pp. 170–189.

19. K Hara. Pharmaceutical application of eicosapentaenoic acid. *Yushi* 46(1):91–99, 1993 (in Japanese).

20. ML Garg, EB Bosschieter, CDL Coulander. The inverse relation between fish consumption and 20-year mortality from coronary heart disease. *New Engl J Med* 312:1205–1209, 1985

21. T Kanayasu, I Morita, J Nakano-Hayashi, N Asuwa, N Fujisawa, T Ishii, M Ito, S Murota. Eicosapentaenoic acid inhibits tube formation of vascular endothelial cells *in vitro*. *Lipids* 26:271–276, 1991.

22. BE Phillipson, DW Rothrock, WE Connor, WS Harris, DR Illingworth. Reduction of plasma lipids, lipoproteins, and apoproteins by dietary fish oils in patients with hypertriglyceridemia. *New Engl J Med* 312:1210–1216, 1985.

23. T Lee, RL Hoover, JD Williams, RI Sperling, J Ravalese, BW Spur, DR Robinson, EJ Corey, RA Lewis, KF Austen. Effect of dietary enrichment with eicosapentaenoic and docosahexaenoic acids on *in vitro* neutrophil and monocyte leukotriene generation and neutrophil function. *New Engl J Med* 312:1217–1224, 1985.

24. MG Bravo, RJ Antueno, J Toledo, ME Tomas, OF Mercuri, C Quintans. Effect of an eicosapentaenoic and docosahexaenoic acid concentrate on a human lung carcinoma grown in nude mice. *Lipids* 26:866–870, 1991.

25. W Stillwell, W Ehringer, LJ Jenski. Docosahexaenoic acid increases permeability of lipid vesicles and tumor Cell. *Lipids* 28:103–108, 1993.

26. R Uauy, I De Andraca. Human milk and breast feeding for optimal mental development. *J Nutr* 125:2278–2280, 1995.

27. M Nuringer, WE Conner, CV Petten, L Barstad. Dietary omega-3 fatty acid deficiency and visual loss in infant rhesus monkeys. *J Clin Invest* 73:272–276, 1984.

28. SE Carlson, SH Werkman, JM Peeples, RJ Cooke, EA Tolley. Arachidonic acid status correlate with first year growth in preterm infants. *Proc Natl Acad Sci USA* 90:1073–1077, 1993.

29. CI Lanting, V Fidler, M Huisman, BCL Touwen, ER Boersma. Neurological differences between 9-year-old children fed breast-milk as babies. *Lancet* 344:1319–1322, 1994.

30. K Maruyama, M Nishikawa. Function of fish oil and its application to foods. *Food Chemicals* 1995(4):31–37, 1995 (in Japanese).

31. P Borgeat, M Nadeau, H Salari, P Poubelle, BF DeLaclos. Leukotrienes: biosynthesis, metabolism, and analysis. In: R Paoletti, D Kritchevsky, Eds. *Advancesin Lipid Research*. Vol. 21. New York: Academic Press, 1985, pp. 47–77.

32. GD Minno, AM Cerbone, A Postiglione. Lipids in platelet function: platelet and vascular prostaglandins in thromboembolic disease. In: R Paoletti, D Kritchevsky, Eds. *Advances in Lipid Research*. Vol. 22. New York: Academic Press, 1987, pp. 63–82.

33. Y Shinmen, S Shimizu, K Akimoto, H Kawashima, H Yamada. Production of arachidonic acid by *Mortierella* Fungi. *Appl Microbiol Biotechnol* 31:11-16, 1989.

34. M Certik, S Shimizu, Biosynthesis and regulation of microbial polyunsaturated fatty acid production. *J Biosci Bioeng* 87:1–14, 1999.

35. MS Manku, DF Horrobins, NL Morse, S Wright, JL Burton. Essential fatty acids in the plasma phospholipids of patients with atopic eczema. *Br J Dermatol* 110:643–648, 1984.

36. A Fiocchi, M Sala, P Signoroni, G Banderai, C Agostoni, E Riva. The efficacy and safety of γ-linolenic acid in the treatment of infantile atopic dermatitis, *J Int Med Res* 22:2–32, 1994.

37. RB Zurier, P Deluca, D Rothman. γ-linolenic acid, inflammation immune responses, and rheumatoid arthritis. In Y-S Huang, DE Mills, Eds. *γ-Linolenic Acid: Metabolism and Its Roles in Nutrition and Medicine*, Champaign, IL: AOCS Press, 1996, pp. 129–136.

38. J Jantti, E Seppala, H Vapaatalo, H Isomaki. Evening primrose oil and olive oil in treatment of rheumatoid arthritis. *Clin Rheumatol* 8:238–244, 1989.

39. T Hoshino, T Yamane, S Shimizu. Selective hydrolysis of fish oil by lipase to concentrate n-3 polyunsaturated fatty acids. *Agric Biol Biochem* 54:1459–1467, 1990.

40. Y Tanaka, J Hirano, T Funada. Concentration of docosahexaenoic acid in glyceride by hydrolysis of fish oil with *Candida cylindracea* lipase. *J Am Oil Chem Soc* 69:1210–1214, 1992.

41. MSK Syed Rahmatullah, VKS Shukla, KD Mukherjee. Enrichment of g-linolenic acid from evening primrose oil and borage oil via lipase-catalyzed hydrolysis. *J Am Oil Chem Soc* 71:569–573, 1994.

42. Y Shimada, K Maruyama, S Okazaki, M Nakamura, A Sugihara, Y Tominaga. Enrichment of polyunsaturated fatty acids with *Geotrichum candidum* lipase. *J Am Oil Chem Soc* 71:951–954, 1994.

43. Y Shimada, A Sugihara, K Maruyama, T Nagao, S Nakayama, H Nakano, Y Tominaga. Enrichment of arachidonic acid: selective hydrolysis of a single-cell oil from *Mortierella* with *Candida cylindracea* lipase. *J Am Oil Chem Soc* 72:1323–1327, 1995.

44. Y Shimada, K Maruyama, M Nakamura, S Nakayama, A Sugihara, Y Tominaga. Selective hydrolysis of polyunsaturated fatty acid-containing oil with *Geotrichum candidum* lipase. *J Am Oil Chem Soc* 72:1577–1581, 1995.

45. Y Shimada, N Fukushima, H Fujita, Y Honda, A Sugihara, Y Tominaga. Selective hydrolysis of borage oil with Candida rugosa lipase: two factors affecting the reaction. *J Am Oil Chem Soc* 75:1581–1586, 1998.

46. A Sugihara, Y Shimada, N Takada, T Nagao, Y Tominaga. *Penicillium abeanum* lipase: purification, characterization, and its use for docosahexaenoic acid enrichment of tuna oil. *J Ferment Bioeng* 82:498–501, 1996.

47. Y Shimada, A Sugihara, Y Tominaga. Enzymatic enrichment of polyunsaturated fatty acids. In: TM Kuo, HW Gardner, Eds. *Lipid Biotechnology*, New York: Marcel Dekker, 2002, pp. 493–515.

48. K Higashiyama, T Yaguchi, K Akimoto, S Fujikawa, S Shimizu. Effect of mineral addition on the growth morphology of and arachidonic acid production by *Mortierella alpina* 1S-4. *J Am Oil Chem Soc* 75:161–165, 1998.

49. Y Shimada, T Nagao, A Kawashima, A Sugihara, S Komemushi, Y Tominaga. Enzymatic purification of n-6 polyunsaturated fatty acids. *Kagaku to Kogyo* 73:125–130, 1993 (in Japanese).

50. H Noda, Y Noda, K Hata, T Fujita. *Kagakukogaku* 55:623–625, 1991 (in Japanese).

51. Y Yamamura, Y Shimomura. Industrial high-performance liquid chromatography purification of docosahexaenoic acid ethyl ester and docosapentaenoic acid ethyl ester from single-cell oil. *J Am Oil Chem Soc* 74:1435–1440, 1997.

52. M Yamaguchi, I Tanaka, Y Ohtsu. New method for separation and purification of polyunsaturated fatty acids using a silver ion-exchanged spherical clay mineral. *Yukagaku* 40:959–964, 1991 (in Japanese).

53. GG Haraldsson, B Kristinsson, R Sigurdardottir, GG Gudmundsson, H Breivik. The preparation of concentrates of eicosapentaenoic acid and docosahexaenoic acid by lipase-catalyzed transesterification of fish oil with ethanol. *J Oil Chem Soc* 74:1419–1424.

54. MJ Hills, I Kiewitt, KD Mukherjee. Enzymatic fractionation of fatty acids: enrichment of γ-linolenic acid and docosahexaenoic acid by selective esterification catalyzed by lipases. *J Am Oil Chem Soc* 67:561–564, 1990.

55. TA Foglia, RE Sonnet. Fatty acid selectivity of lipases: γ-linolenic acid from borage oil. *J Am Oil Chem Soc* 72:417–420, 1997.

56. F-C Huang, Y-H Ju, C-W Huang. Enrichment of γ-linolenic acid from borage oil via lipase-catalyzed reactions. *J Am Oil Chem Soc* 74:977–981, 1997.

57. Y Shimada, A Sugihara, H Nakano, T Kuramoto, T Nagao, M Gemba, Y Tominaga. Purification of docosahexaenoic acid by selective esterification of fatty acids from tuna oil with *Rhizopus delemar* lipase. *J Am Oil Chem Soc* 74:97–101, 1997.

58. Y Shimada, A Sugihara, M Shibahiraki, H Fujita, H Nakano, T Nagao, T Terai, Y Tominaga. Purification of γ-linolenic acid from borage oil by a two-step enzymatic method. *J Am Oil Chem Soc* 74:1465–1470, 1997.

59. Y Shimada, N Sakai, A Sugihara, H Fujita, Y Honda, Y Tominaga. Large-scale purification of γ-linolenic acid by selective esterification using *Rhizopus delemar* Lipase. *J Am Oil Chem Soc* 75:1539–1534, 1998.

60. Y Shimada, K Maruyama, A Sugihara, S Moriyama, Y Tominaga. Purification of docosahexaenoic acid from tuna oil by a two-step enzymatic method: hydrolysis and selective esterification. *J Am Oil Chem Soc* 74:1441–1446, 1997.

61. Y Shimada, A Sugihara, Y Minamigawa, K Higashiyama, K Akimoto, S Fujikawa, S Komemushi, Y Tominaga. Enzymatic enrichment of arachidonic acid from *Mortierella* single-cell oil. *J Am Oil Chem Soc* 75:1213–1217, 1998.

62. S Jareonkitmongkol, H Kawashima, S. Shimizu, H Yamada. Production of 5,8,11-*cis*-eicosatrienoic acid by a Δ12-desaturase-defective mutant of *Mortierella alpina* IS-4. *J Am Oil Chem Soc* 69:939–944, 1992.

63. H Kawashima, M Nishihara, Y Hirano, N Kamada, K Akimoto, K Konishi, S Shimizu. Production of 5,8,11-*cis*-eicosatrienoic acid (Mead acid) by a Δ6-desaturation activity-defective mutant of an arachidonic acid-producing fungus, *Mortierella alpina* IS-4. *Appl Environ Microbiol* 63:1820–1825, 1997.

64. Y Shimada, Y Watanabe, A Kawashima, K Akimoto, S Fujikawa, Y Tominaga, A Sugihara. Enzymatic fractionation and enrichment of n-9 PUFA. *J Am Oil Chem Soc* 80:37–42, 2003.

65. MW Pariza. CLA, a new cancer inhibitor in dietary products. *Bull Int Dairy Fed* 257:29–30 (1991).

66. C Ip, SF Chin, JA Scimeca, MW Pariza. Mammary cancer prevention by conjugated dienoic derivatives of lioleic acid. *Cancer Res* 51:6118–6124, 1991.

67. YL Ha, NK Grimm, MW Pariza. Anticarcinogens from field ground beef: heat-altered derivatives of linolenic acid. *Carcinogenesis* 8:1881–1887, 1987.

68. Y Park, KJ Albright, W Liu, JM Storkson, ME Cook, MW Pariza. Effect of conjugated linoleic acid on body composition in mice. *Lipids* 32:853–858, 1997.

69. E Ostrowska, M Muralitharan, RF Cross, DE Bauman, FR Dunshea. Dietary conjugated linoleic acids increase lean tissue and decrease fat deposition in growing pigs. *J Nutri* 129:2037–2042, 1999.

70. SM Rahman, Y-M Wang, H Yotsumoto, J-Y Cha, S-Y Han, S Inoue, T Yanagita. Effect of conjugated lonoleic acid on serum leptin concentration, body-fat accumulation, and β-oxidation of fatty acid in OLETF rats. *Nutrition* 17:385–390, 2001.

71. KN Lee, D Kritchevsky, MW Pariza. Conjugated linoleic acid and atherosclerosis in rabbits, *Atherosclerosis* 108:19-25, 1994.

72. RJ Nicolosi, EJ Rogers, D Kritchevsky, JA Scimeca, PJ Huth. Dietary conjugated linoleic acid deduces plasma lipoproteins and early aortic atherosclerosis in hypercholesterolemic hamsters. *Artery* 22:266–277, 1997.

73. M Sugano, A Tsujita, M Yamasaki, M Noguchi, K Yamada. Conjugated linoleic acid modulates tissue levels of chemical mediators and immunoglobulins in rats. *Lipids* 33:521–527, 1998.

74. YL Ha, JM Storkson, MW Pariza. Inhibition of benzo(α)pyrene-induced mouse forestomach neoplasia by conjugated dienoic derivatives of linoleic acid. *Cancer Res* 50:1097–1101, 1990.

75. Y Park, KJ Albright, JM Storkson, W Liu, ME Cook, MW Pariza. Change in body composition in mice during feeding and withdrawal of conjugated linoleic acid. *Lipids* 34:243–248, 1999.

76. Y Park, KJ Albright, JM Storkson, W Liu, MW Pariza. Evidence that the *trans*-10,*cis*-12 isomer of conjugated linoleic acid induces body composition changes in mice. *Lipids* 34:235–241, 1999.

77. EA De Deckere, JM Van Amelsvoort, GP McNeill, P Jones. Effect of conjugated linoleic acid (CLA) isomers on lipid levels and peroxisome proliferation in the hamster. *Br J Nutr* 82:309–317, 1999.

78. K Nagao, N Inoue, Y-M Wang, J Hirata, Y Shimada, T Nagao, T Matsui, T Yanagita. The 10*trans*,12*cis* isomer of conjugated linoleic acid suppresses the development of hypertension in Otsuka Long-Evens Tokushima fatty rats. *Biochem Biophys Res Commun* 306:134–138, 2003.

79. MJ Haas, JKG Kramer, G McNeill, K Scott, TA Foglia, N Sehat, J Fritsche, MM Mossoba, MP Yurawecz. Lipase-catalyzed fractionation of conjugated linoleic acid isomers, *Lipids* 34: 979–987, 1999.

80. GP McNeill, C Rawlins, AC Peilow. Enzymatic enrichment of conjugated linoleic acid isomers and incorporation into triglycerides. *J Am Oil Chem Soc* 76:1265–1268, 1999.

81. T Nagao, Y Shimada, Y Yamauchi-Sato, T Yamamoto, M Kasai, K Tsutsumi, A Sugihara, Y Tominaga. Fractionation and enrichment of CLA isomers by selective esterification with *Candida rugosa* lipase. *J Am Oil Chem Soc* 79:303–308, 2002.

82. T Nagao, Y Yamauchi-Sato, A Sugihara, T Iwata, K Nagao, T Yanagita, S Adachi, Y Shimada. Purification of conjugated linoleic acid isomers through a process including lipase-catalyzed selective esterification. *Biosci Biotechonol Biochem* 67:1429–1433, 2003.

83. Y Shimada, Y Watanabe, A Sugihara, T Baba, T Ooguri, S Moriyama, T Terai, Y Tominaga. Ethyl esterification of docosahexaenoic acid in an organic solvent-free system with immobilized *Candida antarctica* lipase. *J Biosci Bioeng* 92:19–23, 2001.

84. Y Watanabe, Y Shimada, A Sugihara, Y Tominaga. Stepwise ethanolysis of tuna oil using immobilized *Candida antarctica* lipase. *J Biosci Bioeng* 88:622–626, 1999.

85. Y Shimada, A Sugihara, S Yodono, T Nagao, K Maruyama, H Nakano, S Komemushi, Y Tominaga. Enrichment of ethyl docosahexaenoate by selective alcoholysis with immobilized *Rhizopus delemar* lipase. *J Biosci Bioeng* 84:138–143, 1997.

86. Y Shimada, K Maruyama, A Sugihara, T Baba, S Komemushi, S Moriyama, Y Tominaga. *J Am Oil Chem Soc* 75:1565–1571, 1998.

87. K Maruyama, Y Shimada, T Baba, T Ooguri, A Sugihara, Y Tominaga, S Moriyama. Purification of ethyl docosahexaenoate through selective alcoholysis with immobilized *Rhizomucor miehei* lipase. *J Jpn Oil Chem Soc* 49:793–799, 2000.

88. Y Shimada, Y Watanabe, A Sugihara, Y Tominaga. Enzymatic alcoholysis for biodiesel fuel production and application of the reaction to oil processing. *J Mol Catal B: Enzym* 17:133–142, 2002.

89. MJ Haas, GJ Piazza, TA Foglia. Enzymatic approach to the production of biodiesel fuels. In: TM Kuo, HW Gardner Eds. *Lipid Biotechnology*, New York: Marcel Dekker, 2002, pp. 587–598.

90. M Mittelbach. Lipase catalyzed alcoholysis of sunflower oil. *J Am Oil Chem Soc* 67:168–170, 1990.

91. Y-Y Linko, M Lämsä, A Huhtala, O Rantanen. Lipase biocatalysis in the production of esters. *J Am Oil Chem Soc* 72:1293–1299, 1995.

92. LA Nelson, TA Foglia, WN Marmer. Lipase-catalyzed production of biodiesel. *J Am Oil Chem Soc* 73:1191–1195, 1996.

93. S-B Park, Y Endo, K Maruyama, K Fujimoto. Enzymatic synthesis of ethyl ester of highly unsaturated fatty acids from fish oils using immobilized lipase. *Food Sci Technol Res* 6:192–195, 2000.

94. B Selmi, D Thomas. Immobilized lipase-catalyzed ethanolysis of sunflower oil in a solvent-free medium. *J Am Oil Chem Soc* 75:691–695.

95. H Breivik, GG Haraldsson, B Kristinsson. Preparation of highly purified concentrates of eicosapentaenoic acid and docosahexaenoic acid. *J Am Oil Chem Soc* 74:1425–1429, 1997.

96. Y Shimada, Y Watanabe, T Samukawa, A Sugihara, H Noda, H Fukuda, Y Tominaga. Conversion of vegetable oil to biodiesel using immobilized *Candida antarctica* lipase. *J Am Oil Chem Soc* 76:789–793, 1999.

97. R Irimescu, K Furihata, K Hata, Y Iwasaki, Y Yamane. Utilization of reaction medium-dependent regiospecificity of *Candida antarctica* lipase (Novozym 435) for synthesis of 1,3-dicapryloyl-2-docosahexaenoyl (or eicosapentaenoyl) glycerol. *J Am Oil Chem Soc* 78:285–289, 2001.

98. R Irimescu, Y Iwasaki, CT Hou. Study of TAG ethanolysis to 2-MAG by immobilized *Candida antarctica* lipase and synthesis of symmetrically structured TAG. *J Am Oil Chem Soc* 79:879–883, 2002.

99. Y Watanabe, Y Shimada, A Sugihara, H Noda, H Fukuda, Y Tominaga. Continuous production of biodiesel fuel from vegetable oil using immobilized *Candida antarctica* lipase. *J Am Oil Chem* 77:355–360, 2000.

100. Y Watanabe, Y Shimada, A Sugihara, Y Tominaga. Enzymatic conversion of waste edible oil to biodiesel fuel in a fixed-bed bioreactor. *J Am Oil Cem Soc* 78:703–707, 2001.

101. WW Nawar. Chemical changes in lipids produced by thermal processing. *J Chem Educ* 61:299–302, 1984.

102. C Cuesta, FJ Sánchez-Muniz, C Garrido-Polonio, S López-Varela, R Arroyo. Thermoxidative and hydrolytic changes in sunflower oil used in fryings with a fast turnover of fresh oil. *J Am Oil Chem Soc* 70:1069–1073, 1993.

102. DW Peterson. Effect of soybean sterols in the diet on plasma and liver cholesterol in chicks. *Proc Soc Exp Biol Med* 78:219–225, 1951.

103. I Ikeda, Y Tanabe, M Sugano, Effect of sitosterol and sitostanol on micellar solubility of cholesterol. *J Nutr Sci Vitaminol* 35:361–369, 1989.

104. PJ Jones, M Raeini-Sarjaz, FY Ntanois, CA Vanstone, JY Feng, WE Parsons. Modulation of plasma lipid levels and cholesterol kinetics by phytosterol versus phytosterol esters. *J Lipid Res* 41:689–705, 2000.

105. Y Shimada, S Nakai, M Suenaga, A Sugihara, M Kitano, Y Tominaga. Facile purification of tocopherols from soybean oil deodorizer distillate in high yield using lipase. *J Am Oil Chem Soc* 77:1009–1013, 2000.

106. Y Watanabe, Y Shimada, Y Hirota, S Nakai, M Kitano, A Sugihara. Two-step *in situ* reaction with *Candida rugosa* lipase for purification of tocopherols from soybean deodorizer distillate. *Abstracts of Annual Meeting of the Society for Biotechnology*, Japan, Osaka, 2002, pp. 91.

107. Y Hirota, T Nagao, Y Watanabe, M Suenaga, S Nakai, M Kitano, A Sugihara, Y Shimada. Purification of steryl esters from soybean oil deodorizer distillate. *J Am Oil Chem Soc* 80:341–346, 2003.

108. Y Shimada, T Nagao, Y Watanabe, Y Takagi, A Sugihara. Enzymatic conversion of steryl esters to free sterols. *J Am Oil Chem Soc* 80:243–247, 2003.

109. T Matsuno. Xanthophylls as precursors of retinoids. *Pure Appl Chem* 63:81–88, 1991.

110. P Palozza, NI Krinsky. Astaxanthin and canthaxanthin are potent antioxidants in a membrane model. *Arch Biochim Biophys* 297:291–295, 1992.

111. YMA Naguib. Antioxidant activities of astaxanthin and related carotenoids. *J Agric Food Chem* 48:1150–1154, 2000.

112. BP Chew, JS Park, MW Wong, TS Wong. A comparison of the anticancer activities of dietary β-carotene, canthaxanthin and Astaxanthin in mice *in vivo*. *Anticancer Res* 19:1849–1854, 1999.

113. H Jyonouchi, S Sun, K Iijima, MD Gross. Antitumor activity of astaxanthin and its mode of action. *Nutr Cancer* 36:59–65, 2000.

114. H Jonouchi, L Zhand, Y Tomita. Studies of immunomodulating actions of carotenoides. Astaxanthin enhances in vitro antibody production to T-dependent antigens without facilitating polyclonal B-cell activation. *Nutr Cancer* 19:269–280, 1993.

115. X Wang, R Willen, T Wadstrom. Astaxanthin-rich algal meal and vitamin C inhibit *Helicobacter pylori* infection in BALB/cA mice. *Antimicrob Agents Chemother* 44:2452–2457, 2000.

116. T Nagao, A Sugihara, A Kawashima, T Fukami, H Ikushima, S Komemushi, Y Shimada. Enzymatic enrichment of astaxantin from *Haematococcus pluvialis* cell extracts. *J Am Oil Chem Soc* 80: 975–981, 2003.

9

Applications of Lipases in Modifications of Food Lipids

Subramani Sellappan

Casimir C. Akoh

9.1   Introduction

Lipases (E. C. 3.1.1.3) are important hydrolytic enzymes that catalyze the hydrolysis of fats and oils. Unlike other enzymes, oil/water or air/water interfaces activate the lipase.[1] In a natural environment, water acts as a secondary substrate for the hydrolysis of lipids. With the recent advancement in reaction medium engineering and enzyme technology, it is now possible to use lipases in synthetic applications. Introduction of a microaqueous environment with decreased and controlled water activity drives the hydrolytic reactions toward the synthetic side. The reversal of hydrolysis reaction is due to the change in the equilibrium constant in low water activity systems.[2] Such a possibility opened a new branch of biotechnology for the synthesis of a variety of food, pharmaceutical, and chemically important compounds with greater specificity than chemical catalysis. There are many advantages of using enzymes for these purposes, including reactions at mild temperature, atmospheric pressure, low energy consumption, and cleaner products with minimum or no by-products. Lipases can be applied to regio- and stereospecific synthesis of drugs in pharmaceutical industries, enantiometrically pure compounds synthesis in chemical industries, modification of fats and oils, and flavor esters synthesis in food industries.[3,4]

The diversity of fatty acid composition of natural lipids gives them a variety of physiochemical, nutritional, and functional properties. Fatty acid carbon chain length, one or more unsaturations, and distribution of fatty acids in a triacylglycerol molecule give each fat or oil its unique properties, and hence their application. Lipids are derived from both plant and animal sources; however, not all lipids are suitable for oral consumption, for instance, castor oil. The majority of the lipids from these sources are edible. Lipids are part of the human diet and essential for healthy living. A high concentration of lipids is found in neuronal tissues and cell membranes as insulators and barriers for compartmentalization of cellular organelles. Nutritionally, lipids are classified based on their importance as essential or nonessential with respect to their fatty acid composition. Essential fatty acids are those that cannot be synthesized by human biochemical machineries such as linoleic, linolenic, and arachidonic acids.

9.2   Lipids and Health

The role of dietary fats and oils in human nutrition is currently a focus of many researchers. Nutritional fats and oils contain both saturated and unsaturated fatty acids, mostly of cis-configuration. However, the type of fatty acids and their ratio in oil determines the nutritional and commercial importance. The major diseases caused by the imbalance or excessive intake of fats includes coronary heart disease, cancer, hypertension, high cholesterol, and diabetes.[5-7]

Lipids in general are a storehouse of energy to supply on demand as and when the body needs. The lipid in the body comes from external as well as internal sources. Dietary lipid is the external source and is obtained from plant and animal origins. Internal sources of lipids are synthesized within the body. Excess calories from carbohydrates also contribute to the internal production of lipids and their storage. Generally, long-chain fatty acids (LCFAs) are not easily metabolized unless there is great demand for energy supply. Consumed lipids are hydrolyzed during digestion and resynthesized before storage.

Several decades of research shows that reduction in fat intake has been linked to a decreased risk of coronary heart disease. Several lines of evidence, however, have indicated that types of fat have a more important role in determining risk of coronary heart disease than the total amount of fat in the diet. Metabolic studies have shown that the type of fat, but not the total amount of fat, predicts serum cholesterol levels. Clinical studies further indicates that replacing saturated fat with unsaturated fat is more effective in lowering risk of coronary heart disease than simply reducing total fat consumption. Increasing intake of *n*-3 fatty acids from fish or plant sources significantly reduces the risk of cardiovascular disease.

9.3   Trans Fatty Acids

Trans fatty acids are present in very low concentration in most nutritional fats. The majority of the fatty acids in edible oils are in cis-configuration. In nature, trans fatty acids are found in ruminants and trace amounts in plants. The dietary source of trans fatty acids comes mainly from hydrogenated or partially hydrogenated oils, specifically from margarines and shortenings. The concept of hydrogenation emerged in 1900s to improve the consistency and stability of dietary fats. During hydrogenation of vegetable oils mostly C18 acids of monoene and diene are formed with trans elaidic acid as one of the major trans-fatty acids.[8] Hydrogenation of fish oil produces mainly monoene type trans-fatty acids of 20 and 22 carbons. However, the main source of trans-fatty acids comes from dairy products, margarine, cooking fat, shortening and salad oils.[9] Although the health effects of trans fatty acids are highly debated, a positive correlation between the intake of trans fatty acids and the risk of coronary heart disease has been established through epidemiological studies.[10]

Unlike other major nutrients, lipids are very sensitive to processing, cooking, and storage. Consumption of processed foods can incorporate considerable amounts of oxidized lipid products in the diet.

However, the extent of deteriorated lipids in foods depends on many factors, including source, composition, presence of antioxidants, and type of processing.

9.4 Importance of Essential Fatty Acids

As stated earlier, essential fatty acids are those that cannot be synthesized by humans, and therefore should be supplied through diet. Specifically, the long-chain omega-3 and omega-6 fatty acids are considered essential fatty acids for humans. Arachidonic acid (20:4n-6) formed through the elongation of linoleic acid (18:2n-6) is an important essential fatty acid. The omega-3 fatty acids, such as linolenic acid (18:3n-3), eicosapentaenoic acid, 20:5n-3 (EPA), and docosahexaenoic acid 22:6n-3 (DHA), are a group of essential fatty acids. Human beings evolved with a diet containing about equal amounts of n-3 and n-6 essential fatty acids.[11] In recent times there has been an increase in the consumption of oils rich in n-6 fatty acids such as oils from corn, sunflower seeds, safflower seeds, cottonseed, and soybeans. Currently, the ratio of n-6 to n-3 fatty acids ranges from <20 to 30:1 in typical western diets.[11] High intake of n-6 fatty acids affects the physiological function with increase in blood viscosity, vasospasm, and vasoconstriction and decreased bleeding time.[11] N-3 fatty acids such as EPA and DHA have antiinflammatory, antiarrhythmic, hypolipidemic, and vasodilator properties.[12] These beneficial effects of n-3 fatty acids have been shown to prevent coronary heart disease, hypertension, type 2 diabetes, renal disease, rheumatoid arthritis, ulcerative colitis, crohn disease, and chronic obstructive pulmonary disease.[13,14] α-Linolenic acid (ALA), found in flaxseed, rapeseed, and walnuts, desaturates and elongates in the human body to EPA and DHA to a limited extent.[11] N-3 fatty acids are essential for normal growth and development.[15] The n-3 polyunsaturated fatty acids (PUFAs) inhibit tissue eicosanoid biosynthesis and reduce inflammation. Diets rich in n-3 PUFAs also increase high-density lipoprotein (HDL) cholesterol, while decreasing low-density lipoprotein (LDL) and very low-density lipoprotein (VLDL) cholesterol levels.[11]

Unsaturated fatty acids such as linoleic acid, ALA, and their derivatives are essential components of cell membranes. These fatty acids affect the membrane fluidity and functionality. Ingestion of EPA and DHA from fish or fish oil reduces production of prostaglandin E2 metabolites and the concentrations of thromboxane A2, a potent platelet aggregator and vasoconstrictor. Further, they prevent the formation of leukotriene B4, an inducer of inflammation and a powerful inducer of leukocyte chemotaxis and adherence. These fatty acids also increase the concentrations of thromboxane A3, a weak platelet aggregator and vasoconstrictor, concentrations of prostacyclin PGI3, and concentrations of leukotriene B5, a weak inducer of inflammation and chemotactic agent.[16,17]

9.5 Role of Saturated Fats

The link between excessive consumption of a diet rich in saturated fatty acids and coronary heart disease is now well established.[18] Primarily, C14.0 is thought to be atherogenic. Long-chain saturated fatty acids are known to increase serum cholesterol levels except stearic acid (18:0), which is neutral and can be desaturated to oleic acid. On the other hand, consumption of fats rich in medium-chain triacylglycerols (MCTs) might be advantageous because they are absorbed intact, highly stable, and not re-esterified. As a result, MCTs provide a ready source of energy and may be useful in diet therapy or specific food formulations. Nevertheless, there is a possible negative effect of consuming saturated fatty acids-rich oils without the essential fatty acids such as linolenic and linoleic acids. Higher consumption of saturated fats are responsible for many diseases such as Parkinson's disease,[19] heart disease,[20,21] cancer,[22–24] and cell apoptosis.[25,26] Oxidized LDL play a major role in the development of atherosclerosis. Saturated fatty acids, especially fatty acids with 12–16 carbon atoms, are the most important determinants of the LDL cholesterol level. Recent evidence shows how the content of saturated fats in the brain can influence numerous behaviors, including body temperature regulation, pain sensitivity, feeding behavior including macronutrient selection, and cognitive performance.[27] Importantly, saturated fatty acids might be the important component of dietary fat mediating macronutrient selection and cognition. However, a direct role of saturated fatty acids in modulating brain functions has not been elucidated.[27] Dietary saturated fatty acids, especially lauric (12:0), myristic (14:0), and palmitic (16:0) acids, which are hypercholesterolemic, influence cell membrane fatty acid composition and affect LDL receptor function. Further, fatty acid changes may modulate mid-bilayer fluidity.[28]

9.6   Availability of Specific Fatty Acids

Fatty acid carbon chain length and the number of unsaturations influence fat absorption. It is well known that MCFAs are better absorbed than LCFAs because they are more polar and can be solubilized in the intestinal contents. After absorption they are transported to the liver by the portal vein.[29] Dietary triacylglycerol stereochemical structure influences the bioavailability of its component fatty acids. For instance, a better absorption of human milk fat due to presence of palmitic acid in the *sn*-2 position was reported.[30] Dietary fats are mainly composed of triacylglycerols with 3–6% of phospholipids.[31]

The availability of specific fatty acids or type of fatty acids in diet is determined by the choice of oil used for cooking or types of meat selected for consumption. In general, oils originating from temperate region have a high amount of saturated fatty acids, for instance, coconut oil, palm oil, and cocoa butter. In contrast, oils originating from colder regions are high in unsaturated fatty acids such as fish and borage oils. Furthermore, economics and affordability also have an impact on the type of fats chosen for edible purposes. Geographically, the climate determines the type of oil seeds that can be cultivated and hence the availability of the specific oils for the people living in that part of the globe.

In recent years, there has been an increasing awareness that long-chain PUFAs such as arachidonic and docosahexaenoic acids are essential for the development of central nervous system during fetal development and early childhood brain development. The association between dietary long-chain fatty acids and visual development has been reported.[32] A number of polyunsaturated dietary lipid sources are currently available for supplementing with long-chain polyunsaturates such as egg yolk lipids, fish oils, and oils from fungi and algae. These lipid sources differ in their structure and fatty acid composition, and that can affect their absorption, distribution, and tissue uptake.

Unsaturated fatty acids and MCFAs are more efficiently absorbed than LCFAs. MCFAs can be absorbed in the stomach, after hydrolysis of medium-chain triacylglycerol (MCT) by gastric lipase.[33] Because of efficient absorption, the MCT has been used as an energy source in pancreatic lipase deficiency conditions such as cystic fibrosis.[34] With increasing chain length of saturated fatty acids, an increasing proportion is absorbed through the lymphatic pathway and a decreasing proportion is absorbed through the portal venous blood.[31]

The positional distribution of fatty acids in dietary triacylglycerol determine whether fatty acids are absorbed as *sn*-2-monoacylglycerol or as free fatty acid, and hence influences the composition of the newly formed chylomicrons because triacylglycerols are resynthesized in the intestinal mucosa using *sn*-2 monoacylglycerols from dietary lipids. However, little specificity is shown toward the unesterified fatty acids, which are re-esterified to the *sn*-1 and *sn*-3 positions.[35] Long-chain saturated fatty acids such as 16:0 are not well absorbed from the lumen as free fatty acid, because their melting point is substantially above body temperatures and they have a strong tendency to form insoluble calcium soap with divalent cations at the alkaline pH of the small intestine.[30] Palm oil and its derivatives are mixed with other vegetable oils to increase the content of 16:0 in infant formulas up to the percentage found in human milk. However, infants fed formulas containing 50% of total fat as palm olein showed a lower fat and calcium absorption than infants fed a soy-based formula.[36,37] The use of dietary TAG with 16:0 mainly at the *sn*-2 position (obtained by enzymatic interesterification) has highlighted the importance of fatty acid positional distribution and *sn*-2 position of TAG on fat absorption both in experimental animals and in newborn infants.[38–42] Moreover, the presence of 16:0 in the *sn*-2 position may influence the composition and size of chylomicrons after the digestion process and also the metabolism of cholesterol esters and long-chain PUFAs.[41,43] Likewise, the position of long-chain PUFAs in the lipid structure influences their absorption and metabolism. The absorption of 20:5*n*-3 and 22:6*n*-3 was higher when those fatty acids were predominantly at the *sn*-2 position than when they were distributed at random between the three positions of triacylglycerol molecule.[33] Furthermore, the incorporation of both fatty acids into lipids of plasma lipoprotein fractions was related to their distribution between the inner and outer positions of triacylglycerol.[44]

9.7 Triacylglycerol Structure and Composition

Oils from animal and plant origin exclusively consist of triacylglycerol, except minor quantities of free fatty acids and in some cases phospholipids. The structure of naturally occurring triacylglycerols is genetically determined. Chemically, triacylglycerols consist of glycerol esterified with three fatty acids with asymmetric structures. The distribution of fatty acids in triacyglycerols varies considerably from natural fats and oils. The positional distribution of fatty acids in selected fats and oils is shown in Table 9.1.

In plant oils, linoleic and linolenic acids are primarily distributed in the *sn*-2 position, while the saturated fatty acids are distributed mainly at the *sn*-1 position. However, monounsaturated fatty acids are evenly distributed in all three positions. There are exceptions to this generalization, such as in cocoa butter where the oleic acid is mainly present at the *sn*-2 position. Longer-chain fatty acids are more concentrated in the *sn*-1 position and small percentages at the *sn*-3 position. In animal fats such as tallow or lard, saturated fatty acids are distributed primarily at the *sn*-1 position along with significant amounts of oleic acids. Unsaturated fatty acids tend to be present at the *sn*-2 position along with small amounts of short-chain fatty acids. Longer-chain fatty acids are predominantly distributed at the *sn*-3 position. Fats obtained from marine animals, especially from fish, show some discrepancies in the distribution of fatty acids. Oleic and longer chain monoenoic fatty acids are primarily located at the *sn*-1 position. Increase in chain length of monoenoic changes the preference of distribution to the *sn*-3 position. Myristic, palmitic, and palmitoleic acids are primarily distributed at the *sn*-1 and *sn*-3 positions. PUFAs are mostly distributed at the *sn*-2 position with significant amounts also being distributed at position *sn*-3. However, there are significant differences in triacylglycerol structure between fish and other marine animals.

9.8 Position-Specific Triacylglycerols

9.8.1 Fatty Acid Sources and Digestion

Absorption of fat is influenced by the chain length of fatty acids and unsaturation. MCFAs are easily absorbed compared to LCFAs because of their better solubility.[29] The structure of triacylglycerols influences the bioavailability of fatty acids, specifically at the *sn*-2 position. The fatty acids present at the *sn*-1 and 3 positions are released during digestion and reformulated as part of chylomicrons before absorption. Released fatty acids can easily combine with other ions such as calcium to form soap and could easily be lost before absorption. However, the fatty acid present at the *sn*-2 position is retained during digestion as *sn*-2 monoacylglycerol and absorbed more easily than hydrolyzed free fatty acids. The presence of specific fatty acid at the *sn*-2 position also affects the overall absorption of digested fats. For instance, palmitic acid present at the *sn*-2 position is believed to be responsible for enhanced absorption of human milk fat than fats from formula fed to infants.[30]

The type of fatty acids present at the *sn*-2 position can affect the emulsification property of released monoacylglycerols, which can further affect both their digestion and absorption. S*n*-2 monoacylglycerols having short or medium-chain fatty acids can be more surface active than LCFAs because of higher hydrophilic to lipophilic ratio. Pancreatic lipase specifically cleaves fatty acids at the *sn*-1 and 3 positions, leaving sn-2 monoacylglycerols during digestion. In contrast, diacylglycerol is produced during gastric lipase action.[55] Pancreatic lipase is relatively inefficient in digesting marine oils and arachidonic acid-containing TAG.[56]

Ionized FFA, *sn*-2 monoacylglycerols, bile salts, and phospholipids together form mixed micelles, which help apolar lipids to go through the unstirred water layer and reach the microvillous membrane, where they are absorbed.[57] Absorbed lipids are re-esterified to form new TAG and phospholipids (PL) in the

TABLE 9.1 Sn-1,2 and 3 Positional Distribution of Fatty Acids in Selected Fats and Oils[45-54]

Oil Sources	Sn-Position	C4:0	C6:0	C8:0	C10:0	C12:0	C14:0	C16:0	C16:1	C18:0	C18:1	C18:2	C18:3	C20+
								Plant oils						
Cocoa butter	1	—	—	—	—	—	—	34	—	50	12	1	1	1
	2	—	—	—	—	—	—	2	—	2	87	9	—	—
	3	—	—	—	—	—	—	37	—	53	9	<0.5	—	2
Olive	1	—	—	—	—	—	—	13	—	3	72	10	1	—
	2	—	—	—	—	—	—	1	—	—	83	14	1	—
	3	—	—	—	—	—	—	17	—	4	74	5	—	—
Soybean	1	—	—	—	—	—	—	14	—	6	23	48	9	—
	2	—	—	—	—	—	—	1	—	<0.5	22	70	7	—
	3	—	—	—	—	—	—	13	—	6	28	45	9	—
Peanut	1	—	—	—	—	—	—	14	—	5	59	19	—	4
	2	—	—	—	—	—	—	2	—	<0.5	59	39	—	1
	3	—	—	—	—	—	—	11	—	5	57	10	—	15
Flaxseed	1	—	—	—	—	—	—	10	—	6	15	16	53	—
	2	—	—	—	—	—	—	2	—	1	16	21	60	—
	3	—	—	—	—	—	—	6	—	4	17	13	59	—
Rapeseed	1	—	—	—	—	—	—	4	—	2	23	11	6	53
	2	—	—	—	—	—	—	1	—	—	37	36	20	6
	3	—	—	—	—	—	—	4	—	3	17	4	3	70
Corn	1	—	—	—	—	—	—	8	—	3	28	50	1	—
	2	—	—	—	—	—	—	2	—	<0.5	27	70	1	—
	3	—	—	—	—	—	—	14	—	3	31	52	1	—
						Animal Fats—Milk								
Human	1	—	—	—	<0.5	1	3	16	4	15	46	11	<0.5	—
	2	—	—	—	<0.5	2	7	58	5	3	13	7	1	—
	3	—	—	—	<0.5	6	7	6	8	2	50	15	1	—
Cow	1	—	—	1	2	5	10	34	2	10	30	2	1	—
	2	—	1	1	3	6	18	32	4	10	19	4	—	—
	3	35	13	4	6	1	6	5	1	1	23	2	—	—
Pig	1	—	—	—	—	—	2	22	7	7	50	11	1	—
	2	—	—	—	—	—	7	58	11	1	14	8	1	—
	3	—	—	—	—	—	4	15	10	6	52	12	2	—
Rat	1	—	—	3	10	10	10	20	2	5	24	14	1	—
	2	—	—	6	20	16	18	29	2	1	3	5	1	—
	3	—	—	10	26	15	9	13	2	2	12	12	1	—

Animal—Adipose Tissue Fats

Oil Sources	Sn-position	C4:0	C6:0	C8:0	C10:0	C12:0	C14:0	C16:0	C16:1	C18:0	C18:1	C18:2	C18:3	C20+
Human Oil Sources	1	—	—	—	—	—	4	42	3	15	27	6	1	—
	2	—	—	—	—	—	6	10	12	2	55	4	2	—
	3	—	—	—	—	—	4	19	6	6	57	11	1	—
Chicken	1	—	—	—	—	—	1	47	7	8	31	5	1	—
	2	—	—	—	—	—	<0.5	13	5	6	55	19	1	—
	3	—	—	—	—	—	1	31	7	3	49	8	1	—
Chicken—egg	1	—	—	—	—	—	—	71	5	4	17	2	—	—
	2	—	—	—	—	—	—	4	3	3	63	26	—	—
	3	—	—	—	—	—	—	12	6	14	67	1	—	—
Beef	1	—	—	—	—	—	4	41	6	17	20	4	1	—
	2	—	—	—	—	—	9	17	6	9	41	5	1	—
	3	—	—	—	—	—	1	22	2	24	37	5	1	—
Sheep	1	—	—	—	—	—	1	35	2	47	4	—	—	—
	2	—	—	—	—	—	4	14	1	15	52	5	—	—
	3	—	—	—	—	—	3	16	9	42	26	2	—	—
Rabbit	1	—	—	—	—	—	3	34	12	6	25	14	2	—
	2	—	—	—	—	—	6	25	7	1	26	23	5	—
	3	—	—	—	—	—	1	24	5	3	35	22	5	—
Rat	1	—	—	—	—	—	2	32	4	9	32	15	1	—
	2	—	—	—	—	—	1	10	5	1	37	45	—	—
	3	—	—	—	—	—	2	27	2	7	37	17	1	—
Pig	1	—	—	—	—	—	1	10	2	30	51	6	—	—
	2	—	—	—	—	—	4	72	5	2	13	3	—	—
	3	—	—	—	—	—	—	<0.5	2	7	17	18	—	—
Fish Oil Cod	1	—	—	—	—	—	6	15	14	6	28	2	—	22
	2	—	—	—	—	—	8	16	12	1	9	2	—	47
	3	—	—	—	—	—	4	7	14	1	23	2	—	44
Herring	1	—	—	—	—	—	6	12	13	1	16	3	—	44
	2	—	—	—	—	—	10	17	10	1	10	3	—	45
	3	—	—	—	—	—	4	7	5	5	8	1	—	76
Skate	1	—	—	—	—	—	2	19	12	1	30	1	—	30
	2	—	—	—	—	—	3	15	7	3	9	1	—	63
	3	—	—	—	—	—	1	6	6	1	28	2	—	54
Mackerel	1	—	—	—	—	—	6	15	11	3	21	2	—	34
	2	—	—	—	—	—	10	21	6	1	9	1	—	45
	3	—	—	—	—	—	2	5	4	2	21	2	—	59

smooth endoplasmic reticulum for which the *sn*-2 monoacylglycerol can be a template. Triacylglycerols, phospholipid, cholesterol, and apoproteins are used to synthesize chylomicroms (CM), which are secreted to the lymph, and then to the general blood stream through the thoracic duct. In the peripheric tissues, they are cleaved by lipoprotein lipase releasing TAG and forming CM remnants. Chylomicron remnants interchange components with other plasma lipoproteins and finally, are taken up by the liver.

9.9 Properties of Fats and Oils

Naturally occurring fats and oils are very different from each other in terms of fatty acid composition, fatty acid distribution, ratio of saturated to unsaturated fatty acids, melting point, crystallization temperature, processing, cooking and storage stability, physical state at ambient temperature (liquid or solid), nutritional value, caloric value, health-promoting effects (such as fish oil), health-deteriorating effects (such as tallow), cost, and their application in various food product formulations (such as cocoa butter). Different kinds of food preparations demand a range of physico-chemical characteristics from fats and oils to achieve particular functionality. Cost is also considered an important factor at large-scale industrial operations. For instance, cocoa butter is costlier than many other lipids due to their unique melting property and application in confectionery industry. Nutritional quality is an important factor for direct consumer consumption.

9.10 Why Modify Lipids

Fats and oils are modified mainly for two reasons: (1) to change or alter physico-chemical properties, and (2) to change or alter their nutritional value. The properties in the first category include: melting point, smoking point, crystallization behavior, slip melting point, spreadability, pourability, solid fat content, oxidative stability, cooking and frying properties, flavor and off-flavor development, emulsification ability, emulsion stability, fatty acid composition, fatty acid positional distribution within triacylglycerol, saturated-to-unsaturated fatty acid ratio, and composition as short-, medium-, and long-chain fatty acids. The nutritional properties of importance are based on essential or nonessential fatty acids content, *n*-3 to *n*-6 ratio, saturated-to-unsaturated ratio, contents of monounsaturates, contents of polyunsaturates, contents of medium- or short-chain fatty acids, hypo- or hypercholesterolmic effect, and if directly metabolized or not. These two properties can be modified by (1) physical blending of different fats or oils, (2) hydrogenation, (3) chemical interesterification, and (4) enzymatic interesterification. The modified properties provide new possibilities for designing special lipids for particular purposes in human nutrition. For instance, triacylglycerols containing MCFAs at the *sn*-1 and *sn*-3 positions and essential fatty acids or LCFAs at the *sn*-2 position of the same TAG molecule can be used in malabsorption and cystic fibrosis syndromes to provide energy as well as essential fatty acids in a more absorbable manner.[34]

9.11 Sources of Lipase

Lipases are found in all living organisms and broadly classified as intracellular and extracellular. They are also classified based on the sources from which they are obtained, such as microbial, animal, and plant. The selection of a lipase for lipid modification is based on the nature of modification sought, for instance, position-specific modification of triacylglycerol, fatty acids specific modification, modification by hydrolysis, and modification by synthesis. The literature survey shows the use of lipase from the following sources. The microbial lipases are derived from *Aspergillus niger*, *Bacillus thermoleovorans*, *Candida cylindracea*, *Candida rugosa*, *Chromobacterium viscosum*, *Geotrichum candidum*, *Fusarium heterosporum*, *Fusarium oxysporum*, *Humicola lanuginose*, *Mucor miehei*, *Oospora lactis*, *Penicillium cyclopium*, *Penicillium roqueforti*, *Pseudomonas aeruginosa*, *Pseudomonas cepacia*, *Pseudomonas fluorescens*, *Pseudomonas putida*, *Rhizopus arrhizus*, *Rhizopus boreas*, *Rhizopus thermosus*, *Rhizopus usamii*, *Rhizopus stolonifer*, *Rhizopus fusiformis*, *Rhizopus*, *Rhizopus circinans*, *Rhizopus delemar*, *Rhizopus chinensis*, *Rhizopus japonicus* NR400, *Rhizopus microsporus, Rhizomucor miehei, Rhizopus nigricans, Rhizopus niveus, Rhizopus oryzae, Rhizopus rhizopodiformis, Rhizopus stolonifer* NRRL 1478, *Rhodotorula rubra,* and *Staphylococcus hyicus.*

Animal sources are from pancreatic lipases, and plant lipases are from papaya latex, oat seed lipase, and castor seed lipase.

9.12   Lipase Reaction Mechanisms

Enzymes, including lipases, have specific active three-dimensional structure in aqueous environment with polar groups exposed and nonpolar groups buried inside. Unlike other enzymes, the nature of lipolytic reaction catalyzed by lipases is very complex in which the lipid substrates are water insoluble. The need for water to maintain and activate lipase and the immiscibility of lipids in water makes the reaction media heterogeneous by forming a liquid-liquid interface. The interface is the point where the lipase can access the substrate and catalyze the reaction. Lipase activity can be easily influenced by the nature of interface, interfacial property, and interfacial area. Interface activates the enzyme by adsorption, which aids the opening of the lid on the catalytic site. All types of interfaces such as solid-liquid, liquid-liquid, or liquid-gas can influence the activity due to the interfacial hydrophobicity. Such an effect of adsorption on activity has been demonstrated.[58] An increase in interfacial area increases the amount of enzyme adsorbed onto the interface and that is why increases in interfacial area increase the activity of enzyme in a lipid/water heterogeneous system. Adsorption of enzyme onto the interface initiates a sequence of events before complete catalysis can be achieved. Adsorption leads to activation and substrate binding followed by catalysis. The accumulation of reaction products on the interface reduces interfacial pressure, which in turn corresponds to high surface energy. These effects are undesirable because they exert denaturing effect on the enzyme molecule, although it is well tolerated by lipase.

9.13   Reversible Reaction of Lipases

Biologically, both intra- and extracellular lipases are designed to catalyze hydrolytic reactions since the living cells are made up of and surrounded by a water-rich environment. Water plays an important role as a medium to disperse the enzyme molecule and participate as a co-substrate in hydrolysis. Reduction in water content may not affect the direction of hydrolysis as long as the water activity, a_w, is maintained at 1. Reduction in water activity below 1 affects the equilibrium constant of the system and the direction of hydrolysis is changed to synthesis, in which water molecules will be produced to shift the system toward higher water activity. The application of water immiscible solvents such as *n*-hexane in lipid modification serves two purposes: (1) the ability to control the water content and therefore the water activity, and (2) the possibility to modify high-temperature melting lipids at low temperature by solubilizing them. Other advantages of using water immiscible solvents in lipid modification include: (a) easy process control in large scale production by reducing viscosity of the oil, (b) the ability to keep the enzyme in insoluble form, (c) ease of enzyme recovery and reuse, (d) easy recovery of products, and (e) increased enzyme stability due to low water content and hence increased productivity and low cost of final product.

9.14   Modification of Lipids by Lipases

Unlike other enzymes, lipases are active at the oil-water interface. Lipases have to be activated by interface and work better in heterogeneous reaction systems such as in emulsions. Lipases can be used to carry out a variety of reactions in microaqueous systems such as direct esterification, acidolysis, and alcoholysis reactions, as described next.

1. Hydrolysis:
   $$R\text{-COOR'} + H_2O \leftrightarrow R\text{-COOH} + R'\text{-OH}$$

2. Synthesis:
    (a) Direct esterification:
        R-COOH + R'-OH ↔ R-CO-OR' + H$_2$0
    (b) Transesterification
        (i) Interesterification:
            R-COOR' + R''-COOR* ↔ R-COOR* + R''-COOR'
        (ii) Acidolysis:
            R-COOR' + R''-COOH ↔ R''-COOR' + R-COOH
        (iii) Alcoholysis:
            R-COOR' + R''-OH ↔ R-COOR'' + R'-OH

9.14.1 Hydrolysis

Partial or controlled hydrolysis is used to produce modified fats with enriched desirable characteristics. The reaction parameter has to be precisely controlled to achieve the maximum yield of products. Products like monoacylglycerols, diacylglycerols, lysolecithins, polyunsaturated fatty acid-enriched mono- and diacylglycerols, and free fatty acids are produced through hydrolytic reactions of both position-specific and-nonspecific lipases. A combination of hydrolysis, esterification, and interesterification of oils using lipases from various sources with a range of specificities and activities can yield position-specific and enriched fatty acid products. For instance, heat-labile long-chain PUFAs can be enriched under mild reaction conditions. Regioselective lipase is used for selective hydrolysis of erucic acid-rich oils from *sn*-1,3-positions of triacylglycerols. Similarly, fatty acid-specific lipases can be used to enrich α-linolenic acid from evening primrose oil or borage oil, docosahexaenoic acid from fish oils, and petroselinic acid from coriander oil.[59] Production of monoacylglycerols using lipases by partial hydrolysis, esterification of glycerol with fatty acids, and protected group reactions that range from organic solvents to reverse micelles was reviewed.[60] Diacylglycerols containing a variety of unsaturated fatty acids such as oleic, linoleic, erucic, ricinoleic, and hydroxystearic acids were used to produce a variety of compounds on a preparative scale (>150 mmol) in good yield (ca. 85%) and with high regioisomerical purities (>95%).[61] *Sn* 1,3-diacylglycerols and dicaprylin were synthesized by direct esterification of glycerol with free fatty acids in a solvent-free system with stoichiometric ratios of the reactants and water removal by evaporation to yield 98% of products. Such diacylglycerol was further used to synthesize 1,3-dicaproyl-2-eicosapentaenoylglycerol and 1,2-dicaprylol-3-eicosapentaenoylglycerol.[62] Mono- and diacylglycerols containing PUFAs up to 28.7% were prepared using lipase in selective hydrolysis reaction.[63] Lipases have also been used to modify phospholipids by hydrolysis.[64] More examples are given in the following section under specific lipases.

9.14.2 Synthesis

Unlike hydrolysis, there are many ways by which lipids can be modified. In synthetic reactions a variety of lipid derivatives such as acids, esters, and alcohols can be used to modify the lipids. However, synthesis reaction is only possible in water-restricted environment as discussed earlier. Except for direct esterification, all other reaction mechanisms are highly preferred for modification of lipids. Formation of water in direct esterification makes it difficult to keep the reaction running in one direction, toward synthesis, compared to other types of reactions where no water is produced or consumed during the reaction. Modifications can be performed by solublizing the lipids in solvents such as *n*-hexane or without solvent. Addition of solvent is helpful in reducing viscosity and solublizing high melting fats so as to allow reactions below melting point.

9.14.2.1 By Direct Esterification

Lipase-catalyzed hydrolysis and esterification reactions occur sequentially even in synthetic reactions. That means a small quantity of water is needed for the reactions and for the enzyme activity. However, excessive presence of water inhibits or reverses the synthetic reactions toward hydrolytic side. In direct esterification reactions, water is produced as a by-product and should be removed from the system in order to drive the reaction toward synthesis. A variety of techniques are applied to solve this problem, including adsorption,[65] azeotropic distillation,[66–68] pervaporation,[69,70] the addition of molecular-sieves,[71] saturated salt solutions,[72,73] and salts.[74] These control mechanisms work better at a smaller scale, but mass transfer limitations affect the efficient control or removal of water from reactors at large-scale operations. The use of saturated salt solutions seems to be ideal in these situations.[72,73,75,76] Direct esterification catalyzed by lipases was used to synthesize a variety of flavor esters and waxes from fatty acids and fatty acid derivatives in organic solvents. Some examples are butyl butyrate,[77] ethyl butyrate, isoamyl butyrate, isoamyl acetate,[78] geranyl acetate,[79] terpene esters,[80,81] geraniol and citronellol esters,[82] citronellyl acetate,[83] butyl laurate,[84] methyl acetate,[85] methyl propionate,[86] l-menthyl esters,[87] ethyl butyrate,[88] geraniol esters,[89] citronellyl butyrate, geranyl caproate,[90] short-chain fatty acid esters,[91] methyl benzoate,[92] isoamyl isovalerate,[93] isoamyl isobutyrate,[94] and isoamyl isovalerate.[95]

9.14.2.2 By Transesterification

In transesterification reactions it is possible to carry out the reaction between ester and acid (acidolysis), ester and an alcohol (alcoholysis), ester and an ester (interesterification), and ester and an amine (aminolysis). Transesterification is the most favored and studied reaction system in the modification of lipids due to better control of water activity and enhanced stability of enzymes.

9.14.2.3 Acidolysis

Acidolysis is an effective way to incorporate or remove a specific or a range of fatty acids from a particular oil. Using the *sn*-1,3 position-specific enzyme, it is possible to incorporate fatty acids of interest at these positions. Examples are discussed in the following section. Acidolysis also allows easy purification or removal of unwanted fatty acids by different methods such as saponification and short-path distillation, whereas use of fatty acid derivatives can only allow the latter method.

9.14.2.4 Alcoholysis

An alcoholysis reaction between an alcohol and ester, in this case triacylglycerols, produces corresponding alcohol esters. However, alcohol can inhibit the enzyme activity at high water content. It is essential to keep the water content low in order to carry out this type of reaction. Many products are synthesized using this technique (Table 9.2 and Table 9.3).

9.14.3 Lipases Used in Lipid Modifications

9.14.3.1 Lipase from *Rhizopus arrhizus* (RAL)

The mycelium of *Rhizopus arrhizus* was used directly in organic solvent for the hydrolysis and interesterification of triacylglycerols, and the synthesis of esters, acyglycerols, and natural esters. The advantage of using mycelium directly is that it acts as a natural carrier for the enzyme in immobilized form. No purification and immobilization are needed, which keeps the cost of mycelium preparation low, and makes it suitable for industrial adaptation.[96] Enrichment of γ-linolenic acid from commercial fungal oil derived from *Mucor* sp. by selective esterification with *n*-butanol or by selective hydrolysis of the oil are most effective with this enzyme.[97] Immobilized RAL was effective in releasing short-chain fatty acids such as butyric to capric acids from milk fat.[98] In the treatment of bovine and caprine milk fats, RAL produced more medium-chain and branched-chain fatty acids than other lipases.[99] The interesterification of tripalmitin with triolein in canola lecithin-hexane reverse micelles was successfully performed to yield a fat with intermediate improved rheological properties between the two initial substrates. The modified fat had its crystallization temperature reduced from 47.7 to 37.5°C and melting point from 61 to 57°C. Such modification was recommended for use in edible plastic fat applications.[100] Modification of palm oil midfraction with stearic acid by immobilized enzyme in *n*-hexane yielded interesterified product with fatty acid composition similar to that of cocoa butter. Significantly increased substrate conversion with defatted soy lecithin reverse micelles was also noted for this enzyme.[101] Isomerically pure 2-monoacylglycerols by alcoholysis of triacylglycerols with ethanol by celite-immobilized RAL was obtained in more than 97% yield.[102] Sorbitan ester surfactant activated this enzyme in the interesterification of triacylglycerols with

TABLE 9.2   Modification of Lipids by Lipases in Various Types of Reactors

Substrates	Product	Enzyme	References
	Stirred Tank Batch Reactor		
Peanut, sunflower, safflower, soybean, and linseed oils + oleic acid	High oleic oils	*Mucor miehei* lipase	[212]
1,2-Isopropylidene glycerol + oleic acid	1,2-Isopropylidene-3-oleoyl glycerol	*Mucor miehei* lipase	[199]
Myristic acid + ethanol	Ethyl myristate	Immobilized *Mucor miehei* lipase	[213]
n-Butanol + oleic acid	n-Butyl oleate	Immobilized *Mucor miehei* lipase	[214]
Rapeseed oil + 2-ethyl-1-hexanol	2-Ethyl-1-hexyl ester	*Candida rugosa* lipase	[215]
Oleic acid + oleyl alcohol	Oleyloleate	*Mucor miehei* lipase	[216]
Lauric acid + geraniol	Geranyl laurate	Lipozyme (*Mucor miehei* lipase)	[217,218]
Monoacylglycerols of erucic acid + caprylic acid	Caprucin	*Pseudomonas cepacia, Geotrichum candidum, Candida rugosa* lipases	[219]
Oleic acid + various primary alcohols	Oleic esters	Lipozyme (*Rhizomucor miehei* lipase)	[201]
Peanut oil + caprylic acid	Structured lipids	Lipozyme (*Mucor miehei* lipase)	[220]
Glycerol + ethyl docosahexaenoate	Tri- docosahexaenoylglycerol	Novozym SP 435 (*Candida antarctica*)	[221]
Ferulic acid + ethanol and octanol	Ethyl and octyl ferulate	Novozym SP 435 (*Candida antarctica*)	[222]
Oleic acid + cetyl alcohol	Oleic ester	*Candida antarctica* lipase	[223]
Tristearin + lauric and oleic acids	Structured lipid	Lipozyme (*Mucor miehei* lipase)	[177]
Palm stearin + coconut oil	Margarine fats	Lipozyme (*Mucor miehei* lipase)	[179,224]
High oleate sunflower oil	Free fatty acids	*Candida rugosa* lipase	[225]
	Stirred Tank Continuous Reactor		
Olive oil + glycerol	Acylglycerols	*Chromobacterium viscosum* lipase	[226]
Olive oil and milk fat hydrolysis	Free fatty acids	Immobilized *Rhizopus delemar* lipase	[227]
	Packed-bed Reactor		
Trilaurin and myristic acid	Structured lipid	*Rhizopus arrhizus* lipase	[228]
Sal (Shorea robusta), kokum (Garcinia indica), mahua (Madhuca latifolia), dhupa (Vateria indica) and mango (Mangifera indica) fats + methyl palmitate and/or stearate	Cocoa butter substitutes	Lipozyme (*Mucor miehei* lipase)	[189]
Triolein + stearic acid	Structured lipid	*Mucor miehei* lipase	[229]
Oleic acid + lauryl alcohol	Oleic ester	Lipozyme (*Mucor miehei* lipase)	[200]
Tallow + glycerol	Monoacylglycerols	Lipozyme (*Mucor miehei* lipase)	[159]
Acetic acid + ethanol	Ethyl acetate	*Mucor miehei* lipase	[230]
Phospholipids + alcohols	Lysophospholipids	Lipozyme (*Mucor miehei* lipase)	[231]
Palm kernel olein	Free fatty acids	*Rhizopus arrhizus* lipase	[232]
Butter oil + conjugated linoleic acid	Modified fat	*Candida antarctica* lipase	[233]
Triacylglycerol + PUFA	Structured lipid	Lipozyme (*Mucor miehei* lipase)	[234]
Medium-chain triacylglycerols + oleic acid	Structured lipid	Lipozyme (*Mucor miehei* lipase)	[235]

(continued)

TABLE 9.2   Modification of Lipids by Lipases in Various Types of Reactors (continued)

Substrates	Product	Enzyme	References
High oleic sunflower oil	Butyl ester + glycerol	Lipozyme (*Mucor miehei* lipase)	[236]
Isopropyl alcohol + palmitic acid	Isopropyl palmitate	Lipozyme (*Mucor miehei* lipase)	[204]
Borage oil + caprylic acids	Structured lipid	*Rhizopus delemar* lipase	[120]
Cetyl alcohol + palmitic acid	Cetyl palmitate	*Candida antarctica* lipase	[237]
Glycerol + PUFA	Mono-, di-, and triacylglycerols	Lipozyme (*Mucor miehei* lipase)	[238]
Oleic acid + ethanol	Oleic ester	Lipozyme (*Mucor miehei* lipase)	[239]
Palm olein	Fatty acids	*Candida rugosa* lipase	[240]
Chicken fat + caprylic acid	Structured lipid	Papaya latex lipase	[241]
Tallow	High oleic tallow fraction	Novozyme (*Candida antarctica* lipase)	[242]
Menhaden oil + caprylic acid	Structured lipid	Lipozyme (*Mucor miehei* lipase)	[178]
Canola oil + caprylic acid	Structured lipid	Lipozyme (*Mucor miehei* lipase)	[243,244]
Butteroil + conjugated linoleic acid	Structured lipid	*Candida antarctica* and *Mucor miehei* lipase	[245]
Palm oil midfraction + stearic acid	Cocoa butter equivalent	Lipozyme (*Mucor miehei* lipase)	[192]
Olive oil + caprylic acid	Structured lipid	Lipozyme (*Mucor miehei* lipase)	[181]
Borage oil + caprylic acid	Structured lipid	*Rhizopus oryzae* lipase	[132]
Cod liver oil + caprylic acid	Structured lipid	Lipozyme (*Mucor miehei* lipase)	[183]
Tristearin + conjugated linoleic acid	Structured lipid	Lipozyme (*Mucor miehei* lipase)	[246]
Menhaden oil + conjugated linoleic acid	Structured lipid	Lipozyme (*Mucor miehei* lipase)	[247]
Medium-chain triacylglycerol + fish oil	Structured lipid	Lipozyme (*Mucor miehei* lipase), Novozym (*Candida antarctica* lipase)	[248]
Membrane Reactor			
Triacetin	Fatty acids	*Candida cylindracea* lipase	[249]
Butter oil	Fatty acids	*Aspergillus niger* lipase	[250]
Oilve oil + glycerol	Monoacylglycerol	*Candida rugosa*, *Chromobacterium viscosum* and *Rhizopus* sp. lipases	[147]
Cetyl alcohol + palmitic acid	Wax ester	*Pseudomonas* sp. lipase	[251]
Medium-chain triacylglycerol + fish oil	Structured lipid	Lipozyme (*Mucor miehei* lipase)	[252]
Rapeseed oil + capric acid	Structured lipid	Lipozyme (*Mucor miehei* lipase)	[253]
Salmon oil	PUFA-enriched oil	Novozyme (*Candida antarctica* lipase)	[254]

fatty acids in *n*-hexane.[103] High yields of 1,3-distearoyl-2-monooleyl glycerol (36%) and 1(3)-2-dioleyl-1(3)-monostearoyl glycerol (27%) with incorporation of stearic acid at the 2-position were produced by immobilized RAL on polypropylene powder by transesterification of high oleic acid rapeseed oil with stearic acid or methyl stearate.[104] Partial hydrolysis of high erucic acid seed oils from white mustard (*Sinapis alba*), oriental mustard (*Brassica juncea*) and honesty (*Lunaria annua*), catalyzed by RAL yielded fatty acids with substantially higher levels of very long-chain monounsaturated fatty acids compared to

TABLE 9.3 Fats and Oils Modified Using Lipases

Source Oil	Modified with	Final Product	Reference
Borage oil	By partial hydrolysis	α-Linolenic acid	[150]
	By partial hydrolysis and esterification	α-Linolenic acid	[115,116]
	Capric and eicosapentaenoic acids (EPA)	Structured lipids	[142]
	Partial hydrolysis	α-Linolenic acid containing triacylglycerols	[255]
	Docosahexaenoic acid (DHA)	Modified lipid	[256]
	EPA and DHA	Modified lipid	[257,258]
	Caprylic acid	Structured lipid	[120]
	Crystallization and butanol	α-Linolenic acid	[188]
Canola oil	Methanol	Fatty acid methyl ester	[259]
	Anhydrous milk fat	C42-C50 and C54 triacylglycerols	[156]
	Palm oil	High fluidity oil	[260]
	Milk fat	Modified fat	[261]
	Butter fat	Spreads	[262]
	Butter fat	Modified fat	[106]
	Caprylic acid	Structured lipid	[243]
	Stearic acid	Margarine fats	[263]
Chicken fat	Caprylic acid	Structured lipids	[241,264]
	Monounsaturated fatty acids	Structured lipids	[265]
Coconut oil	Medium chain fatty acid methyl esters	Medium-chain triacylglycerols	[266]
	Oleic acid	Structured lipids	[267]
	Palm stearin	Margarine	[179,224]
Corn oil	Glycerol	Monoacylglycerols	[268]
	Conjugated linoleic acid	Modified fat	[269]
	By hydrolysis	Linoleic acid	[270]
	Tristearin	Modified oil	[271]
Cottonseed oil	Hydrogenated soybean oil	Blended, modified fat	[136]
	Hydrogenated soybean oil	Margarine fats	[272]
	Hydrogenated soybean oil and olive oil	Zero-*trans* margarine	[273]
	Fatty acids	Low melting and crystallizing fats	[274]
	Caprylic acid	Structured lipids	[130]
	Methanol	Methyl esters	[275]
Fish oil	Soy phospholipid	Polyunsaturated phospholipid	[108]
	By hydolysis	Docosahexaenoic acid enrichment	[276,277]
	By acidolysis	Eicosapentaenoic acid and docosahexaenoic acid enrichment	[278]
	By hydrolysis and interesterification	EPA and DHA containing structured lipids	[279]
	Isopropanol	*n*-3 PUFA	[280]
	Glycerol and free fatty acids	PUFA triacylglycerols	[281]
	n-3 Fatty acids	Mixed acylglycerols	[282]
	By hydrolysis	PUFA-rich acylglycerols	[283]
	Hydrolysis and interesterification with fatty acids	DHA-rich acylglycerols	[284]
	Tributyrin and glycerol	Homogeneous triacylglycerols of EPA or DHA	[138]
	By hydrolysis	Arachidonic acid	[151]
	By hydrolysis	Enriched EPA and DHA	[285]

(continued)

TABLE 9.3   Fats and Oils Modified Using Lipases (continued)

Source Oil	Modified with	Final Product	Reference
	By hydrolysis and glycerol	PUFA enriched triacylglycerols	[286]
	Caprylic acid	Structured lipids	[114]
	Lauryl alcohol	DHA	[287]
	Lauryl alcohol	Ethyl docosahexaenoate	[288]
	Ethanol	EPA and DHA	[187]
	By hydrolysis and glycerol	EPA and DHA rich triacylglycerols	[289]
	By hydrolysis	*n*-3 PUFA	[153]
	Capric acid	Structured lipids	[178,290–292]
	By hydrolysis	*Sn*-2 PUFA monoacylglycerol	[171]
	Lauric acid	Structured lipids	[172]
	Ethanol	Fatty acid ethyl esters	[293]
	Conjugated linoleic acid (CLA)	CLA-enriched fish oil	[294]
	Caprylic acid	EPA and DHA containing structured lipids	[295]
	Caprylic acid	Symmetrically structured lipid	[296]
	By saponification and glycerol	Structured lipids	[297]
	Conjugated linoleic acid	Structured lipids	[298]
	Glycerol	Mixed acylglycerols	[299]
Milk fat	Glycerol	Monoacylglycerol	[268]
	Oleic acid	Low melting milk fat	[126]
	Oleyl alcohol	Wax esters	[198]
	Palm stearin	Transesterified product	[176]
	Conjugated linoleic acid	Modified butter fat	[245]
Lard	Glycerol	Mixed acylglycerols	[268]
	High-oleic sunflower oil	Plastic fats	[143]
Linseed or flax seed oil	Oleic acid	High oleic oils	[212]
	Caprylic acid	Structured lipids	[113]
Olive oil	Glycerol	Mixed acylglycerols	[268]
	n-3 PUFA	*n*-3 PUFA containing oils	[300]
	Glycerol	Monoacylglycerols	[147,301]
	Lauryl and palmityl alcohols	Wax esters	[125]
	Caprylic acid	Structured lipids	[181]
Rapeseed oil	Tallow	Low melting fat	[302]
	Lauric acid	Modified fat	[303]
	Stearic acid or methyl stearate	Confectionery fat	[104]
	Capric acid	Structured lipid	[304]
	Glycerol	Monoacylglycerols	[211]
	1-Hexadecanol	Wax esters	[197]
	Caproic acid	Structured lipids	[305]
Peanut oil	Oleic acid	High oleic oils	[212]
	EPA and DHA	Vegetable oil with *n*-3 PUFA	[300]
	Tricaprylin	Structured lipids	[141]
	Caprylic acid	Structured lipids	[220]
	Caprylic, oleic or linoleic acid	Structured lipids	[131]
Palm oil	Glycerol	Monoacylglycerol	[268,306,307]
	Stearic acid	Cocoa butter equivalent	[101,192]
	Canola or soybean oil	Modified oil	[260]
	Coconut oil	Low melting fats	[161]
	Methanol or propanol	Alkyl esters	[127]
Palm kernel oil	Glycerol	Monoacylglycerol	[307]
	Palm stearin	Margarine	[308]

(continued)

TABLE 9.3 Fats and Oils Modified Using Lipases (continued)

Source Oil	Modified with	Final Product	Reference
	Glycerol	Mixed acylglycerols	[309]
	Ethanol	Ethyl esters	[310]
Palm olein	Glycerol	Monoacylglycerols	[306]
	Selected fatty acids	Structured lipids	[267]
	Palm stearin	Solid frying shortening	[311]
Palm stearin	Glycerol	Monoacylglycerols	[268,312]
	Coconut oil and other vegetable oils	Low slip melting point lipids	[161]
	Sunflower oil	Plastic fats	[313]
	Palm kernel olein	Plastic fats	[166]
	Palm kernel olein	Table margarine	[173,308,314,315]
	Sunflower oil	Table margarine	[169]
	Milk fat	Low melting fat blends and margarines	[176,316,317]
	Coconut oil	Margarine fats	[179,224]
	Palm kernel olein	Frying oil	[311,318]
	Sunflower oil	Low slip melting point lipids	[319]
Primrose oil	By hydrolysis	α-Linolenic acid	[150,320]
	n-Butanol	α-Linolenic acid	[184]
	EPA	Structured lipids	[152,258]
Rice bran oil	Capric acid	Structured lipids	[321]
Safflower oil	Caprylic acid	Structured lipids	[113]
Sesame oil	Capric acid	Structured lipids	[322]
Soybean oil	Oleic acid	High oleic oils	[212]
	Myristic acid, trimyristin, or fully hydrogenated soybean oil	Zero *trans* margarine and table spreads	[323]
	Palm oil	Improved fluidity oil blends	[260]
	EPA and DHA	*N*-3 containing fats	[300,324]
	Cottonseed oil	Modified fat	[136]
	Cottonseed oil	*Trans* free margarine fats	[272]
	Ethyl caprylate	Structured lipids	[325]
Sunflower oil	Tallow	Modified solid fat content fat	[326]
	Palm stearin	Lower slip melting point products	[161,313]
	Ethyl caprylate	Structured lipids	[325]
	Glycerol	Partial acylglycerols	[327–329]
	Ethanol	Ethyl esters	[330]
	Lard	Plastic fats	[143]
	By partial hydrolysis	Very low saturate oils	[331]
	Palm stearin	Table margarine	[169]
	Palm stearin	Low slip melting point fats	[319]
Phospholipids	Soy phospholipid and Sardine oil	Polyunsaturated phospholipids	[108]
	Phosphatidylcholine	Modified phospholipids	[209]
	Phosphatidylcholine and *n*-3 PUFA	*n*-3 PUFA containing phospholipids	[205]
	By hydrolysis	Modified phospholipids	[332]
	Oleic acid	Modified phospholipids	[64,333]
	Alcohols	Lysolecithin	[231]
	PUFA	Modified phospholipids	[334,335]
	By hydrolysis	Lysophospholipids	[336]
	Heptadecanoic acid	Modified phospholipids	[206]
	Methyl esters of fatty acids	Modified phospholipids	[337]
	Alcohols	Lysophospholipids	[338]
	Soy phospholipids	Structured phospholipids	[339]

the starting material, while the C-18 fatty acids are enriched in the acylglycerol fraction.[105] Position-specific interesterification with butter fat/canola oil blends produced a modified fat with unique fatty acid composition.[106] In an acidolysis reaction between digalactosyldiacylglycerol and heptadecanoic acid in toluene, RAL gave the highest reaction rate compared to other enzymes, with a yield of 24%.[107]

9.14.4 Lipases from *Rhizopus delemar* (RDL)

Transesterifications were investigated by preparing polyunsaturated phospholipids from soy phospholipid and sardine oil. Lipases from *Candida cylindracea* and *Rhizopus delemar* produced approximately 45% polyunsaturated phospholipids under optimal conditions. The resulting phospholipid was rich in polyunsaturated fatty acids and linoleic acid with total percentage of polyunsaturated fatty acids incorporation of 8.4%.[108] Regio-isomerically pure 1,3-monoacylglycerols were prepared in multigram scale with high yields (>75%) by esterification of glycerol with a variety of different acyl donors, such as free fatty acids, fatty acid alkyl esters, vinyl esters and triacylglycerols, as well as natural fats and oils in the presence of various lipases, including RDL.[109] A lipid-coated RDL catalyzed di- and triacylglycerol syntheses from monoacylglycerols and aliphatic acids in homogenous and dry benzene solution. Such a coated lipase could also catalyze ester exchange reactions and had a higher stability than uncoated enzyme.[110] In the production of mono- or dipalmitoylglycerol, RDL showed specificity toward monopalmitoylglycerol with 60 wt% yield.[111] For the synthesis of MCTs, RDL showed higher production for dicaprin than other products.[112]

Structured lipids with essential fatty acids at the 1,3-position could be prepared by modifying safflower or linseed oil with caprylic acid using immobilized RDL. By repeated acidolysis with safflower oil as a starting material and purification of triacylglycerol, the only products obtained were 1,3-capryloyl 2-linoleoylglycerol and 1,3- capryloyl-2-oleoyl-glycerol, at a ratio of 86:14 (w/w). The products from linseed oil were 1,3-capryloyl-2-alpha-linolenoyl-glycerol, 1,3-capryloyl-2-linoleoyl-glycerol, and 1,3-capryloyl-2-oleoyl-glycerol (60:22:18, w/w/w). All fatty acids at the 1,3-positions in the original oils were exchanged for caprylic acids by the repeated acidolyses.[113] When tuna oil was interesterified with caprylic acid by 1,3-position-specific RDL, structured lipids with caprylic acids at the 1- and 3- positions and functional fatty acid, docosahexaenoic acid, at the 2-position were produced. Approximately 65% of the fatty acids at the 1- and 3-positions were exchanged for caprylic acid. However, no exchange was reported for docosahexaenoic acid at 1,3-position due to low activity of the lipase toward this fatty acid.[114] γ-Linolenic acid was purified from borage oil by a two-step enzymatic reaction. The oil was hydrolyzed with *Pseudomonas* sp. lipase to release free fatty acids followed by selective esterification with lauryl alcohol by RDL. In this modification γ-linolenic acid was purified to 93.7 wt% with a recovery of 67.5% of its initial content.[115] The large-scale purification of γ-linolenic acid by a combination of enzymatic reactions using RDL and distillation yielded oil containing 45% γ-linolenic acid. Selectively hydrolyzed borage oil with *Pseudomonas* sp. lipase was film distilled after dehydration with recovery of 94.5% free fatty acids. The free fatty acids were selectively esterified with two molar equivalents of lauryl alcohol using RDL with repeated purification and re-esterification and elimination of lauryl esters by urea adduct fractionation to give 98.6% γ-linolenic acid with a recovery of 49.4%.[116] Similarly, DHA was purified by a two-step enzymatic method that consisted of hydrolysis of tuna oil with *Pseudomonas* sp. lipase and selective esterification of the resulting free fatty acids with RDL. As a result, the DHA content was raised to 91 wt% in 88% yield by the repeated esterification.[117] In acidolysis reactions, 1,3-positional specific RDL acted strongly on myristic, palmitic, palmitoleic, stearic, oleic, linoleic, and α-linolenic acids, and moderately on arachidonic and eicosapentaenoic acids. On the other hand, the lipase acted moderately on γ-linolenic acid and DHA in the hydrolysis, but only very weakly on these fatty acids in the early stage of acidolysis.[118]

Structured lipids important in infant nutrition, 1,3-oleoyl-2-palmitoylglycerol was synthesized by a two-step enzymatic process in high yields and purity using *sn*-1,3-regiospecific lipases. First, tripalmitin was alcoholyzed to yield corresponding 2-monopalmitin with 85% yield and 95% purity. In the second step, 2-monopalmitin was esterified with oleic acid to produce structured triacylglycerol containing oleic-palmitic-oleic acids up to 78% yield with 96% palmitic acid in the *sn*-2 position. Best results were reported with lipases from *Rhizomucor miehei* and RDL immobilized on EP 100.[119] Similarly, structured lipids containing γ-linolenic acid were synthesized from borage oil first by alcoholysis and then esterification with caprylic acid using immobilized 1,3-specific *Rhizopus delemar* lipase as a catalyst.[120] Structured lipids having similar *sn*-2 positional distribution of palmitic acid along with arachidonic acid for possible use in infant foods have been synthesized using 1,3-arachidonoyl-2-palmitoyl-glycerol by acidolysis of tripalmitin with arachidonic acid using 1,3-specific RDL.[121] To avoid heat and oxidation, PUFAs such as EPA, DHA, γ-linolenic acid, and arachidonic acid were purified using lipases from *Candida antarctica* and RDL by selective esterification with lauryl alcohol. Such processes increased the purity of DHA, γ-linolenic acid, and arachidonic acid to 91, 98, and 96 wt%, respectively. Selective alcoholysis was also effective for increasing the purity of ethyl docosahexaenoate to 90 wt%.[122]

9.14.5 Lipase from *Rhizopus niveus* (RNL)

A commercial preparation from *Rhizopus niveus* was used to concentrate the omega-3 fatty acid, DHA from cod liver oil to 9.64% (w/w) DHA. The DHA content in the free fatty acid, triacylglycerol, diacylglycerol, and monoacylglycerol components were 5.72%, 9.95%, 15.16%, and 29.17% (w/w) of total fatty acids, respectively. The same enzyme could synthesize phosphatidylcholine containing long-chain PUFAs whereas phospholipase A$_2$ failed to catalyze such transesterification in microemulsions.[123] Interesterified butter fat with increased amounts of palmitic acid at *sn*-2 position was produced by RNL in phosphatidylcholine reverse micellar system in *n*-hexane. Such modification also yielded small quantities of free fatty acids.[124] Alcoholysis of olive oil with lauryl and palmityl alcohols to produce wax esters were catalyzed efficiently by cell-bound lipase from *Rhizopus niveous* cells immobilized within cellulose biomass support particles.[125]

9.14.6 Lipases from Other *Rhizopus* sp.

Immobilized commercial lipase from *Rhizopus oryzae* catalyzed the modification of milk fat with oleic acid. The interesterified milk fat had about 50% more oleic acid and a significantly lower palmitic acid content than those of the original milk fat. The crystallization and melting curves analysis showed that the transition temperature of the major milk fat peaks decreased by 7.6°C and 5.4°C, respectively. The reaction induced specific interesterification between oleic acid and palmitic acid in the milk fat triacylglycerols without loss of the short-chain fatty acids.[126] A mycelium of *Rhizopus rhizopodiformis* was used as lipase for alcoholysis of palm oil mid-fraction and resulted in high-percentage conversions of palm oil mid-fraction to alkyl esters with methanol or propanol. In this reaction, palmitic acid seemed to be preferred over oleic acid in the formation of methyl and propyl esters.[127] However, lipases from *Rhizopus javanicus* and *Rhizopus niveus* showed less than 20% catalytic activity in the synthesis of wax esters using triolein and stearyl alcohol.[128] Structured triacylglycerols of the ABA-type, containing one type of fatty acid (A) in the *sn*-1 and *sn*-3 positions and a second type of fatty acid (B) in the *sn*-2 position of the glycerol, were synthesized using *Rhizomucor miehei*, *Rhizopus delemar*, and *Rhizopus javanicus* lipases.[129] In the two-step process, a triacylglycerol of the B-type was subjected to an alcoholysis reaction catalyzed by *sn*-1,3-regiospecific lipases yielding the corresponding 2-monoacylglycerol. Using this strategy, 2-monopalmitin was obtained up to 88% yield at > 95% purity by crystallization. Esterification of 2-monopalmitin with oleic acid resulted in the formation of 1,3-oleyl-2-palmitoyl-glycerol in up to 72% yield and containing 94% palmitic acid in the *sn*-2 position. Also 2-monoacylglycerols from fish oil were produced by alcoholysis with 84% yield at > 95% purity.[129] Structured triacylglycerols containing MCFAs at *sn*-1 and *sn*-3 positions and an unsaturated LCFA at *sn*-2 position were synthesized using 1,3-regiospecific lipases. First, the 2-monoacylglycerols were prepared by alcoholysis with ethanol from pure triacylglycerols (triolein and trilinolein) or natural oil (cottonseed oil) with lipase from *Rhizopus delemar*. The purified 2-monoacylglcerols with 71.8% yield was esterified with caprylic acid in the second step using lipase from *Rhizomucor miehei* (Lipozyme). Such modification resulted in a product containing more than 94% caprylic acid at *sn*-1 and *sn*-3 positions with *sn*-2 position composed of 78% of unsaturated LCFAs.[130] Similarly, structured triacylglycerols with caprylic acid at *sn*-1 and *sn*-3 positions and oleic or linoleic acid at *sn*-2 position were synthesized from a combination of pure 2-monoacylglycerol produced from triolein, trilinolein, or peanut oil with 1,3-regiospecific lipases (from *Rhizomucor miehei, Rhizopus delemar,* and *Rhizopus javanicus*) and re-esterified with caprylic acid yield a final product of more than 90% caprylic acid in the *sn*-1 and *sn*-3 positions and 98.5% unsaturated LCFAs at *sn*-2 position.[131] Structured lipids rich in 1,3-dicapryloyl-2-gamma-linolenoyl glycerol from γ-linolenic acid rich oil was prepared by hydrolysis of borage oil with *Candida rugosa* lipase and alcoholysis by immobilized *Rhizopus oryzae* with caprylic acid. The distilled and purified product yielded 52.6 mol% of 1,3-dicapryloyl-2-gamma-linolenoyl glycerol.[132] *Sn*-1,3 positional specific intracellular lipase from *Rhizopus japonicus* NR400 was applied to interesterification of trioleoylglycerol with palmitic acid to prepare highly pure 1,3-dipalmitoyl-2-oleoylglycerol.[133]

9.14.7 Lipases from *Candida* sp.

9.14.7.1 Modification and Structured Lipid Synthesis

Candida cylindracea lipase (CCL) synthesized acylglycerols using individual free fatty acids (C18:1, C18:2, C18:3, C18:4, C20:4, C20:5, and C22:6) with glycerol at levels greater than 70% along with docosahexaenoic acid incorporation of 63%.[134] *Candida antarctica* lipase (CAL) was used for esterification of free fatty acids or transesterification of fatty acid methyl esters with isopropylidene glycerols. Mono- and diacylglycerols were obtained by acid catalyzed cleavage of the isopropylidene groups. Oleic and eicosapentaenoic acid were successfully incorporated in the range of 46.9–96.9% with monoacylglycerol content of up to 88.5%.[135] Interesterification reaction between blends of refined cottonseed oil and fully hydrogenated soybean oil catalyzed by CAL produced fats with reduced triunsaturated and trisaturated triacylglycerols and increased the amounts of mono- and disaturated triacylglycerols in the blends. The content of 1,3- diacyglycerols formed exceeded that of 1,2-diacyglycerols.[136] In the modification of the fatty acid composition of crude melon seed (*Citrullus colocynthis* L) oil by incorporating *n*-3 polyunsaturated fatty acid, CAL incorporated higher EPA when EPA ethyl ester (97% pure) was used than by using free acid (45% pure). By this modification *n*-6 fatty acid content of melon seed oil was significantly lowered and the *n*-3 PUFA content increased in the modified oil.[137]

Homogeneous triacylglycerols of pure eicosapentaenoic acid and/or docosahexaenoic acid were synthesized by CAL. The interesterification of tributyrin and direct esterification of glycerol with stoichiometric amount of 99% EPA or DHA as ethyl esters and free fatty acids was reported.[138] The synthesis of structured lipid using tricaprin and trilinolein by CAL showed some degree of preference for the triacylglycerol form (tricaprin).[139] Different structural forms of the capric acid-containing substrate (triacylglycerol vs. ethyl ester) and different chain lengths of triacylglycerols were studied to gain knowledge of the specificity of the enzyme. CAL had some degree of preference for the triacylglycerol form (tricaprin) but no chain length selectivity from tricaprin, trilinolein, tristearin, and trilinolein.[139] Randomization of fats and oils in super-critical carbon dioxide by CAL produced a fat suitable for margarine applications from palm olein and high stearate soybean oil.[140] Structured lipids containing MCFAs and LCFAs were synthesized by many enzymes including *Candida* sp. from tricaprylin and peanut oil with 71% yield.[141] Borage oil was restructured by incorporation of capric acid and EPA catalyzed by CAL and other enzymes. CAL incorporated 8.8% and 15.5% EPA at the *sn*-2 position.[142] Lard and high oleic sunflower oil were interesterified by CAL to produce plastic fats. Increased amount of high oleic sunflower oil produced fats with increased ratio of unsaturated fatty acid/saturated fatty acid, oxidizability, and the amount of 18:1 found at the *sn*-2 position of triacylglycerol products. A 60:40 (w/w) ratio of lard to high oleic sunflower resulted in a product that had 60.1% 18:1 at the *sn*-2 position compared to 44.9% for the physical blend. The solid fat content of the 60:40 interesterified mixture resembled soft-type margarine oil.[143] In a similar study, the interesterification of triolein and tristearin by immobilized CAL produced more of the high-melting oleoyl-distearoyl triacylglycerols than dioleoyl-stearoyl triacylglycerols (OSS, SOO). As the proportion of tristearin was increased, the production of SOO and OSS decreased, and the melting profile of the interesterified triacylglycerols shifted toward higher melting forms to produce hard fats.[144]

9.14.7.2 Hydrolysis Reactions

The lipases from *Candida rugosa* lipase (CRL) can hydrolyze triacylglycerols in an organic solvent. The presence of secondary amines, i.e., diethyl amine, N-methylbutylamine, or the tertiary amine, triethylamine, greatly increased the extent of hydrolysis. Such enhanced hydrolytic activity enabled complete lipolysis of tallow within 20 h at 45°C.[145] CRL was also effective in enhancing the release of short-chain fatty acids from milk fat.[98] High-melting animal fats were hydrolyzed below their melting points by CRL in organic solvents. Edible pork lard was a better substrate than inedible beef tallow, yielding up to 96% hydrolysis.[146] Approximately 46% γ-linolenic acid (GLA)-containing oil was produced by selective hydrolysis of borage oil with CRL. Similarly, CRL produced highest monoacylglycerols (70%) from olive oil.[147]

9.14.7.3 Enrichment

The hydrolysis of tuna oil by *Candida cylindracea* lipase resulted in acylglycerol mixture enriched with DHA.[148] *Candida rugosa* has been shown to discriminate against erucic acid and was used to produce dierucin followed by synthesis of trierucin.[149] Selective and partial hydrolysis of evening primrose (*Oenothera biennis* L.) seed oil and borage (*Borago officinalis* L.) seed oil by *Candida cylindracea* lipase increased the level of γ-linolenic acid in unhydrolyzed acylglycerols. γ-Linolenic acid content in evening primrose oil was raised from 9.4% in the starting material to 46.5% in the unhydrolyzed acylglycerols and in borage oil, the γ-linolenic acid content of the acylglycerols was increased to 47.8%.[150] Arachidonic acid was enriched with *Candida cylindracea* effectively from *Mortierella alpina* single-cell oil. The resulting oil had 60% arachidonic acid.[151] *Candida antarctica* was used to modify the composition of evening primrose oil by incorporating eicosapentaenoic acid, which resulted in increased EPA content up to 43%.[152] N-3 polyunsaturated fatty acid concentrates from seal blubber oil and menhaden oil were prepared at 43.5% and 44.1%, respectively, with *Candida cylindracea* lipase by hydrolytic reactions. However, other lipases from *Aspergillus niger*, *Chromobacterium viscosum*, *Geotrichum candidum*, *Mucor miehei*, *Pseudomonas sp.*, *Rhizopus oryzae, and Rhizopus niveus* were found to be less efficient.[153] Enrichment of EPA and DHA from fish oil by transesterification of ethanol with oil by *Pseudomonas sp.* lipase yielded 46% EPA and DHA as mono-, di-, and triacylglycerols. These partial acylglycerols were converted to ethyl esters and urea-fractionated to give EPA and DHA content up to 85%.[154] Selective hydrolysis of borage oil by immobilized *Candida rugosa* lipase yielded γ-linolenic acid in unhydrolyzed acylglycerols up to 51.7 mol% from an initial content of 23.6 mol% with a yield of 59%.[155] CCL was most active in the hydrolysis of seal blubber and menhaden oil, which increased EPA and DHA content in the nonhydrolyzed portion to produce EPA and DHA-rich acylglycerols.[153]

9.14.7.4 Ester Synthesis

Ethyl oleate was synthesized from oleic acid and ethanol by using *Candida cylindracea* lipase in supercritical carbon dioxide with faster reaction rate than in organic solvents.[156] Wax esters from fatty alcohols and uncommon fatty acids were synthesized up to 90% yield with *Candida antarctica* lipase.[157]

9.14.8 Lipozyme

Lipozyme is a trade name for *sn*-1,3 position specific or selective enzyme produced from *Rhizomucor miehei* by Novo Nordisk, Inc. This enzyme has been extensively used both for research and commercial applications.

9.14.8.1 Modification and Structured Lipid Synthesis

Lipozyme (LZ) catalyzed transesterification of castor and coconut oils in batch and tubular recycle reactors effectively.[158] Monoacylglycerols were synthesized up to 50% yield by LZ from melted tallow.[159] Acidolysis of babassu fat and palmitic acid gave a modified fat with unique properties.[160] Transesterification of palm stearin, and palm stearin with coconut oil and other vegetable oils produced lower slip melting point products than the respective unreacted mixtures. The solid fat content of an equal mixture of the stearin feedstock and coconut oil revealed it was softer after transesterification.[161] Improved nutritional quality of soybean oil with decreased linoleoyl moiety was prepared with oleic acid by *sn*-1,3-specific LZ, in which triacylglycerols with 50.8% oleoyl, 38.8% linoleoyl, and 5.4% alpha-linolenoyl moieties, were produced.[162] Triacylglycerol with conjugated linoleic acid (CLA) were prepared by LZ in a solvent-free reaction medium. Incorporation of at least 95% of the original CLA into the product acylglycerols was achieved.[163] Surface active 1(3)-monooleoyl-rac-glycerol (83% yield) was produced from glycerol and 2-methyl-2-butanol by LZ, and was effectively recovered at 100%.[164] Modified borage oil enriched with *n*-3 polyunsaturated fatty acids was synthesized using LZ. The content of γ-linolenic acid, EPA, and DHA were 26.5, 19.8, and 18.1%, respectively. Further, controlling the reaction parameters yielded 70–72% PUFA with a ratio of *n*-3 PUFA to *n*-6 PUFA of 0–1.09.[165] An improved physical and/or melting characteristic of palm stearin was obtained by transesterification with palm olein in a solvent-free system catalyzed by various lipases including Lipozyme. The resulting product had substantially lower melting points. The repositioning of the fatty acids of triacylglycerols in the higher melting range to form lower- or middle-melting components was responsible for this change.[166]

Structured triacylglycerols resembling human milk fat that contain palmitoyl (16:0) moieties predominantly at the *sn*-2- position of the glycerol backbone and oleoyl (18:1) and linoleoyl (18:2) moieties at the *sn*-1,3-positions have been prepared by transesterification of tripalmitin with fatty acids of low erucic rapeseed oil using Lipozyme.[167] Medium-chain triacylglycerols such as tricaprylin, tricaprin, trilaurin, and trimyristin were synthesized in a solvent-free system using LZ with glycerol and corresponding fatty acids. Appreciable levels of medium-chain triacylglycerols were achieved, except for tricaprylin. Fatty acid/glycerol molar ratio was the most significant variable affecting the synthesis of triacylglycerols.[168] Palm stearin-sunflower oil blends transesterified by lipases from *Pseudomonas* sp. and LZ gave products with reduced slip melting point of 37–40°C in table margarine formulations.[169] A structured triacylglycerol, 1,3-dilauroyl-2-oleoyl-glycerol was obtained by enzymatic transesterification between triolein and lauric acid using LZ with final purity of 95%.[170] The preparation of *sn*-2 monoacylglycerol containing *sn*-2 eicosapentaenoyl glycerol and *sn*-2 docosahexaenoyl glycerol by the hydrolysis of fish oil with LZ was achieved. Such monoacylglycerides could be used as substrates for the synthesis of structured triacylglycerols containing long-chain polyunsaturated fatty acids at specific positions.[171] Position-specific structured triacylglycerols *sn*-1, *sn*-3 dilauryl, *sn*-2 eicosapentaenoyl glycerol and *sn*-1, *sn*-3 dilauryl, *sn*-2 docosahexaenoyl glycerol were prepared by LZ-catalyzed interesterification of lauric acid and *sn*-2 eicosapentaenoyl glycerol and *sn*-2 docosahexaenoyl glycerol. Such structured triacylglycerols containing MCFAs at the *sn*-1 and *sn*-3 positions and long-chain PUFA from marine origin at the *sn*-2 glycerol position were prepared from previously hydrolyzed fatty acids and monoacylglycerols from coconut and fish oil, respectively.[172]

Margarine fats were synthesized from palm stearin and palm kernel oil blends by enzymatic interesterification using LZ. The thermal characteristics of interesterified blend ratios from 30:70 to 70:30 were comparable to those of commercial margarines with less than 0.5% *trans* fatty acids.[173] Shortening, margarine fat bases, and fat products like cocoa butter substitute were synthesized from stearin fraction of tallow and sunflower, soybean, and rice bran oils by LZ-catalyzed reactions. The olein fractions of tallow were interesterified with sat (*Shore robust*) fat, sale olein, and acidolyzed karanja (*Pongamia glabra*) stearin. The interesterified final products had slip melting point and solid fat index suitable for shortening, margarine fat bases, and vanaspati substitute.[174] 3-Dicapryloyl-2-eicosapentaenoylglycerol was synthesized by interesterification of trieicosapentaenoylglycerol with ethyl caprylate catalyzed by a combination of hydrolysis and esterification by Novozyme and LZ. The total yield was over 88%, and no purification of the intermediates was necessary. The regioisomeric purity of the product was found to be 100% by silver-ion high-pressure liquid chromatography.[175] Modified fats with reduced slip melting point were produced by interesterification of palm stearin and anhydrous milk fat catalyzed by LZ.[176] Coating lipids with enhanced moisture restriction properties were prepared by acidolysis of tristearin with lauric and oleic acids using LZ. The prepared fat was compared with cocoa butter for melting point and fatty acid composition.[177] Structured lipids from menhaden oil were produced by enzymatic acidolysis with caprylic acid as acyl donor. Incorporated caprylic acid did not replace DHA, but the content of EPA decreased somewhat with an increase in caprylic acid incorporation.[178] LZ-catalyzed interesterification of palm stearin and coconut oil (75/25%, w/w) was used to produce margarine fats.[179] Acidolysis of rapeseed oil with capric acid was carried out to obtain structured lipids. The transesterified triacylglycerols contained mainly oleic, linoleic and linolenic acids (about 90%) in the internal *sn*-2 position, whereas capric acid was mostly in the external *sn*-1,3 positions (approximately 40%).[180] Structured lipids containing 7.2% caprylic acid, 69.6% oleic acid, 21.7% linoleic acid, and 1.5% palmitic acid at the *sn*-2 position were synthesized from LZ-catalyzed acidolysis of olive oil and caprylic acid.[181] Structured lipids were also synthesized by acidolysis of perilla oil and caprylic acid using LZ.[182] Structured TAGs enriched with EPA and DHA at position 2 of the triacylglycerol backbone were synthesized by acidolysis of cod liver oil with caprylic acid catalyzed by LZ. The structured lipid product contained caprylic acid 57%, EPA 5.1%, DHA 10.0%, and palmitic acid 6.3%. The proportion of EPA and DHA at position 2 of the triacylglycerol was 13.5%, which represented 44% of the total fatty acids in the 2-position.[183]

9.14.8.2 Enrichment

Various lipases, including LZ, have been evaluated for the enrichment of γ-linolenic acid from commercial fungal oil by selective esterification with *n*-butanol or by selective hydrolysis of the oil. LZ was found to be more effective than the other lipases because it increased γ-linolenic acid content from 10.4% in the starting material to 68.8% in the final product. Selective hydrolysis of the fungal oil triacylglycerols using LZ resulted in about 1.5-fold enrichment of γ-linolenic acid in the unhydrolyzed acylglycerols.[97] γ-Linolenic acid was enriched from fatty acids of borage (*Borago officinalis* L.) seed oil to 93% from the initial concentration of 20% by LZ-catalyzed selective esterification of the fatty acids with *n*-butanol.[184] Selective hydrolysis of borage oil in isooctane was performed with immobilized *Candida rugosa* lipase followed by selective esterification of free fatty acids with *n*-butanol by LZ combined with acidolysis of unhydrolyzed acylglycerols and unesterified free fatty acids.[185] The possibility of enriching lauric acid from coconut oil has also been demonstrated.[186] Transesterification of various fish oil triacylglycerols with a stoichiometric amount of ethanol catalyzed by LZ under anhydrous solvent-free conditions resulted in a good separation of EPA and DHA. When free fatty acids from the various fish oils were directly esterified with ethanol under similar conditions, greatly improved results were obtained. When tuna oil comprising 6% EPA and 23% DHA was transesterified with ethanol, 65% conversion into ethyl esters was obtained. The residual acylglycerol mixture contained 49% DHA and 6% EPA (8:1), with 90% DHA recovery into the acylglycerol mixture and 60% EPA recovery into the ethyl ester product. When the corresponding tuna oil free fatty acids were directly esterified with ethanol, 68% conversion was obtained. The residual free fatty acids comprised 74% DHA and only 3% EPA. The recovery of both DHA into the residual free fatty acid fraction and EPA into the ethyl ester product was reported to be very high at 83% and 87%, respectively.[187] High-purity γ-linolenic acid was obtained by employing a modified low-temperature solvent crystallization process, followed by a lipase-catalyzed esterification with borage oil fatty acids. After the esterification of γ-linolenic acid-rich fatty acids with *n*-butanol catalyzed by LZ, γ-linolenic acid content in the fatty acids was further raised from 92.1% to 99.1%. The overall yield of the combined process was reported at 72.8%.[188]

9.14.8.3 Cocoa Butter Substitutes

Solid and semi-solid fats obtained from trees originating from India, namely sal (*Shorea robusta*), kokum (*Garcinia indica*), mahua (*Madhuca latifolia*), dhupa (*Vateria indica*), and mango (*Mangifera indica*), were interesterified with methyl palmitate and/or stearate to produce cocoa butter substitutes. The interesterified fat compared well with cocoa butter in total fatty acid composition, the *sn*-2 position of triacylglycerols, and triacylglycerol composition.[189] Modified kokum fat closely resembled cocoa butter in solid fat content and peak melting temperature.[189] Interesterification of palm olein with stearic acid in solvent-free system catalyzed by LZ IM20 resulted in the formation of 39.3% of the desired cocoa butter-like triacylglycerols, distearoyl-oleoyl-glycerol (SOS), palmitoyl-oleoyl-stearoyl-glycerol (POS), and dipalmitoyl-oleoyl-glycerol (POP). Purification of the reaction mixture gave a fat, whose triacylglycerol composition and melting profile were comparable to cocoa butter. The yield of the cocoa butter-like fat was reported at approximately 25%.[190] Transesterification between triolein and ethylbehenate catalyzed by 1,3-regiospecific LZ in supercritical carbon dioxide produced an anti-blooming agent, 1,3-dibehenoyl-2-oleoyl glycerol.[191] Enzymatic interest-erification of palm oil midfraction with stearic acid in a solvent-free system using LZ produced cocoa butter equivalent. Thermograms of the products obtained by scanning differential calorimetry were similar to cocoa butter (CB), but exhibited several distinct peaks.[192]

9.14.8.4   Synthesis of Wax

Wax esters of absolute purity with no additional purification requirements were synthesized in a solvent-free medium with fatty acid and stearyl alcohol. In another method, alcoholysis reactions involving triolein with stearyl alcohol were carried out to produce 1,2-diolein, 2-monoolein, and the wax ester of oleic acid. Synthesis of waxes from high erucic acid rapeseed oil was also reported.[193] Wax esters from fatty alcohols and uncommon fatty acids were synthesized in yields up to 90% with LZ in nonpolar solvents under mild conditions.[157] Lesquerolic acid wax and alpha, omega-diol esters were synthesized by LZ-catalyzed esterification of lesquerolic acid and alcoholysis of lesquerella oil. Several lesquerolic acid esters were synthesized on a preparative scale.[194] A series of saturated esters, from 20 to 28 carbon atoms, were synthesized in more than 80% yield. Monounsaturated wax esters from 28 to 36 carbon atoms were obtained in yields greater than 70%. In these reactions, primary esters were found to be the sole reaction products (yields from 79% to 81.5%). It was also possible to prepare diesters from octanediol and suberic acid with good yields (84–92%). However, the synthesis of monoesters was moderate.[195] Long-chain waxes of high molecular mass with low solubility and high viscosity have been synthesized from methyl oleate and stearyl alcohol in high yield of 95%. However, substrate inhibition of stearyl alcohol was observed at high concentration.[196] Wax esters of fatty acid methyl esters of rapeseed and a fatty alcohol (1-hexadecanol, 16:0) were synthesized by LZ. The reaction reached equilibrium at 83% conversion in 20 minutes. However, continuous evaporation of methanol increased the yield to 90%.[197] Wax esters from oleyl alcohol and long-chain fatty acid fraction of milk fat were synthesized by various enzymes, including LZ. The yield of esters for LZ-catalyzed reaction was less than 20%, which was lower than other lipases.[198] Alcoholysis of triolein with oleyl alcohol catalyzed by LZ and Novozyme was used to produce oleyl oleate, a wax ester. Synthesis in hexane and heptane yielded more than 75% of oleyl oleate.

9.14.8.5   Ester Synthesis

Fatty ester synthesis with 1,2-isopropylidene glycerol and oleic acid to obtain 1,2-isopropylidene-3-oleoyl glycerol with 80% conversion was reported.[199] LZ- catalyzed esterification of oleic acid with lauryl alcohol and water removal by adsorption with silica gel was performed.[200] The strategies for removal and control of water generated during the esterification of citronellol with butyric acid using LZ were investigated. Modification of the hydration state of the LZ during the reaction was found to be the most important factor in inhibiting the ester synthesis in consecutive batch runs. Removal of water restored the activity in consecutive batch reactions.[80] Oleic acid esters with various primary alcohols have been synthesized at a pressure of 300 bars. The esterification rate was higher at high pressure and the enzyme was very stable.[201] Terpene esters of short- and medium-chain fatty acids using geraniol and citronellol were synthesized by LZ in a solvent-free system with yields of 96% to 99% molar conversion.[202] The syntheses of geranyl acetate and citronellyl acetate by alcoholysis with ethyl and butyl acetates as acyl donors produced molar conversion of 75–77%.[203] Commercially important isopropyl palmitate ester was obtained using LZ with 0.15M palmitic acid and isopropyl alcohol in 1:1 stoichiometric ratio. Excess isopropyl alcohol seemed to inhibit the enzyme activity.[204] Apple flavor ester, isoamyl isovalerate was synthesized using LZ by esterification of isoamyl alcohol and isovaleric acid. A maximum yield of >85% was obtained using solvents such as cyclohexane, hexane, and heptane/isooctane.[93]

9.14.8.6   Modification of Phospholipids

Phospholipids were modified by transesterification with *n*-3 PUFA such as EPA and DHA by LZ. Eicos-apentaenoic acid was incorporated to 17.7 mol%.[205] Transesterification of phospholipids with heptade-canoic acid by acetone-dried cells of a *Rhizopus* species were performed on immobilized biomass support particles. Higher reaction rate for enzymes from immobilized cells of *R. niveus, R. delemar,* and *R. javanicus* were obtained or were equal to that of a commercially available immobilized lipase preparation, LZ.[206] LZ was employed to catalyze acidolysis reaction of 1,2-diacyl-sn-glycero-3-phosphatidylcholine with n-3 PUFAs under nonaqueous solvent-free conditions. In the final product, the phospholipids with 32% EPA and 16% DHA content were obtained as a mixture of phosphatidylcholine and lysophosphatidylcholine.[207] Lysophosphatidylcholine was synthesized from L-alpha- glycerophosphatidylcholine by LZ-catalyzed esterification with more than 90% conversion.[208]

9.14.8.7 Hydrolysis

Soybean phosphatidylcholine was hydrolyzed by LZ in organic media. Solvent polarity had a profound effect on the degree of hydrolysis along with water content.[209] Hydroxy fatty acids from lesquerella oil were extracted using LZ by combined saponification and extraction after enzymatic hydrolysis, which yielded 85–90% free fatty acids with 75–80% of hydroxy free fatty acids in it.[210] Synthesis of monoacylglycerols by enzyme-catalyzed glycerolysis of rapeseed oil using LZ produced only 17.4% of monoacylglycerols.[211]

9.15   Modification of Lipids in Large Scale

Large-scale modification of lipids can be economically achieved with the use of bioreactors. The design and operation of bioreactors and their type determine the productivity and final cost of the product. There are mainly four configurations of reactors available for this purpose: (1) plug flow and packed-bed reactors, (2) continuous stirred tank reactors, (3) stirred-tank reactors, and (4) membrane reactors. Packed-bed reactors are easy to construct and operate and relatively low in cost. The lower ratio of enzyme to substrate in these reactors results in shorter retention time than the batch reactors. Packed-bed reactors are more advantageous for time-sensitive reactions such as reducing acyl migration. Some of the lipid modifications using lipases in various types of reactors are listed in Table 9.2.

References

1. T Maruyama, M Nakajima, S Ichikawa, H Nabetani, S Furusaki, M Seki. Oil–water interfacial activation of lipase for interesterification of triacylglycerol and fatty acid. *J Am Oil Chem Soc* 77:1121–1126, 2000.
2. LF Garcia-Alles, V Gotor. Lipase-catalyzed transesterification in organic media: Solvent effects on equilibrium and individual rate constants. *Biotechnol Bioeng* 59:684–694, 1998.
3. A Kovac, H Scheib, J Pleiss, RD Schmid, F Paltauf. Molecular basis of lipase stereoselectivity. *Eur J Lipd Sci Technol* 102:61–77, 2000.
4. NN Gandhi, NS Patil, SB Sawant, JB Joshi, PP Wangikar, D Mukesh. Lipase-catalyzed esterification. *Catal Rev-Sci Eng* 42:439–480, 2000.
5. M Kratz, P Cullen, U Wahrburg. The impact of dietary mono- and poly-unsaturated fatty acids on risk factors for atherosclerosis in humans. *Eur J Lipid Sci Technol* 104:300–311, 2002.
6. T Sasaki, Y Kobayashi, J Shimizu, M Wada, S In'nami, Y Kanke, T Takita. Effects of dietary n-3-to-n-6 polyunsaturated fatty acid ratio on mammary carcinogenesis in rats. *Nutr Cancer-An Intl J* 30:137–143, 1998.
7. G Fernandes, JT Venkatraman. Role of omega-3-fatty acids in health and disease. *Nutr Res* 13:S19–S45, 1993.
8. J Fritsche, H Steinhart. Analysis, occurrence, and physiological properties of trans fatty acids (tfa) with particular emphasis on conjugated linoleic acid isomers (cla)—a review. *Fett-Lipid* 100: 190–210, 1998.
9. D Kritchevsky. The effects of dietary trans fatty acids. *Chem Ind* 565–567, 1996.
10. WC Willett, MJ Stampfer, JE Manson, GA Colditz, FE Speizer, BA Rosner, LA Sampson, CH Hennekens. Intake of trans fatty-acids and risk of coronary heart disease among women. *Lancet* 341:581–585, 1993.
11. AP Simopoulos. Essential fatty acids in health and chronic disease. *Am J Clin Nutr* 70:560S–569S, 1999.

12. AP Simopoulos. Omega-3 fatty acids in inflammation and autoimmune diseases. *J Am Coll Nutr* 21:495-505, 2002.

13. AP Simopoulos. Omega-3-fatty-acids in health and disease and in growth and development. *Am J Clin Nutr* 54:438–463, 1991.

14. AP Simopoulos. Summary of the conference on the health effects of polyunsaturated fatty-acids in seafoods. *J Nutr* 116:2350–2354, 1986.

15. AP Simopoulos. Human requirement for n-3 polyunsaturated fatty acids. *Poultry Sci* 79:961-970, 2000.

16. PC Weber, S Fischer, C von Schacky, R Lorenz, T Strasser. Dietary omega-3 polyunsaturated fatty acids and eicosanoid formation in man. In: AP Simopoulos, RR Kifer, RE Martin, eds. *Health Effects of Polyunsaturated Fatty Acids in Seafoods*. Orlando, FL: Academic Press, 1986, pp. 49–60.

17. RA Lewis, TH Lee, KF Austen. Effects of omega-3 fatty acids on the generation of products of the 5-lipoxygenase pathway. In: AP Simopoulos, RR Kifer, RE Martin, eds. *Health Effects of Polyunsaturated Fatty Acids in Seafoods*. Orlando, FL: Academic Press, 1986, pp. 227–38.

18. DJ Pehowich, AV Gomes, JA Barnes. Fatty acid composition and possible health effects of coconut constituents. *West Ind Med J* 49:128–133, 2000.

19. H Chen, SMM Zhang, MA Hernan, WC Willett, A Ascherio. Dietary intakes of fat and risk of Parkinson's disease. *Am J Epidemiol* 157:1007–1014, 2003.

20. AH Lichtenstein. Dietary fat and cardiovascular disease risk: Quantity or quality? *J Womens Health Gender-Based Med* 12:109–114, 2003.

21. A Ascherio. Epidemiologic studies on dietary fats and coronary heart disease. *Am J Med* 113:9–12, 2002.

22. M Huncharek, B Kupelnick. Dietary fat intake and risk of epithelial ovarian cancer: A meta-analysis of 6,689 subjects from 8 observational studies. *Nutr Cancer* 40:87–91, 2001.

23. C Byrne, H Rockett, MD Holmes. Dietary fat, fat subtypes, and breast cancer risk: Lack of an association among postmenopausal women with no history of benign breast disease. *Cancer Epidemiol Biomarkers Prev* 11:261–265, 2002.

24. F Levi, C Pasche, F Lucchini, C La Vecchia. Macronutrients and colorectal cancer: A Swiss case-control study. *Ann Oncol* 13:369–373, 2002.

25. K Eitel, H Staiger, MD Brendel, D Brandhorst, RG Bretzel, HU Haring, M Kellerer. Different role of saturated and unsaturated fatty acids in beta-cell apoptosis. *Biochem Biophys Res Commun* 299:853–856, 2002.

26. YM Mu, T Yanase, Y Nishi, A Tanaka, M Saito, CH Jin, C Mukasa, T Okabe, M Nomura, K Goto, H Nawata. Saturated free fatty acids, palmitic acid and stearic acid, induce apoptosis in human granulosa cells. *Endocrinology* 142:3590–3597, 2001.

27. RJ Kaplan, CE Greenwood. Dietary saturated fatty acids and brain function. *Neurochem Res* 23:615–626, 1998.

28. E Berlin, JS Hannah, K Yamane, RC Peters, BV Howard. Fatty acid modification of membrane fluidity in chinese hamster ovary (tr715-19) cells. *Int J Biochem Cell Biol* 28:1131–1139, 1996.

29. EA Decker. The role of stereospecific saturated fatty acid positions on lipid nutrition. *Nutr Rev* 54:108–110, 1996.

30. RM Tomarell, BJ Meyer, JR Weaber, FW Bernhart. Effect of positional distribution on absorption of fatty acids of human milk and infant formulas. *J Nutr* 95:583–590, 1968.

31. ABR Thomson, M Keelan, ML Garg, MT Clandinin. Intestinal aspects of lipid absorption—in review. *Can J Physiol Pharmacol* 67:179–191, 1989.

32. WC Heird, TC Prager, RE Anderson. Docosahexaenoic acid and the development and function of the infant retina. *Curr Opin Lipidology* 8:12–16, 1997.

33. MS Christensen, CE Hoy, CC Becker, TG Redgrave. Intestinal-absorption and lymphatic transport of eicosapentaenoic (EPA), docosahexaenoic (DHA), and decanoic acids—dependence on intramolecular triacylglycerol structure. *Am J Clin Nutr* 61:56–61, 1995.

34. MM Jensen, MS Christensen, CE Hoy. Intestinal-absorption of octanoic, decanoic, and linoleic acids—effect of triacylglycerol structure. *Ann Nutr Metab* 38:104–116, 1994.

35. TG Redgrave, DR Kodali, DM Small. The effect of triacyl-sn-glycerol structure on the metabolism of chylomicrons and triacylglycerol-rich emulsions in the rat. *J Biol Chem* 263:5118–5123, 1988.

36. SE Nelson, RR Rogers, JA Frantz, EE Ziegler. Palm olein in infant formula: Absorption of fat and minerals by normal infants. *Am J Clin Nutr* 64:291–296, 1996.

37. SE Nelson, JA Frantz, EE Ziegler. Palm olein in infant formula affects absorption of fat and calcium. *Faseb J* 10:1321–1321, 1996.

38. EL Lien, FG Boyle, R Yuhas, RM Tomarelli, P Quinlan. The effect of triacylglycerol positional distribution on fatty acid absorption in rats. *J Pediatr Gastroenterol Nutr* 25:167–174, 1997.

39. NJ Defouw, GAA Kivits, PT Quinlan, WGL Vannielen. Absorption of isomeric, palmitic acid-containing triacylglycerols resembling human-milk fat in the adult-rat. *Lipids* 29:765–770, 1994.

40. SM Innis, RA Dyer, EL Lien. Formula containing randomized fats with palmitic acid (16:0) in the 2-position increases 16:0 in the 2-position of plasma and chylomicron triacylglycerols in formula-fed piglets to levels approaching those of piglets fed sow's milk. *J Nutr* 127:1362–1370, 1997.

41. SM Innis, R Dyer. Dietary triacylglycerols with palmitic acid (16:0) in the 2- position increase 16:0 in the 2-position of plasma and chylomicron triacylglycerols, but reduce phospholipid arachidonic and docosahexaenoic acids, and alter cholesteryl ester metabolism in formula-fed piglets. *J Nutr* 127:1311–1319, 1997.

42. VP Carnielli, IHT Luijendijk, JB Vangoudoever, EJ Sulkers, AA Boerlage, HJ Degenhart, PJJ Sauer. Feeding premature newborn infants palmitic acid in amounts and stereoisomeric position similar to that of human-milk-effects on fat and mineral balance. *Am J Clin Nutr* 61:1037-1042, 1995.

43. S Aoe, J Yamamura, H Matsuyama, M Hase, M Shiota, S Miura. The positional distribution of dioleoyl-palmitoyl glycerol influences lymph chylomicron transport, composition and size in rats. *J Nutr* 127:1269–1273, 1997.

44. H Sadou, CL Leger, B Descomps, JN Barjon, L Monnier, AC Depaulet. Differential incorporation of fish-oil eicosapentaenoate and docosahexaenoate into lipids of lipoprotein fractions as related to their glyceryl esterification—a short-term (postprandial) and long-term study in healthy humans. *Am J Clin Nutr* 62:1193–1200, 1995.

45. H Brockerh, M Yurkowsk. Stereospecific analyses of several vegetable fats. *J Lipid Res* 7:62–64, 1966.

46. H Brockerh, RJ Hoyle, N Wolmark. Positional distribution of fatty acids in triacylglycerols of animal depot fats. *Biochimica et Biophysica Acta* 116:67–69, 1966.

47. WW Christie, JH Moore. Structures of adipose tissue and heart-muscle triglycerides in domestic chicken (gallus-gallus). *J Sci Food Agric* 23:73–77, 1972.

48. WW Christie, JH Moore, AR Lorimer, TDV Lawrie. Structures of triglycerides from atherosclerotic plaques and other human tissues. *Lipids* 6:854–856, 1971.

49. WW Christie, JH Moore. Structures of triglycerides isolated from various sheep tissues. *J Sci Food Agric* 22:120–124, 1971.

50. WW Christie, JH Moore. A comparison of structures of triglycerides from various pig tissues. *Biochimica et Biophysica Acta* 210: 46–56, 1970.

51. WW Christie. Structure of the triacyl-sn-glycerols in the plasma and milk of the rat and rabbit. *J Dairy Res* 52:219–222, 1985.

52. WW Christie, JL Clapperton. Structures of the triglycerides of cows milk, fortified milks (including infant formulas), and human milk. *J Soc Dairy Technol* 35:22–24, 1982.

53. Breckenr.Wc, L Marai, A Kuksis. Triglyceride structure of human milk fat. *Can J Biochem* 47:761–769, 1969.

54. RW Walker, H Barakat, JGC Hung. Positional distribution of fatty acids in phospholipids and triglycerides of mycobacterium-smegmatis and m-bovis bcg. *Lipids* 5:684–691, 1970.

55. H Carlier, A Bernard, C Caselli. Digestion and absorption of polyunsaturated fatty-acids. *Reprod Nutr Dev* 31:475–500, 1991.

56. NR Bottino, Vandenburg GA, R Reiser. Resistance of certain long-chain polyunsaturated fatty acids of marine oils to pancreatic lipase hydrolysis. *Lipids* 2:489–493, 1967.

57. A Ganem-Quintanar, D Quintanar-Guerrero, P Buri. Monoolein: A review of the pharmaceutical applications. *Drug Dev Ind Pharm* 26:809–820, 2000.

58. B Benzonan, P Desnuell. Etude cinetique de laction de la lipase pancreatique sur des triglycerides en emulsion essai dune enzymologie en milieu heterogene. 105:121–136, 1965. (see webofscience).

59. KD Mukherjee. Production of added-value lipids using enzymatic reactions. *Fett Wiss Technol-Fat Sci Technol* 94:542–546, 1992.

60. UT Bornscheuer. Lipase-catalyzed syntheses of monoacylglycerols. *Enzyme Microb Technol* 17:578–586, 1995.

61. C Waldinger, M Schneider. Enzymatic esterification of glycerol. 3. Lipase-catalyzed synthesis of regioisomerically pure 1,3-sn-diacylglycerols and 1(3)-rac-monoacylglycerols derived from unsaturated fatty acids. *J Am Oil Chem Soc* 73:1513–1519, 1996.

62. R Rosu, M Yasui, Y Iwasaki, T Yamane. Enzymatic synthesis of symmetrical 1,3-diacylglycerols by direct esterification of glycerol in solvent-free system. *J Am Oil Chem Soc* 76:839–843, 1999.

63. M Linder, E Matouba, J Fanni, M Parmentier. Enrichment of salmon oil with n-3 PUFA by lipolysis, filtration and enzymatic re-esterification. *Eur J Lipid Sci Technol* 104:455–462, 2002.

64. A Mustranta, P Forssell, AM Aura, T Suortti, K Poutanen. Modification of phospholipids with lipases and phospholipases. *Biocatalysis* 9:181–194, 1994.

65. P Mensah, JL Gainer, G Carta. Adsorptive control of water in esterification with immobilized enzymes: I. Batch reactor behavior. *Biotechnol Bioeng* 60:434–444, 1998.

66. F Monot, F Borzeix, M Bardin, JP Vandecasteele. Enzymatic esterification in organic media—role of water and organic-solvent in kinetics and yield of butyl butyrate synthesis. *Appl Microbiol Biotechnol* 35:759–765, 1991.

67. YC Yan, UT Bornscheuer, LQ Cao, RD Schmid. Lipase-catalyzed solid-phase synthesis of sugar fatty acid esters-removal of byproducts by azeotropic distillation. *Enzyme Microb Technol* 25:725–728, 1999.

68. L Gubicza, A Szakacsschmidt. Online water removal during enzymatic asymmetric esterification in organic media. *Biotechnol Tech* 9:687–690, 1995.

69. A Vanderpadt, JJW Sewalt, K Vantriet. Online water removal during enzymatic triacylglycerol synthesis by means of pervaporation. *J Membr Sci* 80:199–208, 1993.

70. SJ Kwon, KM Song, WH Hong, JS Rhee. Removal of water produced from lipase-catalyzed esterification in organic-solvent by pervaporation. *Biotechnol Bioeng* 46:393–395, 1995.

71. HF Decastro, SS Jacques. Influence of molecular-sieves on the performance of citronellyl butyrate synthesis using immobilized lipase. *Arq Biol Technol* 38:339–344, 1995.

72. CM Rosell, AM Vaidya, PJ Halling. Continuous *in situ* water activity control for organic phase biocatalysis in a packed bed hollow fiber reactor. *Biotechnol Bioeng* 49:284–289, 1996.

73. E Wehtje, J Kaur, P Adlercreutz, S Chand, B Mattiasson. Water activity control in enzymatic esterification processes. *Enzyme Microb Technol* 21:502–510, 1997.

74. HP He, LR Yang, ZQ Zhu. Control of water activity of enzymatic reaction in organic solvent media using salt/salt hydrate pairs. *Chin J Org Chem* 21:376–379, 2001.

75. Z Ujang, AM Vaidya. Stepped water activity control for efficient enzymatic interesterification. *Appl Microbiol Biotechnol* 50:318–322, 1998.

76. J Kaur, E Wehtje, P Adlercreutz, S Chand, B Mattiasson. Water transfer kinetics in a water activity control system designed for biocatalysis in organic media. *Enzyme Microb Technol* 21:496–501, 1997.

77. F Fayolle, R Marchal, F Monot, D Blanchet, D Ballerini. An example of production of natural esters-synthesis of butyl butyrate from wheat flour. *Enzyme Microb Technol* 13:215–220, 1991.

78. FW Welsh, RE Williams, SC Chang, CJ Dicaire. Production of low-molecular-weight esters using vegetable oils or butter oil as reaction media. *J Chem Technol Biotechnol* 52:201–209, 1991.

79. W Chulalaksananukul, JS Condoret, D Combes. Kinetics of geranyl acetate synthesis by lipase-catalyzed transesterification in normal hexane. *Enzyme Microb Technol* 14:293–298, 1992.

80. HF Decastro, WA Anderson, RL Legge, M Mooyoung. Process-development for production of terpene esters using immobilized lipase in organic media. *Indian J Chem Sect B-Org Chem Incl Med Chem* 31:891–895, 1992.

81. HF deCastro, EB Pereira, WA Anderson. Production of terpene ester by lipase in non-conventional media. *J Braz Chem Soc* 7:219–224, 1996.

82. PA Claon, CC Akoh. Enzymatic-synthesis of geraniol and citronellol esters by direct esterification in n-hexane. *Biotechnol Lett* 15:1211–1216, 1993.

83. HF deCastro, PC deOliveira, EB Pereira. Evaluation of different approaches for lipase catalysed synthesis of citronellyl acetate. *Biotechnol Lett* 19:229–232, 1997.

84. NN Gandhi, SB Sawant, JB Joshi. Studies on the lipozyme-catalyzed synthesis of butyl laurate. *Biotechnol Bioeng* 46:1–12, 1995.

85. R Perraud, L Moreau, E Krahe. Optimization of the enzyme-catalyzed biosynthesis of methyl acetate. *Dtsch Lebensm-Rundsch* 91:219–221, 1995.

86. R Perraud, F Laboret. Optimization of methyl propionate production catalysed by *Mucor miehei* lipase. *Appl Microbiol Biotechnol* 44:321–326, 1995.

87. Y Shimada, Y Hirota, T Baba, S Kato, A Sugihara, S Moriyama, Y Tominaga, T Terai. Enzymatic synthesis of l-menthyl esters in organic solvent-free system. *J Am Oil Chem Soc* 76:1139–1142, 1999.

88. JP Chen. Production of ethyl butyrate using gel-entrapped *Candida cylindracea* lipase. *J Ferment Bioeng* 82:404–407, 1996.

89. M KarraChaabouni, S Pulvin, D Touraud, D Thomas. Enzymatic synthesis of geraniol esters in a solvent-free system by lipases. *Biotechnol Lett* 18:1083–1088, 1996.

90. LN Yee, CC Akoh, RS Phillips. Lipase PS-catalyzed transesterification of citronellyl butyrate and geranyl caproate: Effect of reaction parameters. *J Am Oil Chem Soc* 74:255–260, 1997.

91. D Leblanc, A Morin, D Gu, XM Zhang, JG Bisaillon, M Paquet, H Dubeau. Short chain fatty acid esters synthesis by commercial lipases in low-water systems and by resting microbial cells in aqueous medium. *Biotechnol Lett* 20:1127–1131, 1998.

92. JP Leszczak, C Tran-Minh. Optimized enzymatic synthesis of methyl benzoate in organic medium. Operating conditions and impact of different factors on kinetics. *Biotechnol Bioeng* 60:356–361, 1998.

93. GV Chowdary, MN Ramesh, SG Prapulla. Enzymic synthesis of isoamyl isovalerate using immobilized lipase from *Rhizomucor miehei*: A multivariate analysis. *Process Biochem* 36:331–339, 2000.

94. SH Krishna, AP Sattur, NG Karanth. Lipase-catalyzed synthesis of isoamyl isobutyrate—optimization using a central composite rotatable design. *Process Biochem* 37:9–16, 2001.

95. GV Chowdary, S Divakar, SG Prapulla. Modelling on isoamyl isovalerate synthesis from *Rhizomucor miehei* lipase in organic media: Optimization studies. *World J Microbiol Biotechnol* 18:179–185, 2002.

96. C Gancet. Utilization of a mycelium lipase for the production of free fatty-acids and fatty esters. *Revue Francaise Des Corps Gras* 38:79–84, 1991.

97. KD Mukherjee, I Kiewitt. Enrichment of gamma-linolenic acid from fungal oil by lipase catalyzed reactions. *Appl Microbiol Biotechnol* 35:579–584, 1991.

98. JP Chen, BK Yang. Enhancement of release of short-chain fatty acids from milk fat with immobilized microbial lipase. *J Food Sci* 57:781–782, 1992.

99. JK Ha, RC Lindsay. Release of volatile branched-chain and other fatty-acids from ruminant milk fats by various lipases. *J Dairy Sci* 76:677–690, 1993.

100. AG Marangoni, RD McCurdy, ED Brown. Enzymatic interesterification of triolein with tripalmitin in canola lecithin-hexane reverse micelles. *J Am Oil Chem Soc* 70:737–744, 1993.

101. L Mojovic, S Silermarinkovic, G Kukic, G Vunjaknovakovic. Rhizopus-arrhizus lipase-catalyzed interesterification of the midfraction of palm oil to a cocoa butter equivalent fat. *Enzyme Microb Technol* 15:438–443, 1993.

102. A Millqvist, P Adlercreutz, B Mattiasson. Lipase-catalyzed alcoholysis of triglycerides for the preparation of 2-monoglycerides. *Enzyme Microb Technol* 16:1042–1047, 1994.

103. S Basheer, K Mogi, M Nakajima. Surfactant-modified lipase for the catalysis of the interesterification of triglycerides and fatty acids. *Biotechnol Bioeng* 45:187–195, 1995.

104. T Gitlesen, I Svensson, P Adlercreutz, B Mattiasson, J Nilsson. High-oleic-acid rapeseed oil as starting material for the production of confectionary fats via lipase-catalyzed transesterification. *Ind Crop Prod* 4:167–171, 1995.

105. KD Mukherjee, I Kiewitt. Enrichment of very long chain mono-unsaturated fatty acids by lipase catalysed hydrolysis and transesterification. *Appl Microbiol Biotechnol* 44:557–562, 1996.

106. D Rousseau, AG Marangoni. Tailoring the textural attributes of butter fat canola oil blends via *Rhizopus arrhizus* lipase-catalyzed interesterification. 1. Compositional modifications. *J Agric Food Chem* 46:2368–2374, 1998.

107. M Persson, I Svensson, P Adlercreutz. Enzymatic fatty acid exchange in digalactosyldiacylglycerol. *Chem Phys Lipids* 104:13–21, 2000.

108. Y Totani, S Hara. Preparation of polyunsaturated phospholipids by lipase- catalyzed transesterification. *J Am Oil Chem Soc* 68:848–851, 1991.

109. M Berger, MP Schneider. Enzymatic esterification of glycerol. 2. Lipase-catalyzed synthesis of regioisomerically pure 1(3)-rac-monoacylglycerols. *J Am Oil Chem Soc* 69:961–965, 1992.

110. Y Okahata, K Ijiro. Preparation of a lipid-coated lipase and catalysis of glyceride ester syntheses in homogeneous organic solvents. *Bull Chem Soc Jpn* 65:2411–2420, 1992.

111. SJ Kwon, JJ Han, JS Rhee. Production and *in situ* separation of monoacylglycerol or diacylglycerol catalyzed by lipases in n-hexane. *Enzyme Microb Technol* 17:700–704, 1995.

112. DY Kwon, HN Song, SH Yoon. Synthesis of medium-chain glycerides by lipase in organic solvent. *J Am Oil Chem Soc* 73:1521–1525, 1996.

113. Y Shimada, A Sugihara, H Nakano, T Yokota, T Nagao, S Komemushi, Y Tominaga. Production of structured lipids containing essential fatty acids by immobilized *Rhizopus delemar* lipase. *J Am Oil Chem Soc* 73:1415–1420, 1996.

114. Y Shimada, A Sugihara, K Maruyama, T Nagao, S Nakayama, H Nakano, Y Tominaga. Production of structured lipid containing docosahexaenoic and caprylic acids using immobilized *Rhizopus delemar* lipase. *J Ferment Bioeng* 81:299–303, 1996.

115. Y Shimada, A Sugihara, M Shibahiraki, H Fujita, H Nakano, T Nagao, T Terai, Y Tominaga. Purification of gamma-linolenic acid from borage oil by a two-step enzymatic method. *J Am Oil Chem Soc* 74:1465–1470, 1997.

116. Y Shimada, N Sakai, A Sugihara, H Fujita, Y Honda, Y Tominaga. Large-scale purification of gamma-linolenic acid by selective esterification using *Rhizopus delemar* lipase. *J Am Oil Chem Soc* 75:1539–1544, 1998.

117. Y Shimada, K Maruyama, A Sugihara, S Moriyama, Y Tominaga. Purification of docosahexaenoic acid from tuna oil by a two-step enzymatic method: Hydrolysis and selective esterification. *J Am Oil Chem Soc* 74:1441–1446, 1997.

118. Y Shimada, A Sugihara, H Nakano, T Nagao, M Suenaga, S Nakai, Y Tominaga. Fatty acid specificity of *Rhizopus delemar* lipase in acidolysis. *J Ferment Bioeng* 83:321–327, 1997.

119. U Schmid, UT Bornscheuer, MM Soumanou, GP McNeill, RD Schmid. Highly selective synthesis of 1,3-oleoyl-2-palmitoylglycerol by lipase catalysis. *Biotechnol Bioeng* 64:678–684, 1999.

120. Y Shimada, M Suenaga, A Sugihara, S Nakai, Y Tominaga. Continuous production of structured lipid containing γ-linolenic and caprylic acids by immobilized *Rhizopus delemar* lipase. *J Am Oil Chem Soc* 76:189–193, 1999.

121. Y Shimada, T Nagao, Y Hamasaki, K Akimoto, A Sugihara, S Fujikawa, S Komemushi, Y Tominaga. Enzymatic synthesis of structured lipid containing arachidonic and palmitic acids. *J Am Oil Chem Soc* 77:89–93, 2000.

122. Y Shimada, A Sugihara, Y Tominaga. Enzymatic purification of polyunsaturated fatty acids. *J Biosci Bioeng* 91:529–538, 2001.

123. K Holmberg. Lipase and phospholipase catalyzed transformations in microemulsions. *Indian J Chem Sect B-Org Chem Incl Med Chem* 31:886–890, 1992.

124. M Safari, S Kermasha, L Lamboursain, JD Sheppard. Interesterification of butterfat by lipase from *Rhizopus niveus* in reverse micellar systems. *Biosci Biotechnol Biochem* 58:1553–1557, 1994.

125. JP Chen, JB Wang, HS Liu. Alcoholysis of olive oil for producing wax esters by intracellular lipase in immobilized fungus cells. *Biotechnol Lett* 17:1177–1182, 1995.

126. T Oba, B Witholt. Interesterification of milk fat with oleic acid catalyzed by immobilized *Rhizopus oryzae* lipase. *J Dairy Sci* 77:1790–1797, 1994.

127. M Basri, AC Heng, CNA Razak, W Yunus, M Ahmad, RNA Rahman, K Ampon, AB Salleh. Alcoholysis of palm oil mid-fraction by lipase from *Rhizopus rhizopodiformis*. *J Am Oil Chem Soc* 74:113–116, 1997.

128. B Decagny, S Jan, JC Vuillemard, C Sarazin, JP Seguin, C Gosselin, JN Barbotin, F Ergan. Synthesis of wax ester through triolein alcoholysis: Choice of the lipase and study of the mechanism. *Enzyme Microb Technol* 22:578–582, 1998.

129. U Schmid, UT Bornscheuer, MM Soumanou, GP McNeill, RD Schmid. Optimization of the reaction conditions in the lipase-catalyzed synthesis of structured triglycerides. *J Am Oil Chem Soc* 75:1527–1531, 1998.

130. MM Soumanou, UT Bornscheuer, U Schmid, RD Schmid. Synthesis of structured triglycerides by lipase catalysis. *Fett-Lipid* 100:156–160, 1998.

131. MM Soumanou, UT Bornscheuer, RD Schmid. Two-step enzymatic reaction for the synthesis of pure structured triacylglycerides. *J Am Oil Chem Soc* 75:703–710, 1998.

132. A Kawashima, Y Shimada, T Nagao, A Ohara, T Matsuhisa, A Sugihara, Y Tominaga. Production of structured TAG rich in 1,3-dicapryloyl-2-linolenoyl glycerol from borage oil. *J Am Oil Chem Soc* 79:871–877, 2002.

133. M Kimura, K Hasegawa, H Takamura, T Matoba. Preparation of triacylglycerol molecular-species by interesterification using endocellular lipase in normal hexane. *Agric Biol Chem* 55:3039–3043, 1991.

134. K Osada, K Takahashi, M Hatano. Polyunsaturated fatty glyceride syntheses by microbial lipases. *J Am Oil Chem Soc* 67:921–922, 1990.

135. CC Akoh. Lipase-catalyzed synthesis of partial glyceride. *Biotechnol Lett* 15:949–954, 1993.

136. HMA Mohamed, S Bloomer, K Hammadi. Modification of fats by lipase interesterification. 1. Changes in glyceride structure. *Fett Wiss Technol-Fat Sci Technol* 95:428–431, 1993.

137. KH Huang, CC Akoh, MC Erickson. Enzymatic modification of melon seed oil—incorporation of eicosapentaenoic acid. *J Agric Food Chem* 42:2646–2648, 1994.

138. GG Haraldsson, BO Gudmundsson, O Almarsson. The synthesis of homogeneous triglycerides of eicosapentaenoic acid and docosahexaenoic acid by lipase. *Tetrahedron* 51:941–952, 1995.

139. KT Lee, CC Akoh. Effects of selected substrate forms on the synthesis of structured lipids by two immobilized lipases. *J Am Oil Chem Soc* 74:579–584, 1997.

140. MA Jackson, JW King, GR List, WE Neff. Lipase-catalyzed randomization of fats and oils in flowing supercritical carbon dioxide. *J Am Oil Chem Soc* 74:635–639, 1997.

141. MM Soumanou, UT Bornscheuer, U Menge, RD Schmid. Synthesis of structured triglycerides from peanut oil with immobilized lipase. *J Am Oil Chem Soc* 74:427–433, 1997.

142. CC Akoh, CO Moussata. Lipase-catalyzed modification of borage oil: Incorporation of capric and eicosapentaenoic acids to form structured lipids. *J Am Oil Chem Soc* 75:697–701, 1998.

143. V Seriburi, CC Akoh. Enzymatic interesterification of lard and high-oleic sunflower oil with *Candida antarctica* lipase to produce plastic fats. *J Am Oil Chem Soc* 75:1339–1345, 1998.

144. V Seriburi, CC Akoh. Enzymatic interesterification of triolein and tristearin: Chemical structure and differential scanning calorimetric analysis of the products. *J Am Oil Chem Soc* 75:711–716, 1998.

145. A Bilyk, RG Bistline, MJ Haas, SH Feairheller. Lipase-catalyzed triglyceride hydrolysis in organic solvent. *J Am Oil Chem Soc* 68:320–323, 1991.

146. M Derenobales, I Agud, JM Lascaray, JC Mugica, LC Landeta, R Solozabal. Hydrolysis of animal fats by lipase at temperatures below their melting points. *Biotechnol Lett* 14:683–688, 1992.

147. S Ferreiradias, MMR Dafonseca. Production of monoglycerides by glycerolysis of olive oil with immobilized lipases—effect of the water activity. *Bioprocess Eng* 12:327–337, 1995.

148. Y Tanaka, T Funada, J Hirano, R Hashizume. Triglyceride specificity of *Candida-cylindracea* lipase —effect of docosahexaenoic acid on resistance of triglyceride to lipase. *J Am Oil Chem Soc* 70:1031–1034, 1993.

149. M Trani, R Lortie, F Ergan. Enzymatic-synthesis of trierucin from high erucic acid rapeseed oil. *J Am Oil Chem Soc* 70:961–964, 1993.

150. M Rahmatullah, VKS Shukla, KD Mukherjee. Enrichment of gamma-linolenic acid from evening primrose oil and borage oil via lipase-catalyzed hydrolysis. *J Am Oil Chem Soc* 71:569–573, 1994.

151. Y Shimada, A Sugihara, K Maruyama, T Nagao, S Nakayama, H Nakano, Y Tominaga. Enrichment of arachidonic-acid—selective hydrolysis of a single-cell oil from Mortierella with *Candida-cylindracea* lipase. *J Am Oil Chem Soc* 72:1323–1327, 1995.

152. CC Akoh, BH Jennings, DA Lillard. Enzymatic modification of evening primrose oil: Incorporation of n-3 polyunsaturated fatty acids. *J Am Oil Chem Soc* 73:1059–1062, 1996.

153. UN Wanasundara, F Shahidi. Lipase-assisted concentration of n-3 polyunsaturated fatty acids in acylglycerols from marine oils. *J Am Oil Chem Soc* 75:945–951, 1998.

154. H Breivik, GG Haraldsson, B Kristinsson. Preparation of highly purified concentrates of eicosapentaenoic acid and docosahexaenoic acid. *J Am Oil Chem Soc* 74:1425–1429, 1997.

155. FC Huang, YH Ju, CW Huang. Enrichment in gamma-linolenic acid of acylglycerols by the selective hydrolysis of borage oil. *Appl Biochem Biotechnol* 67:227–236, 1997.

156. ZR Yu, SSH Rizvi, JA Zollweg. Enzymatic esterification of fatty-acid mixtures from milk fat and anhydrous milk fat with canola oil in supercritical carbon dioxide. *Biotechnol Prog* 8:508–513, 1992.

157. R Multzsch, W Lokotsch, B Steffen, S Lang, JO Metzger, HJ Schafer, S Warwel, F Wagner. Enzymatic production and physicochemical characterization of uncommon wax esters and monoglycerides. *J Am Oil Chem Soc* 71:721–725, 1994.

158. D Mukesh, AA Banerji, R Newadkar, HS Bevinakatti. Lipase catalyzed transesterification of vegetable oils—A comparative-study in batch and tubular reactors. *Biotechnol Lett* 15:77–82, 1993.

159. DE Stevenson, RA Stanley, RH Furneaux. Glycerolysis of tallow with immobilized lipase. *Biotechnol Lett* 15:1043–1048, 1993.

160. LA Gioielli, RNM Pitombo, M Vitolo, R Baruffaldi, MN Oliveira, MS Augusto. Acidolysis of babassu fat catalyzed by immobilized lipase. *J Am Oil Chem Soc* 71:579–582, 1994.

161. HM Ghazali, A Maisarah, S Yusof, M Yusoff. Triglyceride profiles melting properties of lipase-catalyzed transesterified palm stearin and coconut oil. *Asia Pac J Mol Biol* 3:280–289, 1995.

162. M Akimoto. Nutritional improvement of soybean oil via lipase-catalyzed interesterification. *Appl Biochem Biotechnol* 74:31–41, 1998.

163. JA Arcos, C Otero, CG Hill. Rapid enzymatic production of acylglycerols from conjugated linoleic acid and glycerol in a solvent-free system. *Biotechnol Lett* 20:617–621, 1998.

164. C Edmundo, D Valerie, C Didier, M Alain. Efficient lipase-catalyzed production of tailor-made emulsifiers using solvent engineering coupled to extractive processing. *J Am Oil Chem Soc* 75:309–313, 1998.

165. YH Ju, FC Huang, CH Fang. The incorporation of n-3 polyunsaturated fatty acids into acylglycerols of borage oil via lipase-catalyzed reactions. *J Am Oil Chem Soc* 75:961–965, 1998.

166. LO Ming, HM Ghazali, CC Let. Effect of enzymatic transesterification on the fluidity of palm stearin-palm kernel olein mixtures. *Food Chem* 63:155–159, 1998.

167. KD Mukherjee, I Kiewitt. Structured triacylglycerols resembling human milk fat by transesterification catalyzed by papaya (carica papaya) latex. *Biotechnol Lett* 20:613–616, 1998.

168. MAP Langone, GL Sant'Anna. Enzymatic synthesis of medium-chain triglycerides in a solvent-free system. *Appl Biochem Biotechnol* 77-9:759–770, 1999.

169. LO Ming, HM Ghazali, CC Let. Use of enzymatic transesterified palm stearin sunflower oil blends in the preparation of table margarine formulation. *Food Chem* 64:83–88, 1999.

170. S Miura, A Ogawa, H Konishi. A rapid method for enzymatic synthesis and purification of the structured triacylglycerol, 1,3-dilauroyl-2-oleoyl-glycerol. *J Am Oil Chem Soc* 76:927–931, 1999.

171. S Nieto, J Gutierrez, J Sanhueza, A Valenzuela. Preparation of sn-2 long-chain polyunsaturated monoacylglycerols from fish oil by hydrolysis with a stereo-specific lipase from *Mucor miehei*. *Grasas Aceites* 50:111–113, 1999.

172. S Nieto, J Sanhueza, A Valenzuela. Synthesis of structured triacylglycerols containing medium-chain and long-chain fatty acids by interesterification with a stereoespecific lipase from *Mucor miehei*. *Grasas Aceites* 50:199–202, 1999.

173. Z Zainal, MSA Yusoff. Enzymatic interesterification of palm stearin and palm kernel olein. *J Am Oil Chem Soc* 76:1003–1008, 1999.

174. S Bhattacharyya, DK Bhattacharyya, BK De. Modification of tallow fractions in the preparation of edible fat products. *Eur J Lipid Sci Technol* 102:323–328, 2000.

175. R Irimescu, M Yasui, Y Iwasaki, N Shimidzu, T Yamane. Enzymatic synthesis of 1,3-dicapryloyl-2-eicosapentaenoylglycerol. *J Am Oil Chem Soc* 77:501–506, 2000.

176. OM Lai, HM Ghazali, F Cho, CL Chong. Physical properties of lipase-catalyzed transesterified blends of palm stearin and anhydrous milk fat. *Food Chem* 70:215–219, 2000.

177. S Sellappan, CC Akoh. Enzymatic acidolysis of tristearin with lauric and oleic acids to produce coating lipids. *J Am Oil Chem Soc* 77:1127–1133, 2000.

178. XB Xu, LB Fomuso, CC Akoh. Modification of menhaden oil by enzymatic acidolysis to produce structured lipids: Optimization by response surface design in a packed bed reactor. *J Am Oil Chem Soc* 77:171–176, 2000.

179. H Zhang, XB Xu, HL Mu, J Nilsson, J Adler-Nissen, CE Hoy. Lipozyme im-catalyzed interesterification for the production of margarine fats in a 1 kg scale stirred tank reactor. *Eur J Lipid Sci Technol* 102:411–418, 2000.

180. E Ledochowska, A Jewusiak, M Szymczak. Preparation of structured lipids with special functional properties. *J Food Lipids* 8:239–250, 2001.

181. LB Fomuso, CC Akoh. Lipase-catalyzed acidolysis of olive oil and caprylic acid in a bench-scale packed bed bioreactor. *Food Res Int* 35:15–21, 2002.

182. IH Kim, H Kim, KT Lee, SH Chung, SN Ko. Lipase-catalyzed acidolysis of perilla oil with caprylic acid to produce structured lipids. *J Am Oil Chem Soc* 79:363–367, 2002.

183. BC Paez, AR Medina, FC Rubio, PG Moreno, EM Grima. Production of structured triglycerides rich in n-3 polyunsaturated fatty acids by the acidolysis of cod liver oil and caprylic acid in a packed-bed reactor: Equilibrium and kinetics. *Chem Eng Sci* 57:1237–1249, 2002.

184. M Rahmatullah, VKS Shukla, KD Mukherjee. Gamma-linolenic acid concentrates from borage and evening primrose oil fatty-acids via lipase-catalyzed esterification. *J Am Oil Chem Soc* 71:563–567, 1994.

185. FC Huang, YH Ju, CW Huang. Enrichment of gamma-linolenic acid from borage oil via lipase-catalyzed reactions. *J Am Oil Chem Soc* 74:977–981, 1997.

186. R Rodriguez, J Sanhueza, A Valenzuela, S Nieto. Hydrolysis of coconut oil (cocos nucifera l.) by specificity and no positional specificity enzymes. *Grasas Aceites* 48:6–10, 1997.

187. GG Haraldsson, B Kristinsson. Separation of eicosapentaenoic acid and docosahexaenoic acid in fish oil by kinetic resolution using lipase. *J Am Oil Chem Soc* 75:1551–1556, 1998.

188. YH Ju, TC Chen. High purity gamma-linolenic acid from borage oil fatty acids. *J Am Oil Chem Soc* 79:29–32, 2002.

189. R Sridhar, G Lakshminarayana, TNB Kaimal. Modification of selected indian vegetable fats into cocoa butter substitutes by lipase-catalyzed ester interchange. *J Am Oil Chem Soc* 68:726–730, 1991.

190. CN Chong, YM Hoh, CW Wang. Fractionation procedures for obtaining cocoa butter-like fat from enzymatically interesterified palm olein. *J Am Oil Chem Soc* 69:137–140, 1992.

191. SH Yoon, O Miyawaki, KH Park, K Nakamura. Transesterification between triolein and ethylbehenate by immobilized lipase in supercritical carbon dioxide. *J Ferment Bioeng* 82:334–340, 1996.

192. D Undurraga, A Markovits, S Erazo. Cocoa butter equivalent through enzymic interesterification of palm oil midfraction. *Process Biochem* 36:933–939, 2001.

193. M Trani, F Ergan, G Andre. Lipase-catalyzed production of wax esters. *J Am Oil Chem Soc* 68:20–22, 1991.

194. DG Hayes, R Kleiman. Lipase-catalyzed synthesis of lesquerolic acid wax and diol esters and their properties. *J Am Oil Chem Soc* 73:1385–1392, 1996.

195. E Ucciani, M SchmittRozieres, A Debal, LC Comeau. Enzymatic synthesis of some wax esters. *Fett-Lipid* 98:206–210, 1996.

196. N GomaDoncescu, MD Legoy. An original transesterification route for fatty acid ester production from vegetable oils in a solvent-free system. *J Am Oil Chem Soc* 74:1137–1143, 1997.

197. ML Hallberg, DB Wang, M Harrod. Enzymatic synthesis of wax esters from rapeseed fatty acid methyl esters and a fatty alcohol. *J Am Oil Chem Soc* 76:183–187, 1999.

198. L Poisson, S Jan, JC Vuillemard, C Sarazin, JP Seguin, JN Barbotin, F Ergan. Lipase-catalyzed synthesis of waxes from milk fat and oleyl alcohol. *J Am Oil Chem Soc* 76:1017–1021, 1999.

199. S Pecnik, Z Knez. Enzymatic fatty ester synthesis. *J Am Oil Chem Soc* 69:261–265, 1992.

200. D Mukesh, S Jadhav, D Sheth, AA Banerji, R Newadkar, HS Bevinakatti. Lipase-catalyzed esterification reaction in a tubular reactor. *Biotechnol Tech* 7:569–572, 1993.

201. M Habulin, V Krmelj, Z Knez. Synthesis of oleic acid esters catalyzed by immobilized lipase. *J Agric Food Chem* 44:338–342, 1996.

202. T Chatterjee, DK Bhattacharyya. Synthesis of terpene esters by an immobilized lipase in a solvent-free system. *Biotechnol Lett* 20:865–868, 1998.

203. T Chatterjee, DK Bhattacharyya. Synthesis of monoterpene esters by alcoholysis reaction with *Mucor miehei* lipase in a solvent-free system. *J Am Oil Chem Soc* 75:651–655, 1998.

204. CYG Kee, M Hassan, KB Ramachandran. Studies on the kinetics of isopropyl palmitate synthesis in packed bed bioreactor using immobilized lipase. *Artif Cells Blood Substit Immobil Biotechnol* 27:393–398, 1999.

205. LN Mutua, CC Akoh. Lipase-catalyzed modification of phospholipids-incorporation of n-3 fatty acids into biosurfactants. *J Am Oil Chem Soc* 70:125–128, 1993.

206. F Hara, T Nakashima. Transesterification of phospholipids by acetone-dried cells of a rhizopus species immobilized on biomass support particles. *J Am Oil Chem Soc* 73:657–659, 1996.

207. GG Haraldsson, A Thorarensen. Preparation of phospholipids highly enriched with n-3 polyunsaturated fatty acids by lipase. *J Am Oil Chem Soc* 76:1143–1149, 1999.

208. L Kim, BG Kim. Lipase-catalyzed synthesis of lysophosphatidylcholine using organic cosolvent for in situ water activity control. *J Am Oil Chem Soc* 77:791–797, 2000.

209. MJ Haas, DJ Cichowicz, J Phillips, R Moreau. The hydrolysis of phosphatidylcholine by an immobilized lipase-optimization of hydrolysis in organic solvents. *J Am Oil Chem Soc* 70:111–117, 1993.

210. DG Hayes, KD Carlson, R Kleiman. The isolation of hydroxy acids from lesquerella oil lipolysate by a saponification/extraction technique. *J Am Oil Chem Soc* 73:1113–1119, 1996.

211. I Elfman-Borjesson, M Harrod. Synthesis of monoglycerides by glycerolysis of rapeseed oil using immobilized lipase. *J Am Oil Chem Soc* 76:701–707, 1999.

212. R Sridhar, G Lakshminarayana, TNB Kaimal. Modification of selected edible vegetable oils to high oleic oils by lipase-catalyzed ester interchange. *J Agric Food Chem* 39:2069–2071, 1991.

213. T Dumont, D Barth, M Perrut. Continuous synthesis of ethyl myristate by enzymatic reaction in supercritical carbon dioxide. *J Supercrit Fluids* 6:85–89, 1993.

214. M Habulin, Z Knez. Influence of reaction parameters on synthesis of n-butyl oleate by immobilized *Mucor miehei* lipase. *Fett Wiss Technol-Fat Sci Technol* 95:249–252, 1993.

215. YY Linko, M Lamsa, A Huhtala, P Linko. Lipase-catalyzed transesterification of rapeseed oil and 2-ethyl-1-hexanol. *J Am Oil Chem Soc* 71:1411–1414, 1994.

216. Z Knez, V Rizner, M Habulin, D Bauman. Enzymatic-synthesis of oleyl oleate in dense fluids. *J Am Oil Chem Soc* 72:1345–1349, 1995.

217. FV Lima, DL Pyle, JA Asenjo. Factors affecting the esterification of lauric acid using an immobilized biocatalyst-enzyme characterization and studies in a well-mixed reactor. *Biotechnol Bioeng* 46:69–79, 1995.

218. FV Lima, DL Pyle, JA Asenjo. Reaction kinetics of the esterification of lauric acid in iso-octane using an immobilized biocatalyst. *Appl Biochem Biotechnol* 61:411–422, 1996.

219. GP McNeill, PE Sonnet. Low-calorie triglyceride synthesis by lipase-catalyzed esterification of monoglycerides. *J Am Oil Chem Soc* 72:1301–1307, 1995.

220. KT Lee, CC Akoh. Solvent-free enzymatic synthesis of structured lipids from peanut oil and caprylic acid in a stirred tank batch reactor. *J Am Oil Chem Soc* 75:1533–1537, 1998.

221. P Borg, M Girardin, B Rovel, D Barth. Comparison between two processes for the enzymatic synthesis of tri-docosahexaenoylglycerol in a solvent-free medium. *Biotechnol Lett* 22:777–781, 2000.

222. DL Compton, JA Laszlo, MA Berhow. Lipase-catalyzed synthesis of ferulate esters. *J Am Oil Chem Soc* 77:513–519, 2000.

223. T Garcia, A Coteron, M Martinez, J Aracil. Kinetic model for the esterification of oleic acid and cetyl alcohol using an immobilized lipase as catalyst. *Chem Eng Sci* 55:1411–1423, 2000.

224. H Zhang, XB Xu, J Nilsson, HL Mu, J Adler-Nissen, CE Hoy. Production of margarine fats by enzymatic interesterification with silica-granulated *Thermomyces lanuginosa* lipase in a large-scale study. *J Am Oil Chem Soc* 78:57–64, 2001.

225. D Rooney, LR Weatherley. The effect of reaction conditions upon lipase catalysed hydrolysis of high oleate sunflower oil in a stirred liquid-liquid reactor. *Process Biochem* 36:947–953, 2001.

226. PS Chang, JS Rhee, JJ Kim. Continuous glycerolysis of olive oil by chromobacterium-viscosum lipase immobilized on liposome in reversed micelles. *Biotechnol Bioeng* 38:1159–1165, 1991.

227. JP Chen, SD McGill. Enzymatic hydrolysis of triglycerides by *Rhizopus delemar* immobilized on biomass support particles. *Food Biotechnol* 6:1–18, 1992.

228. DA Miller, HW Blanch, JM Prausnitz. Enzyme-catalyzed interesterification of triglycerides in supercritical carbon-dioxide. *Ind Eng Chem Res* 30:939–946, 1991.

229. SW Cho, JS Rhee. Immobilization of lipase for effective interesterification of fats and oils in organic-solvent. *Biotechnol Bioeng* 41:204–210, 1993.

230. AL Paiva, FX Malcata. Process integration involving lipase-catalyzed ester synthesis reactions. *Biotechnol Tech* 8:629–634, 1994.

231. DB Sarney, G Fregapane, EN Vulfson. Lipase-catalyzed synthesis of lysophospholipids in a continuous bioreactor. *J Am Oil Chem Soc* 71:93–96, 1994.

232. FC Huang, YH Ju. Continuous hydrolysis of palm kernel olein with immobilized *Rhizopus arrhizus* lipase in a packed bed reactor. *J Chin Inst Chem Eng* 26:435–439, 1995.

233. HS Garcia, KJ Keough, JA Arcos, CG Hill. Continuous interesterification of butteroil and conjugated linoleic acid in a tubular reactor packed with an immobilized lipase. *Biotechnol Tech* 13:369–373, 1999.

234. HL Mu, XB Xu, J Adler-Nissen, CE Hoy. Production of structured lipids by lipase-catalyzed interesterification in a packed bed reactor: Effect of reaction parameters on the level of diacylglycerols in the products. *Fett-Lipid* 101:158–164, 1999.

235. X Xu, H Mu, CE Hoy, J Adler-Nissen. Production of specifically structured lipids by enzymatic interesterification in a pilot enzyme bed reactor: Process optimization by response surface methodology. *Fett-Lipid* 101:207–214, 1999.

236. V Dossat, D Combes, A Marty. Continuous enzymatic transesterification of high oleic sunflower oil in a packed bed reactor: Influence of the glycerol production. *Enzyme Microb Technol* 25:194–200, 1999.

237. E Wehtje, D Costes, P Adlercreutz. Continuous lipase-catalyzed production of wax ester using silicone tubing. *J Am Oil Chem Soc* 76:1489–1493, 1999.

238. JA Arcos, HS Garcia, CG Hill. Continuous enzymatic esterification of glycerol with (poly)unsaturated fatty acids in a packed-bed reactor. *Biotechnol Bioeng* 68:563–570, 2000.

239. R Goddard, J Bosley, B Al-Duri. Esterification of oleic acid and ethanol in plug flow (packed bed) reactor under supercritical conditions-investigation of kinetics. *J Supercrit Fluids* 18:121–130, 2000.

240. A H-Kittikun, P Prasertsan, C Sungpud. Continuous production of fatty acids from palm olein by immobilized lipase in a two-phase system. *J Am Oil Chem Soc* 77:599–603, 2000.

241. KT Lee, TA Foglia. Synthesis, purification, and characterization of structured lipids produced from chicken fat. *J Am Oil Chem Soc* 77:1027–1034, 2000.

242. AD MacKenzie, DE Stevenson. Production of high-oleic acid tallow fractions using lipase-catalyzed directed interesterification, using both batch and continuous processing. *Enzyme Microb Technol* 27:302–311, 2000.

243. XB Xu, LB Fomuso, CC Akoh. Synthesis of structured triacylglycerols by lipase-catalyzed acidolysis in a packed bed bioreactor. *J Agric Food Chem* 48:3–10, 2000.

244. CC Akoh, CO Moussata. Characterization and oxidative stability of enzymatically produced fish and canola oil-based structured lipids. *J Am Oil Chem Soc* 78:25–30, 2001.

245. HS Garcia, JA Arcos, KJ Keough, CG Hill. Immobilized lipase-mediated acidolysis of butteroil with conjugated linoleic acid: Batch reactor and packed bed reactor studies. *J Mol Catal B-Enzym* 11:623–632, 2001.

246. CF Torres, F Munir, LP Lessard, CG Hill. Lipase-mediated acidolysis of tristearin with CLA in a packed-bed reactor: A kinetic study. *J Am Oil Chem Soc* 79:655–661, 2002.

247. CF Torres, E Barrios, CG Hill. Lipase-catalyzed acidolysis of menhaden oil with CLA: Optimization by factorial design. *J Am Oil Chem Soc* 79:457–466, 2002.

248. XB Xu, T Porsgaard, H Zhang, J Adler-Nissen, CE Hoy. Production of structured lipids in a packed-bed reactor with *Thermomyces lanuginosa* lipase. *J Am Oil Chem Soc* 79:561–565, 2002.

249. RPM Guit, M Kloosterman, GW Meindersma, M Mayer, EM Meijer. Lipase kinetics-hydrolysis of triacetin by lipase from *Candida cylindracea* in a hollow-fiber membrane reactor. *Biotechnol Bioeng* 38:727–732, 1991.

250. FX Malcata, CG Hill, CH Amundson. Use of a lipase immobilized in a membrane reactor to hydrolyze the glycerides of butteroil. *Biotechnol Bioeng* 38:853–868, 1991.

251. Y Isono, H Nabetani, M Nakajima. Wax ester synthesis in a membrane reactor with lipase-surfactant complex in hexane. *J Am Oil Chem Soc* 72:887–890, 1995.

252. XB Xu, A Skands, G Jonsson, J Adler-Nissen. Production of structured lipids by lipase-catalysed interesterification in an ultrafiltration membrane reactor. *Biotechnol Lett* 22:1667–1671, 2000.

253. X Xu, S Balchen, G Jonsson, J Adler-Nissen. Production of structured lipids by lipase-catalyzed interesterification in a flat membrane reactor. *J Am Oil Chem Soc* 77:1035–1041, 2000.

254. M Linder, J Fanni, M Parmentier. Enzyme-catalysed enrichment of n-3 polyunsaturated fatty acids of salmon oil: Optimisation of reaction conditions. *OCL-Ol Corps Gras Lipides* 8:73–77, 2001.

255. FC Huang, YH Ju, JC Chiang. Gamma-linolenic acid-rich triacylglycerols derived from borage oil via lipase-catalyzed reactions. *J Am Oil Chem Soc* 76:833–837, 1999.

256. S Senanayake, F Shahidi. Enzymatic incorporation of docosahexaenoic acid into borage oil. *J Am Oil Chem Soc* 76:1009–1015, 1999.

257. S Senanayake, F Shahidi. Enzyme-assisted acidolysis of borage (*Borago officinalis* l.) and evening primrose (*Oenothera biennis* l.) oils: Incorporation of omega-3 polyunsaturated fatty acids. *J Agric Food Chem* 47:3105–3112, 1999.

258. S Senanayake, F Shahidi. Structured lipids via lipase-catalyzed incorporation of eicosapentaenoic acid into borage (*Borago officinalis* l.) and evening primrose (*Oenothera biennis* l.) oils. *J Agric Food Chem* 50:477–483, 2002.

259. S Ramamurthi, PR Bhirud, AR McCurdy. Enzymatic methylation of canola oil deodorizer distillate. *J Am Oil Chem Soc* 68:970–975, 1991.

260. J Kurashige, N Matsuzaki, H Takahashi. Enzymatic modification of canola palm oil mixtures-effects on the fluidity of the mixture. *J Am Oil Chem Soc* 70:849–852, 1993.

261. S Ainsworth, C Versteeg, M Palmer, MB Millikan. Enzymatic interesterification of fats. *Aust J Dairy Technol* 51:105–107, 1996.

262. AG Marangoni, D Rousseau. Chemical and enzymatic modification of butterfat and butterfat-canola oil blends. *Food Res Int* 31:595–599, 1998.

263. LB Fomuso, CC Akoh. Enzymatic modification of high-laurate canola to produce margarine fat. *J Agric Food Chem* 49:4482–4487, 2001.

264. KT Lee, TA Foglia. Fractionation of chicken fat triacylglycerols: Synthesis of structured lipids with immobilized lipases. *J Food Sci* 65:826–831, 2000.

265. KT Lee, TA Foglia, MJ Oh. Lipase-catalyzed synthesis of structured lipids with fatty acids fractionated from saponified chicken fat and menhaden oil. *Eur J Lipid Sci Technol* 103:777–782, 2001.

266. S Ghosh, DK Bhattacharyya. Medium-chain fatty acid-rich glycerides by chemical and lipase-catalyzed polyester-monoester interchange reaction. *J Am Oil Chem Soc* 74:593–595, 1997.

267. K Long, HM Ghazali, A Ariff, C Bucke. Acidolysis of several vegetable oils by mycelium-bound lipase of *Aspergillus flavus* link. *J Am Oil Chem Soc* 74:1121–1128, 1997.

268. GP McNeill, S Shimizu, T Yamane. High-yield enzymatic glycerolysis of fats and oils. *J Am Oil Chem Soc* 68:1–5, 1991.

269. CE Martinez, JC Vinay, R Brieva, CG Hill, HS Garcia. Lipase-catalyzed interesterification (acidolysis) of corn oil and conjugated linoleic acid in organic solvents. *Food Biotechnol* 13:183–193, 1999.

270. PS Sehanputri, CG Hill. Biotechnology for the production of nutraceuticals enriched in conjugated linoleic acid: I. Uniresponse kinetics of the hydrolysis of corn oil by a pseudomonas sp. Lipase immobilized in a hollow fiber reactor. *Biotechnol Bioeng* 64:568–579, 1999.

271. CF Torres, F Munir, RM Blanco, C Otero, CG Hill. Catalytic transesterification of corn oil and tristearin using immobilized lipases from *Thermomyces lanuginosa*. *J Am Oil Chem Soc* 79:775–781, 2002.

272. HMA Mohamed, K Larsson. Modification of fats by lipase interesterification. 2. Effect on crystallization behavior and functional properties. *Fett Wiss Technol-Fat Sci Technol* 96:56–59, 1994.

273. HMA Mohamed, MH Iskandar, B Sivik, K Larsson. Preparation and characterization of a zero-trans margarine. *Fett Wiss Technol-Fat Sci Technol* 97:336–340, 1995.

274. DJ Sessa, WE Neff, GR List, MAM Zeitoun. Melting and crystalline properties of enzyme catalysed interesterified vegetable oil-soapstock fatty acid blends. *Food Sci Technol-Lebensm-Wiss Technol* 29:581–585, 1996.

275. O Kose, M Tuter, HA Aksoy. Immobilized *Candida antarctica* lipase-catalyzed alcoholysis of cotton seed oil in a solvent-free medium. *Bioresour Technol* 83:125–129, 2002.

276. VB Yadwad, OP Ward, LC Noronha. Application of lipase to concentrate the docosahexaenoic acid (DHA) fraction of fish oil. *Biotechnol Bioeng* 38:956–959, 1991.

277. Y Tanaka, J Hirano, T Funada. Concentration of docosahexaenoic acid in glyceride by hydrolysis of fish oil with *Candida cylindracea* lipase. *J Am Oil Chem Soc* 69:1210–1214, 1992.

278. T Yamane, T Suzuki, Y Sahashi, L Vikersveen, T Hoshino. Production of n-3 polyunsaturated fatty acid-enriched fish oil by lipase-catalyzed acidolysis without solvent. *J Am Oil Chem Soc* 69:1104–1107, 1992.

279. GG Haraldsson, BO Gudmundsson, O Almarsson. The preparation of homogeneous triglycerides of eicosapentaenoic acid and docosahexaenoic acid by lipase. *Tetrahedron Lett* 34:5791–5794, 1993.

280. ZY Li, OP Ward. Lipase-catalyzed alcoholysis to concentrate the n-3 polyunsaturated fatty-acid of cod liver oil. *Enzyme Microb Technol* 15:601–606, 1993.

281. Y Kosugi, N Azuma. Synthesis of triacylglycerol from polyunsaturated fatty-acid by immobilized lipase. *J Am Oil Chem Soc* 71:1397–1403, 1994.

282. H Maehr, G Zenchoff, DL Coffen. Enzymatic enhancement of n-3 fatty-acid content in fish oils. *J Am Oil Chem Soc* 71:463–467, 1994.

283. Y Sahashi, H Ishizuka, S Koike, K Suzuki. Separation and concentration of polyunsaturated fatty acids by a combined system of liquid-liquid-extraction and membrane separation. *Kag Kog Ronbunshu* 20:148–155, 1994.

284. Y Tanaka, J Hirano, T Funada. Synthesis of docosahexaenoic acid-rich triglyceride with immobilized *Chromobacterium viscosum* lipase. *J Am Oil Chem Soc* 71:331–334, 1994.

285. GP McNeill, RG Ackman, SR Moore. Lipase-catalyzed enrichment of long-chain polyunsaturated fatty acids. *J Am Oil Chem Soc* 73:1403–1407, 1996.

286. SR Moore, GP McNeill. Production of triglycerides enriched in long-chain n-3 polyunsaturated fatty acids from fish oil. *J Am Oil Chem Soc* 73:1409–1414, 1996.

287. Y Shimada, A Sugihara, H Nakano, T Kuramoto, T Nagao, M Gemba, Y Tominaga. Purification of docosahexaenoic acid by selective esterification of fatty acids from tuna oil with *Rhizopus delemar* lipase. *J Am Oil Chem Soc* 74:97–101, 1997.

288. Y Shimada, A Sugihara, S Yodono, T Nagao, K Maruyama, H Nakano, S Komemushi, Y Tominaga. Enrichment of ethyl docosahexaenoate by selective alcoholysis with immobilized *Rhizopus delemar* lipase. *J Ferment Bioeng* 84:138–143, 1997.

289. VJ Robles, HS Garcia, JA Monroy, O Angulo. Lipase-catalyzed esterification of glycerol with n-3 polyunsaturated fatty acids from winterized fish oil. *Food Sci Technol Int* 4:401–405, 1998.

290. X Xu, S Balchen, CE Hoy, J Adler-Nissen. Pilot batch production of specific-structured lipids by lipase-catalyzed interesterification: Preliminary study on incorporation and acyl migration. *J Am Oil Chem Soc* 75:301–308, 1998.

291. BH Jennings, CC Akoh. Enzymatic modification of triacylglycerols of high eicosapentaenoic and docosahexaenoic acids content to produce structured lipids. *J Am Oil Chem Soc* 76:1133–1137, 1999.

292. BH Jennings, CC Akoh. Lipase catalyzed modification of fish oil to incorporate capric acid. *Food Chem* 72:273–278, 2001.

293. Y Watanabe, Y Shimada, A Sugihara, Y Tominaga. Stepwise ethanolysis of tuna oil using immobilized *Candida antarctica* lipase. *J Biosci Bioeng* 88:622–626, 1999.

294. HS Garcia, JA Arcos, DJ Ward, CG Hill. Synthesis of glycerides containing n-3 fatty acids and conjugated linoleic acid by solvent-free acidolysis of fish oil. *Biotechnol Bioeng* 70:587–591, 2000.

295. DQ Zhou, XB Xu, HL Mu, CE Hoy, J Adler-Nissen. Lipase-catalyzed production of structured lipids via acidolysis of fish oil with caprylic acid. *J Food Lipids* 7:263–274, 2000.

296. R Irimescu, K Furihata, K Hata, Y Iwasaki, T Yamane. Two-step enzymatic synthesis of docosahexaenoic acid-rich symmetrically structured triacylglycerols via 2- monoacylglycerols. *J Am Oil Chem Soc* 78:743–748, 2001.

297. CF Torres, HS Garcia, JJ Ries, CG Hill. Esterification of glycerol with conjugated linoleic acid and long-chain fatty acids from fish oil. *J Am Oil Chem Soc* 78:1093–1098, 2001.

298. CF Torres, CG Hill. Lipase-catalyzed acidolysis of menhaden oil with conjugated linoleic acid: Effect of water content. *Biotechnol Bioeng* 78:509–516, 2002.

299. C Torres, B Lin, CG Hill. Lipase-catalyzed glycerolysis of an oil rich in eicosapentaenoic acid residues. *Biotechnol Lett* 24:667–673, 2002.

300. ZY Li, OP Ward. Enzyme catalyzed production of vegetable-oils containing omega-3 polyunsaturated fatty acid. *Biotechnol Lett* 15:185–188, 1993.

301. U Bornscheuer. Lipase-catalyzed synthesis of monoglycerides. *Fett Wiss Technol-Fat Sci Technol* 97:241–249, 1995.

302. P Forssell, R Kervinen, M Lappi, P Linko, T Suortti, K Poutanen. Effect of enzymatic interesterification on the melting-point of tallow-rapeseed oil (lear) mixture. *J Am Oil Chem Soc* 69:126–136, 1992.

303. P Forssell, P Parovuori, P Linko, K Poutanen. Enzymatic transesterification of rapeseed oil and lauric acid in a continuous reactor. *J Am Oil Chem Soc* 70:1105–1109, 1993.

304. XB Xu, ARH Skands, J Adler-Nissen, CE Hoy. Production of specific structured lipids by enzymatic interesterification: Optimization of the reaction by response surface design. *Fett-Lipid* 100:463–471, 1998.

305. DQ Zhou, XB Xu, HL Mu, CE Hoy, J Adler-Nissen. Synthesis of structured triacylglycerols containing caproic acid by lipase-catalyzed acidolysis: Optimization by response surface methodology. *J Agric Food Chem* 49:5771–5777, 2001.

306. GP McNeill, RG Berger. Enzymatic glycerolysis of palm oil fractions and a palm oil based model mixture-relationship between fatty-acid composition and monoglyceride yield. *Food Biotechnol* 7:75–87, 1993.

307. M Tuter, B Babali, O Kose, S Dural, HA Aksoy. Solvent-free glycerolysis of palm and palm kernel oils catalyzed by a 1,3-specific lipase and fatty acid composition of glycerolysis products. *Biotechnol Lett* 21:245–248, 1999.

308. OM Lai, HM Ghazali, DM Hashim, F Cho, CL Chong. Viscoelastic properties of table margarine prepared from lipase-catalyzed transesterified mixtures of palm stearin and palm kernel olein. *J Food Lipids* 6:25–46, 1999.

309. M Tuter, HA Aksoy. Solvent-free glycerolysis of palm and palm kernel oils catalyzed by commercial 1,3-specific lipase from *Humicola lanuginosa* and composition of glycerolysis products. *Biotechnol Lett* 22:31–34, 2000.

310. D Oliveira, JV Oliveira. Enzymatic alcoholysis of palm kernel oil in n-hexane and supercritical carbon dioxide. *J Supercrit Fluids* 19:141–148, 2001.

311. BS Chu, HM Ghazali, OM Lai, YBC Man, S Yusof, SB Tee, MSA Yusoff. Comparison of lipase-transesterified blend with some commercial solid frying shortenings in Malaysia. *J Am Oil Chem Soc* 78:1213–1219, 2001.

312. GP McNeill, T Yamane. Further improvements in the yield of monoglycerides during enzymatic glycerolysis of fats and oils. *J Am Oil Chem Soc* 68:6–10, 1991.

313. OM Lai, HM Ghazali, CL Chong. Effect of enzymatic transesterification on the melting points of palm stearin sunflower oil mixtures. *J Am Oil Chem Soc* 75:881–886, 1998.

314. OM Lai, HM Ghazali, F Cho, CL Chong. Flow properties of table margarine prepared from lipase-catalysed transesterified palm stearin palm kernel olein feedstock. *Food Chem* 64:221–226, 1999.

315. OM Laia, HM Ghazalia, F Cho, CL Chong. Physical and textural properties of an experimental table margarine prepared from lipase-catalysed transesterified palm stearin: Palm kernel olein mixture during storage. *Food Chem* 71:173–179, 2000.

316. IN Hayati, A Aminah, S Mamot, IN Aini, HMN Lida, S Sabariah. Melting characteristic and solid fat content of milk fat and palm stearin blends before and after enzymatic interesterification. *J Food Lipids* 7:175–193, 2000.

317. OM Lai, HM Ghazali, F Cho, CL Chong. Enzymatic transesterification of palm stearin: Anhydrous milk fat mixtures using 1,3-specific and non-specific lipases. *Food Chem* 70:221–225, 2000.

318. BS Chu, HM Ghazali, OM Lai, YBC Man, S Yusof, MSA Yusoff. Performance of a lipase-catalyzed transesterified palm kernel olein and palm stearin blend in frying banana chips. Food Chem 74: 21-33, 2001.

319. LP Lim, HM Ghazali, OM Lai, BS Chu. Comparison of transesterified palm stearin/sunflower oil blends catalyzed by *Pseudomonas* and *Mucor javanicus* lipase. *J Food Lipids* 8:103–114, 2001.

320. MJ Hills, I Kiewitt, KD Mukherjee. Enzymatic fractionation of evening primrose oil by rape lipase—enrichment of gamma-linolenic acid. *Biotechnol Lett* 11:629–632, 1989.

321. BH Jennings, CC Akoh. Lipase-catalyzed modification of rice bran oil to incorporate capric acid. *J Agric Food Chem* 48:4439–4443, 2000.

322. BH Jennings, CC Akoh, JB Eun. Lipase-catalyzed modification of sesame oil to incorporate capric acid. *J Food Lipids* 7:21–30, 2000.

323. MAM Zeitoun, WE Neff, TL Mounts. Interesterification of soybean oil using a sn-1,3- triacylglycerol lipase. *Revue Franncaise Des Corps Gras* 39:85–90, 1992.

324. KH Huang, CC Akoh. Lipase-catalyzed incorporation of n-3 polyunsaturated fatty acids into vegetable oils. *J Am Oil Chem Soc* 71:1277–1280, 1994.

325. KH Huang, CC Akoh. Optimization and scale-up of enzymatic synthesis of structured lipids using RSM. *J Food Sci* 61:137–141, 1996.

326. TA Foglia, K Petruso, SH Feairheller. Enzymatic interesterification of tallow-sunflower oil mixtures. *J Am Oil Chem Soc* 70:281–285, 1993.

327. N El, L Dandik, HA Aksoy. Solvent-free glycerolysis catalyzed by acetone powder of *Nigella sativa* seed lipase. *J Am Oil Chem Soc* 75:1207–1211, 1998.

328. M Tuter, F Arat, L Dandik, HA Aksoy. Solvent-free glycerolysis of sunflower oil and anchovy oil catalyzed by a 1,3-specific lipase. *Biotechnol Lett* 20:291–294, 1998.

329. FA Zaher, SM Aly, OS El-Kinawy. Lipase-catalyzed glycerolysis of sunflower oil to produce partial glycerides. *Grasas Aceites* 49:411–414, 1998.

330. B Selmi, D Thomas. Immobilized lipase-catalyzed ethanolysis of sunflower oil in a solvent-free medium. *J Am Oil Chem Soc* 75:691–695, 1998.

331. RMM Diks, MJ Lee. Production of a very low saturate oil based on the specificity of *Geotrichum candidum* lipase. *J Am Oil Chem Soc* 76:455–462, 1999.

332. MJ Haas, K Scott, W Jun, G Janssen. Enzymatic phosphatidylcholine hydrolysis in organic-solvents—an examination of selected commercially available lipases. *J Am Oil Chem Soc* 71:483–490, 1994.

333. A Mustranta, T Suortti, K Poutanen. Transesterification of phospholipids in different reaction conditions. *J Am Oil Chem Soc* 71:1415–1419, 1994.

334. M Liljahallberg, M Harrod. Enzymatic and nonenzymatic esterification of long polyunsaturated fatty-acids and lysophosphatidylcholine in isooctane. *Biocatal Biotransform* 12:55–66, 1995.

335. M Hosokawa, M Ito, K Takahashi. Preparation of highly unsaturated fatty acid-containing phosphatidylcholine by transesterification with phospholipase A$_2$. *Biotechnol Tech* 12:583–586, 1998.

336. A Mustranta, P Forssell, K Poutanen. Comparison of lipases and phospholipases in the hydrolysis of phospholipids. *Process Biochem* 30:393–401, 1995.

337. M Ghosh, DK Bhattacharyya. Soy lecithin-monoester interchange reaction by microbial lipase. *J Am Oil Chem Soc* 74:761–763, 1997.

338. M Ghosh, DK Bhattacharyya. Enzymatic alcoholysis reaction of soy phospholipids. *J Am Oil Chem Soc* 74:597–599, 1997.

339. LF Peng, XB Xu, HL Mu, CE Hoy, J Adler-Nissen. Production of structured phospholipids by lipase-catalyzed acidolysis: Optimization using response surface methodology. *Enzyme Microb Technol* 31:523–532, 2002.

10

Lipase-Catalyzed Condensation in an Organic Solvent

Shuji Adachi

10.1 Introduction

Enzymatic synthesis of fatty acid esters of saccharides,[1–16] sugar alcohols,[6,17–22] ascorbic acid,[23–29] and other alcohols through both transesterification and condensation reaction has been extensively studied. The esters have been of much interest for use in industries for food, cosmetics, and pharmaceuticals. Compared to the conventional chemical synthesis of the esters, enzymatic preparation has some advantages such as the direct use of unmodified substrates, moderate reaction conditions, and high regiospecificity of the enzyme. Because the lipase-catalyzed reaction in a conventional aqueous system thermodynamically favors the hydrolysis, the reaction was carried out in an organic medium with a low water content, in a solvent-free system and under reduced pressure and/or in the presence of a desiccant to shift the reaction toward synthesis.

Although the reaction equilibrium constant is an important factor for predicting the equilibrium yield of a desired product, only a few studies have reported the factors affecting the constant for the lipase-catalyzed esterification (condensation) in organic solvents. Knowledge of the kinetics of the enzymatic reactions in an organic solvent also seems to be insufficient. The lipase-catalyzed synthesis of the esters in an organic solvent has usually been performed using a batch reaction. However, a continuous reaction would be preferred for large-scale production. Characterization of the products is also important for their utilization.

In this context, factors affecting the reaction equilibrium constant, the substrate selectivity of a lipase for various carboxylic acids, and continuous production of the esters will be described. Properties of acyl ascorbates will also be mentioned. An immobilized lipase from *Candida antarctica*, Novozym® 435 or Chirazyme® L-2 C2, was used throughout this chapter.

10.2   Factors Affecting the Reaction Equilibrium Constant

10.2.1 Reaction Equilibrium Constant

Lipase-catalyzed condensation of an alcohol, A, and a fatty (or carboxylic) acid, F, can be described as follows:

$$A + F \rightleftharpoons P + W \tag{10.1}$$

where P and W represent a product (an ester) and water, respectively. A reaction equilibrium constant K_a for the above reaction is thermodynamically defined based on the activities of substrates and products by Equation 10.2:

$$K_a = \frac{a_{Pe} a_{We}}{a_{Ae} a_{Fe}} \tag{10.2}$$

where the subscript e indicates equilibrium. The K_a value for a reaction should be intrinsic under specific temperature and pressure. However, we usually define an apparent reaction equilibrium constant K_C based on the equilibrium concentrations C of the substrates and products as follows:

$$K_C = \frac{C_{Pe} C_{We}}{C_{Ae} C_{Fe}} \tag{10.3}$$

The K_C value is conveniently used for estimating the equilibrium yield of a desired product under any conditions. The K_C value can be related to the K_a value by the following equation:

$$K_C = \frac{\gamma_A \gamma_F}{\gamma_P \gamma_W} K_a \tag{10.4}$$

where γ is the activity coefficient.

10.2.2   Correlation of the Apparent Equilibrium Constant with a Solvent Parameter[30]

Mannose was condensed with lauric acid using the immobilized lipase in four water-miscible organic solvents, acetonitrile, acetone, 2-methyl-2-propanol, and 2-methyl-2-butanol, with various initial water contents. Figure 10.1 shows the equilibrium conversions of lauroyl mannose. In every solvent, the equilibrium conversion was higher at the lower initial water contents. This dependence of the conversion on the water content would be reasonable because water is one of the products in the condensation. However, the conversion largely depended on the kind of solvent. Among the solvents, acetonitrile gave the highest conversion at any initial water content. In order to elucidate whether the reaction equilibrium constant K_C itself was different among the solvents or other factors affected the yield, the K_C values were calculated.

The concentrations of product and water at equilibrium, C_{Pe} and C_{We}, were experimentally observed. To obtain C_{Ae}, we measured the solubility of mannose in each solvent. Figure 10.1 also shows the solubility of mannose at 50°C in the solvents with different water contents. The solubility depended on the kind of solvent. Mannose was the most soluble in 2-methyl-2-propanol and 2-methyl-2-butanol, followed by acetone and acetonitrile. The solubility could empirically be expressed as an exponential function of the water content for each solvent, as shown by the dotted curve in the figure. In this reaction system, mannose was not fully dissolved in the solvent because of the limited solubility. Only the mannose dissolved in the solvent would be effective as a substrate for the condensation. When $C_{A0} - C_{Pe}$, where C_{A0} is the overall initial concentration of mannose, was lower than the solubility C_A at C_{We}, the concentration of mannose at equilibrium C_{Ae} was equal to $C_{A0} - C_{Pe}$. On the other hand, when $C_{A0} - C_{Pe} > C_A$, the solubility C_A at C_{We} was regarded as C_{Ae}.

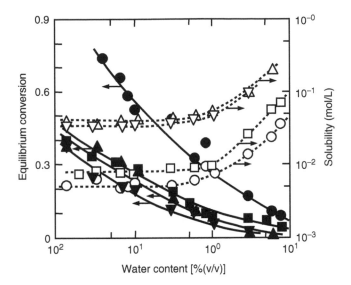

FIGURE 10.1 Effect of the initial water content on the equilibrium conversion of lauroyl mannose and solubility of mannose at 50°C in (●, ○) acetonitrile, (■, ▢) acetone, (▲, △) 2-methyl-2-propanol and (▼, ▽) 2-methyl-2-butanol.[30] The closed and open symbols represent the conversion and the solubility, respectively.

Lauric acid was completely dissolved in any case; therefore, its concentration at equilibrium C_{Fe} was calculated from $C_{F0} - C_{Pe}$, where C_{F0} is the initial concentration of lauric acid.

By substituting the concentrations of substrates and products at equilibrium into Equation 10.3, the K_C value was determined. The K_C values were different by about two orders of magnitude among the solvents. The K_C value was the highest in acetonitrile, intermediate in acetone, and small in 2-methyl-2-propanol and in 2-methyl-2-butanol.

As mentioned earlier, the K_C value evaluated here is an apparent one because it was defined based not on the activities but on the concentrations. Because the K_C can be related to the K_a by Equation 10.4, the K_C would be inversely proportional to γ_W if the water activity is only a parameter affecting the K_C value. The reaction system consisted of five components even if the immobilized enzyme was assumed to be inert, and it was too complicated to estimate the γ_W value. Therefore, we estimated the γ_W value under the assumption of a binary system consisting of water and organic solvent according to the UNIFAC.[31] However, the γ_W did not correlate with the K_C value. This might indicate that the water activity was not the sole factor affecting the reaction equilibrium in an organic solvent. Another possibility for the failure was that the γ_W was not adequately evaluated because the assumption of a binary system was too simple.

We tried to find a solvent parameter that could correlate with the K_C value to obtain a criterion for selection of the solvent. The correlations of the K_C values with log$P$, where P is the partition coefficient between 1-octanol and water phases, the Dimroth-Reichardt parameter for polarity of the solvents $E_T(30)$,[32] and the relative dielectric constant of the solvent were examined. There was a tendency for the K_C to be smaller in a more hydrophobic solvent, but log$P$ did not seem to be a satisfactory parameter to correlate with the K_C value. The $E_T(30)$ did not correlate with the K_C value at all. As shown in Figure 10.2, the relative dielectric constant correlated best with the K_C (ln K_C) value among the parameters tested. However, the reason for the correlation remains unclear.

10.2.3 Apparent Reaction Equilibrium Constants for the Synthesis of Lauroyl Hexoses[33]

The K_C values for the synthesis of lauroyl glucose, galactose, mannose, and fructose through the immobilized-lipase-catalyzed condensation in acetonitrile at 50°C were estimated. The K_C significantly depended on the kind of hexose. There would be two possible reasons for the dependency: one is that

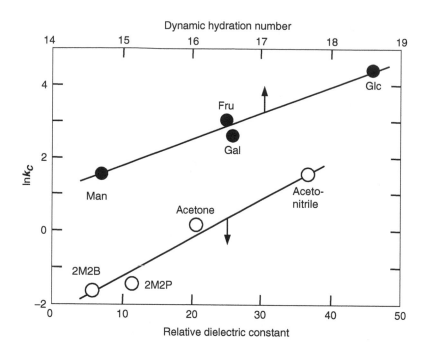

FIGURE 10.2 Correlations of (O) the apparent equilibrium constant K_C for lauroyl mannose synthesis with the relative dielectric constant of various solvents,[30] and of (●) that for lauroyl hexose synthesis in acetonitrile with the dynamic hydration number of hexoses[33] 2M2P: 2-methyl-2-propanol; 2M2B: 2-methyl-2-butanol; Fru: fructose, Gal: galactose; Glc: glucose; Man: mannose.

the free energy change is different for each ester formation, and the other is that activities of the substrates and products depend on the kind of hexose. Because the bond formed through the condensation is common for every ester, the free energy change seems to be almost the same for every ester.

When a hexose hydrates, the water activity decreases. Therefore, it would be supposed that a hexose with stronger binding of water gives a larger K_C value although the activities of other components would also be affected by the presence of the hexose. The dynamic hydration number of hexose[34,35] was selected as a measure of the extent of hydration, and the correlation of the K_C with the dynamic hydration number was examined (Figure 10.2). As expected, there was a positive correlation. This indicates that water activity plays an important role for the condensation in microaqueous organic solbent.

10.2.4 Interaction of a Substrate with a Solvent[36]

The K_C values for the formation of fatty acid butyl esters by the lipase-catalyzed condensation were estimated in nitriles, tertiary alcohols, and their mixtures at 50°C. The K_C values for the formation of butyl decanoate in 2-methyl-2-butanol and 2-methyl-2-propanol were 1.9 and 1.4, while those in the acetonitrile, propionitrile, and butyronitrile were from 21 to 38. The values in the nitriles were almost 1 order greater than those in tertiary alcohols. The alkyl chain length of the nitriles or tertiary alcohols scarcely affected the constant.

The K_C value for the synthesis of butyl decanoate was estimated in the mixtures with nitriles and tertiary alcohols at various molar ratios. Figure 10.3 shows the relationship between the K_C value and the molar fraction of the nitrile in the mixtures. All the plots could be connected by a curve, and the K_C value increased as the molar fraction of the nitrile in the mixture increased. The alkyl chain length of the nitrile or tertiary alcohol only slightly affected the K_C value. These results suggest that the equilibrium is controllable to a certain degree using a mixture of a nitrile and a tertiary alcohol as the reaction medium.

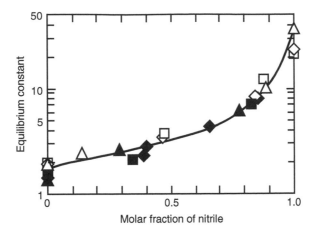

FIGURE 10.3 Dependence of the apparent equilibrium constant at 50°C on the molar fraction of nitrile in mixtures of alcohols and nitriles.[36] The open and closed symbols represent the mixtures of 2-methyl-2-propanol and 2-methyl-2-butanol, respectively, with (◊, ◆) acetonitrile, (■, ❑) propionitrile and (▲, △) butyronitrile.

To elucidate the effect of a reaction medium on the K_C value for the synthesis of butyl decanoate, the IR spectra of decanoic acid and butyl decanoate were measured in various solvents. Although four reactants were included in the reaction system, the spectra of decanoic acid and butyl decanoate were measured because they have a carbonyl double bond (C=O), which absorbs in the IR range of 1700 to 1800 cm[1], and the reaction media used do not have a C=O double bond. The wave number at the absorption peak of a C=O double bond did not depend on the alkyl chain length of the medium for both the nitriles and tertiary alcohols, but the peak shifted to a higher wave number when the volumetric fraction of acetonitrile in the mixture with 2-methyl-2-propanol was higher.

Figure 10.4 shows the relationship between the wave number at the absorption peak of a C=O double bond and the molar fraction of acetonitrile in the mixture with 2-methyl-2-propanol for decanoic acid. The wave number for decanoic acid rose with an increase in the fraction of acetonitrile. The change in

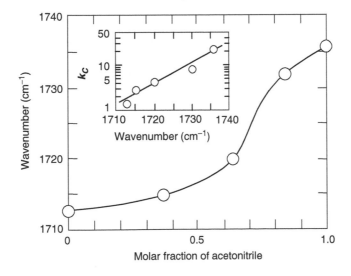

FIGURE 10.4 Relationship between the molar fraction of acetonitrile in its mixture with 2-methyl-2-propanol and the wave number where decanoic acid showed a peak for C=O.[36] Inset: Correlation of the apparent equilibrium constant K_C for the butyl decanoate synthesis with the wave number for the C=O peak of decanoic acid.[36]

wave number for butyl decanoate was smaller than that for decanoic acid. These results suggest that the hydroxyl group of decanoic acid interacts with the hydroxyl group of 2-methyl-2-propanol via the hydrogen bond, while the hydrogen bonding between butyl decanoate and the hydroxyl group of 2-methyl-2-propanol is weak because butyl decanoate does not have a hydroxyl group. Thus, the stronger interaction of decanoic acid with 2-methyl-2-propanol than that of butyl decanoate allows the reaction to shift toward the hydrolysis in the mixture with higher fraction of 2-methyl-2-propanol.

The K_C value estimated in the mixture of acetonitrile and 2-methyl-2-propanol could be linearly, on a semi-logarithmic scale, correlated to the wavelength at the peak of the C=O double bond of decanoic acid, as shown in the inset of Figure 10.4. The correlation would allow us to predict the K_C value for the synthesis of a fatty acid ester in a mixture of a tertiary alcohol and nitrile by measuring the IR spectrum of the C=O double bond of the fatty acid in the mixture. However, this is rather limited because the other reactants, water and 1-butanol, are not taken into account. It is possible that the reactants also more strongly interact with the tertiary alcohol than with the nitrile because they have a hydroxyl group.

10.2.5 Prediction of the Equilibrium Conversion in the Presence of a Desiccant[37]

Because the condensation is thermodynamically controlled, removal of one of the products, usually water, is effective for increasing the conversion. For this purpose, addition of a desiccant such as molecular sieve to the reaction system has often been adopted. However, the criterion for the amount of the desiccant to be added to achieve the desired conversion seems not to be elucidated. We proposed a method for predicting the equilibrium conversion for the synthesis of a monoacyl hexose through the lipase-catalyzed condensation of a fatty acid and a hexose in water-miscible organic solvent in the presence of molecular sieve.

The mass balance equation with respect to water in the presence of molecular sieve is given at equilibrium by

$$VC_{W0} + VC_{Pe} = VC_{We} + wq_{We} \tag{10.5}$$

where V is the volume of solvent, w is the amount of molecular sieve, and q_{We} is the amount of water adsorbed onto molecular sieve, which was expressed by the Langmuir equation (10.6):

$$q_{We} = \frac{bC_{We}}{1 + aC_{We}} \tag{10.6}$$

where a and b are the constants. Under the assumptions that the adsorption of water onto the molecular sieve is fast and that the molecular sieve acts only as a desiccant, the equilibrium product concentration C_{Pe} for the synthesis of an acyl hexose in the presence of the molecular sieve can be estimated by solving Equation 10.3, Equation 10.5, and Equation 10.6 simultaneously, and the equilibrium conversion x_e is then given by $x_e = C_{Pe}/C_{A0}$. In the numerical estimation of C_{Pe}, the relationships of $C_{Fe} = C_{F0} - C_{Pe}$ and of $C_{We} = C_{W0} + C_{Pe}$ were used and the C_{Ae} was expressed by an exponential function in terms of C_{We}, as mentioned above, or by $C_{Ae} = C_{A0} - C_{Pe}$ depending on the quantity of $C_{A0} - C_{Pe}$.

The inset of Figure 10.5 shows the adsorption isotherms at 50°C of water onto molecular sieve 3A in 2-methyl-2-propanol and 2-methyl-2-butanol. The isotherms could be expressed by the Langmuir equation (10.6), and the constants a and b were evaluated. The K_C values for lauroyl mannose formation in 2-methyl-2-propanol and 2-methyl-2-butanol were evaluated in the absence of molecular sieve according to the above-mentioned procedures.

Lauroyl mannose was synthesized in 2-methyl-2-propanol and 2-methyl-2-butanol in the presence of various amounts of molecular sieve 3A. The apparent equilibrium conversions to lauroyl mannose in the solvents are plotted versus the concentration of molecular sieve 3A in Figure 10.5. The conversions were comparable to those previously reported for the synthesis of other esters.[2,5,38] The solid curves in the figure were calculated by the method mentioned above from the K_C value, the adsorption isotherm of water, and the solubility of mannose in each solvent. The calculated results expressed well the experimental results.

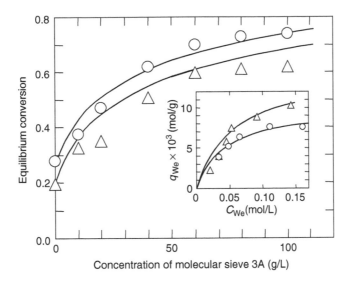

FIGURE 10.5 Equilibrium conversion at 50°C for the lauroyl mannose synthesis in (O) 2-methyl-2-propanol and (Δ) 2-methyl-2-butanol at various amount of molecular sieve 3A.[37] Inset: Adsorption isotherm at 50°C of water onto molecular sieve 3A in 2-methyl-2-propanol and (Δ) 2-methyl-2-butanol.[37] C_{We} and q_{We} are the concentration of water in the solvent and the amount of water adsorbed onto molecular sieve, respectively, at equilibrium.

This fact indicates the appropriateness of the method for predicting the equilibrium conversion in the presence of a desiccant.

The molecular sieve acts as a catalyst for a reaction as well as an adsorbent. The catalytic ability, in some cases, brings about an undesirable phenomenon. 6-O-Oleoyl ascorbate was synthesized through the lipase-catalyzed condensation in acetone in the presence of molecular sieve 4A.[39] The addition of molecular sieve into the reaction system increased both the initial reaction rate and the conversion. However, the conversion gradually decreased in prolonged reaction time. When an excess amount of molecular sieve was added into the system, the conversion was lowered because of adsorption and subsequent degradation of the product on molecular sieve.

10.3 Substrate Selectivity for Carboxylic Acids[40]

10.3.1 Characterization of Carboxylic Acids

Although there have been many studies on the evaluation of substrate selectivity for various lipases, many of them deal with alcohol substrates.[41–43] Studies about acid substrates are fewer than those about alcohols. In many of the investigations about the acid substrates, long-chain fatty acids were usually used as substrates.[44] Short-chain carboxylic acids and their derivatives are also substrates for lipase-catalyzed esterification and they would be adequate substrates for investigating the effect of the nearby structure of the substrate on the selectivity. Some reports[41,45,46] indicated the possibility that both the steric and electrical effects have a great influence on the substrate selectivity.

We assessed the substrate selectivity for various short-chain carboxylic acids for their condensation with p-methoxyphenethyl alcohol catalyzed by the immobilized lipase by the maximum reaction rate V and the Michaelis constant K_m. The selectivity was discussed from the viewpoints of the steric and electrical properties of the substrate molecules, which were estimated from the optimized structure of a carboxylic acid molecule and the electron density of a carboxylic carbon using the molecular orbital calculation software, MOPAC 2000. Carboxylic acids used and their estimated properties are listed in Table 10.1. The acids can be divided into for groups: linear and conjugated acids, linear and nonconjugated ones, branched and conjugated ones, and branched and nonconjugated ones. The conjugated acids

TABLE 10.1 Carboxylic Acid Used and Their Estimated Properties

No.	Carboxylic Acid	Characteristic	Electro Density of Carboxyl Carbon	Projection Area [2]
1	Propionic acid	Straight, nonconjugated	3.696	18.3
2	Butyric acid		6.695	19.9
3	Vinylacetic acid		3.691	17.9
4	Isobutyric acid	Branched, nonconjugated	3.695	24.1
5	Isolaveric acid		3.693	25.7
6	Cyclohexanecarboxylic acid		3.694	28.5
7	Acrylic acid	Straight, conjugated	3.665	14.9
8	Crotonic acid		3.661	16.1
9	Methacrylic acid	Branched, conjugated	3.662	20.7
10	Benzoic acid		3.648	19.2

had lower electron densities than the nonconjugated ones. The projection area of an acid molecule becomes larger when a branched chain is present in the molecule.

10.3.2 Effects of Steric and Electrical Properties on the Kinetic Parameters

Figure 10.6 shows the relationship between the V/K_m value and the electron density of a carboxyl carbon of an acid. For both the straight- and branched-chain carboxylic acids, a carboxylic carbon of the conjugated acids has an electron density lower than that of the nonconjugated acids and the V/K_m values of the conjugated acids were about 10 times lower than those of the nonconjugated acids. The V/K_m values of the acids with a branched chain were also about 1/10 of those of the acid with a straight chain, although the electron density of the acids with a branched chain was not so different from that of the acids with a straight chain for each of the conjugated and nonconjugated acids. This would suggest that the structure of an acid molecule also affects the substrate selectivity of the enzyme.

The relationship between the projection area of the noncarboxylic region of an acid and the K_m value of the acid is shown in Figure 10.7. The K_m value increased with an increase in the projection area and could be linearly related to the projection area. The linear relationship between them indicates that a more bulky acid exhibits a larger steric hindrance for the formation of the enzyme-substrate complex. Both the projection areas and the K_m values for the branched-chain carboxylic acids tended to be larger

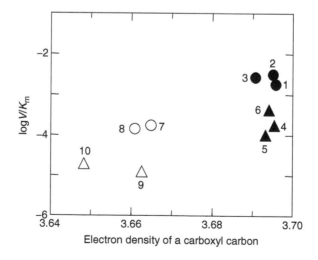

FIGURE 10.6 Relationship between the electron density of a carboxyl carbon of an acid and the logV/K_m value for the acid.[40] The unit of V/K_m is min^{-1}. Labels in the graph correspond to the substrate numbers in Table 10.1.

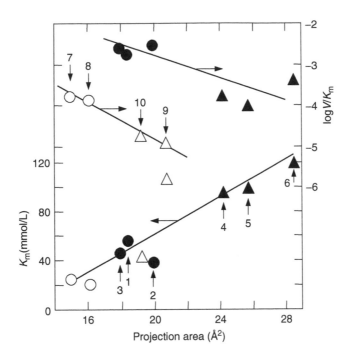

FIGURE 10.7 Effects of the projection area of the noncarboxylic region of an acid molecule on the K_m and the $\log V/K_m$ value of the acid.[40] The unit of V/K_m is min^{-1}. The labels are the same as in Figure 10.6.

than those of straight-chain ones. As for the K_m value, the values of the conjugated acids were slightly smaller than those of nonconjugated ones, while whether an acid had a straight- or a branched-chain significantly affected the K_m value. The large K_m value for the branched-chain acids means that their binding affinity to the enzyme is smaller than that of the straight-chain acids.

Figure 10.7 also shows the relationship between the logarithm of the V/K_m value and the projection area of the noncarboxylic region of an acid. The V/K_m values for the branched-chain acids were about 1/10 of those for the straight-chain acids. The conjugated acids had much lower V/K_m values than the nonconjugated ones. A straight line could connect the plots for the conjugated acids with straight and branched chains. The plots for the nonconjugated acids also lay on a straight line. The values of the slopes were not so different. Therefore, the lowering of the V/K_m value by the presence of conjugation in an acid seems to be independent of the lowering of the value by the presence of a branched chain in the acid. Roughly speaking, the presence of conjugation and a branched chain independently lowers the order of the V/K_m value by 1. The V value was also usable instead of the V/K_m value for estimation of the bulkiness of an acid on the selectivity for the molecule because the difference in the order of the V values was more than 1 but that for the K_m values was less than 1. Although the bulkiness of an acid molecule was estimated by the projection area, the maximum width of the projection image was also effective for the estimation.

10.4 Continuous Synthesis of Fatty Acid Esters

10.4.1 Continuous Stirred Tank Reactors

The lipase-catalyzed synthesis of fatty acid esters in an organic solvent has mostly performed using a batch reactor. However, a continuous reaction would be preferred for large-scale production. A continuous stirred tank reactor (CSTR) and a plug flow reactor (PFR) are typical types of continuous reactors based on the liquid mixing in the reactors.

Because an alcohol substrate such as saccharides, sugar alcohols, and ascorbic acid is hydrophilic and its solubility in a dehydrated organic solvent is relatively low, it remains undissolved in the solvent when an excess amount of the substrate is added into the solvent. Only the substrate solubilized in the solvent would be condensed with a fatty acid to produce an ester, and the consumption of the substrate would be supplemented by dissolution of the substrate. Therefore, CSTR seems to be a promising candidate for continuous production of an ester through lipase-catalyzed condensation in the solvent. We realized continuous or semicontinuous synthesis of fatty acid esters of kojic acid,[47] maltose,[48] and ascorbic acid.[49]

10.4.2 Plug Flow Reactors

An alcohol substrate dissolves in a dehydrated organic solvent although its concentration is, in many cases, from 0.001 to 0.05 mol/L. Let us assume that the conversion is 0.7, the molecular mass of a product is 300, and the concentration of the product is 0.21–10.5 g/L. The concentration is not high but at the same time not extremely low. Based on the hypothetical calculation, we tried to construct a PFR system for producing fatty acid esters of erythritol,[20,50] mannose,[51] and ascorbic acid.[52]

A schematic diagram of the reactor system for continuously producing acyl ascorbate is shown in the inset of Figure 10.8. L-Ascorbic acid powders were packed into a column and immobilized-lipase particles were packed into another column. The columns were connected in series. A fatty acid solution dissolved in acetone was fed to the column packed with ascorbic acid through the preheating coil and then pumped to the immobilized-enzyme column at a specified flow rate. The preheating coil and columns were installed in a thermoregulated chamber at 50°C. After a steady state was achieved, the effluent was sampled and the product concentration in it was determined.

A fatty acid solution (200 mmol/L) was continuously fed to the system for 11 days at a flow rate of 0.5 mL/min, which corresponded to the superficial residence time of 5 min. The acid solution was changed in the order of arachidonic, oleic, linoleic, capric, lauric, and myristic acids. The arachidonic acid solution was fed for 1 day, and the other acids for 2 days. At appropriate intervals, the effluent was sampled and the concentration of the product in it was quantified (Figure 10.8). The system could be stably operated

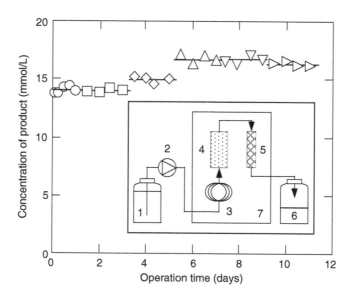

FIGURE 10.8 Continuous production of (O) arachidonoyl, (▢) oleoyl, (◊) linoleoyl, (Δ) decanoyl, (∇) lauroyl and (▷) myristoyl L-ascorbates using the reactor packed with immobilized-lipase particles at a flow rate of 0.5 mL/min and 50°C.[52] The concentration of each fatty acid in the feed was 200 mmol/L. Inset: A reactor system for the continuous synthesis of the acyl ascorbates.[52] 1: feed reservoir; 2: pump; 3: preheating coil; 4: column packed with ascorbic acid; 5: column packed with immobilized lipase; 6: effluent reservoir; 7: thermoregulated chamber.

for 11 days, and no loss in the enzyme activity was observed. The product concentrations in the effluent were in the range of 14 to 17 mmol/L. The lower concentration of unsaturated acyl ascorbates compared to those of the saturated acyl ascorbates would be ascribed to the lower purity of the unsaturated fatty acids. These product concentrations corresponded to the productivity of 1.6 to 1.9 kg/L-reactor•day, depending on the molecular mass of the product.

10.5 Properties of Acyl Ascorbates

10.5.1 Antioxidant Property

L-Ascorbic acid (vitamin C) is a hydrophilic antioxidant with a strong reducing ability. Because its derivatives with fatty acids are expected to possess both antioxidative ability, emulsifying one and oil-solubility, we synthesized the derivatives through immobilized-lipase-catalyzed condensation of ascorbic acid and saturated or unsaturated fatty acids[29,39,49,52–54] and examined the properties. The radical scavenging activities of ascorbic acid and saturated acyl ascorbates against DPPH free radical were measured.[49] The high activity of the ascorbic acid was not affected by the introduction of a saturated acyl group to the hydroxyl group at the C-6 position of the ascorbic acid.

Figure 10.9 shows the oxidation processes of the unmodified arachidonic and docosahexaenoic acids and the acyl moiety of their ascorbates at 65°C and nearly 0% relative humidity.[54] The unmodified fatty acids were almost completely oxidized within 2 h, whereas all the unsaturated acyl ascorbates were significantly resistant to oxidation. The unsaturated acyl moiety of 90% or more remained in the unoxidized state during the test period for the ascorbates.

Linoleoyl ascorbate was mixed with linoleic acid at various molar ratios and the oxidation process of linoleic acid was measured at 65°C and *ca.* 0% relative humidity (Figure 10.10). The oxidation of linoleic acid was almost completely suppressed at the molar ratios of 0.2 and 1.0. These results suggested that the addition of saturated acyl ascorbate to a polyunsaturated fatty acid would also suppress the oxidation of the polyunsaturated fatty acid. We measured the oxidation processes of docosahexaenoic acid, which is very susceptible to oxidation, mixed with lauroyl ascorbate at various molar ratios. As shown in Figure 10.10, the oxidation of docosahexaenoic acid was significantly suppressed at the molar ratios of 0.2.

Acyl ascorbates were surface active,[55] and were also effective for suppressing the oxidation of linoleic acid encapsulated with a polysaccharide by spray drying[56] and of the membrane of intestinal epithelial cells, Caco-2.[57]

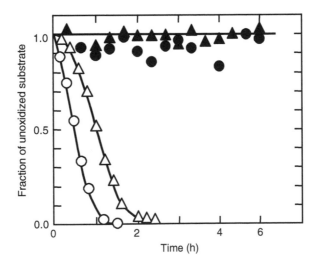

FIGURE 10.9 Oxidation of (Δ) arachidonic and (O) docosahexaenoic acids, and (▲) arachidonoyl and (●) docosahexaenoyl ascorbates at 65°C and nearly 0% relative humidity.[54]

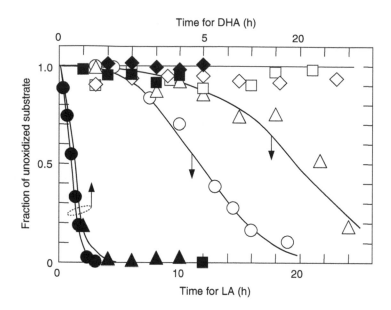

FIGURE 10.10 Effect of the addition of (open symbols) linoleoyl ascorbate to linoleic acid (LA) or (closed symbols) lauroyl ascorbate to docosahexaenoic acid (DHA) on oxidation of linoleic or docosahexaenoic acid at 65°C and nearly 0% relative humidity.[54] The molar ratios of linoleoyl or lauroyl ascorbate to linoleic or docosahexaenoic acid were (O, ●) 0, (Δ, ▲) 0.05, (□, ■) 0.2 and (◊, ◆) 1.0.

10.5.2 Solubility in Water and Soybean Oil

The solubilities of the saturated acyl L-ascorbates in water or soybean oil were measured at various temperatures (Figure 10.11).[49] Although L-ascorbic acid is insoluble in soybean oil and highly soluble in water, the acylation of ascorbic acid significantly improved its solubility in soybean oil but decreased the solubility in water. The solubilities of the acyl ascorbates in both soybean oil and water were higher for those with a shorter acyl chain. The temperature dependence of the solubilities of the acyl ascorbates in

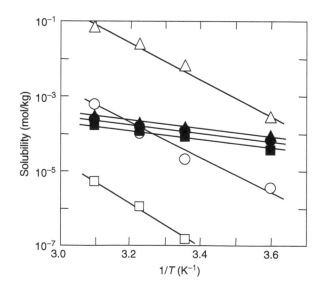

FIGURE 10.11 Temperature dependence of the solubility of (Δ, ▲) decanoyl, (O,●) lauroyl and (□, ■) myristoyl ascorbates in (open symbols) water and (closed symbols) soybean oil.[49]

water or soybean oil could be expressed by the vanít Hof equation. The dissolution enthalpies were about 20 kJ/mol for soybean oil and about 90 kJ/mol for water.

References

1. N Khaled, D Montet, M Farines, M Pina, J Graille. Synthesis of sugar mon-esters by biocatalysis. *Oléagineux* 47: 181–190, 1992.
2. G. Ljunger, P Adlercreutz, B Mattiasson. Lipase catalyzed acylation of glucose. *Biotechnol Lett* 16: 1167–1172, 1994.
3. DV Sarney, EV Vulfson. Application of enzymes to the synthesis of surfactants. *Trends Biotechnol* 13: 164–172, 1995.
4. C Scheckermann, A Schlotterbeck, M Schmidt, V Wray, S Lang. Enzymatic monoacylation of fructose by two procedures. *Enzyme Microb Technol* 17: 157–162, 1995.
5. D Coulon, M Girardin, B Rovel, M Ghoul. Comparision of direct esterification and transesterification of fructose by Candida antartica lipase. *Biotechnol Lett* 17: 183–186, 1995.
6. A Ducret, A Giroux, M Trani, R Lortie. Characterization of enzymatically prepared biosurfactants. *J Am Oil Chem Soc* 73: 109–113, 1996.
7. OP Ward, J Fang, Z Li. Lipase-catalyzed synthesis of a sugar ester containing arachidonic acid. *Enzyme Microb Technol* 20: 52–56, 1997.
8. L Cao, A Fischer, UT Bornsceuer, RD Schmid. Lipase-catalyzed solid phase synthesis of sugar fatty acid esters. *Biocatal Biotransform* 14: 269–283, 1997.
9. JA Arcos, M Bernabé, C Otero. Quantitative enzymatic production of 6-O-acylglucose esters. *Biotechnol Bioeng* 57: 505–509, 1998.
10. JA Arcos, M Bernabé, C Otero. Quantitative enzymatic production of 1,6-diacyl fructofuranose. *Enzyme Microb Technol* 22: 27–35, 1998.
11. JE Kim, JJ Han, JH Yoon, JS Rhee. Effect of salt hydrate pair on lipase-catalyzed regioselective monoacylation of sucrose. *Biotechnol Bioeng* 57: 121–125, 1998.
12. P Degn, LH Pedersen, JØ Duus, W Zimmermann. Lipase-catalysed synthesis of glucose fatty acid esters in tert-butanol. *Biotechnol Lett* 21: 275–280, 1999.
13. M Ferrer, MA Cruces, M Bernabé, A Ballesteros, FJ Plou. Lipase-catalyzed regioselective acylation of sucrose in two-solvent mixtures. *Biotechnol Bioeng* 65: 10–16, 1999.
14. X. Zhang, DG Hayes. Increased rate of lipase-catalyzed saccharide-fatty acid esterification by control of reaction medium. *J Am Oil Chem Soc* 76: 1495–1500, 1999.
15. F Chamouleau, D Coulon, M Girardin, M Ghoul. Influence of water activity and water content on sugar esters lipase-catalyzed synthesis in organic media. *J Mol Catal B: Enzymatic* 11: 949–954, 2001.
16. S Soultani, J-M Engasser, M Ghoul. Effect of acyl donor chain length and sugar/acyl donor molar ratio on enzymatic synthesis of fatty acid fructose esters. *J Mol Catal B: Enzymatic* 11: 725–731, 2001.
17. AEM Jassen, A van der Padt, K van't Riet. Solvent effects on lipase-catalyzed esterification of glycerol and fatty acids. *Biotechnol Bioeng* 42: 953–962, 1993.
18. A Ducret, A Girous, M Trani, R Lortie. Enzymatic preparation of biosurfactants from sugars or sugar alcohols and fatty acids in organic media under reduced pressure. *Biotechnol Bioeng* 48: 214–221, 1995.
19. RP Yadav, RK Saxena, R Gupta, S Davidson. Production of biosurfactant from sugar alcohols and natural tryglicerides by Aspergillus terreus lipase. *J Sci Ind Res* 56: 479–482, 1997.
20. S Adachi, K Nagae, R Matsuno. Lipase-catalysed condensation of erythritol and medium-chain fatty acids in acetonitrile with low water content. *J Mol Catal B: Enzymatic* 6: 21–27, 1999.
21. AR Medina, LE Cerdán, AG Giménez, BC Páez, MJL González, EM Grima. Lipase-catalyzed esterification of glycerol and polyunsaturated fatty acids from fish and microalgae oils. *J Biotechnol* 70: 379–391, 1999.

22. J Piao, T Kobayashi, S Adachi, K Nakanishi, R Matsuno. Synthesis of mon- and dioleoyl erythritols through immobilized-lipase-catalyzed condensation of erythritol and oleic acid in acetone. *Biochem Eng J* 14: 79–84, 2003.

23. C Humeau, M Girardin, D Coulon, A Miclo. Synthesis of 6-*O*-palmitoyl L-ascorbic acid catalyzed by Candida antarctica lipase. *Biotechnol Lett* 17: 1091–1094, 1995.

24. C Humeau, M Girardin, B Rovel, A Miclo. Enzymatic synthesis of fatty acid ascorbyl esters. *J Mol Catal B: Enzymatic* 5: 19–23, 1998.

25. C Humeau, M Girardin, B Rovel, A Miclo. Effect of the thermodynamic water activity and the reaction medium hydrophobicity on the enzymatic synthesis of ascorbyl palmitate. *J Biotechnol* 63: 1–8, 1998.

26. S Bradoo, RK Saxena, R Gupta. High yields of ascorbyl palmitate by thermostable lipase-mediated esterification. *J Am Oil Chem Soc* 76: 1291–1295, 1999.

27. H Stamatis, V Sereti, FN Kolisis. Studies on the enzymatic synthesis of lipophilic derivatives of natural antioxidants. *J Am Oil Chem Soc* 76: 1505–1510, 1999.

28. Y Yan, UT Bornscheuer, RD Schmid. Lipase-catalyzed synthesis of vitamin C fatty acid esters. *Biotechnol Lett* 21: 1051–1054, 1999.

29. Y Watanabe, S Adachi, R Matsuno. Condensation of L-ascorbic acid and medium-chain fatty acids by immobilized lipase in acetonitrile with low water content. *Food Sci Technol Res* 5: 188–192, 1999.

30. Y Watanabe, Y Miyawaki, S Adachi, K Nakanishi, R Matsuno. Equilibrium constant for lipase-catalyzed condensation of mannose and lauric acid in water-miscible organic solvents. *Enzyme Microb Technol* 29: 494–498, 2001.

31. A Fredenslund, RL Jones, JM Prausnitz. Group-contribution estimation of activity coefficients in nonideal liquid mixtures. *AIChE J* 21: 1086–1099, 1975.

32. C Reichardt. Empirical parameters of solvent polarity as linear free-energy relationships. *Angew Chem Int Ed Engl* 18: 98–110, 1979.

33. Y Watanabe, Y Miyawaki, S Adachi, K Nakanishi, R Matsuno. Synthesis of lauroyl saccharides through lipase-catalyzed condensation in microaqueous water-miscible solvents. *J Mol Catal B: Enzymatic* 10: 241–247, 2000.

34. H Uedaira, M Ikura, H Uedaira. Natural-abundance oxygen-17 magnetic relaxation in aqueous solutions of carbohydrates. *Bull Chem Soc Jpn* 62: 1–4, 1989.

35. H Uedaira, M Ishimura, S Tsuda, H Uedaira. Hydration of oligosaccharides. *Bull Chem Soc Jpn* 63: 3376–3379, 1990.

36. T Kobayashi, W Furutani, S Adachi, R Matsuno. Equilibrium constant for the lipase-catalyzed synthesis of fatty acid butyl ester in various organic solbents. *J Mol Catal B: Enzymatic* 24,25: 61–66, 2003.

37. X Zhang, S Adachi, Y Watanabe, T Kobayashi, R Matsuno. Prediction of the equilibrium conversion for the synthesis of acyl hexose through lipase-catalyzed condensation in water-miscible solvent in the presence of molecular sieve. *Biotechnol Prog* 19: 293–297, 2003.

38. J Giacometti, F Giacometti, Milin, Vasi-Raki. Kinetic characterization of enzymatic esterification in a solvent system: adsorptive control of water with molecular sieves. *J Mol Catal B: Enzymatic* 11: 921–928, 2001.

39. K Kuwabara, Y Watanabe, S Adachi, K Nakanishi, R Matsuno. Synthesis of 6-*O*-unsaturated acyl L-ascorbates by immobilized lipase in acetone in the presence of molecular sieve. *Biochem Eng J* 16: 17–23, 2003.

40. T Kobayashi, S Adachi, R Matsuno. Lipase-catalyzed condensation of *p*-methoxyphenethyl alcohol and carboxylic acids with different steric and electrical properties in acetonitrile. *Biotechnol Lett* 25: 3–7, 2003.

41. HS Bevinakatti, AA Banerji. Lipase catalysis: factors governing transesterification. *Biotechnol Lett* 6: 397–398, 1988.

42. G Langrand, N Rondot, C Triantaphylides, J Baratti. Short chain flavour esters synthesis by microbial lipases. *Biotechnol Lett* 12: 581–586, 1990.

43. QL Chang, CH Lee, KL Parkin. Comparative selectivities of immobilized lipases from *Pseudomonas cepacia* and *Candida antarctica* (fraction B) for esterification reactions with glycerol and glycerol analogues in organic media. *Enzyme Microb Technol* 25: 290–297, 1999.

44. R Borgdorf, S Warwel. Substrate selectivity of various lipases in the esterification of cis- and trans-9-octadecanoic acid. *Appl Microb Biotechnol* 51: 480–485, 1999.

45. G Kichner, MP Scollar, AM Klivanov. Resolution of racemic mixtures via lipase catalysis in organic solvents. *J Am Chem Soc* 107: 7072–7076, 1985.

46. M Charton. Contribution of steric, electrical, and polarizability effects in enantioselective hydrolyses with *Rhizopus nigricans*: a quantitative analysis. *J Org Chem* 52: 2400–2403, 1987.

47. T Kobayashi, S Adachi, K Nakanishi, R Matsuno. Semi-continuous production of lauroyl kojic acid through lipase-catalyzed condensation in acetonitrile. *Biochem Eng J* 9: 85–89, 2001.

48. X Zhang, T Kobayashi, Y Watanabe, T Fujii, S Adachi, K Nakanishi, R Matsuno. Lipase-catalyzed synthesis of monolauroyl maltose through condensation of maltose and lauric acid. *Food Sci Technol Res* 9: 110–113, 2003.

49. Y Watanabe, K Kuwabara, S Adachi, K Nakanishi, R Matsuno. Production of saturated acyl L-ascorbate by immobilized lipase using a continuous stirred tank reactor. *J Agric Food Chem* 51: 4628–4632, 2003.

50. J Piao, T Kobayashi, S Adachi, K Nakanishi, R Matsuno. Continuous synthesis of lauroyl or oleoyl erythritol by a packed-bed reactor with an immobilized lipase. *Process Biochem* 39:681–686, 2004.

51. Y Watanabe, Y Miyawaki, S Adachi, K Nakanishi, R Matsuno. Continuous production of acyl mannose by immobilized lipase using a packed-bed reactor and their surfactant properties. *Biochem Eng J* 8: 213–216, 2001.

52. K Kuwabara, Y Watanabe, S Adachi, K Nakanishi, R Matsuno. Continuous production of acyl L-ascorbates using a packed-bed reactor with immobilized lipase. *J Am Oil Chem Soc* 80:895–899, 2003.

53. Y Watanabe, Y Minemoto, S Adachi, K Nakanishi, Y Shimada, R Matsuno. Lipase-catalyzed synthesis of 6-O-eicosapentaenoyl L-ascorbate in acetone and its autoxidation. *Biotechnol Lett* 22: 637–640, 2000.

54. Y Watanabe, S Adachi, K Nakanishi, R Matsuno. Lipase-catalyzed synthesis of unsaturated acyl L-ascorbates and their ability to suppress the autoxidation of polyunsaturated fatty acids. *J Am Oil Chem Soc* 78: 823–826, 2001.

55. Y Watanabe, S Adachi, T Fujii, K Nakanishi, R Matsuno. Surface activity of 6-O-hexanoyl, octanoyl, decanoyl and dodecanoyl ascorbates. *Jpn J Food Eng* 2: 73–75, 2001.

56. Y Watanabe, X Fang, Y Minemoto, S Adachi, R Matsuno. Suppressive effect of saturated acyl L-ascorbate on the oxidation of linoleic acid encapsulated with maltodextrin or gum arabic by spray-drying. *J Agric Food Chem* 50: 3984–3987, 2002.

57. Y Kimura, H Kanatani, M Shima, S Adachi, R Matsuno. Anti-oxidant activity of acyl ascorbates in intestinal epithelial cells. *Biotechnol Lett* 25:1723–1727, 2003.

11

Enzymatic Production of Diacylglycerol and Its Beneficial Physiological Functions

Naoto Yamada

Noboru Matsuo

Takaaki Watanabe

Teruyoshi Yanagita

11.1 Introduction

Fats and oils are not only essential nutrients but also contribute to the flavor and aroma of foods. However, an excessive intake of fat can result in obesity. The importance of preventing the accumulation of body fat, which also prevents certain lifestyle-related diseases, has been noted by many investigators and surveys. Clinical studies suggest that weight loss in the range of 5–10% of initial weight can confer a significant reduction in obesity-related disorders. Although energy restriction and the limitation of total and saturated fat intake may be the primary measures for the treatment of obesity, these goals are difficult to achieve. Therefore, numerous studies on dietary fats that influence the accumulation of body fat have been reported. While changing the fatty acid composition of triacylglycerol (TAG) is one approach, this study focused on the structure of other acylglycerols.

Diacylglycerol (DAG) is a natural component of various edible oils and is frequently used in foods as emulsifiers. We have investigated an efficient process to produce a DAG that is mainly composed of the 1,3-isomer and found that the immobilized lipases can be used to produce the 1,3-isomer on an industrial scale. Furthermore, the nutritional characteristics of dietary DAG in comparison with triacylglycerol (TAG) have been evaluated and evidence is available to show that DAG, particularly in the 1,3-isomer,

TABLE 11.1   Contents of Acylglycerols (in Weight %) in the Edible Oils of Various Origins

	Soybean	Cottonseed	Palm	Corn	Safflower	Olive	Rapeseed
Monoacylglycerol	—	0.2	—	—	—	0.2	0.1
Diacylglycerol	1.0	9.5	5.8	2.8	2.1	5.5	0.8
Triacylglycerol	97.9	87.0	93.1	95.8	96.0	93.3	96.8
Others	1.1	3.3	1.1	1.4	1.9	2.3	2.3

Data from Abdel-Nabey et al. (1992), D'alonzo et al. (1982)

has metabolic characteristics that are distinct from TAG and that it has beneficial effects with regard to the prevention and management of postprandial lipemia and obesity.[1]

A cooking oil product containing 80% (w/w) or greater of DAG has been marketed in Japan since 1999 as "Food for Specified Health Use" approved by the Ministry of Health, Labor, and Welfare. In this chapter, based on recent reports, we report on the efficient enzymatic production of DAG and its beneficial effects. The clinical studies are discussed and possible mechanisms are proposed.

11.2 Occurrence of DAG

TAG is the major constituent of edible oils. However, edible oils generally contain approximately 10% DAG, with the relative content depending on the origin of the oil. The DAG content of various edible oils has been reported.[2,3] Table 11.1 shows the content of DAG and other acylglycerols in edible oils from a variety of sources. For example, olive oil contains 5.5% DAG and cottonseed oil contains 9.5% DAG. These values may depend on the storage conditions or the variety of oil plant.

Approximately 70% of the DAG in cooking oil is converted to 1,3-diacyl-sn-glycerol (1,3-DAG) as the result of acyl group migration during the manufacturing process that involves the heating of the oil.

11.3 Enzymatic Production of DAG

Mixtures of monoacylglycerol (MAG) and diacylglycerol (DAG) have been used as emulsifiers and stabilizers in the food industry for a long time. Mono- and diacylglycerols are generally synthesized from fats and oils with glycerol via a chemical glycerolysis reaction at temperatures in excess of 200°C using an alkaline catalyst. The distribution of components such as mono-, di-, and tri-acylglycerol depends on the mole ratio of glycerol to acyl groups used in the reaction.[4] Chemical glycerolysis is a mature process used to produce mono- and diacylglycerol. However, it is not ideal for producing oils with a high DAG content because the main product of chemical glycerolysis is MAG. Our target product is a cooking oil high in DAG that is intended to replace conventional oils. The requirements are as follows:

- High DAG content to promote health beneficial effects
- Mild processing conditions to minimize the deterioration and trans fatty acids
- Physical properties comparable to those of conventional oils that contribute to the taste and flavor of cooked foods

The focus of our efforts was enzymatic reactions using lipases developed since the 1980s. Some researchers reported DAG production using lipases through alternative pathways.[5–15] Hirota et al. patented a method for preparation of an oil containing a high purity of DAG by the esterification of glycerol with fatty acids, using 1,3-regioselective lipases, combined with the simultaneous removal of water.[13] Namely 1,3-DAG is synthesized by esterification of fatty acid with glycerol using immobilized 1,3-regioselective lipases (Figure 11.1). Similarly Rosu et al. carried out esterification reactions of some fatty acids and glycerol in a nonsolvent system with water removal using Lipozyme IM. As a result, they achieved an 85% yield of 1,3-DAG when caplylic acid was used.[14] Thus these esterification methods are widely available for the production of oils that are high in DAG. Considering the production of DAG in an industrial scale, the reuse of the enzyme is necessary to reduce enzyme cost. For this purpose, an immobilized

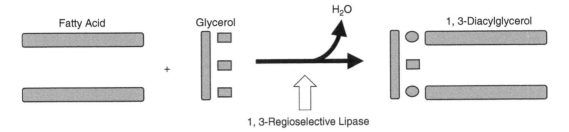

FIGURE 11.1 Esterification reaction by using 1,3-regioselective lipase.

enzyme is suitable. Only a few commercially immobilized lipases are available. The kinetics and reaction conditions of Lipozyme RM IM *(Rhizomucor meihei* 1,3-regioselective lipase immobilized on ion exchange resin), produced by Novozymes, was investigated by Watanabe et al.[15]

11.3.1 Diacylglycerol Production in a Stirred-Batch Reactor

The esterification of fatty acids with glycerol in a solvent-free system using Lipozyme RM IM in a stirred-batch reactor has been investigated previously.[15] The fatty acids used as the esterification substrate were obtained from rapeseed oil and soybean oil, which was reduced to saturated fatty acids. The composition of this product is shown in Table 11.2. Because the melting point of DAG is higher than TAG with the same fatty acid composition, it is necessary to use reduced saturated fatty acids as the substrate for an oil high in DAG. An example of stirred-batch reactor for producing DAG is shown in Figure 11.2. Figure 11.3 shows a typical result of the esterification using a stirred-batch reactor at 50°C and 3 mmHg.

11.3.1.1 Treatment of Immobilized Lipase

It is generally thought that the lipase reaction takes place at the oil-water interface, but in a reaction using an immobilized lipase, the reaction of fatty acids with glycerol would occur in the oil phase and not in the glycerol phase.[15] Experimental data indicate that the esterification reaction does not take place when the immobilized enzyme first comes into contact with glycerol followed by the fatty acid. Therefore an immobilized lipase initially should be saturated in fatty acids (oil phase) prior to adding the glycerol.[15] Baerger et al.[10] and Castillo et al.[17] reported that the use of absorbed glycerol on the silica gels was an effective method for the esterification of fatty acid with glycerol in organic solvents. These methods also avoid the direct contact of glycerol with the immobilized lipase.

11.3.1.2 Removal of Water

The removal of water generated by esterification is an important point, since the equilibrium might be shifted in favor of esterification with a decrease in water content. Several methods for the removal of water are available, such as the use of a stream of nitrogen gas, desiccating agents such as silica gel or molecular

TABLE 11.2 Fatty Acid Composition

Fatty Acid Component	Content (wt%)
Palmitic acid (C16:0)	3.1
Stearic acid (C18:0)	1.3
Oleic acid (C18:1)	38.2
Linoleic acid (C18:2)	47.5
Linolenic acid (C18:3)	7.5
Arachidic acid (C20:0)	0.7
Gadoleic acid (C20:1)	0.6
Behenic acid (C22:0)	0.1
Erucic acid (C22:1)	0.2

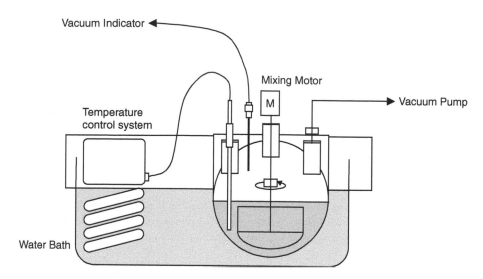

FIGURE 11.2 A diagram of the stirred-batch reactor used for esterification reaction for diacylglycerol synthesis.

sieves, and the use of a vacuum pump. A vacuum system would be preferable for the commercial production of DAG. Various vacuum conditions between 1 and 10 mmHg have been investigated, and as a result, the maximum yield of 1,3-DAG was obtained at a concentration of 1.09M at 1 mmHg (Figure 11.4). These results suggest that a higher vacuum (higher water removal rate) would be advantageous in the 1,3-DAG production. It is known that water plays an important role in maintaining enzyme structure and catalytic activity. Thus, the removal of excess water might lead to deactivation of the enzyme. In experiments at a vacuum of 3 mmHg, the water content including that in the immobilized enzyme was around 3% at the end of the reaction, similar to that before the reaction. Thus, the amount of water essential for an enzyme reaction can be maintained during the esterification reaction under vacuum conditions, when an immobilized lipase is used.

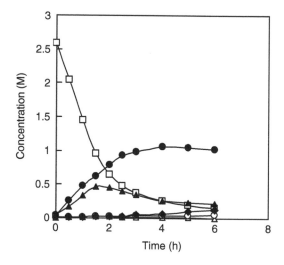

FIGURE 11.3 Courses for the esterification reaction using an immobilized lipase starting with 2.59M fatty acid and 1.29M glycerol. The reaction was performed at 50°C and the molar ratio of FA to GLY was 2.0 and an immobilized enzyme resin concentration of 5% (dry weight basis). Water removal was performed under 3 mmHg vacuum. (□),FA; (▲),1-MAG; (△), 2-MAG;(●),1,3-DAG; (○),1,2(2,3)-DAG; (◆),TAG. (Data from Reference 15, simulation data not shown).

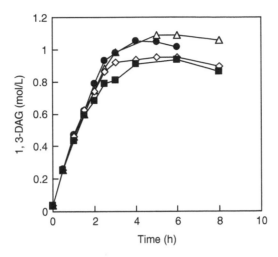

FIGURE 11.4 Effect of vacuum condition on 1,3-DAG synthesis. The reaction was performed at 50°C and the molar ratio of FA to GLY was 2.0 and an immobilized enzyme resin concentration of 5% (dry weight basis). Water removal was performed under 1mm Hg ($\triangle$), 3 mm Hg($\bullet$), 6 mm Hg ($\diamond$), 10 mm Hg ($\blacksquare$). (Data from Reference 15; simulation data not shown).

11.3.1.3 Acyl Migration

TAG generation is unfavorable for DAG production since it causes a decrease in DAG purity. Since the separation of TAG from the product oil is difficult, TAG, produced during the esterification reaction, may be retained in the final product oil. Thus, lowering the TAG content is important in obtaining an oil high in DAG. In the esterification reaction using Lipozyme RM IM, which is a 1,3-regioselective lipase, 1,3-DAG is produced in good yield with negligible amount of 1,2-DAG. Acyl migration, in which 1,3-DAG converted to 1,2-DAG, takes place during the reaction, and once 1,2-DAG is generated, it is easily converted to TAG by lipases. Consequently, low levels of TAG lowers the extent of acyl migration. The effect of various factors such as water content, temperature, enzyme load, and reaction time on acyl migration in transesterification reactions was investigated by Xu et al.[18,19] The results showed that all factors had a positive influence on the acyl migration. The effects of various factors on the extent of acyl migration were also investigated for esterification reactions using Lipozyme RM IM.[15] As a result, in the production of 1,3-DAG using Lipozyme RM IM a higher reaction temperature and a higher enzyme load led to an increase in TAG concentration, instead of higher reaction rate. These results suggest that a lower temperature and smaller enzyme load would have advantages in terms of higher 1,3-DAG production. The establishment of the optimal reaction temperature and enzyme load is important for the effective production of DAG in an industrial scale from the viewpoints of both the rate of production of 1,3-DAG and DAG purity.

11.3.1.4 Model Study

The kinetics of 1,3-DAG production from fatty acids and glycerol were investigated on a basis of a simplified model.[15] The model used for the reaction scheme shown in Figure 11.5 is based on the triolein synthesis model by Lortie et al.[16] This model takes into consideration the acyl migration reactions, glycerol dissolution into the oil phase and water removal from the reaction mixture in addition to the esterification reactions. Reaction rate constants were determined by fitting the simplex method on this model. Data generated by this model were in good agreement with experimental data. Arrhenius plots of the reaction rate constants determined by this model are shown in Figure 11.6. The temperature dependency of acyl migration rate is stronger than that of the other esterification reaction rates. This result supports the experimental data that the purity of DAG becomes higher when the reaction temperature is decreased. This simplified model can be used for the study of DAG production.

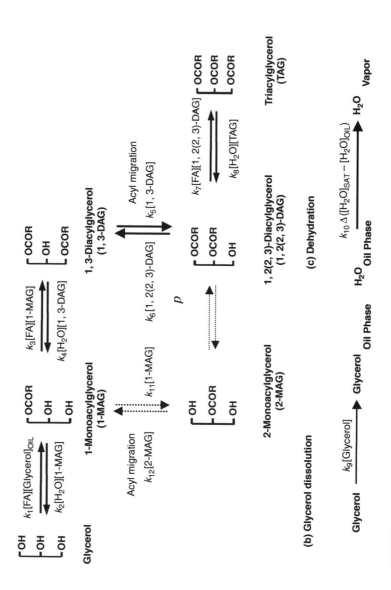

FIGURE 11.5   Reaction scheme for the 1,3-DAG synthetic reactions (data from Reference 15).

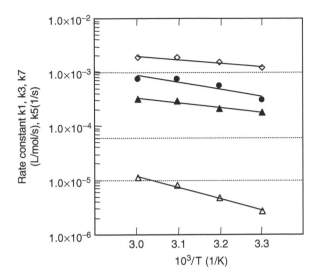

FIGURE 11.6 Arrhenius plots for various reaction rate constants. (●),k1; (▲),k3; (△),k5; (◇), k7. (data from Reference15).

11.3.2 DAG Production in a Packed-Bed Bioreactor

A packed-bed reactor is frequently used as an immobilized enzyme reactor on an industrial scale since it has the advantage of being compact. Other benefits include the separation of the product and continuous production.

Enzymatic reactions using lipases in packed-bed reactors have been investigated by a number of investigators for transesterification[20–22] and for alcoholisis of fatty acid with alcohol.[23,24] Reactions in a packed-bed reactor for the esterification of fatty acid and glycerol have not been studied extensively.[25] Water is generally crucial to maintain the structure and stability of the enzyme. However, in the case of esterification, excess water is generated in the reaction. Thus a system for the removal of water generated in the reaction is necessary to obtain a high yield in an esterification reaction using a packed-bed reactor. The accumulation of water in the packed-bed column also led to a decrease in enzyme stability. Thus methods to prevent the accumulation of water in a reactor were reported in the case of an esterification reaction in an organic solvent, such as using a polar solvent,[23] drying periodically by airflow,[23] or by a pure solvent.[24] Acros et al. studied the nonsolvent esterification of fatty acid with glycerol in a continuous packed-bed reactor containing immobilized lipase from *Mucor miehei*.[25] They used an excess of glycerol to remove the water. Namely, excess glycerol was used to dissolve water generated by the esterification. Although the conversion of fatty acids was 90%, the DAG concentration in the resulting oil was only 48%. An effective DAG production system using a packed-bed reactor was investigated by Sugiura et al.[26] Namely the system consisted of a packed-bed column and a water removal vessel external to the packed-bed column (Figure 11.7). In this system, the removal of water was achieved by the use of a vacuum. This DAG production system from fatty acid and glycerol using Lipozyme RM IM involved a repeated batch operation, as shown in Table 11.3. This system has the advantage in controlling the water content by vacuum. The yield of DAG prepared from soybean and rapeseed fatty acids were around 60%. A DAG purity of 80% or greater was attained using Lipozyme RM IM in this system.

11.3.3 Conclusion of Enzymatic DAG Production

The esterification of fatty acids and glycerol using a 1,3-regioselective lipase, which has high 1,3-selectivity, could lead to the efficient production of 1,3-DAG. Lipozyme RM IM, which is immobilized on the resin, has significant performance not only with respect to selectivity but is also stable. High-purity 1,3-DAG was obtained in a stirred-batch reactor and also in a packed-bed reactor. A packed-bed reactor with a

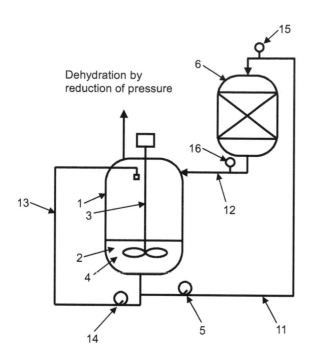

FIGURE 11.7 A diagram of the packed-bed reactor used for esterification reaction for diacylglycerol synthesis (data from Reference 26).

TABLE 11.3 Diacylglycron Production Results in Packed-Bed Reactors

	1	2	3
Batch size (kg)	100	100	100
Oleic acid (kg)	86	86	86
Glycerol (kg)	14	14	14
Immobilized enzyme	Lipozyme IM	Lipozyme IM	Lipozyme IM
Average particle diameter d (mm)	0.43	0.43	0.43
Amount (kg)	5	5	20
Packing thickness (m)	0.18	0.18	0.7
Superficial velocity U (mm/s)	4.4	2.2	3.7
Residence time (s)	40	79	190
Pressure loss P (kg/cm^2)	2.6	1.5	9.5
Spray nozzle	Used	Used	Used
Droplet diameter (mm)	1	1	1
Circulation on spray side (m^3/hr)	1.2	1.2	1.2
Reaction time (hr)	3.5	3.5	3.5
Reaction Product (wt%)			
Oleic acid	14.1	15.4	11.6
Glycerol	0.3	0.7	0.4
Monoglyceride (M)	14.1	18.3	15.0
Diglyceride (D)	65.6	58.1	55.7
Triglyceride (T)	5.9	7.5	17.3
Total	100.0	100.0	100.0
Yield of reaction (D + T) (wt%)	71.5	65.6	73.0
Purity of diglyceride (D/D + T) (wt%)	91.7	88.6	76.3

Data from Reference 26

vacuum system could be used for the industrial production of oils high in DAG content. The 1,3-DAG content in the total DAG was greater than 90% by this esterification reaction. However, a part of 1,3-DAG was converted into 1,2-DAG during the manufacturing process that involved heating of the oil. Finally, it would be converted into a mixture of approximately 70% 1,3-DAG and 30% 1,2-DAG as a result of equilibration.

The water content in the reaction mixture has a considerable effect on 1,3-DAG yields in the esterification reaction. The water content is controlled by the balance of the generation water by the reaction and the removal water by vacuum. Thus, the choice of suitable vacuum conditions is an important point for the efficient production of DAG on an industrial scale.

Although the enzymatic production of DAG thus appears to be effective, further development of enzyme techniques in this area will be necessary for the improvement of the production process.

11.4 Beneficial Functions of DAG Oils

11.4.1 Less Postprandial Lipemia after the Ingestion of DAG in Humans

The most remarkable acute effect of DAG can be observed in the postprandial state. When a fat emulsion is ingested after overnight fasting, the blood TAG concentration elevates and reaches a peak level at around 4 hr after ingestion of the oil and then gradually decreases in humans. This postprandial elevation of blood TAG concentration has been shown to be lower when a DAG oil is used, compared to a TAG oil. This postprandial hypertriglyceridemia reducing effect is one of the nutritional characteristics of DAG. TAG in the serum is a source of fat to be accumulated in the body and postprandial hyperlipidemia is a risk factor of cardiovascular diseases. Figure 11.8 shows changes in serum TAG concentrations after a single ingestion of lipid emulsion in healthy men.[27] The two test oils had the same fatty acid composition. We have shown in other experiments that chylomicron TAG concentration after ingestion of DAG was reduced by half compared to TAG ingestion.[28] The lower serum TAG levels after DAG ingestion may be the result of the slower rate of chylomicron formation after DAG ingestion compared with that of TAG ingestion, as discussed in the section on animal studies (Section 11.4.6).

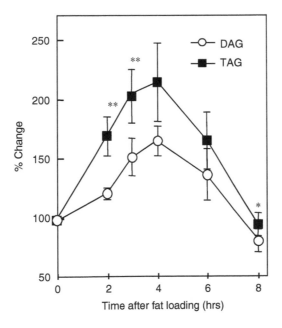

FIGURE 11.8 Differential effects of diacylglycerol (DAG) and triacylglycerol (TAG) on postprandial serum TAG concentration. Serum TAG was measured after a single ingestion of lipid emulsion (dose of test oil = $30 g/m^2$) in healthy men (M ± SD, n = 6, *: $P < 0.05$, **: $P < 0.01$). (Data from Reference 27.)

TABLE 11.4 Changes of Anthropometric Values and Body Composition of the Subjects[†1]

	Wk	Diacylglycerol Group	Change[†2]	Triacylglycerol Group	Change
Body weight, kg	0	72.1 ± 1.8	—	68.1 ± 1.3	—
	16	69.5 ± 1.7##[3]	−2.6 ± 0.3**	67.0 ± 1.5#	−1.1 ± 10.4
Waist circumference, cm	0	85.0 ± 1.4	—	82.0 ± 1.0	—
	16	80.6 ± 1.3##	−4.4 ± −0.6*	79.5 ± 1.2##	−2.5 ± 0.6
Visceral fat area, cm²	0	79 ± 7.0*	—	56 ± 6.0	—
	16	63 ± 7.0##	−6 ± 2.0**‡	51 ± 6.0	−5.0 ± 3.0
Subcutaneous fat area, cm²	0	148 ± 11	—	126 ± 10	—
	16	126 ± 10##	−22 ± 3.0**	118 ± 13	−8.0 ± 4.0

[†1] Values are mean ± SEM (n = 19).

[†2] 16 wk value minus 0 wk value.

[†3] Significantly different from the initial value by Student's t-test:

# $p < 0.05$,

$p < 0.01$.

[†4] Significantly different from triacylglycerol diet group by Student's t-test for paired value:

* $p < 0.05$,

** $p < 0.01$.

[†5] Significantly different from triacylglycerol diet group by analysis of covariance no,

‡ $p < 0.05$.

11.4.2 Repeated DAG vs. TAG Consumption Reduces Body Fat in Humans

To observe the chronic effect of DAG, we conducted a long-term comparative study of DAG oil and TAG oil in 38 healthy Japanese men.[29] Ten grams of 50 g fat in the daily diet were replaced with the test oil for a period of 16 weeks in a double-blind parallel study. As shown in Table 11.4, significant decreases were observed in the DAG group compared with the TAG group in body weight, visceral fat, subcutaneous fat, and waist circumference. Abdominal fat was evaluated by the area of computed tomography (CT) scan cross-section images.

This Japanese study was expanded using male and female subjects who were categorized as either obese or overweight. The Chicago Center for Clinical Research conducted a double-blind parallel study in which 131 overweight or obese men and women ingested DAG and TAG diets for 24 weeks.[30] In this study, 15% of the energy intake was ingested as the test oil under mild hypo-calorie conditions established by subtracting 500–800 kcal from the energy requirement calculated based on body weight, activity level, and age. The food products included muffins, crackers, instant soup mix, sugar cookies, and granola bars. On the basis of the energy-requirement calculations, the number of study food products incorporated into the diet ranged from 2 to 5 serving/d corresponding to 16 to 45g of test oil/d. The decreases in body weight and body fat were significantly higher in the DAG group than in the TAG group (Figure 11.9).

The effect of DAG oil has been shown in a practical study conducted in Japan. After the consumption of approximately 10g of DAG oil/day for 9 months, physical characteristics such as waist circumference and subcutaneous fat thickness significantly decreased from the initial values.[31]

11.4.3 Beneficial Effects of DAG Consumption in Pathological Conditions

Type 2 diabetes, abnormal lipid metabolism, hypertension, hyperuricemia/gout, arteriosclerotic diseases, and fatty liver are some of the known complications of obesity. The importance of diet therapy as a treatment of these lifestyle diseases is widely recognized. DAG oil has also been tested in pathological states such as dialysis patients and type 2 diabetic patients.

Using type 2 diabetic patients, a randomized, single-blind, controlled parallel trial was conducted. In this study, the influence of the long-term ingestion of DAG on blood lipids in type 2 diabetics was examined

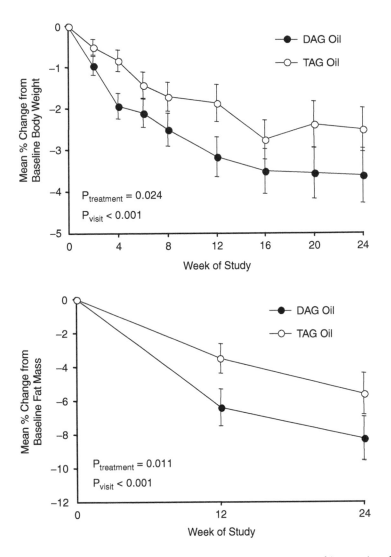

FIGURE 11.9 Mean ± SEM percent changes in body weight from baseline among subjects assigned to DAG oil or TAG oil groups. P-values represent results of repeated measures ANOVA. Mean ± SEM percent changes in fat mass from baseline among subjects assigned to DAG oil or TAG oil groups. P-values represent results of repeated measures ANOVA. (Data from Reference 30.)

using patients with hypertriglyceridemia whose serum TAG levels were persistently increased despite continuous nutritional counseling at an outpatient clinic.[32] The subjects were 16 patients with diabetes. Mean body mass index (BMI) was 26.3 ± 2.9 kg/m^2. Serum TAG levels of these patients were persistently increased despite continuous nutritional counseling at the outpatient clinic over a period of 14.4 ± 12.7 mo.

The baseline serum TAG, total serum cholesterol, and HDL cholesterol levels did not differ significantly between the DAG and the control group. Changes in these parameters during the test period were not significant in the control TAG group. The serum TAG level in the DAG group, in contrast, decreased significantly (from 2.51 ± 0.75 to 1.52 ± 0.28 mmol/L, $p < 0.01$) after a three-month treatment (Table 11.5). No significant changes in blood sugar levels during the study period were observed in either group. In the DAG group, the glycohemoglobin (HbA$_{1c}$) level significantly decreased from 6.41 ± 1.15 to 5.79 ± 0.85% (9.7% reduction, $p < 0.05$). The normal range of HbA$_{1c}$ established by the Japan Diabetes Society is 4.3–5.8%. No significant changes in HbA$_{1c}$ levels were observed in the control group (Table 11.5).

TABLE 11.5 Serum Parameters in Diabetic Patients before and 12 Weeks after the Substitution of Ordinary Cooking Oil with Diacylglycerol Oil

	Diacylglycerol Group (n = 8)		Control Group (n = 8)	
	Before	After[1]	Before	After
Total Cholesterol, mmol/L	5.82 ± 1.32[2]	5.87 ± 0.80	6.00 ± 0.98	5.74 ± 0.70
Triacylglycerol, mmol/L	2.51 ± 0.75	1.52 ± 0.28**†	3.22 ± 2.13	3.59 ± 1.70
HDL Cholesterol, mmol/L	1.27 ± 0.23	1.34 ± 0.36	1.09 ± 0.20	1.22 ± 0.36
Glucose, mmol/L	6.72 ± 0.72	6.94 ± 0.89	7.83 ± 1.78	7.77 ± 2.50
HbA$_{1c}$, %	6.41 ± 1.15	5.79 ± 0.85[3]	6.88 ± 0.53	6.65 ± 0.73

[1]Measured at 12 wk after the substitution of the oil.
[2]Values are mean ± SD.
[3]Significantly different from before ($p < 0.05$; **$p < 0.01$).
[4]Significantly different from control group ($p < 0.05$).

11.4.4 Energy Value and Absorption Coefficient of DAG

The combustion heat of cooking oil containing 87% DAG was 2% less than that of TAG.[28] This difference in energy value between DAG oil and TAG oil may be negligible, in terms of the practical consumption of oils. Furthermore, the absorption coefficients of the DAG and TAG oils determined in rats were similar (96.3%).[28] These results indicate that the physiological differences between DAG and TAG observed in humans and animals described in this chapter are likely to be caused by the different metabolic fates after absorption by gastrointestinal epithelial cells.

11.4.5 Digestion and Absorption of DAG Compared to TAG

The digestion and absorption process of DAG was investigated in an experiment involving time-course changes in lipid composition after the perfusion of triolein and diolein (1,2-:1,3-diolein = 3:7) in the intestinal tract of rats.[33] In the diolein perfusion, unlike the triolein perfusion, 65% of monoolein was found to be 1(3)-monoolein at 60 min after the initiation of perfusion (Figure 11.10). Kondo et al.[34] reported that, when DAG was intraduodenally infused in the form of an emulsion, TAG was digested to

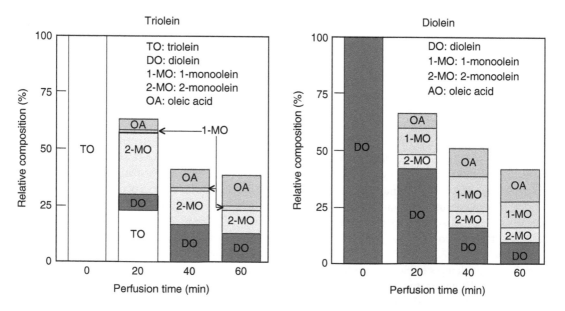

FIGURE 11.10 Analysis of digestion products (data from Reference 33).

1,2-DAG, 2-MAG, and FFA, whereas 1,3-DAG was digested to 1(3)-MAG and FFA. Thus, the production of 1(3)-MAG instead of 2-MAG may be one of the characteristics of DAG metabolism.

The TAG resynthesis pathway in small intestinal epithelial cells includes the 2-MAG pathway and the glycerophosphate pathway, with the former being predominant. Whereas 2-MAG is a good substrate in reactions of the 2-MAG pathway, the reactivity of 1(3)-MAG is lower. Free glycerol is a substrate in the glycerophosphate pathway, but the reaction rate is slower than that in the 2-MAG pathway, and its contribution to TAG resynthesis is small. The incorporation of ^{14}C-labeled linoleic acid into TAG was significantly retarded in the intestinal mucosa of rats in which a DAG oil emulsion had been infused as compared to TAG oil emulsion.[34] Murata et al. reported that the rate of release of resynthesized TAG into the intestinal lymph chylomiceron was lower after the administration of a DAG emulsion than that of a TAG emulsion.[35] A slower lymphatic secretion of ^{14}C-labeled triolein after intra-gastric infusion of ^{14}C-1,3-diolein has also been demonstrated in rats.[36] This may be the result of the slower reesterification rate of fatty acids after DAG ingestion. Figure 11.11 illustrates the metabolic characteristics of DAG, as compared with TAG.

In terms of beneficial effects other than nutritional effect, we found that the stomach-emptying time was reduced when DAG oil is used compared with TAG oil in scrambled eggs. The subjects ingested scrambled eggs cooked with or without the test oil. The stomach-emptying time was monitored by using technetium labeled albumin as a probe of a scintigram. The stomach-emptying time for scrambled eggs cooked with DAG oil was shorter than that cooked with control TAG oil.[37]

11.4.6   Antiobesity Effects of DAG in Animal Studies

Animal models have been used to further confirm the effects of DAG and to elucidate the mechanism of its action. When C57BL/6J mice (obesity- and diabetes-prone model mice) were fed a high fat (30%, w/w) and sucrose (13%, w/w) diet for five months, body weight and body fat (white adipose tissue, WAT) mass increased, compared to a control low-fat diet. Insulin and leptin (a peptide product of the OB gene that is associated with obesity in mice as well as in humans) concentrations also increased in these mice. Substituting DAG for TAG prevented the increases in body weight and fat accumulation associated with the high fat and sucrose diet (Figure 11.12).[38] Increases in insulin and leptin concentrations were also prevented by substituting TAG with DAG. Mice consuming a DAG diet were able to maintain smaller fat stores suggesting that DAG consumption produced an increase in energy expenditure.

Although the mechanism for the suppression of body fat accumulation by a DAG diet is yet to be elucidated, it may be due, at least in part, to the increased beta-oxidation in the liver induced by a DAG diet. When food containing 10% DAG was given to rats for 2–3 weeks, the enzyme activity for fatty acid

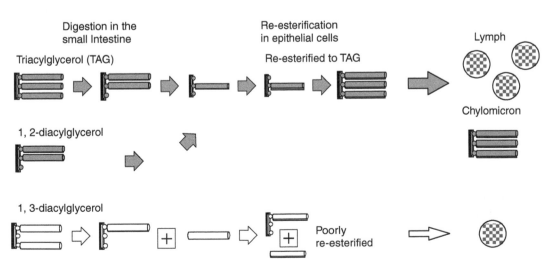

FIGURE 11.11   Metabolic characteristics of diacylglycerol (DAG) in comparison with triacylglycerol (TAG).

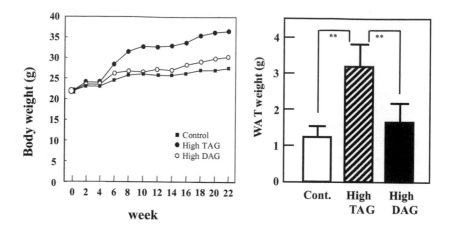

FIGURE 11.12 Effect of diacylglycerol ingestion on body weight (panel A) and total body fat (panel B) in C57BL/6J mice. Panel A, body weight changes during the study period; Panel B, intra-abdominal white adipose tissue (WAT) weight. Sum of epididymal, mesenteric, retroperitoneal and perirenal WAT at the end of the study (5 months) are shown. Mean ± SD. Cont, low fat control diet; high DAG, high fat DAG diet; high TAG, high fat TAG diet. Data from ref. 38.

synthesis in the liver was decreased and that for beta-oxidation of fatty acids was increased.[39] In another experiment with high-fat induced obesity model mice (C57BL/6J), a high DAG diet, as compared to a high TAG diet, increased hepatic acyl-coenzyme A oxidase activity and mRNA for acyl-coenzyme A synthase, suggesting a higher capacity for hepatic lipid oxidation.[38] Using the same animal model, they showed that, within the first 10 days (before onset of obesity), DAG consumption stimulated beta-oxidation and lipid metabolism-related gene expression, including acyl-CoA oxidase, medium-chain acyl-CoA dehydrogenase and uncoupling protein-2 in the small intestine but not in the liver, skeletal muscle, or brown adipose tissue, suggesting the predominant contribution of intestinal lipid metabolism to the effects of DAG.[40]

11.4.7   Further Application of DAG

We examined the effects of the difference in the acylglycerol structure with the same fatty acid composition and found significant beneficial effects of DAG over TAG oil. However, n-3 polyunsaturated fatty acids in foods, such as eicosapentaenoic acid (EPA), docosahexaenoic acid (DHA) and alpha-linolenic acid (ALA) are of interest, due to various physiological functions such as for their anti-arteriosclerotic, antihypertensive antihyperlipidemic, and anti-obesity effects. We pursued more potent beneficial effects of DAG by incorporating n-3 polyunsaturated fatty acid into the DAG structure. In this context, the effects of alpha linolenic acid rich diacylglycerol (ALA-DAG) on body fat were investigated in obese model mice C57BL/6J. By incorporating a small amount (1–4 wt%) of ALA-DAG in a high-fat and sucrose diet (30 wt% and 13 wt%, respectively), body weight gain and visceral fat gain were significantly reduced.[41] These results indicate the antiobesity function of DAG is potentiated by incorporating ALA into the DAG structure, and also suggest the effectiveness of using ALA-DAG for the prevention and treatment of lifestyle-related diseases in which obesity is a risk factor.

In addition to obesity, hypercholesterolemia is another health concern since it has been shown to be associated with various lifestyle-related diseases. Clinical investigations have revealed that the administration of phytosterols in human subjects reduces plasma total cholesterol and LDL cholesterol levels. Meguro et al. found that DAG was a good solvent for phytosterols as compared to TAG. They conducted a study to investigate the difference in the serum-cholesterol-lowering activities between phytosterols dissolved in DAG and those dispersed in TAG. The administration of 500 mg/day of phytosterols dissolved in DAG oil for two weeks significantly reduced the total and LDL cholesterol levels from 5.57 to 5.31 mmol/l (4.7% reduction) and 3.69 to 3.39 mmol/l (7.6% reduction), respectively, whereas the same amount of phytosterols dispersed in TAG oil had no significant effect.[42] DAG oil containing phytosterols has been on the market in Japan since 2001 as "Food for Specified Health Use" approved by the Ministry of Health, Labour, and Welfare.

11.5   Conclusion

In this chapter, we focused on the structure of acylglycerols in the edible oil that influence body fat accumulation, and described 1,3-DAG production using immobilized 1,3-regioselective lipase and its nutritional functions in comparison with conventional TAG oil with the same fatty acid composition. Consequently we can conclude that the efficient enzymatic DAG production is possible and that the product has beneficial functions in human health as compared to the conventional TAG oil.

A cooking oil product containing 80% (w/w) DAG or greater has been used in Japan since 1999 as "Food for Specified Health Use" approved by the Ministry of Health, Labour, and Welfare. The approved claims are: 1) a lower elevation in postprandial TAG concentrations in the blood after DAG ingestion, as compared to TAG with the same fatty acid composition, and 2) less body fat accumulation by DAG oil consumption as compared to TAG oil. We conclude that this oil will contribute to human health by reducing risk factors for lifestyle-related diseases.

These effects are probably caused by metabolic differences during or after the absorption of the oils into mucosal cells, and not by the difference in the bioavailability of the oils. Although mechanisms of the digestion, absorption, and metabolic process are yet to be fully elucidated, the ingestion of DAG oil has been shown to be beneficial in lipid metabolism in various conditions.

Clinical studies are still in progress to further confirm the efficacy and safety of DAG oil consumed under various conditions.

References

1. N Tada, H Yoshida. Diacylglycerol on lipid metabolism. *Curr Opin Lipidol* 14:29–33, 2003.
2. AA Abdel-Nabey, AA Shehata, MH Y, Ragab, JB Rossell. Glycerides of cottonseed oils from Egyptian and other varieties. *Riv Ital Sostanze Grasse* 69:443–447, 1992.
3. RP D'alonzo, WJ Kozarek, RL Wade. Glyceride composition of processed fats and oils as determined by glass capillary gas chromatography. *J Am Oil Chem Soc* 59:292–295, 1982.
4. N O V Sonntag. Glycerolysis of fats and methyl esters-status. *J Am Oil Chem Soc* 59:795A–802A, 1982.
5. J F Plou M Barandiarn, V M Calvo, A Ballesteros, E Paster. High-yield production of mono- and di-oleylglycerol by lipase-catalyzed hydrolysis of triolein. *Enzyme Microb Technol* 18:66–71, 1996.
6. A M Fureby, L Tian, P Adlercreutz, B Mattiasson. Preparation of diglycerides by lipase-catalyzed alcoholysis of triglycerides. *Enzyme Microb Technol* 20:198–206, 1997.
7. B Yang, W J Harper, K L Parkin, J Chen. Screening of commercial lipase for production of mono- and diacylglycerols from butteroil by enzymic glycerolysis. *Int Dairy Journal* 4:1–13, 1994.
8. T Yamane, S T Kang, K Kawahara, Y Koizumi. High-yield DAG formation by solid-phase enzymatic glycerolysis of hydrogenenated beef tallow. *J Am Oil Chem Soc* 71:339–342, 1994.
9. Y Ota, T Takasugi, M Suzuki. Synthethic of either mono- or diacylglycerols from high-oleic sunflower oil by lipase-catalyzed glycerolysis. *Food Sci Technol* 3:384–387, 1997.
10. M Berger, K Laumen, M Schneider. Enzymatic esterification of glycerol. I. Lipase-catalyzed synthesis of resioisomerically pure 1,3-sn-DAGs. *J Am Oil Chem Soc* 69:955–960, 1992.
11. C Waldinger, M Schneider. Enzymatic Esterification of Glycerol III. Lipase-catalyzed synthesis of regioisomerically pure 1,3-sn-DAGs and 1(3)-rac-monoacylglycerols derived from unsaturated fatty acids. *J Am Oil Chem Soc* 73:1513–1519, 1990.
12. F Ergan, M Trani, G André. Production of glycerides from glycerol and fatty acid by immobilized lipase in non-aqueous media. *Biotechnol Bioeng* 35:195–200, 1990.
13. Y Hirota, J Kohori, Y Kawahara. Preparation of diglycerides. Europe Patent No. 0307154, 1988.

14. R Rosu, M Yasui, Y Iwasaki, T Yamane. Enzymatic synthesis of symmetrical 1,3-DAGs by direct esterification of glycerol in solvent-free system. *J Am Oil Chem Soc* 76:839–843, 1999.

15. T Watanabe, M Shimizu, M Sugiura, M Sato, J Kohori, N Yamada, K Nakanishi. Optimization of reaction conditions for production of DAG using immobilized 1,3-regiospecific lipase Lipozyme RM IM. *J Am Oil Chem Soc,* 80:1201–1207, 2003.

16. R Lortie, M Trani, F Ergan. Kinetic study of the lipase-catalyzed synthesis of triolein. *Biotechnol Bioeng* 41:1021–1026, 1993.

17. E Castillo, V Dossat, A Marty, J S Condoret, D Combes. The role of silica gel in lipase-catalyzed esterification reactions of high-polar substrates. *J Am Oil Chem Soc* 74:77–85, 1997.

18. X Xu, A R H Skands, C E Høy, H Mu, S Balchen, J A Nissen. Production of specific-structured lipids by enzymatic interesterification: elucidation of acyl migration by response surface design. *J Am Oil Chem* Soc 75:1179–1186, 1998.

19. X Xu, A R H Skands, C E Høy, J A Nissen. Parameters affecting DAG formation during the production of specific-structured lipids by lipase catalyzed interesterification. *J Am Oil Chem Soc* 76:175–181, 1999.

20. X Xu, S Balchen, C E Høy. Production of specific-structured lipids by enzymatic interseterification in a pilot continuous enzyme bed reactor. *J Am Oil Chem Soc* 75:1573–1579, 1998.

21. X Xu, L B Fomuso, C C Akoh. Synthesis of structured triacylglycerols by lipase-catalyzed acidlysis in a packed-bed bioreactor. *J Agric Food Chem* 48:3–10, 2000.

22. Y Shimada, A Sugihara, H Nakano, T Yokota, T Nagao, S Komemushi, Y Tominaga. Production of structured lipids containing essential fatty acids by immobilized rhizopus delemar lipase. *J Am Oil Chem Soc* 73:1415–1420, 1996.

23. S Colombie, R J Tweddell, J Condoret, A Marty. Water activity control: a way to improve the efficiency of continuous lipase esterification. *Biotech. Bioeng* 60:362–368, 1999.

24. P Mensah, G Catra. Adsorptive control of water in esterification with immobilized enzymes. continuous operation in a periodic counter-current reactor. *Biotech. Bioeng* 66:137–146, 1999.

25. J A Arcos, H S Garcia, C G Hill. Continuous enzymatic esterification of glycerol with (poly)unsaturated fatty acid in a packed reactor. *Biotech Bioeng* 68:563–570, 2000.

26. M Sugiura, H Kohori, N Yamada. Preparation Process of Diglycerides. US Patent No. 6361980B2, 2002.

27. N Tada, H Watanabe, N Matsuo, I Tokimitsu, M Okazaki. Dynamics of postprandial remnant-like lipoprotein particles in serum after loading of diacylglycerols. *Clin Chim Acta* 311:109–117, 2001.

28. H Taguchi, T Nagao, H Watanabe, K Onizawa, N Matsuo, I Tokimitsu, H Itakura. Energy value and digestibility of dietary oil containing mainly 1,3-diacylglycerol are similar to those of triacylglycerol. *Lipids* 36:379–382, 2001.

29. T Nagao, H Watanabe, N Goto, K Onizawa, H Taguchi, N Matsuo, T Yasukawa, R Tsushima, H Shimasaki, H Itakura. Dietary diacylglycerol suppresses accumulation of body fat compared to triacylglycerol in men in a double-blind controlled trial. *J Nutr* 130:792–797, 2000.

30. KC Maki, MH Davidson, R Tsushima, N Matsuo, I Tokimitsu, DM Umporowicz, MR Dicklin, GS Foster, KA Ingram, BD Anderson, SD Frost, M Bell. Consumption of diacylglycerol oil as part of a mildly reduced-energy diet enhances loss of body weight and fat compared with a triacylglycerol control oil. *Am J Clin Nutr* 76:1230–1236, 2002.

31. Y Katsuragi, T Toi, T Yasukawa. Effects of dietary diacylglycerols on obesity and hyperlipidemia. *J Jpn Soc of Human Dry Dock* 14:258–262, 1999.

32. K Yamamoto, H Asakawa, K Tokunaga, H Watanabe, N Matsuo, I Tokimitsu, N Yagi. Long-term ingestion of dietary diacylglycerol lowers serum triacylglycerol in type II diabetic patients with hypertriglyceridemia. *J Nutr* 131:3204–3207, 2001.

33. H Watanabe, K Onizawa, H Taguchi, M Kobori, H Chiba, Y Naito, N Matsuo, T Yasukawa, M Hattori, H Shimasaki. Nutritional characterization of diacylglycerols in rats. *J Jpn Oil Chem Soc* 46:301–307, 1997.

34. H Kondo, T Hase, T Murase, I Tokimitsu. Digestion and assimilation features of dietary DAG in the rat small intestine. *Lipids* 38:25–30, 2003.

35. M Murata, K Hara, T Ide. Alteration by diacylglycerols of the transport and fatty acid composition of lymph chylomicrons in rats. *Biosci Biotech Biochem* 58:1416–1419, 1994.

36. T Yanagita, I Ikeda, Y-M Wang, H Nagakiri. Comparison of the lymphatic transport of radio labeled 1,3-diacylglycerol and triacyglycerol in rats, *Lipids* 39:827–832, 2002.

37. K Yasunaga, Y Seo, Y Katsuragi, N Oriuchi, H Ootake, K Endo, T Yasukawa. 54th Annual meeting of the Japanese Society of Nutrition and Food Science, 2000.

38. T Murase, T Mizuno, T Omachi, K Onizawa, Y Komine, H Kondo, T Hase, I Tokimitsu. Dietary diacylglycerol suppresses high fat and high sucrose diet-induced body fat accumulation in C57BL/6J mice. *J Lipid Res* 42:372–378, 2001.

39. M Murata, T Ide, K Hara. Reciprocal responses to dietary DAG of hepatic enzymes of fatty acid synthesis and oxidation in the rat. *Br J Nutr* 77:107–121, 1997.

40. T Murase, A Nagasawa, J Suzuki, T Wakisaka, T Hase, I Tokimitsu. Dietary α-linolenic acid-rich diacylglycerol reduces body weight gain accompanying stimulation of intestinal β-oxidation and related gene expression in C57BL/KsJ-*db/db* mice. *J Nutr* 132:3018–3022, 2002.

41. T Hase, T Mizuno, K Onizawa, K Kawasaki, H Nakagiri, Y Komine, T Murase, S Meguro, I Tokimitsu, H Shimasaki, H Itakura. Effect of α-linolenic acid-rich diacylglycerol on diet-induced obesity in mice. *J Oleo Sci* 50:701–710, 2001.

42. S Meguro, K Higashi, T Hase, Y Honda, A Otsuka, I Tokimitsu, H Itakura. Solubilization of phytosterols in diacylglycerol versus triacylglycerol improves the serum cholesterol-lowering effect. *Eur J Clin Nutr* 55:513–517, 2001.

12

The Use of Nonimmobilized Lipase for Industrial Esterification of Food Oils

Satoshi Negishi

12.1   Introduction

Nowadays, most food oils are isolated from milk or fatty meat of animals or from sarcocarps, seeds or germs of plants, and vegetable oils commonly produced by purifications of their crude extracts. Consequently, the characteristics of food oils, such as physicochemical properties, nutritious efficacy, suitability of cooking properties, and taste directly depend on the kind of oil sources. Therefore, some efforts to modify the structure of triacylglycerides have been made to obtain new food oils with valuable characteristics. The structural modifications include hydrogenation, transesterification, and fractionation, which are the commonly used techniques in many countries.

Lipase catalyzed transesterification is one of the methods employed to obtain functional food oils. This method is used to produce food oils under mild conditions at normal temperatures and pressures, to reduce the deterioration of products as well as side reactions, and to maintain the substrate specificity. Nevertheless, the current status of this technique has been put into practice only in limited areas compared to the other techniques mentioned above. One of the reasons for this is that although the cost performance of the lipase method is low, it requires complicated equipment for industrial production.

Based on this background, our studies on the transesterification reaction using lipase were addressed to reduce the cost and to develop it for versatile use. Currently, the method we developed has been put into practice in the industrial production of several kinds of functional food oils. I wish to present here some findings obtained in the course of our studies on the methodology for industrial transesterification using lipase.

12.2 Transesterification Using Nonimmobilized Lipase

To reduce the cost of the transesterification process, it is necessary to design the reaction system as simply as possible. Usually, immobilized lipase is used for transesterification reactions, but, in principle, lipase can be reused without immobilization processing because the enzyme is practically insoluble in oil. For the immobilization processing, a third substance is required as a carrier, by which the specific activity of the enzyme is inevitably decreased. Thus, we investigated the transesterification reaction given that a compact reaction system can be designed by using powder lipase with which the specific activity per unit volume or weight can be increased and the cost can be reduced as a result.[1]

12.2.1 Relationship between Water Content and Transesterification Reactivity

Studies on enzymatic transesterification reactions in organic media have a long history, and Dastoli et al. had already reported on the reactions in nonpolar solvents.[2] After that, in the 1980s, enzymatic reactions in organic solvents were intensively studied, and reviewed by Drodick in 1989.[3] In these studies, it was found that water was an important factor for the transesterification reaction. Water plays major roles in two contradictory ways; the first is the induction of enzymatic activity and the second is the thermal deactivation of the enzyme.[4] Namely, when powder lipase is used, water is required to induce the catalytic activity of transesterification, whereas too much water leads to the thermal deactivation of the enzyme or the hydrolysis of the products. Therefore, a very precise control of water content is required for the transesterification reaction using powder lipase, this being a bottleneck in the industrialization of this technique. The use of immobilized lipase aimed, of course, at the recovery or recycle of the enzyme, but, as a matter of fact, it also served as an effective means to solve the problems mentioned above.

To put the transesterification method using powder lipase into practice for industrial utilization, it is necessary to select an enzyme with which the content of water in the reaction system can be simply controlled. Usually, the water content in a reaction system is examined by measuring the concentrations of water in the reaction medium, but we examined the effects of water content on the transesterification activity using an index of water quantity relative to powder lipase. The optimal amount of water for the maximum transesterification activity differs depending on the kind of lipase used. In addition, the optimal temperature for the maximum reactivity of a transesterification reaction using a lipase with optimal water content also differs depending on the lipase used. We examined the optimal water content and temperature for a maximum transesterification reactivity for tens of different lipases. The results of some typical lipases are shown in Figure 12.1, in which most lipases were categorized as high concentration of water region group. The high concentration of water region lipases are characterized by showing no transesterification reactivity unless water is added to some extent, and the addition of water, of course, eventually leads to the hydrolysis of esters and the thermal deactivation of the enzyme. Therefore, a very precise control of water content is required when high concentrations of water region lipases are used for the transesterification reaction, its practical use being considered therefore difficult.

However, some of the lipases were categorized as a low concentration of water region group (Figure 12.1). In the case of low concentration of water region lipases, the volume of water required for enzyme activation is very little. Some low concentration of water region lipases can be activated by the water present in oil in as little amount that can be considered mere moisture without addition of water. Namely, the negative factors such as hydrolysis and thermal deactivation of the enzyme caused by

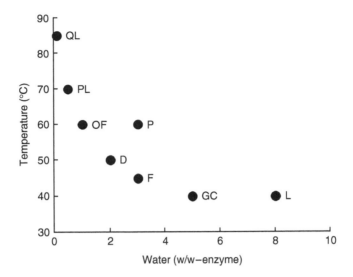

FIGURE 12.1 The optimal temperature of various lipases versus the percentage of added water. The following enzymes were used: lipase OF (*Candida cylindracea*; Meito Sangyo Co., Ltd.), lipase PL (*Alcaligenes* sp.; Meito), Lipase QL (*Alcaligenes* sp.; Meito), lipase D (*Rhizopus delemar*; Amano Pharmaceutical Co., Ltd.), lipase F (*Rhizopus javanicus*; Amano), lipase GC (*Geotorichum candidum*; Amano), lipase L (*Candida lipolytica*; Amano), and lipase P (Pseudomonas sp.; Nagase Sangyo & Co., Ltd) (From Reference 1.)

externally added water are avoidable by using lipases of this category because the transesterification activity can be induced even in a condition such as the apparent absence of water. By using the powder form of these lipases in place of immobilized enzymes, the construction of a simple reaction system for the transesterification process became possible.

12.2.2   Changes in the Property of Water Dissolved in Oil

As for the difference between lipases categorized as high and low concentration of water regions (Figure 12.1), a trace amount of water can be dissolved in oil as inherent moisture. According to the report by Takasago et al. on the water present in oil, the water can be classified roughly into two types: the first exists in a hydrogen-bonded form with the polar part of triacylglyceride, and the second exists as cluster water in a constrained form to some extent but not in a hydrogen-bonded form.[5] The physical state of the two types of water can be detected by near-infrared spectroscopy.

Figure 12.2 shows the near-infrared spectra of the oils treated with respective lipases by stirring at room temperature for 30 min, followed by removal of the lipases by filtration. The absorption peak at the position 5170 cm^{-1} in Figure 12.2 can be assigned to the hydrogen-bonded water, while that at the position 5260 cm^{-1} to the cluster water. When low concentration of water region lipases were added to oil, the cluster water in the oil decreased. This suggested the possibility that the cluster water present in oil was consumed for the activation of low concentration of water region lipases. In any case, the water dissolved in oil is not free water, which is considered not to cause hydrolysis or thermal deactivation of the enzyme. Therefore, transesterification reactions can be conducted by directly using low concentration of water region powder lipases, thereby avoiding hydrolysis and thermal deactivation.

12.2.3   Stability of Powder Lipases

As mentioned above, a very simple reaction system for the transesterification process could be constructed by using a powder lipase. However, even though a simple reaction system using powder lipase can be constructed, it cannot directly be put into practice unless the stability of the process is guaranteed. For the

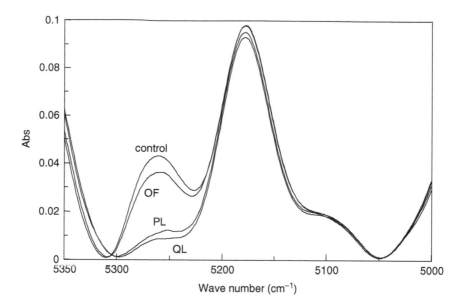

FIGURE 12.2 Influence of lipase addition on near-infrared spectra of the dissolved water in triolein. Conditions: 100 mg of lipase was stirred for 30 min in 10 g of triolein at room temperature and filtrated. (From Reference 1.)

stable production process, the stability of lipase is an essential factor. Meanwhile, an enzyme can be generally stabilized by immobilization; thus some immobilized lipases are already commercially available. However, in view of the cost performance, the use of immobilized lipases for industrial production of food oils is unacceptable. In this context, we investigated the stability of the synthetic process using powder lipase.

Figure 12.3 shows the results of a continuous transesterification reaction between triolein and tricaprylin using a column in which powder lipase QL was directly packed without any carriers. The reaction was carried out without adding water into the reaction medium. As a result, the reaction

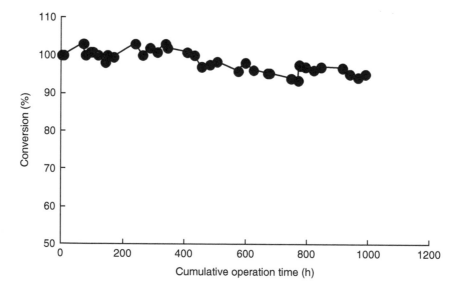

FIGURE 12.3 Transesterification using a 1 cm diameter column with filter (ADVANTEC,5A). A mixture of triolein and tricaprylin was continuously pumped at a flow rate of 5 g/h. The depth of the bed of lipase QL(0.3 g) was 1.9 cm. The temperature of the reaction system was 40°C. Water was not added to the reaction. (From Reference 1.)

could be continued for a long period of 1000 hours maintaining the overall conversion rate over 95%. The total quantity of oil reacted in a batch amounted to 16,000-fold of the lipase used, indicating sufficient stability of the process for practical use.

As described above, enzymatic transesterification could be conducted in a simple and convenient system without requiring any complicated procedures such as enzyme immobilization. By using this method, the transesterification reaction with lipase, which has not yet commonly been used in industrial production, could become widely used in the fat industry.

12.3 Applications to Food Oils Consisting of Medium-Chain Fatty Acids

Most fatty acids derived from conventional foodstuffs have long-chain C_{14}–C_{18} hydrocarbon residues. In contrast, human milk, bovine milk, and coconut oil contain fatty acids with medium-chain C_8–C_{10} hydrocarbon residues. Medium-chain fatty acids have the characteristics that they are directly metabolized in the liver after absorption through epithelial cells in the intestine, followed by transportation to the portal vein without resynthesis.[6] Many studies on the physiological functions of medium-chain fatty acids have been documented in reviews or printed books.[7,8] Furthermore, medium-chain fatty acids are expected to provide the nutritious advantage that they are not only safer but also more readily converted to energy than long-chain fatty acids, some of which have been utilized as fat absorption inhibitors for medical purposes.

However, when we try to impart these nutritious advantages of medium-chain fatty acids to food oils for general use, a great problem regarding their cooking properties arises. The problem is due to their following chemical properties: when MCT, a triacylglyceride consisting solely of medium-chain fatty acids, is used in frying, it smokes markedly; furthermore, when it is used in combination with triacylglyceride consisting of long-chain fatty acids, it foams violently, causing bubbling over or a fire. With this background in mind, we tried to develop versatile food oils with a superior cooking property and a nutritious value by designing the triglyceride structure of food oils.[9,10]

12.3.1 Improvement in Smoking Tendency on Frying

When MCT is directly used in frying, it inevitably leads to smoking, which may result not only in a reduced cooking performance but also in dangerous problems such as causing a fire. This is probably due to the chemical composition of the oil, the molecular weight of which is smaller than that of rape oil or soybean oil. To reduce the smoking tendency, we must convert the triglyceride structure of MCT to that of a greater molecular weight consisting of both medium-chain and long-chain hydrocarbon residues (MLCT). The structural modification includes two possibilities of fatty acid composition of triglyceride in which two molar medium-chain and one molar long-chain fatty acid or one molar medium-chain and two molar long-chain fatty acid residues are bound as glycerol esters, and we confirmed that neither of these MLCT compositions caused smoking on frying. Thus, the conversion of MCT to MLCT by means of the lipase-catalyzed transesterification with conventional vegetable oils (LCT: triacylglyceride with long-chain hydrocarbon residues) could solve the problem of smoking.

12.3.2 Improvement in Foaming Tendency on Frying

When MCT is mixed with LCT, violent foaming develops on frying, which appears to be dangerous, and also can cause a firelike smoking. But the foaming is not caused solely due to the low molecular weight of MCT, because the foaming tendency was not improved by converting it to MLCT by the lipase-catalyzed transesterification with LCT. Therefore, it was necessary to examine in more detail the relationship between the structure of triacylglyceride and foaming tendency on frying.

12.3.2.1 Quantitative Determination of Foaming on Frying

To examine the relationship between the structure of triacylglyceride and foaming on frying, first it was necessary to establish a method for quantitative determination of the amount of foam generated on frying.

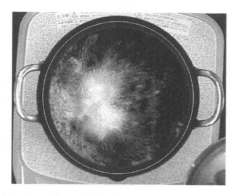 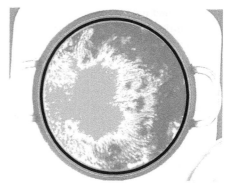

Photograph   Image analyzed

FIGURE 12.4 Example of image analysis. The foaming spot was defined by a fixed luminance criterion. (From Reference 9.)

For the quantitative analysis of foaming on frying, we introduced an image analysis method. As foaming on frying is continuously changing, it cannot be grasped from a momentary image; therefore we shot it continuously at intervals of 1 sec. The proportion of bubbling area to the total area of the frying pan was calculated for each image and it was integrated for all images to obtain an integrated value, which was defined as the amount of foam generated on frying.

Figure 12.4 shows examples of the photographs used for the image analysis. A total of 517 images were shot at intervals of 1 sec in each experiment and the relative area of the gray portion to the area encircled by a black line was assigned as the foam in each image. The bubbling area in each image varied greatly, but the IF value (integration of foam), which is the integrated proportions of foaming area of all 517 images, was practically invariable. Thus, the quantitative determination of foam generated on frying became possible by using the index IF.

12.3.2.2 Relationship between the Structure of Triacylglyceride and Foaming Tendency

Several kinds of triacylglycerides consisting of medium-chain fatty acids were prepared by means of the lipase reactions, and they were used for frying to determine the IF values. As a result, it was revealed that the amount of foam generated on frying depends on the molecular weight distribution of the triacylglyceride used. Namely, when triacylglycerides with different molecular weights were mixed into cooking oil, it readily bubbled on frying, while the foaming tendency was slight when the oil was composed of triacylglycerides with comparable molecular weights. We set a new index, the foam index of triacylglyceride (FIT), as a parameter relevant to the molecular weight distribution of triacylglycerides composition, which was calculated according to the following equation:

$$FIT = \Sigma \; RM - ARM \times (Mol\%)/100\}$$

$$RM = \{(MW - 470)/(976 - 470) \times 100$$

where
    MW: molecular weight
    RM: relative molecular weight
  ARM: average relative molecular weight
    470:   molecular weight of tricapryloylglycerol
    976:   molecular weight of triarachidonylglycerol

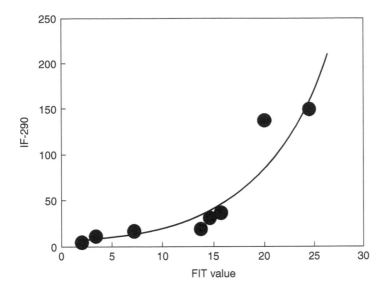

FIGURE 12.5 The correlation between the FIT and IF (from Reference 9).

Figure 12.5, where the IF value was deducted from 290 as the blank value of the frying object, shows the relationship between IF and FIT values. As shown, the IF values increased with the increase of FIT values, indicating that the foaming tendency increased in relation to the molecular weight of tricylglycerides. Considering the whole situation, i.e., foaming and the risk of fire, it was suggested that food oils consisting of medium-chain fatty acids with better cooking properties can be obtained when the triacylglyceride composition is designed so that the FIT value is less than 15 (Figure 12.6).

12.3.3 Functionalization with Nutritious Efficacy

By designing the structure of triacylglycerides, as described above, a favorable cooking property can be imparted to the structured lipids consisting of medium-chain fatty acids. In the design, many structural possibilities are considered, and the following points should be kept in mind: continuity of the synthetic process before and after the enzymatic transesterification reaction;

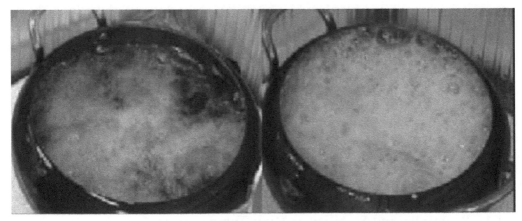

Structured Lipid   Mixed

FIGURE 12.6 Photograph of foaming.

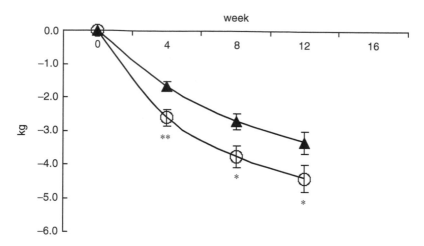

FIGURE 12.7 Change in the body fat composition of men and women consuming either long-chain triacylglycerols or medium- and long-chain triacylglycerols for 12 weeks (from M Kasai et al. Effect of dietary medium- and long-chain triacylglycerols (MLCT) on accumulation of body fat in healthy humans. *Asia Pac J Clin Nutr* 12:151–160, 2003): -▲- ; long-chain triacylglycerols, -O- ; medium- and long-chain triacylglycerols significantly different from long-chain triacylglycerols diet group, *, $P < 0.05$; **, $P < 0.01$.

equipment occupancy; required manpower and costs. Considering these factors, we decided to commercialize a food oil product containing ca. 13% of a medium-chain fatty acid. Figure 12.7 shows the results of a study on the nutritious efficacy of long-term intake of the product in healthy volunteers with BMI indices over 24.7 kg/m². In addition to the favorable cooking property, it was expected that the product would provide a more effective nutritious advantage to reduce body fat during long-term intake compared to conventional food oils, which was suggested from the data shown in Figure 12.7. This product has been approved in Japan as a food for specified health use and marketed already.

12.4 Applications of the Method to General-Purpose Food Oils Consisting of Phytosterols

Phytosterols and their fatty acid esters have been known to suppress cholesterol absorption.[11] Therefore, the development of food oils consisting of phytosterols has been a goal of researchers. However, phytosterols have melting points higher than 100°C, and are not soluble in food oil forming precipitates, this being a great reason for the difficult commercialization of food oils containing phytosterol. To solve the problem, it is necessary to convert phytosterols to an ester of fatty acid with a low melting point such as oleic acid. For the synthesis of a fatty acid ester of phytosterols, chemical methods are considered, but some problems such as the formation of 3,5-diene as a by-product and coloring of the product are involved. Alternatively, the lipase-catalyzed transesterification method is considered useful to avoid these problems. However, the reaction requires a high temperature to dissolve phytosterols in oil; therefore enzyme deactivation should be taken into consideration. We took into account that the high-temperature performance of powder lipase could be fitted for the reaction and therefore examined the transesterification method in an attempt to develop a method for industrial production of phytosterol containing food oils.[12]

12.4.1 Reactivity and Stability at High Temperatures

For the synthesis of phytosterol esters by the transesterification reaction between phytosterols and vegetable oils, as described above, the reaction has to be carried out at very high temperatures because of

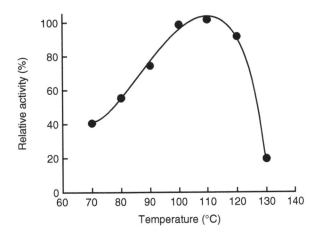

FIGURE 12.8 Dependence of the relative activity of transesterification of phytosterol and sunflower oil by lipase QLM on temperature. Conditions: 0.1 g of lipase powder was added to 9 g of sunflower oil and 1 g of phytosterol. (From S Negishi et al. Transesterification of phytosterols and edible oil by lipase powder at high temperature. *J Am Oil Chem Soc* 80:805–907, 2003.)

the high melting points of phytosterols. Figure 12.8 shows the results of the examination of the effect of the temperature on transesterification reactions between phytosterols and sunflower oil using powder lipase. As shown in the Figure, the maximum reactivity was observed at temperatures in the range of 100–120°C, and 20% of the maximum activity was reserved even at 130°C. The results suggested that the method could be efficiently applied to the transesterification of phytosterols for the industrial production of phytosterol containing food oils.

Usually, it is believed that the three-dimensional stereo-structure of an enzyme protein can be stabilized by immobilization, by which the enzyme can be protected from thermal deactivation, this being an advantage for its use in high-temperature reactions. But it was demonstrated that transesterification proceeded at adequately high temperatures by using powder lipase without immobilization.

When we consider a practical industrial production, the enzyme has to be reused repeatedly. In this context, we examined repeated use of powder lipase (10 times) by consecutively carrying out transesterification at 90°C for 20 hr. Conversion rates of transesterification were determined at intervals of 2 hr for each reaction and a logarithmic plot of the data is shown in Figure 12.9. From the slope of the graph, a considerably long half-life (260 hr) of the enzyme activity was estimated, although the activity slightly declined by repetition. Compared to the aforementioned reaction at 40°C (Figure 12.3), the costs of enzyme used in this reaction were more expensive; nevertheless, we considered that the industrial production of phytosterol-containing food oils using powder lipase might be practically possible by devising a synthetic process.

12.4.2 Improvement in Cold Resistance

Figure 12.10 shows the photographs of the bottled products of phytosterol-containing food oil, in which "mixed" is a mixture of phytosterol and a vegetable oil, while "structured lipid" is the product obtained by transesterification reaction using powder lipase. As it is evident in the photographs, considerable precipitates are seen in the mixture oil, whereas the structured lipid shows practically no precipitates, indicating an increased solubility of phytosterol, thus the commercialization of the product became possible. Furthermore, this product was demonstrated to reduce blood cholesterol levels, and it was also approved as a food for specified health use, like the aforementioned structured food oil consisting of medium-chain fatty acids.

In addition, we have further commercialized some kinds of solid palm oils with improved taste for business use, which were also produced by means of high-temperature transesterification using powder lipase.

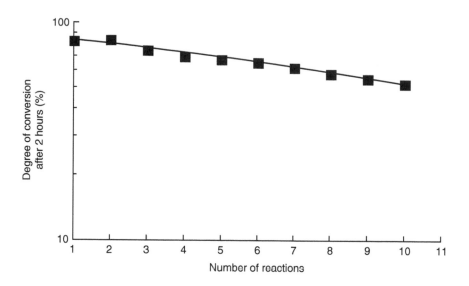

FIGURE 12.9 Stability of lipase QLM on repeated use. Conditions: 0.1 g of lipase powder was added to 9 g of sunflower oil and 1 g of phytosterol at 90°C. (From S Negishi et al. Transesterification of phytosterols and edible oil by lipase powder at high temperature. *J Am Oil Chem Soc* 80:805–907, 2003.)

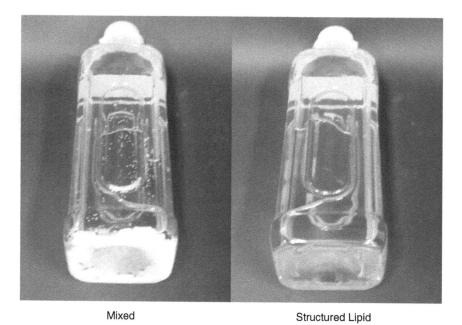

Mixed   Structured Lipid

FIGURE 12.10 Photograph of structured oil.

12.5   Synthesis of Structured Lipids Consisting of DHA

In recent years, attention has focused on the high nutritious functions of polyunsaturated fatty acids (PUFA) such as eicosapentaenoic acid (EPA) or docosahexaenoic acid (DHA).[13] Among the structured lipids consisting of these fatty acids, the lipids in which a PUFA is bound at the sn-2-position and medium-chain fatty acids at the sn-1,3-positions have attracted special attention because of their excellent absorption profiles.[14]

For the synthesis of lipids of this type, the lipase reaction is essential not only because of the thermal instability of PUFA, but also because of the fact that the fatty acid binding positions in the glycerol skeleton have to be specified. For example, the structured lipid in which DHA is located at the sn-2-position and capric acid at the sn-1,3-positions can be synthesized *via* a lipase catalyzed, sn-1,3 regioselective hydrolysis using triacylglyceride of DHA as the starting material, followed by the esterification of the resulting mono-acylglyceride intermediate with capric acid at the sn-1,3-positions. From the practical point of view, however, this method is considered inefficient for industrial production, and all relevant processes such as the separation, recovery, or recycle of large amounts of fatty acids produced in the reaction system have to be improved.

Considering the improvement in efficiency of the recovery and recycling process of the reactants, we examined a synthetic flow in which a specific reaction and a nonspecific reaction are conducted in combination.[15,16]

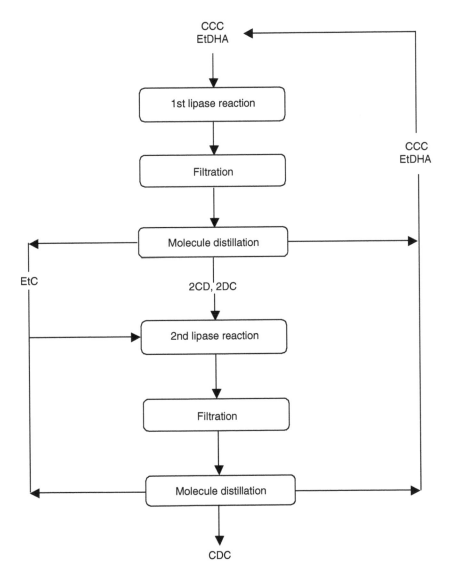

FIGURE 12.11 Schematic diagram of the reaction process showing combination of different enzyme reactions. (From S Negishi et al. Synthesis of 1,3-dicapryloyl-2-docosahexaenoylglycerol by a combination of non-selective and sn-1,3-selective lipase reactions for industrial utilization. *J Am Oil Chem Soc* 80:971–974, 2003.)

12.5.1 Combination of Specific Reaction and Nonspecific Reaction

Figure 12.11 shows the flow chart of the production processes examined. In the first lipase reaction, random transesterification is performed between tricaprylin (CCC) and ethyl docosahexaenoate (EtDHA), and in the second lipase reaction, sn-1,3 regioselective transesterification is performed in the presence of ethyl caprylate (EtC), giving sn-1,3-dicapryloyl-sn-2-docosahexaenoylglycerol (CDC). In the synthetic process, CCC, EtDHA and EtC are removed by molecular distillation and the first two are recycled as the substrates for the first lipase reaction, while the last one is also recycled as the substrate for the second lipase reaction. In the molecular distillation, EtC can be separated from CCC and EtDHA because of its lower molecular weight, and the residual mixture of CCC and EtDHA needs not to be separated and can be directly recycled for the first reaction together. The thin film molecular distillation method was used, by which the distillation can be carried out within a short heating time to avoid thermal deterioration of the recovered substrates.

Figure 12.12 shows the time courses of the change in triacylglycerides composition of the first lipase reaction. The proportion of DHA-containing triacylglycerol in the final product was over 50%. DHA-triacylglyceride (DDD) was not produced in this reaction. Furthermore, the formation of diacylglyceride was limited to 1.5%.

Figure 12.13 shows the time courses of the changes in triacylglyceride composition and the proportion of CDC to CCD produced in the second lipase reaction. As a result, CCC increased with decreases of dicapryloyl-docosahexaenoylglycerol (2CD) and didocosahexaenoyl-caprytoylglycerol (2DC), and the final product composition of DHA-containing triacylglycerides was mostly 2CD. CCC is removable by distillation, but 2DC is not because of its high boiling point; therefore the production of only a little amount of 2DC is considered a great advantage of this synthetic process.

In conclusion, CDC could finally be obtained with a purity of ca. 70% by the synthetic process depicted in the flow chart (Figure 12.11). In addition, no deterioration of DHA was observed in GLC not in HPLC

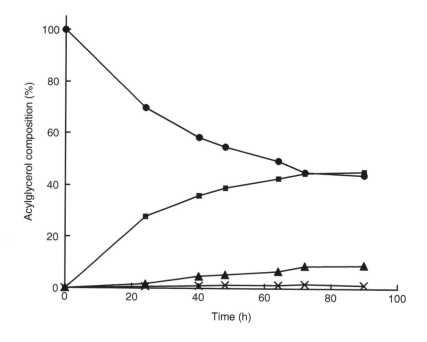

FIGURE 12.12 Time course of acylglycerol composition during the first enzyme reaction. The reaction was performed with 2% of lipase QLM at 50°C. Acylglycerol content [CCC(-●-), 2CD(-■-), 2DC(-▲-), DAG(-×-)].(From S Negishi et al. Synthesis of 1,3-dicapryloyl-2-docosahexaenoylglycerol by a combination of non-selective and *sn*-1,3-selective lipase reactions for industrial utilization. *J Am Oil Chem Soc* 80:971–974, 2003.)

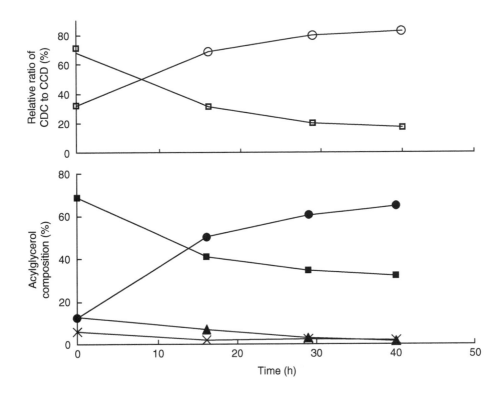

FIGURE 12.13 Time course of acylglycerol composition and relative ratio of CDC and CCD during the second enzyme reaction. The reaction was performed with 2% of Novozyme 435 at 40°C. (A) Relative ratio of CDC and CCD[CDC(-O-), CCD(-□-)]. (B) Acylglycerol content [CCC(-●-), 2CD(-■-), 2DC(-▲-), DAG(-×-)]. (From S Negishi et al. Synthesis of 1,3-dicapryloyl-2-docosahexaenoylglycerol by a combination of non-selective and *sn*-1,3-selective lipase reactions for industrial utilization. *J Am Oil Chem Soc* 2003, in press.)

chromatograms during the overall process using recycled substrates, EtC, EtDHA and CCC, according to the synthetic flow.

12.5.2 Future Developments

The synthetic process has not yet been put into practice for industrial production. Considering the thermal history, it is thought to be desirable to renew all materials, including the enzyme, after some repetitions. Moreover, for the industrialization of the synthetic process, it is considered necessary to confirm the safety of CDC produced using recycled substrates.

As the production process of CDC presented in this chapter is thought to be useful for the synthesis of symmetrical triacylglycerides with a functional fatty acid residue at the sn-2-position, using it is considered to facilitate the industrial production of structured lipids with various functions that has been investigated at the laboratory levels.

References

1. S Negishi, S Shirasawa, Y Arai, J Suzuki, S Mukataka. Activation of powdered lipase by cluster water and the use of lipase powders for commercial esterification of food oils. *Enzyme Microb Technol* 32:66–70, 2003.
2. FR Dastoli, NA Musto, S Price. Reactivity of active sites of chymotrypsin suspended in an organic medium. *Arch Biochem Biophys* 115: 44–47, 1966.

3. JS Drodick. Enzymatic catalysis in monophasic organic solvents. *Enz Microb Technol* 11: 194–211, 1989.

4. A Zaks, AM Klibanov. Enzymatic catalysis in organic media at 100°C. *Science* 224: 1249–1251, 1984.

5. M Takasago, K Takaoka. Analysis of the state of dissolved water in methyldecanoate and safflower oil by FT-near infrared spectrometry. *J Jpn Oil Chem Soc* 33: 772–775, 1984.

6. B Bloom, L L Chaikoff, WO Reinhardt. Intestinal lymph as pathway for transport of absorbed fatty acids of different chain lengths. *Am J Physiol* 166:451–455, 1951.

7. JR Senior (ed.) *Medium Chain Triglycerides*. Pennsylvania Press, Philadelphia, 1967

8. AC Bach, VK Babayan. Medium-chain triglycerides: an update. *Am J Clin Nutr* 36:950–962, 1982.

9. S Negishi, M Itakura, S Arimoto, T Nagasawa, K Tsuchiya. Measurement of foaming of frying oil and effect of the composition of triglycerides on foaming. *J Am Oil Chem Soc* 80:471–474, 2003.

10. M Kasai, N Nosaka, H Maki, S Negishi, T Aoyama, M Nakamura, Y Suzuki, H Tsuji, H Uto, M Okazaki, K Kondo. Effect of dietary medium- and long-chain triacylglycerols (MLCT) on accumulation of body fat in healthy humans. *Asia Pac J Clin Nutr* 12:151–160, 2003.

11. F Mattson, R A Volpenhein, B A Erickson. Effect of plant sterol esters on the adsorption of dietary cholesterol. *J Nutr* 107:1139–1146, 1977.

12. S Negishi, I Hidaka, I Takahashi, S Kunita. Transesterification of phytosterols and edible oil by lipase powder at high temperature. *J Am Oil Chem Soc* 80:805–907, 2003.

13. W E Connor. Importance of n-3 fatty acids in health and disease. *Am J Clin Nutr* 71:171s–175s, 2000.

14. M S Christensen, C E Hoy, C C Becker, T G Redgrave. Intestinal adsorption and lymphatic transport of eicosapentaenoic(EPA), docosahexaenoic (DHA), and decanoic acids: Dependence on intramolecular triacylglycerol structure.[1-3] *Am J Clin Nutr* 61:56–61, 1995.

15. S Negishi, Y Arai, S Arimoto, K Tsuchiya, I Takahashi. Synthesis of 1,3-dicapryloyl-2-docosa-hexaenoylglycerol by a combination of non-selective and *sn*-1,3-selective lipase reactions for industrial utilization. *J Am Oil Chem Soc* 80:971–974, 2003.

16. S Negishi, Y Arai, S Shirasawa, S Arimoto, T Nagasawa, H Kouziu, k Tsuchiya. Analysis of regiospecific distribution of FA of TAG using the lipase-catalyzed ester-exchange. *J Am Oil Chem Soc* 80:353–365, 2003.

13

Preparation of Polyunsaturated Phospholipids and Their Functional Properties

Masashi Hosokawa

Koretaro Takahashi

13.1 Introduction

Polyunsaturated fatty acids (PUFAs), such as eicosapentaenioc acid (EPA) and docosahexaenoic acid (DHA), have many physiological functions. Recent studies have shown that phospholipids containing PUFA as an acyl chain (PUFA-PL) exhibit novel functions other than the functionalities of PUFA itself.[1–6] It has been reported that phosphatidylcholine (PC) with DHA in the *sn*-2 position exhibits increased cell membrane permeability,[7] antitumor activity,[8] cytotoxicity,[9,10] and modulation of class I major histocompatibility complex (MHC I) structure.[11] Dietary phospholipids containing EPA was shown to decrease the weight of adipose tissue of rats.[12] These properties of PUFA-PLs have resulted in exploring their role in nutraceutical, pharmaceutical, and medical fields. It has been shown through those studies that molecular species, polar head group, and fatty acid-binding position are also important in expressing the biological functions of PUFA-PLs. Increased understanding of the functions of PUFA-PLs at the molecular level would progress the exploitation and application of PUFA-PLs for health benefits.

To understand the biological functions in detail and to design an industrial production process of PUFA-PLs, it is important to develop the preparation methods of PUFA-PL molecular species. Since lipase and phospholipase show specificities on PL, enzymatic reactions are one of the key techniques to prepare tailor-made PUFA-PLs. Enzymatic reaction are usually undertaken under milder conditions than chemical reactions. For this reason, toxic side reactions can be avoided. And in general, enzymatic techniques require less process than chemical methods to purify the desired products. There is no doubt that these reactions could be applied for the modification of natural PLs such as those from soybean, egg yolk, and marine organisms. Enzymatic preparation of tailor-made PLs including PUFA-PLs has been extensively reviewed by some researchers.[13–15]

In this chapter, we briefly review the enzymatic preparation of n-3 PUFA-PLs from convenient industrial phospholipids, then evaluate their functional properties with respect to health applications.

13.2   Production of Tailor-Made n-3 Polyunsaturated Phospholipids

13.2.1   Techniques Available to Incorporate PUFA into the *sn*-2 Position of PL

Phospholipse A$_2$ (PLA$_2$) catalyzes esterification of lysophospholipid (lysoPL) with PUFA under strictly moisture limited reaction system. PLA$_2$ mediated esterification, shown in Figure 13.1a, is very useful in incorporating specific PUFA such as EPA and DHA into the *sn*-2 position of PL. The esterification of lysoPL with PUFA by porcine pancreatic PLA$_2$ has been achieved in organic solvents, as shown in Table 13.1. Na et. al.[17] reported esterification reaction of soy lysophosphatidylcholine (lysoPC) with PUFA in isooctane-water microemulsion. Harrod and Elfman[20] have shown that the reaction could be carried out in a high-pressure reactor using propane and carbon dioxide, although it was difficult to adjust the optimal water content in PLA$_2$-mediated system. To improve the yield of PUFA-PL, approaches using a water miscible solvent were effective.[21] By adding formamide as a water mimic, a yield of 40% was obtained from the esterification between lysoPC and PUFA. And a combinational device of decompression with water mimics improved the yield to more than 70% by removing the moisture produced during the esterification in the reaction system.[23] Additional PLA$_2$ in the latter half of the esterification reaction was also effective to increase the yield of PUFA-PL.[23] In another reaction system proposed by Egger et. al.[24] a decreasing order step-wise water–activity reaction system was effective in improving the yield of PC from lysoPC and oleic acid to 60%.

Alternatively, a one-step acyl exchange at the *sn*-2 position on PC mediated by PLA$_2$ was also reported.[25–27] PLA$_2$ catalyzed acidolysis between soy PC and PUFA in glycerol as solvent under low-moisture conditions. Incorporation of PUFA into PC was 35% in this system. But the yield of the desired PC remained around 18% of the substrate soy PC.

13.2.2   Exchange of Fatty Acid at the *sn*-1 Position in Phospholipid

Some lipases exhibit positional specificity towards PL. This specificity is useful in exchanging the fatty acids on position *sn*-1 (Figure 13.1b). Acidolysis and transesterification of PL has been widely used for exchanging the fatty acids on *sn*-1 with a desired fatty acid.[28–34] Svensson et. al.[30] almost completely exchanged (97.7%) the fatty acid in *sn*-1 position with immobilized *Rhizopus arrhizus* lipase under controlled water activity limiting to 0.11. They attained 60% yield of the desired PC. We observed that polar solvents with high dielectric constant and low log *P* (solvent hydrophobicity value), such as propylene glycol in the reaction system, was found to be effective for the acidolysis between soy PC and PUFA, as these act as "water mimics".[32] An 80% incorporation of EPA in the *sn*-1 position was achieved with water and propylene glycol as water mimics. The yield of the desired PC was approximately 80%.[32]

13.2.3   Exchanging Polar Head Group on Phospholipid

Phosphalipase D (PLD) is known to catalyze transphosphatidylation between a polar base and a desired alcohol. Applications involving PLD have been reviewed by Iwasaki and Yamane.[35] Juneja et. al.[36–39] synthesized phosphatidylglycerol, phosphatidylethanolamine (PE), and phoshatidylserine from PC or

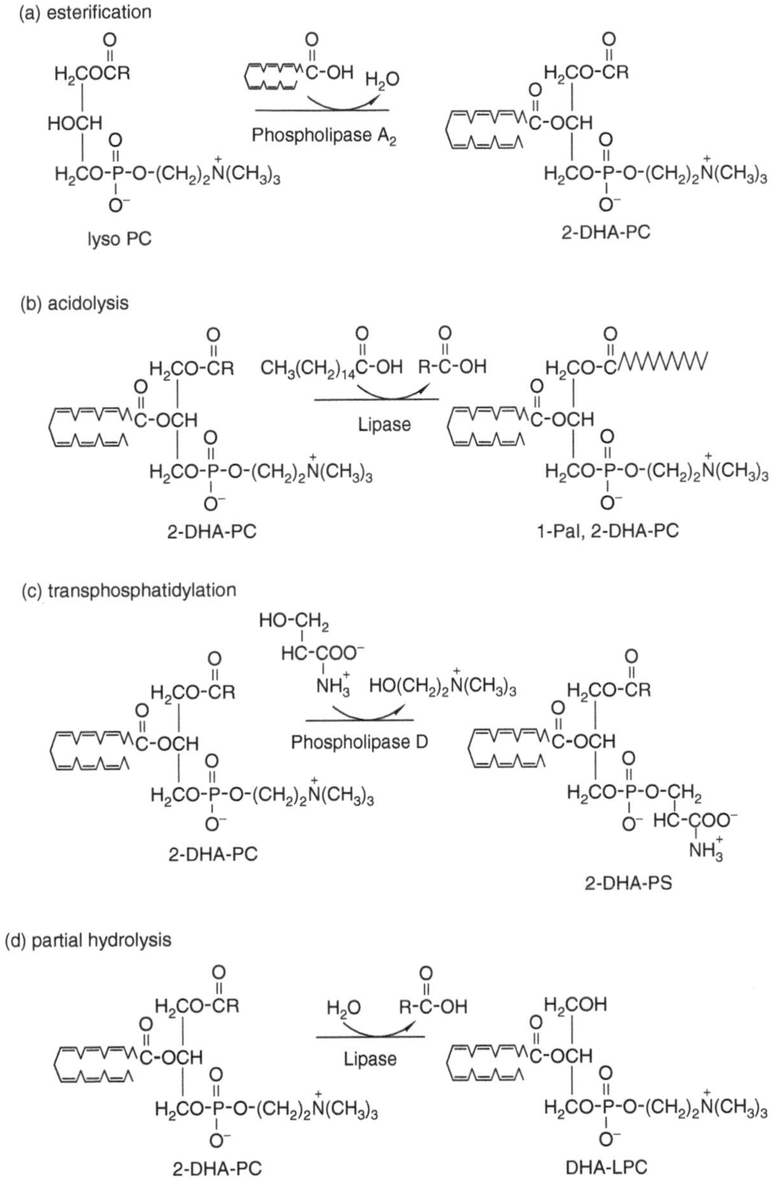

(a) esterification

(b) acidolysis

(c) transphosphatidylation

(d) partial hydrolysis

FIGURE 13.1 Reaction scheme for preparation of polyunsaturated phospholipids using phospholipase and lipase.

lecithin. This reaction is useful in modifying the polar head group into various PUFA-PLs prepared by enzymatic reactions as above mentioned, and naturally occuring PUFA-PL (Figure 13.1c).[40] Esterification, acyl exchange reaction, and transphosphatidylation with lipase and/or phospholipase is no doubt a useful tool for preparing tailor-made PL molecules (Table 13.2).

13.2.4 Partial Hydrolysis of Polyunsaturated Phospholipid

Partial hydrolysis of DHA-PL is very effective in preparing lysoPL with exclusive DHA content (Figure 13.1d). In this reaction system, a_w regulation is desirable for the enrichment of DHA and also for obtaining the substantial yield of lysoPL.[41,42] For instance, partial hydrolysis of DHA-enriched egg yolk PL with Lipozyme IM (immobilized *Rhizomucor meihei* lipase) increased the DHA content from approximately 11% to more

TABLE 13.1 Phospholipase-Mediated Esterification, Acidolysis and Transesterification

Product	Acyl donor	Reaction	Solvent	Yield (incorporation)	Technique	Reference
PC	Oleic acid	Esterification	Toluene	6.5%	Microemulsion	16
PC	PUFA	Esterification	Iisooctane	6%		17
PC	Oleic acid etc.	Esterification	Benzene	35%		18
PC	PUFA	Esterification	Isooctane	21%	Immobilized PLA_2	19
PC	PUFA	Esterification	Isooctane, Carbone dioxide, Propane	25%	High pressure reactor	20
PC	EPA	Esterification	Glycerol	40%	Formamide used as water mimic	21
PE	PUFA	Esterification	Glycerol	18%		22
PC	Oleic acid etc.	Esterification	Toluene	60%	Water activity controlled	25
PC	DHA	Esterification	Glycerol	71%	Formamide used as water mimic + Decompression + Molecular seive trap	23
PC	DHA	Esterification	Glycerol	76%	Formamide used as water mimic + Decompression + Additional PLA_2	23
PC	Lauric acid	Acidolysis	non solvent	20% (6%)	Immobilized PLA_2	24
PC	EPA	Acidolysis	Glycerol	18% (35%)		26
PC	EPA ethyl ester	Transesterification	Toluene	(14.3%)	Water activity controlled	27

TABLE 13.2 Fatty Acid Composition of Polyunsaturated Phospholipids Prepared by Enzymatic Reaction

Fatty acid	EDPC[*1]	ODPE[*2]	PDPS[*3]	Squid skin PC[*4]	Squid skin PS[*5]
16:0	0.4	1.4	46.6	36.9	38.0
18:0	nd	nd	0.6	1.6	1.7
18:1	0.4	45.8	0.8	2.0	1.9
20:4	nd	nd	nd	0.9	1.1
20:5 (EPA)	45.8	1.4	2.7	8.0	8.9
22:6 (DHA)	49.3	49.1	46.6	44.2	43.3

*1 EDPC: 1-EPA, 2-DHA-phosphatidylcholine (PC), *2 ODPE: 1-Oleoyl, 2-DHA-phosphatidylethanolamine,
*3 PDPS: 1-Palmitoyl, 2-DHA-phosphatidylserine (PS), *4 Squid skin PC: PC extracted from squid skin
*5 Squid skin PS: PS transphosphatidylated from Squid skin PC

TABLE 13.3 Fatty Acid Composition of Substrate Phospholipids and Polyunsaturated Lysophospholipids Produced by Lipase-Mediated Partial Hydrolysis

Fatty acid	DHA-enriched egg yolk PC[*1]	DHA-enriched egg yolk lysoPC[*2]	Squid skin PC[*3]	Squid skin lysoPC[*4]
16:0	39.1	8.3	31.4	1.0
18:0	9.6	3.7	4.3	nd
18:1	28.9	38.7	3.4	0.4
20:4	1.4	2.9	1.4	1.4
20:5 (EPA)	0.4	1.1	8.9	10.8
22:6 (DHA)	6.7	29.2	44.9	83.1

*1 DHA-enriched egg yolk PC: Phosphatidylcholine (PC) prepared from egg yolk of fish oil fed hens.
*2 DHA-enriched egg yolk lysoPC: lysoPC prepared from DHA-enriched egg yolk PC through lipase-mediated partial hydrolysis.
*3 Squid skin PC: PC prepared from squid skin.
*4 Squid skin lysoPC: lyso PC prepared from DHA-enriched egg yolk PC through lipase-mediated partial hydrolysis.

than 35% in the recovered total PL fraction. Under the same conditions, DHA increased from 33% to nearly 60% when squid skin PL was used as the substrate. In this process, the sum of EPA and DHA in the obtained lysoPC reached greater than 93% (Table 13.3).

13.3 Functional Properties of n-3 Polyunsaturated Phospholipids

13.3.1 Effect of Polyunsaturated Phospholipids on Cell Differentiation

Retinoic acid (RA)[43] and dibutyryl cyclic AMP (dbcAMP)[44] induce human leukemia HL-60 cells to differentiate into granulocytes. These differentiation-inducing agents have been recently used in combination with other compounds to alleviate the side effects without losing their therapeutic effects.[45]

2-DHA-PL (PL with DHA bound on *sn*-2 position) exhibited synergistic effect on differentiation of HL-60 cells induced by RA or dbcAMP.[46,47] When 2-DHA-PL was treated with RA or dbcAMP, the NBT reducing activity, which is an indicator of cell differentiation, was enhanced compared to RA or dbcAMP alone (Figure 13.1). The enhancing effect of 2-DHA-PL on cell differentiation was more potent than DHA in free fatty acid chemical form. In contrast, di-oleoyl PC and di-oleoyl PE did not enhance cell differentiation induced by RA or dbcAMP. Focusing acyl residues binding at the *sn*-1 position of 2-DHA-PL, palmitic acid and oleic acid showed high potency. In addition, the effect was much more prominent in 2-DHA-PE than in 2-DHA-PC (Figure 13.2). These indicate that PUFA at the *sn*-2 position of PL is crucial for the enhancing effect. Furthermore, the molecular structure including the polar head of PUFA-PL is important for the enhanced expression of cell differentiation induced by RA and dbcAMP. Membrane modification by 2-PUFA-PL may be medically beneficial because it may make the HL-60 cells more sensitive to differentiation-inducing compounds.

13.3.2 Oncogenes Regulation by n-3 Polyunsaturated Phospholipids

Studies undertaken by Hosokawa et al.[47] found that a combination of 2-PUFA-PL and dbcAMP would regulate expression of functional genes in HL-60 cells. When 1-oleoyl, 2-DHA-PE was combined with dbcAMP to induce HL-60 cell differentiation, the expression of *c-jun* mRNA was up-regulated (Figure 13.3). In particular, the expression level of *c-jun* mRNA was higher in HL-60 cells treated with the combination of 2-DHA-PE and dbcAMP as compared to dbcAMP alone. The c-*jun* gene product associates to form hetero/homodiametric transcription factor AP-1, which is reported to be a modulator of cell differentiation.[48] 2-DHA-PE in combination with dbcAMP may enhance cell differentiation through up-regulation of the transcriptional factor AP-1.

The c-*myc* mRNA is one of the oncogenes that is expressed excessively in many tumor cells and affects disordered cell cycle.[49] The combination of 2-DHA-PE with dbcAMP is suggested to induce cell growth inhibition through down-regulation of c-*myc* oncogene. Combination of 2-DHA-PE and dbcAMP

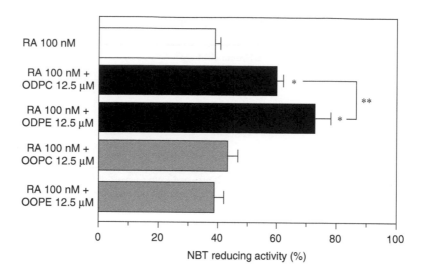

FIGURE 13.2 Effect of phospholipids on HL-60 cell differentiation induced by retinoic acid. HL-60 cells (5×10^4 cells/ml) were incubated with 100 nM retinoic acid (RA) for 24 h after preincubation with 12.5 μM phospholipids for 24 h. Values present means $\pm$ SD (n-3). *$P < 0.01$ vs. RA 100 nM, **$P < 0.01$ vs. ODPC 12.5 μM. ODPC: 1-Oleoyl, 2-docosahexaenoyl (DHA)-phosphatidylcholine, ODPE: 1-Oleoyl, 2-DHA-phosphatidylethanolamine, OOPC: dioleoyl-PC, OOPE: dioleoyl-PE. (From Reference 45.)

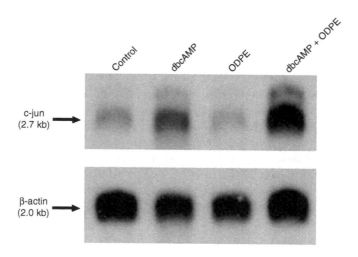

FIGURE 13.3 Expression of *c-jun* mRNA in HL-60 cells treated with dibutyryl cAMP and 2-DHA-phosphatidyle-thanolamine. HL-60 cells were incubated with 200 μM dibutyryl cAMP (dbcAMP) and 0.5 mM IBMX for 24 h after preincubation with 50 μM 1-oleoyl, 2-docosahexaenoyl-phosphatidylethanolamine (ODPE) for 24 h.

remarkably decreased the expression of *c-myc* mRNA, while 2-DHA-PE alone did not have an effect (Figure 13.4). Growth inhibition in correspondence to decreased c-*myc* mRNA was also observed in HL-60 cells treated by 2-DHA-PE and dbcAMP.

13.3.3 Characteristics and Antitumor Effect of n-3 Polyunsaturated Phospholipid Liposomes

PL forms liposome in aquatic media. Liposomes have been studied extensively for designing a drug-delivery system. The liposomes prepared from 2-PUFA-PL may be beneficial as they should have many

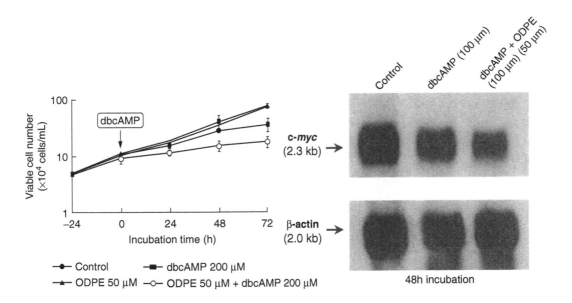

FIGURE 13.4 Growth inhibition and down-regulation of c-*myc* mRNA in HL-60 cells treated with dibutyryl cAMP and 2-DHA-phosphatidylethanolamine. HL-60 cells were incubated with dibutyryl cAMP (dbcAMP) after preincubation with 50 μM 1-oleoyl, 2-docosahexaenoyl-phosphatidylethanolamine (ODPE) for 24 h.

desirable functions as described above. However, in contrast to the wealth of information pertaining to the characteristics of soy PL and egg yolk PL liposomes, none has been reported on the characteristics and functionalities of n-3 PUFA-PL liposomes. Liposomes (2-DHA-PC/PS liposome) prepared from 2-DHA-PC and 2-DHA-PS (with a molar ratio of 7:3) by the Proliposome method[50] showed 6–9 multilamella vesicles, with 0.7–1.2 μm diameter size when filtered through a 1 μm pore sized filter.[51] 2-DHA-PC/PS liposomes were more stable than soy PC/PS liposomes from the point of view of particle size distribution stability and liposome suspension turbidity. 2-DHA-PC/PS liposomes retained a single particle size distribution over at least a week, whereas a peak of another particle size distribution appeared in soy PC/PS liposomes stored under the same period.[51]

2-DHA-PC/PS liposomes exerted an antitumor activity against mouse fibrosarcoma *in vivo*. When mice were intraperitoneally injected with 2-DHA-PC/PS liposomes at the dosage of 1 mg/mouse on 2nd, 4th, and 6th days after tumor implantation, tumor growth of fibrosarcoma was suppressed by approximately 50% of the control tumor size[51] (Figure 13.5). In contrast, soy PC/PS liposomes did not show any antitumor activity under the same condition. *In vitro* experiments undertaken on fibrosarcoma also showed that 2-DHA-PC/PS liposomes suppress the growth of Meth-A cells more than the soy PC/PS liposomes. Neither liposomes suppressed the cell growth of macrophage-like J744-1. It was notable that 2-DHA-PC/PS liposomes enhanced the phagocytic activity of the individual macrophage-like J744-1 cell more than the soy PC/PS liposomes, and also increased the ratio of cells undergoing phogocytosis. In other words, these results indicate that 2-DHA-PC/PS liposomes would suppress cell growth of fibroarcoma *in vivo* at least in part by direct growth inhibition on tumor cells and by enhancing phagocytic activity on macrophages.

13.3.4 Improvement of Erythrocytes Deformability by n-3 Polyunsaturated Lysophospholipids

LysoPL rich in n-3 PUFA (PUFA-lysoPL) can be produced through selective lipase-mediated partial hydrolysis of 2-DHA-PL, as described in section 13.2.4. This reaction results in a dominant form of 2-acyl-lysoPL, though 1-acyl-lysoPC may be produced in part by acyl migration. LysoPC (designated as "2-DHA-lysoPL" in the text) prepared from DHA-enriched egg yolk PC with Lipozyme IM effectively improved the deformability of human erythrocytes *in vitro*.[52] Under the same molar amount of DHA,

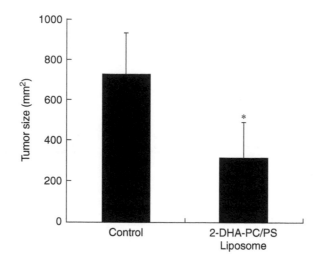

FIGURE 13.5 Antitumor effect of 2-DHA-PC/PS liposome on Meth-A fibrosarcoma-bearing BALB/c mice. The Meth-A fibrosarcoma (2.7×10^6 cells/mouse) suspended in Hanks' balance salt solution were implanted intraperitoneally into the Female BALB/c mice. The liposomes were injected intraperitoneally at dosages of 1 mg/mouse on days 2, 4, and 6 after tumor implantation. Mice receiving similar treatment with PBS alone served as controls. The mice were observed for 3 weeks and tumor size of those mice was measured. Data are shown as means ± S.D. (n = 6), * $P < 0.01$ vs. control. (From A Fujimoto et al. Fibrosarcoma may be suppressed by liposomal *sn*-2 DHA-inserted phospholipid therapy. In: YS Huang, SJ Lin, PC Huang, Eds. *Essential Fatty Acids and Eicosanoids: Invited Papers from the Fifth International Congress.* Champaign, IL: American Oil Chemists' Society, 2003.)

the deformability of erythrocytes incubated for 1 h with 2-DHA-lysoPC was higher than that of the erythrocytes treated with 2-DHA-PC (Figure 13.6). Incorporation of DHA into erythrocyte phospholipids was more rapid when erythrocytes were incubated with 2-DHA-lysoPC than 2-DHA-PC. However, after 3 h incubation, the deformability of erythrocytes and the incorporation of DHA became comparable between the 2-DHA-lysoPC and the 2-DHA-PC treatments. Improvement of deformability within 1 h was attributed to a rapid intake of the 2-DHA-lysoPC into the erythrocytes compared to 2-DHA-PC.

The functions of PUFA-lysoPC were different between the PUFA-binding positions on lysoPC backbone. We prepared 1-EPA-lysoPC and 2-EPA-lysoPC with lipase and PLA_2 mediated reactions to compare those effects on erythrocytes deformability. 2-EPA-lysoPC had a more beneficial effect on erythrocytes deformability than 1-EPA-lysoPC. On the other hand, 2-soy-lysoPC (prepared from soy PC) treated erythrocytes did not show any improvement in deformability, unlike the 2-EPA-lysoPC. Hemolysis of the erythrocytes after incubation with 2-EPA-lysoPC was also much lower compared to 1-EPA-lysoPC (Figure 13.7). Thus, the PUFA-binding position and the species of fatty acid on lysoPC backbone seem to be very important in increasing the erythrocyte deformability and decreasing the hemolysis. Bioavailability of 2-EPA-lysoPC may be much higher than that of 1-EPA-lysoPC.

Preferential incorporation of 2-DHA-lysoPC over unesterified DHA has been reported in the brain of young mice.[53,54] Morash et. al.[55] found that 2-acyl-lysoPC was taken up and acylated to PC faster than 1-acyl-lysoPC in neuroblastoma and glionoma cells. The ability to take up and acylate 2-acyl-lysoPC is suggested to be a general feature among many cell types. Thus, 2-acyl-lysoPC was likely to be an efficient delivery form of PUFA to some extrahepatic tissues. Study on utilization of 2-PUFA-lysoPC to express their functions *in vivo* seems to be a promising area of research.

13.3.5 Other Functionalities

PUFA-PLs have received attention because of their novel physiological and nutraceutical functions. Matsumoto et. al.[3] showed that 2-DHA-PC with DHA at the *sn*-2 position inhibits 5-lipoxygenase. 5-Lipoxygenase is an important enzyme that catalyzes the first step in leukotriene production in the

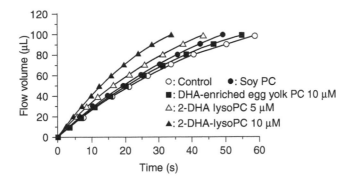

FIGURE 13.6 Flow curves of human erythrocytes treated with DHA-enriched egg yolk PC and 2-DHA-lysoPC for 1 h. Data were obtained through evaluation of deformability with artificial capillary model.

arachidonic acid cascade. In their study, 1-oleoyl, 2-DHA-PC (ODPC) was the most potent inhibitor of 5-lipoxygenase and had no effect on 12-, 15-lypoxygenase and cyclooxygenase. Thus, a specific DHA-PC molecular species (ODPC) affects 5-lipoxygenase activity and regulates leukotriene biosynthesis. Izaki et. al.[4] observed an enhancement of discriminatory shock-avoidance learning in rats injected intraperitoneally with ODPC.

On the other hand, nutraceutical effects of dietary DHA-PL have also been reported recently. Morizawa et. al.[56] reported that dietary DHA-PL extracted from fish roe exhibited an anti-inflammatory effect on the contact hypersensitivity reaction in the ears of mice sensitized with 2,4-dinitro-1-fluorobenzene. In mice that were administered with DHA-PL, the expression of IFN-γ, IL-6, and IL-1β mRNA, which are related with inflammatory reaction, were suppressed. A notable feature was that anti-inflammatory activity by DHA-PL was stronger than that of DHA-triacylglycerol. These results indicate that the lipid structure of DHA affects the anti-inflammatory effect *in vivo*. Furthermore, it was also reported that

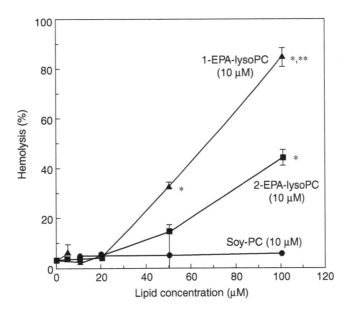

FIGURE 13.7 Effect of 1-EPA-lysoPC and 2-EPA-lysoPC on hemolysis of human erythrocytes. Human erythrocytes were incubated with EPA-lysoPC and soy PC at 37°C for 1 h. *$P < 0.01$ vs. Soy PC, **$P < 0.01$ vs. 2-EPA-lysoPC. (From M Hosokawa et al. Increase in deformability of human erythrocyter through the action of b-lysophospholipid rich in n-3 polyunsaturated fatty acid content. *J Jpn Oil Chem Soc* 47:1313–1318, 1998.)

dietary PUFA-PL normalizes urinary melatonin excretion in n-3 PUFA deficient adult rats[57] and delayed survival time of stroke-prone spontaneously hypersensitive rats (SHR-SP).[58] Thus, PUFA-PL has an effective chemical structure to improve *in vivo* functions of PUFA.

13.4   Concluding Remarks

Enzymes are useful tools to convert natural PLs into PUFA-PLs with desired fatty acids on the *sn*-1 and/ or *sn*-2 positions, coupled with a specific polar head group. Lipase, PLA$_2$, and PLD have been used for these reactions. To expand the uses of PUFA-PL in nutraceutical, pharmaceutical, and medical applications, it is important to increase the reaction yield by employing novel techniques such as water activity modulation and water mimics enzyme activation. Furthermore, investigations of PUFA-PL functions in relation to those molecular structures would presumably result in a discovery of a novel functional PUFA-PL molecule. The preparation and modification of PUFA-PL using lipase and phospholipase will be very important techniques to produce highly functional PUFA-PL.

Acknowledgments

Permissions granted from AOCS Press (Figure 13.4 and Table 13.1, "Fibrosarcoma may be suppressed by liposomal *sn*-2 DHA-inserted phospholipid therapy." In YS Huang, SJ Lin, PC Huang, Eds. *Essential Fatty Acids and Eicosanoids: Invited Papers from the Fifth International Congress*. Champaign, IL: American Oil Chemists' Society, 2003) and Japan Oil Chemists' Society (Figure 13.2, *J. Jpn. Oil Chem. Soc.* Vol. 46[4]: 383–390, 1997; Figure 13.4, *J. Jpn. Oil Chem. Soc.* Vol. 47[12]: 1313–1318, 1998.) Original sources are gratefully acknowledged.

References

1. JS Parks, TY Thuren, JD Schmitt. Inhibition of lecithin:cholesterol acyltransferase activity by synthetic phosphatidylcholine species containing eicosapentaenoic acid or docosahexaenoic acid in the *sn*-2 position. *J Lipid Res* 33:879–887, 1992.
2. M Suzuki, K Asahi, K Isono, A Sakurai, N Takahashi. Differentiation inducing phosphatidylcholine from the embryos of rainbow trout (Salmo gairdneri): Isolation and structural elucidation. *Develop Growth & Differ* 34:301–307, 1992.
3. K Matsumoto, I Morita, H Hibino, S Murota. Inhibitiory effect of docosahexaenoic acid-containing phospholipids on 5-lipoxygenase in rat basophilic leukemia cells. *Prostagla Leukoto Essen Fatty Acids* 49:861–866, 1993.
4. M Hashimoto, X-W Gong, Y Izaki, M Iriki, H Hibino. 1-Oleoyl-2-docosahexaenoyl phosphatidylcholine increased paradoxical sleep in F344 rats. *Neurosci Lett* 158:29–32, 1993.
5. Y Izaki, M Hashimoto, J Arita, m Iriki, H Hibino. Intraperitoneal injection of 1-oleoyl-2-docosahexaenoyl phosphatidylcholine enhances discriminatory shock avoidance learning in rats. *Neurosci Lett* 167:171–174, 1994.
6. DC Mitchell, SL Niu, BJ Litman. Enhancement of G protein-coupled signaling by DHA phospholipids. *Lipids* 38:437–443, 2003.
7. W Stillwell, W Ehringer, LJ Jenski. Docosahexaenoic acid increases permeability of lipid vesicles and tumor cells. *Lipids* 28:103–108, 1993.
8. LJ Jenski, M Zerouga, W Stillwell. ω-3 Fatty acid-containing liposomes in cancer therapy. *Proc Soc Exp Biol Med* 210:227–233, 1995.
9. M Zerouga, W Stillwell, J Stone, A Powner, LJ Jenski. Phospholipid class as a determinant in docosahexaenoic acid's effect on tumor cell viability. *Anticancer Res* 16:2863–2868, 1996.
10. O Kafrawy, M Zerouga, W Stillwell, LJ Jenski. Docosahexaenoic acid in phosphatidylcholine mediates cytotoxicity more effectively than other ω-3 and ω-6 fatty acids. *Cancer Lett* 132:23–29, 1998.

11. LJ Jenski, PK Nanda, P Jiricko, W Stillwell. Docosahexaenoic acid-containing phosphatudylcholine affects the binding of monoclonal antibodies to purified Kb reconstituted into liposomes. *Biochim Biophys Acta* 1467:293–306, 2000.

12. K Yazawa, K Watanabe, C Ishikawa, K Kondo, S Kimura. Production of eicosapentaenoic acid from marine bacteria. In: DJ Kyle, C Ratledge, ed. *Industrial Application of Single Cell Oils.* Champaign, IL: American Oil Chemists Society, 1992, pp. 29–51.

13. P D'Arrigo, S Servi. Using phospholipases for phospholipid modification. *Trends in Biotechnol* 15:90–96, 1997.

14. P Adlercreutz. Enzymatic conversion of glycerophospholipids. In: UT. Bornscheuer, ed. *Enzymes in Lipid Modification:* Wiley-VCH, 2000, pp 293–306.

15. K Takahashi, M Hosoawa. Production of tailor-made polyunsaturated phospholipids through bioconversion. In: TM Kuo, HW Gardner, ed. *Lipid Biotechnology.* New York: Marcel Dekker, 2002, pp. 517–526.

16. P Pernas, JL Olivier, MD Legoy, G Bereziat. Phospholipid synthesis by extracellular phospholipase A2 in organic solvents. *Biochem Biophys Res Commun* 168:644–650, 1990.

17. A Na, C Eriksson, E. Osterberg, K Holmberg. Synthesis of phosphatidylcholine with (n-3) fatty acids by phospholipase A_2 in microemulsion. *J Am Oil Chem Soc* 67:766–770, 1990.

18. I Mingarro, C Abad, L Braco. Characterization of acylating and deacylating activities of an extracellular phospholipase A_2 in a water–restricted environment. *Biochemistry* 33:4652–4660, 1994.

19. M. Lija-Hallberg, M Harrod. Enzymatic and non-enzymatic esterification of long polyunsaturated fatty acids and lysophosphatidylcholine in isooctane. *Biocatal Bitransform* 12:55–66, 1995.

20. M Harrod, I Elfman. Enzymatic synthesis of phosphatidylcholine with fatty acids, isooctane, carbon dioxide, and propane as solvents. *J Am Oil Chem Soc* 72:641–646, 1995.

21. M Hosokawa, K Takahashi, Y Kikuchi, M Hatano. Preparation of therapeutic phospholipids through porcine pancreatic phospholipase A_2-mediated esterification an lipozyme-mediated acidolysis. *J Am Oil Chem Soc* 72:1287–1291, 1995.

22. M Hosokawa, K Takahashi, M Hatano, M Egi. Phospholipase A_2-mediated synthesis of phosphatidylethanolamine containing highly unsaturated fatty acids. *Int J Foos Sci and Technol* 29:721–725, 1995.

23. K Takahashi, M Hosokawa, T Ueno, S Norinobu, M Mankura. Novel functions of DHA-phospholipids and their preparation through bioconversions. 92nd AOCS annual Meeting & Expo Abstracts S18, 2001.

24. D. Egger, E Wehtje, P Adlercreutz. Characterization and optimization of phospholipase A_2 catalyzed synthesis of phosphatidylcholine. *Biochim Biophys Acta* 1343:76–84, 1997.

25. AM Aura, P Forssell, A Mustranta, K Poutanen. Transesterification of soy lecithin by lipase and phospholipase. *J Am Oil Chem Soc* 72:1375–1379, 1995.

26. M Hosokawa, M Ito, K Takahashi. Preparation of highly unsaturated fatty acid-containing phosphatidylcholine by transesterification with phospholipase A_2. *Biotechnol Technique,* 12:583–586, 1998.

27. CW Park, SJ Kwon, JJ Ham, JS Rhee. Transesterification of phosphatidylcholine with eicosapentaenoic acid ethyl ester using phospholipase A_2 in organic solvent. *Biotechnol Lett* 22:147–150, 2000.

28. T Yagi, T Nakanishi, Y Yoshizawa, F Fukui. The enzymatic acyl exchange of phospholipids with lipase. *J Ferment Technol* 69:23–25, 1990.

29. I Svensson, P Adlercreutz, B Mattiasson. Interesterification of phosphatidylcholine with lipases in organic media. *Appl Microbiol Biotechnol* 33:255–258, 1990.

30. I Svensson, P Adlercreutz, B Mattiasson. Lipase-catalyzed transesterification of phosphatidylcholine at controlled water activity. *J Am Oil Chem Soc* 69:986–991, 1992.

31. GG Haraldsson, A Thorarensen. Preparation of phospholipids highly enriched with n-3 polyunsaturated fatty acids by lipase. *J Am Oil Chem Soc* 76:1143–1149, 1999.

32. M Hosokawa, K Takahashi, N Miyazaki, k Okamura, M Hatano. Application of water mimics on preparation of eicosapentaenoic and docosahexaenoic acids-containing glycerolipids. *J Am Oil Chem Soc* 72:421–425, 1995.

33. L Peng, X Xu, H. Mu, CE Hoy, J Adler-Nissen. Production of structured phospholipids by lipase-catalyzed acidolysis: Optimization using response surface methodology. *Enzyme Microb Technol* 31:523–532, 2002.

34. D Adlercreutz, H Budde, E Wehtje. Synthesis of phosphatidylcholine with defined fatty acid in the *sn*-1 position by lipase-catalyzed esterification and transesterification reaction. *Biotechnol Bioeng* 78:403–411, 2002.

35. Y Iwasaki, T Yamane. Phospholipases in enzyme engineering pf phospholipids for food, cosmetics, and medical applications. In: TM Kuo, HW Gardner, Ed. *Lipid Biotechnology*. New York: Marcel Dekker, pp. 417–431, 2002.

36. LR Juneja, N Hibi, N Inagaki, T Yamane, S Shimizu. Comparative study on conversion of phosphatidylcholine to phosphatidylglycerol by cabbage phospholipase D in micelle and emulsion systems. *Emzyme Microb Technol* 9:350–354, 1987.

37. LR Juneja, T Kazuoka, N Inagaki, T Yamane, S Shimizu. Kinetic evaluation of conversion of phosphatidylcholine to phosphatidylethanolamine by phospholipase D from different sources. *Biochem Biophys Acta* 960:334–341, 1988.

38. LR Juneja, T Kazuoka, N Goto, N Inagaki, T Yamane, S Shimizu. Conversion of phosphatidylcholine to phosphatidylserine by various phospholipase D in the presence of L- or D-serine. *Biochem Biophys Acta* 1003:277–283, 1989.

39. LR Juneja, E Taniguchi, S Shimizu, T Yamane. Increasing productivity by removing choline in conversion of phosphatidylcholine to phosphatidylserine by phospholipase D. *J Ferment Bioeng* 73:357–361, 1992.

40. M Hosokawa, T Shimatani, T Kanada, Y Inoue, K Takahashi. Conversion to docosahexaenoic acid-containing phosphatidylserine from squid skin lecithin by phospholipase D-mediated transphosphatidylation. *J Agric Food Chem* 48:4550–4554, 2000.

41. M Ono, m Hosokawa, Y Inoue, K Takahashi. Concentration of docosahexaenoic acid-containing phospholipid through Lipozyme IM-mediated hydrolysis. *J Jpn Oil Chem Soc* 46:867–872, 1997.

42. M Ono, M Hosokawa, Y Inoue, K Takahashi. Water activity-adjusted enzymatic partial hydrolysis of phospholipids to concentrate polyunsaturated fatty acids. *J Am Oil Chem Soc* 74:1415–1417, 1997.

43. TR Breitman, SE Selonics, SJ Collins. Induction of differentiation of the human promyelicytic leukemia cell line (HL-60) by retinoic acid. *Proc Natl Acad Sci USA* 77:2936–2940, 1980.

44. SS McCachren, J Jr Nichols, RE Kaufman, J Niedel. Dibutyryl cyclic adenosine monophosphate reduces expression of c-myc during HL-60 differentiation. *Blood* 68:412–416, 1986.

45. H Hemmi, TR Breitman. Combinations of recombinant human interferons and retinoic acid synergistically induce differentiation of the humanpromyelocytic leukemia cell line HL-60. *Blood* 69:501–507, 1987.

46. K Tochizawa, M Hosokawa, H Kurihara, H Kohno, S Odashima, K Takahashi. Effects of phospholipids containing docosahexaenoic acid on differentiation and growth of HL-60 human promyelocytic leukemia cells. *J Jpn Oil Chem Soc* 46:383–390, 1997.

47. M Hosokawa, A Sato, H Ishigamori, H Kohno, T Tanaka, K Takahashi. Synergistic effects of highly unsaturated fatty acid-containing phosphatidylethanolamine on differentiation of human leukemia HL-60 cells by dibutyryl cyclic adenosine monophosphate. *Jpn J Cancer Res* 92:666–672, 2001.

48. J Nichols, SD Nimer. Transcription factors, translocations, and leukemia. *Blood* 80:2953–2963, 1992.

49. LE Grosso, HC Pitot. Transcriptional regulation of c-myc during chemically induced differentiation of HL-60 cultures. *Cancer Res* 45:847–850, 1985.

50. Anonymous. *The Lecithin People.* Lucas Meyer, Hamburg. 1993.

51. A Fujimoto, J Sasaki, M Hosokawa, K Takahashi. Fibrosarcoma may be suppressed by liposomal *sn*-2 DHA-inserted phospholipid therapy. In: YS Huang, SJ Lin, PC Huang eds. *Essential Fatty Acids and Eicosanoids: Invited Papers from the Fifth International Congress.* Champaign, IL: American Oil Chemists' Society, 2003.

52. M Hosokawa, M Ono, K Takahashi, Y Inoue. Increase in deformability of human erythrocytes through the action of β-lysophospholipid rich in n-3 polyunsaturated fatty acid content. *J Jpn Oil Chem Soc* 47:1313–1318, 1998.
53. F Thies, C Pillon, P Moliere, M Lagarde, J Lecerf. Preferential incorporation of *sn*-2 lysoPC DHA over unesterified DHA in young rat brain. *Am J Physiol* 267:R1273–1279, 1994.
54. N Bernoud, L Fernart, P Moliere, MP Dehouck, M Lagarde, R Cecchelli, J Lecerf. Preferential transfer of 2-docosahexaenoyl-1-lysophosphatidylcholine through an *in vitro* blood-brain barrier over unesterified docosahexaenoic acid. *J Nurochem* 72:338–345, 1999.
55. SC Morash, HW Cook, MW Spence. Lysophosphatidylcholine as an intermediate in phosphatidylcholine metabolism and glycerophosphocholine synthesis in cultured cells: An evaluation of the 1-acyl and 2-acyl lysophosphatidylcholine. *Biochim Biophys Acta* 1004:221–229, 1989.
56. K Morizawa, Y I. Tomobe, M Tsuchida, Y Nakano, H Hibino, Y Tanaka. Dietary oils and phospholipids containing n-3 highly unsaturated fatty acids suppress 2,4-dinitro-1-fluorobenzene-induced contact dermatitis in mice. *J Jpn Oil Chem Soc* 49:59–65, 2000.
57. M Zaouali-Ajina, A Gharib, G. Durand, N Gazzah, B Claustrat, C Gharib, N Sarda. Dietary docosahexaenoic acid-enriched phospholipids normalize urinary melatonin excretion in adult (n-3) polyunsaturated fatty acid-deficient rats. *J Nutr* 129:2074–2080, 1999.
58. Y Inoue, T Kanada. Manufacturing and physiological properties of phospholipids containing DHA. *Oleoscience* 2:67–74, 2002 (in Japanese).

14

Production of Biosurfactants by Fermentation of Fats, Oils, and Their Coproducts

Daniel K.Y. Solaiman

Richard D. Ashby

Thomas A. Foglia

14.1 Introduction

Fats, oils, and their coproducts (FOC) represent a class of substrates that can assume a uniquely important role in industrial fermentation processes. First of all, the structural diversity afforded by this class of substrates can potentially impart a variety of chemical composition and physical properties to the final products. The majority of fats and oils derived from biological sources are triacylglycerols containing fatty acids ranging from 8- to 22-carbon chain length. In some instances, these fatty acids may contain one or more functional groups, notably the carbon-carbon double bond and hydroxyl group. These functional groups are often incorporated into the structure of the fermentation products. Moreover, when triacylglycerols are hydrolyzed either during fermentation or in other industrial processes to release the fatty acids, glycerol is generated. This coproduct is a viable carbon source in many microbial fermentation systems to support cell growth and the synthesis of bioproduct, and can also play a role in modulating the structure and properties of the fermentation products.

FOCs are also attractive substrates from an economics point of view. The prices of the major fats and oils are comparable to those of refined carbohydrate-based substrates such as glucose and starch. Fats and oils are also large-volume commodities readily available to the users. The world production of animal fats according to a report from the Food and Agriculture Organization of the United Nations (FAO) reached 30 million metric tons (MMT) in 1999, and the global production of the nine major vegetable oils (soybean, palm, sunflower, rapeseed, cottonseed, groundnut, coconut, palm kernel and olive oils) is projected at 100 million metric tons in the 2003–2004 harvest year.[1] Together these commodities represent a viable resource of renewable materials that can be utilized for the production of biobased products.

Major areas in which fats and oils have been utilized in commercial fermentation processes are the production of pharmaceuticals, the manufacture of lipases, and the control of excessive foaming in bioprocesses. The addition of oil to antibiotic-producing fermentation systems is often accompanied by increased yields. As early as 1959, Pan et al.[2] reported the use of fatty oils to increase penicillin yields and noted the antifoaming effect of the oils. Subsequently, other studies ensued that validated the beneficial roles of fats and oils in the fermentation of various antibiotics. For example, when soybean oil was added to the medium, the production of cephalosporin C by *Acremonium chrysogenum* increased by ~30%.[3] The production of erythromycin and triketide lactone by *Saccharopolyspora erythraea* was shown to increase significantly in the presence of rapeseed oil.[4] The production of cephalosporin C by *Cephalosporium acremonium* increased 54% with the inclusion of sesame oil.[5] Jones and Porter[6] found that the addition of a small amount of vegetable oil enhanced the productivity of tetracyline fermentation by increasing the rates of synthesis. Numata et al.[7] showed that olive oil and other oleic acid-rich oils selectively stimulated the productivity of the antibiotic glidobactins C by *Polyangium brachysporum*. The productivity of josamycin increased when oils were added to the fermentation system,[8] and a requirement for triacylglycerols in the production of efrotomycin by *Nocardia lactamdurans* also was reported.[9]

In the second major area of oil utilization in fermentation processes, numerous reports have documented the necessity or effectiveness of oil addition to induce and increase the production of lipases.[10–16] Animal fats and vegetable oils also are widely used as antifoaming agents. In studies with simulated system containing only fermentation media without microorganism, Vardar-Sukan[17] showed that various vegetable oils dramatically increased the rates of foam collapse, thus effectively suppressing foam formation. Vidyarthi et al.[18] reported that the inhibitory effects of chemical antifoaming agents on cell growth and biopesticide (endotoxin) production in *Bacillus thuringiensis* fermentation could be avoided by using vegetable oils as the antifoaming agents. Accordingly, fats, oils, and fatty acids often are key components of antifoaming agents.[19–20]

Biobased products are receiving increasing attention due to the interplay of various technical, societal, and geopolitical factors. With a continually growing world population, an urgent need exists to further develop the utilization of the earth's renewable resources to sustain the increasing demands of goods consumption. A heightened global awareness of ecological and environmental conservation further exacerbates this need to develop sustainable production systems with minimal environmental impacts to meet the consumer demands of the world populace. A National Research Council report[21] cited among the benefits of biobased industry the minimizing of adverse economic impacts due to the disruption of petrochemical supplies resulting from volatile geopolitical circumstances. Coinciding with the need for a biobased industry is the advent of the biotechnology era. Quantum-leaped advances over the past decades in the basic knowledge and applied technologies of genomics, proteomics, and metabolomics have presented new potential for the design and implementation of economically viable bioprocesses. As a result, various biobased research initiatives have been forwarded in the last ten years, both in the public and private sectors. For example, the DuPont Corporation in 1999 set the goals of deriving 10% of its energy needs and 25% of its revenues from renewable resources by year 2010. The U.S. Department of Agriculture and U.S. Department of Energy (DOE) in 2003 announced the funding of 19 projects totaling $23 million for research, development, and demonstration projects on biomass research. With this background in mind, this chapter presents a survey of the current advances in the production of biosurfactants from the renewable raw materials, specifically the animal fats, vegetable oils, and their coproducts.

14.2 Microbial Glycolipids as Biosurfactants

Many microorganisms such as bacteria, yeasts and fungi synthesize glycolipids that have surface-active properties due to their amphiphilic structure.[22–24] These microbial products are usually synthesized during the stationary phase of cell growth and are thus classified as secondary metabolites. The use of biosurfactants has been proposed in many industrial applications, including enhanced oil recovery, bioremediation, herbicide and pesticide formulations, antimicrobial uses, manufacture of detergents and cosmetics, and the formulation of lubricants.[25] Improvement in their production cost and surface-active properties, however, are needed to help advance the commercialization of these biosurfactants. This chapter will focus on the fermentative production of sophorolipids and rhamnolipids from fats, oils, and their coproducts in an effort to contain the production costs and to impart compositional variation to these materials as a means of improving their surfactant properties.

14.2.1 Sophorolipids

Sophorolipids (SLs) are glycolipids produced and secreted by *Candida* and related yeast species. The structures of these glycolipids consist of a disaccharide sophorose (2-*O*-β-D-glucopyranosyl-β-D-glucopyranose) unit linked through a glycosidic bond to a hydroxy fatty acid. These microbial products possess surface-active properties because of their amphiphilic structure. The most studied SL-producing organism is *Candida bombicola*, which produces copious amounts of extracellular SLs. SL yields as high as 400 g/L of culture have been reported for this organism.[23] The glycolipids produced by *C. bombicola* are a mixture of SLs with varying degrees of acetylation on the sophorose group and different compositions of the hydroxy fatty acyl moiety. The sophorose group of the SLs is largely acetylated at the 6' and 6" positions (Figure 14.1), with 6"-monoacetylated and nonacetylated species also being minor species. The fatty acids of the SLs from *C. bombicola* are predominantly hydroxylated at the (ω-1)-carbon, with the ω-hydroxy derivatives being the minor components. The major SL species of the mixture contain a C18- or C16-fatty acyl group that may have one or more double bonds. The majority of the SLs of *C. bombicola* exist in the lactone form in which the carboxyl group of the fatty acyl chain forms a hemi-acetal-type bond with the 4" hydroxy group of the sophorose unit (Figure 14.1, SL-1). Other *Candida* species that synthesize similar sophorolipid mixtures include *C. apicola*[26] and *C. magnoliae*.[27] On the other hand, *Rhodotorula* (formerly *Candida*) *bogoriensis* produces sophorolipids containing a 13-hydroxydocosanoic acid as the major lipid constituent.[28–29]

   Production of SL by *C. bombicola* is greatly stimulated by the addition of a lipid feedstock along with a carbohydrate substrate such as glucose.[30–31] Production levels of 300–400 g/L of fermentation culture have been claimed under these cosubstrate culture conditions.[23] Various lipid substrates have been evaluated for their effects on the yield and composition of the fatty-acyl moiety of SL. An early study showed that the addition of soybean oil to a *C. bombicola* batch culture resulted in the production of the glycolipids in yields of 80 g/L of culture.[32] A carbon mass balance calculation showed that the input carbon was channeled to the cell mass (13%), SLs (37%), and CO_2 (50%). Switching to fed-batch culture with intermittent oil feeding improved the carbon flow to the desired SL products (60%) and reduced the conversion to CO_2 (30%). These results showed that based on total carbon input, the substrate-conversion rates in SL production could be increased from 0.37 g/g-substrate in a batch-culture setting to 0.6 g/g-substrate by instituting a fed-batch culture with intermittent feeding of soybean oil. Davila et al.[33] also showed that a controlled feeding of rapeseed ethyl esters to the culture markedly increased the yields of SL production to as high as 317 g/L. Presumably, the instituting of a controlled feeding of the lipid cosubstrate prevents the growth inhibitory effect of fatty acids. Daniel et al.[34] studied the production of SLs from *C. bombicola* grown in deproteinized whey concentrate supplemented with rapeseed oil. Product yields of 280 g SL/L

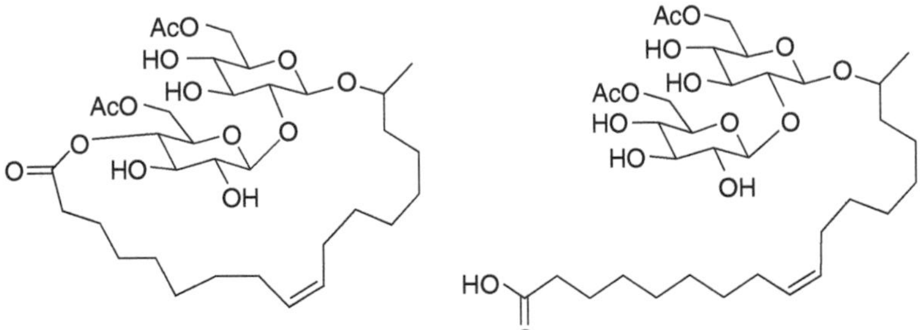

SL-1   SL-1A

FIGURE 14.1 Sophorolipid (17-L-([2'-*O*-β-glucopyranosyl-β-D-glucopyranosyl]-oxy)-9-*cis*-octadecenoic acid 1',4"-lactone 6',6" -diacetate (SL-1) and its free-acid form (SL-1A)).

culture were obtained with intermittent feeding of the oil substrate. Interestingly, lactose in the whey concentrate was not consumed, nor was galactosidase activity detected in the culture. A high lipase activity, however, was observed, suggesting that the cells used the fatty acid and glycerol generated from the hydrolysis of rapeseed oil as the carbon sources. The same research group[35] devised a two-stage fed batch fermentation process to improve SL production using whey concentrate and rapeseed oil as substrates. In the first fermentation stage, the oleaginous yeast *Cryptococcus curvatus* was used to consume the lactose in the whey concentrate to produce single-cell oil in the organism. In the second-stage fermentation, the sterilized cell lysate prepared from the first-stage culture was inoculated with *C. bombicola*, and intermittent feeding of rapeseed oil was instituted to result in a production level of 422 g SL/L of culture. Safflower oil was also studied as a lipid substrate for SL production by Zhou et al.[36] to achieve a product yield of 135 g SL/L culture. Compositional analysis indicated that 50% of the products were the diacetylated lactone form of the SL, but the chain length of the hydroxy fatty acyl unit was not determined. Animal fats (i.e., white choice hog grease) also have been used as lipid substrates for SLs production.[37] SLs levels of 120 g/L culture, as estimated by an anthrone-based colorimetric assay, were achieved with glucose as a cosubstrate. The production rate of 2.4 g SL/L/hr, however, was as high as twice those observed with oil substrates. Rau et al.[38] reported that when oleic acid was used as the lipid cosubstrate, a product yield of 180 g SL/L culture could be achieved. They noted, however, that excess lipid substrate led to a paste-like SL product that required additional washing whereas controlled addition of the lipid substrate produced a microcrystalline SL precipitate that was relatively easy to collect and purify.

In a systematic analysis of the effects of medium composition and culture condition on the production of sophorolipids by *C. bombicola*, Casas and Garcia-Ochoa[39] surveyed six lipid substrates (i.e., palmitic acid, coconut oil, corn oil, grape seed oil, olive oil, and sunflower oil) and various concentrations of the oil, glucose, and yeast extract (YE). Their results indicated that optimal production of SLs of 120 g/L culture was achieved using the following production medium: glucose (100 g/L), sunflower oil (100 g/L), and YE (1 g/L). They observed also that when the concentration of YE was 5 g/L or higher, the SL products were mainly of the free-acid form, whereas at YE concentrations lower than 5 g/L, the lactone form of the SLs predominated. In a similar study, Ogawa and Ota[40] compared the production of SL from coconut oil, olive oil, rapeseed oil, and soybean oil. They reported that olive oil and soybean oil gave yields of SL at 14 g/L, followed by rapeseed oil at 10 g/L. Krivobok et al.[41] compared methyl oleate, olive oil, and rapeseed oil in the production of SLs. Their results showed that methyl oleate gave the highest SL yields (46–53 g SL/L), followed by olive oil (40-48 g SL/L) and rapeseed oil (7-10 g SL/L). Davila et al.[42] performed an extensive comparison of the influence of rapeseed, sunflower and palm oils and esters, linseed esters, fish oil, and n-alkanes (C12, C14, C16, and C18) on the yields and compositions of SLs obtained from *C. bombicola*. They concluded that C18 alkane and oils and esters rich in C18:1 and C18:0 fatty acids were the preferred substrates for SL production. The best feedstock in their study was rapeseed ester; SL production levels of 340 g/L culture were reached using this lipid cosubstrate. The effect of lipid cosubstrates containing fatty acids of varying degrees of unsaturation on the fatty acyl composition of SL product mix was studied using tallow oil, soybean oil, and linseed oil as cofeedstock.[43–44] The data (based on LC/MS analyses) showed that the predominant hydroxy acyl chains of the SLs obtained from tallow oil cosubstrate were the C18:1-(51.3%) and the C16:0-(39.5%) containing SL species. The SLs from the soybean oil fermentation primarily contained as their hydroxy fatty acyl units the unsaturated C18:1 (42.3%) and C18:2 (36.5%) species.

In addition to triacylglycerols and fatty acids and esters, hydrocarbons and alcohols were tested as substrates for SL production. Hu and Ju[45] compared the effects of hexadecane to soybean oil on SL production, and found that in conjunction with glucose substrate, the hexadecane supported SL synthesis at a conversion rate of 0.84 g/g substrate. The major constituents of the hexadecane-derived SL mixture were diacetylated SL lactones (80%) containing either an ω- or ω-1 hydroxy palmitoyl acyl group. Soybean oil, on the other hand, gave substrate conversion rate of 0.20 g SL/g substrate, with the diacetylated lactone SLs containing ω- or ω-1 hydroxy oleoyl or linoleoyl acyl groups constituting 50% of the product mix. Brakemeier et al.[46–48] succeeded in producing alkyl glycosides by culturing *C. bombicola* in the presence of glucose and an alcohol or ketone cosubstrate. Product yields range from 12 to 22 g alkyl glycosides/L culture. Like sophorolipids, these sophorose alkyl-glycosides were shown to reduce the surface tension of water from 72 mN/m to 31-38 mN/m. The coproduct streams of industrial processes, such as the glycerol-rich coproduct of biodiesel manufacture and the carbohydrate-rich molasses from soybean processing also were studied as potential inexpensive substrates for SL production. Ashby et al.[49] succeeded in producing SLs consisting mainly of the acid form at a yield of 60 g/L culture using the crude glycerol coproduct stream from biodiesel production as the feedstock. Solaiman et al. (personal communication, 2003) demonstrated the feasibility of producing SLs at a yield of 21 g/L by using a soy molasses coproduct stream in place of glucose as the carbohydrate cosubstrate in conjunction with oleic acid as the lipid substrate.

The SL-producing *Candida* species also were used as biocatalysts for the transformation of fatty acids. Prabhune et al.[50] cultured *C bombicola* and *C. apicola* on glucose (100 g/L) and arachidonic acid (5Z, 8Z, 11Z, 14Z-eicosatetraenoic acid or AA) (1.25 g/L) and showed that up to 0.93 g/L of sophorolipid could be produced. Acid hydrolysis of this SL product yielded 19-hydroxy-5Z, 8Z, 11Z, 14Z-eicosatetraenoic acid (19-HETE) (73%), and 20-hydroxy-5Z, 8Z, 11Z, 14Z-eicosatetraenoic acid (20-HETE; 27%) as the lipid components from the SLs. This demonstrated that the organism effectively transformed AA into its ω- and (ω-1)-hydroxyl derivatives.

14.2.2 Rhamnolipids

Rhamnolipids (RLs) are glycolipids produced by certain Gram-negative bacteria. The most studied species for RL biosynthesis is *Pseudomonas aeruginosa* because of its high level of RL production[51] in comparison to other RL-synthesizing organisms such as *Burkholderia* (previously *Pseudomonas*) *pseudomallei*,[52] *P. putida*,[53] and a *Pseudomonas* sp.[54] Genetically engineered organisms have been constructed in an attempt to improve RL production.[55]

As with sophorolipids, the amphiphilic structure of RLs imparts to them their surface-active properties. The sugar moiety of rhamnolipids is the hydrophilic rhamnose (Rh) molecule, and the lipid portion of the glycolipid is a lipophilic 3-hydroxyalkanoic-acid (3-HAA) unit. Similar to the sophorolipids, the sugar and fatty acyl group of RLs are linked through a glycosidic bond, but the sugar residue is not acetylated. The most commonly found lipid moiety in RLs is the dimer of 3-hydroxydecanoic acid (3-HD) or 3-hydroxydodecanoic acid (3-HDD) linked by an estolide ester bond (Figure 14.2). In *P. aeruginosa*, a mixture of $(Rh)_{1-2}(3\text{-}HD)_{1-2}$ was produced.[56] RLs have potential uses in a wide range of applications, such as agricultural antifungal treatments, enhanced oil recovery, heavy metal remediation, cosmetic emulsification, and as food thickening agents.[25,57–60]

Various feedstocks including fats, oils, and their coproducts, have been used as potential low-cost feedstocks for the fermentative production of RLs.[61–64] Unlike the results of the studies with SLs in which the important role of lipid substrate on SL productivity is unequivocally established, the influence of fat

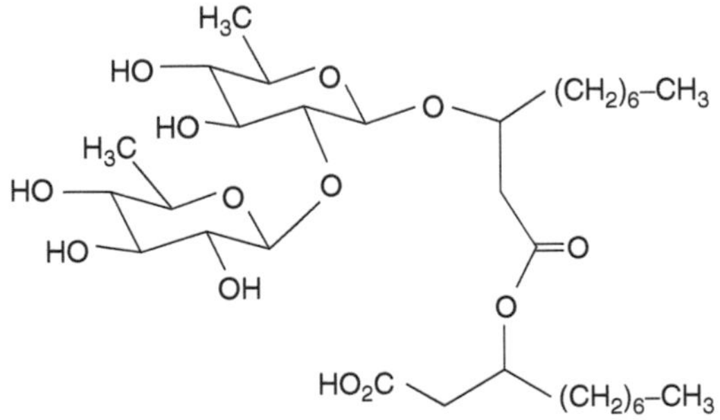

FIGURE 14.2 Rhamnolipid (3-L-([2-O-α-L-rhamnopyranosyl-α-L-rhamnopyranosyl]-oxy,-β-hydroxydecanoyl-β-hydroxydecanoate).

and oil substrate on the yields of RLs has not been systematically investigated. Nevertheless, a comparison of the yields of RLs reported by various researchers showed that oil feedstocks (e.g., olive oil, soybean oil, corn oil, and canola oil) often supported higher levels of RL production than the non-lipid substrates such as whey, sucrose, and glycerol did.[57] Giani et al.[51] described in a patent the construction of *P. aeruginosa* DSM 7107 and DSM 7108 mutant strains for high-efficiency production of RLs from soybean oil. Using these mutants, these authors achieved production levels of 70–120 g RLs/L of batch culture. Haba et al.[62] demonstrated that low-cost waste frying oils could be utilized by *P. aeruginosa* 47T2 NCIB 40044 to produce RLs at 2.7 g rhamnose-equivalent/L of culture. Trummler et al.[54] recently demonstrated that growing cells of *Pseudomonas* sp DSM 2874 under nitrogen limiting conditions as well as the resting cells could use rapeseed oil as the sole carbon source to produce a mixture of $(Rh)_1(3-HD)_2$ and $(Rh)_2(3-HD)_2$ at levels of 45 g RLs/L culture. These results collectively show that fats, oils, and their coproducts are suitable substrates for the fermentative production of RLs.

14.3 Conclusion

Animal fats, vegetable oils, and the coproducts derived either from their production or utilization processes are suitable and in some cases preferred or even indispensable substrates in industrial fermentation processes. This chapter specifically highlights their uses in the fermentative production of biosurfactants. Their roles in other important fermentation processes such as those for the production of poly(hydroxyalkanoates[65–73] (this volume, Chapter 19) cannot be overemphasized. Continued research and development efforts in this field should further expand the usefulness of these renewable resources for the cost-effective production of many important bulk and specialty chemicals and industrial materials.

References

1. F Gunstone. Early forecasts for world supplies of oilseeds and vegetable oil in 2003–04. *INFORM* 14:668–668, 2003.
2. SC Pan, S Bonanno, GH Wagman. Efficient utilization of fatty oils as energy sources in penicillin fermentation. *Appl Microbial* 7:176–180, 1959.
3. E Sandor, A Szentirmai, GC Paul, CR Thomas, I Pocsi, L Karaffa. Analysis of the relationship between growth, cephalosporin C production, and fragmentation in *Acremonium chrysogenum*. *Can J Microbiol* 47:801–806, 2001.
4. N Mirjalili, V Zormpaidis, PF Leadlay, AP Ison. The effect of rapeseed oil uptake on the production of erythromycin and triketide lactone by *Saccharopolyspora erythraea*. *Biotechnol Prog* 15:911–918, 1999.
5. S Paul, RL Bezbaruah, RS Prakasham, MK Roy, AC Ghosh. Enhancement of growth and antibiotic titre in *Cephalosporium acremonium* induced by sesame oil. *Folia Microbiol* 42:211–213, 1997.
6. AM Jones, MA Porter. Vegetable oils in fermentation: beneficial effects of low-level. *J Ind Microbiol Biotechnol* 21:203–207, 1998.
7. K Numata, T Murakami, M Oka, H Yamamoto, M Hatori, T Miyaki, T Oki, H Kawaguchi. Enhanced production of the minor components of glidobactins in *Polyangium brachysporum*. *J Antibiot* 41:1358–1365, 1988.
8. J Eiki, H Gushima, T Saito, H Ishida, Y Oka, T Osono. Product inhibition and its removal on josamycin fermentation by *Streptomyces narbonensis* var *josamyceticus*. *J Ferment Technol* 66:559–565, 1988.
9. M Chartrain, G Hunt, L Horn, A Kirpekar, D Mathre, A Powell, L Wassel, J Nielsen, B Buckland, R Greasham. Biochemical and physiological characterization of the efrotomycin fermentation. *J Ind Microbiol* 7:293–300, 1991.
10. S Benjamin, A Pandey. Optimization of liquid media for lipase production by *Candida rugosa*. *Bioresourc Technol* 55:167–170, 1996.

11. S Benjamin, A Pandey. Mixed-solid substrate fermentation. A novel process for enhanced lipase production by *Candida rugosa*. *Acta Biotechnol* 18:315–324, 1998.

12. S-J Chen, C-Y Cheng, T-L Chen. Production of an alkaline lipase by *Acinetobacter radioresistens*. *J Ferment Bioeng* 86:308–312, 1998.

13. A Hiol, MD Jonzo, D Druet, L Comeau. Production, purification and characterization of an extracellular lipase from *Mucor hiemalis f. hiemalis*. *Enzyme Microb Technol* 25:80–87, 1999.

14. ND Mahadik, US Puntambekar, KB Bastawde, JM Khire, DV Gokhale. Production of acidic lipase by *Aspergillus niger* in solid state fermentation. *Process Biochem* 38:715–721, 2002.

15. D Ozer, M Elibol. Effects of Some Factors on Lipase Production by *Rhizopus arrhizus*. *J Food Sci Technol* (Mysore) 37:661–664, 2000.

16. CB Tamerler, AT Martinez, T Keshavarz. Production of lipolytic enzymes in batch cultures of *Ophiostoma piceae*. *J Chem Technol Biotechnol* 76:991–996, 2001.

17. F Vardar-Sukan. Effects of natural oils on foam collapse in bioprocesses. *Biotechnol Lett* 13:107–112, 1991.

18. AS Vidyarthi, M Desrosiers, RD Tyagi, JR Valero. Foam control in biopesticide production from sewage sludge. *J Ind Microbiol Biotechnol* 25:86–92, 2000.

19. M Hayashi; Y Hioki, M Shonaka, T Moriyama. Antifoaming agent for fermentation, L-amino acid-producing medium and production process of L-amino acids. US Patent 5567606, 1996.

20. M Shonaka, K Hasebe, M Hayashi. Antifoaming agent for fermentation and fermentation production process using the same. US Patent 5843734, 1998.

21. National Research Council, Committee on Biobased Industrial Products (2000) Biobased Industrial Products: Research and Commercialization Priorities. The National Academies Press.

22. D Kitamoto, H Isoda, T Nakahara. Functions and potential applications of glycolipid biosurfactants—from energy-saving materials to gene delivery carriers. *J Biosci Bioeng* 94:187–201, 2002.

23. S Lang. Biological amphiphiles (microbial biosurfactants). *Curr Opin Colloid Interface Sci* 7:12–20, 2002.

24. S-C Lin. Biosurfactants: Recent advances. *J Chem Tech Biotechnol* 66:109–120, 1996.

25. IM Banat, RS Makkar, SS Cameotra. Potential commercial applications of microbial surfactants. *Appl Microbiol Biotechnol* 53:495–508, 2000.

26. JP Tulloch. Structures of extracellular glycolipids produced by yeasts. In: *Glycolipid Methodology*. Champaign: American Oil Chemists' Society, 1976, pp. 329–344.

27. PA Gorin, JFT Spencer, AP Tulloch. Hydroxy fatty acid glycosides of sophorose from *Torulopsis magnoliae*. *Can J Chem* 39:846–855, 1961.

28. TW Esders, RJ Light. Glucosyl- and acetyltransferases involved in the biosynthesis of glycolipids from *Candida bogoriensis*. *J Biol Chem* 247:1375–1386, 1972.

29. AP Tulloch, JFT Spencer, MH Deinema. A new hydroxy fatty acid sophorosides from *Candida bogoriensis*. *Can J Chem* 46:345–348, 1968.

30. H-J Asmer, S Lang, F Wagner, V Wray. Microbial production, structure elucidation and bioconversion of sophorose lipids. *J Am Oil Chem Soc* 65:1460–1466, 1988.

31. DG Cooper, DA Paddock. Production of a biosurfactant from *Torulopsis bombicola*. *Appl Environ Microbiol* 47:173–176, 1984.

32. KH Lee, JH Kim. Distribution of substrates carbon in sophorose lipid production by *Torulopsis bombicola*. *Biotechnol Lett* 15:263–266, 1993.

33. A-M Davila, R Marchal, J-P Vandecasteele. Sophorose lipid fermentation with differentiated substrate supply for growth and production phases. *Appl Microbiol Biotechnol* 47:496–501, 1997.

34. H-J Daniel, RT Otto, M Reuss, C Syldatk. Sophorolipid production with high yields on whey concentrate and rapeseed oil without consumption of lactose. *Biotechnol Lett* 20:805–807, 1998.

35. H-J Daniel, M Reuss, C Syldatk. Production of sophorolipids in high concentration from deproteinized whey and rapeseed oil in a two stage fed batch process using *Candida bombicola* ATCC 22214 and *Cryptococcus curvatus* ATCC 20509. *Biotechnol Lett* 20:1153–1156, 1998.

36. QH Zhou, V Klekner, N Kosaric. Production of sophorose lipids by *Torulopsis bombicola* from safflower oil and glucose. *J Am Oil Chem Soc* 69:89–91,1992.

37. M Deshpande, L Daniels. Evaluation of sophorolipid biosurfactant production by *Candida bombicola* using animal fat. *Biotechnol Lett* 54:143–150, 1995.

38. U Rau, C Manzke, F Wagner. Influence of substrate supply on the production of sophorose lipids by *Candida bombicola* ATCC 22214. *Biotechnol Lett* 18:149–154, 1996.

39. JA Casas, F García-Ochoa. Sophorolipid production by *Candida bombicola*: Medium composition and culture methods. *J Biosci Bioeng* 88:488–494, 1999.

40. S Ogawa, Y Ota. Influence of exogenous natural oils on the ω-1 and ω-2 hydroxy fatty acid moiety of sophorose lipid produced by *Candida bombicola*. *Biosci Biotechnol Biochem* 64:2466–2468, 2000.

41. S Krivobok, P Guiraud, F Seigle-Murandi, R Steiman. Production and toxicity assessment of sophorosides from *Torulopsis bombicola*. *J Agric Food Chem* 42:1247–1250, 1994.

42. A-M Davila, R Marchal, J-P Vandecasteele. Sophorose lipid production from lipidic precursors: predictive evaluation of industrial substrates. *J Ind Microbiol* 13:249–257, 1994.

43. A Nuñez, R Ashby, TA Foglia, DKY Solaiman. Analysis and characterization of sophorolipids by liquid chromatography with atmospheric pressure chemical ionization. *Chromatographia* 53:673–677, 2001.

44. DKY Solaiman, RD Ashby, TA Foglia, A Nuñez, WN Marmer. Production of biodegradable polymers and surfactants by fermentation of agricultural coproducts. Proceedings of the 30th Annual Meeting of United States-Japan Cooperative Program in Natural Resources (UJNR) Protein Panel, Tsukuba (Japan), 2001, pp. 161–164.

45. Y Hu, L-K Ju. Sophorolipid production from different lipid precursors observed with LC-MS. *Enz Microbiol Technol* 29:593–601, 2001.

46. A Brakemeier, S Lang, D Wullbrandt, L Merschel, A Benninghoven, H Buschmann, F Wagner. Novel sophorose lipids from microbial conversion of 2-alkanols. *Biotechnol Lett* 17:1183–1188, 1995.

47. A Brakemeier, D Wullbrandt, S Lang. Microbial alkyl-sophorosides based on 1-dodecanol or 2-, 3- or 4-dodecanones. *Biotechnol Lett* 20:215–218, 1998.

48. A Brakemeier, D Wullbrandt, S Lang. *Candida bombicola*: production of novel alkyl glycosides based on glucose/2-dodecanol. *Appl Microbiol Biotechnol* 50:161–166, 1998.

49. RD Ashby, A Nuñez, DKY Solaiman, TA Foglia. Synthesis of biopolymers and biosurfactants from glycerol and glycerol-based waste streams. Proceedings of the 32nd Annual Meeting of United States-Japan Cooperative Program in Natural Resources (UJNR) Food and Agricultural Panel, Tsukuba (Japan), 2003, pp. 269–278.

50. A Prabhune, SR Fox, C Ratledge. Transformation of arachidonic acid to 19-hydroxy- and 20-hydroxy-eicosatetraenoic acids using *Candida bombicola*. *Biotechnol Lett* 24:1041–1044, 2002.

51. C Giani, D Wullbrandt, R Rothert, J Meiwes. *Pseudomonas aeruginosa* and its use in a process for the biotechnological preparation of L-rhamnose. US Patent 5658793, 1997.

52. S Haussler, M Rohde, N von Neuhoff, M Nimtz, I Steinmetz. Structural and functional cellular changes induced by *Burkholderia pseudomallei* rhamnolipid. *Infect Immun* 71:2970–2975, 2003.

53. N Hoffmann, AA Amara, BB Beermann, Q Qi, H-J Hinz, BHA Rehm. Biochemical characterization of the *Pseudomonas putida* 3-hydroxyacyl ACP:CoA transacylase, which diverts intermediates of fatty acid de novo biosynthesis. *J Biol Chem* 277:42926–42936, 2002.

54. K Trummler, F Effenberger, C Syldatk. An integrated microbial/enzymatic process for production of rhamnolipids and L-(+)-rhamnose from rapeseed oil with *Pseudomonas* sp DSM 2874. *Eur J Lipid Sci Technol* 105:563–571, 2003.

55. UA Ochsner, J Reiser, A Fiechter, B Witholt. Production of *Pseudomonas aeruginosa* rhamnolipid biosurfactants in heterologous hosts. *Appl Environ Microbiol* 61:3503–3506, 1995.

56. NB Rendell, GW Taylor, M Somerville, H Todd, R Wilson, J Cole. Characterization of *Pseudomonas* rhamnolipids. *Biochem Biophys Acta* 1045:189–193, 1990.

57. S Lang, D Wullbrandt. Rhamnose lipids—biosynthesis, microbial production and application potential. *Appl Microbiol Biotechnol* 51:22–32, 1999.

58. RM Maier, G Soberón-Chávez. *Pseudomonas aeruginosa* rhamnolipids: biosynthesis and potential applications. *Appl Microbiol Biotechnol* 54:625–633, 2000.

59. E Rosenberg, EZ Ron. High- and low-molecular-mass microbial surfactants. *Appl Microbiol Biotechnol* 52:154–162, 1999.

60. ME Stanghellini, RM Miller, SL Rasmussen, DH Kim, Y Zhang. Microbially produced rhamnolipids (biosurfactants) for the control of plant pathogenic zoosporic fungi. US Patent 5767090, 1998.

61. PS Babu, AN Vaidya, AS Bal, R Kapur, A Juwarkar, P Khanna. Kinetics of biosurfactant production by *Pseudomonas aeruginosa* strain BS2 from industrial wastes. *Biotechnol Lett* 18:263–268, 1996.

62. E Haba, MJ Espuny, M Busquets, A Manresa. Screening and production of rhamnolipids by *Pseudomonas aeruginosa* 47T2 NCIB 40044 from waste frying oils. *J Appl Microbiol* 88:379–387, 2000.

63. ME Mercade, MA Manresa. The use of agroindustrial by-products for biosurfactant production. *J Am Oil Chem Soc* 71:61–64, 1994.

64. RM Patel, AJ Desai. Biosurfactant production by *Pseudomonas aeruginosa* GS3 from molasses. *Lett Appl Microbiol* 25:91–94, 1997.

65. HM Alvarez, A Steinbüchel. Triacylglycerols in prokaryotic microorganisms. *Appl Microbiol Biotechnol* 60:367–376, 2002.

66. RD Ashby, TA Foglia. Poly(hydroxyalkanoate) biosynthesis from triglyceride substrates. *Appl Microbiol Biotechnol* 49:431–437, 1998.

67. A-M Cromwick, TA Foglia, RW Lenz. The microbial production of poly(hydroxyalkanoates) from tallow. *Appl Microbiol Biotechnol* 46:464–469, 1996.

68. G Eggink, P de Waard, GNM Huijberts. Formation of novel poly(hydroxyalkanoates) from long-chain fatty acids. *Can J Microbiol* 41(Suppl 1):14–21, 1995.

69. W He, W Tian, G Zhang, G-Q Chen, Z Zhang. Production of novel polyhydroxyalkanoates by *Pseudomonas stutzeri* 1317 from glucose and soybean oil. *FEMS Microbiol Lett* 169:45–49, 1998.

70. P Kahar, T Tsuge, K Taguchi, Y Doi. High yield production of polyhydroxyalkanoates from soybean oil by *Ralstonia eutropha* and its recombinant strain. *Polym Degrad Stabil* 83:79–86, 2004.

71. H Salehizadeh, MCM van Loosdrecht. Production of polyhydroxyalkanoates by mixed culture: recent trends and biotechnological importance. *Biotechnol Adv* 22:261–279, 2004.

72. DKY Solaiman, RD Ashby, TA Foglia. Production of polyhydroxyalkanoates from intact triacylglycerols by genetically engineered *Pseudomonas*. *Appl Microbiol Biotechnol* 56:664–669, 2001.

73. DKY Solaiman, RD Ashby, TA Foglia. Physiological characterization and genetic engineering of *Pseudomonas corrugata* for medium-chain-length polyhydroxyalkanoates synthesis from triacylglycerols. *Curr Microbiol* 44:189–195, 2002.

15

Fatty Acid-Modifying Enzymes: Implications for Industrial Applications

Tsunehiro Aki

Seiji Kawamoto

Seiko Shigeta

Kazuhisa Ono

15.1 Introduction

Hydrophobic moieties in lipid molecules are often composed of a fatty acid consisting of a hydrocarbon chain and carboxylic acid. Fatty acids generally combine with glycerin to form acylglycerol, which accumulates in cells and/or plant seeds as an energy storage. Fatty acids are also constituents of the phospholipid hydrophobic layer of the cell membrane. The various species of fatty acids differ with regard to the length of the carbon chain, the degree of desaturation, and type of modification such as epoxidation, hydroxylation, and conjugation.[1] The physicochemical properties of fatty acids, ranging from regulation of membrane functions to maintenance of cell homeostasis, differ with respect to structure. Recent studies demonstrated that some kinds of fatty acids are participants in the intracellular signaling pathways. They act as ligands for nuclear receptors regulating cell responses and modulate eicosanoid metabolism in cells of the immune system.[2] Fatty acids also have nonphysiological uses, and demand for fatty acids for use as industrial raw materials for detergents, lubricating oils, coating agents, and cosmetics is rising.

At present, more than 75% of the edible and industrially useful fatty acids are obtained from seeds of oleageneous higher plants such as soybean (*Glycine max*), parm (*Elaeis guineensis*), rapeseed (*Brassica napus*; canola), sunflower (*Helianthus annuus*), and peanut (*Arachis hypogaea*), while most of the remaining fatty acids are obtained from animals such as cattle, pig, and fish. In addition, microbial lipids, termed single cell oils, containing polyunsaturated fatty acids (PUFA) have been developed and continue to gain greater market share.[3–5] Oils produced by transgenic organisms are becoming more prevalent due to the continual accumulation of genetic information about fatty acid biosynthetic enzymes. This chapter presents an overview of the current metabolic engineering and the important issues concerning the development and industrialization of "transgenic oils."

15.2 Biosynthetic Pathways of Modified Fatty Acids

A large proportion of naturally available long-chain (C16-18) and very-long-chain (C20 or more) fatty acids contain one or more double bonds, an acetylene bond, a hydroxyl group, an epoxy group, and/or conjugated double bonds on their acyl moiety. These molecules, considered "modified fatty acids," have a basic, saturated fatty acid backbone such as palmitic acid (C16:0) or stearic acid (C18:0). Palmitic acid is biosynthesized from acetyl-CoA by fatty acid synthase complex (FAS) associated with plastids (higher plants) or cytosol (lower eukaryotes and animals).[6] Stearic acid is a chain-elongated product with a C2-unit from palmitic acid. These saturated fatty acids that esterified with an acyl carrier protein (ACP) or coenzyme A (CoA) are used to biosynthesize acylglycerols, phospholipids, steryl esters, and modified fatty acids.

15.2.1 Fatty Acid Desaturases

Modified fatty acids are biosynthesized by catalytic reactions employing appropriate "fatty acid–modifying enzymes" along the pathways shown in Figure 15.1. The most abundant unsaturated fatty acids in natural oils are produced by fatty acid desaturases that act on acyl-ACP, glycerolipids, or acyl-CoA.[7] The glycerolipid and acyl-CoA type of desaturases, but not the soluble acyl-ACP type desaturases, are associated with membranes and have similar primary structures in part.[8] Desaturases are the terminal enzymes in the enzyme complex system and use ferredoxin or cytochrome b_5 as coenzymes. The position of introduction of a double bond on the acyl chain depends on the specificity of the desaturase. The yeast *Saccharomyces cerevisiae* has only $\Delta 9$ desaturase and produces palnitic acid (C16:1n-7) and oleic acid (C18:1n-9).[9] Most higher plants that produce α-linolenic acid (C18:3n-3) have at least three desaturases, each of which catalyzes desaturation at the $\Delta 9$, $\Delta 12$, or $\Delta 15$ position. Animals and some fungal species of *Mucor* and *Mortierella* have $\Delta 6$ and $\Delta 5$ desaturases that are involved in the synthesis of γ-linolenic acid and arachidonic acid.[10–12] Note that these fungi also have $\Delta 9$ and $\Delta 12$ desaturases and therefore produce all the fatty acids *de novo*. Recently, a $\Delta 4$ desaturase was identified in *Thraustochytrium*, a marine heterokont algae that produces docosapentaenoic acid (C22:5n-6) and docosahexaenoic acid (C22:6n-3).[13]

15.2.2 Desaturase Family

Some higher plants also produce other types of fatty acids such as 9-octadecen-12-ynoic acid (crepenynic acid), D-12-hydroxyoctadec-*cis*-9-enoic acid (ricinoleic acid), 12,13-epoxy-9-octadecenoic acid (vernolic acid), or 5t,7t,9c-octadecatrienoic acid (α-eleostearic acid), where characteristic structures are underlined. Formation of these modified structures is catalyzed by acetylenase,[14] hydroxylase,[15] epoxygenase,[14] or conjugase,[16] respectively. Genes for all of these enzymes have been isolated. These enzymes along with the membrane-bound desaturases share partial primary structures, three histidine cluster motifs (amino acids HXXXH, HXXHH, and QXXHH) typical of some redox enzymes,[8] and two long hydrophobic domains (Figure 15.2). In particular, the oleate hydroxylase from crucifer (*Lesquerella fendleri*) has about 81% sequence identity to the oleate $\Delta 12$ desaturase from *Arabidopsis thaliana*.[15] Interestingly, only four amino acid substitution could convert the $\Delta 12$ desaturase to a hydroxylase.[17]

15.2.3 Fatty Acid Elongases

Fatty acid elongases are important in the biosynthesis of fatty acids of more than 20 carbons. The elongation reaction consists of four step reactions.[7] The first one, in which an acyl substrate and a C2-unit from malonyl-CoA are condensed, is considered the limiting step and is substrate specific. The enzyme that catalyzes this particular reaction is usually called elongase. The very-long-chain plant fatty acids found

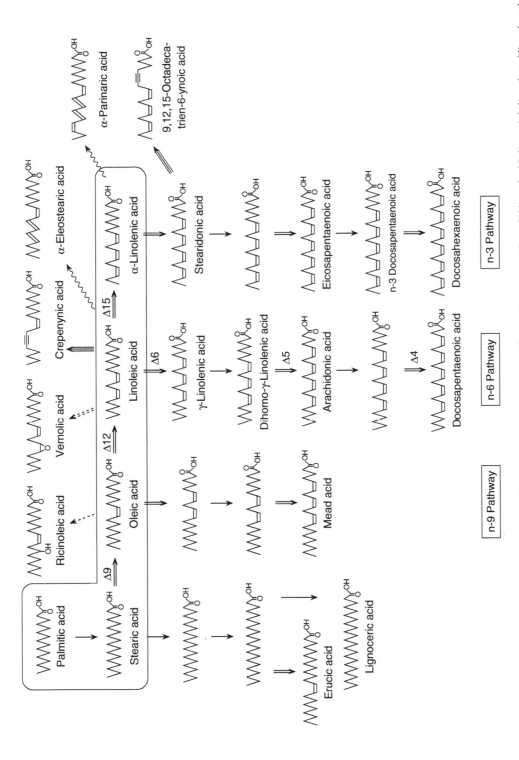

FIGURE 15.1 Biosynthetic pathways of polyunsaturated fatty acids and very-long-chain saturated fatty acids. Arrows with a solid line, double line, triple line, dotted line, dotted double line, or waved line indicate the steps of elongation, desaturation, acetylation, hydroxylation, epoxidation, or conjugation reactions, respectively.

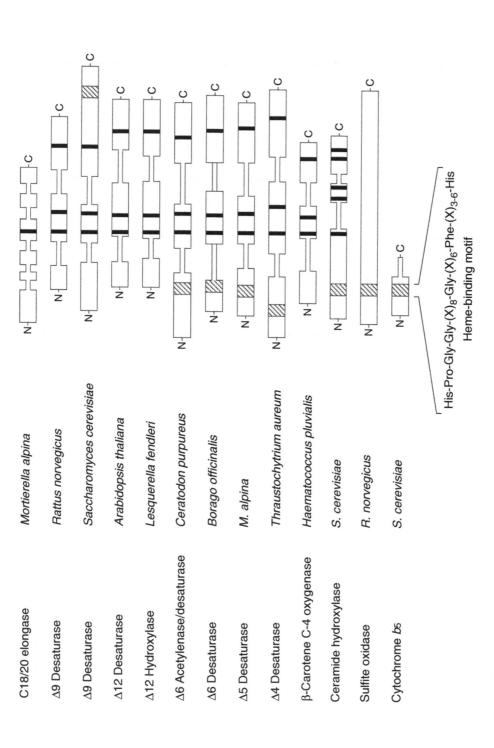

C18/20 elongase — *Mortierella alpina*

Δ9 Desaturase — *Rattus norvegicus*

Δ9 Desaturase — *Saccharomyces cerevisiae*

Δ12 Desaturase — *Arabidopsis thaliana*

Δ12 Hydroxylase — *Lesquerella fendleri*

Δ6 Acetylenase/desaturase — *Ceratodon purpureus*

Δ6 Desaturase — *Borago officinalis*

Δ5 Desaturase — *M. alpina*

Δ4 Desaturase — *Thraustochytrium aureum*

β-Carotene C-4 oxygenase — *Haematococcus pluvialis*

Ceramide hydroxylase — *S. cerevisiae*

Sulfite oxidase — *R. norvegicus*

Cytochrome *b5* — *S. cerevisiae*

His-Pro-Gly-Gly-(X)$_8$-Gly-(X)$_6$-Phe-(X)$_{3-6}$-His
Heme-binding motif

FIGURE 15.2 Structural comparisons of fatty acid modifying enzymes and other redox enzymes. Structures of the fatty acid Δ12 acetylenase, epoxygenase, conjugase, and Δ15 desaturase are similar to that of Δ12 desaturase and not shown. The potential heme-binding regions and histidine clusters are indicated by hatched and shaded boxes, respectively. The narrow areas in the horizontal bars indicate the long hydrophobic regions. Some other redox enzymes (β-carotene C-4 oxygenase, ceramide hydroxylase, and sulfite oxidase) are also shown.

in wax and sphingolipids are biosynthesized by C18/20-specific elongases such as FAE1 and KCS.[18,19] Fungi and animals have other types of elongases to biosynthesize C20/22 PUFA.[20,21] Unlike FAE1 and KCS, the primary structures of the unsaturates-specific elongases contain at least one histidine cluster motif (Figure 15.2).

15.2.4 Other Fatty Acid-Related Enzymes

Genetic manipulation of enzymes that change the lipid forms of fatty acids is a useful means of producing lipids with desirable fatty acid composition. The lipid accumulation in microorganisms is considered to be controlled by the activity of malic enzyme, which acts as a source of cofactor (NADPH) for fatty acid synthase.[22] Acyltransferase, involved in the biosynthesis of glycerides, is the key enzyme determining the fatty acid composition of storage lipids.[23] Lipase and esterase are involved in the release and transposition of fatty acids in glycerides, and thus control the lipid composition and the level of the intracellular acyl-CoA pool (see Section 15.5). In addition, acyl-CoA synthetase plays an important role not only in the activation with CoA but also in the transportation of fatty acids across membranes.[24]

15.3 Genetically Manipulated Improvement of Fatty Acid Compositions

The introduction, disruption, modification, or regulated expression of genes of the above mentioned fatty acid modifying enzymes and other related enzymes enables oleageneous organisms to produce oils with novel compositions.[25] This includes the generation of new species of fatty acids, not produced by the unaltered organism. Some representative cases of industrial and laboratory scale production of transgenic oils are briefly described below.

15.3.1 Lauric Acid

High lauric acid (C12:0) containing rapeseed oil was the first transgenic oil available on the market. Lauric acid is used as an industrial raw material for the detergent production. Natural rapeseed oil contains very low levels of this fatty acid but introduction into rapeseed of the C12:0-specific acyl-ACP thioesterase gene from the laurel tree (*Laurus nobilis*) increases the lauric acid content up to ~60% of the total fatty acid.[26]

15.3.2 Oleic Acid

Oleic acid is a major constituent of olive oil and has received much attention recently, as it reduces LDL-cholesterol levels in blood. To increase the oleic acid content in soybean and rapeseed oils that naturally contains a high level of linoleic acid, the expression of $\Delta 12$ desaturase was down-regulated by silencing the gene. This modification increased the oleic acid content up to ~90%.[27,28]

15.3.3 γ-Linolenic Acid

γ-Linolenic acid is put to practical use as a therapeutic agent for the treatment of atopic dermatitis and is known to reduce tumor transposition and blood cholesterol levels.[29] Although seeds from evening primrose (*Oenothera biennis*) and borage (*Borago officinalis*) have been a source of γ-linolenic acid,[30] the total oil content is not high in these plant seeds. Thus, a $\Delta 6$ desaturase gene was introduced into rapeseed or other plants to produce oil rich in γ-linolenic acid.[31–33] This fatty acid has also been targeted for single cell oil production by *Mucor* and *Mortierella* fungi.[34] Since these fungi have at least two types of $\Delta 6$ desaturases with distinct expression patterns,[12] the accurate regulation of their gene expression will contribute to increase the content of γ-linolenic acid.

15.3.4 Erucic Acid

Erucic acid (C22:1), an industrial material used to produce plastics, detergents, and lubricating oils, has been produced from a rapeseed oil at levels of ~30% of the glyceride fraction. Attempts to increase the erucic acid content in this oil were initially unsuccessful due to the substrate specificity of lysophosphatidic acid acyltransferase (LPAAT), which introduces an acyl group on glycerides at the *sn*-2 position. However, transformation with a gene coding for a yeast LPAAT with lower substrate specificity resulted in production of glycerides composed largely (nearly 50%) of erucic acid.[35]

15.3.5 Miscellaneous

At the laboratory scale, crepenynic acid,[14] 9,12,15-Octadecatrien-6-ynoic acid,[14] ricinoleic and other hydroxy fatty acids,[36] vernolic acid,[14] α-eleostearic acid,[16] α-parinaric acid,[16] and other very-long-chain fatty acids[37] have been successfully produced in seeds or other tissues of soybean, tobacco, or in *Arabidopsis* by gene manipulation.

Protein engineering is also a powerful tool to alter enzyme functions for the purpose of producing target fatty acids. The 18:0-ACP desaturase is a soluble protein, and its crystal structure has been elucidated. Substitution of an amino acid residue at the substrate binding site with a larger residue increased the specificity of the enzyme to C16:0.[38] The shallower, modified substrate binding pocket may be suitable for binding to smaller substrates.

15.4 Potential PUFA Production by Transgenic Organisms

Among the modified fatty acids, PUFA are probably the most valuable for use in edible oils and pharmaceuticals. Single-cell oils have recently been expected to become a practical source of PUFA. The fungus *Mortierella* has been used to ferment arachidonic acid (C20:4n-6).[4,39] Dihomo-γ-linolenic acid (C20:3n-6) and mead acid (C20:3n-9) are produced by mutant species of *Mortierella*.[40,41] *Schizochytrium*, a marine algae similar to *Thraustochytrium*, accumulates docosahexaenoic acid to ~40% of its total fatty acid content[42] and is used as an industrial producer. To reconstitute biosynthetic pathways in higher plants, the introduction of more than one gene-encoding fatty acid modifying enzyme is necessary.

The fungus *Mortierella* has a pathway similar to that in higher plants for linoleic acid production and, in addition, a Δ6 desaturase to produce γ-linolenic acid. Certain species of this genus also produce the PUFA-specific elongase and Δ5 desaturase proteins, which are involved in the biosynthesis of dihomo-γ-linolenic acid and arachidonic acid.[4] Thus, in order to produce arachidonic acid in soybeans for instance, at least three genes for Δ6 and Δ5 desaturases and elongase must be transferred. A model study was performed in the yeast *S. cerevisiae* in which homologous genes from the nematode *Caenohabiditis elegans* were introduced.[21] In this case, arachidonic acid was successfully produced in the transformed yeast cells by the addition of excess linoleic acid, although the conversion efficiency from linoleic acid to arachidonic acid was lower than 1%. Production of eicosapentaenoic acid (C20:5n-3) from α-linolenic acid (n-3 pathway) was also inefficient. Thus, fatty acid production is anticipated to be more difficult when multiple genes are expressed in plants.

The PUFA-specific elongases in animals act on both C18 and C20 PUFA and are involved in the production of C22 PUFA, such as docosatetraenoic acid (C22:4n-6) and n-3 docosapentaenoic acid (C22:5n-3).[20,43] Therefore, production of docosapentaenoic acid and docosahexaenoic acid may be possible in higher plants transformed with the animal elongase and *Thraustochytrium* Δ4 desaturase genes. However, acquiring sufficient amounts of fatty acid substrates may be problematic.

The enterobacteria *Shewanella* and *Vibrio* found in marine fish produce eicosapentaenoic acid and docosahexaenoic acid, respectively.[44,45] A gene cluster containing five open reading frames required for the biosynthesis of eicosapentaenoic acid has been isolated and functionally expressed in *Escherichia coli*.[46] Homologous genes have also been found in *Vibrio*.[47] The amino acid sequences deduced from these new genes contain some sequences characteristic to polyketide biosynthetic enzymes, suggesting novel PUFA biosynthetic pathways. Recently, gene segments substantially similar to the *Shewanella* and *Vibrio* genes have been isolated from *Schizochytrium*.[48] Therefore, the polyketide-related pathway is also functional in eukaryotes, although the mechanism underlying this novel system has not yet been elucidated. Further research is necessary to determine if industrial production of docosahexaenoic acid will be possible using *Schizochytrium* genes.

15.5   The Efficient Production of Transgenic Oils

Molecular cloning of a number of fatty acid-modifying enzymes enabled the production of transgenic oils in plants and microorganisms. The timing of gene expression and the tissues (in plants) in which genes are expressed as well as the accumulation of the target lipid form must be accurately regulated for efficient production. In addition, in the case of multiple gene expression for the PUFA production, substrate supply as well as the efficiency of each enzymatic reaction will be critical for the efficient production of target fatty acids.

A model experiment was done to investigate which factor should be regulated to improve the activity of fatty acid desaturase, using a reconstituted system in the yeast *S. cerevisiae*.[49] The desaturation system in fungi and animals is considered to consist of two electron transporters, NADH-dependent cytochrome b_5 reductase and cytochrome b_5, and the catalytic enzyme desaturase.[7] Up-regulation of the coenzymes was expected to boost the activity of the entire enzymatic reaction. However, cloning of the Δ6 and Δ5 desaturases revealed that these proteins contain an amino acid sequence characteristic to cytochrome *b*5 (Figure 15.2; the potential heme-binding region),[10,33,50–52] suggesting that the free cytochrome *b*5 molecules are not necessary. To address this, a Δ6 desaturase gene was overexpressed in a cytochrome *b*5 deficient yeast strain. Interestingly, Δ6 desaturation activity was detected even in this system but was significantly lower than that in wild-type yeast (Figure 15.3). The activity was then increased two- to threefold when the free cytochrome b_5 was overexpressed. These results indicate that coenzyme(s) may be the limiting factor(s) and should be coordinately expressed to achieve optimal production.

Until transgenic plant oils rich in PUFA become readily available and affordable, production systems for the single cell oils should be further improved. For example, the fractions into which target lipids accumulate need to be controlled. The conventional fungal strains used in industrial production accumulate lipids in their cells. When lipids are released in culture medium, product recovery is efficient,

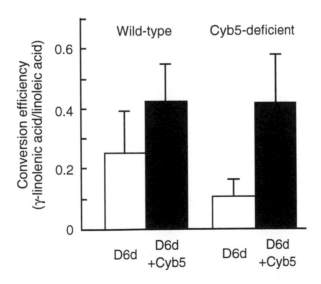

FIGURE 15.3   The role of cytochrome b_5 in the Δ6 desaturation system reconstituted in the yeast *S. cerevisiae*.[49] A rat Δ6 desaturase gene (D6d) or D6d and the cytochrome b_5 gene (D6d + Cyb5) was overexpressed in the wild-type or cytochrome b_5-deficient yeast strain in the presence of linoleic acid. The conversion efficiency was calculated as the ratio of γ-linolenic acid produced to the substrate linoleic acid.

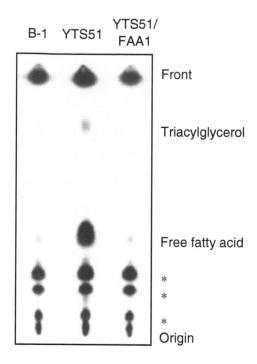

FIGURE 15.4 Characterization of a free fatty acid secreting mutant from the yeast *S. cerevisiae.*[54] Analysis of extracellular lipids by thin-layer chromatography. Spots indicated by asterisks are derived from the detergent present in the medium. B-1, wild-type; YTS51, a free fatty acid-secreting mutant; YTS51/FAA1, YTS51 carrying a FAA1 gene.

and promotion of the lipogenesis due to alteration of intracellular lipid homeostasis is also expected. In the development of genetically engineered lipid secretion systems, some microorganisms that secrete acylglycerols (yeast *Tricosporon*[53]) or free fatty acids (*S. cerevisiae*[54]) have been isolated. Genetic analysis of *S. cerevisiae* has revealed that a gene that complements the free fatty acid-secreting phenotype encodes acyl-CoA synthetase (Figure 15.4). This enzyme is involved not only in the activation of fatty acids but also in the incorporation of fatty acids across membranes in cooperation with a fatty acid transporter protein.[55] Free fatty acids might accumulate and be secreted due to the inactivation of acyl-CoA synthetase and the disordered transport system in this mutant. Application of this system to the oleageneous microorganisms will be effective.

15.6   Future Prospects

Manipulation of isolated fatty acid-modifying enzyme genes will enable efficient and affordable production of the currently expensive PUFA and unusual fatty acids. However, some critical problems mentioned in this chapter must be overcome to fully employ this technology. Extensive research, including the elucidation of biosynthetic mechanisms, on fatty acid modification is essential.

References

1. SF O'Keefe. Nomenclature and classification of lipids. In: CC Akoh, DB Min, eds. *Food Lipids.* New York: Marcel Dekker, 1998, pp. 1–36.
2. P Yaqoob. Lipids and the immune response: from molecular mechanisms to clinical applications. *Curr Opin Clin Nutr Metab Care,* 6:133–150, 2003.
3. W Yongmanitchai, OP Ward. Omega-3 fatty acids: Alternative sources of production. *Process Biochem,* 24:117–125, 1989.

4. M Certik, S Shimizu. Biosynthesis and regulation of microbial polyunsaturated fatty acid production. *J Biosci Bioeng*, 87:1–14, 1999.

5. C Ratledge. Microbial lipids: Commercial realities or academic curiosities. In: D Kyle, C Ratledge, eds. *Industrial Applications of Single Cell Oils*. Champaign, IL: American Oil Chemists' Society, 1992, pp. 1–15.

6. LM Salati, AG Goodridge. Fatty acid synthesis in eukaryotes. In: DE Vance, JE Vance, eds. *Biochemistry of Lipids, Lipoproteins and Membranes*. Amsterdam: Elsevier, 1996, pp. 101–127.

7. HW Cook. Fatty acid desaturation and chain elongation in eukaryotes. In: DE Vance, JE Vance, eds. *Biochemistry of Lipids, Lipoproteins and Membranes*. Amsterdam: Elsevier Science, 1996, pp. 129–152.

8. J Shanklin, E Whittle, BG Fox. Eight histidine residues are catalytically essential in a membrane-associated iron enzyme, stearoyl-CoA desaturase, and are conserved in alkane hydroxylase and xylene monooxygenase. *Biochemistry*, 33:12787–12794, 1994.

9. JE Stukey, VM McDonough, CE Martin. Isolation and characterization of OLE1, a gene affecting fatty acid desaturation from *Saccharomyces cerevisiae*. *J Biol Chem*, 264:16537–16544, 1989.

10. LV Michaelson, CM Lazarus, G Griffiths, JA Napier, AK Stobart. Isolation of a Delta 5-fatty acid desaturase gene from *Mortierella alpina*. *J Biol Chem*, 273:19055–19059, 1998.

11. E Sakuradani, M Kobayashi, S Shimizu. Δ6-fatty acid desaturase from an arachidonic acid-producing *Mortierella* fungus. Gene cloning and its heterologous expression in a fungus, *Aspergillus*. *Gene*, 238:445–453, 1999.

12. Y Michinaka, T Aki, T Shimauchi, T Nakajima, S Kawamoto, S Shigeta, O Suzuki, K Ono. Differential response to low temperature of two Δ6 fatty acid desaturases from *Mucor circinelloides*. *Appl Microbiol Biotechnol*, 2003.

13. X Qiu, H Hong, SL MacKenzie. Identification of a Delta 4 Fatty Acid Desaturase from *Thraustochytrium* sp. Involved in the Biosynthesis of Docosahexanoic Acid by Heterologous Expression in *Saccharomyces cerevisiae* and *Brassica juncea*. *J Biol Chem*, 276:31561–31566, 2001.

14. M Lee, M Lenman, A Banas, M Bafor, S Singh, M Schweizer, R Nilsson, C Liljenberg, A Dahlqvist, PO Gummeson, S Sjodahl, A Green, S Stymne. Identification of non-heme diiron proteins that catalyze triple bond and epoxy group formation. *Science*, 280:915–918, 1998.

15. FJ van de Loo, P Broun, S Turner, C Somerville. An oleate 12-hydroxylase from *Ricinus communis* L. is a fatty acyl desaturase homolog. *Proc Natl Acad Sci USA*, 92:6743–6747, 1995.

16. EB Cahoon, TJ Carlson, KG Ripp, BJ Schweiger, GA Cook, SE Hall, AJ Kinney. Biosynthetic origin of conjugated double bonds: production of fatty acid components of high-value drying oils in transgenic soybean embryos. *Proc Natl Acad Sci USA*, 96:12935–12940, 1999.

17. P Broun, J Shanklin, E Whittle, C Somerville. Catalytic plasticity of fatty acid modification enzymes underlying chemical diversity of plant lipids. *Science*, 282:1315–1317, 1998.

18. MW Lassner, K Lardizabal, JG Metz. A jojoba beta-Ketoacyl-CoA synthase cDNA complements the canola fatty acid elongation mutation in transgenic plants. *Plant Cell*, 8:281–292, 1996.

19. J Todd, D Post-Beittenmiller, JG Jaworski. KCS1 encodes a fatty acid elongase 3-ketoacyl-CoA synthase affecting wax biosynthesis in *Arabidopsis thaliana*. *Plant J*, 17:119–130, 1999.

20. JM Parker-Barnes, T Das, E Bobik, AE Leonard, JM Thurmond, LT Chaung, YS Huang, P Mukerji. Identification and characterization of an enzyme involved in the elongation of n-6 and n-3 polyunsaturated fatty acids. *Proc Natl Acad Sci USA*, 97:8284–8289, 2000.

21. F Beaudoin, LV Michaelson, SJ Hey, MJ Lewis, PR Shewry, O Sayanova, JA Napier. Heterologous reconstitution in yeast of the polyunsaturated fatty acid biosynthetic pathway. *Proc Natl Acad Sci USA*, 97:6421–6426, 2000.

22. C Ratledge. Regulation of lipid accumulation in oleaginous micro-organisms. *Biochem Soc Trans*, 30:1047–1050, 2002.

23. N Murata, Y Tasaka. Glycerol-3-phosphate acyltransferase in plants. *Biochem Biophys Acta*, 1348:10–16, 1997.

24. LJ Knoll, DR Johnson, JI Gordon. Biochemical studies of three *Saccharomyces cerevisiae* acyl-CoA synthetases, Faa1p, Faa2p, and Faa3p. *J Biol Chem*, 269:16348–16356, 1994.

25. AJ Kinney, EB Cahoon, WD Hitz. Manipulating desaturase activities in transgenic crop plants. *Biochem Soc Trans,* 30:1099–1103, 2002.

26. TA Voelker, TR Hayes, AM Cranmer, JC Turner, HM Davies. Genetic engineering of a quantitative trait: metabolic and genetic parameters influencing the accumulation of laurate in rapeseed. *Plant J,* 9:229–241, 1996.

27. AJ Kinney. Development of genetically engineered soybean oils for food applications. *J Food Lipids,* 3:273–292, 1996.

28. PA Stoutjesdijk, C Hurlestone, SP Singh, AG Green. High-oleic acid Australian *Brassica napus* and *B. juncea* varieties produced by co-suppression of endogenous Delta 12-desaturases. *Biochem Soc Trans,* 28:938–940, 2000.

29. DF Horrobin. Nutritional and medical importance of gamma-linolenic acid. *Prog Lipid Res,* 31:163–194, 1992.

30. AM Galle, M Joseph, C Demandre, P Guerche, JP Dubacq, A Oursel, P Mazliak, G Pelletier, JC Kader. Biosynthesis of gamma-linolenic acid in developing seeds of borage (*Borago officinalis* L.). *Biochem Biophys Acta,* 1158:52–58, 1993.

31. AS Reddy, TL Thomas. Expression of a cyanobacterial delta 6-desaturase gene results in gamma-linolenic acid production in transgenic plants. *Nat Biotechnol,* 14:639–642, 1996.

32. H Hong, N Datla, DW Reed, PS Covello, SL MacKenzie, X Qiu. High-level production of gamma-linolenic acid in *Brassica juncea* using a delta 6 desaturase from *Pythium irregulare. Plant Physiol,* 129:354–362, 2002.

33. O Sayanova, MA Smith, P Lapinskas, AK Stobart, G Dobson, WW Christie, PR Shewry, JA Napier. Expression of a borage desaturase cDNA containing an N-terminal cytochrome b5 domain results in the accumulation of high levels of delta 6-desaturated fatty acids in transgenic tobacco. *Proc Natl Acad Sci USA,* 94:4211–4216, 1997.

34. O Suzuki. Recent trends of oleochemicals by biotechnology. Proceedings of World Conference on Oleochemicals into the 21st Century, Champaign, IL, 1990, pp. 221–230.

35. J Zou, V Katavic, EM Giblin, DL Barton, SL MacKenzie, WA Keller, X Hu, DC Taylor. Modification of seed oil content and acyl composition in the brassicaceae by expression of a yeast *sn*-2 acyl-transferase gene. *Plant Cell,* 9:909–923, 1997.

36. P Broun, C Somerville. Accumulation of ricinoleic, lesquerolic, and densipolic acids in seeds of transgenic *Arabidopsis* plants that express a fatty acyl hydroxylase cDNA from castor bean. *Plant Physiol,* 113:933–942, 1997.

37. KD Lardizabal, JG Metz, T Sakamoto, WC Hutton, MR Pollard, MW Lassner. Purification of a jojoba embryo wax synthase, cloning of its cDNA, and production of high levels of wax in seeds of transgenic arabidopsis. *Plant Physiol,* 122:645–655, 2000.

38. EB Cahoon, J Shanklin. Substrate-dependent mutant complementation to select fatty acid desaturase variants for metabolic engineering of plant seed oils. *Proc Natl Acad Sci USA,* 97:12350–12355, 2000.

39. T Aki, Y Nagahata, K Ishihara, Y Tanaka, T Morinaga, K Higashiyama, K Akimoto, S Fujikawa, S Kawamoto, S Shigeta, K Ono, O Suzuki. Production of arachidonic acid by filamentous fungus, *Mortierella alliacea* strain YN-15. *J Am Oil Chem Soc,* 78:599–604, 2001.

40. S Jareonkitmongkol, H Kawashima, N Shirasaka, S Shimizu, H Yamada. Production of dihomo-γ-linolenic acid by a Δ5-desaturase-defective mutant of *Mortierella alpina* 1S-4. *Appl Environ Microbiol,* 58:2196–2200, 1992.

41. E Sakuradani, N Kamada, Y Hirano, M Nishihara, H Kawashima, K Akimoto, K Higashiyama, J Ogawa, S Shimizu. Production of 5,8,11-eicosatrienoic acid by a delta 5 and delta 6 desaturation activity-enhanced mutant derived from a delta 12 desaturation activity-defective mutant of *Mortierella alpina* 1S-4. *Appl Microbiol Biotechnol,* 60:281–287, 2002.

42. T Yokochi, D Honda, T Higashihara, T Nakahara. Optimization of docosahexaenoic acid production by *Schizochytrium limacinum* SR21. *Appl Microbiol Biotechnol,* 49:72–76, 1998.

43. K Inagaki, T Aki, Y Fukuda, S Kawamoto, S Shigeta, K Ono, O Suzuki. Identification and expression of a rat fatty acid elongase involved in the biosynthesis of C18 fatty acids. *Biosci Biotechnol Biochem,* 66:613–621, 2002.

44. K Yazawa, K Araki, N Okazaki, K Watanabe, C Ishikawa, A Inoue, N Numao, K Kondo. Production of eicosapentaenoic acid by marine bacteria. *J Biochem (Tokyo),* 103:5–7, 1988.

45. Y Yano, A Nakayama, H Saito, K Ishihara. Production of docosahexaenoic acid by marine bacteria isolated from deep sea fish. *Lipids,* 29:527–528, 1994.

46. K Yazawa. Production of eicosapentaenoic acid from marine bacteria. *Lipids,* 31:S297–300, 1996.

47. M Tanaka, A Ueno, K Kawasaki, I Yumoto, S Ohgiya, T Hoshino, K Ishizaki, H Okuyama, N Morita. Isolation of clustered genes that are notably homologous to the eicosapentaenoic acid biosynthesis gene cluster from the docosahexaenoic acid-producing bacterium *Vibrio marinus* strain MP-1. *Biotechnol Lett,* 21:939–945, 1999.

48. JG Metz, P Roessler, D Facciotti, C Levering, F Dittrich, M Lassner, R Valentine, K Lardizabal, F Domergue, A Yamada, K Yazawa, V Knauf, J Browse. Production of polyunsaturated fatty acids by polyketide synthases in both prokaryotes and eukaryotes. *Science,* 293:290–293, 2001.

49. Y Michinaka, T Aki, K Inagaki, H Higashimoto, Y Shimada, T Nakajima, T Shimauchi, K Ono, O Suzuki. Production of polyunsaturated fatty acids by genetic engineering of yeast. *J Oleo Sci,* 50:359–365, 2001.

50. LV Michaelson, JA Napier, M Lewis, G Griffiths, CM Lazarus, AK Stobart. Functional identification of a fatty acid delta 5 desaturase gene from *Caenorhabditis elegans. FEBS Lett,* 439:215–218, 1998.

51. HP Cho, MT Nakamura, SD Clarke. Cloning, expression, and nutritional regulation of the mammalian Delta-6 desaturase. *J Biol Chem,* 274:471–477, 1999.

52. T Aki, Y Shimada, K Inagaki, H Higashimoto, S Kawamoto, S Shigeta, K Ono, O Suzuki. Molecular cloning and functional characterization of rat Δ-6 fatty acid desaturase. *Biochem Biophys Res Commun,* 255:575–579, 1999.

53. Y Nojima, T Yagi, T Miyakawa, H Matsuzaki, T Hatano, S Fukui. Extracellular formation of triglycerides from glucose by a mutant strain of *Tricosporon. J Ferment Bioeng,* 80:88–90, 1995.

54. Y Michinaka, T Shimauchi, T Aki, T Nakajima, S Kawamoto, S Shigeta, O Suzuki, K Ono. Extracellular secretion of free fatty acids by disruption of a fatty acyl-CoA synthetase gene in *Saccharomyces cerevisiae. J Biosci Bioeng,* 95:435–440, 2003.

55. NJ Færgeman, PN Black, XD Zhao, J Knudsen, CC DiRusso. The Acyl-CoA synthetases encoded within *FAA1* and *FAA4* in *Saccharomyces cerevisiae* function as components of the fatty acid transport system linking import, activation, and intracellular utilization. *J Biol Chem,* 276:37051–37059, 2001.

16

Low-Calorie Fat Substitutes: Synthesis and Analysis

Ki-Teak Lee

Thomas A. Foglia

Jeung-Hee Lee

16.1 Lipid Digestion/Absorption/Energy

Most dietary fat calories come from the ingestion of triacylglycerols (TAGs) since they are the most abundant lipid forms in both plants and animals. Because most lipids, including TAGs, are practically water-insoluble, fat digestion must be aided by bile salts to form a coarse emulsion in the small intestine. As digestion proceeds, the triacylglycerol lipid droplets form micelles, which renders them more accessible to digestion by lipases secreted into the intestine from the pancreas. Hydrolysis of TAGs by pancreatic lipases liberates free fatty acids (FFA) from the sn-1, 3 positions of the TAGs leaving 2-monoacylglycerols as remnants.[1] For the body to use these lipids they must first be absorbed from the small intestine. At the surface of the intestinal tract, the liberated free fatty acids and monoacylglycerols diffuse into the epithelial cells where TAGs re-synthesis occurs. The newly synthesized TAGs are transported as chylomicra into the blood stream via the lymphatic system and transported to the liver, adipose tissue, and the muscular and other organs for oxidation or storage.[2,3] Because lipids in general have higher caloric content (~9 Kcal/gram) than proteins or carbohydrates, they are the main fuel for the body. For fatty acids with carbon-chain length <12, however, their metabolism is quite different. Short- and medium-chain fatty acids of C2 to C12 in carbon chain length have a caloric value of ~5 Kcal/gram or less due to their shorter chain length. This characteristic makes such fatty acids attractive for use in low-calorie applications. Because their adsorption occurs by transport through the portal system, they are metabolized more rapidly than long-chain fatty acid and because of this are utilized also in dietary energy supplements.[4–6]

16.2 Trends in Low-Calorie Fats

Generally, fats and oils contribute palatability, flavor, creaminess, and mouthfeel to foods, transmit heat rapidly and uniformly, evaporate moisture, and provide crisp texture in fried foods. Despite these attractive properties, researchers have sought ways to reduce the lipid content of foods with other materials as a way of reducing their caloric content. The ideal fat substitute should impart most attributes of a fat while reducing the calorie content of the food, which is not an easy task. Several types of fat substitutes have been developed, that chemically and physically resemble fats and oils and generally fall into three broad categories: carbohydrates, proteins, and fat analogs. Carbohydrates and proteins are fat mimetics that interact with water to provide some of the functionality of a fat. Carbohydrate-based fat substitutes can be made from gums, algins, starches, or fibers. They form a gel with water and provide the mouthfeel (creaminess) and body of fat but are not stable when heated. Oatrim (Beta-Trim®, TrimChoice®) is one example and is made from enzyme-treated oat flour. It can be used in place of conventional fats in food manufacturing for baked products, processed meat products, spreads, soups, and frozen desserts.[7]

Protein-based fat substitutes can be prepared by heating and blending milk and egg white protein or from a mixture of egg whites, whey protein, and xanthan gum. Mixtures of such materials when subjected to high shear with heat form micro spheroidal particles of <3.0 μm diameter. Such shaped particles imitate the texture characteristics of fats, creating a creamy feeling. One example is Simplesse®, which is made from whey and egg white. Their usefulness is similar to carbohydrate products and they are suited for use in dairy products such as yogurt, cheese, and cream products, but not for baking and frying applications.[7–9]

True fat subanalogs are hydrophobic substances with molecular structures similar to conventional fats and oils, and can replace the full functionality of fat. The fat-based substitutes that have received attention include the sucrose polyester known as Olestra™ (now known as Olean™). Olestra is made by esterifying from 6 to 8 fatty acids to a sucrose molecule to produce a mixture of the hexa to octa fatty acyl esters of sucrose. This family of sugar esters has similar taste and texture characteristics of a fat or oil and performs like a conventional fat in baking and frying applications. Pancreatic lipases, however, are not able to hydrolyze sucrose esters due to their relatively large molecular size, and thus the esters are nondigestible and, therefore, provide no calories.

Another class of lipid-based fat substitutes is a family of tailored fats that are characterized by being composed of a combination of short- and long-chain fatty acyl groups. The most familiar class of these low-calorie type fats is the so-called Benefat™/Salatrim™ family, which is obtained by base-catalyzed interesterification of a highly hydrogenated vegetable oil with triacetin, tripropriorin and/or tributyrin.[10] Such low-calorie fats are mainly designed for use in selected nutritional applications,[11] and as such are characterized by a combination of short-chain (C_{2-4}) and long-chain (C_{16-22}) acyl residues into a single triacylglycerol structure. Interest in these classes of lipids stems from the fact that they contain only 5 cal/g[10,11] compared with the 9 cal/g of natural fats and oils because of the lower caloric content of short-chain acyl residues (C_{2-4}) compared with their long-chain (C_{16-22}) counterparts. Such triacylglycerols have the functional properties of ordinary fats but are metabolized differently since the short-chain fatty acids are transported to the liver directly through the portal vein and metabolized while the saturated long-chain fatty acids are less bioavaiable.[10,11] These tailored fats contain randomly distributed acyl groups on the glycerol backbone and are composed of a mixture of two types of triacylglycerol (TAG) structures; the first contain two short-chain and one long-chain acyl residues (SSL-TAG), and the second contain two long-chain and one short-chain acyl residues (LLS-TAG). By predetermining the fatty acid composition and ratio of SSL- and LLS-TAG, it is possible to produce a range of products intended for use in baking chips, coatings, dips, baked products, or as cocoa butter substitutes.[7–9]

16.3 Lipase-Catalyzed Production of Structured Lipids (SLs) as Low-Calorie Fats

16.3.1 Production of Low-Calorie SLs

Structured lipids (SLs) are triacylglycerols (TAG) that have been chemically or enzymatically modified by rearrangement of the native fatty acid triacylglycerol distribution, by the introduction of new fatty acids, and/or by the redistribution of fatty acids on the glycerol backbone. For nutritional purposes, SLs are synthesized for use as functional foods, as infant formulas, as dietary supplements, and for the treatment for disease or maintenance of good health (nutraceuticals). Physical properties of SL are important when intended for applications such as spreads, cooking and baking fats, frying oils, creams, etc.[12,13]

To attain the maximum benefits of SLs, the structural/compositional modification of TAG often is carried out with a selective lipase, since chemical modification does not selectively replace fatty acyl residues because of its random nature. After modification, the nutritional, physical, and chemical characteristics, such as digestibility, titre, solid fat content, or iodine value are improved or changed from the starting TAG. Thus, SLs with defined characteristics can be used to provide specific metabolic effects for nutritive or pharmaceutical purposes, and hence hold promise for wider usage in nutritional, medical, or food applications.[14] A variety of fatty acids are used in the synthesis of SL, taking advantage of the functions and properties of each to obtain maximum benefits for a given SL. These fatty acids include short, medium, saturated, monounsaturated, and polyunsaturated fatty acids (PUFA).[15] SLs produced from short- and long-chain fatty acids are designated as low-calorie fats with intended use as coatings and confectionary fats.[16] The enzymatic combination of a medium-chain and saturated or unsaturated long-chain fatty acid into a single SL species has received recent interest because of their advantages in parenteral and enteral nutrition.[17] Another active area of SL research is the synthesis of SLs containing n-3 PUFA and medium-chain triacylglycerols (MCT), which have been shown to have the ability to improve immune function and reduce cholesterol concentration.[18]

Our goal was to construct SLs that are intended for applications as low-calorie fats that have been modified by enzymatic enhancement alone or in combination with other processes such as fat fractionation or interesterification. These lipase-mediated processes are conducted under mild conditions and exclude toxic chemicals in their syntheses. Hence, such products could be considered more natural than those obtained by chemical routes, making lipase-catalyzed routes potentially advantageous processes.[12]

To produce SLs by lipase-catalyzed reaction, the fatty acyl substrates to be placed on the glycerol backbone should be selected depending on the desired purposes of the SL. As acyl moieties for low-calorie SLs, short-, medium-chain, and saturated long-chain acyl moieties that are longer than C18 are suitable. It is known that saturated short-chain fatty acids provide about 3.5–6 Kcal/g whereas unsaturated long-chain fatty acids provide about 9–9.5 Kcal/g. Saturated long-chain fatty acids, C18, however, are considered less bioavailable due to their low absorbability by the body.[19–21] For example, Mattson et al.[22] reported that the absorbability of stearic acid (C18) was dependent on its positional distribution in the TAG structure. They found that the highest rate of absorption was observed when the stearoyl residue was located at the *sn*-2 position of the TAG structure. Accordingly, positional distribution as well as degree of acyl saturation should be considered when low-calorie TAG SLs are designed. In addition, stearic acid is not thought to raise blood cholesterol levels (unlike palmitic acid) because it is rapidly converted to oleic acid *in vivo*.[23] Recently, there has been much interest in low-calorie fats as functional food ingredients. Thus, TAG containing short-, medium-, or saturated long-chain acyl moieties are ideal components for low-calorie TAG molecules. As mentioned, short (C2–C6), and medium- (C8–C12) chain acyl moieties usually provide less energy than long- (≥C18) chain acyl moieties. Saturated medium-chain fatty acids (MCFA) containing 8 to 12 carbon atoms are often used because of their lower caloric content compared to saturated long-chain fatty acids (LCFA). Because the adsorption of MCFA occurs by transport through the portal system, they are metabolized more rapidly than LCFA and because of this are utilized in dietary energy supplements. Examples of the dietary use of MCFA include the treatment of fat malabsorption disorders, hyperlipidemia, obesity, and diabetes.[21]

One example of SL lipids as low-calorie fats is the SALATRIM™ family of TAG, which is produced by the chemical interesterification of highly hydrogenated vegetable oils with short-chain triacylglycerols. Thus, the major fatty acid components of SL are saturated short-chain fatty acids (C2: acetic acid—C4: butyric acid) and saturated long-chain fatty acids (C16: palmitic acid; C18: stearic acid).[10,11] It is possible that low-calorie SL molecules composed of these short- and long-chain acyl species can be produced by enzymatic reaction.[13] In the lipase-catalyzed reaction, acidolysis was conducted because lipase-catalyzed reaction products from short-chain TAG and long-chain TAG provide LLS (1,2-long-3-short), SLL, and SLS-TAG species similar to those produced in the chemical reaction. Through acidolysis, stearoyl groups are selectively incorporated into the 1 and/or 3 positions in triacetin glycerol backbone using a 1,3 positional lipase to produce low-calorie SL species. Similar type SLs also can be produced by interesterification of the medium-chain triacylglycerols (MLCT) tributyrin, tricaproin, and tricaprylin with stearic acid or hydrogenated soybean oil to produce a series of MLCT triacylglycerols.[24] For these reactions, lipases from *Mucor miehei* (immobilized on a macroporous anion exchange resin [Novozyme IM60]) or *Carica papaya* latex were used because they exhibit a high 1,3 selectivity in lipid esterification and acidolysis reactions.

Other low-calorie SLs can be produced from edible fats and oils. The literature abounds in the production of mixed medium-chain long-chain triacylglycerols (MLCT) produced from medium-chain fatty acids or triglycerides with naturally occurring fats and oils and is extensively reviewed elsewhere.[12,13,24] On the other hand, animal fats in general are of dietary concern because of their relatively high long-chain (C16 and C18 carbon atoms) saturated fatty acid (SFA) content and have received limited attention for the production of structured lipids as low-calorie fats.[25–27] Poultry fats such as chicken fat, however, can be considered a source of monounsaturated fatty acids (MUFA) since they constitute 45–50% of its constitutive fatty acids, while tallow contains only 30–40% MUFA.[25,27] Monounsaturated fatty acids such as oleic acid (*cis*-9-octadecenoic acid) are known to reduce blood cholesterol levels in nonhypertriglycerimic individuals.[28] Among vegetable oils, those of olive, sunflower, and canola being identified as rich sources of MUFA, with the latter type fatty acids constituting from 50% to 80% of their fatty acid composition. Because of the importance placed on dietary MUFA, it has been recommended that MUFA intake be as high as half of the total recommended dietary intake of calories from fat (30%) as a means for reducing the risk of coronary artery disease.[29] Therefore, when low-calorie SLs are designed with animal fats, it is desirable to reduce the saturated long-chain fatty acid (LCFA) content of the starting fat and increase the MUFA and PUFA content as one means of improving the nutritional quality of the targeted structured lipid. Thus, chicken fat can be fractionated to produce MUFA-enriched triacylglycerol (TAG) fractions (MUFA-TAG). Selected liquid MUFA-TAG fractions were then subjected to enzyme-catalyzed acidolysis with a medium-chain fatty acid.[30] This was done to produce SLs that combine the beneficial dietary effects of the MUFA and MCFA classes of fatty acids. It is well known that vegetable oil has a high content of unsaturated fatty acids such as oleic and linoleic acid. Because oleic acid is one of the major fatty acids in peanut oil, peanut oil with medium-chain fatty acid was esterified to produce the low calorie SL.

16.3.2 Analysis of Mixed Medium-, Long-Chain Containing Triacylglycerols

Once synthesized SLs can be separated, quantified, and characterized by normal-phase HPLC (cyanopropyl [$NP_{CN)}$ or silica NP_{Sil}] columns) using evaporative light scattering detection (ELSD) or mass spectrometric (MS) detection and/or high-temperature GC.[31,32] Typical HPLC and GC chromatograms for an MLCT are shown in Figure 16.1a and Figure 16.1b, respectively. Figure 16.1a shows the normal phase cyanopropyl (NP_{CN}-HPLC) chromatogram for the crude MLCT product obtained by lipase-catalyzed acidolysis of tributyrin with stearic acid. This HPLC method, which utilizes ELSD, was developed to separate mixtures of mixed short- and long-chain TAG molecular species.[31] In this instance, the MLCT product was composed of two TAG types, corresponding to medium-long-long chain (MLL) and medium-medium-long (MML)-TAG species, peaks 2 and 3 in Figure 16.1a, that were readily separated from each other and the starting materials. After removing the starting substrates by column chromatography, the composition and homogeneity of this MLCT product was verified by high-temperature GC (Figure 16.1b), which also established the presence of MML and MLL-TAG species. It is interesting to note that the elution orders of the two MLCT species are reversed in Figure 16.1a and Figure 16.1b because in HPLC the separation is based on polarity of the TAG species, whereas in GC the separation is based on molecular weight. Neither the NP_{CN}-HPLC nor GC method, however, resolved the two TAG species into their respective 1,3- or 1,2-TAG isomers. Because regioisomeric MLCT can have different physical and nutritional properties, it is important that the isomeric TAG composition present in an MLCT product be identified and characterized.

Figure 16.2a shows the normal phase silica (NP_{Si}-HPLC) chromatogram of the same MLCT product discussed above. This HPLC method[33] not only resolves the LML- and MML-TAG (peaks 1 and 4, respectively, in Figure 16.2a) but also separates the LML- and LLM–TAG isomers, peaks 1 and 2 in Figure 16.2a. Because the lipase used in the acidolysis reaction is regarded as being *sn*-1,3-regioselective, only

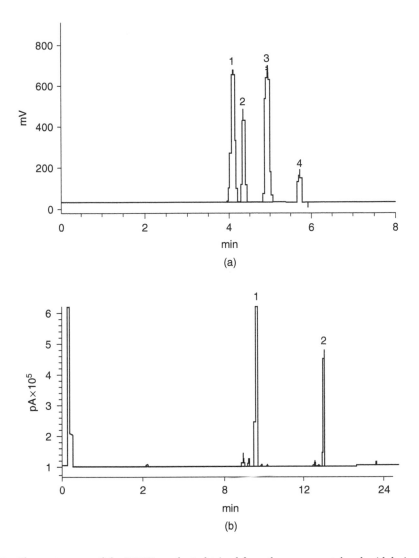

FIGURE 16.1 Chromatograms of the MLCT product obtained from the enzyme-catalyzed acidolysis of tributyrin with stearic acid at a 1:1.2 mole ratio. Panel A is the NP_{CN}-HPLC chromatogram of the crude MLCT product. Peak number: 1 = stearic acid; 2 = LML-TAG; 3 = LMM-TAG; 4 = tributyrin, respectively. Panel B is the HTGC chromatogram of the purified MLCT product. Peak number: 1 = LMM-TAG; 2 = LML-TAG. MML-TAG contain two-butyroyl and one-stearoyl acyl residues; LML-TAG contain two-stearoyl and one-butyroyl acyl residues.

MML- and LML-TAG structures were expected (Scheme 16.1). Coupling the HPLC to a mass detector (LC-MS), however, confirmed that peak 2 in Figure 16.2a was indeed an LLM-TAG species. The single ion chromatogram for peak 2 in Figure 16.2b produced two major ion fragments m/z 411 and 607 confirming that peak 2, Figure 16.2a, was indeed the 1,2-LLM-TAG isomer.[33] Formation of this isomer suggests that either the lipase is not totally rigorous in its selectivity or that acyl migration occurred during synthesis. Since there was no indication of the corresponding MLM positional isomer for the MML-TAG (peak 4), it is concluded that the LLM-TAG (peak 2, Figure 16.1b) resulted from acyl migration. Suppression of acyl migration can be diminished by adjusting several reaction parameters, such as substrate type, mole ratio of reactants, water activity of the reaction medium, shorter reaction times, higher enzyme loading, and lower reaction temperatures.[34]

Figure 16.3 shows the time courses of MLCT production for the enzyme-catalyzed reactions conducted in this study, which were obtained by using NP_{Si}-HPLC. At a 1:0.3 mole ratio of stearic acid:tributyrin

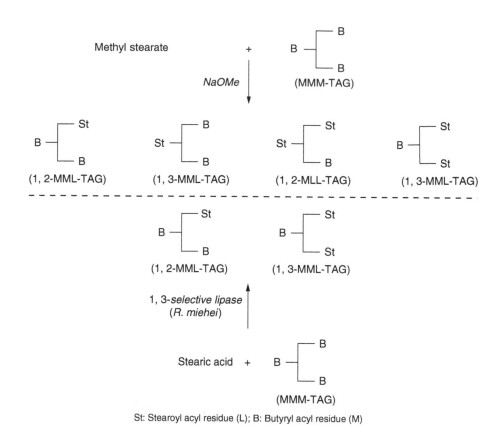

St: Stearoyl acyl residue (L); B: Butyryl acyl residue (M)

SCHEME 16.1 Triacylglycerols produced by lipase-catalyzed acidolysis of tributyrin with stearic acid or by chemical interesterification of tributyrin with methyl stearate. Abbreviations used are: (M) medium-chain; (B) butyryl acyl residue; (L) long-chain; (St) stearoyl acylresidue.

the bulk of TAG production occurred within the first 3 h of reaction (Figure 16.3a) with the reaction reaching equilibrium after 5 h reaction. During the 2–4 h reaction period, the amount of the initially formed MML-TAG decreased while there was a concomitant increase in the amount of LML-TAG in the reaction system. Time course curves for the reaction conducted using a 1:0.6 mole ratio of reactants (Figure 16.3b) were similar to those of the 1:0.3 mole ratio reaction with the exception that about 20% more MML-TAG species were produced. For the 1:1.2 stearic acid:tributyrin reaction (Figure 16.3c), the population of TAG in the MLCT were predominately MML-TAG species. These results show that by changing the mole ratio of reactants one can manipulate the TAG distribution (MML- or LML-TAG) in the final MLCT product and accordingly tailor its final properties to suit its intended application.

After separating unreacted starting substrates from the MLCT products by column chromatography, the distribution of TAG types (MML- and LML-TAG) present in the purified MLCT products obtained by enzyme-catalyzed reaction using various starting substrate mole ratios was determined (Table 16.1). At a 1:0.3 mole ratio (stearic acid:tributyrin) the MLCT product contained 45% MML-TAG, which increased with increased amounts of tributyrin in the starting substrate mixture. For example, at a 1:1.8 mole ratio (stearic acid: tributyrin) the final MLCT product contained 84% MML-TAG. In parallel, the percentage of LML-TAG in the final MLCT product decreased with increased amounts of tributyrin since larger molar amounts of tributyrin increases the probability of producing MML-TAG during reaction. Overall, the production of LML-TAG decreased from 55% to 15% in the MLT product over the mole ratio of reactants used in this study (Table 16.1).

The distribution of TAG species in the MLCT products produced by chemical transesterification is listed in Table 16.1. Transesterification of tributyrin with methyl stearate initially produces an MML-

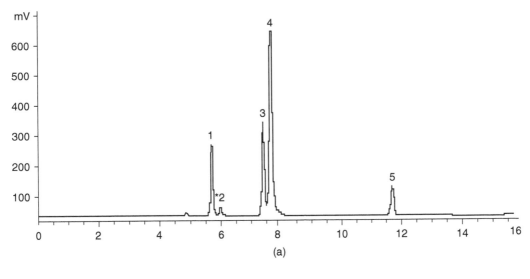

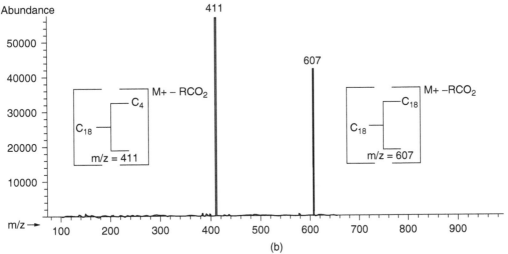

FIGURE 16.2 Chromatogram of the MLCT product obtained from the enzyme-catalyzed acidolysis of tributyrin with stearic acid at a 1:1.2 mole ratio. Panel A is the NP_{Si}-HPLC chromatogram of the crude MLCT product. Peak: 1) LML-TAG; 2) LLM-TAG; 3) stearic acid; 4) MML-TAG; 5) tributyrin. MML-TAGs contain two-butyroyl and one-stearoyl acyl residues; LML-TAG contain two-stearoyl and one-butyryl acyl residues. Panel B is the single ion mass spectrum of peak 2 in the NP_{Si}-HPLC chromatogram of Figure 16.2a. The two major fragmentation ions at *m/z* 411 and 607 represent the $[M-RCO_2]^+$ ions that contain one-butyryl and one stearoyl acyl residue and two stearoyl acyl residues, respectively.

TAG species, which is subsequently converted to an MLL-TAG species.[13] The final distribution of TAG types, however, is determined from the initial mole ratio of reactants used, which was as found in the enzyme-catalyzed reactions. At a methyl stearate:tributyrin mole ratio of 1:3, the purified MLCT product was composed of >99% and <1% of MML- and MLL-TAG species, respectively, as determined by NP_{CN}-HPLC. In contrast, when the mole ratio of methyl stearate:tributyrin used was 3:1 the MLCT product was dominated by MLL-TAG species (Table 16.2). Analysis of the latter MLCT product by NP_{Si}-HPLC showed that the distribution of 1,3 and 1,2-MLL-TAG isomers was 66.6% and 33.3%, respectively. This result is in agreement with that which would be theoretically predicted based on the random nature of chemical transesterification.[12,13,15] Because of the higher bioavailability of stearic acid at the 2-position

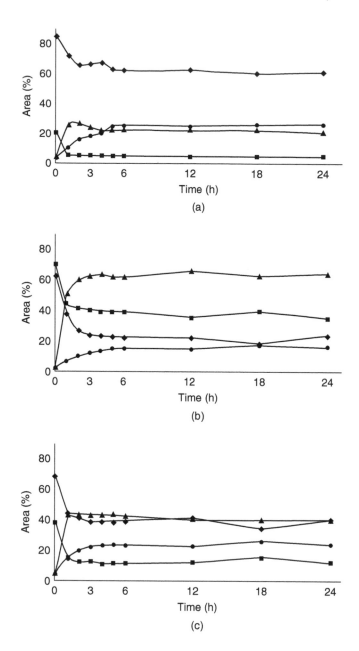

FIGURE 16.3 Progress curves for the production of MML-TAG and MLL-TAG in the enzyme-catalyzed acidolysis of tributyrin with stearic acid at various mole ratios. Panel A: acidolysis at a mole ratio of 1.0:0.3; Panel B: acidolysis at a mole ratio of 1.0:0.6; and Panel C: acidolysis at a mole ratio of 1:1.2. Symbols are: ■, stearic acid; ♦, tributyrin; ▲, MML-TAG; and, ●, MLL-TAG. MML-TAGs contain of two-butyryl and one-stearoyl acyl residues; LML-TAGs contain two-stearoyl and one-butyryl acyl residues.

of TAG, it is best to minimize the presence of MLM- and MLL-TAG species in MLCT that are intended for use in low-calorie applications.

Representative DSC melting profiles for the MLCT prepared in this study are presented in Figure 16.4. For the MLCT product obtained from chemical transesterification that contained 98% of its TAG species as MML-TAG (mole ratio of methyl stearate to tributyrin of 1:3, Table 16.2), the major exothermic peak is located around 17°C (Figure 16.4a). This exothermic peak is broadened since the MLCT is a 2:1 mixture

TABLE 16.1 Distribution of MML-TAG (Triacylglycerol with Two Butyryl- and One Stearoyl Acyl Groups) and LML-TAG (Triacylglycerol with Two Stearoyl- and One Butyryl Acyl Groups) in Mixed Medium- and Long-Chain Triacylglycerol (MLCT) Products Produced by Enzymatic Acidolysis of Tributyrin with Stearic Acid at Different Substrate Mole Ratios

Stearic Acid: Tributyrin Mole Ratio		Reactant Area %[a]	Crude MLCT Area %[a]	Purified MLCT Area %[b]
1:0.3	Tributyrin	16.5	0.8	nd[c]
	MML–TAG[d]	—	18.7	45
	LML–TAG[d]	—	22.1	55
	Stearic acid	83.5	58.4	nd
1:0.6	Tributyrin	35.4	10.8	0.1
	MML–TAG	—	36.8	55.2
	LML–TAG	—	21.7	44.7
	Stearic acid	65.5	30.7	0.02
1:1.2	Tributyrin	51.8	28.7	1.4
	MML–TAG	—	47.2	84.8
	LML–TAG	—	11.5	13.8
	Stearic acid	48.2	12.6	nd
1:1.8	Tributyrin	65.4	42.3	0.8
	MML–TAG	—	43.6	83.7
	LML–TAG	—	6.3	15.5
	Stearic acid	34.6	7.7	nd

[a]Evaporative light scattering detector (ELSD) response of crude MLCT mixture after separation by NP_{CN}-HPLC.
[b]ELSD response of MLCT product after Florisil column chromatography and separation by NP_{CN}-HPLC.
[c]nd: not detected.
[d]MML-TAG: triacylglycerols composed of two butyryl- and one stearoyl acyl groups; LML-TAG: triacylglycerols composed of two stearoyl- and one butyryl acyl groups.

of 1,3- and 1,2-MML-TAG isomers. In contrast, the MLCT obtained by lipase-catalyzed acidolysis of tributyrin with stearic acid (1:0.3 substrate mole ratio, Table 16.1), which was a 45:55 mix of MML- and LM-TAG (Table 16.1), showed two distinct melting exothermic peaks at 43°C and 33°C (Figure 16.4b), which suggests the presence of two distinct TAG species in this MLCT product. Since butyric acid has a lower melting point (m.p −7.9°C) than stearic acid (m.p 69.6°C),[25] it is thought that the higher exothermic peak represents melting of the LML-TAG species and the lower exothermic peak represents melting of the MML-TAG species. For the purified MLCT products containing increased amounts of MML-TAG species, however, the observed exothermic peaks were not distinctly separated (Figure 16.4c and Figure 16.4d). For example, the melting profile for the MLCT product obtained from the reaction run at a 1:1.2 mole ratio of reactants, which contained 80% MML-TAG species (Table 16.1), two overlapping exothermic peaks are observed at 15°C and 18°C.

The distribution of TAG species within a fat or oil can affect its melting/solidification profile. When pure solid tripalmitin (DCS crystallization exothermic peak at ≈45°C) is mixed in sesame oil, it co-crystallizes with the solid TAG species present in sesame oil, causing a shifting and broadening of the tripalmitin exothermic peak to a lower temperature. Because of this, it was suggested that tripalmitin was solubilized into the sesame oil to form a homogeneous tripalmitin-sesame oil solution.[35] As the concentration of tripalmitin in sesame oil is increased, however, tripalmitin tends to crystallize independently from the sesame oil TAG, forming two new shifted and broadened exothermic peaks. Thus, the melting/crystallization behaviors of mixed TAG are affected by both the variation in and concentration of individual TAG species that may be present in the fat or oil. Because the distribution of MML- and LML-TAG in the MLCT products obtained from the 1:1.2 and 1:1.6 mole ratio reactions was similar (Table 16.1), only the MLCT product from the 1:1.2 mole ratio reaction was analyzed by DSC (Figure 16.4d). As expected, this MLCT had a lower exothermic peak than the MLCT product form the 1:0.6 mole ratio

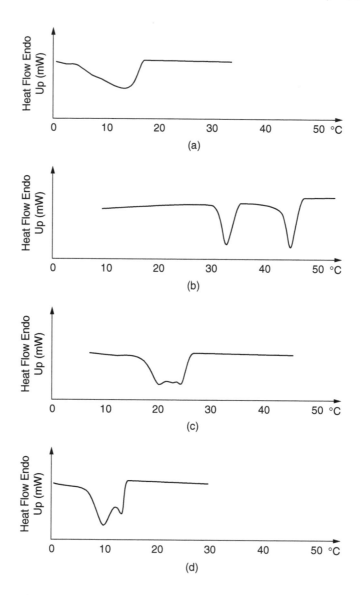

FIGURE 16.4 DSC cooling curves of purified structured MLCT products. Panel A: MLCT product from chemical transesterification of methyl stearate and tributyrin at a mole ratio of 1:3. Panel B, C, and D: MLCT product from enzymatic acidolysis of tributyrin with stearic acid at mole ratio of 1:0.3, 1:0.6, and 1:1.2, respectively.

reaction because of its higher content of MML-TAG. Therefore, it appears that the higher LML-TAG content of the MLCT product obtained from the 1:0.3 mole ratio reaction results in two distinct exo-thermic peaks, which gradually shift to lower temperatures and tend to overlap as the LML-TAG content in the MLCT decreases.

Melting profiles of the low-calorie SL TAG were determined by differential scanning calorimetry (DSC). The samples were tempered at 100°C for 10 min and the cooling curves obtained by cooling at 20°C/min until reaching –20°C. Purified SL-TAG molecules from lipase reaction (1:0.3 substrate mole ratio, stearic acid:tributyrin) showed two distinct crystallization occurrences (Figure 16.4b) at 43°C and 33°C, which suggested that co-crystallization did not occur. It was thought that the peak with higher crystallization temperature represented the SLL-TAG species because butyric acid (butanoic acid, m.p = –7.9°C) has a lower melting point than stearic acid (octadecanoic acid, m.p = 69.6°C).[25] Thus, SSL-TAG species have lower crystallization temperature than SLL-TAG. However, the purified SL-TAG product from the 1:0.6 and 1:1.2 reactions, Figure 16.4c and Figure 16.4d, respectively, had exotherm peaks that were not distinctly separated. The crystallization pattern of the 1:0.6 SL product gave two exotherm peaks at 25°C and 20°C. Therefore, low-calorie TAG molecules produced from lipase-catalyzed reaction should be less bioavailable, allowing their use as a low-calorie functional ingredient in foods.

TAG species also were prepared by interesterification of either tricaproin (a C_{18}-TAG) or tricaprylin (a C_{24}-TAG) with stearic acid (C_{18} or hydrogenated soybean oil (a C_{52-54}-TAG) as low-calorie SLs. The reactions were conducted for 24 hr at 65°C. A screw-cap vial in water-jacketed beaker with magnetic stirring at 200 rpm was used as the reactor. After the reaction, each resulting product was mixed with 100 mL hexane and 3 mL aliquot was filtered through disposable Fluoropore PTFE membrane filter. The products after reaction were applied to a florisil chromatography column to obtain the TAG molecules. An aliquot was applied to the plate and developed by hexane: diethyl ether: acetic acid (50:50:1, v/v/v). The bands were visualized with ultraviolet light after spraying with 0.2% 2,7-dichlorofluorescein in methanol. The fractions containing TAG were pooled, evaporated under nitrogen, and the purified TAG analyzed by HPLC. A Hewlett-Packard Model 1050 HPLC equipped with a Varex ELSD II mass detector was used for analysis. A Phenomenex cyanopropyl (IB-Sil, 5 micron, 250 × 4.6 mm i.d.) column with accompanying guard column (30 × 4.6 mm i.d.) was used for the NP_{CN}-HPLC method and the Chromegasphere (Si60, 3 micron, 250 × 4.6 mm i.d., ES Industries) was used for the NP_{SIL}-HPLC method. For the NP_{CN}-HPLC method, the binary mobile phase gradient of hexane and MTBE (each fortified with 0.4% acetic acid) was used at a flow rate of 1.0 mL min^{-1}. The gradient conditions used were as reported by Foglia and Jones.[36] For the NP_{SIL}-HPLC method, a binary solvent system of 15% methyl-t-butyl ether and 85% hexane (each fortified with 0.4% acetic acid) was used as mobile phase. The mobile phase flow was 0.75 mL min^{-1}. Nonaqueous reverse-phase HPLC was conducted using a Beckman/Altex Ultrasphere ODS 5 μm (4.6 mm × 25 cm) column on a Hewlett-Packard Model 1050 HPLC equipped with a Varex (Burtonville, MD) ELSD MK III. The mass spectrometer (HP Model HP5989 A quadrupole) coupled to a HP1050 HPLC via direct liquid APCI interface HP5998A was operated in the positive ion mode.[31,36]

Theoretically, in this study, MML (1,2-medium-3-long-triacylglycerols) and LML (1-long-2-medium-3-long) SL-TAG molecules can be synthesized by lipase-catalyzed esterification of S(M)CT and long-chain fatty acid such as stearic acid. When hydrogenated soybean oil and S(M)CT are esterified, a more complex combination of SL-TAG molecules can be expected such as S(M)S(M)L, LS(M)S(M), LS(M)L, LLS(M), S(M)LL, and S(M)LS(M), as shown in Scheme 16.2. The chromatogram showing the separation of the SL-TAG species in the SL product obtained from the acidolysis of tributyrin with stearic acid is shown in Figure 16.5. In Figure 16.5a, the SL-TAG species containing one short (butyroyl moiety; C4) with two long (stearoyl moieties; C18) acyl residues (peak # 2), and SL-TAG containing two short and one long acyl residues (peak # 3) were readily resolved. Previously, NP_{CN} with binary mobile phase gradient using hexane and MTBE (both fortified by 0.4% acetic acid) was successfully applied to separate partially purified SL-TAG molecules and unreacted TAG substrates. However, this system could not further separate SL positional isomers. Because each SL-TAG species generally has different nutritional properties, it is important to separate all SL-TAG isomers. Accordingly, a more detailed separation of these isomers could be obtained using a normal-phase silica column (NP_{SIL}). The NP_{SIL} system using an isocratic mobile phase of 85% hexane: 15% MTBE, both fortified with 0.4% acetic acid that successfully separated isomers containing short- (C4) and long-chain (C18) fatty acid (Figure 16.5b). Positional isomers, LSL (1-stearoyl-2-butyroyl-3-stearoyl) and LLS/ (1,2 distearoyl-3-butyroyl glycerol), could be well separated by silica column under the HPLC conditions used. A minor peak designated by an asterisk on the chromatograph in Figure 16.5b was LLS/SLL isomers, which might be produced due to acyl migration.

Comparing the tributyrin and stearic acid SL product chromatogram in Figure 16.5b with chromatogram of the SL product obtained from lipase-catalyzed interesterification of tributyrin and hydrogenated soybean oil in Figure 16.6a, Figure 16.5b indicated the presence of less LLS/SL-TAG species than those of hydrogenated soybean oil, suggesting that lipase-catalyzed reaction successfully allowed the positional (1,3) selective esterification (Scheme 16.2). In addition, SL-TAG products from tributyrin and hydroge-

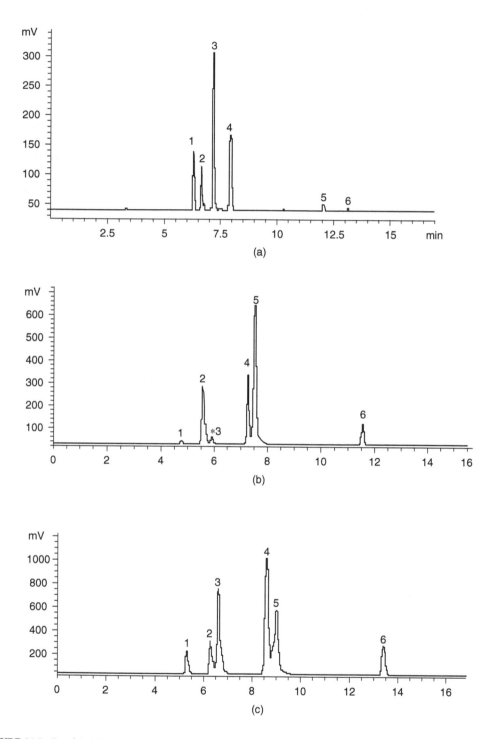

FIGURE 16.5 Panel A: NP_{CN} HPLC chromatograph of crude products of SL. SL was produced from tributyrin and stearic acid. 1 = stearic acid, 2 = LSL, 3 = SSL/LSS, 4 = tributyrin, 5 = 1,3 diacylglycerols, 6 = 1,2 diacylglycerols. Panel B: NP_{SIL} HPLC chromatography of crude products of SL. SL was produced from tributyrin and stearic acid. 1 = tristearin, 2 = LSL, 3 = LLS/SLL, 4 = stearic acid, 5 = SSL/LSS, 6 = tributyrin. L and S represent stearoyl and butroyl acyl residues, respectively. Panel C: NP_{SIL} HPLC chromatogram of partially purified SL product produced by the lipase-catalyzed interesterification of tributyrin and hydrogenated soybean oil. Peak number 1 = hydrogenated soybean oil, 2 = LSL, 3 = LLS/SLL, 4 = SSL/LSS, 5 = SLS, 6 = tributyrin.

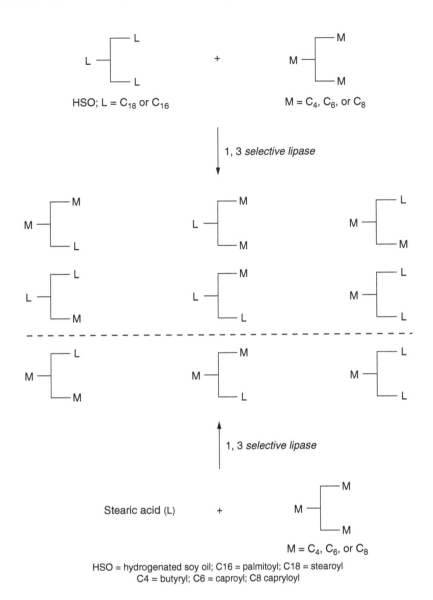

SCHEME 16.2 Triacylglycerols produced by lipase-catalyzed interesterification of medium-chain triacylglycerols [tributyrin (T4), tricaproin T6, or tricaprylin T8] with hydrogenated soybean oil (HSO) with or by acidolysis with stearic acid. L = stearoyl or palmitoyl acyl residue, S/M = butyryl, caproyl, or capryloyl acyl residue, St = stearoyl acyl residue.

nated soybean oil also were well separated into SSL and SLS isomers by HPLC with the silica type column (Figure 16.5c).

Partially purified SL-TAG from tricaproin with hydrogenated soybean oil, tricaprylin with stearic acid, and tricaprylin with hydrogenated soybean oil, in Figures 16.6a, Figure 16.6b, and Figure 16.6c, respectively, also were analyzed for SL isomer content using this NP$_{SI}$ protocol. However, The SL isomers (LSL and LLS or SSL and SLS were not fully resolved (Figure 16.6a) or co-eluted with similar retention times (Figure 16.6b and Figure 16.6c) under the analysis conditions used. These results indicated that the NP$_{SIL}$ HPLC method can successfully separate SL-TAG isomers composed of short- (C4 acyl groups) and long-chain acyl groups (C16-C18) but does not fully resolve the isomers of SL-TAG containing medium- (C6 or C8) and long chain (C18) fatty acids. In addition, the separation of the SL by silica-based HPLC sometimes

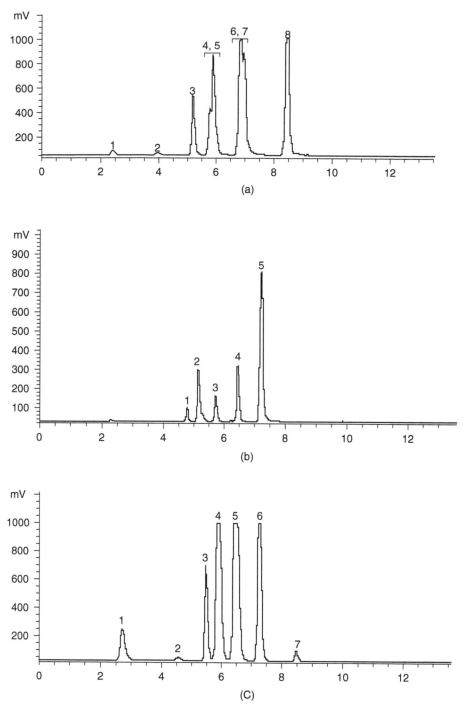

FIGURE 16.6 Panel A) NP_{SIL} HPLC chromatogram of partially purified SL product produced by the lipase-catalyzed interesterification of tricaproin and hydrogenated soybean oil. Peak number 1, 2 = unknown, 3 = hydrogenated soybean oil, 4 = LML, 5 = LLM/SLL, 6 = SSM/LMM, 7 = MLM, 8 = tricaproin. L = stearoyl/palmitoyl acyl residuesand M = butroyl and capoyl acyl residues, respectively. Panel B: NP_{SIL} HPLC chromatogram of crude SL product obtained from the lipase-catalyzed acidolysis of tricaprylin with stearic acid. Peak number: 1 = tristearin, 2 = LML/LLM; 3 = MML/ MLM; 4 = tricaprylin, 5 = stearic acid. Panel C: NP_{SIL} HPLC chromatograph of crude SL product produced from tricaprylin and hydrogenated soybean oil. Peak number: 1, 2 = unknown; 3 = hydrogenated soybean oil; 4 = LML/LLM; 5 = MML/MLM; 6 = tricaprylin; and 7 = stearic acid. L and M represent stearoyl and capryloyl acyl residues, respectively.

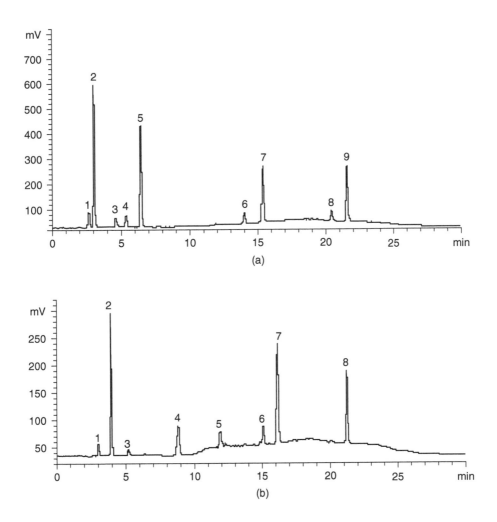

FIGURE 16.7 Panel A: Reverse-phase HPLC chromatogram of partially purified SL produced from tricaproin and hydrogenated soybean oil. Peak number: 1 = diacylglycerols; 2 = tricaproin; 3 = 6-18-6 (1-caproyl-2-stearyl-3-caproyl glycerol); 4 = 6-6-16 (1,2-dicaproyl-3-palmitoyl glycerol); 5 = 6-6-18; 6 = 6-18(16)-18(16); 7 = 18(16)-6-18(16); 8 = 18-16-16/16-18-18; and 9 = tristearin. Panel B: Reverse-phase HPLC chromatograph of partially purified SL produced from tricaprylin and stearic acid. Peak number: 1 = diacylglycerols; 2 = tricaprylin; 3 = 8-18-8 (1-capryloyl-2,3-stearoyl glycerol); 4 = 18-8-8; 5 = unknown; 6 = 18-18-8; 7 = 18-8-18; and 8 = tristearin.

generated unreproducible separations since small amounts of water advantageously introduced into the system altered the silica surface can result in changed elution orders for the SL-TAG species.

Reverse-phase HPLC separated the TAG species according to theoretical carbon number (TCN) and double bonds with relatively polar solvents such as acetonitrile, acetone, propionitrile or water. Although the analysis by RP-HPLC of TAG species containing long-chain saturated acyl groups is limited due to their limited solubility in commonly used mobile phase solvents, reverse-phase HPLC has been widely used to analyze TAG molecular species.[34–37] As seen in Figure 16.7, SL-TAG produced by interesterification of tricaproin and hydrogenated soybean oil (Figure 16.7a) and SL-TAG produced by acidolysis of tricaprylin with stearic acid (Figure 16.7b) were resolved using reverse-phase HPLC. Positional isomers (LSL and LLS/SSL or SSL/LSS and SLS), which were not resolved adequately by NPSIL HPLC, were well resolved. In Figure 16.7a, the peaks from SL-TAG were well grouped into two classes, TAG containing one short-chain (caproic) fatty acid and TAG containing two short-chain fatty acids. When short-chain fatty acids positioned at the *sn*-1 or *sn*-3 position of the TAG molecule, it makes TAG more polar than

TAG containing short-chain fatty acid at the *sn*-2 position, resulting in earlier elution than their positional isomers. The retention times of the 6-18-6 and 6(18)-6-18(6) TAG are 4.6 and 6.4, respectively. S(L)SL(S)-TAG were further separated according to the chain length of the long-chain fatty acid (C16 or C18) because hydrogenated soybean oil contains both palmitoyl (C16:0) and stearoyl (C18:0) acyl residues. SL-TAG with two long-chain fatty acids, however, could not be further separated based on acyl chain length. Rather, they were separated according to the position of short-chain fatty acid in SL-TAG species. In the reaction with stearic acid, less SL-TAG molecular species were expected than from hydrogenated soybean oil acidolysis reactions. In Figure 16.7, the TAG peaks were separated into two groups of SL-TAG molecules, depending on fatty acid composition in acyl group. Besides, SL-TAG isomers from acyl migration (described previously) from 18-8-18 also were fully separated. Accordingly, reverse-phase HPLC is a more suitable method for analysis of SL-TAG species that are produced from medium-chain (C6-C10) and long-chain (C16-C18) fatty acids.

Acidolysis reactions also were conducted by reacting a chicken fat fraction with caprylic acid (approximately 1:2 molar ratio) in the presence of immobilized lipase. Chicken fat before and after fractionation was transesterified. Temperature fractionation of a fat or oil can be regarded as a thermomechanical separation process wherein individual TAG species characteristic for a given fat or oil are selectively crystallized from the melt or liquid phase. During cooling of the liquid oil or melted fat, TAG species with the highest melting points preferentially crystallize, resulting in a solid phase within the liquid phase.[37,38] Most natural fats and oils are a complex mixture of individual TAG that can contain from one to three different fatty acyl residues on their glycerol backbone. Because of this, there is a large variation in the melting points of the TAG species, which complicates the fractionation process.

To produce low-calorie fats from chicken fat a series of acidolysis reactions were run at 35°, 45°, 55°, and 65°C. To determine enzyme stability, the 55°C acidolysis sample was centrifuged, the immobilized lipase recovered, and fresh substrates added for subsequent reactions at 55°C. Immobilization provides several benefits in enzyme-catalyzed reactions.[39,40] Reaction mixtures were separated by TLC and the purified TAG products isolated from the plates with ether. The solvent was removed under nitrogen and the triacylglcyerol converted to FAME for determination of fatty acid composition. From the results of the various fractionation steps, chicken fat and enriched MUFA-TAG fractions were obtained and used for the production of SLs. This was done by lipase-catalyzed acidolysis of the TAG with caprylic acid. After acidolysis, 5.8–8.1% of caprylic acid was incorporated into the TAG molecules of the synthesized

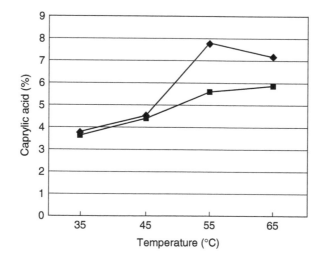

FIGURE 16.8 Effect of temperature on the lipase-catalyzed acidolysis of fractionated chicken fat (fraction-38L20, Table 16.2) with caprylic acid. Mole percent of caprylic acid incorporated into triacylglycerols. Symbols: ◆ immobilized *G. candidum* lipase; ■ immobilized *C. rugosa* lipase.

TABLE 16.2 Fatty Acid Composition (area %) of SL Produced by the Lipase-Catalyzed Acidolysis of Neat or Fractionated Chicken Fat (Liquid TAG fraction isolated at −38°C) with Caprylic Acid

			C. rugosa		G. candidum	
Fatty Acyl Group	Chicken Fat	Fractionated Chicken Fat	SL Product from Chicken Fat	SL Product from Fractionated Chicken Fat	SL Product from Chicken Fat	SL Product from Fractionated Chicken Fat
C8:0	nd	nd	7.6	5.8	8.1	7.8
C14:0	0.7	0.6	0.5	0.5	1.0	0.6
C14:1	0.2	0.5	0.4	0.5	0.3	0.3
C16:0	25.2	6.3	23.5	14.2	23.9	8.2
C16:1	7.8	12.6	7.1	7.9	6.1	9.6
C18:0	5.9	1.4	6.8	3.6	6.2	2.1
C18:1	40.5	44.2	36.3	42.2	37.1	44.4
C18:2	18.4	32.4	16.1	23.8	15.9	25.4
C18:3	0.7	1.4	0.9	0.8	1.0	1.0
C20:0	0.1	—	0.03	nd	nd	nd
C20:1	0.5	0.4	0.7	0.6	0.3	0.5
C20:3	0.1	0.2	0.05	0.1	0.05	0.1

SLs (Table 16.2). The incorporation of caprylic acid into the chicken fat and chicken fat fractions was intentionally kept low (>10%) since the goal was to prepare a SL wherein the long-chain SFA acyl groups at the 1,3 positions of the original TAG species were replaced preferentially with capryloyl residues. The low efficiency of incorporation of caprylic acid (approximately 8%) was not of concern since it can be readily recovered for reuse. The data in Table 16.2 show that the content of several fatty acids in the SL products differed compared to the starting fats. When acidolysis was catalyzed with the immobilized lipase from *G. candidum*, the products from the fractionated chicken fat contained 25.3% more MUFA and 65.7% less palmitic acid, the major SFA in chicken fat, than that of unfractionated chicken fat. Of the 18–24% total SFA in the product from the fractionated chicken fat, caprylic acid constituted 24–42% of the SFA. This is to be compared to the products prepared from unfractionated chicken fat where total SFA content ranged between 38% and 40% with caprylic acid constituting only 20–21% of total SFA. The acidolysis of chicken fat with caprylic acid catalyzed by the immobilized lipases in general was favored at higher temperature (Figure 16.8). The *G. candidum* lipase gave the highest incorporation of caprylic acid at 55°C (7.8%) while for the *C. rugosa* lipase 65°C (5.9% incorporation) was the optimal temperature. At 35°C, the incorporation of caprylic acid decreased by 52.1% and 38.3% compared to the reaction at 55°C for the *G. candidum* and 65°C for the *C. rugosa* lipases, respectively.

16.4 Conclusion

This chapter described several approaches to the production of SLs involving the directed enzymatic interesterification of fats and oils with medium-chain triacylglycerols or selected fatty acids/esters. Other approaches to the synthesis of these predefined lipids include the dry fractionation combined of natural fats and oils so as to provide triacylglycerol fractions rich in nutritionally important fatty acids combined with the directed enzymatic interesterification of these fractions with selected fatty acids. An understanding of the functional properties of and metabolic fate of the component fatty acids incorporated into will aid in the synthesis of new SLs species with beneficial properties. The processes described allow scientists to design SLs for various applications that may include medical applications, foods, and nutritional supports as well as continuing to play a role in enteral and parenteral applications. When Sls are synthesized, the products must first be separated from unreacted substrates, usually by distillation or chromatography, its TAG molecular species composition determined and its physical properties measured for the intended application. Several high-performance liquid chromatograph methods are presented in this chapter that can aid researchers in completing the latter tasks.

Acknowledgments

Ki-Teak Lee thanks the Ministry of Health & Welfare, Republic of Korea, Korea Health 21 R&D Project 02-PJ1-PG1-CH15-0001, for their support of portions of the work described.

References

1. Brody, T. Digestion and Absorption. In *Nutritional Biochemistry*, Academic Press, San Diego, CA, 1999, pp. 57–131.
2. Ober, W.C., Garrison, C.W., Silverthorn, A.C. Cellular Metabolism. In *Human Physiology*, Prentice-Hall Inc., Upper Saddle River, NJ, 1998, pp. 72–98.
3. Kritchevsky, D. Fats and Oils in Human Health. In *Food Lipids, Chemistry, Nutrition, and Biotechnology*, Akoh, C.C. and Min, D.B., Eds., Marcel Dekker Inc., New York, NY, 1998, pp. 449–462.
4. Brisson, G.J. Lipids in Human Nutrition, Jack K. Burgess, Inc., Englewood, NJ, 1981, pp. 65–90.
5. Mead, J.F., Alfin-Slater, R.B., Howton, D.R., Popjak, G. Digestion and Absorption of Lipids. In *Lipids: Chemistry, Biochemistry, and Nutrition*, Plenum Press, New York, 1986, pp. 225–272.
6. Clarke, S.D., Ramsos, D.R., Leveille, G.A. Differential effects of dietary methyl esters of long chain saturated and polyunsaturated fatty acids on rat liver and adipose tissue lipogenesis, *J. Nutr. 107*: 1170–1181 (1977).
7. Akoh, C.C. Lipid-based synthetic fat substitutes. In *Food Lipids, Chemistry, Nutrition, and Biotechnology*, Akoh, C.C. and Min, D.B., Eds., Marcel Dekker Inc., New York, NY, 1998, pp. 559–588.
8. Artz, W.E., Hansen, S.L. Other Fat Substitutes, In *Carbohydrate Polyesters as Fat Substitutes*, Akoh, C.C. and Swanson, B.G., Eds., Marcel Dekker, Inc., New York, NY, 1994, pp. 197–236.
9. Morrison, R.M. Fat Substitutes in Foods: Growing Demand and Potential Markets, In *Carbohydrate Polyesters as Fat Substitutes*, Akoh, C.C. and Swanson, B.G., Eds., Marcel Dekker, Inc., New York, NY, 1994, pp. 237–250.
10. Smith, R.E., Finley, J.W., Leveille, G.A. Overview of SALATRIM, a family of low-calorie fats, *J. Agric. Food Chem. 42*: 432–434 (1994).
11. Klemann, L.P., Finley, J.W., Leveille, G.A. Estimation of the absorption coefficient of stearic acid in SALATRIM. *J. Agric. Food Chem. 42*: 484–488 (1994).
12. Gunstone F.D., *Food Applications of Lipids in Food Lipids*, Akoh, C.C., and Min D. B., Eds. Marcel Decker, Inc. New York, NY, pp. 729–750, 2002.
13. Akoh, C. C., *Structured Lipids in Food Lipids*, Akoh, C.K., and Min D.B., Eds. Marcel Decker, Inc. New York, NY, pp. 877–908, 2002.
14. Fomuso, L.B., Akoh, C.C. Structured lipids: their food applications and physical property testing methods. *Food Sci. Biotechnol. 10*: 690–698 (2001).
15. Akoh, C.C., Lee, K-T., Fomuso, L.B. Synthesis of Positional Isomers of Structured Lipids with Lipases as Biocatalysts. In *Structural Modified Food Fats: Synthesis, Biochemistry, and Use*, Christophe, A.B., Ed.; AOCS Press: Champaign, IL, 1998, pp. 46–72.
16. Fomuso, L.B., Akoh, C.C. Enzymatic modification of Triolein: Incorporation of caproic and butyric acids to produce reduced-calorie structured lipids. *J. Am. Oil Chem. Soc., 74*: 269–272 (1997).
17. Akoh, C.C. Structured lipids-enzymatic approach. *INFORM. 6*: 1055–1061 (1995).
18. Lee, K-T., Akoh, C.C., Dawe, D.L. Effects of structured lipid containing omega-3 and medium chain fatty acids on serum lipids and immunological variables in mice. *J. Food Biochem. 23*: 197–208 (1999).
19. Ikeda, I., Tomari, Y., Sugano, M., Watanabe, S., Nagata, J. Lymphatic absorption of structured glycerolipids containing medium-chain fatty acids and linoleic acid and their effects on cholesterol absorption in rats, *Lipids 26*: 369–373 (1991).
20. Jandacek, R.J., Whiteside, J.A., Holcombe, B.N., Volpenhein, R.A. Taulbee, J.D. The rapid hydrolysis and efficient absorption of triglycerides with octanoic acid in the 1 and 3 position and long chain fatty acid in the 2 position, *Am. J. Clin. Nutr. 45*: 940–945 (1987).

21. Bach, A.C., Babayan. V.K. Medium-chain triglycerides: An update, *Am. J. Clin. Nutr. 36:* 950–961 (1982).

22. Mattson, F.H., Nolen, G.A., Webb, M.R. The absorbability by rats of various triglycerides of stearic and oleic acid and the effect of dietary calcium and magnesium, *J. Nutr. 109:* 1682–1687 (1979).

23. Grundy, S.M., Which Saturated Fatty Acids Raise Plasma Cholesterol Levels? In: *Health Effects of Dietary Fatty Acids;* Nelson, G.J., Ed.; AOCS Press: Champaign IL, 1991, pp. 83–93.

24. Willis, W.M., Lencki, R.W., Marangoni, A.G. Lipid modification strategies in the production of nutritionally functional fats and oils. *Crit. Rev. Food Sci. Nutr. 38(8):* 639–674 (1998).

25. Belitz, H-D., Grosch, W. Lipids. In: *Food Chemistry;* Springer-Verlag Press, New York, 1986, pp. 128–198.

26. Foglia, T.A., Petruso, K., Feairheller, S.H. Enzymatic interesterification of tallow-sunflower oil mixtures, *J. Am. Oil Chem. Soc. 70:* 281–285 (1993).

27. Brockerhoff, h., hoyle, r.j., wolmark, n. positional distribution of fatty acids in triglycerides of animal depot fats, *Biochim. Biophys. Acta. 116:* 67–72 (1966).

28. Mattson, F.H., Grundy, S.M. Comparison of effects of dietary saturated, monounsaturated, and polyunsaturated fatty acids on plasma lipids and lipoproteins in man, *J. Lipid Res. 26:* 194–202 (1985).

29. Nicolosi, R.J., Stucchi, A.F., Loscalzo, J. Effect of dietary fat saturation on low-density lipoprotein metabolism. In: *Health Effects of Dietary Fatty Acids,* G.J. Nelson Ed., AOCS Press, Champaign, IL, 1991, pp. 77–82.

30. Lee, K-T., Foglia, T.A. Fractionation of chicken fat triacylglycerols: Synthesis of structured lipids with immobilized lipases. *J. Food Sci. 65:* 826–831 (2000).

31. Mangos, T.J.; Jones, K. C.; Foglia, T.A. Lipase-catalyzed synthesis of structured low-calorie triacylglycerols. *J. Am. Oil Chem. Soc. 76:* 1127–1132 (1999).

32. Lee, K-T, Foglia, T.A., Chang, K-S. Production of alkyl ester as biodiesel from fractionated lard and restaurant grease. *J. Am. Oil Chem. Soc. 79:* 191–195 (2002).

33. Lee, J-H, Jones, K.C., Lee, K-T., Kim. M-R, Foglia, T.A. High-performance liquid chromatographic separation of structured lipids produced by interesterfication of macadamia oil with tributyrin and tricaprylin. *Chromatographia 58:* 653–658 (2003).

34. Bloomer, S., Adlercreutz, P., Mattiasson, B. Triglyceride interesterification by lipases; Reaction parameters for the reduction of trisaturate impurities and diglycerides in batch reactions. *Biocatalysis 5:* 145–162 (1991).

35. Dibildox-Alvarado, E., Toro-Vazquez J.F. Isothermal crystallization of tripalmitin in sesame oil. *J. Am. Oil Chem. Soc. 74:* 69–76 (1997).

36. Foglia, T.A., Jones, K.C. Quantitation of neutral lipid mixtures using high performance liquid chromatography with light scattering detection, *J. Liq. Chrom. Rel. Technol. 20:* 1829–1838 (1997).

37. Bull, W.C., Wheeler, D.H. Low-temperature solvent crystallization of soybean oil and soybean oil fatty acids, *Oil and Soap 24:* 137–141 (1943).

38. Yokochi, T., Usita, M.T., Kamisaka, Y., Nakahara, T., Suzuki, O. Increase in the γ-linolenic acid content by solvent winterization of fungal oil extracted from *Mortierella Genus, J. Am. Oil Chem. Soc. 67:* 846–851 (1990).

39. Hsu, A.F., Shen, S., Wu, E., Foglia, T.A. Characterization of soybean lipoxygenase immobilized in cross-linked phyllosilicates, *Biotechnol. Appl. Biochem. 28:* 55–59 (1998).

40. Lee, K-T., Akoh, C.C. Immobilization of lipases on clay, celite 545, diethylaminoethyl and carboxymethyl-sephadex and their interesterification activity, *Biotechnol. Techniques 12:* 381–384 (1998).

17

Microbial Polyunsaturated Fatty Acid Production

Toro Nakahara

17.1 Introduction

Polyunsaturated fatty acids (PUFA) have been the most important target among microbial lipids, what is called single-cell oils (SCO).[1] Among PUFAs, SCOs including gamma-linolenic acid (GLA, 18:3n-6) became the first industrial scale product about 18 years ago in Japan[2] and the UK. This was caused by a rather scarce distribution of GLA in the main plant oil resources, and its function of anti-inflammation, antithrombosis, and antiallergy. Recent attention has focused more on PUFAs with longer carbon chains (C20 and C22), such as docosahexaenoic acid (DHA), eicosapentaenoic acid (EPA), and arachidonic acid (AA), than GLA.

In this chapter, reports related to microbial PUFA production from bacteria to thraustochytrids are described mostly in chronological order.

When we discuss PUFA productivity by microbial culture, attention must be paid to whether it is a batch or a chemostat culture. In batch culture, growth rate and PUFA productivity are kind of integrated (or average) values during the culture time; on the other hand, in chemostat culture (or continuous culture) growth rate converges to dilution rate (D h^{-1}) and so can be kept constant (Figure 17.1). It is thought, therefore, that continuous culture gives more accurate values than batch, but the operation of chemostat culture is generally more complicated and prone to be contaminated than batch culture.

17.2 Bacterial Lipid

17.2.1 Unsaturated Fatty Acid

Bacteria do not principally accumulate lipid in their cells as triacylglycerol (TG), which is the biggest weak point as the resource of oil-producing microbes, though their growth rates are higher than other

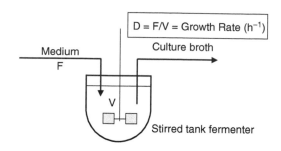

FIGURE 17. 1 Chemostat (continuous) vulture.

eukaryotic microbes. The most abundant lipid in most bacteria is phospholipid (PL), which is generally composed of phosphatidylethanolanine (PE), phosphatidylglycerol (PG), and diphosphatidylglycerol (DPG). Glycolipid (GL) as well as PL is main component of bacterial cell membrane. The effect of temperature on the FA composition in *Escherichia coli* was tested in batch and chemostat cultures of *Escherichia coli.*[3] In batch culture, monounsaturated fatty acid (MFA) increased as the temperature decreased and cyclopropane fatty acid (CPA) increased at the stationary growth phase. In chemostat culture, longer saturated fatty acid (SFA) with N-limited and slightly longer MFA with C-limited than batch were observed. Shaw et al.[4] reported that, by temperature shift, FA composition of *Escherichia coli* ML30 was changed during lag phase when glucose was present but not without glucose, though FA composition of the cells did not determine the minimal temperature of growth. In PL of *Escherichia coli*, Cronan[5] also showed that MFA was converted to CPA from the logarithmic growth phase to the stationary phase. Gill[6] showed that, in chemostat of *Pseudomonas fluorescens*, 16:1, 18:1 decreased and 17-CPA increased as the dilution rate (D) decreased, or as the growth rate decreased, which coincide with an increase in CPA at the stationary phase in batch culture. As the temperature was decreased, 16:1 and 18:1 increased but 16:0 and 17-CPA decreased. Minnikin et al.[7] reported that, in chemostat of *Pseudomonus diminuta* at 30°C, D = 0.2 h^{-1} and pH 7.0, between P- and Mg-limited, PL and GL content changed, and at P-limited, acidic PL (PG, DPG) was replaced by acidic GL. DeLong et al.[8] showed that the unsaturation index (UI) of a deep-sea bacterium, *Vibrio* sp. CNPT3, whose main FAs were 16:1, 16:0, 14:1, and 18:1, increased according to pressure, suggesting adaptations in the membrane lipids to environmentally relevant pressure. At 2°C, the strain grew optimally at pressures of 300 to 500 bars. Intrago et al.[9] showed that, in *Flexibacter* sp. strain Inp, whose main FAs were iso15, 16:1n-5, 18:1n-9 (OA), and 18:2n-6 (LA), UFA content was increased as temperature decreased and especially LA content increased as salinity increased and that the inverse relationship between 16:1n-5 and 18:1n-9 indicated two different pathways of UFA synthesis.

17.2.2 PUFA

Johns et al.[10] reported high content of EPA in *Flexibacter polymorphus*, marine origin, Gram-negative, nonphotosynthetic, nonfruiting, filamentous bacteria capable of gliding motility at the first specimen of bacterial PUFA in 1977, suggesting the membrane flexibility was necessary for its gliding motility. Na_2S inhibited production of EPA. DeLong and Yayanos[11] found that a deep-sea psychrophilic bacterium, *Vibrio marinus* MP-1, produced long-chain PUFA. The content of PUFA increased at 2°C as pressure was increased, and as temperature was decreased at atmospheric pressure. They suggested PUFAs acted to keep optimal membrane fluidity. They also suggested that this microbe has a specific role to produce essential fatty acid (EFA) for deep-sea bacteria in abyssal food webs.

Yazawa et al.[12] tried to isolate EPA-producing bacteria from mainly fish intestines, and isolated 112 strains out of some 7000 screened. They reported the highest EPA productivity of 3.8–10.8 mg/g-DCW (dry cell weight) and EPA content of 27.8–36.3% in isolates from the Pacific mackerel, *Pneumatophorus japonicus*. Yazawa et al.[13] also found 88 EPA-producing strains among 5000 isolates. One isolate, SCRC-8132, was obligate aerobic, Gram-negative, motile, short rod-shaped bacterium, with 30 min doubling time at 25°C, and accumulated 20 mg-EPA/L (4 mg-EPA/g-DCW) for 18 h, when EPA content in total fatty acid (TFA) was 24%. At 4°C for 5 days, the strain accumulated 26 mg/L (15 mg/g-DCW) with 40% of EPA. Akimoto et al.[14] reported that a higher EPA accumulation at 45.6 mg-EPA/L-culture broth after 8 h (at 25°C, pH 7) was found from an isolate from mackerel intestines, SCRC-2738. They[15] also reported that EPA and other cellular FAs were *in vivo* synthesized from acetyl coenzyme A by the usual de novo synthesis route in this marine bacterium, SCRC-2738. They[16] also tried to perform a large-scale EPA production by *Shewanella* sp. SCRC-2738 using 16 L and 200 L fermenters and obtained 376 mg-EPA/L at 8°C and 145 mg/L at 15°C. As for salinity, 75% salt concentration of seawater was optimum. Watanabe et al.[17] tested the incorporation of exogenous DHA into various bacterial PLs using *Shewanella* sp. strain SCRC-2738 and three other strains. SCRC-2738 incorporated the largest amount of DHA into its PLs. Contents of DHA in PE and PG were 16% and 29%, respectively.

Nichols et al.[18] reported *Shewanella putrefaciens*, strain ACAM342, produced 18:2n-3, 18:3n-3 (ALA), EPA (total 2–4%) under both aerobic and anaerobic condition at 15 and 25°C, suggesting that it possesses both pathways for PUFA synthesis. Ringo et al.[19] found 14 isolates from turbot larvae produced EPA, and *Vibrio pelagius* contained a high proportion of EPA. At a lower temperature (5°C), EPA concentration of the bacterium increased. *Vibrio* from freshwater origin was also found to produce EPA. They[20] also found 3.5% EPA in the intestinal liquor of arctic charr (fed a pellet diet without n-3PUFA and 10,000 cfu/mL aerobic bacteria in intestinal liquor). Among the 17 bacterial isolates producing EPA, four among them contained a high EPA proportion and these belong to *Vibrio* sp. One of them, *Vibrio* sp. strain no. 5710, showed 35, 9, 6, and 10 h-generation times at −5, 0, 5, and 10°C, respectively. Content of DHA decreased as growth temperature increased, indicating that membrane fluidity is adjusted by the amount of DHA and 16:0.[21] Among 22 psychrophilic and psychrotrophic Vibrios isolated from deep-sea sediments, 12 isolates contained DHA and five contained EPA. Optimum growth temperature for DHA producers was lower than 20°C and that for EPA producers, 20–25°C, but that of other bacteria was higher than 25°C. Hamamoto et al.[22] suggested that these PUFAs may play a role in low-temperature adaptation of marine Vibrios, and that PUFAs could be used to discriminate psychrophilic and psychro-tolerant marine bacteria. EPA producing *Vibrio* sp. was also isolated from sea urchin, which contained 17.5% EPA and less than 1% of other PUFAs.[23] Conversion of [1-C14]OA to EPA in the strain was oxygen-dependent and cerulenin-sensitive. With cerulenin, trienoic FA increased and tetra- and pentaenoic FA decreased, suggesting OA was aerobically desaturated and elongated to EPA via ALA. Both radio labeled 20:4n-6 (AA) and ALA were converted to EPA, showing that the desaturation of n-6 and n-3 PUFAs occurs in C18 and C20 chain length. Nichols et al.[24] showed that in the psychrophilic EPA producer, *Shewanella gelidimarina* ACAM 456T, the EPA yield was relatively constant at all temperatures within and below the optimum growth temperature, but decreased at higher temperatures. Association of EPA with 17:1 and 18:0 acyl chain in PL was specific to PG, but with iso13/13 and 14/iso14 was specific to PE. The EPA yield was 1–16 mg/g-DCW. Alvarez et al.[25] showed that among 40 psychrophile or psy-chrotrophic crude-oil-utilizing marine bacteria, 73% were able to accumulate specialized lipids like polyhydroxyalkanoic acids (PHA), while two strains were able to accumulate wax esters. Accumulation of PHA occurred predominantly at low temperatures (4–20°C). Bowman et al.[26] showed that *Psychroflexus gondwanense* gen. nov., *comb.* nov., a psychrophilic species from anarctic sea ice, produced EPA and AA. They[27] also showed that *Colwellia demingiae* sp. nov., and other psychrophillic antarctic species had the ability to synthesize DHA. Eight strains associated with antarctic sea-ice diatom assemblages were identified as genus *Colwellia* by 16S rRNA sequence analysis. Watanabe et al.[28] isolated a novel DHA-producing marine bacterium, SCRC-21406, from the intestine of a marine fish, *Glossanodon semifasciatus*, which produced DHA at 23% mol/mol of TFA at 12°C under atmospheric pressure. A marine bacterium Pseudomonas/Alteromonas, associated with the sponge, *Dysidea fragilis*, DF-1, was proved to produce 16:3 and 16:4.[29] Data on bacterial PUFA are summarized in Table 17.1.

TABLE 17.1 Bacterial PUFA

Strain	Mode of Growth	PUFA Productivity/ Growth Rate	Reference	No
Flexibacter polymorphus		High EPA content	Johns et al. (1977)	10
Vibrio marinus MP-1		PUFAs ↑ (at 2°C as pressure ↑ as Temp ↓ at atmosphilic pressure)	DeLong & Yayanos (1986)	11
EPA producers(112 from 7,000)	batch	4–11 mg-EPA/g-DCW, 28-36% EPA/TFA	Yazawa et al. (1988)	12
EPA producers(SCRC-8132)	batch	24 mg-EPA/g-DCW for 18 h, 24% EPA/TFA, At 4°C for 5 days, 26 mg-EPA/L (15 mg/g-DCW), 40% EPA	Yazawa et al. (1988)	13
SCRC-2783	batch	46 mg-EPA/L after 8 h	Akimoto et al. 1990	14
SCRC-2738	fermenter	380 mg-EPA/L at 8C, 150 mg/L at 15C	Suzuki et al. (1992)	16
Shewanella sp. strain SCRC-2738		DHA incorporated in PL, 16% DHA in PE, 29% in PG	Watanabe et al. (1994)	17
Shewanella putrefaciens strain ACAM342		18:2n-3, ALA, EPA (total 2ñ–4%) were produced under aerobic/anaerobic conditions at 15/25°C	Nichols et al. (1992)	18
Vibrio pelagius		At 5°C, EPA ↑	Ringo et al. (1992)	19
Vibrio sp. Strain no. 5710		DHA ↓ as Temp ↑, membrane fluidity adjusted by DHA	Hamamoto et al. (1994)	22
22 psychrophilic/ psychrotrophic Vibrios		Max growth temp of DHA producers: lower than 20°C, EPA producers: 20-25°C, others:higher than 25°C	Hamamoto et al. (1995)	23
Vibrio sp. (from sea urchin)		18% EPA and less than 1% of other PUFAs	Iwanami et al. (1995)	23
Shewanella gelidimarina ACAM 456T	batch	EPA yield: 1–16 mg/g-DCW	Nichols et al. (1997)	24
Colwellia demingiae sp. nov.		DHA producr (Antarctic sea-ice diatom assemblages)	Bowman et al. (1998)	26
SCRC-21406		23% DHA/TFA at 12°C under atmospharic pressure, Cell yield: 0.43 g/L	Watanabe et al. (1997)	28
DF-1, associated with the sponge		16:3 and 16:4 were produced	Rosa et al. (2000)	29

: ↑ increase, ↓: decrease, DCW: dry cell weight, TFA: total fatty acid, PL: phospholipid, PE: phosphatidyletha-nolamine, PG: phosphatidylglycerol

17.3 Yeast Lipid

17.3.1 Nutrient Effect

Enebo et al.[30] found, in batch cultures of *Rhodotorula gracilis*, that lipid content was increased up to 50–60% with N and P-limitation, when the fat coefficient was 16–18 g from 100 g C-source, but long generation time at 15–20 h. Sufficient supply of N and P resulted in low lipid content with 2.8 h-generation time. On the other hand in *Cryptococcus terricolus*, Pedersen[31] reported that lipid content was independent on N quantity, and even with high N content (C/N ratio = 1.9), lipid content was as much as 55% in flask culture. The different behavior of lipid accumulation to N content (C/N ratio) may be caused by the presence of the regulatory system called the Crabtree effect[32] in yeast strains. Using *Candida utilis* (Crabtree-negative strain), Babij et al.[33] reported that lipid content was increased as glucose was increased in chemostat, and that PUFA was increased as DO was increased. When both glucose and DO were high, the ALA amount was the highest. Brown et al.[34] also reported in chemostat of *Candida utilis* that the lower temperature resulted in the increase of UFA and decreased dissolved oxygen (DO) caused increased palmitic acid (16:0). Johnson et al.[35] examined the influence of glucose concentration on lipid composition of six Crabtree-positive and -negative yeast strains in chemostat (D = 0.1 h^{-1}, 25°C, pH 4.5). The content of FA decreased with increased glucose concentration in Crabtree-positive yeasts, but FA content increased in Crabtree-negative yeasts. Johnson et al.[36] also reported the effect of P-limitation on the lipid of *Saccharomyces cerevisiae* (Crabtree positive) and *Candida utilis* (Crabtree negative) in chemostat cultures. In *S. cerevisiae*, TFA and TG increased but PL did not change; on the other hand, in *C. utilis*, PL decreased and FA composition changed. Brown et al.[37] also tested the influence of C-source concentration in chemostat of *Saccharomyces cerevisiae* (D = 0.1 h^{-1}, 25°C, pH 4.5). At glucose more than 5 g/L and galactose more than 10 g/L, TFA content decreased and 16:1 decreased with a corresponding decrease of total amount of UFA. Kessell et al.[38] showed that lipid content of *Rhodotorula gracilis* increased after N exhaustion owing to the absence of proliferation but that the lipid production rate was unchanged. At low pH, growth rate decreased but the lipid production rate increased (though final lipid content remained the same) and UI decreased (18:1 decreased). Wilson et al.[39] tested the effect of growth condition on sterol, FA content, and viability in *Saccharomyces cerevisiae*. Aerobic cells contained more UFA (16:1 and 18:1), but anaerobic cells contained more SFA (e.g., 10:0 and 14:0) and less sterol. Viability after 72-h starvation was 75% (aerobic) and 17% (anaerobic) of intact cells. Gill et al.[40] reported 40% lipid content at D = 0.06 h^{-1} (N-limited) and 14% at D = 0.21 h^{-1} (C-limited) in *Candida* 107. In C-limited condition, UI was decreased as D increased (opposite to N-limited condition). Neutral lipid (NL) content was 66% (C-limited) and 92% (N-limited). Yoon et al.[41] reported 40–62% of lipid content and 8–22 of fat coefficient in the flask culture of *Rhodotorula gracilis* NRRL Y-1091, which contained 84% NL, 14% GL, 2% PL. Using a oleaginous yeast, *Candida curvata* D, Evans et al.[42] showed maximum lipid content of 49% with xylose in batch and 37% at optimum D (= 0.05 h^{-1}) in chemostat. Highest biomass and lipid yield at 60 g-DCW and 18.6 g-TL (total lipid) from 100 g-lactose were obtained at D = 0.04 h^{-1}. Zhelifornia et al.[43] proposed a characterization of lipid-forming yeast by the energy yield of lipids based on substrate (ethanol) consumed by the result of high lipid content (67%) of *Lipomyces* and *Cryptococcus* sp. Boulton et al.[44] reported, during 32 h-transition from steady-state C-limiting to N-limiting of *Lipomyces starkeyi* CBS 1809, DCW and lipid content increased; AMP rapidly decreased, though ADP and ATP increased but specific activity of APT citrate lyase did not significantly change. Naganuma et al.[45] showed at lower NH^{4+} and Zn^{2+}, and at higher Ca^{2+} lipid content was increased in *Lipomyces starkeyi*, but decreased at K$^+$ deficiency, Zn^{2+} and Mn^{2+} sufficiency, without Fe^{3+} and Mn^{2+}, at very low pH (1.8), low temperature (15 and 19°C), and low DO. They[46] also reported that by adding Zn^{2+}, Mn^{2+}, and Na$_2$HPO$_4$, *Lipomyces starkeyi* started its second log growth phase but decreased its lipid content. Ykema et al.[47] tried mathematical modeling of lipid production by oleaginous yeast (*Apitotrchum curvatum*) in continuous cultures and found lipid yield increased with increasing C/N ratio but maximum growth rate was obtained at relatively low C/N ratio. Dedyukhina et al.[48] reported that the lipid content of *Candida valida* VKM Y-2328 was increased with Mg-limiting from 3.4% to 15%, and the specific rate of lipid synthesis was increased by a factor of 3.9 in comparison with EtOH-limiting, though lipid content and FA composition were not significantly changed in EtOH-, N-, and P-limiting chemostat. Simultaneous utilization of glucose and xylose by *Candida curvata* was examined and similar biomass yield, lipid content, and FA composition were obtained, showing the capability of growing hydrolyzed wood and straw wastes.[49] Ykema et al.[50] realized lipid production rate of 1.0 g/Lh in a partial recycling culture of *Apiotrichum curvatum* with whey permeate as C-source. Lipid content remained at 22% up to 25 C/N ratio and the maximum lipid content at 50% was obtained at C/N = 70. Prapulla et al.[51] showed inoculum level was important for lipid accumulation using *Rhodotorula gracilis* CRF-1, i.e., at higher level of inoculum, the strain was found to be more tolerant to higher sugar concentration. The flask culture with 20% of inoculum level (10% sugar, 0.37 g-N/L) resulted in the maximum biosynthesis of lipids. Andlid. et al.[52] showed that energy content increased during the lipid-accumulating phase (TFA/DCW from 8.3% to 46.2%) in *Rhodotorula glutinis*.

17.3.2   Temperature and Dilution Rate (D)

Hunter et al.[53] showed that decreased temperatures in batch cultures of *Saccharomyces cerevisiae* NCYC 366 resulted in the increase of TL, TFA, TG, and PL. In chemostat cultures, a decrease in D resulted in an increase of TG and PL but little changes of FA composition and UI. Moon et al.[54] showed that more UFA (18:1) was produced at decreased temperature in *Candida curvata* and *Trichosporon cutaneum*. Dedyukhina et al.[55] showed that, as temperature decreased, UI of *Hansenula polymorpha* increased in chemostat on methanol. Lipid-accumulation yeasts *Candida* 107 and *Rhodotorula gracillis*, showed lower specific O_2 uptake rate at lower D in chemostat C- and N-limited culture.[56] Choi et al.[57] reported the effect of growth rate and DO on *Rhodotorula gracilis* NRRL Y-1091 in N-limited continuous culture. As growth rate increased, protein content increased but DCW, lipid content, and lipid productivity decreased. The specific lipid production rate remained constant (0.012 g/g-DCW h) and the maximum lipid content was 49.9% at $D = 0.02$ h^{-1}. Contents of LA and ALA were increased with the increase in D, lipid content was increased at increased DO, and UI was increased according to the increase in specific O_2 uptake rate. Dedyukhina et al.[58] reported the effect of D on lipid synthesis with ethanol-limited growth of *Trichosporon pullulans*, and obtained 16.6% highest lipid content at $D = 0.05$ h^{-1}. Higher content of LA and ALA in PL fraction but OA in TG fraction occurred and specific OA synthesis rate rose with an increase of D. Granger et al.[59] showed that N-limited conditions (C/N ratio > 0.14) resulted in enhanced lipid content but a reduced amount of PUFA in *Rhodotorula glutinis* NRRL Y 1091. A higher PUFA content and maximum ALA productivity was obtained at 25°C lower than the optimum growth temperature (30°C). Yoon et al.[60] showed the maximum lipid content at 58% for N-limited and 19% for C-limited in *Rhodotorula glutinis* batch culture. Maximum specific growth rate for C- and N-limited cultures were 0.13 and 0.24 h^{-1}. In continuous culture, maximum lipid yield at 16.4 was obtained at lowest D (0.02 h^{-1}). The higher D resulted in the higher UI. Hansson et al.[61] showed that in both N-limited and excess-N conditions, lipid content was increased and UI was decreased using *Cryptococcus albidus* var. *albidus*. Lipid content and UI were not changed between 20 and 25°C of culture temperature.

17.4   Fungal Lipid

17.4.1   Character of Fungal Lipid

Sumner et al.[62] reported that spores contained less lipids than vegetative mycelium of Mucorales, and that FA composition of spore lipid was more highly saturated, though qualitatively similar, in their flask stationary culture. Lowering temperature resulted in an increase of UFA synthesis in thermotolerant and thermophilic fungi. Mumma et al.[63] compared FA composition between thermophilic and mesophilic fungi, and reported PA (16:0) and OA were the main FA and LA content varied between 0% and 18.5% in mesophiles but less than 0.5% in thermophiles. Some Mucor contained GLA. Fatty acid of thermophiles was more saturated than the corresponding mesophiles. Rogers et al.[64] reported that, in chemostat of *Cokeromyces poitrasii* (NRRL 2845)(D = 0.06 h^{-1}, pH 5.6, 30°C), UFA content increased according to increase in DO. Konova et al.[65] reported that higher C/N ratio and higher content of these elements increased lipid content, but decreased UFA content in *Entomophthora conica*. Kendrick et al.[66] reported that temperature but not DO was the principal regulation factor of lipid unsaturation in chemostat culture (D = 0.04 h^{-1}, 20–30°C) of *Entomophthora exitalis*. As temperature was decreased, PUFA increased from 18% to 27%, particularly AA, but TG did not change.

As for GLA, Shaw[67] reported the occurrence of GLA only in the Phycomycetes among orders Phycomycetes, Ascomycetes and Basidiomycetes and among nine Phycomycetes. Only six species of the order Mucorales contained GLA, but no ALA, which suggested a different evolution of Phycomycetes from other fungi. With starch as C-source, *Mortierella ramanniana* CBS 112.08 gave the highest GLA content (25.7%) though lower DCW than glucose.[68] and the maximum lipid content (66%) at high C/N ratio (80), higher UI at lower temperature. Hansson et al.,[69] obtained 39.3 mg-GLA/Lh and 99 mg-TL/Lh (GLA 39.7%) in fed-batch culture of *Mucor rouxii* CBS 416.77 and 37 mg-GLA/Lh and 95 mg-TL/Lh in continuous culture (at high 4.5 g-NH_4Cl/L, D = 0.10 h^{-1}). This strain also showed higher GLA production with starch and its hydrolysate than glucose at 0.33 g-GLA/L.[70] Torlanova et al.[71] reported the increase in lipid content with Cu and Zn, and GLA yield with Zn in flask cultures of *Mucor lusitanicus* INMI. Emelyanova[72] reported high GLA content in the lipid (37%) of *Mucor inaquisporus* IBFM-1 grown on ethanol as C-source.

17.4.2 GLA Production by Genus *Mortierella* (by Suzuki and Yokochi)

Suzuki et al.[73] examined lipid accumulation of 13 species of Deuteromycetes, and found 51.3% of lipid content in *Penicillium lilacinum*. They[74] also examined lipid composition of six species of Mucorales, including GLA but not ALA, and found that the lipid contents were 86.1% and 63.3% at 20°C and 30°C, respectively, with *Mortierella isabellina* IFO 7884. They[75] found high LA contents at 60–70% in NL (neutral lipid) and 70–80% in PL in 5 species of Pellicularia in Bacidiomycetes. The maximum TL was obtained at 197 mg/200 with *Pellicularia filamentosa* sp. sasakii IFO 8985[76] with LA content at 70% in NL and 80% in PL. With 2 strains of *Mortierella isabellina* (IFO 7884 and 7824),[77] the increase in lipid content occurred with the increase in C/N ratio and the highest lipid content was 83.5% at 343 C/N ratio (Glucose/NH_4NO_3). Ammonium sulfate was the best N-source and the maximum lipid coefficient was 14. These fungi can grow well with n-decane as C-source and their maximum lipid amount was 0.375 g/400 ml, when lipid coefficient was 10.3, lower than with glucose as C-source.[78] Yokochi et al.[79] examined 33 strains of genus *Mortierella* and found that GLA content was the highest in *Mortierella ramanniana* var. *angulispora* IFO 8187, the values of which were 7.7% in NL and 23.6% in PL with glucose (30°C) and 28.2% in NL and 35.9% in PL with n-decane (20°C). As to the GLA content, n-decane gave much higher content than glucose. Kamisaka et al.[80] examined the effect of metal ions and temperature on *Mortierella ramanniana* var. *anglispora* IFO 8187 and reported that increase in Zn^{2+} and Fe^{2+} caused the increase of GLA in PL and NL, respectively, indicative of different enzymatic systems modulating FA composition of PL and NL.

Higher C-source concentration in growth media may be beneficial in obtaining higher DCW, though it is often inhibitory to microbial growth. Yokochi et al.[81] showed that *Mortierella isabellina* IFO 7884 can grow at over 200 g-glucose/L and at a higher initial glucose concentration and C/N ratio (up to 40) gave higher lipid content. The maximum DCWs at 156.4 g/L and TL at 83.1 g/L were obtained with a 390 g-glucose/L (C/N ratio = 20). And the maximum lipid productivity at 0.69 g/Lh was obtained with a 270 g-glucose/L (C/N ratio = 20). In these conditions the shape of the cell became round rather than mycerial form.

Using 3 strains of *M. isabellina* (7884, 8183, and 8308), Yokochi et al.[82] found that lipid was highly accumulated at lower pH but the growth rate was also decreased, and pH shift from 4.5 to 3 was suitable for lipid production, a kind of separation with growth phase (pH 4.5) and lipid accumulation phase (pH 3). The maximum lipid accumulation at 5.5 g/L was obtained at higher C/N ratio (62.9) using the pH shift, when lipid content was 66.6% and lipid coefficient was 22. Then they[83] attempted to separate the optimum growth and lipid accumulation conditions of *Mortierella ramanniana* var. *angulispora* IFO 8187 by using two-fermenters system, and obtained the maximum lipid production of 13.9 g/Ld with a repeated batch mode.

Further Hiruta et al.[84] tried to isolate a low temperature-resistant mutant of *M. ramanniana*, and obtained *Mortierella ramanniana* mutant MM 15-1. The GLA content was 16.5%, almost twice that of its parent strain. Content of GLA was increased to 18.3% in a 600-L fermenter. Hiruta et al.[85] also performed a scale-up optimization of the mutant and obtained a higher GLA content by pellet formation than filamentous forms and higher lipid content with higher glucose concentration at 300 g/L. Higher rotation at 800 rpm and relatively less amount of innoculant helped forming pellet. Yokochi et al.[86]

reported the production of GLA-containing PL by *Mortierella ramanniana* var. *angulispora* IFO 8187 in fed-batch cultures with n-decane as C-source. Production of GLA-containing PC and PE were 300 and 125 mg/L. Data on fungal GLA production are summarized in Table 17.2.

17.4.3 AA and PUFA Production by Genus *Mortierella* (Shimizu et al.)

Totani et al.[87] reported a solid fermentation on 6% glucose at 20°C for 20 days using *Mortierella alpina* IFO 8568 and obtained AA content at 13.1 g/kg-agar medium. They[88] also reported 68.5–78.8% AA in TFA and 25% AA in DCW in the 4 strains of *Mortierella alpina*.

Yamada et al.[89] isolated 7 AA producers from 464 strains. One soil isolate, *Mortierella elongata* 1S-5, showed 0.99 mg-AA/mL (22 mg/g-DCW) in a shake flask for 4 days. They[90] also isolated *Mortierella alpina* 1S-4 and 2O-17. Strain 1S-4 produced 3.6 g-AA/L (147 mg/g-DCW) for 7 d and 2O-17 produced 0.5 g-EPA/L (27 mg/g-DCW) for 7 d at 12°C. Shimizu et al.[91,92] found that EPA was produced in their mycelia only when grown at a low temperature (12°C), not at physiological growth temperatures (20–28°C) and that maximum EPA productivity was obtained by the culture at 20°C at the early growth phase followed by a shift to 12°C. *Mortierella alpina* 1S-4 then produced 0.3 g-EPA/L (27 mg/g-DCW). Shimizu et al.[93] further reported that *Mortierella alpina* 1S-4 accumulated DGLA by the addition of nonoil fraction of sesame oil and produced 2.17 g-DGLA/L (107 mg/g-DCW) in 50 L- fermenter culture for 7 days. The factor in the sesame oil was later identified as (+)-sesamin, which proved to be a potent and specific inhibitor of Δ5-desaturase in PUFA biosynthesis.[94] Shimizu et al.[95] reported *Mortierella alpina* 2O-17 converted ALA (added in the medium) to EPA and accumulated large amounts of EPA at 1.35 g-EPA/L (41.5 mg/g-DCW) for 6-day cultures and 7-day aging in 4% glucose and 1% linseed oil (60% ALA). Addition of soybean oil to *Mortierella alpina* similarly increased AA production to 3.6 g-AA/L for 7 days in 5-L fermenter, which was 2.8-fold higher than without the addition. In a 2000-L fermenter for 10 days, 22.5 kg-AA/kL (9.9 kg-lipid, AA 31% of TFA) was produced.[96]

The fungus strain had such a flexible FA synthetic pathway, as *Mortierella alpina* 1S-4 accumulated an odd chain PUFA, 5,8,11,14-19:4, with 5% n-heptadecane (C15) and 1% yeast extract, and further addition of sasamin caused an increase in C19:3. *Mortierella* producing 18-PUFAs also produced C17-PUFAs with n-heptadecane.[97] When grown on 1-alkenes, *Mortierella alpina* 1S-4 produced 5,8,11,14,19-EPA (ω1-EPA) together with ω1-16:1 and ω1-18:1.[98] Concentration of EPA and DHA in *Mortierella alpina* 1S-4, grown with fish oil, was also performed. Contents of EPA and DHA reached 29.2% and 20.0% of TFA in this strain grown on salmon oil as the main C-source in a 5-L fermenter. Both EPA and DHA were incorporated into both PL and TG fraction.[99]

17.4.4 PUFA Production by the Mutant from *Mortierella* (Shimizu et al.)

Jareonkitmongkol et al.[100] tried to isolate FA desaturation-defective mutant from an AA-producing fungus, M. alpina 1S-4, using MNNG (nitrosoguanidine). They obtained three mutants with low Δ5, Δ12 and Δ6-desaturase.

Using the Δ5-desaturase defective mutant of *M. alpina* 1S-4, Jareonkitmongkol et al.[101] reported DGLA production at 3.2 g-DGLA/L (123 mg/g-DCW) (23.4% of TFA) in a 10-L fermenter at 28°C for 6 days. Using the Δ12-desaturase defective mutant, mead acid (20:3n-9) was produced at 0.8 g/L (56 mg/g-DCW)(15% of TFA) in a 5-L fermenter at 20°C for 10 days.[102] The conversion of ALA to EPA was studied using the same Δ12 desaturase-defective mutant and obtained 1 g-EPA/L (64 mg/g-DCW) (20% of TFA) with a medium comprising 1% glucose, 1% yeast extract, and 3% linseed oil.[103] Data on fungal PUFA (AA) are summarized in Table 17.3.

17.5 Thraustochytrids' Lipid

The ALA synthesis by a nonphotosynthetic zooflagrllate, *Leishmania enriettii* whose FA composition were 22% OA, 17% LA, 23% ALA, and no absolute correlation between chlorophyll-dependent photosynthesis, was first reported by Korn and Greenblatt.[104] Goldstein[105] first observed in 1963 species of thraustochytrids,

TABLE 17.2 Fungal GLA

Strain	Mode of Growth	PUFA Productivity/ Growth Rate	Reference	No.
31 species of fungi		GLA was found only in phycomycetes, in which 6 sp of the order Mucorales contained GLA, but no ALA.	Shaw (1965)	67
Mortierella ramanniana CBS 112.08		Glu & fructose: 12 g-DCW/L, LC = ~24%, starch: 26% GLA, Max LC (66%) at C/N = 80, UI as Temp ↓	Hansson et al (1988)	68
M. rouxii CBS 416.77	fed-batch	39 mg-GLA/Lh, 99 mg-TL/Lh (40% GLA)	Hansson et al (1989)	69
	chemostat	37 mg-GLA/Lh, 95 mg-TL/Lh		
Mucor lusitanicus INMI	batch	Cu & Zn increased LC, Zn increased GLA yield	Torlanova et al. (1995)	71
Mucor inaquisporus IBFM-1	batch (EtOH)	36% GLA at highest growth rate	Emelyanova (1997)	72
	chemostat (EtOH)	37% GLA in the lipid, 3% GLA in the cell		
6 sp of Mucorales		*M. isabellina* IFO 7884: 86% LC at 20C, 63% at 30C/ GLA instesd of ALA	Suzuki et al. (1981)	74
2 strains of *Mortierella isabellina*		LC ↑ with C/N ratio ↑ (84% LC), Lipid Coef = 14	Suzuki et al. (1982)	77
2 strains of *M. isabellina*		0.94 g-TL/L on n-decane, Lipid Coef = 10.3	Suzuki et al. (1982)	78
33 strains of genus *Mortierella*	Glu	5 g-TL/L (50% LC) in IFO 8308, 8%-GLA in IFO 8187	Yokochi et al. (1986)	79
	n-decane	0.85 g-TL/L(32% LC) in IFO 6739, 28%-GLA in IFO 8187		
Mortierella ramanniana var. *anglispora* IFO 8187		Zn ↑: LA & GLA in PL (not NL)↑/ Fe↑: GLA in NL↑/ Temp↓:GLA in PL↑	Kamisaka et al. (1988)	80
M. isabellina IFO 7884		↑Glu & ↑C/N (<40) gave↑LC, 160-DCW/L & 83 g-TL/L with a 390 g/L Glu (C/N = 20), 0.7 g-TL/Lh with a 270 g/L Glu	Yokochi et al. (1989)	81
3 strains of *M. isabellina*		5.5 g-TL/L at C/N = 63 (67% LC), Lipid Coef = 22	Yokochi et al. (1987)	82

(Continued)

TABLE 17.2 Fungal GLA (Continued)

Strain	Mode of Growth	PUFA Productivity/ Growth Rate	Reference	No.
M. ramanniana var. *angulispora* IFO 8187	batch/repeated batch	14 g-TL/Ld with 2 stage culture	Yokochi et al. (1993)	83
M. ramanniana mutant MM 15-1	batch (fermenter)	18% GLA in a 600-L fermenter	Hiruta et al. (1996)	84
M. ramanniana mutant MM 15-1	batch (fermenter)	Glu↑ (300 g/L) gave↑ LC, Pellet gave ↑ GLA content than filamentous forms, ↑rotation (800 rpm) helped forming pellet	Hiruta et al. (1996)	85

Glu: glucose, TL: total lipid, LC: lipid content, C/N: the ratio of C-source/N-source, NL: neutral lipid, Lipid Coef: lipid (g) produced from 100 g-glucose, GLA: gamma-linolenic acid, ALA: alpha-linolenic acid, LA: linoleic acid

Thraustochytrium motivum and *Thra. multirudimentale*, which were obligatory marine microbes requiring 2.5–3.0% NaCl. Goldstein[106] also found a new monocentric chytrid, *Thraustochytrium aureum*, which was primarily a saprophyte of vascular plant debris and obligatory marine (euryhaline). He[107] also reported a new species of thraustochytrid, *Thraustochytrium Roseum*, that displays light-stimulated growth. Optimum and available C-source were different among these thraustochytrids. Some 25 years after Goldstein's findings, these thraustochytrids gathered only taxonomical interest. As these organisms, a kind of marine fungi, began to be known to produce docosahexaenoic acid (22:6n-5, DHA) as well as n-6 docosapentaenoic acid (22:5n-6, n-6DPA), trials to grow thraustochytrids for the SCO including DHA were started.

In 1991 Bajpai et al.[108] obtained lipid content from 2.7% to 16.5%, DHA yield from 26 to 270 mg/L in flask culture of *Thraustochytrium aureum* ATCC 34304 under light at 25°C for 6 days. Content of DHA was 48.5% of TFA. They[109] also tried optimization of DHA production by the same strain and obtained 20% lipid content and DHA yield at 511 mg/L in light-exposed cultures containing 2.5% starch at 28°C. Kendrick and Ratledge[110] obtained less than 10% lipid content from *Thra. aureum, Thra. roseum*, and *Schizochytrium aggregatum* and found that there was no activity of ATP:citrate lyase when lipid content was less than 10%. They obtained 4 g-DCW/L, 10% lipid content, 30% DHA, or 120 mg-DHA/L in a flask culture of *Thra. aureum* at 28°C for 3 days. Li & Word[111] cultivated three thraustochytrids on 2.5% starch and 0.2% yeast extract at initial pH 6.0 under light for 5 days. Content of DHA were 46.7%, 32.9%, 0.75% of TFA in *Thra. roseum, aureum*, and *stratum*, respectively. The maximum DHA yield at 0.85 g/L was obtained in *Thra. Roseum*.

An important US patent (US Patent 5,130.24) was issued in 1992 that claimed n-3PUFA production by the genera *Thraustochytrium* and *Schizochytrium*. Thraustochytrids have been morphologically thought to comprise seven genera and labyrinthulids one genus (principally by Porter[112]). Among seven genera only *Thraustochytrium* and *Schizochytrium* are practically important owing to their growth rate and wide distribution. The claims of the patent will affect novel strains in the two genera because of the disagreement on phylogenetic analysis and the morphological taxonomy. Therefore, the currently used taxonomic criteria of Labyrinthulomycota need a serious reconsideration (Honda et al.[113]).

Iida et al.[114] tried an improvement of DHA production in a culture of *Thra. aureum* ATCC 34304 and obtained 5.7 g-DCW/L and 460 mg-TL/L in flask culture at 25°C for 69 h. DHA content was 40% of TFA. Fermenter culture showed a lower growth than flask indicative of growth inhibition by stirrer. In a fermenter culture of *Schizochytrium* sp. strain SR21 isolated from seawater near Yap Islands, Nakahara et al.[115] realized the productivities at 2.0 g-DHA/Ld and 0.44 g-DPA/Ld at 28°C and pH4 in the medium contained 60 g-glucose/L and corn steep liquor and $(NH_4)_2SO_4$ in 50%-salt concentration of artificial seawater. This is the first report on lipid accumulative thraustochytrids with DHA content of 34% and

TABLE 17.3 Fungal PUFA (AA)

Strain	Mode of Growth	PUFA Productivity/ Growth Rate	Reference	No.
Mortierella alpina IFO 8568	solid fermentation	13 g-AA/kg for 20 days	Totani et al. (1987)	87
4 strains of *M. alpina*	solid fermentation	69-79% AA in TFA, 25% AA in DCW for 20 days	Totani et al. (1987)	88
Mortierella elongata 1S-5		1.0 mg-AA/mL (22 mg-AA/g-DCW) for 4 days	Yamada et al. (1987)	89
M. alpinaa 1S-4		3.6 g-AA/L(150 mg-AA/g-DCW) for 7 days	Yamada et al. (1988)	90
M. alpinaa 2O-17		At ↓ Temp in AA producing *Mortierella*, EPA was produced	Shimizu et al. (1988)	91
M. alpina 1S-4		0.3 g-EPA/L at 12C	Shimizu et al. (1988)	92
M. alpina 1S-4	batch (fermenter)	2.2 g-DGLA/L(110 mg-DGLA/g-DCW) for 7 days (by the addition of sesamin)	Shimizu et al. (1989)	93
M. alpina 2O-17		1.4 g-EPA/L (42 mg-EPA/g-DCW) for 6 days culture + 7 days aging with 1%-linseed oil	Shimizu et al. (1989)	95
M. alpina 1S-4	batch (fermenter)	23 kg-DCW/kL (9.9 kg-lipid, AA 31% of TFA)	Shinmen et al. (1989)	96
M. alpina 1S-4		44 mg-C19:4/g-DCW (0.7 mg/mL) with 5% n-heptadecane, Addition of sasamin caused an increase in C19:3	Shimizu et al. (1991)	97
M. alpina 1S-4		ω1-EPA was produced (0.13 mg/mL) with 4% 1-hexadecene	Shimizu et al. (1991)	98
M. alpina 1S-4	batch (fermenter)	29% EPA & 20% DHA grown with fish oil	Shinmen et al. (1992)	99
D5-desaturase defective mutant of *M. alpina* 1S-4	batch (fermenter)	3.2 g-DGLA/L(120 mg-DCW)(23% of TFA) for 6 days	Jareonkitmongkol et al. (1992)	101
D12-desaturase defective mutant		0.8 g-MA/L(56 mg/g-DCW)(15% of TFA) for 10 days	Jareonkitmongkol et al. (1992)	102
D12 desaturase-defective mutant		1 g-EPA/L(64 mg/g-DCW)(20% of TFA) with 3%-linseed oil	Jareonkitmongkol et al. (1993)	103

AA: arachidonic acid, EPA: icosapentaenoic acid, DGAL: dihomogamma-linolenic acid, MA: Mead acid

lipid contents of 50%. Characterization of the strain was performed and the strain appeared to be widely tolerant to salinity, high C/N ratio resulted in high lipid content up to 50%, maximum DHA yield of 4 g/L were obtained on 9% glucose or 12% glycerol in flask culture for 5 days.[116] In a fermenter with the same strain, Yaguchi et al.[117] showed that an increase in glucose content (6, 10, 12%) resulted in an increase in DHA productivity (2.0, 2.7, 3.3 g/Ld). At 12% glucose, the productivities were 48.1 g-DCW/L and 13.3 g-DHA/L in 4 days and lipid content at 77.5%, DHA at 35.6% of TFA and neutral lipid (NL) at 95% of TL. The strain was able to grow well at higher glucose content as well as in the shear stress caused by stirrer in fermenter. After observing the isolated *Schizochytrium* sp. strain SR21, Honda et al.[118] identified the strain to be a novel one because of its limacinum amoeboid cells, the size of zoospores, zoospores released from both an amoeboid and ovoid cells, and the assimilation profile of C-source. They designated the strain as *Schizochytrium limacinum* sp. nov.

TABLE 17.4   Thraustochytrids PUFA

Strain	Mode of Growth	PUFA Productivity/Growth Rate	Reference	No.
Thraustochytrium Motivum and T. Multirudimentale		Obligatory marine 2.5–3.0% NaCl, first description of thraustochytrids	Goldstein S. (1963)	105
T. aureum		Primarily a saprophyte of vascular plant debris, obligatory marine 2% NaCl, best N-source was glutamate	Goldstein (1963)	106
T. Roseum		Light accelerated the log growth phase/1.5–4.0% NaCl	Goldstein (1963)	107
T. aureum ATCC 34304	batch under light	Glu 5–20 g/L caused 2.7–17% LC, 26–270 mg-DHA/L for 6 days	Bajpai et al. (1991)	108
T. aureum ATCC 34304	batch	510 mg-DHA/L in light-exposed cultures containing 2.5% starch at 28°C (20% LC)	Bajpai et al. (1991)	109
T. Aureum	batch	4 g-DCW/L, 10% LC, 120 mg-DHA/L for 3 days	Kendrick & Ratledge (1992)	110
T. Roseum	batch under light	47%-DHA, 0.85 g-DHA/L for 5 days	Li & Word (1994)	111
DHA in Palestinian freshwater sponges	batch	(5,8,11,14-20:4) up to 10% of TFA, EPA up to 12% and DHA up to 12%	Dembitsky et al. (1996)	119
T. aureum ATCC 34304	batch	5.7 g-DCW/L & 460 mg-TL/L in 69 h-shake flask (DHA 40% of TFA), fermenter culture showed a lower growth	Iida et al. (1996)	114
Schizochytrium sp. strain SR21	batch (fermenter)	2.0 g-DHA/Ld, DHA 34%, 50% LC, TG 93% of TL	Nakahara et al. (1996)	115
Schizochytrium sp. strain SR21	batch	C/N resulted in LC (<50%), 4 g-DHA/L were obtained on 9% Glu or 12% glycerol	Yokochi et al. (1998)	116
Schizochytrium sp. strain SR21	batch (fermenter)	WithGlu & N-sources, DHA productivity, 48 g-DCW/L, 13 g-DHA/L for 4 days/At 12% Glu, 78% LC, 36% DHA	Yaguchi et al. (1997)	117
57 isolates from three different location	batch (fermenter)	Isolates from cold habitat had DHA content (<50%), but low growth, 14 g-DCW/L, 78% LC, 2.2 g-DHA/L for 107 h.	Bowles et al. (1999)	120

DHA: docosahexaenoic acid, TG: triacylglycerol

Fatty acid composition of three freshwater sponge were studied and revealed high levels of (5,8,11,14-20:4) up to 10.1% of TFA, EPA up to 11.6% and DHA up to 11.8%, which may be produced by coexisting thraustochytrids.[119] Bowles et al.[120] isolated thraustochytrids from three different locations, cold temperate littorals, cool temperate littoral and subtropical mangroves, and found that average DHA content were 35.9, 13.7%, and 16.6% for cold, cool, and subtropical location. Growth rate was highest in subtropical location, but isolates from the cold one had higher DHA content up to 50%, though with very low growth. Data on thraustochytrids' lipid are summarized in Table 17.4.

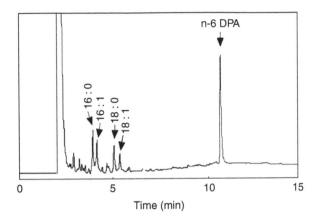

FIGURE 17. 2 GC chromatogram showing fatty acid profile of labyrinthulid L59.

Trial of PUFA production by labyrinthulids was reported by Kumon et al.[121] in monoxenic growth with *Psychlobacter phenylpyruvicus* on oil-dispersed agar medium and the maximum productivities were 0.59 g-PUFA/L and 4.93-DCW/L for 14 days. They[122] also isolated an labyrinthulid isolate from the northern part of Japan, strain L59, which contained only n-6DPA among all PUFAs (Figure 17.2), and obtained n-6DPA productivity at 0.53 g/L at 20°C for 7 days.

References

1. DJ Kyle, and C Ratledge, ed. Industrial Applications of Single Cell Oils. *Am Oil Chem Soc*, Champaign IL 1992
2. O Suzuki, T Yokochi, K Amano, T Sano, S Seto, Y Ohtsu, S Ishida, S Iwamoto, K Morioka, A Satoh, and K Uotani. Development in production of fat containing gamma-linolenic acid by fungi and its industrialization. *J Jpn Oil Chem Soc* 37:1081–1096, 1988 (in Japanese).
3. AG Marr, and JL Ingraham. Effect of temperature on the composition of fatty acids in *Escherichia coli, J Bacteriol* 84:1260–1267, 1962.
4. MK Shaw, and JL Ingraham. Fatty acid composition of *Escherichia coli* as a possible controlling factor of the minimal growth temperature. *J Bacteriol* 90:141–146, 1965.
5. JE Cronan, JR. Phospholipid alterations during growth of *Escherichia coli, J Bacteriol* 95:2054–2061, 1968.
6. CO Gill. Effect of growth temperature on the lipids of *Pseudomonas fluorescens. J Gen Microbiol* 89:293–298, 1975.
7. DE Minnikin, H Abdolrahimzadeh, and J Baddiley. Replacement of acidic phospholipids by acidic glycolipids in *Pseudomonus diminuta Nature* 249:268–269, 1974.
8. EF DeLong, and AA Yayanos. Adaptation of the membrane lipids of a deep-sea bacterium to changes in hydrostatic pressure. *Science* 228:1101–1103, 1985.
9. P Intriago, and GD Floodgate. Fatty acid composition of the estuarine Flexibacter sp. Strain Inp: effect of salinity, temperature and carbon source for growth. *J Gen Microbiol* 137:1503–1509, 1991.
10. RB Johns, and GJ Perry. Lipids of the marine bacterium *Flexibacter polymorphus. Arch Microbiol* 114:267–271, 1977.
11. EF DeLong, and AA Yayanos. Biochemical function and ecological significance of novel bacterial lipids in deep-sea procaryotes. *Appl Environ Microbiol* 51:730–737, 1986.
12. K Yazawa, K Araki, K Watanabe, C Ishikawa, A Inoue, K Kondo, S Watabe, and K Hashimoto. Eicosapentaenoic acid productivity of the bacteria isolated from fish intestines, *Nippon Suisan Gakkaishi* 54:1835–1838, 1988 (in Japanese).

13. K Yazawa, K Araki, N Okazaki, K Watanabe, C Ishikawa, A Inoue, N Numao, and K Kondo. Production of eicosapentaenoic acid by marine bacteria. *J Biochem* 103:5–7, 1988.

14. M Akimoto, T Ishii, K Yamagaki, K Ohtaguchi, K Koide, and K Yazawa. Production of eicosapentaenoic acid by a bacterium isolated from mackerel intestines. *J Am Oil Chem* Soc 67:911–915, 1990.

15. M Akimoto, K Yamagaki, K Ohtaguchi, and K Koide. Metabolism of L-amino acids in a marine bacterium isolated from mackerel intestines in relation to eicosapentaenoic acid biosynthesis. *Biosci Biotech Biochem* 56:1640–1643, 1992.

16. N Suzuki, K Yazawa, K Watanabe, Y Akahori, C Ishikawa, K Kondo, and K Takada. Culture condition of marine bacterium SCRC-2738 for production of eicosapentaenoic acid (EPA). *Nippon Suisan Gakkaishi* 58:323–328, 1992.

17. K Watanabe, C Ishikawa, H Inoue, D Cenhua, K Yazawa, and K Kondo. Incorporation of exogenous docosahexaenoic acid into various bacterial phospholipids. *J Am Oil Chem* Soc 71:325–330, 1994.

18. DS Nichols, PD Nichols, and TA Mcmeekin. Anaerobic production of polyunsaturated fatty acids by *Shewanella putrefaciens* strain ACAM342. *FEMS Microbiol Lett* 98:117–122, 1992.

19. E Ringo, PD Sinclair, H Birkbeck, and A Barbour. Production of eicosapentaenoic acid (20:5 n-3) by *Vibrio pelagius* isolated from turbot (*Scophthalmus maximus* (L.)) larxae, *Appl Environ Microbiol* 58:3777–3778, 1992.

20. E Ringo, JP Jostensen, and RE Olsen. Production of eicosapentaenoic acid by freshwater Vibrio. *Lipids* 27:564–566, 1992.

21. T Hamamoto, N Takata, T Kudo, and K Horikoshi. Effect of temperature and growth phase on fatty acid composition of the psychrophilic Vibrio sp. Strain no. 5710. *FEMS Microbiol Lett* 119:77–82, 1994.

22. T Hamamoto, N Takata, T Kudo, and K Horikoshi. Characteristic presence of polyunsaturated fatty acids in marine psychrophilic vibrios. *FEMS Microbiol Lett* 129:51–56, 1995.

23. H Iwanami, T Yamaguchi, and M Takeuchi. Fatty acid metabolism in bacteria that produce eicosapentaenoic acid isolated from sea urchin *Strongylocentrotus nudus*. *Nippon Suisan Gakkaishi* 61:205–210, 1995.

24. DS Nichols, PD Nichols, NJ Russell, NW Davies, and TA McMeekin. Polyunsaturated fatty acids in the psychrophilic bacterium *Shewanella gelidimarina* ACAM 456T: molecular species analysis of major phospholipids and biosynthesis of eicosapentaenoic acid. *Biochim Biophys Acta* 1347:164–176, 1997.

25. HM Alvarez, OH Pucci, and A Steinbuchel. Lipid storage compounds in marine bacteria. *Appl Microbiol Biotecnol* 47:132–139, 1997.

26. JP Bowman, SA McCammon, T Lewis, JH Skerratt, JL Brown, DS Nichols, and TA McMeekin. *Psychroflexus torquis* gen. nov., sp. nov., a psychrophilic species from Anatarctic sea ice, and reclassification of *Flavobacterium gondwanensa* (Dobson et al. 1993) as *psychroflexus gondwanense* gen. nov., comb. nov. *Microbiology* 144:1601–1609, 1998.

27. LP Bowman, JJ Gosink, SA McCammon, TE Lewis, DS Nichols, PD Nichols, JH Skerratt, JT Staley, and TA McMeekin. *Colwellia demingiae* sp. nov., and *Colwellia hornerae* sp. nov., *Colwellia rossensis* sp. nov., and *Colwellia psychrotropica* sp. nov.: Psychrophilic antarctic species with the ability to synthesize docosahexaenoic acid (22:6ω3). *Int J Sys Bacteriol* 48:1171–1180, 1998.

28. K Watanabe, C Ishikawa, Y Ohtsuka, M Kamata, M Tomita, K Yazawa, and H Muramatsu. Lipid and fatty acid compositions of a novel docosahexaenoic acid-producing marine bacterium. *Lipids* 32:975–978, 1997.

29. S De Rosa, A Milone, A Kujumgiev, K Stefanov, I Nechev, and S Popov. Metabolites from a marine bacterium Pseudomonas/Alteromonas, associated with the sponge *Dysidea fragilis*. *Comparative Biochem Physiol*, Part B 126:391–396, 2000.

30. L Enebo, G Anderson, and H Lundin. Microbiological fat synthesisi by Rhodotorula yeast. *Arch Biochem* 11:383–395, 1946.

31. TA Pedersen. Lipid formation in *Cryptococcus terricolus* I. Nitrogen nutrition and lipid formation. *Acta Chemica Scand* 15:651–662, 1961.

32. RH De Deken. The Crabtree effect: A regulatory system in yeast. *J Gen Microbiol* 44:149–156, 1966.

33. T Babij, FJ Moss, and BJ Ralph. Effects of oxygen and glucose levels on lipid composition of yeast Candida utilis grown in continuous culture. *Biotech Bioeng* 11:593–603, 1969.

34. CM Brown, and AH Rose. Fatty-acid composition of *Candida utilis* as affected by growth temperature and dissolved-oxygen tension. *J Bacteriol* 99:371–378, 1969.

35. B Johnson, SJ Nelson, and CM Brown. Influence of glucose concentration on the physiology and lipid composition of some yeasts. *Antonie van Leeuwenhoek* 38:129–136, 1972.

36. B Johnson, CM Brown and DE Minnikin. The effect of phosphorus limitation upon the lipids of *Saccharomyces cerevisiae* and *Candida utilis* grown in continuous culture. *J Gen Microbiol* 75: proceedings of Society for General Microbiology, 1973.

37. CM Brown, and B Johnson. Influence of the concentration of glucose and galactose on the physiology of *Saccharomyces cerevisiae* in continuous culture. *J Gen Microbiol* 64:279–287, 1970.

38. RHJ Kessell. Fatty acids of *Rhodotorula gracilis*: Fat production in submerged culture and the particular effect of pH value. *J Appl Bact* 31:220–231, 1968.

39. K Wilson, and BJ McLeod. The influence of conditions of growth on the endogenous metabolism of *Saccharomyces cerevisiae*: Effect on protein, carbohydrate, sterol and fatty acid content and on viability. *Antonie van Leeuwenhoek* 42:397–410, 1976.

40. CO Gill, MJ Hall, and C Ratledge. Lipid accumulation in an oleaginous yeast (Candida 107) growing on glucose in single-stage continuous culture, *Appl Environ Microbiol* 33:231–239, 1977.

41. SH Yoon, JW Rhim, SY Choi, DDY Ryn, and JS Rhee. *J Ferm Tech* 60:243–246, 1982.

42. CT Evans, and C Ratledge. A comparison of the oleaginous yeast, *Candida curvata*, grown on different carbon sources in continuous and batch culture. *Lipids* 18:623–629, 1983.

43. VP Zhelifonova, NI Krylova, EC Dedyukhina, and VK Eroshin. Investigation of lipid-forming yeasts growing on a medium with ethanol. *Mikrobiologiya* 52:219–224, 1983.

44. CA Boulton, and C Ratlege. Use of transition studies in continuous cultures of *Lipomyces starkeyi*, an oleaginous yeast, to investigate the physiology of lipid accumulation. *J Gen Microbiol* 129:2871–2876, 1983.

45. T Naganuma, Y Uzuka, and K Tanaka. Physiological factors affecting total cell number and lipid content of the yeast, *Lipomyces starkeyi*. *J Gen Appl Microbiol* 31:29–37, 1985.

46. T Naganuma, Y Uzuka, and K Tanaka. Using inorganic elements to control cell growth and lipid accumulation in *Lipomyces starkeyi*. *J Gen Appl Microbiol* 32:417–424, 1986.

47. A Ykema, EC Verbree, HW Van Verseveld, and H Smit. Mathematical modelling of lipid production by oleaginous yeasts in continuous cultures. *Antonie van Leeuwenhoek* 52:491–506, 1986.

48. EG Dedyukhina, NV Feoktistova, and VK Eroshin. Influence of specific growth rate on lipid synthesis by *Trichosporon pullulans* with growth limited by ethanol. *Microbiology* 56:346–349, 1987.

49. L Heredia, and C Ratledge. Simultaneous utilization of glucose and xylose by *Candida curvata* D in continuous culture. *Biotech Let* 10:25–30, 1988.

50. A Ykema, EC Verbree, MM Kater, and H Smit. Optimization of lipid production in the oleaginous yeast *Apiotrichum curvatum* in whey permeate. *Appl Microbiol Biotechnol* 29:211–218, 1988.

51. SG Prapulla, Z Jacob, N Chand, D Rajalakshmi, and NG Karanth. Maximization of lipid production by *Rhodotorula gracilis* CFR-1 using response surface methodology. *Biotech Bioeng* 40:965–970, 1992.

52. T Andlid, C Larsson, C Liljenberg, I Marison, and L Gustafsson. Enthalpy content as a function of lipid accumulation in *Rhodotorula glutinis*. *Appl Microbiol Biotechnol* 42:818–825, 1995.

53. K Hunter, and AH Rose. Lipid composition of *Saccharomyces cerevisiae* as influenced by growth temperature. *Biochem Biophys Acta* 260:639–653, 1972.

54. NJ Moon, and EG Hammond. Oil production by fermentation of lactose and the effect of temperature on the fatty acid composition, *J Am Oil Chem Soc* 55:683–688, 1978.

55. EG Dedyukhina, LP Dudina, and VK Eroshin. Composition of the lipids of *Hansenula Polymorpha* as a function of the conditions of continuous culturing *Microbiol* 49:30–33, 1979.

56. C Ratledge, and MJ Hall. Oxygen demand by lipid-accumulating yeasts in continuous culture. *Appl Environ Microbiol* 34:230–231, 1977.

57. SY Choi, DY Ryu, and JS Rhee. Production of microbial lipids: effects of growth rate and oxygen on lipid synthesis and fatty acid composition of *Rhodotorula gracilis, Biotech Bioen* 24:1165–1172, 1982.

58. EG Dedyukhina, NV Feoktistova, and VK Eroshin. Influence of specific growth rate on lipid synthesis by *Trichosporon pullulans* with growth limted by ethanol, *Microbiology* 56:346–349, 1987.

59. LM Granger, P Perlot, G Goma, and A Pareilleux. Kinetics of growth and fatty acid production of *Rhodotorula glutinis. Appl Microbiol Biotechnol* 37:13–17, 1992.

60. SH. Yoon, and JS Rhee. Lipid from yeast fermentation: Effects of cultural conditions on lipid production and its characteristics of *Rhodotorula glutinis. J Am Oil Chem Soc* 60:1281–1286, 1983.

61. L Hansson, and M Dostalek. Effect of culture conditions on fatty acid composition in lipids produced by the yeast *Cryptococcus albidus* var. abidus. *J Am Oil Chem Soc* 63:1179–1184, 1986.

62. JL Sumner, and ED Morgan. The fatty acid composition of sporangiospores and vegetative mycelium of temperature-adapted fungi in the order Mucorales. *J Gen Microbiol* 59:215–221, 1969.

63. RO Mumma, CL Fergus, and RD Sekura. The lipids of thermophilic fungi: Lipid composition comparisons between thermophilic and mesophilic fungi. *Lipids* 5:100–103, 1969.

64. PJ Roger, and FH Gleason. Metabolism of *Cokeromyces poitrasii* grown in glucose-limited continuous culture at controlled oxygen concentrations. *Mycologia* 66:919–925, 1974.

65. IV Konova, LM Rudakova, OI Pan'kina, and LV Orekhova. Lipogenesis in mycerial fungi in relation to their culture conditions. *Mikrobiologiya* 56:783–791, 1987.

66. A Kendrick, and C Ratledge. Lipid formation in the oleaginous mould *Entomophthora exitalis* grown in continuous culture: Effects of growth rate, temperature and dissolved oxygen tension on polyunsaturated fatty acids. *Appl Microbiol Biotech* 37:18–22, 1992.

67. R Shaw. The occurrence of γ-linolenic acid in fungi. *Biochem Biophys Acta* 98:230–237, 1965.

68. L Hansson, and M Dostalek. Effect of culture conditions on mycelial growth and production of γ-linolenic acid by the fungus *Mortierella ramanniana. Appl Microbiol Biotechnol* 28:240–246, 1988.

69. L Hansson, M Dostalek, and B Sorenby. Production of γ-linolenic acid by the fungus *Mucor rouxii* in fed-batch and continuous culture. *Appl Microbiol Biotechnol* 31:223–227, 1989.

70. A-M Lindberg, and L Hansson. Production of γ-linolenic acid by the fungus *Mucor rouxii* on cheap nitrogen and carbon sources. *Appl Microbiol Biotecnol* 36:26–28, 1991.

71. BO Torlanova, NS Funtikova, IV Konova, and NK Babanova. Synthesis of the lipid complex containing γ-linolenic acid and carotenoids by a Mucorous fungus under various cultivation conditions. *Microbiol* 64:417–421, 1995.

72. EV Emelyanova. Lipid and γ-linolenic acid production by *Mucor inaguisporus. Process Biochemistry* 32:173–177, 1997.

73. O Suzuki, T Yamashina, and T Yokochi. Studies on production of lipids in fungi 1: Lipid compositions of 13 species of Deuteromycetes. *J Jpn Oil Chem Soc* 30:854–862, 1981 (in Japanese).

74. O Suzuki, T Yokochi, and T Yamashina. Studies on production of lipids in fungi 2: Lipid compositions of six species of Mucorales in Zygomycetes. *J Jpn Oil Chem Soc* 30:863–868, 1981 (in Japanese).

75. O Suzuki, T Yokochi, and T Yamashina. Studies on production of lipids in fungi 3: Lipid compositions of five species of Pellicularia in Bacidiomycetes. *J Jpn Oil Chem Soc* 30:869–876, 1981 (in Japanese).

76. O Suzuki, T Yokochi, and T Yamashina. Studies on production of lipids in fungi 6: Change of lipid compositions in three species of the genus Pellicularia of Bacidiomycetes fungi by cultural conditions. *J Jpn Oil Chem Soc* 31:494–502, 1982 (in Japanese).

77. O Suzuki, T Yokochi, and T Yamashina. Studies on production of lipids in fungi 8: Influence of cultural conditions on lipid compositions of two strains of *Mortierella isabellina J Jpn Oil Chem Soc* 31:921–931, 1982 (in Japanese).

78. T Yokochi, and O Suzuki. Studies on production of lipids in fungi 10: Influence of cultural conditions on lipid compositions of two strains of *Mortierella isabellina* from n-paraffin as a carbon source. *J Jpn Oil Chem Soc* 31:993–1003, 1982 (in Japanese).

79. T Yokochi, and O Suzuki. Studies on production of lipids in fungi 16: Lipid composition of 33 strains of genus Mortierella by using glucose or decane as a carbon source. *J Jpn Oil Chem Soc* 35:929–936, 1986 (in Japanese).

80. Y Kamisaka, H Kukutsugi, T Yokochi, T Nakahara, and O Suzuki. Studies on production of lipids in fungi. XX. Effect of metal ions in cultural media and temperature on fungal lipids. *J Jpn Oil Chem Soc* 37:344–348, 1988.

81. T Yokochi, Y Kamisaka, T Nakahara, L Enoshita, and O Suzuki. Studies on production of lipids in fungi 21: Effect of cultural conditions on lipid productivity on *Mortierella isabellina* with a culture at high cell mass. *J Jpn Oil Chem Soc* 38:241–248, 1989 (in Japanese).

82. T Yokochi, and O Suzuki. Studies on production of lipids in fungi 17: Influence of cultural conditions of lipid productivity of *Mortierella isabellina*. *J Jpn Oil Chem Soc* 36:413–417, 1986 (in Japanese).

83. T Yokochi, Y Kamisaka, T Nakahara, and O Suzuki. Production of lipid containing gamma-linolenic acid by continuous culture of *Mortierella ramanniana*. *J Jpn Oil Chem Soc* 42:893–898, 1993.

84. O Hiruta, Y Kamisaka, T Yokochi, T Futamura, H Takebe, A Satoh, T Nakahara, and O Suzuki. γ-Linolenic acid production by a low temperature-resistant mutant of *Mortierella ramanniana*. *J Ferm Bioeng* 82:119–123, 1996.

85. O Hiruta, T Futamura, H Takebe, A Satoh, Y Kamisaka, T Yokochi, T Nakahara, and O Suzuki. Optimization and scale-up of γ-linolenic acid production by *Mortierella ramanniana* MM 15-1, a high γ-linolenic acid producing mutant. *J Ferm Bioeng* 82:366–370, 1996.

86. T Yokochi, Y Kamisaka, T Nakahara, and O Suzuki, Production of γ-linolenic acid-containing phospholipids by *Mortierella ramanniana* in a fed-batch culture of decane. *J Jpn Oil Chem Soc* 44:9–15, 1995.

87. N Totani, A Watanabe, and K Oba. An improved method of arachidonic acid production by *Mortierella alpina*. *J Jpn Oil Chem Soc* 36:328–331, 1987.

88. N Totani, and K Oba. The filamentous fungus *Mortierella alpina*, high in arachidonic acid. *Lipids* 22:1060–1062, 1987.

89. H Yamada, S Shimizu, and Y Shinmen. Production of arachidonic acid by *Mortierella elongata* 1S-5. *Agric Biol Chem* 51:785–790, 1987.

90. H Yamada, S Shimizu, Y Shinmen, H Kawashima, and K Akimoto. Production of arachidonic acid and eicosapentaenoic acid by microorganisms. World Conf on Biotech for the Fats and Oil Industry, 1988, pp. 173–177.

91. S Shimizu, H Kawashima, Y Shinmen, K Akimoto, and H Yamada. Production of eicosapentaenoic acid by *Mortierella fungi*. *J Am Oil Chem Soc* 65:1455–1459, 1988.

92. S Shimizu, Y Shinmen, H Kawashima, K Akimoto, H Yamada. Fungal mycelia as a novel source of eicosapentaenoic acid. Activation of enzyme(s) involved in eicosapentaenoic acid production at low temperature. *Biochem Biophys Res Commun* 150:335–341, 1988.

93. S Shimizu, K Akimoto, H Kawashima, Y Shinmen, and H Yamada. Production of dihomo-γ-linolenic acid by *Mortierella alpina* 1S-4. *J Am Oil Chem Soc* 66:237–241, 1989.

94. S Shimizu, K Akimoto, Y Shinmen, and H Kawashima. Sesamin is a potent and specific inhibitor of Δ5 desaturase in polyunsaturated fatty acid biosynthesis. *Lipids* 26:512–516, 1991.

95. S Shimizu, H Kawashima, K Akimoto, Y Shinmen, and H Yamada. Microbial conversion of an oil containing α-linolenic acid to an oil containing eicosapentaenoic acid. *J Am Oil Chem Soc* 66:342–347, 1989.

96. Y Shinmen, S Shimizu, K Akimoto, H Kawashima, and H Yamada. Production of arachidonic acid by Mortierella fungi. *Appl Microbiol Biotechnol* 31:11–16, 1989.

97. S Shimizu, H Kawashima, K Akimoto, Y Shinmen, and H Yamada. Production of odd chain polyunsaturated fatty acids by Mortierella fungi. *J Am Oil Chem Soc* 68:254–258, 1991.

98. S Shimizu, S Jareonkitmongkol, H Kawashima, K Akimoto, and H Yamada. Production of a novel 1-eicosapentaenoic acid by *Mortierella alpina* 1S-4 grown on 1-hexadecene. *Arch Microbiol* 156:163–166, 1991.

99. Y Shinmen, H Kawashima, S Shimizu, and H Yamada. Concentration of eicosapentaenoic acid and docosahexaenoic acid in an arachidonic acid-producing fungus, *Mortierella alpina* 1S-4, grown with fish oil. *Appl Microbiol Biotechnol* 38:301–304, 1992.

100. S Jareonkitmongkol, S Shimizu, and H Yamada. Fatty acid desaturation-defective mutants of an arachidonic-acid-producing fungus, *Mortierella alpina* 1S-4. *J Gen Microbiol* 138:997–1002, 1992.

101. S Jareonkitmongkol, H Kawashima, N Shirasaka, S Shimizu, and H Yamada. Production of dihomo-γ–Linolenic acid by a Δ5-desaturase-defective mutant of *Mortierella alpina* 1S-4. *Appl Environ Microbiol* 58:2196–2200, 1992.

102. S Jareonkitmongkol, H Kawashima, S Shimizu, and H Yamada. Production of 5,8,11-cis-eicosa-trienoic acid by a Δ12-desaturase-defective mutant of *Mortierella alpina* 1S-4. *J Am Oil Chem Soc* 69:939–944, 1992.

103. S Jareonkitmongkol, S Shimizu, and H Yamada. Production of an eicosapentaenoic acid-containing oil by a Δ12 desaturase-defective mutant of *Mortierella alpina* 1S-4. *J Am Oil Chem Soc* 70:119–141, 1993.

104. ED Korn, and CL Greenblatt. Synthesis of α-linolenic acid by *Leishmania enriettii*. *Science* 142:1301–1303, 1963.

105. S Goldstein. Development and nutrition of new species of Thraustochytrium. *Am J Botany* 50:271–279, 1963.

106. S Goldstein. Morphological variation and nutrition of a new monocentric fungus. *Arch Microbiol* 45:101–110, 1963.

107. S Goldstein. Studies of a new species of Thraustochytrium that displays light stimulated growth. *Mycologia* 55:799–811, 1963.

108. P Bajpai, PK Bajpai, and OP Ward. Production of docosahexaenoic acid by *Thraustochytrium aureum*. *Appl Microbiol Biotech* 35:706–710, 1991.

109. PK Bajpai, P Bajpai, and OP Ward. Optimization of production of docosahexaenoic acid (DHA) by Thraustochytrium aureum ATCC 34304. *J Am Oil Chem Soc* 68:509–514, 1991.

110. A Kendrick, and C Ratledge. Lipids of selected molds grown for production of n-3 and n-6 polyunsaturated fatty acids. *Lipids* 27:15–20, 1992.

111. Z Li, and OP Ward. Production of docosahexaenoic acid by *Thraustochytrium roseum*. *J Ind Microbiol* 13:238–241, 1994.

112. D Porter. *Phylum Labyrinthulomycota: Handbook of Protoctista*. Jones and Bartlett Publ., Athens, 1990, pp. 388–398.

113. D Honda, T Yokochi, T Nakahara, S Raghukumar, A Nakagiri, K Schaumann, and T Higashihara. Molecular phylogeny of labyrinthulids and thraustochytrids based on the sequencing of 18S ribosomal RNA gene. *J Eukaryot Microbiol* 46:637–647, 1999.

114. I Iida, T Nakahara, T Yokochi, Y Kamisaka, H Yagi, M Yamaoka, and O Suzuki. Improvement of docosahexaenoic acid production in a culture of *Thraustochytrium aureum* by medium optimization. *J Ferm Bioeng* 81:76–78, 1996.

115. T Nakahara, T Yokochi, T Higashihara, S Tanaka, T Yaguchi, and D Honda. Productioin of docosa-hexaenoic and docosapentaenoic acids by Schizochytrium sp. isolated from Yap Islands. *J Am Oil Chem Soc* 73:1421–1426, 1996.

116. T Yokochi, T Higashihara, and T Nakahara. Optimization of docosahexaenoic acid production by *Schizochytrium limacinum* SR21. *Appl Microbiol Biotech* 49:72–76, 1998.

117. T Yaguchi, S Tanaka, T Yokochi, T Nakahara, and T Higashihara. Production of high yields of docosahexaenoic acid by Schizochytrium sp. strain SR21. *J Am Oil Chem Soc* 74:1431–1434, 1997.

118. D Honda, T Yokochi, T Nakahara, M Erata, and T Higashihara. *Schizochytrium limacinum* sp. nov., A new thraustochytrid from a mangrove area in the west Pacific Ocean. *Mycol Res* 102:439–448, 1998.

119. VM Dembitsky, and T Rezanka, Unusually high levels of eicosateraenoic, eicosapentaenoic, and docosahexaenoic fatty acids in Palestinian freshwater sponges. *Lipids* 31:647–650, 1996.

120. RD Bowles, AE Hunt, GB Bremer, MG Duchars, and RA Eaton. Long-chain n-3 polyunsaturated fatty acid production by members of the marine protistan group the thraustochytrids: Screening of isolates and optimisation of docasahexaenoic acid production. *J Biotech 70*:193–202, 1999.

121. Y Kumon, T Yokochi, T Nakahara, and M Yamaoka. Production of long-chain polyunsaturated fatty acids by monoxenic growth of labyrinthulids on oil-dispersed agar medium. *Appl Microbiol Biotechnol* 60:275–278, 2002.

122. Y Kumon, R. Yokoyama, T Yokochi, D. Honda, and T Nakahara. A new labyrinthulid isolate, which solely produces n-6 docosapentaenoic acid. *Appl Microbiol Biotechnol* 61 Epub 2003 May 15.

18

Structured Triacylglycerols Comprising Omega-3 Polyunsaturated Fatty Acids

Gudmundur G. Haraldsson

18.1 Introduction

The term *structured lipids*[1–4] usually refers to lipids that have a predetermined composition and distribution of fatty acids at the glycerol backbone. In a broader sense, structured triacylglycerol (TAG) means any modified or synthetic oils and fats obtained from transesterification or esterification by chemical or enzymatic means. Structured TAGs constituting certain types of fatty acids at the end positions and different fatty acids at the mid-position of the glycerol backbone have gained increasing attention of scientists as dietary and health supplements. Of particular interest from the human nutritional point of view are structured TAG possessing biologically active long-chain polyunsaturated fatty acids (PUFA) located at the mid-position with medium-chain fatty acids (MCFA) at the end positions.[5,6] The reason is that the MCFA located at the end positions undergo a rapid hydrolysis by pancreatic lipase, are absorbed into the intestines, and are rapidly carried into the liver where they are consumed as a quick source of energy. The remaining 2-monoacylglycerols (2-MAG), on the other hand, become a source of essential fatty acids after being absorbed through the intestinal wall.[7] They are accumulated as TAGs in the adipose tissues or as phospholipids in the cell membranes from where they can be released upon demand for their desired biological functions.

The various beneficial health effects of the long-chain n-3 PUFA are firmly established.[8,9] They are almost exclusively associated with eicosapentaenoic acid (EPA) and docosahexaenoic acid (DHA), the two most prevalent n-3 PUFA in fish oil.[10] The strongest evidence relates to reductions in cardiovascular and heart diseases by both EPA and DHA. Besides that, the beneficial effects of EPA have been linked to various inflammatory disorders and DHA to pregnancy, infants, and brain and nervous system development.[11,12] Currently there is a great interest in the beneficial effects of EPA on various mental disorders[13] and schizophrenia.[14] This has resulted in strong demands for EPA and DHA concentrates by the health food industry as food supplements as well as the pharmaceutical industry for drugs. The n-3 PUFA concentrates may be divided into three classes. The first class consists of monoesters of various concentration levels, highly enriched with EPA or DHA or both, especially as ethyl esters, that have been developed into various health supplements and drugs.[10] Together with free acid PUFA, they can be used as starting material for structured lipids enriched with these fatty acids. The second class consists of n-3 PUFA concentrates in the natural TAG form or of high TAG content, which are also available in various enrichment levels where their fatty acid distribution into defined positions of the glycerol backbone is not of much concern.[10,15] This type of concentrate belongs to the more broadly defined type of structured TAG. Finally, structured TAGs containing EPA or DHA located at the mid-position with MCFA at the end positions of the glycerol moiety are the most sophisticated concentration form. Such positionally labeled structured TAGs of the MLM (medium-long-medium) type are in high demand.[16] They may in turn be divided into structured TAGs comprising various enrichment levels of the defined fatty acids into the glycerol positions, suitable for large-scale neutraceutical and health supplement products, and symmetrically structured TAG fulfilling criteria of fine chemicals and pharmaceuticals, intended as potential drugs, for clinical studies and as analytical standards. That type of product must fulfill the requirement of homogeneous fatty acids in the defined positions, and chemical and regioisomeric purity. Figure 18.1 illustrates such structured TAGs comprising EPA and capric acid (top) and DHA and caprylic acid (bottom).

Lipases are ideally suited as biocatalysts to various esterification and transesterification processes involving the highly labile n-3 PUFA because of the mild conditions under which they act.[17] Based on their fatty acid selectivity, lipases have been widely used to enrichment of n-3 PUFA in fish oil by kinetic resolution in hydrolysis, transesterification, and esterification reactions.[10] They can be used to highly efficiently make concentrates of EPA and DHA in a whole range of compositions. They may be used to concentrate EPA together with DHA[18] or they may offer a strong discrimination between EPA and DHA to concentrate EPA or DHA individually.[19] And, finally, lipases owing to their regioselectivity are perfectly suited as biocatalysts for preparing structured TAGs comprising n-3 PUFA at the mid-position and MCFA at the sn-1 and sn-3 positions.[10] By acting preferably at the primary end positions of the glycerol backbone,

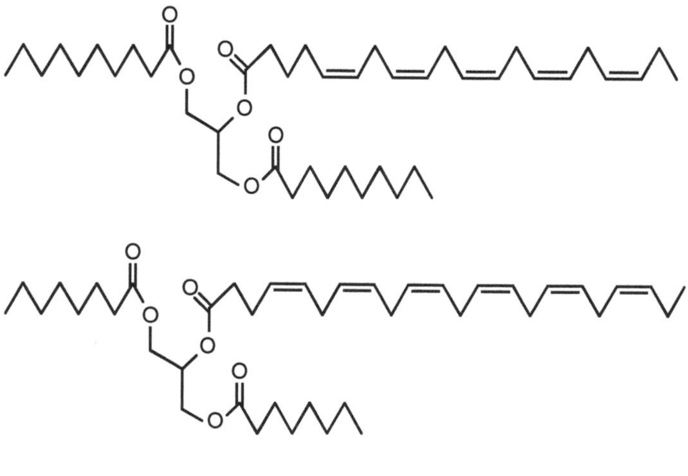

FIGURE 18.1 The structure of homogeneous positionally labeled symmetrically structured TAGs of the MLM type comprising capric acid and EPA (top) and caprylic acid and DHA (bottom).

they can be employed to introduce fatty acids of a certain type or composition at these positions by esterification or transesterification processes. Likewise, they may be used for regioselective hydrolysis or alcoholysis of esters located at these positions. The synthesis of such positionally labeled structured TAGs by traditional synthetic organic chemistry methods requires a full regioselectivity control and can hardly be undertaken without multistep protection-deprotection processes.

In this chapter structured TAGs comprising n-3 PUFA will be discussed and how they can be made by biotransformations involving lipase. First, structured TAGs belonging to the broader definition will be discussed, i.e., TAGs where the aim is high levels of EPA and DHA rather than concern about their distribution within the glycerol backbone. Then, strategies and problems related to preparation of positionally labeled structured TAGs will be described and discussed. This is followed by structurally labeled TAGs where the main concern is enrichment of n-3 PUFA and MCFA into specific positions of the glycerol moiety, especially of the MLM type. Finally, the main emphasis will be put on approaches to the synthesis of positionally labeled symmetrically labeled TAGs of the MLM type comprising pure homogeneous MCFA and EPA or DHA, first by a fully enzymatic approach and then by a detailed description of a chemoenzymatic approach. These methods will be compared and evaluated.

18.2 Strategies and Challenges Related to Structured TAG Synthesis

Various approaches have been developed recently to undertake the synthesis of positionally labeled structured TAGs comprised of n-3 PUFA and MCFA. The simplest method is to treat fish oil, of which the mid-position usually constitutes a significantly higher n-3 PUFA composition than the end positions,[20,21] with a regioselective lipase. The lipase acts preferably or exclusively at the end positions by promoting fatty acid exchange reactions with MCFA as free acids (acidolysis) or monoesters (interesterification). Alternatively, the fish oil TAGs may be alcoholized at the end positions by simple alcohols such as ethanol to accomplish 2-MAG enriched with the n-3 PUFA. A subsequent acylation with MCFA, either by direct esterification of MCFA as free acids, or by transesterification with MCFA as monoesters, will afford the structured TAG. When higher enrichment levels of n-3 PUFA are needed, extra concentration steps must be included. Either TAGs highly enriched with n-3 PUFA need to be prepared or an action must be taken to further enrichment of the 2-MAG, e.q. by low-temperature crystallization.

Positionally labeled symmetrically structured TAGs of the MLM type comprised of pure homogeneous EPA or DHA and MCFA require total synthesis from glycerol with a full regioselectivity control. There are two alternative approaches. One is a fully enzymatic approach involving two or three enzymatic steps. The first step involves the synthesis of homogeneous TAG comprising a single fatty acid that is intended to accommodate the mid-position. This is followed by a second enzymatic step to accomplish the corresponding 2-MAG homogeneous with EPA or DHA. This second enzymatic step is a hydrolysis by a regioselective lipase or, preferably, alcoholysis, for example, ethanolysis. The pure MCFA is then introduced to the end positions by lipase. The second approach is a two-step chemoenzymatic approach. In the first step, a 1,3-regioselective lipase is exploited to prepare 1,3-DAG of a pure MCFA. This is followed by a subsequent chemical coupling reaction of pure EPA or DHA into the free mid-position. Both approaches have their drawbacks and limitations. The full enzymatic approach is obstructed by the need of a threefold excess of the pure EPA or DHA. Also, large excesses of MCFA are often required in order to reach satisfactory results in terms of yields and purity of the final structured TAGs. The main advantage is that the overall process constitutes environmentally friendly processes, where no toxic and hazardous chemicals or organic solvents are involved. This approach is usually hampered by extreme difficulties in affording products of absolute regioisomeric and chemical purity that may need tedious purification processes in where organic solvents eventually will have to be introduced.

Provided that a strict regiocontrol is maintained from a fully regioselective lipase and a suitable coupling agent, the chemoenzymatic approach has the advantages of offering chemically and regioisomerically pure products where only stoichiometric amounts of the pure fatty acids are needed. This approach is ideally suited for synthesizing structured TAGs of absolute regioisomeric and chemical purity of compounds intended for clinical studies, potential drugs, libraries of pure compounds for comparison studies by biological screening, standards for analysis, and isotopic labeling. The drawbacks relate to use of chemicals and organic solvents, but that is widely practiced in pharmaceutical synthesis, and can be justified when the advantages in terms of purity are borne in mind. When homogeneous products of that purity are involved, traditional synthetic organic chemistry methods such as high-field ^{1}H and ^{13}C NMR spectroscopy may be used for analytical purposes, which are ideally suited to monitor the regioisomeric purity and regioselectivity control.

Whatever the approach or methods used, there are always various crucial aspects that must be watched carefully in relation to these processes. One is a strict regioselectivity control requiring a highly regioselective lipase. But it is just as important to keep acyl-migration side-reactions under control, which otherwise may ruin even the best of regioselective lipase. Acyl-migration is a major problem in regioselective acylation of polyhydroxy compounds such as glycerols and carbohydrates.[22] The acyl-migration process is an enzyme-independent intramolecular rearrangement working against regioselectivity. Such processes are speeded up by various factors, including temperature, pH, the presence of acid or base, type of solvent, and immobilized enzyme support material.[23,24] Mildness is one of the major benefits offered by enzymes and the mild conditions offered by lipases retard acyl-migration. Although the acyl-migration is an enzyme-independent process, it may be induced by lipase. This occurs in association with the presence of low quantities of water in the organic reaction medium essential for lipases to retain their optimal activity and consequent hydrolysis side reactions.[25,26] Such hydrolysis of a 1,3-DAG acyl group may easily maintain the presence of 1-MAG in the reaction mixture. It is believed that the acyl-migration rate for 1,3-DAG is lower than for 1-MAG, the latter being more prone to acyl-migration.[24] The extent of such hydrolysis side reaction is related to the optimal water content of the reaction medium, which appears to vary considerably among different lipases.

Synthesis of the positionally labeled structured TAG can hardly be undertaken by traditional synthetic organic chemistry methods without multistep protection-deprotection processes. Scheme 18.1 illustrates a possible approach to the synthesis of structured TAGs of the MLM type based on conventional synthetic organic chemistry methods. This proposed pathway involves a total of six steps: two protection and two deprotection steps. This approach, starting from glycerol, involves blocking of the primary hydroxyl groups of the glycerol moiety, protection of the secondary hydroxyl group, and deblocking to generate the 2-O-benzylglycerol adduct. This is followed by introduction of the MCFA to the end positions.

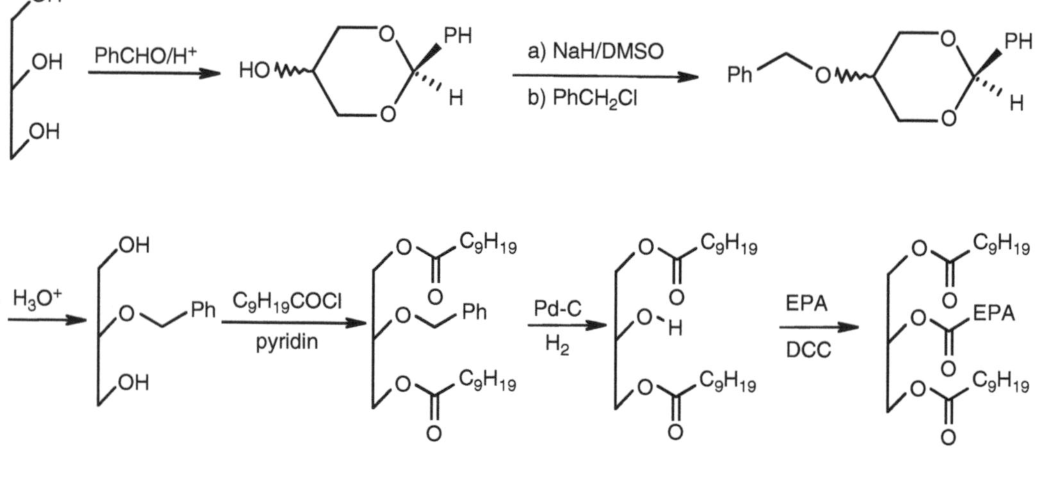

SCHEME 18.1 A proposed pathway for synthesis of an MLM type structured TAG comprising EPA and capric acid by a traditional synthetic organic chemistry approach.

Catalytic hydrogenolysis of the benzyl group affects the 1,3-DAG adduct, which subsequently is esterified with EPA by coupling with the dicyclohexylcarbodiimide (DCC) coupling agent. Assuming average yields of 75% for each step, it is evident that the overall yield of the final structured TAGs will remain very low or approximately 18%. By involving lipase in a two-step chemoenzymatic process, the total number of steps is reduced to two. Basing this on a 1,3-regioselective lipase for generating the 1,3-DAG intermediate and a subsequent coupling of the PUFA into the mid-position, the overall yield may rise to 70–80% provided that the lipase displays satisfactory regioselectivity and that an efficient coupling agent is found. Such a shortcut chemoenzymatic synthesis is illustrated in Section 18.6 (Figure 18.10).

18.3 Fish Oil–Derived Structured TAGs

Although concentrates of n-3 PUFA in the natural TAG form constituting randomly distributed fatty acids do not belong to positionally labeled structured lipids, they may certainly be considered as one category of structured TAG in the broad sense of the definition of such lipids. TAG concentrates of EPA and DHA were introduced as a dietary n-3 supplement to the market in the early 1980s as MaxEPA containing 18% EPA and 12% DHA.[27] It was widely used for various clinical studies for over 15 years. Preparation of TAGs up to the 30–35% EPA+DHA concentration level can be brought about directly on fish oils without splitting the fat by a careful selection of fish oils and simple methods such as winterization, molecular distillation, and solvent crystallization.[28,29] Concentration beyond that level requires splitting of the TAG into free acids or monoesters, concentration of EPA and DHA by various physical methods and combination of methods, and reintroduction of such free acid or monoester concentrates into TAG concentrates.[10,15]

Such resynthesis of TAG highly enriched with EPA and DHA is not easy by traditional chemical esterification methods, and in the late 1980s lipases were introduced to the n-3 field to solve these problems of producing TAG highly enriched with EPA and DHA. Haraldsson and coworkers were the first to report on the preparation of such TAG in the literature.[30,31] They used Lipozyme, an immobilized 1,3-regioselective lipase from the fungus *Rhizomucor miehei* commercially available from Novozyme A/S in Denmark, as a biocatalyst to bring about transesterification reactions of cod liver oil with EPA and DHA concentrates. Both acidolysis and interesterification reactions were conducted without a solvent, using 10% dosage of lipase as based on weight of fat at 60–65°C and a threefold excess of free acids or monoesters as based on number of mol equivalents of esters present in the fish oil TAG. TAG of high purity constituting 60–65% EPA+DHA and well over 70% total n-3 PUFA content were accomplished. This is illustrated in Figure 18.2.

Yamane and coworkers have also reported on a similar solvent-free methodology to enrich cod liver oil TAGs up to similar levels by lipase-catalyzed acidolysis using Lipozyme and a two-stage acidolysis approach.[32] They also designed a bioreactor packed-bed system for performing the acidolysis of cod liver oil with n-3 PUFA.[33] Adachi et al. reported on a similar acidolysis of sardine oil by *Pseudomonas sp.* lipase

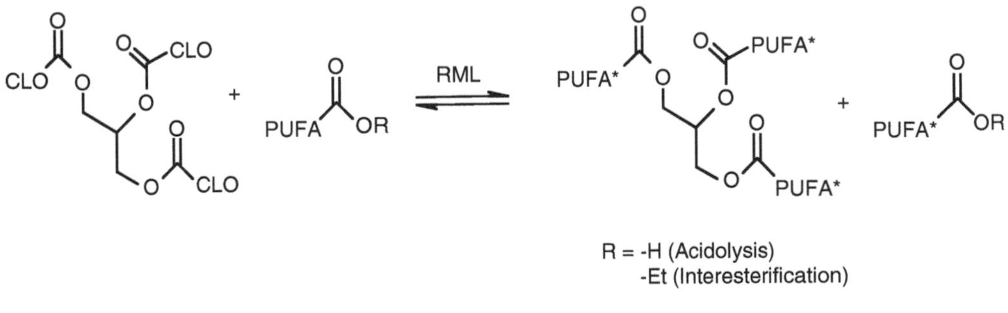

R = -H (Acidolysis)
  -Et (Interesterification)

FIGURE 18.2 Enrichment of cod liver oil (CLO) with PUFA by acidolysis (R = H) or interesterification (R = Et). CLO refers to cod liver oil fatty acid composition, but PUFA* to equilibrium composition.

in organic solvents.[34] The yield of TAGs and enrichment levels of EPA and DHA were strongly dependent on the water content. There are numerous reports describing treatment of various types of TAG oils with n-3 PUFA from both fish and single-cell origin in lipase-catalyzed transesterification reactions. These include incorporation of n-3 fatty acids into vegetable oils,[35,36] melon seed oil,[37] trilinolein,[38] evening primrose oil,[39] borage oil,[40–42] palm oil stearin,[43] and various TAG of medium-chain fatty acids, trilaurin, tricaprin, and tricaprylin.[44,45] It appears that the most efficient lipases used (immobilized *Rhizomucor miehei* and *Candida antarctica* lipases) acted preferably at the end positions. High-incorporation levels of the n-3 fatty acids were obtained into these positions, although high levels of n-3 fatty acids were also incorporated into the mid-position, but were lower depending on the reaction time. High levels of n-6 and n-3 fatty acids were accomplished with n-6 enriched oils such as borage and evening primrose oils. Most of these reactions were conducted in organic solvents and a few without a solvent, but there are also reports on such reactions under supercritical carbon dioxide conditions.[46]

The fish oil TAG transesterification approach was obstructed by the excessive amounts of n-3 PUFA concentrates needed to obtain high enrichment levels of EPA and DHA into fish oil.[15] Haraldsson and coworkers developed a procedure to produce TAGs of composition identical to that of the concentrate being used to avoid the above mentioned limitations.[47–50] This procedure is based on a direct esterification of stoichiometric amount of free fatty acids with glycerol and opened the possibility to synthesize TAG homogeneous with EPA or DHA, i.e., 100% EPA or DHA. An immobilized *Candida antarctica* lipase, now commercially available as Novozym 435 from Novozyme A/S in Denmark, was observed to be superior to Lipozyme in esterifying glycerol with free fatty acids of varying n-3 PUFA content. That lipase was highly efficient in generating TAGs of both 100% EPA and DHA content using only stoichiometric amount of pure EPA and DHA. No solvent was required and the reaction was performed at 65°C under vacuum with 10% dosage of the immobilized lipase as based on substrate weight. The vacuum was applied to pump off the coproduced water that was condensed into a liquid nitrogen-cooled trap as the reaction proceeded, thus shifting the reaction to completion. The reaction is displayed in Figure 18.3 for EPA. The TAGs homogeneous with both EPA and DHA of excellent purity were accomplished in virtually quantitative yields. High-field ^{1}H and ^{13}C NMR spectroscopy was of great use in monitoring the progress of the reaction and to follow EPA and DHA incorporation into glycerol to form the various acylglycerol intermediates participating in the esterification reaction (1- and 2-MAG, 1,2- and 1,3-DAG) and the TAG products (see Scheme 18.2). That technology was also of high value to establish the purity of the final products. Similar products were accomplished by interesterification of tributyrin and pure ethyl esters of EPA and DHA under identical conditions.[47,50] This reaction is illustrated in Figure 18.4. As with the water in the previous case, ethyl butyrate was pumped off from the reaction upon formation to drive the reaction to completion. The interesterification was observed to proceed significantly slower than the esterification reaction.

Kosugi and Azuma used the same methodology to prepare nearly pure TAG (96%) of EPA, DHA, and arachidonic acid under similar conditions using the *Candida antarctica* lipase.[51] Purification on alumina afforded the pure TAG that were analyzed by high-field ^{1}H NMR spectroscopy. There are also reports on a similar direct esterification of glycerol with n-3 PUFA concentrates where the reaction was conducted in an organic solvent. Medina and coworkers compared three commercially available lipases (Lipozyme, Novozym 435, and an immobilized *Pseudomonas cepacia* lipase) under the direct esterification conditions

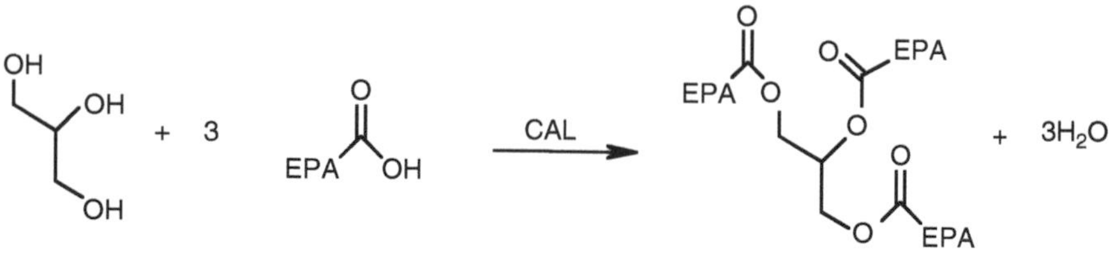

FIGURE 18.3 Direct esterification of glycerol with pure EPA to prepare structured TAGs homogeneous with EPA by immobilized *Candida antarctica* lipase (CAL).

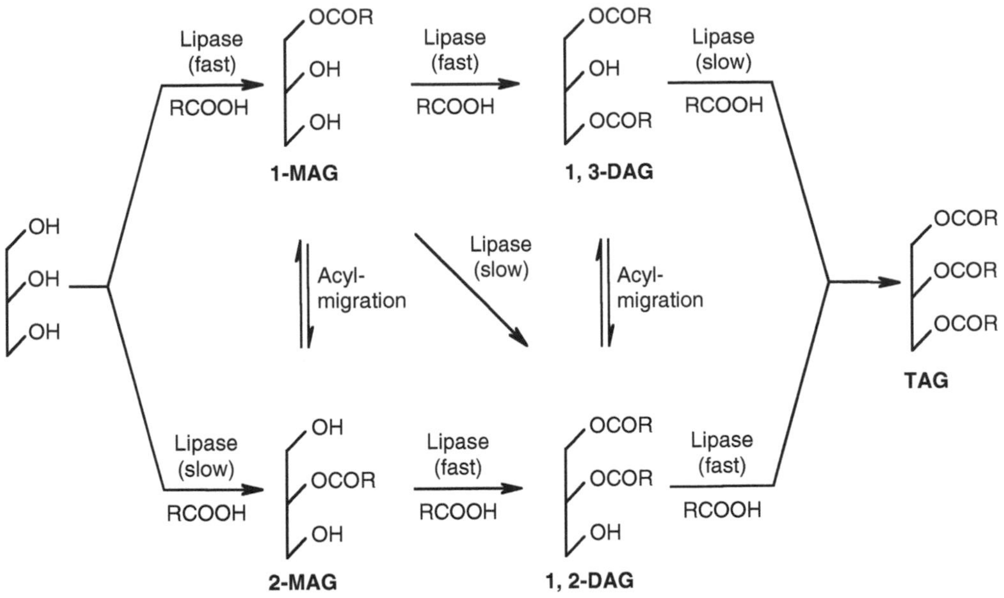

SCHEME 18.2 The overall process for the direct esterification of glycerol with a free fatty acid by *Candida antarctica* lipase showing all potential reactions and intermediates.

using glycerol and stoichiometric amount of an n-3 PUFA concentrate comprising 26% EPA and 48% DHA.[52] Hexane was used as a solvent and the reaction was conducted at 60°C, using 0.5% water as based on volume, but molecular sieves were added as a dehydrating agent to the reaction after 24 h to reduce the water content. The best results were obtained for Novozym 435 where TAGs of 85% purity constituting 27% EPA and 45% DHA were afforded after 48 h. Li and Ward[53] and He and Shahidi[54] investigated numerous lipases under esterification of glycerol with n-3 PUFA in organic solvents. The authors were obviously not much interested in products high in TAG since excessive amounts of glycerol were used and water was not eliminated during these reactions.

There are also reports on TAG synthesis by direct esterification of n-3 PUFA enriched partial acylglycerols obtained from *Candida rugosa* promoted hydrolysis of fish oil with n-3 PUFA as free fatty acids. Tanaka et al. treated such an acylglycerol mixture obtained at 70% hydrolysis level of tuna oil comprising 4% EPA and 53% DHA with n-3 PUFA comprising 23% EPA and 57% DHA to obtain TAGs of higher than 90% TAG purity.[55] The reaction was conducted at 50°C with a threefold molar excess of n-3 PUFA using 40 wt% molecular sieves as a dehydrating agent, using an immobilized *Chromobacterium viscosum* lipase. Similar results were obtained by McNeill and coworkers in their treatment of acylglycerol mixture from fish oil hydrolysis with stoichiometric amount of DHA-enriched fatty acids. TAGs of 95% purity were obtained with both Lipozyme and Novozym with continuous removal of water using vacuum at 55°C.[56,57]

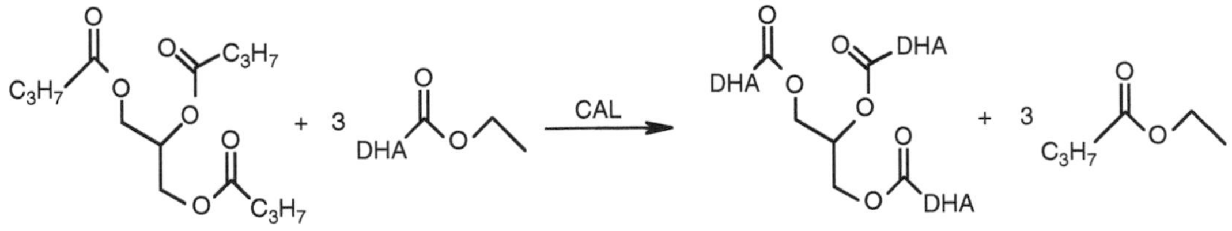

FIGURE 18.4 Interesterification of tributyrin and pure DHA ethyl ester to prepare structured TAGs homogeneous with DHA by immobilized *Candida antarctica* lipase (CAL).

There is no doubt that the immobilized *Candida antarctica* lipase offers superiority over other lipases in terms of TAG synthesis involving n-3 PUFA. That lipase is highly efficient, tolerating the n-3 PUFA very well. Highly pure TAGs were accomplished under the right conditions with very little or no contamination of any MAG or DAG partial acylglycerols. No solvent was needed and only stoichiometric amounts of substrates were required. That lipase is suitable for the production of pure TAG of any composition identical to that of the starting free acids as was demonstrated by Haraldsson and coworkers.[48] This lipase is therefore highly feasible for industrialization.[10]

18.4 Positionally Labeled Structured TAG Derived from Fish Oil

Shimada and coworkers reported on the production of structured TAGs of the MLM type comprising the mid-position enriched with DHA by exchanging fatty acids at the end positions of tuna oil for caprylic acid (CA; C8:0) using an immobilized 1,3-regioselective *Rhizopus delemar* lipase.[58,59] The reaction was conducted at 30°C using approximately sixfold molar excess of CA as based on the fish oil TAG. After 40 h the incorporation level of CA into the fish oil had reached a steady state and remained at 42.5 mol%. The immobilized lipase could be used 15 times (30 days) without a significant decrease in the CA content. Regiospecific analysis indicated that the regioselectivity of the lipase was very high and that the extent of acyl migration was very low. A total of 99% of the CA was confined to the end positions and the mid-position contained only 0.5% of the MCFA. The DHA content remained virtually unchanged during the reaction both at the mid-position (12.4%) and the end positions (7.8%). The fatty acid composition of the mid-position hardly changed during the reaction, whereas the fatty acid composition of the end positions changed dramatically, apart from DHA, which was resistant to the lipase action. High performance liquid chromatography (HPLC) analysis established that all the transesterified oil contained one or two MCFA residues at the end positions. The reaction is shown in Figure 18.5. Likewise, Shimada and coworkers have succeeded in producing structured TAG containing linoleic and α-linolenic acids at the mid-position with CA at the end positions highly successfully, using the same lipase and technology on safflower and linseed oils.[60] They have also produced similar γ-linolenic acid enriched structured TAGs in a continuous fixed-bed reactor from borage oil and CA.[61] Finally, they have reported on the incorporation of arachidonic acid into the end positions of tripalmitin using the same lipase and similar methodology for absorption studies purposes.[62] After 34 h at 40°C, 63 mol% arachidonic acid content was obtained into the structured TAG.

Jennings and Akoh reported on a similar acidolysis to incorporate capric acid (C10:0) into TAGs highly enriched with EPA (40.9%) and DHA (33.0%) with and without organic solvent at 55°C.[63] The highest capric acid incorporation levels of 65.4% were obtained in hexane with a molar ratio of 1:8 between the TAG and capric acid, but 56.1% under the solvent-free conditions with a 1:6 molar ratio. Analysis of the mid-position suggest that some acyl-migration was taking place during this reaction by the presence of capric acid at that position.

Yamane and coworkers enriched single-cell oil (SCO) of high DHA (35%) and docosapentaenoic acid (DPA; 10.2%) content with CA under acidolysis conditions using Lipozyme and *Pseudomonas sp.* lipases.[64]

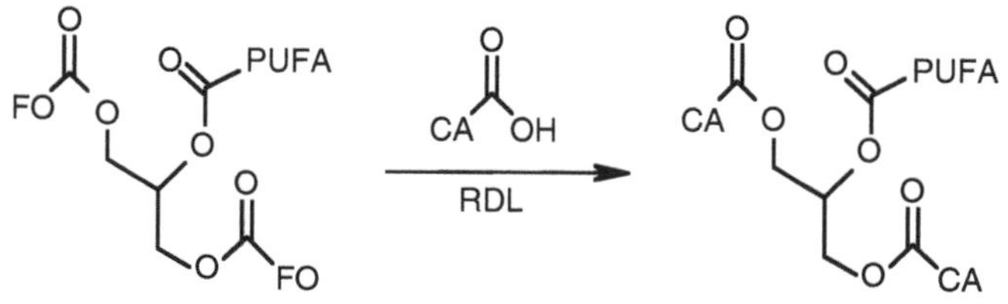

FIGURE 18.5 Production of fish oil–derived positionally labeled structured TAGs comprising caprylic acid (CA) and PUFA by a fish oil acidolysis using immobilized *Rhizopus delamar* lipase (RDL). (FO = fish oil fatty acids located at the end positions of the glycerol moiety of the fish oil TAGs.)

Their objective was to prepare structured TAGs containing CA residues at the end positions and DHA and DPA residues at the mid-position of the glycerol moiety. Both lipases required extended reaction time of several days and high ratios of CA to single-cell oil TAG. Much higher incorporation levels were obtained for the *Pseudomonas* lipase with the final CA content of the TAG reaching 65 mol% after 168 h at 18.8 CA/ SCO molar ratio at 30°C. The results for Lipozyme were strongly affected by low activity of the 1,3-regioselective lipase toward DHA residues located at the end positions of the TAG, which limited the yield of the desired structured TAG. The incorporation stopped at 23 mol% CA content of the TAG for CA/TAG molar ratio of 12.4. The reaction mixture was analyzed with high-temperature gas chromatography (HTGC) and silver-ion HPLC. The analysis indicated that *Pseudomonas* lipase was not nearly as 1,3-regioselective as the *Rhizomucor miehei* lipase since a large quantity of tricaprylin molecular species was formed (29%), indicating that the *Pseudomonas* lipase was acting directly at the mid-position. The elongated reaction time (6–7 days) indicates that both lipases were resistant to the n-3 PUFA. However, it is evident that the *Pseudomonas* lipase tolerated DHA located at the end positions of the TAG moiety much better than the *Rhizomucor miehei* lipase. It has been demonstrated by Haraldsson and coworkers.[19,65] that *Pseudomonas* lipases, unlike most lipases, display preference for DHA over EPA as a substrate. This may partially explain why that lipase offered much higher incorporation levels of CA than the other lipase. On top of that, the *Pseudomonas* lipase appears to have acted directly on the mid-position in accordance with claims to be nonregioselective and it was not obstructed by DHA located at either position.

Xu and coworkers studied the various aspects of structured TAGs of the MLM type extensively as well as their pilot batch production.[66] They investigated the effects of water content and reaction time on production of such positionally labeled TAGs of the MLM type under pilot batch conditions using Lipozyme on fish oil and capric acid (substrate ratio 6:1 FFA/TAG in mole) under solvent-free conditions at 60°C.[67] After 30 h more than 65% incorporation of the MCFA had taken place into the end positions together with 12% acyl-migration levels into the mid-position during the reaction. Distillation under vacuum was used to separate the structured TAG and free acid products under which further acyl-migration was noticed to take place. They also reported on the use of a packed-bed reactor with Lipozyme as a biocatalyst to treat menhaden oil under acidolysis with caprylic acid.[68] Parameters such as flow rate, temperature. and substrate ratio were studied. Structured TAG products containing 40% CA and 30% EPA+DHA were accomplished under optimal conditions with less than 3% CA escaping into the mid-position. Finally, Xu and coworkers have investigated the production of fish oil–related structured TAG in a packed-bed reactor exploiting an immobilized *Thermomyces lanuginosa* lipase recently introduced to the marked by Novozyme A/S in Denmark.[69] This lipase is immobilized on granulated silica as a very economic carrier and is primarily intended for large-scale oil interesterification processes. In their work, fish oil was interesterified with a TAG oil enriched with MCFA. In the bioreactor an equilibrium was reached at 30–40 min residence time at 60°C with the lipase remaining stable for 2 weeks without any adjustment of water content, column activity, or substrate mixture.

Bornscheuer and coworkers approached the task to generate such fish oil–derived structured TAGs differently by proposing a two-step strategy.[70,71] In the first step 2-MAG enriched with n-3 PUFA was generated from fish oil TAGs by lipase catalyzed ethanolysis by a 1,3-regioselective lipase in organic solvent. The resulting 2-MAG were subsequently esterified in a second enzymatic step (see Figure 18.6). This strategy worked well for less unsaturated TAGs, but when fish oils containing n-3 PUFA were used less favorable results were accomplished. This relates to the low yield of the 2-MAG intermediate as a result of low activity of the lipases toward TAGs comprising EPA and especially DHA, but also to complications in isolating and purifying the n-3 PUFA enriched 2-MAG. The first step was improved when lipases that usually are not considered to be 1,3-regioselective, *Pseudomonas sp.* and *Candida antarctica* lipases, were used in the ethanolysis reaction. A further n-3 PUFA enrichment of the 2-MAG from tuna oil was effected by low-temperature crystallization by freezing out the saturated 2-MAG.[72] A question remains as to how the resulting 2-MAG enriched with n-3 PUFA will be purified on a larger preparative scale since the 2-MAG must be highly susceptible to undergo temperature promoted acyl-migration during e.g., short-path distillation. Alternatively, this may be delayed until after the reesterification, which then has to be conducted in the presence of the fatty acid ethyl esters obtained in the first

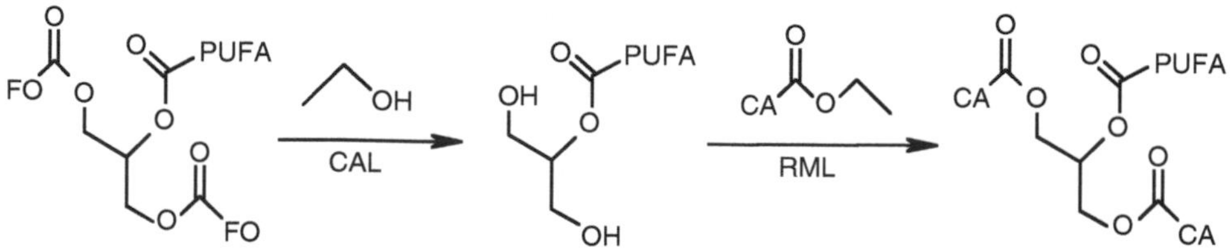

FIGURE 18.6 Production of fish oil–derived positionally labeled structured TAGs comprising caprylic acid (CA) and PUFA by a fish oil ethanolysis using immobilized *Candida antarctica* lipase (CAL) and reesterification with CA by *Rhizomucor miehei* lipase (RML). (FO = fish oil fatty acids located at the end positions of the glycerol moiety of the fish oil TAGs.)

step, but that obviously means a need for excessive amounts of the MCFA. This approach is illustrated in Figure 18.6.

Finally, Yamane and coworkers have recently reported on their modification of the above described methodology of Bornscheuer and coworkers on bonito oil for producing structured TAGs enriched with DHA in the mid-position and caprylic acid residues at the end positions.[73] This approach is illustrated in Figure 18.6. They exploited the immobilized *Candida antarctica* lipase to a highly efficient and regioselective ethanolysis of the initial TAG oil to yield 92.5% of the 2-MAG with 43.5% DHA content in only 2 h at 35°C. The subsequent reesterification was conducted directly on the crude reaction mixture in the presence of the fatty acid ethyl esters produced from the end positions of the original oil, after filtering off the enzyme and stripping off the excessive ethanol. The reaction was conducted for only 1 h under vacuum at 35°C using Lipozyme and excessive amount of ethyl caprylate (seven- to eightfold stoichiometric excess). Structured TAGs comprising well over 40% DHA at the mid-position and well over 90% CA content at the end positions were obtained. This method offers various advantages over other reported methods. It is fast, the regioselectivity is extremely high, and acyl-migration is kept at minimum. The yields of the 2-MAG intermediates are very high with the original fatty acid composition in that position largely preserved. However, an efficient separation, presumably by short-path distillation, and purification of the final product needs to be demonstrated.

18.5 Positionally Labeled Symmetrically Structured TAGs by a Fully Enzymatic Approach

Yamane's group developed a three-step fully enzymatic process starting from pure EPA ethyl ester to fulfill their objective of the production of regioisomerically pure structured TAGs of the MLM type comprising homogeneous EPA and caprylic acid at the defined positions.[74] This was based on a lipase promoted hydrolysis of pure EPA ethyl ester giving rise to EPA as free acid, *in situ* synthesis of TAG homogeneous with pure EPA (EEE), using the same lipase, and a subsequent interesterification with ethyl caprylate by a 1,3-regioselective lipase. This process is illustrated in Figure 18.7. In the first step the

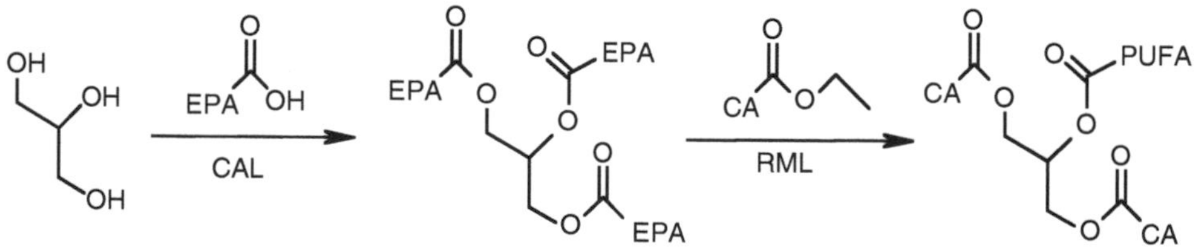

FIGURE 18.7 Synthesis of MLM type symmetrically structured TAGs comprising caprylic acid (CA) and EPA by a fully enzymatic approach using immobilized *Candida antarctica* lipase (CAL) for direct esterification and *Rhizomucor miehei* lipase (RML) for interesterification.

immobilized *Candida antarctica* lipase (Novozym 435) was used to hydrolyse EPA ethyl ester and then the same lipase preparation was subsequently used to esterify the resulting EPA free acid in tandem without any purification with glycerol under the reduced pressure conditions to afford the homogeneous EEE in 90% conversion yield with stoichiometric amount of substrates under solvent-free conditions. The crude reaction product mixture was then subjected to the third enzymatic step without purification, apart from immobilized lipase separation. That step was also conducted without a solvent using the immobilized 1,3-regioselective *Rhizomucor miehei* lipase (Lipozyme). A 100-fold molar excess (50-fold stoichiometric amount) of ethyl caprylate was used presumably to compensate for the presence of two molar equivalents of EPA ethyl ester from the previous step.[16] Water-content control was such that during the first part (10 h) of the reaction water was present to favor the conversion of EEE into CEC and CEOH, but after that vacuum was applied (3 h) to remove water and effect the reesterification part to bring about the CEC formation. Evidently, temperature related acyl-migration side reactions were taking place during the interesterification process, resulting in deterioration of the regioselectivity control. A compromise remained clearly between reaction rate, extent of acyl-migration, and loss of regioselectivity. Acyl-migration was also favored by increased water content. The final reaction product mixture constituted 88.5% of the aimed CEC, with apparently none of the CCE regioisomer present. However, other acyl-migration related products were present together with some key intermediates from the reaction. Apparently, no attempts were made to isolate and purify the desired product from the bulk of the reaction product mixture constituting only 3 wt% of the desired structured TAG. Therefore, it was not fully characterized by conventional analytical methods, and it is evident that the required purification will be tedious by fractional multistep molecular distillation and preparative column chromatography.

Comparison was made between the interesterification and acidolysis reactions by Yamane's group.[75] The interesterification reaction had clear advantages over acidolysis with free acids in terms of reaction rate and CEC production as well as extent of acyl-migration. This time a commercially available EEE was used in a two-step process consisting of transesterification in the presence of 2% water, followed by direct esterification under diminished water content under vacuum as before. The first part resulted in formation of 1,2-DAG and 2-MAG, which were then reesterified with caprylic acid during the second part under water-deficient conditions. The regioisomeric composition of the transesterification products was analyzed by silver ion high-performance liquid chromatography, but high-temperature gas chromatography was used to analyze the products from the lipase-promoted reactions.[76]

Shimada and coworkers made an effort to simplify Yamane's approach somewhat by treating TAGs homogeneous with EPA in three successive acidolysis reactions using 15 mol parts of caprylic acid.[77] The reaction was conducted for 48 h each time at 30°C using their immobilized *Rhizopus delemar* lipase. This was done in a similar manner as previously reported for their work on tuna oil.[58] After the three successive acidolysis reactions, the caprylic acid content of the TAG product reached 66 mol%. The product was still a mixture constituting 86 wt% of the desired CEC structured TAG, but it was contaminated with 2% of the undesired regioisomeric CCE. No attempts were made to isolate and purify the desired product or to fully characterize it. The presence of 2.5% of tricaprylin CCC and 2% of the wrong regioisomer in the reaction mixture indicates that some acyl-migration was taking place under these reaction conditions.

Yamane and coworkers have reported on a modification of their fully enzymatic approach to enable the synthesis of MLM type structured TAGs comprising MCFA such as CA and DHA (CDC).[78] The main impediment in the enzymatic synthesis of MLM type structured CDC is the very low activity of 1,3-regioselective lipase on TAG containing DHA and when DHA residues are located at the end positions. Such TAGs are very resistant to lipase action and DHA remains in place at the end positions, causing the final product to constitute a mixture of regioisomers. Yamane and coworkers managed to solve this by a modification based on a highly 1,3-regioselective lipase catalyzed ethanolysis of the homogeneous EPA and DHA TAGs, and a subsequent lipase promoted esterification of the resulting 2-MAG with a different lipase. The immobilized *Candida antarctica* lipase (Novozym 435) displayed an excellent regioselectivity in ethanolysis of both tridocosahexaenoylglycerol (DDD) and trieicosapentaenoylglycerol (EEE) at 35°C using a 3:1 (w/w; 33-fold stoichiometric excess) ratio of ethanol to TAG. This process is shown in Figure 18.8. For DHA a mixture was afforded comprising 92.7% of the desired 2-MAG together

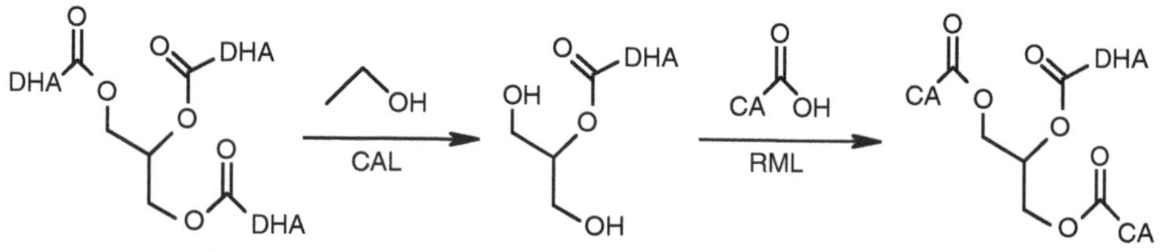

FIGURE 18.8 Synthesis of MLM type symmetrically structured TAGs comprising caprylic acid (CA) and EPA by a fully enzymatic approach using immobilized *Candida antarctica* lipase (CAL) for ethanolysis and *Rhizomucor miehei* lipase (RML) for acidolysis.

with 2% of starting TAG and 5.3% of the 1,2-DAG intermediate. There was no noticeable formation of any of the undesired regioisomers such as 1,2-DAG nor 1(3)-MAG during this step.

Apart from separating the lipase by filtration and removing excess ethanol by evaporation under reduced pressure, the crude product mixture with the DHA ethyl esters present was introduced to the second enzymatic reesterification reaction. This was transesterification with a 20-fold molar excess of ethyl caprylate using the immobilized *Rhizomucor miehei* lipase (Lipozyme) as the biocatalyst under reduced pressure (3–5 mmHg) at 35°C for 1–5 h. That vast excess of MCFA is apparently related to the presence of approximately two molar equivalents of DHA ethyl ester from the previous step, and was necessary in order to diminish reesterification of the 2-MAG with DHA residues present in the reaction mixture. The product mixture comprised 85.4% CDC together with 11.8% of CCOH (mostly as 1,3-TAG), presumably formed by acyl-migration during the second step, 0.8% of the intermediary CDOH and 2% of DDC. There was no contamination with the wrong regioisomer of the desired structured TAGs. Similar results were obtained for the corresponding EPA synthesis. The product constituted not only the structured TAGs contaminated with other acylglycerols that need to be separated, but also there were two molar equivalents of EPA or DHA ethyl esters and excessive amounts of ethyl caprylate that needed to be separated. Therefore, the desired structured TAG product constituted only less than 15% by weight in the crude reaction product mixture. As before, the structured TAGs were not isolated, nor were they fully characterized. It is evident that tedious chromatography procedures will be needed for purifying these positionally homogeneously labeled structured TAGs obtained from the fully enzymatic approach, demanding organic solvents. Apart from that, it is highly probable that this methodology of Yamane and coworkers may become of great use for the synthesis of such structured lipids on a large scale where the presence of various related isomers and intermediates in relatively low quantities will be tolerated. But for pure material both in terms of chemical composition and regioisomeric purity for strict clinical and pharmaceutical purposes, other methods offering a full regiocontrol will be necessary.

It is of interest to note how the regioselectivity of the *Candida antarctica* lipase depends on the reaction type and the reaction conditions. Whether this is a matter of controlling the acyl-migration, regioselectivity of the enzyme or both needs to be clarified unequivocally. That enzyme displayed a strict 1,3-regioselectivity for the ethanolysis reaction at a high excess of ethanol in the reaction mixture, diminishing by lower ethanol content, but no regioselectivity in transesterification and esterification reactions. The first reaction took only 4–6 h, but the subsequent reesterification reaction proceeded even faster and was completed after 1.5 h. Despite its strict regioselectivity in the first reaction, the *Candida* lipase could not be used in the second step. Nor could Lipozyme be used in the first step, despite its reported 1,3-regioselectivity. It is worth noting that only very few lipases have found use in the processes for preparing structured TAGs of the MLM type. These are the *Candida antarctica*, *Rhizomucor miehei*, and *Rhizopus delemar* lipases and to some extent the *Pseudomonas sp.* lipase.

Yamane's method has recently been modified and significantly improved by Hou and coworkers.[79] They managed to improve the regiocontrol and yield in the first step by lowering the temperature to 25°C and carefully controlling the ratio of the substrates to a molar ratio of 77:1 between ethanol and the homogeneous TAGs of DHA and various other PUFA, including EPA. For DHA the final acylglycerol reaction mixture constituted 97.1% of the desired 2-MAG (together with 2.9% of the corresponding 1(3),2-DAG) after 7 h in 93% recovery yield as based on initial TAG, but evidently, some glycerol was formed during the reaction (varying from 3% to 20% depending on type of PUFA). For the corresponding EPA synthesis, the purity of the 2-MAG was 98% in 75% recovery yield. Some 98% pure 2-DAG of DHA was accomplished from the reaction mixture after purification by selective extraction in 87.1% yield. This treatment required the use of organic solvents (acetonitrile, hexane, and chloroform) and was necessary to get rid of excessive amounts of the PUFA, ethanol, and the coproduced glycerol. The subsequent reesterification step was dramatically improved when the purified 2-MAG was directly esterified with stoichiometric amount of caprylic acid at 25°C with Lipozyme under vacuum. Some 96% of the desired DHA structured TAGs adduct was obtained after 8 h in 100% regioisomeric purity, the final reaction product mixture comprising 96.3% CDC together with 3.7% CDOH. Again, isolated yields of the purified structured TAGs from these highly successful processs were not reported and need to be demonstrated together with a full characterization of these products and their 2-MAG intermediates.

18.6 Positionally Labeled Symmetrically Structured TAG by the Chemoenzymatic Approach

In the chemoenzymatic approach, lipase regioselectivity is exploited to synthesize symmetric 1,3-DAG. This is followed by chemical introduction of a different type of fatty acid, usually a long-chain PUFA, into the mid-position, which results in a symmetrically structured TAG of the MLM type. The basis for the enzymatic part of the chemoenzymatic approach was laid by Schneider and coworkers in their lipase promoted 1,3-DAG generation.[80,81] A whole range of 1,3-DAG of pure medium- and long-chain saturated fatty acids was obtained in good yields (74–85%) and regioisomerically pure from glycerol adsorbed on silica gel using several 1,3-regioselective lipases. The reactions were carried out in organic solvents such as hexane, diethyl ether, and t-butyl methyl ether of low water content at room temperature in the presence of molecular sieves. The immobilized *Rhizomucor miehei* lipase provided the best results, but good selectivity was also obtained with 1,3-regioselective lipases from *Rhizopus delemar* and *Chromobacterium viscosum*. The nonregioselective *Pseudomonas sp.* lipase was clearly inferior to the others. Vinyl esters were superior to free acids and methyl esters as acyl donors in terms of reaction rates (24 h) and product yields. The regiocontrol of these methods, however, was by no means perfect since there were clear indications of acyl-migration side reactions taking place under their conditions. Similar methodology was used to generate the corresponding 1(3)-MAG adducts in high yields[82] as well as for the synthesis of 1,3-DAG derivatives of numerous unsaturated fatty acids from their vinyl esters.[83]

Yamane and coworkers were the first to report on a chemoenzymatic synthesis of an MLM type structured TAG comprising pure EPA and caprylic acid.[84] The process is illustrated in Figure 18.9. In the first step, regioisomerically pure 1,3-DAG was prepared by modification of the procedure of Schneider and coworkers using Lipozyme as the biocatalyst. Stoichiometric amount of free acid was used in a solvent-free system at 25°C under vacuum to remove the coproduced water. Yields comparable to those of Schneider and coworkers were obtained (75–80%). The 1,3-DAG content in the reaction mixture reached a maximum

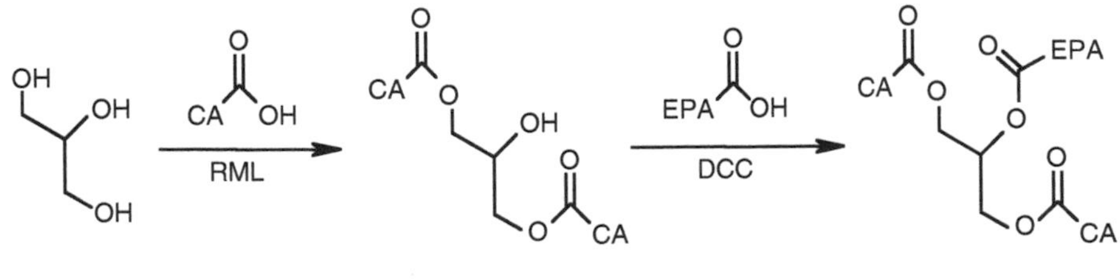

FIGURE 18.9 Chemoenzymatic synthesis of MLM type symmetrically structured TAGs comprising caprylic acid (CA) and EPA by immobilized *Rhizomucor miehei* lipase (RML) and dicyclohexylcarbodiimide coupling agent (DCC) by Yamane's approach.

of 85% after 8 h at 95% conversion. A 98% conversion had been obtained after 12 h at which the 1,3-DAG content had dropped to below 80%. The modification did not fulfill the requirement of a strict regiocontrol since acyl migration was observed to take place during the reaction, as was witnessed by the presence of substantial amount of TAG (approximately 10%). The presence of the TAG precursor, 1,2-DAG, however, remained very low throughout the reaction, but the 1-MAG intermediate content was relatively high at approximately 10% toward the end. No 2-MAG was detected during the reaction. Purified 1,3-DAG (95%) was then subjected to the subsequent chemical esterification step, where dicyclohexylcarbodiimide (DCC) was used as a chemical coupling agent to introduce pure EPA into the mid-position of the 1,3-DAG in the presence of 4-dimethylaminopyridine (DMAP). TAG of 98% purity was obtained, but in only 42% yield, and the product turned out not to be regioisomerically pure. It constituted 10% of the wrong MML type regioisomer as was established by silver-ion HPLC analysis. Thus, the chemoenzymatic approach was unsatisfactory and failed to provide regioisomerically pure structured TAG of the purity and quality anticipated.

Haraldsson and coworkers managed to improve dramatically on the chemical coupling step by using 1-(3-dimethylaminopropyl)-3-ethylcarbodiimide (EDCI) as a coupling medium in the presence of DMAP. This was demonstrated in their synthesis of a structured TAG comprising pure stearic acid residue at the end positions with pure EPA or DHA at the mid-position by the chemoenzymatic approach.[85] The enzymatic step was based on the procedure of Schneider and coworkers in ether to afford regioisomerically pure 1,3-distearoylglycerol in good yields (74%) after purification by crystallization. Exactly two equivalents of stearic acid as based on glycerol were used and the reaction was conducted in ether at r.t. in the presence of silica gel and molecular sieves using Lipozyme as the biocatalyst. Pure EPA and DHA were then subjected to the mid-position using EDCI as a coupling agent in the presence of DMAP in dichloromethane in excellent yields (94% and 91%, respectively, for EPA and DHA), after purification by treatment on silica gel to get rid of chemical waste and a slight excess of the PUFA. All products and intermediates were fully characterized by traditional organic chemistry methods. High field ^{1}H and ^{13}C NMR spectroscopy was of great use to monitor the regiocontrol of both reactions and to establish the regioisomeric purity of all compounds involved. The dramatic improvement in the coupling reaction is worth noting both in terms of the excellent yields and the preserved regiopurity under the reaction conditions. With DCC Yamane and coworkers obtained only low yields with acyl-migration evidently taking place during the reaction that was conducted under similar conditions, but with excessive amounts of EPA, the coupling agent and DMAP.[84]

The synthesis and full characterization of the regioisomeric structured TAG comprising pure EPA and DHA located at one of the outer positions with stearic acid residues at the mid-position and the remaining end position was also reported. This was brought about by a full enzymatic two-step synthesis. In the first step, tristearoylglycerol was prepared in very high yield (88%) by directly esterifying glycerol with stoichiometric amount of stearic acid under vacuum at 70–75°C using the immobilized *Candida antarctica* lipase without a solvent. That adduct was subsequently treated in an acidolysis reaction with two equivalents of EPA or DHA without a solvent at 70–75°C or in toluene at 40°C using Lipozyme as a biocatalyst. The EPA or DHA structured TAGs were afforded in only 44% and 29% yields, respectively, after purification, but chemically and regioisomerically pure. The reason for undertaking this work relates to indications from the literature that location of PUFA such as EPA or DHA in TAG may influence their oxidative stability.[86,87] Preliminary comparative oxidative stability studies performed on these adducts indicated that TAG comprising PUFA at the mid-position are indeed more resistant to oxidations.[88]

Haraldsson and coworkers used the same chemoenzymatic methodology to synthesize various adducts of structured TAG of the MLM type comprising pure MCFA including caprylic, capric, and lauric acid at the end positions and pure EPA or DHA at the mid-position.[89] The 1,3-DAG intermediates were afforded regioisomerically pure in moderate yields, 55–70%, after recrystallization in a highly pure state by the procedure of Schneider and coworkers. As before, EDCI was used to introduce the pure EPA and DHA into the mid-position and all the TAG products were offered in excellent yields after purification.

Very recently, Haraldsson and coworkers have reported on a modification of their chemoenzymatic approach toward synthesis of structured TAG of the above MLM type.[90] A dramatic improvement of the

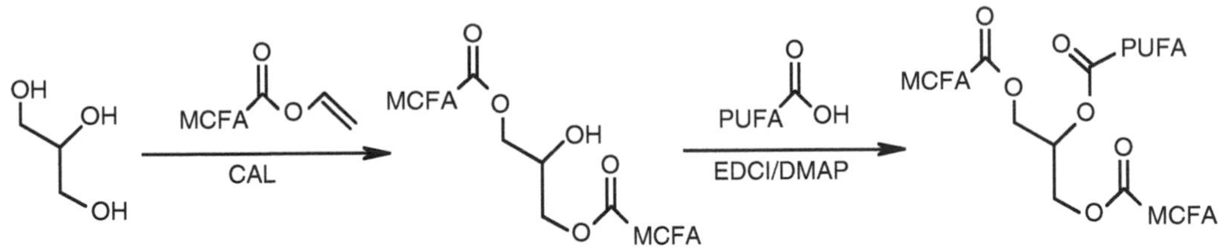

FIGURE 18.10 Chemoenzymatic synthesis of MLM type symmetrically structured TAGs comprising MCFA (C8–C12) and PUFA (EPA or DHA) by immobilized *Candida antarctica* lipase (CAL) and EDCI.

regiocontrol and yields of the enzymatic step was described. This is based on a rapid, irreversible transesterification of glycerol using vinyl esters of the MCFA and the immobilized *Candida antarctica* lipase at 0–4°C. This was followed by coupling of pure EPA and DHA as previously described into the mid-position by EDCI in the presence of DMAP. Their approach is demonstrated in Figure 18.10. A whole series of structured MLM type TAGs was synthesized by this methodology ranging from C8 to C16 saturated fatty acids. These products are listed in Table 18.1 and, as can be noticed, they were all obtained in excellent yields. The yields are based on isolated and purified material, and all compounds were fully characterized and their regioisomeric and chemical purity established by traditional modern organic chemistry methods.

The *Candida antarctica* lipase acted exclusively at the glycerol end positions and no acyl migration took place. The enzymatic reaction was conducted at 0–4°C in dichloromethane or chloroform using 10% dosage of lipase as based on weight of the substrates. Only a minimum amount of solvent needed to dissolve the substrates was used, approximately 0.1 ml per g of substrates. The progress of the reaction was monitored by TLC on silica and ¹H NMR spectroscopy. During the reaction only 1-MAG intermediate was detected in small quantities with the 1,3-DAG product largely dominating the reaction mixture. When the reaction had proceeded to completion, after only 3–5 h virtually quantitative conversion into the desired 1,3-DAG was obtained with only traces of the 1-MAG intermediate left. There were no signs of the unwanted 1,2-DAG regioisomer or 2-MAG, nor was any TAG present and thereby no indications of any acyl-migration side reactions or the lipase acting at the mid-position. However, prolonged reaction time resulted in some TAG formation presumably involving acyl-migration. Also, at higher temperatures (20 and 40°C) there were clear indications of loss of regioselectivity, implying that the temperature is a crucial factor in terms of controlling the regioselectivity of the reaction. To secure yields passing the 90% levels 2.5 equivalents of the vinyl esters were needed, corresponding to 1.25-fold stoichiometric amount, or 25% excess. The excessive amount of the vinyl esters relates to a speed-up of the reaction and hydrolysis

TABLE 18.1 Yields and Type of Products and 2-DAG Intermediates (1a,b,c) from the Chemoenzymatic Synthesis of MLM Type Structured TAGs Comprising Pure MCFA and EPA (2a,b,c) or DHA (3a,b,c)

Compound	MCFA	PUFA	Yield
1a	$-C_7H_{15}$	—	90%
1b	$-C_9H_{19}$	—	92%
1c	$-C_{11}H_{23}$	—	90%
2a	$-C_7H_{15}$	EPA	90%
2b	$-C_9H_{19}$	EPA	93%
2c	$-C_{11}H_{23}$	EPA	92%
3a	$-C_7H_{15}$	DHA	90%
3b	$-C_9H_{19}$	DHA	94%
3c	$-C_{11}H_{23}$	DHA	95%

side reaction competing with glycerol and 1-MAG for the vinyl esters to produce free fatty acids. The yields (90–92%; see Table 18.1) are based on pure material after recrystallization.

In the subsequent coupling reaction EDCI was used as a chemical coupling agent to introduce EPA and DHA into the mid-position of the 1,3-DAG adducts. The reaction was conducted at r.t. in dichloromethane in the presence of 30–50% DMAP (as based on mol) using an exact stoichiometric amount of EPA or DHA as based on the 1,3-DAG adduct. The reactions were completed in 12–15 h. Chemically and regioisomerically pure structured TAGs were afforded as colorless and slightly yellowish oils, respectively, for the EPA and DHA adducts, in yields of 90–95% (see Table 18.1) after chromatography treatment on silica gel. No sign of any acyl-migration side reaction was observed to take place during the coupling reaction.

The excellent results predominantly relate to the elimination of any acyl-migration side reactions and the enzyme displaying a superb regioselectivity. This interrelates to various important factors, including the temperature, apparently the most crucial single parameter; fast and efficient reaction; support material of lipase not inducing acyl-migration; type of reaction; acyl donor; and reaction conditions. Scheme 18.2 may be used to illustrate the complexity involved and all potential reactions, reaction intermediates, and products that may take part in the lipase-promoted reaction. The temperature was maintained low enough to keep the acyl-migration and lipase regioselectivity completely under control. At the same time the enzymatic reaction was fast enough to cope with an adequate reaction rate, since the acyl-migration process is clearly time-dependent. This was brought about by the use of activated vinyl esters, which offered a fast and irreversible reaction, and the lipase tolerated very well as an acylating substrate. The type of reaction and the nature of acyl donor is also very important. This is an irreversible transesterification of vinyl esters and the glycerol adducts. It is irreversible because of the enolic leaving group tautomerizing to a non-nucleophilic acetaldehyde. It is remarkable that Schneider and coworkers observed far lower reaction rates (24 h) under related conditions with vinyl esters as acyl donors in organic solvents at r.t. using the *Rhizomucor miehei* lipase. This clearly demonstrates that the selection of lipase and reaction conditions is of high importance in biotransformation reactions of this type.

Under the described conditions the lipase acted fast enough to eliminate any acyl-migration and also irreversibly to eliminate any equilibrium problems. Prolonged reaction beyond 5 h under the same conditions at 0°C resulted in a noticeable formation of adducts bearing acyl group at the mid-position, 1,2-DAG and TAG. When the *Candida* lipase was used under direct esterification conditions with free acids instead of vinyl esters, the result was a far lower degree of regioselectivity even at 0°C and a much slower reaction. The same lipase was previously used to synthesize homogeneous TAG of EPA and DHA at much higher temperature, 60–65°C, highly efficiently where acyl-migration was believed to play a key role in such synthesis induced by the high temperature and protons from the free fatty acids.[50]

Yamane and coworkers have also observed the regioselectivity of the *Candida antarctica* lipase to be very much dependent on reaction type and conditions.[78] The lipase acted highly regioselectively and effciently in its 1,3-regioselective ethanolysis of homogeneous TAG of EPA or DHA to produce 2-MAG. Interestingly, that lipase did not offer satisfactory results when used to reintroduce MCFA into the free 1,3-positions of the resulting 2-MAG, nor could it be used to incorporate MCFA into the end positions of the homogeneous TAG either by acidolysis or interesterification reactions. The *Rhizomucor miehei* lipase, on the other hand, proved to be highly profitable for that sort of processes, but was far less useful for the TAG ethanolysis approach. When the *Rhizomucor miehei* lipase was substituted for the *Candida antarctica* lipase under identical conditions, losses in 1,3-regioselectivity and yields were obtained. A mixture of 2-MAG, 1,2-DAG, and TAG was clearly noticeable with the latter lipase, as certainly was the case when using the original method of Schneider and coworkers with the same lipase in ether at r.t. for both free acids and vinyl esters as acylating agents. It is clearly evident that the *Rhizomucor* lipase is inferior to the *Candida* lipase to produce 1,3-DAG. When comparing the two lipases, the fact that they are immobilized on different supports should be kept in mind.

The high-field ^{1}H and ^{13}C NMR spectroscopy played a crucial role in monitoring the regioselectivity control in these reactions and to evaluate the purity of the compounds. Previous work revealed that all individual acylglycerols potentially involved in these reactions, 1- and 2-MAG, 1,3- and 1,2-DAG, as well as the TAG adducts (see Scheme 18.2), display quite characteristic ^{1}H NMR spectra in the glyceryl moiety proton region regardless of the type of acyl groups present.[50] This made it quite straightforward to locate all the individual acylglycerol constituents present in the product mixture and to quantify them by reasonable or good accuracy, and, thereby, to monitor the progress of the reaction as it proceeded as well as monitoring the regiocontrol of the reaction. One of the main advantages of this technique is that there is no need to separate the acylglycerol intermediates and they can be measured directly in the sample mixture as soon as samples from the reaction mixture are collected. Similarly, ^{13}C NMR spectroscopy was also of great use to monitor the regiocontrol of the reactions. This is based on two distinctive resonance signals being obtained in the ^{13}C NMR spectra for the carbonyl group carbon of each fatty acid, depending on the location of the acyl groups at the end positions or the mid-position of the glyceryl backbone.[50,85,89] This has also enabled the evaluation of the structured TAG products described in this report in terms of regioisomeric purity. It should, however, be borne in mind that the accuracy of the ^{13}C NMR spectroscopy is not very high, especially when the weak carbonyl signals are involved. Still, it is a good complement to other methods such as ^{1}H NMR spectroscopy and chromatography.

18.7 Concluding Remarks

Finally, it may be of interest to briefly compare and evaluate the two approaches that have been introduced for synthesizing positionally labeled structured TAGs of the MLM type comprising homogeneous fatty acids into the defined positions. The aim is clearly a strictly isomerically and chemically pure MLM type structured TAG of pharmaceutical quality intended for clinical studies, as drug candidates, fine chemicals as standards, and so forth. The main benefit of the fully enzymatic approach is clearly that no toxic chemicals or hazardous organic solvents are involved. However, as is clearly evident from the literature it is hardly possible to accomplish the above stated criteria in terms of chemical and regioisomeric purity without a proper purification after each step for what organic solvents are indeed required. This is clearly implicated from the report of Hou and coworkers and the numerous reports of Yamane and coworkers described earlier in this chapter. The chemoenzymatic approach certainly is dependent on organic solvents, but in relatively low quantities, especially in the enzymatic step, where extremely low volumes of chloroform were involved. The chemical coupling step indeed requires toxic EDCI and DMAP. It should, however, be borne in mind that the nature and qualities of the target products are quite comparable to drugs, that they may become potential drugs, and that the pharmaceutical industry is heavily dependent on multistep synthetic organic chemistry. Therefore, the use of EDCI and DMAP should be easily tolerated and accepted.

From the synthetic organic chemistry point of view, the use of excessive amounts of the highly valuable and expensive pure EPA and DHA is a drawback in the fully enzymatic approach, even though both may be easily recovered and reused after purification. It may also be pointed out that the 2-MAG intermediates involved in the fully enzymatic approach are less stable than the 1,3-DAG intermediates involved in the chemoenzymatic approach and more susceptible to undergo acyl-migration.[24] Finally, the isolation and proper purification of the products and intermediates from the fully enzymatic process is still lacking.

References

1. FD Gunstone, Ed. *Structured and Modified Lipids*. New York: Marcel Decker, Inc., 2001.
2. FD Gunstone. Movements towards tailor-made fats. *Prog Lipid Res* 37: 277–305, 1998.
3. CC Akoh. Structured Lipids. In: CC Akoh and DB Min, Eds. *Food Lipids. Chemistry, Nutrition and Biotechnology*. New York: Marcel Dekker, Inc., 1998, Chapter 26, pp. 699–727.
4. AB Christophe, Ed. *Structural Modified Food Fats: Synthesis, Biochemistry, and Use*. Champaign, IL: AOCS Press, 1998.
5. S Miura, A Ogawa, H Konishi. A rapid method for enzymatic synthesis and purification of the structured triacylglycerol, 1,3-Dilauroyl-2-oleoyl-glycerol. *J Am Oil Chem Soc* 76:927–931, 1999.
6. LB Fomuso, CC Akoh. Structured lipids: lipase-catalyzed interesterification of tricaproin and trilinolein. *J Am Oil Chem Soc* 75:405–410, 1998.

7. MS Christensen, C-E Höy, CC Becker, TG Redgrave. Intestinal absorption and lymphatic transport of eicosapentaenoic (EPA), docosahexaenoic (DHA), and decanoic acids: dependence on intramolecular triacylglycerol structure. *Am J Clin Nutr* 61:56–61, 1995.
8. ME Stansby. Nutritional properties of fish oil for human consumption—early development. In: ME Stansby, Ed. *Fish Oils in Nutrition*. New York: van Nostrand Reinhold, 1990, Chapter 10, pp. 268–288.
9. JA Nettleton. *Omega-3 Fatty Acids and Health*. New York: Chapman and Hall, 1995.
10. GG Haraldsson, B Hjaltason. Fish oils as sources of important polyunsaturated fatty acids. In: FD Gunstone, Ed. *Structured and Modified Lipids*. New York: Marcel Decker, Inc., 2001, pp. 313–350.
11. J Jumpsen, MT Clandinin. *Brain Development: Relationship to Dietary Lipid and Lipid Metabolism*. Champaign, IL: AOCS Press, 1995.
12. BF Haumann. Nutritional aspects of n-3 fatty acids. *INFORM* 8:428–447, 1997.
13. AL Stoll. *The Omega-3 Connection: The Groundbreaking Anti-Depression Diet and Brain Program*, New York: Simon & Schuster, 2001.
14. DF Horrobin. *Schizophr Res* 30:193–208, 1998.
15. GG Haraldsson. Enrichment of lipids with EPA and DHA by lipase. In: UT Bornscheuer, Ed. *Enzymes in Lipid Modification*. Weinheim: Wiley-VCF, 2000, pp. 170–189.
16. T Yamane. Lipase-catalyzed synthesis of structured triacylglycerols containing polyunsaturated fatty acids: monitoring the reaction and increasing the yield. In: UT Bornscheuer, Ed. *Enzymes in Lipid Modification*. Weinheim: Wiley-VCF, 2000, pp. 148–169.
17. GG Haraldsson, B Hjaltason. Using biotechnology to modify marine lipids. *INFORM* 3:626–629, 1992.
18. GG Haraldsson, B Kristinsson, R Sigurdardottir, GG Gudmundsson, H Breivik. The preparation of concentrates of eicosapentaenoic acid and docosahexaenoic acid by lipase-catalyzed transesterification of fish oil with ethanol. *J Am Oil Chem Soc* 74:1419–1424, 1997.
19. GG Haraldsson, B. Kristinsson. Separation of eicosapentaenoic acid and docosahexaenoic acid in fish oil by kinetic resolution using lipase. *J Am Oil Chem Soc* 75:1551–1556, 1998.
20. G Hölmer. Triglycerides. In: RG Ackman, Ed. *Marine Biogenic Lipids, Fats and Oils*. Vol. II, Boca Raton, FL: CRC Press, 1989, pp. 139–173.
21. WW Christie. The positional distributions of fatty acids in triglycerides. In: RJ Hamilton and JB Rossell, Eds. *Analysis of Oils and Fats*. London: Elsevier Science, 1986, pp. 313–339.
22. DR Kodali, A Tercyak, DA Fahey, DM Small. Acyl migration in 1,2-dipalmitoyl-*sn*-glycerol. *Chem Phys Lipids* 52:163–170, 1990.
23. S Bloomer, P Adlercreutz, M Mattiasson. Triglyceride interesterification by lipases. 2. Reaction parameters for the reduction of trisaturated impurities and diglycerides in batch reactions. *Biocatalysis* 5:145–162, 1991.
24. AM Fureby, C Virto, P Adlercreutz, B Mattiasson. Acyl group migrations in 2-monoolein. *Biocat Biotrans* 14:89–111, 1996.
25. JS Dordick. Enzymatic catalysis in monophasic organic solvents. *Enzym Microb Technol* 11:194–211, 1989.
26. AMP Koskinen, AM Klibanov, Eds. *Enzymatic Reactions in Organic Media*. London: Blackie Academic and Professional, 1996.
27. *A History of British Cod Liver Oils. The First 50 Years with Seven Seas*. Cambridge, UK: Martin Books, 1994.
28. RG Ackman. The year of the fish oils. *Chemistry and Industry* March 7:139–145, 1988.
29. H Breivik, KH Dahl. Production and control of n-3 fatty acids. In: JC Frölich, C von Schacky, Eds. *Fish, Fish Oil and Human Health, Clinical Pharmacology*. Vol. 5. München: W. Zuckschwerdt Verlag, 1992, pp. 25–39.
30. GG Haraldsson, PA Höskuldsson, STh Sigurdsson, F Thorsteinsson, S Gudbjarnason. The preparation of triglycerides highly enriched with ω-3 polyunsaturated fatty acids via lipase catalyzed interesterification. *Tetrahedron Lett* 30:1671–1674, 1989.

31. GG Haraldsson, Ö Almarsson. Studies on the positional specificity of lipase from *Mucor miehei* during interesterification reactions of cod liver oil with n-3 polyunsaturated fatty acid and ethyl ester concentrates. *Acta Chemica Scandinavica* 45:723–730, 1991.

32. T Yamane, T Suzuki, Y Sahashi, L Vikersveen, T Hoshino. Production of n-3 polyunsaturated fatty acid-enriched fish oil by lipase-catalyzed acidolysis without a solvent. *J Am Oil Chem Soc* 69:1104–1107 (1992).

33. T Yamane, T Suzuki,T Hoshino. Increasing n-3 polyunsaturated fatty acid content of fish oil by temperature control of lipase-catalyzed acidolysis. *J Am Oil Chem Soc* 70:1285–1287, 1993.

34. S Adachi, K Okumura, Y Ota, M Mankura. Acidolysis of sardine oil by lipase to concentrate eicosapentaenoic and docosahexaenoic acids in glycerides. *J Ferment Bioeng* 75:259–264, 1993.

35. KH Huang, CC Akoh. Lipase-catalyzed incorporation of n-3 polyunsaturated fatty acids into vegetable oils. *J Am Oil Chem Soc* 71:1277–1280, 1994.

36. ZY Li, OP Ward. Enzyme-catalyzed production of vegetable oils containing omega-3 polyunsaturated fatty acid. *Biotechnol Lett* 15:185–188, 1993.

37. KH Huang, CC Akoh, MC Erickson. Enzymatic modification of melon seed oil: incorporation of eicosapentaenoic acid. *J Agric Food Chem* 42:2646–2648, 1994.

38. CC Akoh, BH Jennings, DA Lillard. Enzymatic modification of trilinolein: incorporation of n-3 polyunsaturated fatty acids. *J Am Oil Chem Soc* 72:1317–1321, 1995.

39. CC Akoh, BH Jennings, DA Lillard. Enzymatic modification of evening primrose oil: incorporation of n-3 polyunsaturated fatty acids. *J Am Oil Chem Soc* 73:1059–1062, 1996.

40. CC Akoh, CO Moussata. Lipase-catalyzed modification of borage oil: incorporation of capric and eicosapentaenoic acids to form structured lipids. *J Am Oil Chem Soc* 75:697–701, 1998.

41. Y-H Ju, F-C Huang, C-H Fang. The incorporation of n-3 polyunsaturated fatty acids into acylglycerols of borage oil via lipase-catalyzed reactions. *J Am Oil Chem Soc* 75:961–965, 1998.

42. SPJN Senanayake, F Shahidi. Enzymatic incorporation of docosahexaenoic acid into borage oil. *J Am Oil Chem Soc* 76:1009–1015, 1999.

43. NM Osorio, S Ferreira-Dias, JH Gusmao, MMR da Fonseca. Response surface modelling of the production of ω-3 polyunsaturated fatty acids-enriched fats by a commercial immobilized lipase. *J Mol Cat*. B: Enzymatic 11:677–686, 2001.

44. K-T Lee, CC Akoh. Immobilized lipase-catalyzed production of structured lipids with eicosapentaenoic acid at specific positions. *J Am Oil Chem Soc* 73:611–615, 1996.

45. K-T Lee, CC Akoh. Characterization of enzymatically synthesized structured lipids containing eicosapentaenoic, docosahexaenoic, and caprylic acids. *J Am Oil Chem Soc* 75:495–499, 1998.

46. A Shishikura, K Fujimoto, T Suzuki, K Arai. Improved lipase-catalyzed incorporation of long chain fatty acids into medium-chain triglycerides assisted by supercritical carbon dioxide extraction. *J Am Oil Chem Soc* 71:961–967, 1994.

47. GG Haraldsson, H Svanholm, TT Hansen, B Hjaltason. Preparation of triglycerides of long-chain polyunsaturated fatty acids-e.g. eicosapentaenoic acid, docosahexaenoic acid-by interesterification of long-chain fatty acid with triglyceride of short-chain fatty acid with *Candida antarctica* immobilized lipase. WO9012–858, 1990.

48. GG Haraldsson, B Hjaltason, H Svanholm. Preparation of triglycerides of long-chain polyunsaturated fatty acids-e.g. eicosapentaenoic acid, docosahexaenoic acid-by interesterification of long-chain fatty acid and glycerol with *Candida antarctica* immobilized lipase. WO9116–443, 1991.

49. GG Haraldsson, BÖ Gudmundsson, Ö Almarsson. The preparation of homogeneous triglycerides of eicosapentaenoic acid and docosahexaenoic acid by lipase. *Tetrahedron Lett* 34:5791–5794, 1993.

50. GG Haraldsson, BÖ Gudmundsson, Ö Almarsson. The synthesis of homogeneous triglycerides of eicosapentaenoic acid and docosahexaenoic acid by lipase, *Tetrahedron* 51:941–952, 1995.

51. Y Kosugi, N Azuma. Synthesis of triacylglycerol from polyunsaturated fatty acid by immobilized lipase. *J Am Oil Chem Soc* 71:1397–1403, 1994.

52. LE Cerdan, AR Medina, AG Gimenez, MJI Gonzalez, EM Grima, E.M. Synthesis of polyunsaturated fatty acid-enriched triglycerides by lipase-catalyzed esterification. *J Am Oil Chem Soc* 75:1329–1337, 1998.

53. ZY Li, OP Ward. Lipase-catalyzed esterification of glycerol and n-3 polyunsaturated fatty acid concentrate in organic solvent. *J Am Oil Chem Soc* 70:745–748, 1993.

54. Y He, F Shahidi. Enzyme esterification of ω-3 fatty acid concentrates from seal blubber oil with glycerol. *J Am Oil Chem Soc* 74:1133–1136, 1998.

55. Y Tanaka, J Hirano, T Funada. Synthesis of docosahexaenoic acid-rich triglyceride with immobilized *Chromobacterium viscosum* lipase. *J Am Oil Chem Soc* 71:331–334, 1994.

56. GP McNeill, RG Ackman, SR Moore. Lipase-catalyzed enrichment of long-chain polyunsaturated fatty acids. *J Am Oil Chem Soc* 73:1403–1407, 1996.

57. SR Moore, GP McNeill. Production of triglycerides enriched in long-chain n-3 polyunsaturated fatty acids from fish oil. *J Am Oil Chem Soc* 73:1409–1414, 1996.

.58. Y Shimada, A Sugihara, K Maruyama, T Nagao, S Nakayama, H Nakano, Y Tominaga. Production of structured lipid containing docosahexaenoic and caprylic acids using immobilized *Rhizopus delemar* lipase. *J Ferment Bioeng* 81:299–303, 1996.

59. Y Shimada, A Sugihara, Y Tominaga. Production of functional lipids containing polyunsaturated fatty acids with lipase. In: UT Bornscheuer, Ed. *Enzymes in Lipid Modification*. Weinheim: Wiley-VCF, 2000, pp. 128–147.

60. Y Shimada, A Sugihara, H Nakano, T Yokota, T Nagao, S Komemushi, Y Tominaga. Production of structured lipid containing essential fatty acids by immobilized *Rhizopus delemar* lipase. *J Am Oil Chem* Soc 73:1415–1420, 1996.

61. Y Shimada, M Suenaga, A Sugihara, S Nakai, Y Tominaga. Continuous production of structured lipid containing γ-linolenic and caprylic acids by immobilized *Rhizopus delemar* lipase. *J Am Oil Chem Soc* 76:189–193, 1999.

62. Y Shimada, T Nagao, Y Hamasaki, K Akimoto, A Sugihara, S Fujikawa, S Komemushi, Y Tominaga. Enzymatic synthesis of structured lipid containing arachidonic and palmitic acids. *J Am Oil Chem Soc* 77:89–93, 2000.

63. BH Jennings, CC Akoh. Enzymatic modification of triacylglycerols of high eicosapentaenoic and docosahexaenoic acids content to produce structured lipids. *J Am Oil Chem Soc* 76:1133–1137, 1999.

64. Y Iwasaki, JJ Han, M Narita, R Rosu, T Yamane. Enzymatic synthesis of structured lipids from single cell oil of high docosahexaenoic acid. *J Am Oil Chem Soc* 76:563–569, 1999.

65. A Halldorsson, B Kristinsson, GG Haraldsson. Lipase selectivity toward fatty acids commonly found in fish oil. *Eur. J Lipid Sci Technol* 106:79–87, 2004.

66. X Xu. Production of Specific Structured Lipids by Lipase-Catalyzed Interesterification. Ph.D. Thesis. Technical University of Denmark: Department of Biotechnology, 1999.

67. X Xu, S Balchen, C-E Höy, J Adler-Nielsen. Pilot batch production of specific-structured lipids by lipase-catalyzed interesterification: preliminary study on incorporation and acyl migration. *J Am Oil Chem Soc* 75:301–308, 1998.

68. X Xu, LB Fomuso, CC Akoh. Modification of menhaden oil by enzymatic acidolysis to produce structured lipids: optimization by response surface design in a packed-bed reactor. *J Am Oil Chem Soc* 77:172–176, 2000.

69. X Xu, T Porsgaard, H Zhang, J Adler-Nielsen, C-E Höy. Production of structured lipids in a packed-bed reactor with *Thermomyces lanuginosa* lipase. *J Am Oil Chem Soc* 79:561–565, 2002.

70. MM Soumanou, UT Bornscheuer, RD Schmid. Two-step enzymatic reaction for the synthesis of pure structured triacylglycerols. *J Am Oil Chem Soc* 75:703–710, 1998.

71. U Schmid, UT Bornscheuer, MM Soumanou, GP McNeill, RD Schmid. Optimization of the reaction conditions in the lipase-catalyzed synthesis of structured triglycerides. *J Am Oil Chem Soc* 75:1527–1531, 1998.

72. S Wongsakul, P Prasertsan, UT Bornscheuer, A H-Kittikun. Synthesis of 2-monoglycerides by alcoholysis of palm oil and tuna oil using immobilized lipase. *Eur J Lipid Sci Technol* 105:68–73, 2003.

73. R Irimescu, K Furihata, K Hata, Y Iwasaki, T Yamane. Two-step enzymatic synthesis of docosa-hexaenoic acid-rich symmetrically structured triacylglycerols *via* 2-monoacylglycerols. *J Am Oil Chem Soc* 78:743–748, 2001.

74. R Irimescu, M Yasui, Y Iwasaki, N Shimidzu, T Yamane. Enzymatic synthesis of 1,3-dicaproyl-2-eicosapentaenoylglycerol. *J Am Oil Chem Soc* 77:501–506, 2000.

75. R Irimescu, K Hata, Y Iwasaki, T Yamane. Comparison of acyl donors for lipase-catalyzed production of 1,3-dicaproyl-2-eicosapentaenoylglycero. *J Am Oil Chem Soc* 78:65–70, 2001.

76. JJ Han, Y Iwasaki, T Yamane. Monitoring of lipase-catalyzed transesterification between eicosapentaenoic acid ethyl ester and tricaprylin by silver ion high-performance liquid chromatography and high-temperature gas chromatography. *J Am Oil Chem Soc* 76:31–39, 1999.

77. A Kawashima, Y Shimada, M Yamamoto, A Sugihara, T Nagao, S Komemushi, Y Tominaga. Enzymatic synthesis of high-purity structured lipids with caprylic acid at 1,3-positions and polyunsaturated fatty acid at 2-position. *J Am Oil Chem Soc* 78:611–616, 2001.

78. R Irimescu, K Furihata, K Hata, Y Iwasaki, T Yamane. Utilization of reaction medium-dependent regiospecificity of *Candida antarctica* lipase (Novozym 435) for the synthesis of 1,3-dicaproyl-2-docosahexaenoyl (or eicosapentaenoyl) glycerol. *J Am Oil Chem Soc* 78:285–289, 2001.

79. R Irimescu, Y Iwasaki, CT Hou. Study of TAG ethanolysis to 2-MAG by immobilized *Candida antarctica* lipase and synthesis of symmetrically structured TAG. *J Am Oil Chem Soc* 79:879–883, 2002.

80. M Berger, K Laumen, MP Schneider. Enzymatic esterification of glycerol I. Lipase-catalyzed synthesis of regioisomerically pure 1,3-*sn*-diacylglycerols. *J Am Oil Chem Soc* 69:955–960, 1992.

81. B Aha, M Berger, B Jakob, G Machmüller, C Waldinger, MP Schneider. Lipase-catalyzed synthesis of regioisomerically pure mono- and diglycerides. In: UT Bornscheuer, Ed. *Enzymes in Lipid Modification*. Weinheim: Wiley-VCF, 2000, pp. 100–115.

82. M Berger, MP Schneider. Enzymatic esterification of glycerol II. Lipase-catalyzed synthesis of regioisomerically pure 1(3)-*rac*-monoacylglycerols. *J Am Oil Chem Soc* 69:961–965, 1992.

83. C Waldinger, MP Schneider. Enzymatic esterification of glycerol III. Lipase-catalyzed synthesis of regioisomerically pure 1,3-*sn*-diacylglycerols and 1(3)-*rac*-monoacylglycerols derived from unsaturated fatty acids. *J Am Oil Chem Soc* 73:1513–1519, 1996.

84. R Rosu, M Yasui, Y Iwasaki, N Shimizu, T Yamane. Enzymatic synthesis of symmetrical 1,3-diacylglycerols by direct esterification of glycerol in solvent-free system. *J Am Oil Chem Soc* 76:839–843, 1999.

85. GG Haraldsson, A Halldorsson, E Kulås. Chemoenzymatic synthesis of structured triacylglycerols containing eicosapentaenoic and docosahexaenoic acids. *J Am Oil Chem Soc* 77:1139–1145, 2000.

86. Y Endo, S Hoshizaki, K Fujimoto. Autoxidation of synthetic isomers of triacylglycerol containing eicosapentaenoic acid. *J Am Oil Chem Soc* 74:543–548, 1997.

87. Y Endo, S Hoshizaki, K Fujimoto. Oxidation of synthetic triacylglycerols containing eicosapentaenoic and docosahexaenoic acids: effect of oxidation system and triacylglycerol structure. *J Am Oil Chem Soc* 74:1041–1045, 1997.

88. E Kulås, H Breivik. Unpublished results.

89. A Halldorsson, CD Magnusson, GG Haraldsson. Chemoenzymatic synthesis of structured triacylglycerols. *Tetrahedron Lett* 42:7675–7677, 2001.

90. A Halldorsson, CD Magnusson, GG Haraldsson. Chemoenzymatic synthesis of structured triacylglycerols by highly regioselective acylation. *Tetrahedron* 59:9101–9109, 2003.

19

Biopolyesters Derived from the Fermentation of Renewable Resources

Richard D. Ashby

Daniel K.Y. Solaiman

Thomas A. Foglia

19.1 Introduction

As the chemical industry moves toward a more sustainable future, green chemistry is slowly emerging as an accepted technology to develop and implement chemical products and processes that eliminate or at least ease the generation of substances that are detrimental to human health and the environment. As such, the polymer industry is beginning to focus on the use of renewable resources, waste management options, and life-cycle assessments as new materials are developed and considered for industrial application.

Historically, by strict definition, the term "biopolymer" has referred to any large molecule (macromolecule) that is composed of repeating chemical units from any biological origin. However, recently the definition of biopolymers has broadened to include macromolecules that are synthesized by biological means, such as enzyme-catalyzed polymerization reactions, but whose building blocks may not have originated from biological sources. Some examples include lipase-mediated lactone ring-opening polymerizations, polycondensations of polyols, and transesterification reactions between high molar mass polymers.[1] In addition, some hybrid technologies have been developed where the building blocks of biopolymers are derived from biological means but whose polymerization reactions are conducted through chemical syntheses (e.g., polylactic acid; PLA). In the truest sense, biopolymers are produced in nature by living organisms and are ultimately degraded in a continuous cycling of resources. In biological systems, polyamides seem to be preferred. The most common of the polyamides include proteins that are made up of amino acids and are involved in catalytic (enzymes), regulatory, and structural activities. Other common biopolymers include nucleic acids, which compose the genetic material (DNA and RNA) of the cell, and sugars (polysaccharides), which have structural or storage roles. While these examples represent the majority of natural biopolymers, biopolyesters (produced by native microorganisms or genetically transformed bacteria) have also been described that are functionally diverse and versatile. These polymers have been classified as poly(hydroxyalkanoates) (PHAs) and are naturally produced by a number of diverse bacterial strains as carbon and energy storage materials.

Renewable resources are resources that can be regenerated on a regular basis (usually annually) and are generally derived from agricultural sources (including plants and animals). Today, industry is taking a serious look at the use of renewable resources in the creation of new polymers. The two most noteworthy renewable resources studied in terms of polymer production are polysaccharides (cellulose and starch) and lipids. Starch is the major carbohydrate storage molecule in plants and is present in large amounts in many seeds and roots. The most commercially available sources of starch include corn that is relatively cheap and abundant and to a lesser extent potatoes. Starch is a polysaccharide composed of glucose linked in both unbranched (amylose, consisting of glucose units assembled in α-1,4 linkages) and branched (amylopectin, with approximately 1 α-1,6 linkage per 30 α-1,4 linkages) conformations. Because of its structure, starch can easily be converted either enzymatically or chemically into large amounts of glucose, which can then be used as starting material for the production of a variety of materials including polyols, aldehydes, ketones, acids, and esters.[2] In addition, starch itself has been reacted with some synthetic materials including vinyl or acrylic monomers by graft polymerization to produce natural/synthetic copolymers,[3–5] and polycaprolactone (PCL) to produce starch/PCL copolymers that have been marketed by Novamont SpA (Novara, Italy) under the trade name Mater-Bi[TM6] with a broad range of material properties. These starch-based polymers have been proposed as thickeners, flocculants, and clarification aids, and may be used in structural applications.[7]

Between the years 1999 and 2001, the U.S. produced approximately 10.8 million metric tons of vegetable oil (77% of which was soybean oil) and 5.5 million metric tons of animal fat (47% of which was tallow). With average costs in the range of 12–15¢/lb, these materials have been demonstrated to be excellent candidates for use as precursors in many diverse polymers, including polyolefins, polyesters, polyethers, and polyamides.[8] Two examples include the production of elastomers and rigid plastics through cationic copolymerization of styrene and divinylbenzene with an assortment of vegetable oils,[9,10] and the production of polyurethanes from polyols derived from soybean oil.[11–13] In addition, vegetable oils and animal fats have been documented to be good fermentation feedstocks in the production of PHA polymers.[14–17] This shift in focus to the use of lipids as carbon substrates for fermentation applications stemmed from the notion that the use of an abundant, inexpensive feedstock will reduce production costs involved in PHA polymer synthesis and impart structural control to the resulting polymers based on the fatty acid content of the lipids. In addition, lipid derivatives, including free fatty acids and glycerol, have been utilized as carbon substrates for bacterial fermentations in the synthesis of unique PHA polymers.

19.2 Commercial Biopolyesters from Fermentative Origins

All polyesters, whether derived from biological or synthetic origins, have the structural formula depicted in Figure 19.1. Structural variations in the R group determine whether a particular polyester is classified as an aliphatic polyester (e.g., PCL or PLA) or an aromatic polyester (e.g., polyethylene terephthalate; PET). At present, many of the biopolymers produced from renewable resources are the result of chemical or enzymatic polymerization reactions involving the substrate itself.

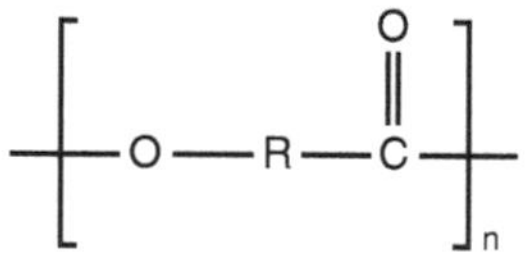

FIGURE 19.1 General formula for all polyesters. R group dictates whether polymer is subdivided into aliphatic or aromatic polyesters.

Recently, however, some of these renewable resources have been found to be good substrates on which microbial fermentations can be performed in the production of the monomeric materials necessary to synthesize cost-effective biopolyesters with good commercial application potential. Two important examples from an industrial perspective include the production of lactic acid (precursor for PLA production) and 1,3 propanediol (precursor for polytrimethylene terephthalate; PTT, production) from the fermentation of cornstarch.

19.2.1 Poly(lactic Acid)

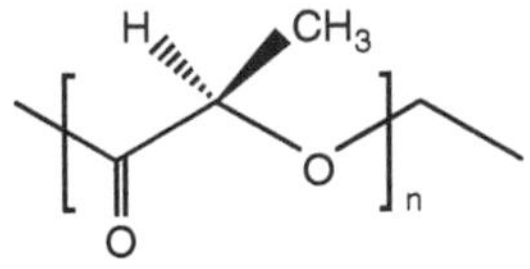

Lactic acid is produced by the optimized batch fermentation of *Lactobacilli*, which naturally synthesizes the L(+) isomer of lactic acid but more recently has been engineered to produce large amounts of the D(-) isomer.[18–21] The ability to differentiate between the syntheses of one isomer versus the other ultimately aids in the property control of the

FIGURE 19.2 Chemical structure of poly(lactic acid).

final product. For example, the crystallinity, crystallization rate, and degradation rate of the final PLA (for structure, see Figure 19.2) product is dictated by the polymerization of different ratios of L(+) isomer and D(-) isomer.[22–25] Because of this property control, PLA has found a number of applications, including packaging applications and as bioresorbable sutures, implants, and drug-delivery systems.[26] Within the last few years Cargill Dow LLC has begun production of PLA at the world's first production-scale plant in Blair, NE, with a capacity of 140,000 tons/yr. The Blair facility uses sugars derived from corn as fermentation substrates in the production of lactic acid, which is subsequently converted into lactide through self-condensation and depolymerization reactions, and then polymerized into PLA by ring-opening polymerization (Figure 19.3) and the use of chain coupling agents. The company markets this material under the Ingeo™ brand for fiber applications or NatureWorks™. Besides the apparent property benefits of PLA, an additional benefit of the use of this polymer is its environmental friendliness. PLA breaks down rapidly in disposal environments and, upon discarding, degrades primarily by hydrolysis rather than enzymatic attack.

19.2.2 Polytrimethylene Terephthalate

More recently, scientists at the Pacific Northwest National Laboratories (PNNL) in conjunction with the National Corn Growers Association (NCGA) developed a fermentation process by which malonic acid (1,3 propanedioic acid) is produced from starch through the use of *Phanerochaete chrysosporium*, a strain of filamentous fungi that is naturally involved in wood degradation.[27] Then, by catalytic hydrogenation, the malonic acid is converted to 1,3 propanediol, a potentially valuable intermediate in the production of polymeric materials. At the same time, E. I. du Pont de Nemours and Co. (Wilmington, DE) in conjunction with Genencor International (Palo Alto, CA) developed a recombinant strain of *Escherichia coli* by transforming the genes encoding glycerol-3-phosphate dehydrogenase, glycerol-3-phosphatase, glycerol dehydratase and 1,3-propanediol oxidoreductase from *Klebsiella* or *Saccharomyces* and by using this newly constructed organism to develop a biobased process to produce 1,3 propanediol from mono-, oligo-, and polysaccharides.[28] This process was subsequently improved by the removal of the transformed 1,3 propanediol oxidoreductase gene (dhaT) and the enhanced nonspecific ability to convert 3-hydroxypropionaldehyde to 1,3 propanediol.[29] The 1,3 propanediol was then polymerized with terephthalic acid to produce aliphatic-aromatic copolyesters of PTT or "3GT" (Figure 19.4).[30–32] This new variation of PET, marketed under the tradename Sorona® 3GT™, has shown promise in both PET and nylon applications. It has excellent

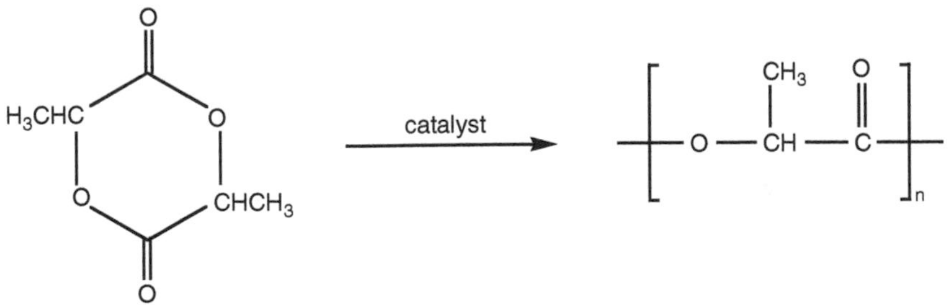

FIGURE 19.3 Ring-opening polymerization of lactide.

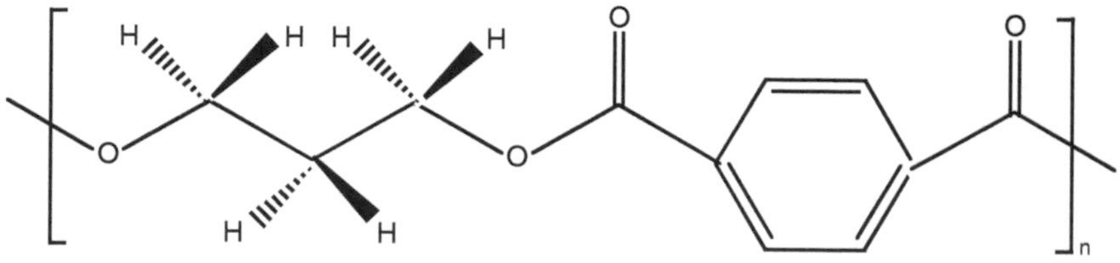

FIGURE 19.4 Chemical structure of poly(trimethylene terephthalate).

stretch-recovery properties and has application potential in textile, upholstery, and other applications where softness, comfort-stretch, and recovery and dyeability are desired. Based on its use in Sorona® 3GT™, 1,3 propanediol has a 2020 market potential of 500 million pounds.

19.2.3 Poly(hydroxyalkanoates)

Poly(hydroxyalkanoates) (PHAs; Figure 19.5) are the only high molecular weight biopolyesters produced from renewable resources entirely by bacteria through whole-cell catalysis. PHAs were discovered in 1925 by Lemoigne,[33] but it was not until the oil crisis of the 1970s that they were seriously considered for potential use in industrial applications. As ecological concerns increased and landfill space started to decline, scientists began to take a hard look at the environmental effects of petroleum-based polymers versus biobased poly-

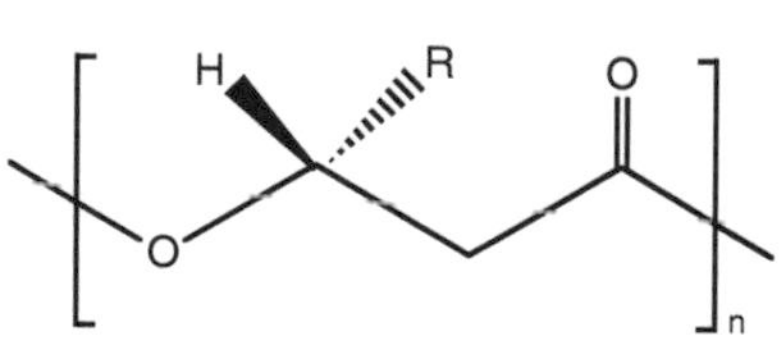

FIGURE 19.5 Chemical structure of poly(hydroxyalkanoates). R group of methyl or ethyl groups = short-chain-length (*scl-*) PHA; R group with chain lengths ≥ C3 = medium-chain-length (*mcl-*) PHA.

mers such as PHA. In some countries, particularly in Europe and Asia where open space is at a minimum, the environmental concerns such as recycling and biodegradation outweighed the steep cost associated with PHA production through fermentation. This focus resulted in the first commercial facility for the manufacture of fully biodegradable biopolyesters by Imperial Chemical Industries (ICI) in England. Poly(3-hydroxybutyrate) (PHB; PHA with a methyl side-chain repeat unit, see Figure 19.5) is the most abundant polymer in the PHA family. It is synthesized from renewable resources by numerous strains of bacteria generally through the sequential action of three enzymes. First, the enzyme β-ketothioloase catalyzes the synthesis of acetoacetyl CoA from two molecules of acetyl CoA. This is followed by the conversion of acetoacetyl CoA to 3-hydroxybutyryl CoA by the action of acetoacetyl CoA reductase and finally the PHB synthase-catalysed polymerization reaction in which the 3-hydroxybutryl-CoA units are condensed to form the PHB polymer. While PHB is the most prevalent biopolyester among many bacterial species, in its native form its properties are not conducive to widespread industrial application because it is stiff and brittle and its degradation temperature is only slightly higher than its processing temperature. To overcome this problem, ICI developed fermentation parameters that allowed the incorporation of valeric acid into these biopolyesters. Poly(3-hydroxybutyrate-*co*-valerate) (PHB/V) was produced from mixed substrate fermentations of glucose and propionic acid by *Alcaligenes eutrophus* (since reclassified as *Ralstonia eutropha*) and marketed under the trade name Biopol™. This structural variation reduced the processing temperature of the polymer and provided a dramatic improvement to the strength and durability of the polymer. In fact, PHB/V exhibits properties comparable to some synthetic polymers (most notably polypropylene) with the added advantage of rapid biodegradation rates in many disposal environments, including soil and compost. However, high costs of production continue to be the largest impedance to a more widespread use and much scientific research has been performed since the marketing of Biopol™ to reduce the cost of PHA synthesis and seek further structural variation and property control.

As researchers continued to search for new methods of property control for PHAs, additional monomers were successfully incorporated into the growing list of short-chain-length (*scl-*) PHA polymers.

In particular was the incorporation of 4-hydroxybutyric acid in the synthesis of the new copolymer of poly(3-hydroxybutyrate-*co*-4-hydroxybutyrate) (P3HB/4HB).[34] Much like the synthesis of PHB/V, the ratios of 3HB to 4HB in the copolymers could be controlled through variations in the types, concentrations, and ratios of the substrates. The physical and mechanical properties of these new copolymers revealed that as the 4HB content increased the tensile strength of the material decreased and the elongation to break increased.[35] Moreover, it was found that the molar mass of *scl*-PHAs could be controlled by the addition of certain polyols to the fermentation media. In particular, small molar mass polyethylene glycols[36–39] and glycerol[40] have been shown to disrupt polymer elongation through end-capping and result in *scl*-PHAs with reduced molar masses, thus providing a simple method to control polymer properties.

In the 1980s a second type of PHA was found and classified as medium-chain-length PHA (*mcl*-PHA) from the fermentation of alkanoic acids with *Pseudomonas oleovorans*.[41–43] These PHAs contained side chains that varied in length from propyl groups to undecanoyl groups, and it was established, at least in the presence of hexanoic acid and octanoic acid, that the carbon chain length of the substrate was highly conserved when polymerized into *mcl*-PHA. As the substrates continued to increase in length, it was evident that substrates with chain lengths of 10 carbon units (decanoic acid) and longer resulted in enhanced β-oxidation and in some cases the addition of 2 carbon subunits. These new polymers exhibited properties that were much different than the *scl*-PHAs of which PHB, PHB/V, and P3HB/4HB belong. The *mcl*-PHAs are elastomeric in nature and have processing temperatures in the vicinity of 40–50°C. While cost continues to be a concern, these polymers opened up an entirely new realm of potential applications.

To date there are over 100 different PHA polymers based on the side-chain substituent.[44] Because of this large structural variation and the properties that each imparts, great efforts have been taken to reduce the cost of PHA production. From a fermentation standpoint, two schools of thought have evolved in an attempt to make PHA production more economical. The first involves the search for more economical feedstocks that result in high polymer yields, while the second includes the development of new techniques to increase PHA yields.

Triacylglycerols (TAGs) are a potentially inexpensive, untapped source of fermentation substrates for PHA production. Recently, successful fermentations using animal fats and vegetable oils as substrates have resulted in unique PHA polymers. In order for a bacterium to utilize TAGs, it is essential that the organism exhibit extracellular esterase activity to liberate the fatty acids from the glycerol backbone. These fatty acids can then be used as substrates for PHA biosynthesis. It has been shown that *mcl*-PHA polymers can be synthesized from TAGs[14–17] and their free fatty acids (FFAs)[45,46] by select bacterial strains (particularly *Pseudomonas*). In general, the absolute fatty acid makeup of the lipid dictates the ratio of medium-chain-length monomers in the final polymer. The final composition of the *mcl*-PHA polymers derived from TAGs range in monomer chain length from C6 (3-hydroxyhexanoic acid) to C14 (3-hydroxytetradecanoic acid). In most cases animal fats and vegetable oils are composed primarily of fatty acyl chains of C16 and higher. This suggests that those bacterial strains that synthesize *mcl*-PHA from TAGs use β-oxidized fatty acids as the precursors for polymerization. Interestingly, many of the olefinic groups present in the unsaturated fatty acids are conserved upon polymerization. These chemical groups provide loci where further chemical and/or physical modification can be performed in the control of their mechanical properties. Studies have shown that *mcl*-PHA can be crosslinked through these olefinic groups by a number of techniques including peroxide crosslinking,[47] sulfur vulcanization,[48] radiation crosslinking,[49–52] and epoxidation and curing.[53] In each case the mechanical properties of the *mcl*-PHA were improved through these reactions.

In addition to *mcl*-PHA synthesis, some bacterial strains synthesize PHB from TAGs or their FFAs. It has been shown that some strains of *P. oleovorans* contain the genes responsible for *scl*-PHA synthesis.[54] As such, in contrast to the *mcl*-PHA producers, these organisms utilize the acetyl-CoA portion of the β-oxidized fatty acids to produce PHB and in some instances the bacterium may synthesize both *scl*-PHA and *mcl*-PHA together. This was reported by Shimamura et al., who were able to synthesize a hybrid PHA copolymer containing 3-hydroxybutyric acid and 3-hydroxyhexanoic acid from olive oil using an isolated strain of *Aeromonas caviae*.[14] This polymer showed improved mechanical properties than PHB or *mcl*-PHA individually and proved to be interesting from an application standpoint.

These efforts resulted in the use of genetic engineering to either stimulate larger polymer yields (overexpression of synthase enzymes), provide the essential enzymes for the utilization of inexpensive carbon substrates that may otherwise be unusable (i.e., imparting extracellular lipase activity to a PHA producing bacterium)[55,56] or catalyze the polymerization of monomeric units that may not be readily possible.[57,58] Recently, scientists at Procter & Gamble (Cincinnati, OH) developed a fermentation process for the synthesis of copolymers of 3-hydroxybutyrate and a broad range of medium-chain-length mono-mers.[59] This material, marketed under the trade name Nodax™, resembles linear low-density polyethylene with the added advantages of biodegradability, hydrolytic stability, and a versatile compatibility with other materials. It can be easily converted to films, fibers, foams, and molded articles[60,61] and with the progression of modern biotechnology is expected to be economically competitive soon.

Attempts to utilize genetic engineering to lower production costs of PHA have also been explored. In one instance, the PHA biosynthetic genes for *scl*-PHA from *Ralstonia eutropha* have been engineered and expressed in *Escherichia coli* (naturally a non-PHA producer),[62,63] which can grow rapidly to very high cell densities. If the PHB cellular productivity of *E coli* can be even moderately high, this may lead to reduced production costs and simplified purification.

Lastly, the use of genetic engineering in PHA biosynthesis has been extended to include the development of genetically altered plants that contain the metabolic pathways for PHA biosynthesis. It was thought that since plants create carbon through photosynthesis they will consistently produce carbon as long as there is access to sunlight. In addition, plants contain large amounts of acetyl-CoA, the precursor for PHA biosynthesis, for fatty acid biosynthesis. The concentration of acetyl-CoA is particularly high in oilseed crops, making them a natural target for genetic modification. Initially, *Arabidopsis thaliana* was chosen as the target crop. The decision to use *Arabidopsis* was based on the fact that it was easily transformed, has many uses as a plant for genetic studies, and is closely related to rapeseed, a commercial, oil-producing crop. Early work proved successful in synthesizing PHB from the leaves of *Arabidopsis*, but the yields continued to be lower than what was needed to make plant production a commercially practical alternative.[64–66] With that in mind, focus shifted to PHB biosynthesis in the seeds of *Brassica napus* with the idea that any plant whose seeds produce large amounts of oil should have high concentrations of acetyl-CoA accessible to initiate PHB synthesis. Once again, success was achieved; however, as mentioned earlier, the properties of PHB are not good for industrial application. It was therefore necessary to devise a procedure that would result in the synthesis of PHB/V.[67,68] After much metabolic engineering, success was achieved and in an attempt to bring this technology to market, Monsanto (who had purchased the patents and marketing rights from ICI) created a plant that was not only 14% plastic but also could be used to produce oil. However, the belief that PHA-producing plants would not be commercially practical unless 20% or more of the plant was plastic caused Monsanto to discontinue further development of PHAs. Today, Metabolix Inc. (Cambridge, MA) continues to pursue the commercialization of PHAs by both fermentative means and in plant crops.

19.3 Conclusion

Increased environmental consciousness has led to the utilization of renewable resources and natural processes to synthesize more benign polyesters. The use of petrochemical-based chemical syntheses generally results in the formation of toxic substances that are difficult to dispose of and may not readily degrade in landfill or compost environments. In contrast, the fermentation of renewable resources in the biosynthesis of polymer precursors (e.g., lactic acid, 1,3-propanediol) and/or biopolyesters themselves (e.g., PHA) has proven to be advantageous with respect to accessibility of the feedstocks and ease of disposal. Presently, starches, lipids (or their components) are widely considered to be good feedstocks for the production of biobased materials by fermentation. However, in many cases, cost continues to be a hindrance. Many methods have been studied in an attempt to reduce the cost of biopolyester synthesis. In the case of PHA synthesis, most of these methods have involved the use of genetic engineering to either induce super-production of the material in natural PHA-producing organisms, allow the utilization of substrates that would otherwise not be possible (e.g., TAGs), induce other producing bacterial strains (primarily *E coli*) to synthesize the polymer or engineering plants to produce these biopolyesters. Further attempts to reduce production costs have used agricultural waste and coproducts such as molasses, whey, and the glycerol fraction of biodiesel production as fermentation feedstocks for PHA production.[69] As the cost problem is conquered and with the capacity to tailor the physical properties of these biopolyesters, additional applications for these biobased materials are being realized.

References

1. RA Gross, A Kumar, B Kalra. Polymer synthesis by in vitro enzyme catalysis. *Chem Rev* 101:2097–2124, 2001.
2. FH Otey, WM Doane. Chemicals from starch. In: RL Whistler, JM BeMiller, EF Paschall, eds. *Starch: Chemistry and Technology 2nd edition*. New York: Academic Press, 1984, pp. 389–416.
3. GF Fanta, EB Bagley. Starch, graft copolymers. In *Encyclopedia of Polymer Science and Technology-Supplement*. New York: Wiley-Interscience, 1977, pp. 665–699.
4. GF Fanta, WM Doane. Grafted starches, modified starches: properties and uses. In: OB Wurzburg, ed. *Grafted Starches: Properties and Uses*. Boca Raton, FL: CRC Press, 1986, pp. 149–178.
5. GF Fanta, FC Felker, JH Salch. Graft polymerization of acrylonitrile onto starch-coated polyethylene film surfaces. *J Appl Polym Sci* 89:3323–3328, 2003.
6. C Bastioli. Properties and applications of Mater-bi starch-based materials. *Polym Degrad Stab* 59:263–272, 1998.
7. C Bastioli. Starch – polymer composites. In: *Degradable Polymers: Principles and Applications, 2nd Edition*. Dordrecht: Kluwer Academic Publishers, 2002, pp. 133–161.
8. S Warwel, F Brüse, C Demes, M Kunz, MR Klaas. Polymers and surfactants on the basis of renewable resources. *Chemosphere* 43:39–48, 2001.
9. F Li, RC Larock. New soybean oil-styrene-divinylbenzene thermosetting copolymers-IV. Good damping properties. *Polym Adv Technol* 13:436–449, 2002.
10. F Li, RC Larock. Novel polymeric materials from biological oils. *J Polym Environ* 10:59–67, 2002.
11. A Guo, I Javni, Z Petrovic. Rigid polyurethane foams based on soybean oil. *J Appl Polym Sci* 77:467–473, 2000.
12. I Javni, ZS Petrovic, A Guo, R Fuller. Thermal stability of polyurethanes based on vegetable oils. *J Appl Polym Sci* 77:1723–1734, 2000.
13. ZS Petrovic, A Guo, W Zhang. Structure and properties of polyurethanes based on halogenated and nonhalogenated soy-polyols. *J Polym Sci Part A-Polym Chem* 38:4062–4069, 2000.
14. E Shimamura, K Kasuya, G Kobayashi, T Shiotani, Y Shima, Y Doi. Physical properties and biodegradability of microbial poly(3-hydroxybutyrate-*co*-3-hydroxyhexanoate). *Macromolecules* 27:878–880, 1994.
15. A-M Cromwick, T Foglia, RW Lenz. The microbial production of poly(hydroxyalkanoates) from tallow. *Appl Microbiol Biotechnol* 46:464–469, 1996.
16. RD Ashby, TA Foglia. Poly(hydroxyalkanoate) biosynthesis from triglyceride substrates. *Appl Microbiol Biotechnol* 49:431–437, 1998.
17. DKY Solaiman, RD Ashby, TA Foglia. Medium-chain-length poly(β-hydroxyalkanoate) synthesis from triacylglycerols by *Pseudomonas saccharophila*. *Curr Microbiol* 38:151–154, 1999.
18. A Demirici, AL Pometto. Enhanced production of D(-)-lactic acid by mutants of *Lactobacillus delbruekii* ATCC 9649. *J Ind Microbiol Biotechnol* 11:23–28, 1992.
19. S Benthin, J Villadsen. Production of optically pure D-lactate by *Lactobacillus bulgaricus* and purification by crystallization and liquid-liquid extraction. *Appl Microbiol Biotechnol* 42:826–829, 1995.
20. K Hofvendahl, B Hahn-Hägerdal. Factors affecting the fermentative lactic acid production from renewable resources. *Enz Microb Technol* 26:87–107, 2000.
21. G Bustos, AB Moldes, JL Alonso, M Vázquez. Optimization of D-lactic acid production by *Lactobacillus coryniformis* using response surface methodology. *Food Microbiol* 21:143–148, 2004.

22. H Tsuji, Y Ikada. Stereocomplex formation between enantiomeric poly(lactic acid)s. XI. Mechanical properties and morphology of solution cast films. *Polymer* 40:6699–6708, 1999.

23. H Tsuji. Autocatalytic hydrolysis of amorphous-made polylactides: effects of L-lactide content, tacticity, and enantiomeric polymer blending. *Polymer* 43:1789–1796, 2002.

24. H Tsuji. In vitro hydrolysis of blends from enantiomeric poly(lactide)s. Part 4: well-homo-crystallized blend and nonblended films. *Biomaterials* 24:537–547, 2003.

25. H Yamane, K Sasai. Effect of the addition of poly(D-lactic acid) on the thermal property of poly(L-lactic acid). *Polymer* 44:2569–2575, 2003.

26. DK Gilding, AM Reed. Biodegradable polymers for use in surgery – polyglycolic / poly(lactic acid) homo- and copolymers: 1. *Polymer* 20:1459–1464, 1979.

27. MT Kingsley, RA Romine, LL Lasure. Effects of medium composition on morphology and organic acid production in *Phanerochaete chrysosporium*. Annual Meeting of the American Society for Microbiology, Poster O-3, 2002.

28. CE Nakamura, AA Gatenby, AK Hsu, RD La Reau, SL Haynie, M Diaz-Torres, DE Trimbur, GM Whited, V Nagarajan, MS Payne, SK Picataggio, RV Nair. Method for the production of 1,3-propanediol by recombinant microorganisms. US Patent 6,013,494, 2000.

29. M Emptage, SL Haynie, LA Laffend, JP Pucci, G Whited. Process for the biological production of 1,3 propanediol with high titer. US Patent 6,514,733, 2003.

30. EN Blanchard, CR Gochanour, KW Leffew, JM Stouffer. Production of poly(trimethylene tereph-thalate). US Patent 5,990,265, 1999.

31. JV Kurian, Y Liang. Preparation of poly(trimethylene terephthalate). US Patent 6,281,325, 2001.

32. CJ Giardino, DB Griffith, CH Ho, JM Howell, MH Watkins. Continuous process for producing poly(trimethylene terephthalate). US Patent 6,538,076, 2003.

33. M Lemoigne. Produits de déshydration et de polymerization de l'acide b-oxobutyrique. *Bull Soc Chem Biol* (Paris) 8:770–782, 1926.

34. M Kunioka, Y Kawaguchi, Y Doi. Production of biodegradable copolyesters of 3-hydroxybutyrate and 4-hydroxybutyrate by *Alcaligenes eutrophus*. *Appl Microbiol Biotechnol* 30:569–573, 1989.

35. Y Doi, A Segawa, M Kunioka. Biosynthesis and characterization of poly(3-hydroxybutyrate-*co*-4-hydroxybutyrate) in *Alcaligenes eutrophus*. *Int J Biol Macromol* 12:106–111, 1990.

36. RD Ashby, F-Y Shi, RA Gross. Use of poly(ethylene glycol) to control the end group structure and molecular weight of poly(3-hydroxybutyrate) formed by *Alcaligenes latus* DSM 1122. *Tetrahedron* 53:15209–15223, 1997.

37. RD Ashby, F-Y Shi, RA Gross. A tunable switch to regulate the synthesis of low and high molecular weight microbial polyesters. *Biotechnol Bioengin* 62:106–113, 1999.

38. RD Ashby, DKY Solaiman, TA Foglia. Poly(ethylene glycol)-mediated molar mass control of short-chain- and medium-chain-length poly(hydroxyalkanoates) from *Pseudomonas oleovorans*. *Appl Microbiol Biotechnol* 60:154–159, 2002.

39. J Zanzig, C Scholz. Effects of poly(ethylene glycol) on the production of poly(β-hydroxybutyrate) by *Azotobacter vinelandii* UWD. *J Polym Environ* 11:145–154, 2003.

40. LA Madden, AJ Anderson, DT Shah, J Asrar. Chain termination in polyhydroxyalkanoate synthesis: involvement of exogenous hydroxy-compounds as chain transfer agents. *Int J Biol Macromol* 25:43–53, 1999.

41. RG Lageveen, GW Huisman, H Preusting, P Ketelaar, G Eggink, B Witholt. Formation of polyesters by *Pseudomonas oleovorans*: effect of substrates on formation and composition of poly(R)-3-hydroxyalkanoates and poly(R)-3-hydroxyalkenoates. *Appl Environ Microbiol* 54:2924–2932, 1988.

42. H Brandl, RA Gross, RW Lenz, RC Fuller. *Pseudomonas oleovorans* as a source of poly(β-hydroxy-alkanoates) for potential applications as biodegradable polyesters. *Appl Environ Microbiol* 54:1977–1982, 1988.

43. RA Gross, C DeMello, RW Lenz, H Brandl, RC Fuller. Biosynthesis and characterization of poly(β-hydroxyalkanoates) produced by *Pseudomonas oleovorans*. *Macromolecules* 22:1106–1115, 1989.

44. A Steinbüchel, HE Valentin. Diversity of bacterial polyhydroxyalkanoic acids. *FEMS Microbiol Lett* 128:219–228, 1995.
45. E Casini, TC de Rijk, P de Waard, G Eggink. Synthesis of poly(hydroxyalkanoate) from hydrolyzed linseed oil. *J Environ Polym Degrad* 5:153–158, 1997.
46. B Hazer, O Torul, M Borcakli, RW Lenz, RC Fuller, SD Goodwin. Bacterial production of polyesters from free fatty acids obtained from natural oils by *Pseudomonas oleovorans. J Environ Polym Degrad* 6:109–113, 1998.
47. KD Gagnon, RW Lenz, RJ Farris, RC Fuller. Chemical modification of bacterial elastomers: 1. peroxide crosslinking. *Polymer* 35:4358–4367, 1994.
48. KD Gagnon, RW Lenz, RJ Farris, RC Fuller. Chemical modification of bacterial elastomers: 2. sulfur vulcanization. *Polymer* 35:4368–4375, 1994.
49. GJM de Koning, HMM van Bilsen, PJ Lemstra, W Hazenberg, B Witholt, H Preusting, JG van der Galien, A Schirmer, D Jendrossek. A biodegradable rubber by crosslinking poly(hydroxyalkanoate) from *Pseudomonas oleovorans. Polymer* 35:2090–2097, 1994.
50. RD Ashby, A-M Cromwick, TA Foglia. Radiation crosslinking of a bacterial medium chain length poly(hydroxyalkanoate) elastomer from tallow. *Int J Biol Macromol* 23:61–72, 1998.
51. RD Ashby, TA Foglia, C-K Liu, JW Hampson. Improved film properties of radiation treated medium-chain-length poly(hydroxyalkanoates). *Biotechnol Lett* 20:1047–1052, 1998.
52. A Dufresne, L Reche, RH Marchessault, M Lacroix. Gamma-ray crosslinking of poly(3-hydroxy-octanoate-co-undecenoate). *Int J Biol Macromol* 29:73–82, 2001.
53. RD Ashby, TA Foglia, DKY Solaiman, C-K Liu, A Nuñez, G Eggink. Viscoelastic properties of linseed oil-based medium chain length poly(hydroxyalkanoate) films: effects of epoxidation and curing. *Int J Biol Macromol* 27:355–361, 2000.
54. RD Ashby, DKY Solaiman, TA Foglia. The synthesis of short- and medium-chain-length poly(hydroxyalkanoate) mixtures from glucose- or alkanoic acid- grown *Pseudomonas oleovorans. J Ind Microbiol Biotechnol* 28:147–153, 2002.
55. DKY Solaiman, RD Ashby, TA Foglia. Production of polyhydroxyalkanoates from intact triacylglycerols by genetically engineered *Pseudomonas. Appl Microbiol Biotechnol* 56:664–669, 2001.
56. DKY Solaiman, RD Ashby, TA Foglia. Physiological characterization and genetic engineering of *Pseudomonas corrugata* for medium-chain-length polyhydroxyalkanoates synthesis from triacylglycerols. *Curr Microbiol* 44:189–195, 2002.
57. K Matsumoto, S Nakae, K Taguchi, H Matsusaki, M Seki, Y Doi. Biosynthesis of poly(3-hydroxybutyrate-co-3-hydroxyalkanoates) copolymer from sugars by recombinant *Ralstonia eutropha* harboring the phaC1(Ps) and the phaG(Ps) genes of *Pseudomonas* sp 61-3. *Biomacromolecules* 2:934–939, 2001.
58. S Taguchi, H Matsusaki, K Matsumoto, K Takase, K Taguchi, Y Doi. Biosynthesis of biodegradable polyesters from renewable carbon sources by recombinant bacteria. *Polym Int* 51:899–906, 2002.
59. PR Green. Medium chain length PHA copolymer and process for producing same. US Patent 6,225,438, 2001.
60. EB Bond. Fiber spinning behavior of a 3-hydroxybutyrate/3-hydroxyhexanoate copolymer. *Macromol Symp* 197:19–31, 2003.
61. I Noda. Production and properties of Nodax™ PHA copolymers. 1st IUPAC International Conference on Bio-based Polymers. Saitama, Japan, 2003.
62. P Schubert, A Steinbüchel, HG Schlegel. Cloning of the *Alcaligenes eutrophus* genes for the synthesis of poly-β-hydroxybutyrate (PHB) and synthesis of PHB in *Escherichia coli. J Bacteriol* 170:5837–5847, 1988.
63. SC Slater, WH Vioge, DE Dennis. Cloning and expression in *Escherichia coli* of the *Alcaligenes eutrophus* H16 poly-β-hydroxybutyrate biosynthetic pathway. *J Bacteriol* 170:4431–4436, 1988.
64. Y Poirier, DE Dennis, K Klomparens, C Somerville. Polyhydroxybutyrate, a biodegradable thermoplastic, produced in transgenic plants. *Science* 256:520–523, 1992.

65. C Nawrath, Y Poirier, C Somerville. Targeting of the polyhydroxybutyrate biosynthetic pathway to the plastids of *Arabidopsis thaliana* results in high levels of polymer accumulation. *Proc Natl Acad Sci USA* 91:12760–12764, 1994.

66. Y Poirier, C Somerville, L Schechtman, MM Satkowski, I Noda. Synthesis of high-molecular-weight poly([R]-(-)-3-hydroxybutyrate) in transgenic *Arabidopsis thaliana* plant cells. *Int J Biol Macromol* 17:7–12, 1995.

67. S Slater, TA Mitsky, KL Houmiel, M Hao, SE Reiser, NB Taylor, M Tran, HE Valentin, DJ Rodriguez, DA Stone, SR Padgette, G Kishore, KJ Gruys. Metabolic engineering of *Arabidopsis* and *Brassica* for poly(3-hydroxybutyrate-*co*-3-hydroxyvalerate) copolymer production. *Nat Biotechnol* 17:1011–1016, 1999.

68. HE Valentin, DL Broyles, LA Casagrande, SM Colburn, WL Creely, PA DeLaquil, HM Felton, KA Gonzalez, KL Houmiel, K Lutke, DA Mahadeo, TA Mitsky, SR Padgette, SE Reiser, S Slater, DM Stark, RT Stock, DA Stone, NB Taylor, GM Thorne, M Tran, KJ Gruys. PHA production, from bacteria to plants. *Int J Biol Macromol* 25:303–306, 1999.

69. G Braunegg, K Genser, R Bona, G Haage. Production of PHAs from agricultural waste material. *Macromol Symp* 144:375–383, 1999.

20

Carbohydrate Active Enzymes for the Production of Oligosaccharides

Hajime Taniguchi

20.1 Introduction

Carbohydrates such as starch and sucrose are abundantly produced as reserve energy by plants and serve as important nutrient sources for almost all organisms. Enzymes acting on this carbohydrate are therefore widely distributed in nature. Amylase, for instance, is one of the enzymes that have the longest history of research and application. In the 1960s a method for producing isomerized sugar (high fructose corn syrup) from starch using amylases and glucose isomerase (xylose isomerase) on an industrial scale was developed in Japan, and this technology rapidly spread in the world. Decades preceding this innovation, active studies on amylases and related enzymes were carried out, mainly to improve the efficiency of starch saccharification, i.e., to get enzymes with higher optimum temperature, with lower transglucosylation activity, with unique substrate specificities, and with varieties of pH optima. Around 1985, following nutrient and sensory functions, physiological function was discovered in Japan as the third function of food. Meanwhile, functional foods that claim special health benefits appeared and the government started an approval system (Foods for Special Health Use, or FOSHU) for these functional foods in 1991. Oligosaccharides were expected as one of the important candidates of these functional foods, and nowadays about 50% of FOSHU contains oligosaccharides as their beneficial ingredients. A variety of oligosaccharides have been developed over the last 20 years in Japan. This development is based on very active research on carbohydrate active-enzymes. As a result, discoveries of new enzymes and of new actions of known enzymes were reported, and certainly will be continued.

In this chapter I will introduce carbohydrate active-enzymes that are used in the production of oligosaccharides commercialized in Japan.

20.2   Oligosaccharides

20.2.1   Industrial Production of Oligosaccharides in Japan

Oligosaccharides are usually defined as sugars consisting of two to ten monosaccharide units connecting through glycosidic linkages. Various kinds of oligosaccharides are present in nature, but their contents in these organisms are usually very low except for sucrose and lactose. For a long time, work on these oligosaccharides was confined to studies on their structures and synthetic pathways, and it was beyond the imagination to produce these oligosaccharides on an industrial scale. However, needs for oligosaccharides as functional food components and accumulation of knowledge and techniques on carbohydrate active-enzymes described above made it possible to produce a wide range of oligosaccharides on an industrial scale. Table 20.1 lists oligosaccharides now commercially available in Japan as foods or food components.

Total production of oligosaccharides in Japan amounts to about 60,000 tons and total sales of oligosaccharides reached to 15 trillion yen in 2001. Of the 13 oligosaccharides listed in Table 20.1, ten saccharides are produced by the actions of carbohydrate active-enzymes. All of these enzymes are hydrolase (EC 3) or transferase (EC 2) of microbial origins.

Techniques of molecular biology such as cloning, sequencing, mutations of enzyme genes, and production of recombinant proteins play an important role in the elucidation of molecular structure and action mechanism of these enzymes as well. Although basic research on recombinant enzymes is being actively carried out, these enzymes are not used in the industrial production of oligosaccharides in Japan, because manufacturers are nervous about consumers' response on this issue.

20.2.2   Physiological Function of Oligosaccharides

It has been well known that oligosaccharides have various levels of sweetness, but the findings that they have various physiological functions greatly concerns both food manufacturers and consumers. Most of these oligosaccharides are resistant to the action of our digestive enzymes and act as a dietary fiber.

TABLE 20.1   Oligosaccharides produced in Japan

Oligosaccharides	Production (ton/year)	Biocatalyst	Price* (Yen/kg)	Manufacturer
Trehalose	25,000	Transglucosidase	300	Hayashibara Biochemical Laboratories Inc.
Isomalto-oligosaccharides	11,000	Transglucosidase	140	Showa Sangyo Co. Ltd.
Galacto-oligosaccharides	6,000	beta-Galactosidase	500	Yakult Pharmaceutical Ind Co. Ltd., Nissin Sugar MFG. Co. Ltd.
Fructo-oligosaccharides	4,000	beta-Fuctofuranosidase	390	MeijZi Seika Kaisha Ltd.
Lactulose	2,800	Chemical isomerization	1,000	Morinaga Milk Industry Co Ltd.
Lactosucrose	2,000	beta-Fructofuranosidase	500	Bioresearch Corporation of Yokohama
Cyclodextrins	1,800	Glucanotransferase	1,450	Bioresearch Corporation of Yokohama
Soy Oligosaccharides	1,000	Extraction	700	Calpis Co. Ltd.
Gentio-oligosaccharides	1,000	beta-Glucosidase	300	Nihon Shokuhin Kako Co. Ltd.
Xylo-oligosaccharides	650	beta-Xylosidase	2,500	Suntory Limited
Nigero-oligosaccharides	300	alpha-Glucosidase	300	NihonnShokuhin Kako Co. Ltd.
Raffinose	230	Extraction	2,000	Nippon Beet Sugar Mfg. Co. Ltd.
Palatinose	150	Glucosyltransferase	1,000	Shin Mitsui Sugar Co. Ltd.

TABLE 20.2   Oligosaccharides and their sources

Source	Oligosaccharides	Calorie	Relative sweetness
Starch	Malto-oligosaccharide	4	25
	Isomalto-oligosaccharide	4	42
	Nigero-oligosaccharide	4	45
	Gentio-oligosaccharide	2	bitter
	Trehalose	4	45
	Cyclodextrin	2	25
Sucrose	Trehalulose	4	50
	Palatinose	4	42
	Glycosylsucrose	4	50
	Fructo-oligosaccharide	2	60
	Lactosucrose	2	55
Others	Galacto-oligosaccharide	2	35
	Xylo-oligosaccharide	2	40
	Chitosan-oligosaccharide		

As a result of vigorous research activities, the following physiological functions have been found with oligosaccharides: (1) enhancement of selective growth of intestinal bifidobacteria helpful for the maintenance of human health, (2) inhibition of dental plaque formation, (3) enhancement of dietary mineral absorption, (4) stimulation of immune response leading to antitumor activity, (5) control of blood glucose level, and (6) reduction of serum cholesterol level. Foods for Special Health use (FOSHU) the regulatory approval system for functional foods in Japan, has categories of functions that can be accepted based on scientific evidence. All above functions are contained in these categories except for category 6 at present.

20.2.3 Classification of Oligosaccharides

Oligosaccharides can be classified in various ways depending on their degree of polymerization, on their sugar composition, or on the method of their production. In this chapter, they are classified according to their source, as shown in Table 20.2.

Starch and sucrose are the cheapest and most abundant carbohydrates produced in pure state annually. Lactose follows these two, as it is obtained as a byproduct of dairy industry. Cellulose is the most abundant and xylan the second most abundant carbohydrate present on the earth, but their utilization as sources of oligosaccharides is still in progress. Enzymes used for oligosaccharides listed in Table 20.1 will be described in this chapter. Enzymes for the productions of cyclodextrins and their derivatives will appear in other chapters and will not be touched on here.

20.3 Enzymes Used for Oligosaccharides Production

20.3.1 Enzyme for Isomalto-Oligosaccharide

Isomalto-oligosaccharide means scientifically a series of oligosaccharides consisting exclusively of alpha-1,6 linkages. But isomalto-oligosaccharide used in this section is a name of commercial products that contain panose and isomaltosylmaltose in addition to isomaltose, isomaltotriose, and isomaltotetraose. The structure of these oligosaccharides is shown in Figure 20.1.

These oligosaccharides have long been known to be present in Japanese sake and to give a special taste and body to this traditional beverage. They had long been troublesome byproducts in the industrial isomerized sugar (high fructose corn syrup) production from starch. Several Japanese companies such as Showa Sangyo Co. changed the idea and have developed industrial production methods for these oligosaccharides starting from starch. Isomalto-oligosaccharides are used as additives in a variety of foods, such as bread and Japanese sweets, to improve their quality. Besides the activity to stimulate bifidobacteria, they have excellent physico-chemical properties that keep the

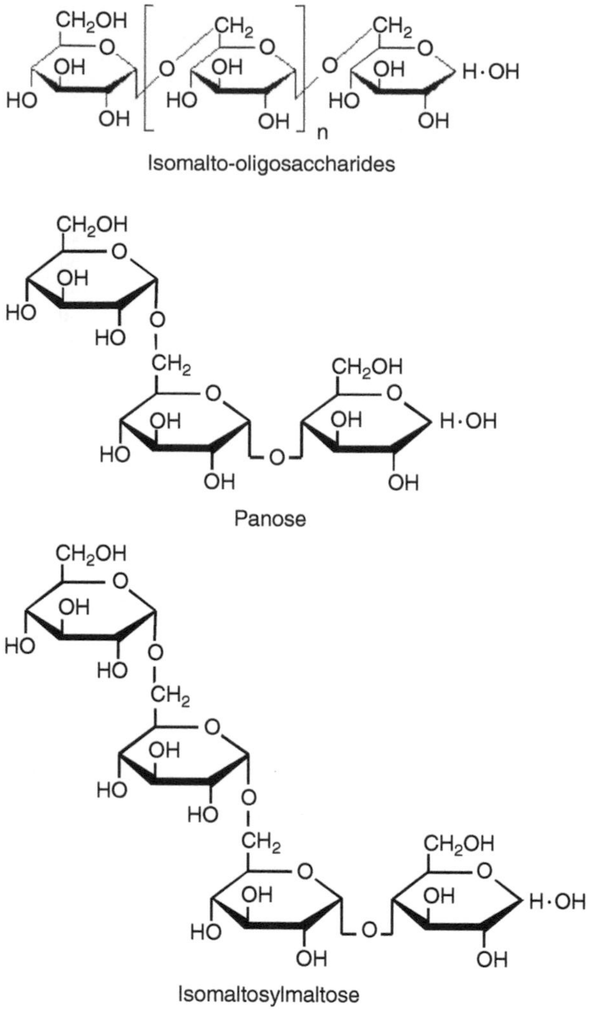

FIGURE 20.1 Structure of isomalto-oligosaccharides.

moisture of added foods and that prevent retrogradation of starch. Alpha-amylase, debranching enzyme such as isoamylase and pullulanase, and alpha-glucosidase are used for the production of isomalto-oligosaccharide from starch. Most of these enzymes have long been known and are used for the industrial production of glucose from starch. Alpha-glucosidases (EC 3.2.1.20) have been found in a wide range of animals, plants, and microorganisms. Some of the alpha-glucosidases catalyze, in addition to hydrolysis reactions, transfer reactions, transferring glucose residue of maltose to the C-2, C-3, C-4, or C-6 OH group of acceptor sugars, depending on the enzyme sources.[1] Alpha-glucosidase from *Aspergillus niger* is known to transfer glucose to C-6 positions of acceptor sugars to form isomalto-oligosaccharides.[2] The enzyme is classified into GH 31 based on its amino acid sequence.[3]

20.3.1.1 Production of Isomalto-Oligosaccharides[4,5]

A slurry of cornstarch (20–40%, w/w) was liquefied with thermophilic alpha-amylase (Termamyl, Novozyme, Japan) to give a dextrin solution of DE 5-15. This dextrin solution was incubated with

0.3–1.2 % (w/w) of beta-amylase (Nagase ChemteX), 0.3–0.8 % (w/w) of pullulanase (Promozyme, Novozyme, Japan) and 0.03–11 U /g of alpha-glucosidase from *Aspergillus niger* at pH 4.0–6.0 and 50–65°C for 1–2 days. The reaction mixture was then subjected to de-coloring and de-salting steps, and was applied to a Dowex 88 (Na salt) column to remove glucose. The obtained solution had the following sugar composition: glucose, 1.4%: maltose, 5.9%: isomaltose, 51.2%: trisaccharides (maltotriose, panose, and isomaltotriose), 24.9%: higher oligosaccharides, 16.7%. This solution was further concentrated to obtain syrup (Isomalto 900) or powdered product (Isomalto 900P). Isomalto-oligosaccharides accounted for more than 85% of total carbohydrates in these products.

20.3.2 Enzyme for Nigero-Oligosaccharide

Nigero-oligosaccharide is a mixture of nigerose, nigerosyl glucose, and nigerosyl maltose, all of which have alpha-1,3 linkage at their nonreducing ends. The structure of these oligosaccharides is shown in Figure 20.2.

Nigerose is known to be present in Japanese sake to give it a special favorable taste and has interesting physiological functions, such as antioxidant activity and enhancement of the human immune system. Recently a unique alpha-glucosidase (EC 3.2.1.20) was found from *Acremonium* sp. S4G13.[6] The crude enzyme preparation had high glucosyl transferring activity to form alpha-1,3 or alpha-1,4 linkages and eventually to accumulate nigero-oligosaccharides in the reaction mixture. It showed the highest transferring activity in the presence of 2% malto-oligosaccharides in pH 4 and at 50°C. A commercial nigero-oligosaccharide product (TasteOligo, Nihon Shokuhinn Kako Co. Ltd.) is produced using this

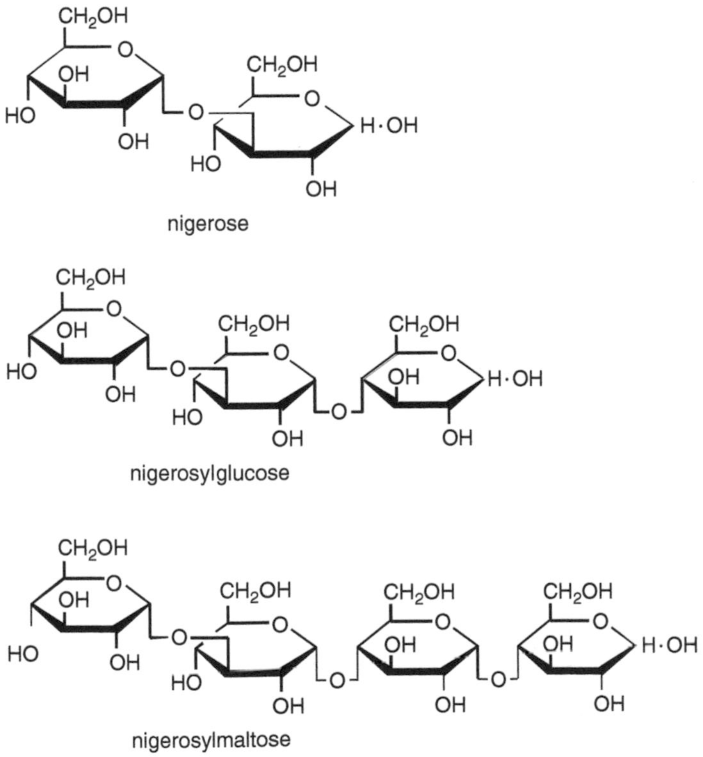

FIGURE 20.2 Structure of nigero-oligosaccharides.

TABLE 20.3 Basic properties of two alpha-glucosidases purified from *Acremonium* sp. S4G13

	Opt pH	Opt Temp	Mol Wt (GPC)	Mol Wt (SDS)	pI
Alpha-glucosidase 1	7–8	50–60	240,000	128,000 + 5,300	12.7
Alpha-glucosidase 2	6–8	50–60	140,000	70,000 + 5,300	13.6

enzyme. It contains nigero-oligosaccharides more than 40% of total carbohydrates. Basic properties of two alpha-glucosidases purified from the culture filtrate of *Acremonium* sp. S4G13 are shown in Table 20.3.[7]

Another unique alpha-glucosidase was found from a fungus, *Paecilomyces lilacinus*.[8] This enzyme catalyzes glucosyl transfer from maltose to form alpha-1,2 or alpha-1,3 linkages, producing nigerose and kojibiose as major products on prolonged incubation.

20.3.3 Enzyme for Gentio-Oligosaccharide

Gentio-oligosaccharide is a mixture of gluco-oligosaccharides connecting through beta-1,6 linkages. Their structure is shown in Figure 20.3.

Gentiobiose, a representative of these oligosaccharides, is known to be present in roots of gentian, one of the raw materials of traditional Chinese medicine. Unlike other sugars, it has a bitter taste. Like many other oligosaccharides recently developed, gentio-oligosaccharides have bifidobacteria-promoting activity in the human intestine. These saccharides find their application in confectioneries, coffee, and cocoa to improve the taste of these foods and drinks. Recently a large-scale preparative method of these oligosaccharides was developed using the transfer reaction of a beta-glucosidase (EC 3.2.1.21) from *Aspergillus niger*,[9] which is commercially available as Novozym 188 (Novo Nordisk Bioindustry Ltd.). The purified enzyme from this product showed condensation reaction from beta-glucose to form gentiobiose and glucosyl transfer reaction from gentiobiose to form gentio-oligosaccharides.[10] Purification and characterization of beta-glucosidases from a variety of plants and microorganisms have been reported. The enzyme from *A. niger* has been reported, for instance, to have an optimum pH of 4.5, an optimum temperature of 60°C, and a molecular weight of 71,000–88,000.[11] Recently, a beta-glucosidae gene from *A. niger* was cloned and sequenced.[12] The enzyme was found to belong to GH3 based on its deduced amino acid sequence and the Asp-261 within the conserved sequence was identified as the catalytic nucleophile of this beta-glucosidase. Beyond its role in forming the covalent glycosyl-enzyme intermediate, this residue is considered to form hydrogen bonds to the sugar 2-OH of the substrate at the transition state.

20.3.3.1 Production of Gentio-Oligosaccharides[13]

In a typical reaction for the industrial production of gentio-oligosaccharides, β-glucosidase was added to 100 kg of glucose solution (solid 65% (w/w), pH 4.0–4.5) at a ratio of 1.5 U/g glucose, and the solution was incubated at 65°C for 3 days. About 50% of added glucose was converted into oligosaccharides. The reaction mixture was heated to inactivate the enzyme, subjected to de-coloring and de-salting processes successively, and then to cation exchange column to remove glucose. The oligosaccharide fractions were

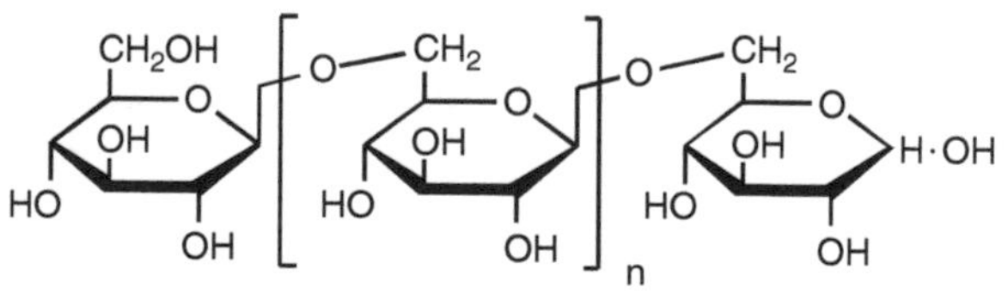

FIGURE 20.3 Structure of gentio-oligosaccharides.

again de-colored with active carbon, de-salted with ion exchange column, and concentrated to give a syrup product (Gentose #80), or spray-dried to give a powdered product (Gentose #80P). Sugar composition of Gentose #80 is: fructose: 1.7%, glucose: 5.8%, gentiobiose: 50.6%, gentiotriose: 28.2%, gentiotetraose and higher: 13.7%.

20.3.4 Enzymes for Trehalose

Trehalose (alpha-D-glucopyranosyl alpha-D-glucopyranoside) is a nonreducing disaccharide widely distributed in nature. It is resistant to heat, pH, and Maillard reaction, and is used as a sweetener and stabilizer for foods, pharmaceuticals, and cosmetics. Biosynthesis of trehalose in microorganisms such as *Escherichia coli* and yeast occurs by the synthesis of trehalose-6-phosphate from UDP-glucose and glucose-6-phosphate followed by its hydrolysis with a specific phosphatase. Production of trehalose has been tried by extraction from trehalose-containing microorganisms such as yeast or by the enzymatic synthesis using maltose phosphorylase and trehalose phosphorylase.[14] However, with these methods economical production of trehalose was difficult.

Recently a group of Hayashibara Biochemical Laboratories found a new enzyme system that produces trehalose directly from starch.[15–22] The system is composed of two enzymes, malto-oligosyl trehalose synthase (MTSase) and malto-oligosyl trehalose trehalohydrolase (MTHase). MTSase catalyzes the conversion of the alpha-1,4-glucosidic linkage between the reducing-end glucose and the next one of malto-oligosaccharides into an alpha alpha-1,1-glucosidic linkage by intramolecular transglycosylation. MTHase catalyzes the hydrolysis of alpha-1,4-glucosidic linkage between the malto-oligosyl and the trehalosyl residues of alpha-malto-oligosyl trehalose, as shown in Figure 20.4.

MTSase and MTHase are classified into isomerase (EC 5.4.99.15) and hydrolase (EC 3.2.1.141), respectively, according to their reaction mechanisms. However, both enzymes are classified into the same GH 13 (alpha-amylase family) according to their primary and secondary structure. Namely, they have four homologous regions in their amino acid sequences characteristic for alpha-amylase family, with two putative catalytic residues, Glu in region 3 and Asp in region 4.[21] Purification and characterization of both enzymes, and cloning and sequencing of the genes coding them, have been carried out on microorganisms such as *Arthrobacter* sp. Q36,[17,18,21] *Sulfolobus acidocaldarius*,[19,20,22] *Sulfolobus solfataricus* KM1,[23,24] *Sulfolobus shibatae*,[25] *Brevibacterium helvolum*,[26] and *Rhizobium* sp. M-11.[27] The enzyme system is present in several bacterial species, as shown in Table 20.4.

No crystallographic study, however, has been reported on these enzymes.

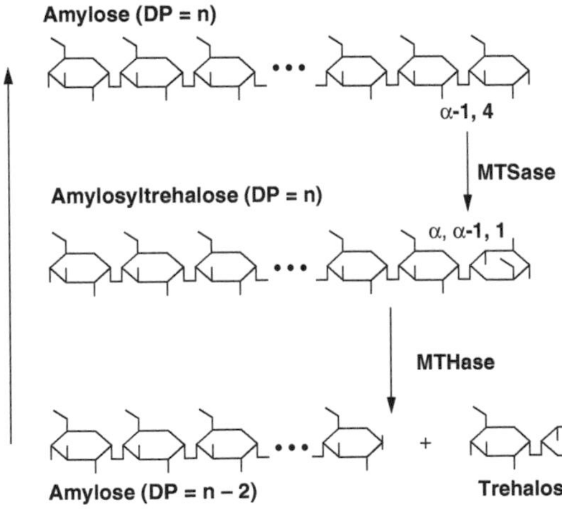

FIGURE 20.4 Conversion of amylose into trehalose by the actions of MTSase and MTHase.

TABLE 20.4 Microorganisms
producing MTSase and MTHase

Arthrobacter sp. Q36
Arthrobacter sp. R190
Arthrobacter sp. ATCC21712
Brevibacterium helvolum ATCC11822
Curtobacterium citreum IFO12677
Curtobacterium luteum IFO15232
Curtobacterium sp. R9
Flavobacterium aquatile IFO3772
Micrococcus luteus IFO3064
Micrococcus roseus ATCC186
Mycobacterium smegmatis ATCC19420
Rhizobium sp. M-11
Sulfolobus acidocaldarius ATCC33909
Sulfolobus solfataricus ATCC35091
Terrabacter tumescens IFO12960

20.3.4.1 Production of Trehalose

Figure 20.5 shows the manufacturing process of trehalose product (TREHA™). Liquefied starch solution was incubated with MTSase, MTHase and isoamylase with an activity ratio of 1:2:200 at 40°C for 24 hr.

Trehalose content of the reaction mixture reached to 85.3%, with glucose, maltose, maltotriose, and others being 1.4%, 5.0%, 6.3%, and 2.0%, respectively. The final product was obtained as a white crystal of trehalose dihydrate. Purity of the commercial products is more than 98%.

Sulfolobus belongs to thermophilic archaebacterium, and MTSase and MTHase from this bacterium showed high optimum temperature and high heat stability, as shown below. From the standpoint of industrial application, these thermophilic enzymes seem to have a great advantage but low productivity of these enzymes (or cells) limits their actual application. Lernia et al.[28] reported expressing of these enzyme activities in *Escherichia coli* cells and immobilization of these recombinant *E. coli* cells after the heat treatment for use as a cell bioreactor for trehalose production at higher temperature. They also constructed the thermophilic fusion enzyme between MTSase and MTHase with an ability to produce trehalose from maltodextrins at 75°C.[29]

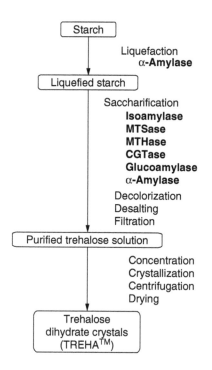

FIGURE 20.5 Flow chart of trehalose production.

20.3.4.2 Basic Properties of Enzymes

Table 20.5 compares basic properties of MTSase and MTHase purified from *Arthrobacter* sp. Q36 and *Sulfolobus acidocaldarius*.

It is clear that MTSase from both origins has higher affinity for oligosaccharides with higher degrees of polymerization. This suggests the enzyme has high affinity toward polysaccharides such as amylose and debranched amylopectin. MTSase does not act on maltotriose and maltose. On the contrary there

TABLE 20.5 Enzymatic properties of MTSase and MTHase produced by *Arthrobacter* sp. Q37 and *Sulfolobus acidocaldarius*

MTSase	Mol. Wt	pI	Opt pH	Opt Temp	Km			
					G4	G5	G6	G7
A. sp. Q36	81,000	4.1	7.0	40	22.9	8.7	1.4	0.9
S. acidocaldarius	74,000	5.9	5.0–6.5	75	41.5	7.1	5.7	1.4
MTHase	Mol. Wt	pI	Opt pH	Opt Temp	Km			
					G2T	G3T	G4T	G5T
A. sp. Q37	62,000	4.1	6.5	45	5.5	4.6	7.0	4.2
S. acidocaldarius	59,000	6.1	5.5–6.0	75	16.7	2.7	3.7	4.9

G2T: maltosyl trehalose, G3T: maltotriosyl trehalose.
G4T: maltotetraosyl trehalose, G5T: maltopentaosyl trehalose.

is no significant difference in Km values of MTHase toward a series of malto-oligosyltrehalose except for Km of *Sufolobus* enzyme for maltosyl trehalose, which was 3–4 times higher than other values. The most remarkable difference between the enzymes from two origins is that enzymes of *Sulfolobus acidocaldarius* have high optimum temperature of 75–80°C whereas enzymes from *Arthrobacter* sp. Q36 showed the optimum temperature of 40–45°C.

20.3.4.3 Trehalose Production from Maltose

Hayashibara Biochemical Laboratories group also found a new enzyme that converts maltose into trehalose in the absence of inorganic phosphate.[30] They purified the enzyme from *Pimelobacter* sp. R48[31] and thermophilic *Thermus aquaticus*[32] and cloned the genes from both organisms.[33,34] The enzyme, trehalose synthase, catalyzes intramolecular conversion of maltose[31,35] and acts only on maltose or trehalose. From these results the enzyme was classified into isomerase (EC 5.4.99.16). It is classified into GH 13 based on its primary and secondary structures. The yield of trehalose from maltose is about 80%, indicating that the equilibrium of the reaction is far in the direction of trehalose. Production of trehalose from maltose using this enzyme seems to be forthcoming. Trehalose-synthesizing enzyme purified from *Pseudomonas* sp. F1[36] appears to have the same catalytic property to the trehalose synthase described above.

Saito et al.[37] purified and characterized a new enzyme from basidiomycete, *Grifola frondosa*. This enzyme, which they called trehalose synthase, catalyzes the synthesis of trehalose from D-glucose and alpha-D-glucose-1-phosphate as shown here:

$$D\text{-glucose} + alpha\text{-D-glucose-1-phosphate} = trehalose + Pi$$

The equilibrium constant of the reaction is 3.5 in favor of trehalose synthesis. Although the reaction mechanism of this enzyme is very close to that of well-known trehalose phosphorylase (EC 2.4.1.64), the distinct difference between them is that the former enzyme recognizes only alpha-D-glucose-1-phosphate whereas the latter recognizes only beta-D-glucose-1-phosphate. Taking advantage of this substrate specificity, the *Grifola frondosa* enzyme can be utilized for industrial trehalose production by coupling with sucrose phosphorylase or starch phosphorylase. The enzyme was cloned and expressed in *Escherichia coli*.[38]

20.3.5 Enzyme for Palatinose (Isomaltulose)

Palatinose is a sucrose isomer having a structure of 6-alpha-D-glucopyranosyl-D-fructofuranose. It is 50% as sweet as sucrose. But unlike sucrose it is noncariogenic and is used as a noncariogenic sweetener. This saccharide is produced from sucrose by microorganisms such as *Protaminobacter rubrum*,[39] *Erwinia rhapontici*,[40] *Serratia plymuthica*,[41] *Pseudomonas mesoacidophila*,[42] *Klebsiella planticola*,[43] and *Agrobacterium radiobacter*.[44] Enzymes responsible for this conversion were purified from several microorganisms and are called alpha-glucosyltransferase,[42] sucrose isomerase,[45] or isomaltulose synthase.[46] This group of enzyme catalyzes conversion of sucrose into a mixture of palatinose and trehalulose, as shown in Figure 20.6.

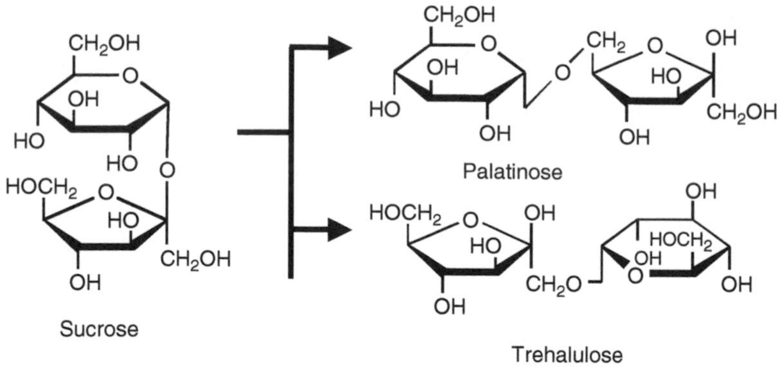

FIGURE 20.6 Conversion of sucrose into palatinose and trehalulose.

The ratio of palatinose to trehalulose varies depending on enzyme sources. These enzymes were classified into isomerase (EC 5.4.99.11) according to its substrate specificity and molecular mechanism, and into GH 13 according to their primary and secondary structures.[47]

20.3.5.1 Production of Palatinose

Protaminobacter rubrum is used for the industrial production of palatinose in Japan.[39] The purified enzyme from this bacterium has a molecular weight of 70,000, pI of 9.9, and produced palatinose (84%) and trehalulose (11%) with a small amount of glucose and fructose when incubated with 50% sucrose in 0.1 N acetate buffer (pH 6.0) at 30°C for 18 hr. In an industrial use, *Protaminobacter rubrum* cells were immobilized in calcium alginate gel and further were cross-linked by treatment with 0.5% glutaraldehyde. The immobilized cells have the same product specificity as the purified enzyme and enough stability. Sucrose solution (40%) adjusted to pH 5.5 was applied to the column packed with the immobilized cells and the flow rate was adjusted so that the sucrose content in the eluate did not exceed 2%. The sugar composition (%) of the eluate determined by HPLC was as follows: palatinose 85.7, trehalulose 8.7, fructose 2.2, glucose 1.8, isomaltose 0.4, isomelezitose 0.2, sucrose 1.1. The eluate was de-salted by passing it through cation and anion exchange columns, successively, and then concentrated by heating to obtain palatinose crystals (90% purity as monohydrate) and palatinose syrup (solid 71%) as commercial products. The sugar composition of palatinose syrup is shown in Table 20.6.

20.3.5.2 Action Mechanism of the Enzyme

Veronese and Perlot[45] studied the action mechanism of sucrose isomerase from *Protaminobacter rubrm* and other two microorganisms using various monosaccharides as inhibitors and acceptors of sucrose conversion reaction. Based on the obtained results they proposed that conformation change of fructose moiety of sucrose must have occurred during the reaction to form trehalulose, while no such conformation change occurred for the conversion to palatinose. Zhang et al.[46,47] carried out biochemical and crystallographic studies on the recombinant isomaltulose synthase from *Klebsiella* sp.LX3 and confirmed the reaction mechanism proposed by Veronese and Perlot. Reaction mechanism presented by Zhang et al.[47] is shown in Figure 20.7.

TABLE 20.6 Sugar Composition of Palatinose Syrup

Sugars	Palatinose syrup-ISN(%)	Palatinose syrup-TN(%)
Trehalulose	53–59	69–75
Palatinose	11–17	11–17
Fructose	12–28	5–10
Glucose	10–16	4–9
Sucrose	<3	<2
Others	<6	<4

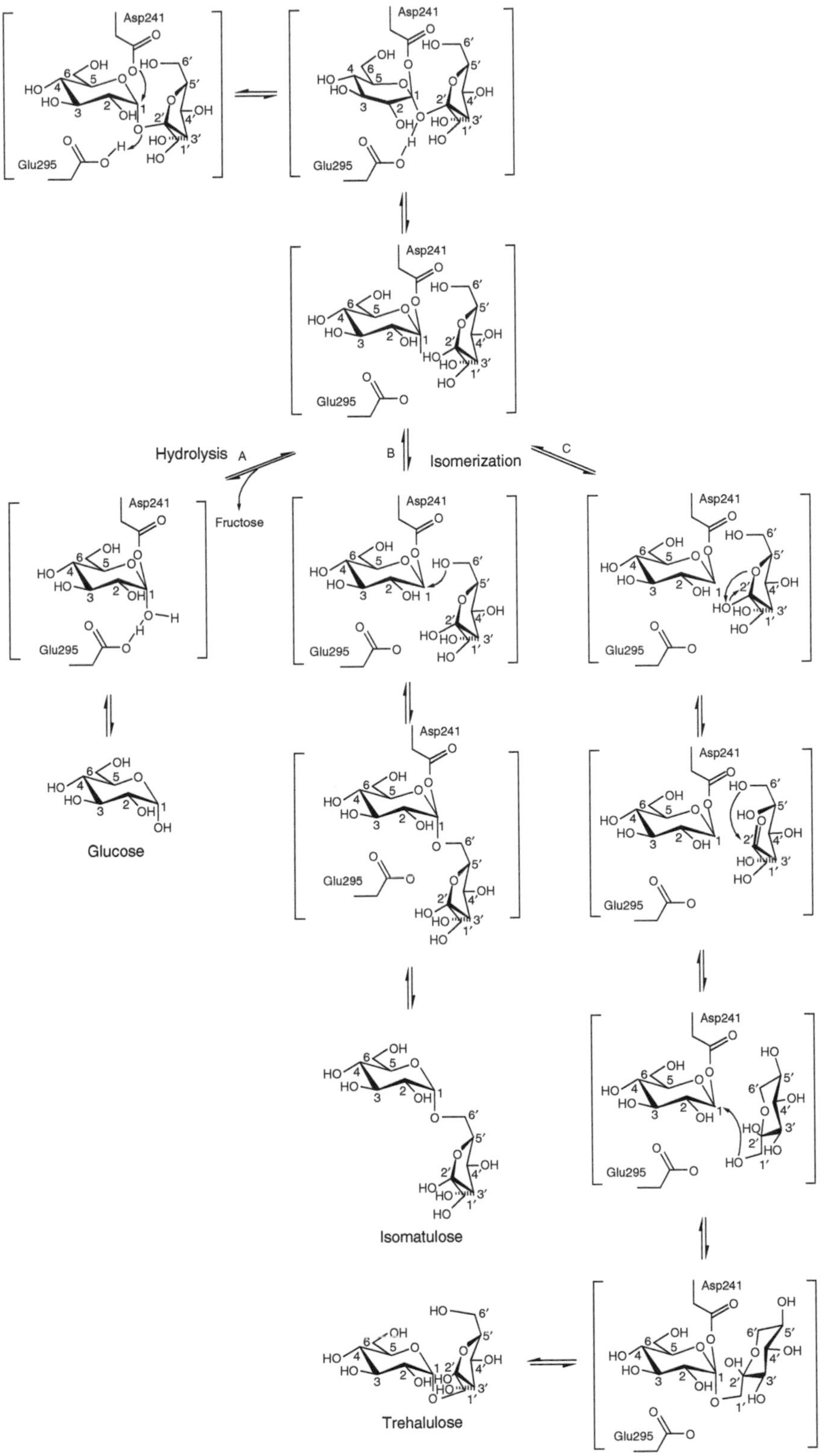

FIGURE 20.7 Reaction mechanism of isomaltulose synthase.

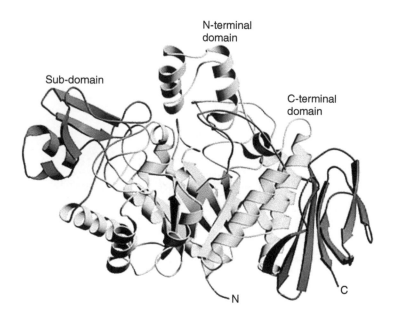

FIGURE 20.8 Three-dimensional model of sucrose isomarase.

From amino acid alignment results they found this enzyme belonged to GH 13. A typical core structure of (beta/alpha)$_8$ barrel characteristic for enzymes of GH13 is also present in this enzyme, as shown in Figure 20.8.

A special motif, 325RLDRD329, in the loop region of isomaltulose synthase plays an important role in keeping the fructose residue released from sucrose in furanose conformation during the catalytic reaction. The purified enzyme produced palatinose (isomaltulose) and trehalulose with a ratio of 82.6% versus 8.3%.

20.3.6 Enzyme for Trehalulose

Trehalulose has a structure of 1-alpha-D-glucopyranosyl-D-fructose and is another structural isomer of sucrose. It is a noncariogenic sugar like palatinose but is more soluble than the latter. This point makes trehalulose more useful as a noncariogenic sweetener. It is produced together with palatinose from sucrose, as shown in Figure 20.6. Enzymes catalyzing this reaction were obtained from various microorganisms as described in Section 20.3.5. Most of these microorganisms, however, produced trehalulose in lower yield (less than 25%) compared to the yield (higher than 75%) of palatinose. So in some cases, trehalulose is produced as a by-product of palatinose production. Nakajima et al. isolated two bacteria, *Pseudomonas mesoacidphila*[48] and *Agrobacterium radiobacter*,[44] both of which converted sucrose into trehalulose with a yield of more than 90%.

20.3.6.1 Production of Trehalulose

Enzyme from *Pseudomonas mesoacidphila* is used for the industrial production of trehalulose. The purified enzyme has a molecular weight of 63,000 and pI of 5.4.[49] It produced trehalulose with an yield of 91% from 20–40% sucrose in pH 5.5–6.5 at 20°C. Sucrose solution (40% w/w) adjusted to pH 5.5 was applied to the column of immobilized cells and the eluate was purified in the same way as described in Section 20.3.5.1, and finally concentrated to obtain trehalulose syrup. The sugar composition of this syrup is shown in Table 20.7.

Similar results are obtained with immobilized *Agrobacterium radiobacter* cells.

TABLE 20.7 Sugar Composition of Trehalulose Syrup

Sugars	%
Trehalulose	85–88
Palatinose	10–12
Fructose	<1
Glucose	<2
Sucrose	<0.5
Others	<3

20.3.6.2 Action Mechanism of the Enzyme

There is no essential difference in action mechanism between enzymes that produce palatinose as a main product and those that produce trehalulose as a main product. Zhang et al.[49] reported that isomaltulose synthase of *Klebisella* sp. LX3, which usually produced palatinose as a main product from sucrose, produced trehalulose with 66.8% yield when Asp residue of 325RLDRD329 motif locating near the active site of this protein was mutated into Asn. This suggests that the charge of 325RLDRD329 determines the product specificity of this enzyme. This motif plays an essential role in keeping the fructose residue in furanose form during the catalysis. For the trehalulose to be produced, fructose residue of sucrose must transform into a pyranose form after it is cleaved out from sucrose at the active site of enzyme and rebinds to glucose through its 1-OH, as shown in Figure 20.7 and Figure 20.8. Very recently Salvucci[50] reported that sucrose isomerase of *Bemisia argentifolii* is distinct from sucrose isomerase of *Erwinia rhapontici* in that the former enzyme produces trehalulose exclusively from sucrose.

20.3.7 Enzyme for Fructo-Oligosaccharide

Fructo-oligosaccharide is a mixture of 1^F-(1-beta-D-fructofuranosyl)$_n$-sucrose, as shown in Figure 20.9.

The occurrence of this oligosaccharide in a variety of plants such as the Jerusalem artichoke has long been known but for its preparation in large-scale, enzymatic methods starting from sucrose is more advantageous than the extraction from plants. Hidaka et al.[51] screened beta-fructofuranosidases

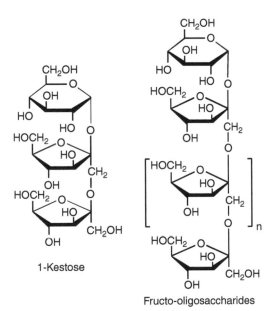

1-Kestose

Fructo-oligosaccharides

FIGURE 20.9 The structure of fructo-oligosaccharides.

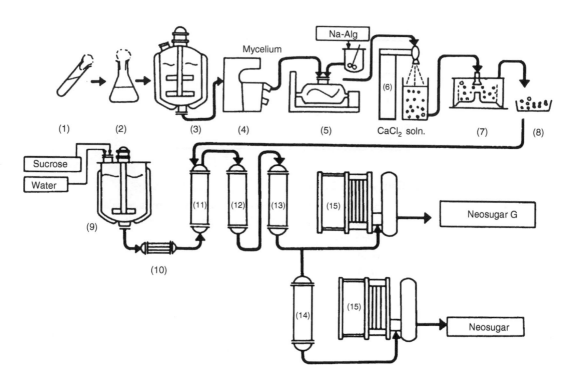

FIGURE 20.10 Process flow of Neosugar production.

(EC 3.2.1.26) from various origins and found that an enzyme produced by *Aspergillus niger* ATCC20611 showed a high fructosyl transferring activity. Other strains of the same *Aspergillus niger* showed little or no transferring activity. The beta-fructofuranosidase from *A.niger* ATCC20611 produced a series of fructo-oligosaccharides with 60% yield when incubated with 50% sucrose. The emergence (commercialization) of fructo-oligosachharide as a bifidobacteria-promoting food component in the 1980s triggered a development rush of a wide range of oligosaccharides in Japan. Fructo-oligosaccharide is widely used not only as food additives but also as feed additives for poultry and swine. Recent studies showed that this sugar has an osteoporosis-preventing activity as well.

20.3.7.1 Production of Fructo-Oligosaccharide[52]

Figure 20.10 shows the process flow of fructo-oligosaccharide product (Neosugar)) developed by Meiji Seika Kaisha Co. Ltd.

As the beta-fructofuranosidase is accumulated inside the cells of *A. niger* ATCC20611, the mycelium was immobilized with sodium alginate and used as an immobilized enzyme. The column (7.0 x 90 cm) of immobilized enzyme was fed with 60% sucrose at 50°C and the reaction mixture was then led to active carbon and ion exchange columns successively to remove colored substance and salts. The obtained solution was concentrated to give Neosugar G preparation, or it was treated with charcoal column to remove glucose and sucrose and then concentrated to give Neosugar preparation. The sugar composition of obtained Neosugar preparations is shown in Table 20.8.

TABLE 20.8 Sugar Composition of Fructo-Oligosaccharides

	G(F)	GF	GF_2	GF_3	GF_4	GF_n
Neosugar G	36-38	10-12	21-28	21-24	3-6	51-53
Neosugar	0	0	28	60	12	100

G: glucose, F: fructose, GF: sucrose, GF2: 1-kestose, GF3: nystose.
GF4: beta-D-fructofuranosyl nystose, GFn: total fructo-oligosaccharides.

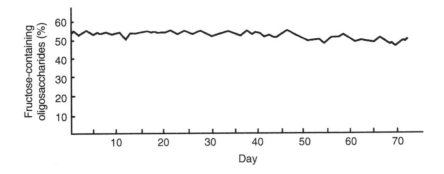

FIGURE 20.11 Continuous production of fructo-oligosaccharides by immobilized fructosyltransferase.

The immobilized enzyme is stable for more than two months, as shown in Figure 20.11.

20.3.7.2 Basic Properties of the Enzyme

Beta-fructofuanosidse of *A. niger* ATCC20611 was extracted from cells and purified.[53] The enzyme showed a high fructosyl transfer activity with optimum pH and temperature being 5.0–6.0°C and 50–60°C, respectively. The molecular weight of the enzyme was 340,000 by gel filtration, 100,000 by SDS-PAGE, and 70,000 after deglycosylation with N-glycosidase F. The last value is coincident with the value of 69,061 estimated from the deduced amino acid sequence of the cloned enzyme gene.[54] The Km value for sucrose was as high as 0.29M. Its catalytic property changes greatly depending on the sucrose concentration, as shown in Figure 20.12.

It showed both hydrolytic and fructosyl transfer reactions in low (0.5%) sucrose solutions, but in 50% sucrose solutions it showed almost exclusively fructosyl transfer reaction. The most remarkable property of this enzyme is its strict regiospecificity to transfer the fructosyl moiety only to the 1-OH of terminal fructofuranosyl residues. The deduced amino acid sequence of beta-fructofuranosidase produced by *A. niger* ATCC20611 showed a high degree of similarity (62% identity) to that[55] of beta-fructofuranosidase produced by *A. niger* NRRL4337, which has low transferring activity.[54] However, the amino acid sequence(s) responsible for the high transferring activity has not yet determined. No crystallographic study on beta-fructofuranosidase has been reported yet. The enzyme belongs to GH32 family based on their primary structure like beta-fructofuranosidases from various origins. A beta-fructofuranosidase secreted by *A. oryzae* ATCC 76080 was purified[56] and found to have high fructosyl transferring activity. But it produced 6^G-(1-beta-fructofuranosyl)$_n$ sucrose in addition to 1^F-(1-beta-fructofuranosyl)$_n$ sucrose indicating lower regiospecificity. A complex biocatalyst system consisting of mycelia of *A. japonicus* CCRC 93007 or *Aureobasidium pullulans* ATCC 9348 with beta-fructofuranosidase activity and *Gluconobacter*

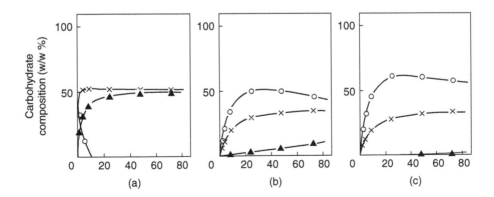

FIGURE 20.12 Effect of sucrose concentrations on the reaction products.

oxydans ATCC 23771 with glucose dehydrogenase activity was developed.[57] The system was run for 7 days to produce 80% fructo-oligosaccharides on a dry weight basis.

20.3.8 Enzyme for Lactosucrose

Lactosucrose is a commercial name given to lactosyl-beta-D-fructofuranoside, the structure of which is shown in Figure 20.13.

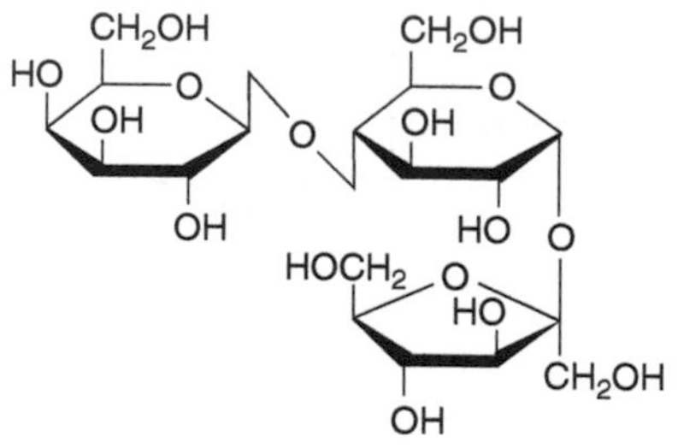

This sugar is 30% as sweet as sucrose and has similar physiological functions as fructo-oligosaccharides described above. Fujita et al.[58] found a unique beta-fructofuranosidase (EC 3.2.1.26) from *Arthrobacter* sp. K-1, which effectively transferred fructosyl residue of sucrose to a wide range of sugars and alcohols.[59] The purified enzyme had a molecular weight of 52,000, pI of 4.3, showed the highest activity at pH 6.5 to 10.0 and at 55°C,

FIGURE 20.13 The structure of lactosucrose.

and was stable up to 45°C at pH 6.5. It showed hydrolytic activity at lower (10 mM) sucrose concentration, but strong transfer activity was observed on incubation with 50% sucrose. Unlike mold enzyme used for the production of fructo-oligosaccharides, this beta-fructofuranosidase showed high transfer activity, especially in the presence of acceptor molecules such as xylose and lactose. Recently, kinetic analysis and modeling of lactosucrose formation by this enzyme was carried out.[60] The transfructosylation reaction was found to proceed according to an ordered bi-bi mechanism in which sucrose bound first to the enzyme and lactosucrose released last.

20.3.8.1 Production of Lactosucrose[61]

Culture supernatant of *Arthroobacter* sp. K-1 was used as the enzyme solution. One hundred kg each of lactose and sucrose was dissolved in water at a total solid concentration of 40%, and this solution was incubated with 750,000 U of beta-fructofuranosidase at 55°C for 10 hr with addition of yeast cells lacking invertase. The solution was heated to stop the enzymatic reaction, subjected to de-coloring and de-salting processes, and then concentrated to syrup (LS-55L) or further, spray-dried to get a powder product (LS-55P). Water content of LS-55L is less than 25% and its sugar composition is: sucrose: less than 25%; lactose: less than 10%; lactosucrose: more than 55%. The sugar composition of LS-55P is: sucrose: less than 10%; lactose: less than 25%; lactosucrose: more than 55%.

20.3.9 Enzyme for Galacto-Oligosaccharide

Galacto-oligosaccharide is a mixture of (beta-1,4 linked-galactosyl)$_n$-lactose, as shown in Figure 20.14.

The occurrence of these oligosaccharides in milk has long been known, and oligosaccharides attract increasing attention because of their prebiotic activity and other activities, such as controlling serum lipid levels and enhancing mineral absorption in the human intestine. Although beta-galactosidase (EC 3.2.1.23) usually catalyzes the hydrolysis of lactose and beta-linked galacto-oligosaccharides, the enzymes from certain microorganisms exhibit high galactosyl transferring activity in some conditions.

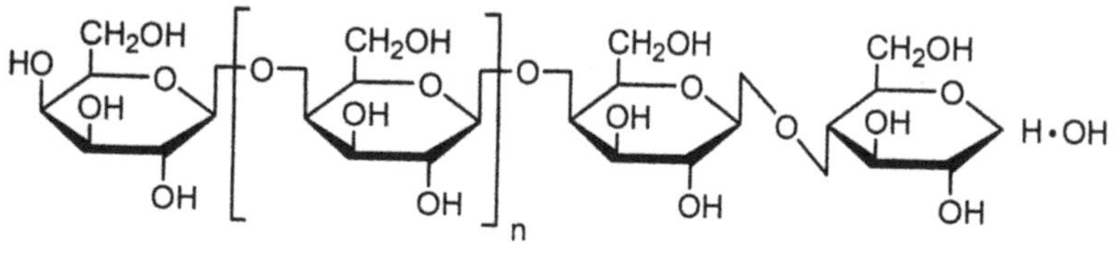

FIGURE 20.14 The structure of galacto-oligosaccharides.

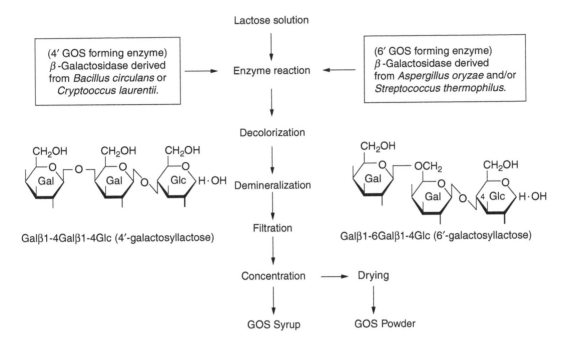

FIGURE 20.15 Industrial production process for galacto-oligosaccharides.

Beta-galacosidase from *Aspergillus oryzae*[62] transfers galactosyl units from lactose to C-6 OH to form a series of 6'-galacto-oligosaccharides, whereas the enzymes from *Bacillus circulans*[63] and *Cryptococcus laurentii*[64] transfer galactosyl units to C-4 OH to form a series of 4'-galacto-oligosaccharides, as shown in Figure 20.15.[65]

20.3.9.1 Production of Galacto-Oligosaccharides[65]

Biolacta, a crude beta-galactosidase preparation from *Bacillus circulans* ATCC 31382, is available from Daiwa Kasei (Osaka, Japan). To 45 kg of lactose solution (5:4, w/w) adjusted to pH 7 was added 150,000 U of the enzyme preparation and the reaction was carried out at 60°C for 12 hr. After heating, the reaction mixture was treated with active carbon, and applied on an ion-exchange column. The eluate was concentrated to obtain a syrup product (water content less than 25%). The sugar composition of this syrup is: tetrasaccharides: 15%; trisaccharides: 23.5%; transferred disaccharides: 20%; and galactose: 1.3%. More than 55% of the added lactose was converted to galacto-oligosaccharides.

20.3.9.2 Basic Properties of the Enzyme

Two beta-galactosidases with molecular weights of 240,000 (galacosidase-I) and 160,000 (galactosidase-II) were purified from the commercial enzyme preparation, Biolacta. They showed similar pI of 4.5 and the same optimum pH of 6.0, but they differed in substrate specificity and galactosyl transferring activity, with galactosidase-II having stronger transferase activity.[63] Ito and Sasaki[66] found that *Bacillus circulans* ATCC 31382 have three beta-galactosidase isoforms and that Biolacta lacked the third one (galactosidase-III). Galacosidase III with the deduced molecular weight of 66,888, catalyzes galactosyl transfer to form 3'-galacto- oligosaccharides.[67] This enzyme was classified into GH 35 based on its amino acid sequence. The *Aspergillus oryzae* beta-galactosidase, which catalyzes the galactosyl transfer to C-6 position, is also classified into GH35.[68] Two nucleotide sequence data of beta-galactosidase gene from *Bacillus circulans* are available in the GenBank (L03424 and L03425). Based on their deduced amino acid sequences they are classified into GH 42, although the relationship between these two genes and two beta-galactosidases present in Biolacta is not known.

20.3.10 Enzyme for Xylo-Oligosaccharide

Xylo-oligosaccharide is a mixture of a series of oligosaccharides consisting of 2 to 7 D-xylose units linked by beta-1,4 linkage as shown in Figure 20.16.

These oligosaccharides are not hydrolyzed by our digestive enzymes and, therefore, are selectively utilized by bifidobacteria when administered. Xylo-oligosaccharides with higher DP are known to act as dietary fiber. They are produced from xylan by the action of xylanase (EC 3.2.1.8). The enzyme is found in plants and a wide range of microorganisms such as *Aspergillus, Bacillus, Clostridium, Penicillium, Pseudomonas,* and *Trichoderma.* For the industrial production of xylo-oligosaccharides, various microbial enzymes were tested and xylanase from *Trichoderma* was selected as the best enzyme.[69] Crude enzyme solution containing 2500 U of xylanase activity was applied on a column packed with a mixture of 6.25 g each of birchwood xylan and celite and the column was irrigated with 50 mM acetate buffer (pH 4.5) at 40°C. The eluted solution contained xylobiose as a main component. Enzyme that was adsorbed on xylan was finally eluted from the column after the elution of xylobiose. The eluted enzyme was then applied to the next column of birchwood xylan. This procedure was repeated four times with more than 90% of packed xylan being hydrolyzed each time. Based on these results the commercial xylo-oligosaccharide product (Xylooligo70) is produced from corncob or bagasse xylan. This syrup contains 75% solid and xylo-oligosaccharides account for 70% of total carbohydrates. Although the detailed information on the source of xylanase used for the commercial production of xylo-oligosaccharides is not available, numerous studies on *Trichoderma* xylanases have been reported. Purification,[70] characterization,[70,71] and crystallographic studies[72,73] of xylanase I and II from *Trichoderma reesei*, and cloning and sequencing[74] of their genes were carried out in the early 1990s. Most of the fungal xylanases including those described above belong to GH11 whereas many plant and bacterial xylanases belong to GH10.

20.3.11 Enzyme for Chitosan-Oligosaccharide

Chitosan-oligosaccharide is a mixture of oligosaccharides consisting of 2 to 7 D-glucosamine (GlcN) units linked by beta-1,4 linkages, as shown in Figure 20.17.

They attract increasing interests as foods or food ingredients because of their physiological activity such as to enhance immune response, to reduce blood glucose level, and to proliferate selectively bifidobacterium. These oligosaccharides are produced by hydrolysis of chitosan either by acid or enzyme. Whereas acid hydrolysis gives a series of chitosan-oligosaccharides including GlcN, enzymatic hydrolysis usually gives chitosan-oligosaccharides with little amounts of GlcN. Enzyme responsible for the hydrolysis of chitosan is called chitosanase (EC 3.2.1.132). Chitosanases have been found in a large number of bacteria, fungi, and plants. They have been purified from various microorganisms such as *Bacillus circulans* MH-K1,[75] *B. coagulans* CK108,[76] *B. pumilus* BN262,[77] *B. Sbutilis* KH1,[78] *B. species* CK4,[79] *Burkholderia gladioli,*[80] *Penicillium islandicum,*[81] and *Streptomyces griseus.*[82] Most enzymes act on partially deacetylated chitosan producing di- and trisaccharides as major products. Based on the specificity of the cleavage positions they are classified into three classes: class I chitosanases cleave GlcNac-GlcN linkage in chitosan, class II chitosanases cleave GlcN-GlcN linkage, and class III chitosanases cleave both GlcN-GlcN and GlcN-GlcNac linkages.

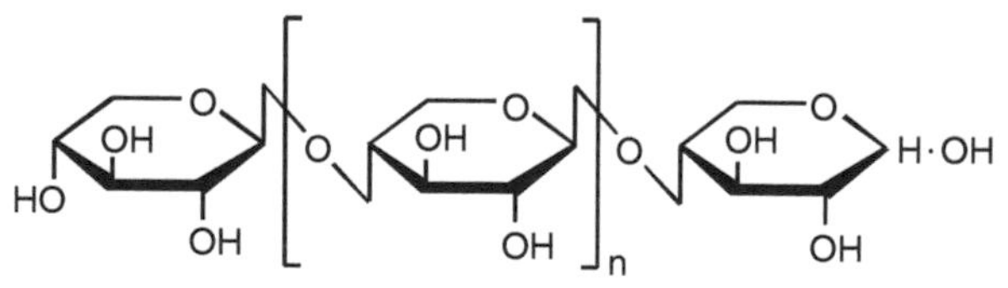

FIGURE 20.16 The structure of xylo-oligosaccharides.

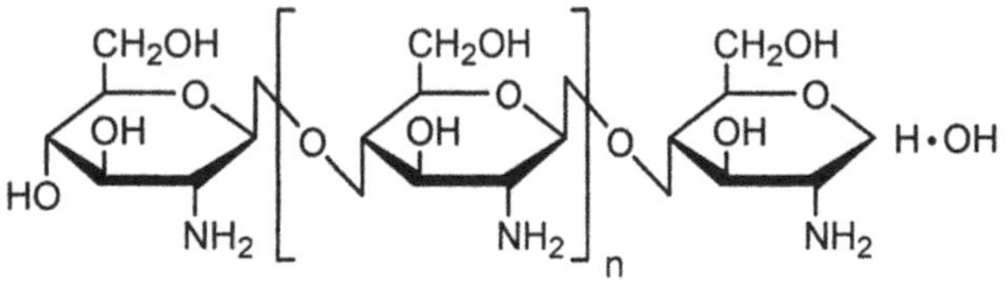

FIGURE 20.17 The structure of chitosan-oligosaccharides.

20.3.11.1 Production of Chitosan-Oligosaccharides

After trying several chitosanases,[83] the enzyme from *Bacillus pumilus* BN262[77] was chosen for the industrial production of chitosan-oligosaccharides because of high enzyme productivity of this bacterium and high oligosaccharides productivity of its enzyme. The purified enzyme showed the maximum activity at 40–55°C and pH 6.5 with a molecular weight of 31,000. Chitin obtained from crab shells was de-acetylated chemically to produce chitosan. A 2% chitosan solution in 50 mM acetate buffer (pH 5.5) was then treated with chitosanase at 40°C for 2 days and the reaction mixture was filtered, de-colorized, and concentrated to yield yellowish powder of chitosan-oligosaccharides. Commercial product (COS-Y, Yaizu Suisan Kagaku Industry Co Ltd.) contains more than 75% of oligosaccharides, and its oligosaccharide composition is shown in Table 20.9.

TABLE 20.9 Oligosaccharide Composition of COS-Y

Sugar	%
Chitobiose	9.8
Chitotriose	23.8
Chitotetraose	27.9
Chitopentaose	23.9
Chitohexaose	9.9
Chitoheptaose	4.7

20.3.11.2 Mechanism of Substrate Specificity

More than 20 chitosanase genes have been cloned and sequenced, and their amino acid sequences have been deduced. Based on their primary structures, these enzymes are grouped into four gluco-hydrolase families: GH46, GH8, GH75, and GH80. The majority of chitosanases belong to GH46. Three-dimensional structures were clarified on enzymes from *B. circulans* MH-K1 and *Streptomyces* sp. N174.[84] Although both enzymes belong to GH46, the former enzyme is classified into class III chitosanase and the latter one is classified into class I enzyme. Crystallographic study of *B. circulans* MH-K1 chitosanase[75] revealed that the shape of substrate binding clefts of this enzyme was significantly different from that of *Streptomyces* sp. N174 enzyme. Furthermore, investigation on the binding model of *B. circulans* MH-K1 chitosanase with chitosan hexamer revealed that its substrate-binding cleft could accommodate only the GlcN-GlcNac motif at the D-E subsite but not the GlcNac-GlcN motif, as shown in Figure 20.18.

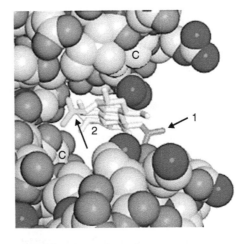

FIGURE 20.18 Substrate fitting at the active site of *B. circulans* MH-K1 chitosanase.

This fact explains why *B. circulans* MH-K1 chitosanase belongs to class III chitosanase.

TABLE 20.10   Sugar Compositions of Fuji-Oligo

	Fuji-oligoG3 (%)		Fuji-oligoG4 (4%)		Fuji-oligoG67 (%)
	#350	#360	#450	#370	
Glucose	2.8	4.0	1.0	2.0	4.4
Maltose	16.8	21.7	7.8	8.5	13.4
Maltotriose	50.0	60.0	10.2	11.0	7.9
Maltotetraose	8.6	7.6	50.5	72.0	7.6
Maltopentaose			2.5	1.0	7.6
Maltohexaose	21.8	6.7			40.0
Maltoheptaose			28.0	5.5	
Others (DP>7)					19.1
Total	100	100	100	100	100

20.3.12 Enzymes for Other Sugars

A series of malto-oligosaccharide products for foods applications are available from Nihon Shokuhin Kako Co Ltd. and Hayasibara Biochemical Laboratories Inc. Sugar compositions of some of these products (Fuji-Oligo) are shown in Table 20.10.

These oligosaccharides are produced by the actions of special malto-oligosaccharide-forming alpha-amylases. For instance, alpha-amylases from *Streptomyces griseus*,[85] *Pseudomonas stutzeli*,[86,87] *Bacillus licheniformis*[88] and *Pseudomonas* sp. KO-8940,[89] and *Aerobacter aerogenes*,[90] are used for the production of maltotriose, maltotetraose, maltopentaose, and maltohxaose, respectively. Basic properties of these malto-oligosacharide-producing enzymes are reviewed in a book.[91] Erythritol is about 70% as sweet as sucrose, a noncalorie sugar alcohol, and is a sole sugar alcohol that is produced by fermentative method. It has been commercially produced by using a yeast strain, *Aureobasidium* sp.[92,93] A commercial production of this sugar alcohol by using *Monilliella yomentosa* var. *pollinis*[94] started recently. Lee et al.[95] are actively studying the erythritol production by *Torula coralline*. 1,5-Anhydro-D-fructose is a newly developed sugar having a structure shown in Figure 20.19.

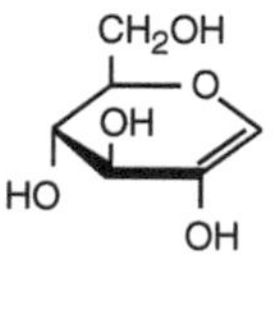

This sugar has a strong antioxidative activity and its application in foods is being researched actively. It is produced from starch by the action of a alpha-1,4-glucan lyase (EC 4.2.2.13) obtained from red algae.[96,97] Another interesting sugar to be introduced is cyclodextrans.[98] They have similar structures to those of cyclodextrins, but are distinctly different from the latter as glucose units are linked

FIGURE 20.19   Structure of 1,5-anhydro-D-fructose.

by alpha-1,6 linkages in cyclodextrans. Cyclodextrans are produced by the action of a unique enzyme, cyclodextran glucanotransferase, from *Bacillus circulans* T-3040.[99] Cellobiose is a disaccharide consisting of two D-glucose units connected through beta-1,4 linkage. Many efforts have been made in vain to produce cellobiose on an industrial scale from cellulose. An efficient conversion of sucrose into cellobiose using sucrose phosphorylase, xylose isomerase, and cellobiose phosphorylase was reported.[100]

References

1. S Chiba, T Shimomura. Transglucosylation of alpha-glucosidase. *Denpun Kagaku*. 26:59–67, 1979.

2. KJ Duan, DC Sheu, CT Lin. Transglucosylation of a fungal alpha-glucosidase. The enzyme properties and correlation of isomaltooligosaccharide production. *Ann NY Acad Sci*. 750:325–328, 1995.

3. A Nakamura, I Nishimura, A Yokoyama, DG Lee, M Hidaka, H Masaki, A Kimura, S Chiba, T Uozumi. Cloning and sequencing of an alpha-amylase gene from *Aspergillus niger* and its expression in *A. nidulans*. *J Biotechnol*. 53:75–83, 1997.

4. E Yasuda, H Takaku, H Matsumoto. Preparative method of branched oligosaccharide syrup. Japanese Patent Registration Number 2131579 (1985).

5. K Kainuma, S Kobayashi, E Yasuda, H Takaku, T Yatake. Preparative method of syrup that contains high amounts of branched oligosaccahrides. Japanese Patent Registration Number 1859017 (1985).

6. Y Konishi, K Shindo. Production of nigerose, nigerosyl glucose, and nigerosyl maltose by *Acremonium* sp. S4G13. *Biosci Biotechnol Biochem* 61:439–442, 1997.

7. Y Konishi, K Shindo. Preparative method of oligosaccharides. Japanese patent Publication Number 1995–059559.

8. I Kobayashi, M Tokuda, H Hashimoto, T Konda, H Nakano, S Kitahata. Purification and characterization of a new type of alpha-glucosidase from *Paecilomyces lilacinus* that has transglucosylation activity to produce alpha-1,3- and alpha-1,2-linked oligosaccharides. *Biosci Biotechnol Biochem* 67:29–35, 2003.

9. T Unno, K Ide, T Yazaki, Y Tanaka, T Nakakuki, G Okada. High recovery purification and some properties of a beta-glucosidase from *Aspergillus niger*. *Biosci Biotechnol Biochem* 57: 2172–2173, 1993.

10. T Unno, T Nakakuki, S Kainuma, K Kataura, G Okada. Production of beta-glucooligosaccharides-containing syrup and its physical properties. *Oyo Toshitsu Kagaku* 41:327–334, 1994.

11. MA Abdel-Naby, MY Osman, AF Abdel-Fattah. Purification and properties of three cellobiases from *Aspergillus niger* A20. *Appl Biochem Biotechnol* 76:33–44, 1999.

12. S Dan, I Marton, M Dekel, BA Bravdo, S He, SG Withers, O Shoseyov. Cloning, expression, characterization, and nuclophile identification of family 3, *Aspergillus niger* beta-glucosidase. *J Biol Chem* 275:4973–4980, 2000.

13. T Unno. Industrial production of gentiooligosaccharides-containing syrup. *Oyo Toshitsu Kagaku* 42:83–89, 1995.

14. M Yoshida, N Nakamura, K Horikoshi. Production and application of maltose phosphorylase and trehalose phosphorylase by a strain of *Plesiomonas*. *Oyo Toshitsu Kagaku (J Appl Glycosci)* 42:19–25, 1995.

15. A Tabuchi, T Mandai, T Shibuya, S Fukuda, T Sugimoto, M Kurimoto. Formation of trehalose from starch by novel enzymes. *Oyo Toshitsu Kagaku* 42:401–406, 1995.

16. K Maruta, T Nakada, M Kubota, H Chaen, T Sugimoto, M Kurimoto, Y Tsujisaka. Formation of trehalose from maltooligosaccharides by a novel enzymatic system. *Biosci Biotechnol Biochem* 59:1829–1834, 1995.

17. T Nakada, K Maruta, K Tsusaki, M Kubota, H Chaen, T Sugimoto, M Kurimoto, Y Tsujisaka Purification and properties of a novel enzyme, maltooligosyl trehalose synthase, from *Arthrobacter* sp. Q36. *Biosci Biotechnol Biochem*. 59:2210–2214, 1995.

18. T Nakada, K Maruta, H Mitsuzumi, M Kubota, H Chaen, T Sugimoto, M Kurimoto, Y Tsujisaka Purification and characterization of a novel enzyme, maltooligosyl trehalose trehalohydrolase, from *Arthrobacter* sp. Q36. *Biosci Biotechnol Biochem*. 59:2215–2218, 1995.

19. T Nakada, S Ikegami, H Chaen, M Kubota, S Fukuda, T Sugimoto, M Kurimoto, Y Tsujisaka. Purification and characterization of thermostable maltooligosyl trehalose synthase from the thermoacidophilic archaebacterium *Sulfolobus acidocaldarius*. *Biosci Biotechnol Biochem* 60:263–266, 1996.

20. T Nakada, S Ikegami, H Chaen, M Kubota, S Fukuda, T Sugimoto, M Kurimoto, Y Tsujisaka. Purification and characterization of thermostable maltooligosyl trehalose trehalohydrolase from the thermoacidophilic archaebacterium *Sulfolobus acidocaldarius*. *Biosci Biotechnol Biochem*. 60:267–270, 1996.

21. K Maruta, K Hattori, T Nakada, M Kubota, T Sugimoto, M Kurimoto. Cloning and sequencing of trehalose biosynthesis genes from *Arthrobacter* sp. Q36. *Biochem Biophys Acta*. 1289:10–13, 1996.

22. K Maruta, H Mitsuzumi, T Nakada, M Kubota, H Chaen, S Fukuda, T Sugimoto, M Kurimoto. Cloning and sequencing of a cluster of genes encoding novel enzymes of trehalose biosynthesis from thermophilic archaebacterium *Sulfolobus acidocaldarius*. *Biochem Biophys Acta* 1291:177–181, 1996.

23. M Kato, Y Miura, M Kettoku, K Shindo, A Iwamatsu, K Kobayashi. Trehalose-producing enzymes isolated from the hyperthermophilic archae, *Sulfolobus solfataricus* KM1. *Biosci Biotechnol Biochem* 60:546–550, 1996.

24. K Kobayashi, M Kato, Y Miura, M Kettoku, T Komeda, A Iwamatsu. Gene cloning and expression of new trehalose-producing enzymes from the hyperthermophilic archaeum *Sulfolobus solfataricus* KM1. Biosci *Biotechnol Biochem* 60:1882–1885, 1996.

25. ID Lernia, A Morana, A Ottombrino, S Fusco, M Rossi, MD Rosa. Enzymes from *Sulfolobus shibatae* for the production of trehalose and glucose from starch. *Extremophiles* 2:409–416, 1998.

26. H Kim, TK Kwon, S Park, HS Seo, JJ Cheong, CH Kim, JK Kim, JS Lee, YD Choi. Trehalose synthesis by sequential reactions of recombinant maltooligosyltrehalose synthase and maltooligosyltrehalose trehalohydrolase from *Brevibacterium helvolum*. *Appl Environ Microbiol* 66:4620–4624, 2000.

27. K Maruta, K Hattori, T Nakada, M Kubota, T Sugimoto, M Kurimoto. Cloning and sequencing of trehalose biosynthesis genes from *Rhizobium* sp. M-11. *Biosci Biotechnol Biochem* 60:717–720, 1996.

28. ID Lernia, C Schiraldi, M Generoso, MD Rosa. Trehalose production at high temperature exploiting an immobilized cell bioreactor. *Extremophiles* 6:341–347, 2002.

29. DD Pascale, ID Lernia, MP Sasso, A Furia, MD Rosa, M Rossi. A novel thermophilic fusion enzyme for trehalose production. *Extremophiles* 6:463–468, 2002.

30. T Nishimoto, M Nakano, S Ikegami, H Chaen, S Fukuda, T Sugimoto, M Kurimoto, Y Tsujisaka. Existence of a novel enzyme converting maltose into trehalose. *Biosci Biotechol Biochem.* 59:2189–2190, 1995.

31. T Nishimoto, M Nakano, T Nakada, H Chaen, S Fukuda, T Sugimoto, M Kurimoto, Y Tsujisaka. Purification and properties of a novel enzyme, trehalose synthase from *Pimelobacter* sp. R48. *Biosci Biotechol Biochem* 60:640–644, 1996.

32. T Nishimoto, T Nakada, H Chaen, S Fukuda, T Sugimoto, M Kurimoto, Y Tsujisaka. Purification and characterization of a thermostable trehalose synthase from *Thermus aquaticus*. *Biosci Biotechol Biochem.* 60:835–839, 1996.

33. K Tsusaki, T Nishimoto, T Nakada, M Kubota, H Chaen, T Sugimoto, M Kurimoto. Cloning and sequencing of trehalose synthase gene from *Pimelobacter* sp. R48. *Biochem Biophys Acta* 1290:1–3, 1996.

34. K Tsusaki, T Nishimoto, T Nakada, M Kubota, H Chaen, S Fukuda, T Sugimoto. Cloning and sequencing of trehalose synthase gene from *Thermus aquaticus* ATCC33923. *Biocim Biophys Acta* 1334:28–32, 1997.

35. S Koh, J Kim, H Shin, D Lee, J Bae, D Kim, D Lee. Mechnistic study of the interamlecular conversion of maltose to trehalose by *Thermus caldophilus* GK24 trehalose synthase. *Carbohydr Res* 338:1339–1343, 2003.

36. M Ohguchi, N Kubota, T Wada, K Yoshinaga, M Uritani, M Yagisawa, K Ohishi, M Yamagishi, T Ohta, K, Ishikawa. Purification and properties of trehalose- synthesizing enzyme from *Pseudomonas* sp. F1. *J Ferment Bioeng* 84:358–360, 1997.

37. K Saito, T Kase, E Takahashi, E Takahashi, S Horinouchi. Purification and characterization of trehalose synthase from basidiomycete *Grifola frondosa*. *Appl Environ Microbiol* 64:4340–4345, 1998.

38. K Saito, H Yamazaki, Y Ohnishi, S Fujimoto, E Takahashi, S Horinouchi. Production of trehalose synthase from a basidiomycete, *Grifola frondosa*, in *Escherichia coli*. *Appl Microbiol Biotechnol.* 50:193–198, 1998.

39. Y Nakajima. Manufacture and utilization of palatinose. *Denpun Kagaku* 35:131–139, 1988.

40. Cheetham PS. The extraction and mechanism of a novel isomaltulose-synthesizing enzyme from *Erwinia rhapontici*, *Biochem J* 220:213–220, 1984.

41. S Fujii, M Kishihara, M Komoto, J Shimizu. Isolation and characterization of oligosaccharides produced from sucrose by transglucosylation action of *Serratia plymuthica*. *Nippon Shokuhin Kogyo Gakkaishi* 30:339–344, 1983.

42. Y Nagai, T Sugitani, K Tsuyuki. Characterization of alpha-glucosyltransferase from *Pseudomonas mesoacidophila* MX-45. *Biosci Biotechnol Biochem* 58:1789–1793, 1994.

43. JH Huang, LH Hsu, YG Su. Conversion of sucrose into isomaltulose by *Klebsiella planticola* CCRC 19112. *J Ind Microbiol Biotechnol* 21:22–27, 1998.

44. J Nagai-Miyata, K Tsuyuki, T Sugitani, T Ebashi, Y Nakajima. Isolation and characterization of a trehalulose-producing strain of *Agrobacterium*. *Biosci Biotechnol Biochem* 57:2049–2053, 1993.

45. T Veronese, P Perlot. Proposition for the biochemical mechanism occurring in the sucrose isomerase active site. *FEBS Lett* 441:348–352, 1998.

46. D Zhang, X Li, LH Xhang. Isomaltulose synthase from *Klebsiella* sp. Strain LX3: gene cloning and characterization and engineering of thermostability. *Appl Environ Microbiol* 68:2676–2682, 2002.

47. D Zhang, N Li, SM Lok, LH Zhang, Swaminathan K. Isomaltulose synthase (PalI) of Klebsiella sp. LX3: Crystal structure and implication of mechanism. *J Biol Chem* 278:35428–35434, 2003.

48. Y Miyata, T Sugitani, K Tsuyuki, T Ebashi, Y Nakajima. Isolation and characterization of *Pseudomonas mesoacidophila* producing trehalulose. *Biosci Biotechnol Biochem* 56:1680–1681, 1992.

49. D Zhang, N Li, K Swaminathan, LH Zhang. A motif rich in charged residues determines product specificity in isomaltulose synthase. *FEBS Lett* 534:151–155, 2003.

50. ME Salvucci. Distinct sucrose isomerases catalyze trehalulose synthesis in whiteflies, *Bemisia argentifolii*, and *Erwinia rhapontici*. *Comp Biochem Physiol B Biochem Mol Biol* 135:385–395, 2003.

51. H Hidaka, M Hirayama, N Sumi. A fructooligosaccharide-producing enzyme from *Aspergillus niger* ATCC 20611. *Agric Biol Chem* 52, 1181–87 (1988).

52. H Hidaka Production and properties of Neosugar. Proceedings of Neosugar Kennkyuukai p. 5–14, (1982).

53. M Hirayama, N Sumi, H Hidaka. Purification and properties of a fructooligosaccharide-producing beta-fructofuranosidase from *Aspergillus niger* ATCC 20611. *Agric Biol Chem* 53:667–673, 1989.

54. K Yanai, A Nakane, A Kawate, M Hirayama. Molecular cloning and characterization of the fructooligosaccharide-producing beta-fructofuranosidase gene from *Aspergillus niger* ATCC 20611. *Biosci Biotechnol Biochem* 65:766–773, 2001.

55. LM Boddy, T Berges, C Barreau, MH Vainstein, MJ Dobson, DC Balance, JF Peberdy. Purification and characterization of an *Aspergillus niger* invertase and its DNA sequence. *Curr Genet* 24:60–66, 1993.

56. CT Chang, YY Lin, MS Tang, CF Lin. Purification and properties of beta-fructofuranosidase from *Aspergillus oryzae* ATCC 76080. *Biochem Mol Biol Int* 32:269–277, 1994.

57. DC Sheu, KJ Duan, CY Cheng, JL Bi, JY Chen. Continuous production of high-content fructooligosaccharides by a complex cell system. *Biotechnol Prog* 18:1282–1286, 2002.

58. K Fujita, K Hara, H Hashimoto, S Kitahata. Purification and some properties of beta-fructofuranosidase I from *Arthrobacter* sp. K-1. *Agric Biol Chem* 54:913–919, 1990.

59. K Fujita, K Hara, H Hashimoto, S Kitahata. Transfructosylation catalyzed by beta-fructofuranosidase I from *Arthrobacter* sp. K-1. *Agric Biol Chem* 54:2655–2661, 1990.

60. A Pilgrim, M Kawase, M Ohashi, K Fujita, K Murakami, K Hashimoto. Reaction kinetics and modeling of the enzyme-catalyzed production of lactosucrose using beta-fructofuranosidase from *Arthrobacter* sp. K-1. *Biosci Biotechnol Biochem* 65:758–765, 2001.

61. K Fujita, T Osawa, K Mikuni, K Hara, H Hashimoto, S Kitahata. Production of lactosucrose by beta-fructofuranosidase and some of its physical properties. *Denpun Kagaku* 38:1–7, 1991.

62. K Matsumoto, Y Kobayashi, N Tamura, T Watanabe, T Kan. Production of galactooligosaccharides with beta-galactosidase. *Denpun Kagaku* 36:123–130, 1989.

63. Z Mozaffar, K Nakanishi, R Matsuno. T Kamikubo. Purification and properties of beta-galactosidases from *Bacillus circulans*. *Agric Biol Chem* 48:3053–3061, 1984.

64. O Ozawa, K Ohtsuka, T Uchida. Production of 4'-galactosyl lactose by mixed cells of *Cryptococcus laurentii* and Baker's yeast. *Nippon Shokuhin Kogyou Gakkaishi*. 36:898–902, 1989.

65. T Sako, K Matsumoto, R Tanaka. Recent progress on research and applications of non-digestible galactooligosaccharides. *International Dairy Journal* 9:69–80, 1999.

66. Y Ito, T Sasaki. Cloning and characterization of the gene encoding a novel beta-galactosidase from *Bacillus circulans*. *Biosci Biotechol Biochem* 61:1270–1276, 1997.

67. H Fujimoto, M Miyasato, Y Ito, T Sasaki, K Ajisaka. Purification and properties of recombinant beta-galactosidase from *Bacillus circulans*. *Glycoconj.* J 15:155–160, 1998.

68. Y Ito, T Sasaki, K Kitamoto, C Kumagai, K Takahashi, K Gomi, G Tamura. Cloning, nucleotide sequencing, and expression of the beta-galactosidase-encoding gene (lacA) from *Aspergillus oryzae*. *J Gen Appl Microbiol* 48:135–142, 2002.

69. S Fujikawa, M Okazaki, N Matsumoto, K Koga. Properties and production of xylo-oligosaccharide. *Denpun Kagaku*, 37:69–77, 1990.

70. A Torronen, RL March, R Meener, G R onzalez, N Kalkkinen, AM Harkki, CP Kubicek. The two major xylanases from *Trichoderma reesei*: characterization of both enzyme and gene. *Biotechnol* 10:1461–1465, 1992.

71. P Biely, L Kremnický, J Alfoldi, M Tenkanen. Stereochemistry of the hydrolysis of glycosidic linakage by endo-beta-1,4-xylanases of Trichoderma reesei. *FEBS Lett* 356:137–140, 1994.

72. A Torronen, A Harkki, J Rouvinen. Three-dimensional structure of endo-1,4-beta-xylanase II from *Trichoderma reesei*: two conformational states in the active site. *EMBO J.* 13:2493–2501, 1994

73. A Torronen, J Rouvinen. Structural comparison of two major endo-1,4-xylanases from *Trichoderma reesei*. *Biochem* 34:847–856, 1995.

74. R Saarelainen, M Palohemio, R Fagerstrom, PL Suominen, KM Nevalainen. Cloning, sequencing and enhanced expression of the *Trichoderma reesei* endoxylanase II (pI 9) gene xln2. Mol *Gen Genet.* 241:497–503, 1993.

75. J Saito, A Kita, Y Higuchi, Y Nagata, A Ando, K Miki. Crystal structure of chitosanase from *Bacillus circulans* MH-K1 at 1.6-A resolution and its substrate recognition mechanism. *J Biol Chem* 274:30818–30825, 1999.

76. HG Yoon, KH Lee, HY Kim, HK Kim, DH Shin, BS Hong, HY Cho. Gene cloning and biochemical analysis of thermostable chitosanase (TCH-2) from *Bacillus coagulans* CK 108. *Biosci Biotechnol Biochem* 66:986–995, 2002.

77. T Matsunobu, O Hiruta, K Nakagawa, H Murakami, S Miyadoh, K Uotani, H Takebe, A Satoh. A novel chitosanase from *Bacillus pumilus* BN 262, properties, production and applications. *Sci Report of Meiji Seika Kaisha* 35:28–59, 1996.

78. CA Omumasaba, N Yoshida, Y Sekiguchi, K Kariya, K Ogawa. Purification and some properties of a novel chitosanase from *Bacillus subtilis* KH1. *J Gen Appl Microbiol* 46:19–27, 2000.

79. HG Yoon, HY Kim, HK Kim, BS Hong, DH Shin, HY Cho. Thermostable chitosanase from *Bacillus* sp. Strain CK4: Its purification, characterization and, reaction patterns. *Biosci Biotechnol Biochem* 65:802–809, 2001.

80. M Shimosaka, Y Fukumori, XY Zhang, NJ He, R Kodaira, M Okazaki. Molecular cloning and characterization of a chitosanase from the chitosanolytic bacterium *Burkholderia gladioli* strain CHB 101. *Appl Microbiol Biotechnol* 54:354–360, 2000.

81. DM Fenton, DE Everleigh. Purification and mode of action of a chitosanase from *Penicillium islandicum*. *J Gen Microbiol* 126:161–165, 1981.

82. A Ohtakara. Chitosanase from *Streptomyces griseus*. *Methods Enzymol* 161:505–510, 1988.

83. K Sakai, F Nanjo, T Usui. Production and utilization of oligosaccharides from chitin and chitosan. *Denpun Kagaku* 37:79–86, 1990.

84. EM Marcotte, AF Monzingo, SR Ernst, R Brzezinski, JD Robertus. X-ray structure of an anti-fungal chitosanase from *Streptomyces* sp. N174. *Nat Struct Biol* 3:155–162, 1996.

85. K Wako, S Hashimoto, S Kubomura, K Yokota, K Aikawa, J Kaneda. Purifications and properties of a maltotriose-producing amylase. *Denpun Kagaku* 26:175–181, 1979.

86. JF Robyt, RK Ackerman. Isolation, purification and characterizationof a maltotetraose-producing amylase from *Pseudomonas stutzeri*. *Arch Biochem Biophys* 145:105–114, 1971.

87. Y Yoshioka, K Hasegawa, Y Matsuura, Y Katsube, M Kubota. Crytstal structures of a mutant maltotetraose-forming exo-amylase cocrystallized with maltopentaose. *J Mol Biol* 271:619–628, 1997.

88. N Saito. A thermophilic extracellular alpha-amylase from *Bacillus licheniformis*. *Arch Biochem Biophys* 155:290–298, 1973.

89. O Shida, T Takano, H Takagi, K Kadowaki, S Kobayashi. Cloning and nucleotide sequence of the maltopentaose-forming amylase gene from *Pseudomonas* sp. KO-8940. *Biosci Biotechnol Biochem* 56:76–80, 1992.

90. K Kainuma, S Kobayashi, T Ito, S Suzuki. Isolation and action pattern of maltohexaose producing amylase from *Aerobacter aerogenes*. *FEBS Lett* 26:281–285, 1972.

91. K Kainuma. Alpha-amylases which produce specific oligosaccharides. In T Yamamoto, ed. *Handbook of Amylases and Related Enzymes*. Pergamon Press, 1988, pp 50–62.

92. H Ishizuka, K Wako, T Kasumi, T Sasaki. Breeding of a mutant of *Aureobasidium* sp. with high erythritol production. *J Ferment Bioeng* 68:310–314, 1989.

93. H Ishizuka, K Tokuoka, T Sasaki, H Taniguchi. Purification and some properties of erythritol reductase from an *Aureobasidium* sp. Mutant. *Biosci Bitechnol Biochem* 56:941–945, 1992.

94. GJ Hanjny, JH Smith, JC Garber. Erythritol production by a yeast like fungus. *Appl Microbiol* 12:240–246, 1964.

95. JK Lee, KW Hong, SY Kim. Purification and properties of a NADPH-dependent erythrose reductase from the newly isolated *Torula corallinea*. *Biotechnol Prog* 19:495–500, 2003.

96. S Yu, L Kenne, M Pedersen. Alpha-1,4 glucan lyase, a new class of starch/glycogen degrading enzyme. I Efficient purification and characterization from red seaweeds. *Biochem Biophys Acta* 1156:313–320, 1993.

97. M Fujise, K Yoshinaga, K Muroya, J Abe, S Hizukuri. Preparation and antioxidative activity of 1,5-anhydrofructose. *J Appl Glycosci* 46:439–444, 1999.

98. T Oguma, T Horiuchi, M Kobayashi. Novel cyclic dextrins, cyclic isomaltooligosaccharides, from *Bacillus* sp. T-3040 culture. *Biosci Biotechnol Biochem* 57:1225–1227, 1993.

99. T Oguma, T Kurokawa, K Tobe, M Kobayashi. Cloning and sequence analysis of the cycloisomaltooligosaccharide glucanotransferase gene from *Bacillus circulans* T-3040 and expression in *Escherichia coli* cells. *Oyo Toshitsu Kagaku (J Appl Glycosci)* 42:415–419, 1995.

100. M Kitaoka, T Sasaki H Taniguchi. Conversion of sucrose into cellobiose using sucrose phosphorylase, xylose isomerase and cellobiose phosphoylase. *Denpun Kagaku (J Jap Soc Starch Sci)* 39:282–283, 1992.

21

Microbial Hemicellulolytic Carbohydrate Esterases

Peter Biely

Gregory L. Côté

21.1 Introduction

In this chapter we introduce and discuss the biocatalytic potential of microbial carbohydrate esterases (CEs). These esterases operate on highly hydrated substrates such as partially acylated polysaccharides occurring in plant cell walls. This is a special group of carboxylic acid esterases (EC 3.1.1.1) discovered relatively recently and therefore less well known than lipases and other esterases, particularly in view of their biocatalytic potential. Microbial hemicellulolytic systems involved in degradation of naturally occurring hemicelluloses include three types of esterases: 1. acetylxylan esterases (AcXEs, EC 3.1.1.72), which deacetylate partially acetylated 4-O-methylglucuronoxylan, the major hardwood hemicellulose (acetylated xylan also occurs in annual plants);[1,2] 2. cinnamoyl esterases, feruloyl esterases (FeEs), and related aryl esterases, which liberate phenolic acids and their dimers (ferulic acid and p-coumaric acid) from plant cell walls, where they mainly occur as esters with L-arabinofuranose–containing polysaccharides, such as L-arabino-D-xylans and L-arabinans;[3–5] 3. acetylgalactoglucomannan esterases, which deacetylate O-acetyl galactoglucomannans, the major softwood hemicellulose.[6,7] In addition to these three basic classes, esterases that deacetylate pectin have also been described.[8,9] In comparison with lipases and some plant esterases (wheat germ, orange peel) that have been used for decades as powerful tools for hydroxyl group protection and deprotection in carbohydrate chemistry,[10] the catalytic properties of hemicellulolytic esterases are relatively unknown beyond their physiological functions in plant cell wall degradation, and therefore remain underutilized as practical biocatalysts.

21.2   Acetylxylan esterase (EC 3.1.1.72) (AcXE)

Acetylxylan esterases (AcXEs) were first described in 1985[11] and have since been recognized as common components of microbial xylanolytic and cellulolytic systems. They are able to remove acetyl ester groups from positions 2 or 3 of D-xylopyranosyl residues in xylan chains[1,2,12,13] (Figure 21.1). The enzyme action on acetylated polysaccharide substrates creates new ester-free regions on the xylan main chain, suitable for productive binding with endo-β-1,4-xylanases. This effect is reflected in a clear cooperativity of the two enzymes during acetylxylan degradation, a typical example of enzyme synergy.[12–14] Acetylxylan degradation by endoxylanases proceeds faster and to a higher degree when AcXEs are also present.[14] Besides acting on polymeric substrates, AcXEs also deacetylate partially acetylated xylooligosaccharides.[12] Deacetylation of xylooligosaccharides renders them fully susceptible to the action of β-xylosidase. Xylooligosaccharides with acetylated nonreducing xylopyranosyl residues do not serve as β-xylosidase substrates.[11] This has been confirmed using 2-, 3-, and 4-mono-O-acetyl derivatives of 4-nitrophenyl β-D-xylopyranosides (Biely P and Mastihubová M, unpublished results). All three monoacetates were resistant to an *Aspergillus niger* β-xylosidase.

The physiological role of AcXEs in connection with biodegradation of plant residues is now clear. The genes coding for AcXEs are coexpressed together with xylanase and cellulase genes,[15–17] and in *Streptomyces* are clustered with xylanase-coding genes.[18] In some bacteria, the catalytic module of AcXE is a part of a multidomain or bifunctional enzyme containing a catalytic module of an endoxylanase as well as the AcXE (see Reference 2 and ref. therein).

The effect of purified, endoxylanase-free AcXEs on physico-chemical properties of acetylated polysaccharides is worth noting. The presence of acetyl residues in polysaccharides influences their physico-chemical properties. Acetylxylan, in comparison to its nonacetylated counterpart, is soluble in water. Acetyl groups on the main chain of the polysaccharide prevent aggregation (or crystallization) of the molecules. Deacetylation, either chemical or enzymatic, leads to diminished solubility.[15,12]

21.2.1   AcXE Substrates and Assays

The natural substrate of AcXEs, which is mainly O-acetyl-4-O-methyl-D-glucurono-D-xylan, cannot be obtained by simple aqueous or alkaline extraction of plant material. However, the polysaccharide, which is most similar to that present in plant cell walls, can be obtained by dimethylsulfoxide (DMSO) extraction

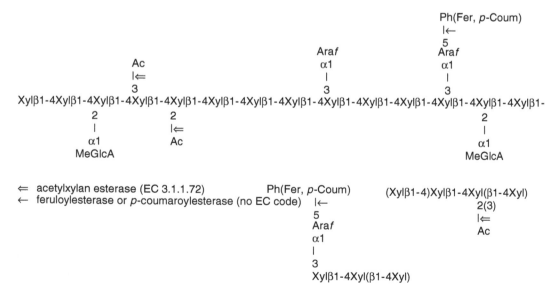

FIGURE 21.1 Hypothetical plant xylan, its acylated fragments and enzymes hydrolyzing two types of ester linkages. Abbreviations: Xyl, D-xylopyranosyl residue; MeGlcA, 4-O-methyl-D-glucuronosyl residue; Ac, acetyl group; Ara*f*, L-arabinofuranosyl residue; Phe (Fer, p-Coum), phenolic acids, ferulic acid and p-coumaric acid esterifying Ara*f* at position 5.

of hardwood hollocellulose, e.g., birchwood or beechwood holocellulose.[19,20] An analogous polysaccharide with a lower degree of polymerization (average DP about 25) and acetyl content of 13% can be obtained as a nondialyzable fraction of a water-soluble, noncellulosic polymer produced by steaming birchwood at 200°C for 10 min.[16] An aqueous thermochemical treatment of hardwood leads to a similar material with a higher acetyl content.[21] Since none of these substrates are commercially available, and their isolation is quite tedious or dependent on availability of special equipment, some authors have replaced natural acetylxylan with a chemically acetylated polysaccharide. One widely used method of chemical acetylation of xylans has been published by Johnson et al.[22] Fully (chemically) acetylated hardwood xylan has also been used as a substrate of AcXE.[23,24] As will be shown below, a majority, but not all, of AcXEs can be assayed on variety of aryl acetates, which also serve as substrates for nonhemicellulolytic acetylesterases and lipases, and for esterases in general.

The most specific AcXE assay employs as a substrate hardwood acetylxylan either extracted by DMSO or obtained by steaming wood. The natural acetylxylan, extracted from plant cell walls, can be replaced by chemically acetylated xylan. In the presence of endoxylanases, both types of polysaccharides are suitable substrates for enzymes that prefer acetylated xylooligosaccharides as substrates. The amount of acetic acid released can be determined chromatographically by HPLC or enzymically using the Boehringer Test Combination No. 148 261. Practically all procedures for determination of acetic acid by HPLC employ a Bio-Rad (Richmond, A) Aminex HPX-87H column eluted with 0.01 N H_2SO_4,[11,22] connected to a refractive index detector. Glycerol is commonly used as an internal standard. The conditions of AcXE assay on acetylxylan vary considerably among different laboratories. The substrate concentration has varied in the range of 0.2–10%, pH of the assay in the range 5.0–7.0, and temperature in the range 30–75°C.

Since most AcXEs exhibit general acetylesterase activity (the exception being family 4[25]), their activity can be determined using chromogenic or fluorogenic acetylesterase substrates such as 4-nitrophenyl acetate, α- or β-naphthyl acetate, or 4-methylumbelliferyl acetate.

Some of the substrates used for AcXE assays can also be employed for detection of the enzyme in gels. The principle of the detection of AcXE using polymeric acetylxylan relies on the precipitation of the polysaccharide due to deacetylation.[15] This approach is applicable only in those cases in which the AcXEs located in the gel are free of endoxylanases, which would hydrolyze the deacetylated polymer, thereby preventing its precipitation. Such a situation may occur, for example, during electrophoretic separation of endoxylanases from AcXEs. This means that precipitation of acetylxylan can be used for selective detection purposes only in the case of strains that do not produce endoxylanases.

The selection of microorganisms and transformants producing AcXEs belonging to families other than family 4 (Table 21.1), may be based on the use of chromogenic and fluorogenic aryl acetates, such as α- or β-naphthyl acetate and 4-methylumbelliferyl acetate.[2] An alternative method is to screen for the production of enzymes that create clear zones in solid media containing insoluble per-O-acetylated carbohydrates, e.g., penta-O-acetyl-D-glucose.[26,27] Subsequent screening of positive acetylesterase or carbohydrate esterase colonies or clones for ability to deesterify acetylxylan would then lead to selection of authentic AcXE-producers. The principles outlined above can be used for detection of AcXEs in electrophoretic gels. The assay and detection procedures are summarized in a recent article.[2]

21.2.2 Properties as Proteins

On the basis of amino acid sequence similarity (frequently established on the basis of a gene sequence), AcXEs have been assigned to 7 families of 13 thus-far recognized families of carbohydrate esterases.[28] This demonstrates that AcXEs show the greatest diversity in molecular organization among all components of microbial xylanolytic systems. Some of the best-known members of individual families are listed in Table 21.1.

Fungal AcXEs from *Aspergillus*, *Penicillium*, *Schizophyllum*, and *Trichoderma* species can be found in carbohydrate esterase families 1 and 5 together with bacterial AcXEs, a majority of which comprise one catalytic domain of bifunctional acetylxylan-degrading enzymes. Diverse AcXEs of the anaerobic fungus

TABLE 21.1 Families of Carbohydrate Esterases Containing Acetylxylan Esterases

Family	Microbial Producer of Acetylxylan Esterase	M.w.	pI	Other Enzymes in the Family
1	Aspergillus niger	30.5	3.1	Feruloyl esterases
	*Schizophyllum commune** Other *Aspergillus* species	30	3.4	
	Penicillium purpurogenum AcXE I	48	7.8	
	Pseudomonas fluorescens	—	—	—
	Catalytic domains of bifunctional enzymes of anaerobic bacteria	—	—	—
2	Ruminal anaerobic bacteria and fungi	—	—	None
3	Ruminal anaerobic bacteria and fungi	—	—	None
4**	*Streptomyces lividans*	34	9.0	Chitin and chito-
	Streptomyces thermoviolaceus	34.3		oligosaccharide deacetylase
	Catalytic domains of bifunctional enzymes of anaerobic bacteria	—	—	—
5	*Trichoderma reesei* AXE1	34	6,8; 7.0	Cutinase
	Trichoderma reesei AXE2	—	—	—
	Penicillium purpurogenum AcXE II	23	7.8	—
6	Anaerobic ruminal fungi	—	—	—
7	*Thermoanaerobacterium*	32, 26	4.2, 4.3	
	Thermotoga neapolitana	—	—	—

*Assignment based on a sequence of 50 NH_2-terminal amino acids.[42]

**This may be the only family whose members do not attack synthetic low molecular mass substrates, such as aryl acetates. For references to individual enzymes see Reference 2.

Source: Part of this information was obtained from http://afmb.cnrs-mrs.fr/~pedro/CAZY/index.html and was presented in Reference 2

Neocallimastix patriciarum[29] can be found in families 2, 3, and 6. AcXEs produced by *Streptomyces* are grouped in family 4 together with similar AcXE modules of bifunctional bacterial enzymes. Their sequences are homologous with nodulating proteins (NodB) from *Rhizobium* species and with some yeast enzymes, identified as chitooligosaccharide and chitin deacetylases.

Fungal AcXEs and some bacterial enzymes are typically monomeric enzymes, while some AcXEs from thermophilic anaerobic bacteria occur as oligomers of smaller subunits.[30] As already mentioned, numerous enzymes have modular architecture. Some AcXEs, such as those from *Pseudomonas fluorescens*[31] or *Trichoderma reesei*,[32] contain cellulose-binding domains (CBDs) separated from the catalytic domain by linker regions. Others, like AcXE from *Streptomyces lividans,* contain a xylan binding domain (XBD).[33] AcXE of *Cellulomonas fimi* is part of a bifunctional enzyme which contains both CBD and XBD.[34] The two-catalytic module architecture, represented by various combinations of endoxylanases linked to AcXEs, is very frequent, particularly among acetylxylan-degrading bifunctional enzymes produced by ruminal anaerobic bacteria.[35] AcXE domains of such bifunctional enzymes can be found in carbohydrate esterase families 1, 2, 3, and 4 (Table 21.1). It has been recently demonstrated that the two catalytic domains of the bifunctional endoxylanase-AcXE proteins from *Clostridium thermocellum*[36] or *Clostridium cellulovorans*[37] act synergistically in the degradation of acetylxylan. This observation suggests that the production of an enzyme having both debranching (or deesterifying) and depolymerizing activity can be a great advantage for a microorganism competing for carbon sources with microorganisms using separate individual enzyme components.

AcXEs existing as single-domain enzymes have molecular masses in the range of 23–48 kDa and acidic or neutral pI values. There does not seem to be a special relation between these two parameters and their assignment into carbohydrate esterase families (Table 21.1).

Three-dimensional structures are known only for AcXEs belonging to family 5. AcXEII from *P. purpurogenum*[38] and AcXE from *T. reesei*[39] have been crystallized and analyzed by X-ray diffraction. Both enzymes show the α/β hydrolase fold, similar to that of cutinase[38,39] (Figure 21.2). Cysteine residues form five S-S bridges, which makes the structure very rigid and compact. The substrate-binding site seems to be small, sufficient to accommodate only an acetyl group, thus giving a relatively high substrate specificity

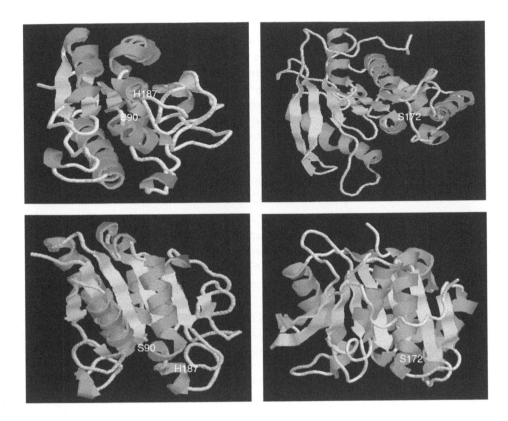

FIGURE 21.2 Ribbon representation of three-dimensional structures of AcXEs from *P. purpurogenum* (AXE II) and *T. reesei*[38,39] belonging to family 5 (left panels, top: top view; bottom: side view), and of *Clostridium* feruloyl esterases belonging to CE family 1[65,66] (right panels). Positions of catalytic serine and histidine are indicated.

to the enzyme. AcXEs of family 1 have not been crystallized, but they may possess protein folding similar to that of FeE classified in CE family 1, the tertiary structure of which has been solved (Figure 21.2).

21.2.3 Properties as Enzymes

21.2.3.1 Substrate Specificity

Substrate specificity of AcXEs is one of the least clarified aspects of these xylanolytic components. There is very limited knowledge on the structure-function relationship, particularly taking into consideration the existence of seven different families in which AcXEs have been placed (Table 21.1). The situation may be complicated by the presence of esterases, which cannot be considered components of xylanolytic systems but which could participate in later stages of acetylxylan degradation, e.g., by deacetylation of xylooligosaccharides. For example, it is known that nonhemicellulolytic esterases and lipases act quite efficiently on low molecular mass acetylated carbohydrates.[40,41,25,10] Consequently, we shall limit the discussion here only to those enzymes that deacetylate the polymeric substrate or which are coinduced with other plant cell wall hydrolyzing enzymes.

Several lines of evidence suggest that there are two major types of acetylesterases that participate in acetylxylan degradation.[12,13] They differ in their relative affinity toward polymeric and oligomeric substrates. The enzymes that perform well on polymeric substrates and are capable of causing precipitation of xylan from solution due to deacetylation are typical acetylxylan esterases.[15,12] This property is very characteristic of AcXEs in family 4 (Table 21.1), which appear to be most specific for acetyl xylan and do not hydrolyze low molecular mass substrates such as α-naphthyl-, 4-nitrophenyl-, or 4-methyl-lumbelliferyl acetate.[25,33,42] The enzymes of other families also exhibit properties similar to those of general acylesterases. AcXEs are active on variety of synthetic aryl acetates and acylates.[25,42,43] However, in contrast to lipases, they precipitate acetylxylan from solutions and their affinity toward synthetic esters decreases with increasing number of carbons of fatty acids in the ester.[25,42]

A special group seems to be formed by enzymes that do not perform so well on polymeric acetylxylan but are more active on acetylated oligosaccharides generated from acetylxylan by endoxylanases.[12,13] The presence of endoxylanases would facilitate the release of acetic acid by this type of acetylesterase, which means that synergy of these enzymes can be expected. Additional work is required to define the specificity of AcXEs relative to the type of polysaccharide that they deacetylate.

21.2.3.2 Specificity for Polysaccharide Type and the Position of the Acetyl Group in the Main Chain

Recent comparative study of AcXEs of families 1, 4, and 5[13,42,43] pointed out that AcXEs differ in ability to deacetylate different polysaccharides. AcXEs of family 1 deacetylate acetylxylan and acetyl galactoglucomannan. AcXEs of family 4 and 5 are more specific as to the type of polysaccharide they require, deacetylating only acetylxylan. Their substrate specificity is not so strict on low molecular mass substrates such as methyl 2,3,4,6-tetra-*O*-acetyl-β-D-mannopyranoside,[42] although this glycoside was not attacked by AcXEs belonging to CE family 4.[44,42]

Xylose residues in typical xylans are acetylated at positions 2 and/or 3 and the acetyl groups are more or less evenly distributed between these positions. The occurrence of the acetyl groups at both positions raises the question of whether the enzymes are capable of releasing acetyl groups bound only at one position or at both positions. The question is further complicated by the possibility of migration of acetyl groups between the 2 and 3 positions. Studies of substrate specificity of three AcXEs on acetylated methyl glycosides showed the regioselectivity of deacetylation to be consistent with their function in acetylxylan degradation, that is, removal of acetyl group from positions 2 and 3.[44–46] AcXE from *S. commune* AcXE exhibited some preference for position 3[45] while the *T. reesei* AcXE deacetylated the model compounds sequentially, first at position 2 and then rapidly at position 3.[46] *S. lividans* AcXE worked almost simultaneously on positions 2 and 3, catalyzing practically a double deacetylation of compounds at both positions.[44] Two methyl xylopyranoside diacetates, which had a free hydroxyl group at position 2 or 3, i.e., the derivatives that most closely mimicked monoacetylated xylopyranosyl residues in acetylxylan, were deacetylated at a rate several orders of magnitude faster than 2,3,4-tri-*O*-acetyl-, or 2,3-di-*O*-acetyl-β-D-xylopyranoside. These observations helped to explain the apparent double deacetylation. The second acetyl group is released immediately after removal of the first one from either of position 2 or 3. An implication of these results is that free hydroxyl group at the vicinal position plays an important role in deacetylation of position 2 or 3. The role of the hydroxyl groups could be envisaged as their participation in the formation of a five-membered transition state intermediate, which may be involved in the deacetylation of position 2 and 3 (Figure 21.3). Such intermediates are believed to be involved in spontaneous migration of acetyl groups along the pyranoside ring in aqueous media.

We have investigated this hypothetical mechanism using deoxy- and deoxy-fluoro-analogues of acetylated methyl β-D-xylopyranoside which do not allow the formation of such an intermediate.[47] When the free hydroxyl group at position 3 or 2 was replaced with a hydrogen or fluorine, the rate of deacetylation was reduced by several orders of magnitude (Table 21.2), but regioselectivity was not affected.[48] The regioselectivity of deacetylation was found to be independent of the prevailing conformation of the substrates in solution as determined by ^{1}H-NMR spectroscopy. These observations confirm the importance of the vicinal hydroxyl group and are consistent with our earlier hypothesis that the deacetylation of positions 2 and 3 may involve a common orthoester intermediate. Another possible role of the free vicinal hydroxyl group could be the activation of the acyl leaving group in the deacetylation mechanism since the low molecular mass substrates could form productive complex with the enzyme in 180° reversed orientation (Figure 21.4), as originally suggested by Hakullinen et al.[39] Involvement of the free hydroxyl group in the enzyme-substrate binding is not supported by the results of inhibition experiments in which methyl 2,4-di-*O*-acetyl β-D-xylopyranoside was used as substrate and its analogues or methyl β-D-xylopyranoside as inhibitors.[48]

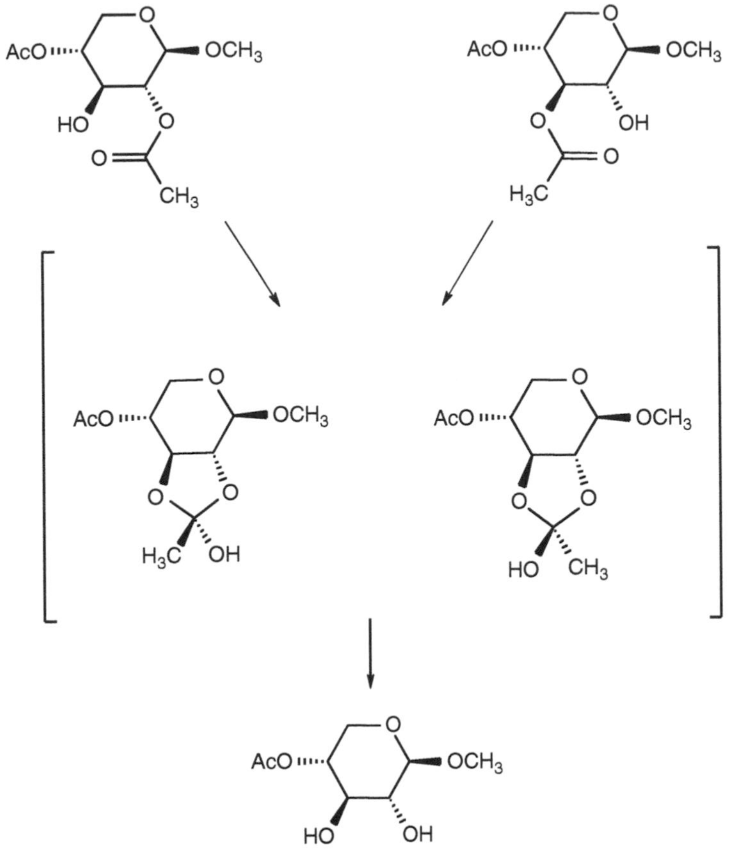

FIGURE 21.3 Hypothetized five-membered *ortho* ester intermediate involved in deacetylation of β-D-xylopyrano-side position 2 and 3 in the presence of the vicinal free hydroxyl group at positions 3 and 2. Notice two possible diastereoisomers of the intermediate. Reproduced from Reference 48 with permission from the publisher.

The enzyme requires for its efficient action the *trans* arrangement of the free and acetylated hydroxyl groups at positions 2 and 3.

The regioselectivity of deacetylation of carbohydrates by AcXEs is complementary to the regioselectivity of deacetylation by lipases, proteases, and other nonhemicellulolytic esterases[25] (Figure 21.5). The regioselectivity exhibited by AcXEs suggests their potential use for targeted modification of acetylated mono-, oligo-, and polysaccharides, and possibly other polyhydroxylated compounds.

21.2.4 Reaction Mechanism

Inactivation of AcXEs with phenyl methane sulfonyl fluoride[32,42] and occurrence of the serine sequence GXSXG in the active site of AcXEs[4] suggest that AcXEs utilize catalytic mechanism similar to lipases and serine proteases. This mechanism involves two major elements: a nucleophilic serine and a general acid-base catalyst, which is usually a histidine. It remains to be established whether all types of AcXEs contain the serine motif (GXSXG) in their catalytic site. Three-dimensional structures of two AcXEs of family 5, the enzymes of *T. reesei* and *P. purpurogenum*, confirmed the serine-type character of these esterases. By contrast, recent data from a Canadian laboratory (C. Dupont and V. Puchart, personal communication) strongly support the view that AcXEs of family 4 are not serine-type esterases. Their evidence is based on mutation and replacement of serine residues in the enzyme.

TABLE 21.2 Action of *S. lividans* AcXE on Acetylated Derivatives of Methyl β-D-xylopyranoside and Its Deoxy and Deoxy-Fluoro Analogues

Derivative of Methyl β-D-xylo-pyranoside	Specific Activity** pmol.min^{-1}.mg^{-1}	Relative Rates
2,3,4-tri-O-Ac*	44.6 ± 28	1.0
2,4-di-O-Ac	175000 ± 4370	3923
3,4-di-O-Ac	34400 ± 9600	771
3-deoxy-2,4-di-O-Ac	63.4 ± 14.6	1.4
3-deoxy-3-fluoro-2,4-di-O-Ac	1616 ± 420	36.2
2-deoxy-3,4-di-O-Ac	55.6 ± 7.8	1.2
2-deoxy-2-fluoro-3,4-di-O-Ac	160 ± 35	4.7

Note: Specific activities of the enzyme were determined at 10 mM substrate concentration in 0.1 M sodium phosphate buffer (pH 6.0) at 40°C on the basis of initial rates of the first deacetylation of the substrates.

Source: P Biely et al. Mode of action of acetylxylan esterase from *Streptomyces lividans*: A study with deoxy and deoxy-fluoro analogues of acetylated methyl β-D-xylopyranoside. *Biochem Biophys Acta* 1622: 82–88, 2003. With permission.

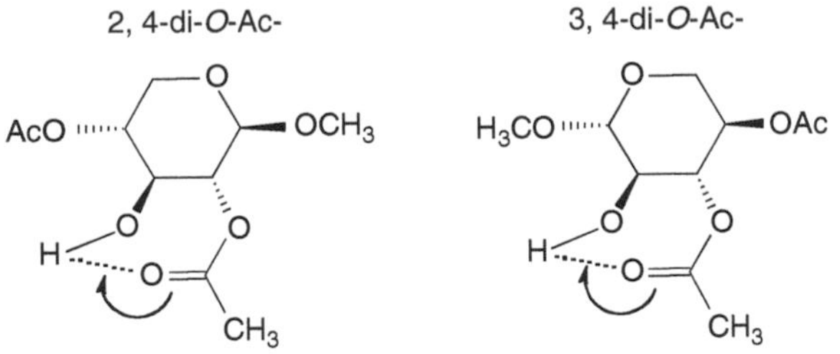

FIGURE 21.4 Activation of the ester carbonyl group via hydrogen bonding with the vicinal hydroxyl group in 2,4-di-O-Ac-Me-β-Xyl*p* and 3,4-di-O-Ac-Me-β-Xyl*p* interacting with the enzyme by 180° reversed orientation. Reproduced from Reference 48 with permission of the publisher.

21.2.5 AcXEs and Other Carbohydrate Esterases in Modification of Cellulose Acetate

The ability of AcXEs to influence solubility and hydrophobicity of acetylxylan by deacetylation has not yet found any commercial applications. Acetylxylan as a raw material or waste is currently not available; its major source would be hemicellulose released from wood during steaming treatment or that present in hardwood sulphite pulp. The action of AcXEs in enzymatic removal of acetylated hemicellulose from eucalyptus suphite pulp was found to be insignificant.[49] From a practical point of view, the ability of AcXEs to deacetylate cellulose acetate seems to be more important.

Carbohydrate esterases can work together with endo-1,4-β-glucanases in degradation of cellulose acetates. This subject was first studied by Reese[50] using cellulase from *T. reesei* and esterases from *Pestalopsiopsis*. He presented the first evidence that the presence of esterases in cellulolytic systems significantly accelerated cellulose acetate depolymerization by cellulases. The important role of an esterase in cellulose acetate biodegradation by cellulase was also reported by Saake et al.[51] However, the first enzymatic deacetylation of cellulose acetate by a purified esterase was reported by Moriyoshi et al.[52] The enzyme was an extracellular protein of *Neisseria sicca* SB, which can assimilate cellulose acetate as a sole carbon source. The purified esterase acted synergistically with endo-1,4-β-glucanases in degradation of water-insoluble cellulose acetate.[53] None of the esterases mentioned in this paragraph have been further classified.

Altaner et al.[54] reported that cellulose acetates of various degrees of substitution (DS) can be deacetylated by several commercial esterase preparations, but the enzyme activity was found to be dependent on the degree of cellulose substitution. None of the tested esterases could deacetylate highly acetylated cellulose acetates (DS = 2.3). The position that was deacetylated was determined by NMR spectroscopy.[55,56] Deacetylation was observed at all three positions—2, 3, and 6—indicating a lack of regioselectivity. Much higher regioselectivity of deacetylation of the same samples was observed with a purified *Aspergillus niger* esterase, which possesses a partial amino acid sequence identical to that of feruloyl esterase from *Aspergillus oryzae*, a member of carbohydrate esterase family 1.[42,57] The enzyme specifically deacetylated positions 2 and 3, while acetyl groups at position 6 remained intact.[58] A later study in which 8 AcXEs and FeEs, classified into 3 CE families, were used[59] led to the following conclusions: All enzymes attacked cellulose acetate at position 2 and/or 3, in consonance with their physiological role in acetylxylan deacetylation. Position 6 was never found to be the enzyme target. Different specific activities of cellulose acetate deacetylation were observed within the same CE family and between members of different CE families. The enzymes of family 1 deacetylated positions 2 and 3, while enzymes of family 5 deacetylated position 2. The enzyme of family 4 deacetylated position 3. The different modes of action suggest that the enzymes could be used for the preparation of regioselectively substituted cellulose acetates, such as 6-O-Ac cellulose or 3,6-di-O-Ac cellulose, which could have special applications.

An unexpected difference in behavior of AcXEs was noticed with cellulose acetates prepared by partial acid and alkaline deacetylation. The enzymes work only on substrates obtained by acid-catalyzed deacetylation. Cellulose acetates of various degrees of substitution (confirmed by IR spectroscopy) obtained by

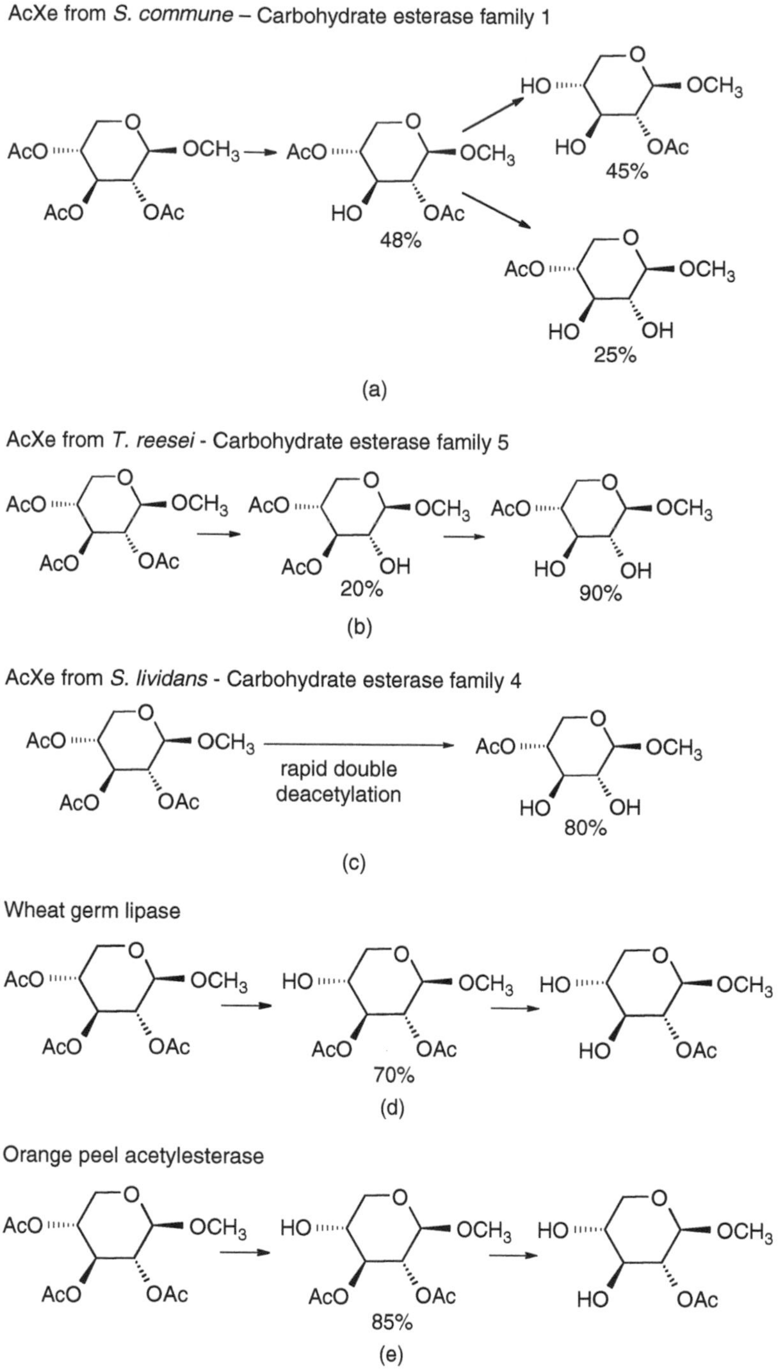

AcXe from *S. commune* – Carbohydrate esterase family 1

AcO''''⟨O⟩—OCH$_3$ → AcO''''⟨O⟩—OCH$_3$

AcO   OAc   HO   OAc
48%

HO''''⟨O⟩—OCH$_3$
HO   OAc
45%

AcO''''⟨O⟩—OCH$_3$
HO   OH
25%

(a)

AcXe from *T. reesei* - Carbohydrate esterase family 5

AcO''''⟨O⟩—OCH$_3$ AcO''''⟨O⟩—OCH$_3$ AcO''''⟨O⟩—OCH$_3$

AcO OAc → AcO OH → HO OH
             20%   90%

(b)

AcXe from *S. lividans* - Carbohydrate esterase family 4

AcO''''⟨O⟩—OCH$_3$ →   AcO''''⟨O⟩—OCH$_3$

AcO OAc   rapid double   HO OH
         deacetylation   80%

(c)

Wheat germ lipase

AcO''''⟨O⟩—OCH$_3$ → HO''''⟨O⟩—OCH$_3$ → HO''''⟨O⟩—OCH$_3$

AcO OAc   AcO OAc   HO OAc
                70%

(d)

Orange peel acetylesterase

AcO''''⟨O⟩—OCH$_3$ → HO''''⟨O⟩—OCH$_3$ → HO''''⟨O⟩—OCH$_3$

AcO OAc   AcO OAc   HO OAc
                85%

(e)

FIGURE 21.5 Differences in regioselectivity of deacetylation of 2,3,4-tri-O-methyl β-D-xylopyranoside by three AcXEs and two nonhemicellulolytic esterases in aqueous medium. Possible yields of some derivatives are indicated in %.

alkali-catalyzed deacetylation did not serve as AcXEs substrates (P Biely, J Puls, GL Cote, unpublished data), suggesting that partial alkaline deacetylation of fully acetylated cellulose leads to block deacetylation. The acetylated part of the substrate is fully acetylated cellulose and therefore is resistant to the enzyme action.

21.2.6 Reverse Reactions of Unclassified CEs

There are only a few examples of transesterifications with carbohydrate acceptors catalyzed by esterases that could be classified as carbohydrate esterases. The first one is the transacetylation to cellobiose and cellulose in aqueous medium in the presence of isopropenyl acetate catalyzed by an intracellular carboxylesterase from *Arthrobacter viscosus*.[60] The physiological role of this enzyme is believed to be the removal of acetyl groups from highly acetylated extracellular polysaccharide secreted by this organism into its culture fluid. The enzyme behaved as a typical carbohydrate esterase, showing decreasing affinity toward 4-nitrophenyl esters with increasing acid chain length. The acetylation of cellulose and cellobiose was clearly demonstrated by NMR spectroscopy. In the case of cellulose, it was a solid-state NMR using crossed polarization magic angle.[61] The position of acetylation of cellulose was not established. NMR spectra of the cellobiose products showed resonances that correspond to acetylation of positions 2 and 3. Surprisingly, further attempts to use this interesting enzyme for specific acetylation of saccharides have not been reported. In terms of carbohydrate esterase families, the enzyme remains unclassified.

An unusual example of reverse reaction of a carbohydrate esterase is the acetylation of the amino groups of chitobiose and chitotetraose in an aqueous solution of 3 M sodium acetate using a chitin deacetylase from *Colletotrichum lindemuthianum*.[62,63] The conditions under which the enzyme operates in a reverse way are simply incredible. The acetylation of chitobiose to β-D-Glc-NAc-(1 → 4)D-GlcN or chitotetraose to β-D-Glc-NAc-(1 → 4) β-D-Glc-NAc-(1 → 4) β-D-Glc-NAc-(1 → 4)-D-GlcN was accomplished in aqueous solution of 3 M sodium acetate adjusted to pH 7.0 with acetic acid. The reaction with the disaccharide was monitored using a β-*N*-acetyl-hexosaminidase-coupled assay with 4-nitrophenyl 2-amino-2-deoxy-β-D-glucopyranosyl-(1 → 4)-2-acetamido-2-deoxy-β-D-glucopyranoside as the substrate. The liberation of 4-nitrophenol was observed as a consequence of enzymatic N-acetylation of the glucosamine residue at the nonreducing end of the substrate. The products of acetylation were purified by amine-adsorption column chromatography and their structures were determined by FABMS and NMR analyses. The reaction product of chitosanobiose after the acetylation reaction was exclusively 2-acetamido-2-deoxy-β-D-glucopyranosyl-(1 → 4)-2-amino-2-deoxy-D-glucose [GlcNAcGlcN]. This work offers a novel method for enzymatic N-acetylation of amino sugars, and especially with chitosanobiose as substrate, a selectively N-acetylated product, GlcNAcGlcN, can be synthesized.

21.2.7 Transesterification and Reverse Reactions of AcXEs and Related Enzymes

The first enzyme classified as an AcXE examined for transesterification and reverse reactions was AcXE from *Schizophyllum commune*,[64] a member of CE family 1, classified on the basis of 50 N-terminal amino acid sequence.[42] By analogy with two feruloyl esterases from the same family, tertiary structures of which have been established,[65,66] *S. commune* AcXE also appears to be a serine-type esterase. However, the serine character of the enzyme still requires experimental confirmation because only a low degree of inhibition of the enzyme was achieved with phenyl methane sulfonyl fluoride.[42]

Various conditions were applied to test the ability of the *S. commune* enzyme to catalyze acetyl group transfer to methyl β-D-xylopyranoside and other substrates.[64] Acetylation of the acceptor was observed under a variety of conditions; however, the reaction did not occur in the absence of water or in a pure aqueous medium such as that used with chitin deacetylase. The best performance of the enzyme was observed in an *n*-hexane-vinyl acetate-sodium dioctylsulfosuccinate (DOSS)-water microemulsion at a molar water-detergent ratio (w_0) of about 4–5. Although the enzyme was found to have a half-life of about 1.5 h in the system, more than 60% conversion of methyl β-D-xylopyranoside to acetylated derivatives was achieved. Under similar reaction conditions the enzyme acetylated other carbohydrates, such as methyl β-D-cellobioside, cellotetraose, methyl β-D-glucopyranoside, 2-deoxy-D-glucose, D-mannose, β-1,4-mannobiose, -mannopentaose, -mannohexaose, β-1,4-xylobiose, and -xylopentaose. Mannohexaose was the largest oligosaccharide that was successfully acetylated. The presence of up to four acetate groups in mannohexaose was demonstrated by electrospray mass spectrometry (Figure 21.6). However, the position of acetyl groups in individual acetylated derivatives has not been established.

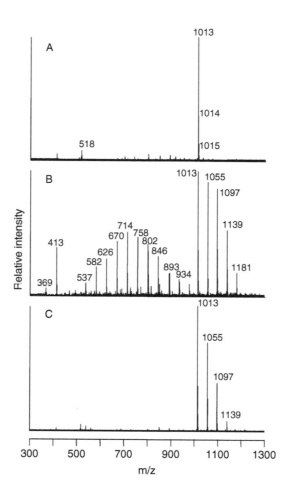

FIGURE 21.6 Electrospray mass spectrometry of mannohexaose used as acetyl group acceptor (part A), and the carbohydrate fraction isolated after its acetylation in the microemulsion system containing vinyl acetate (100 mM) (part B) and triacetin (100 mM) (part C) as the acetyl donor. Reproduced from Reference 64 with permission from the publisher.

Figure 21.6 also shows that the acetylation in the presence of vinyl acetate leads to formation of oligomers of the acetyl group donor (see the masses in the bell-shape curve). This problem can be overcome by using triacetin as the acetyl group donor; however, the efficiency of acetylation is then not as high as with vinyl acetate. This work was the first published example of reverse reactions by AcXE and a carbohydrate esterase belonging to family 1.

Recently, an interesting and potentially useful enzyme-catalyzed acetylation of carbohydrates was described.[67] A partially purified enzyme preparation of the cellulolytic system of *Trichoderma reesei* RUT C-13 was found to catalyze acetyl transfer to carbohydrates not only in organic solvents but even in water. The enzyme efficiently acetylates saccharides in a two-phase system composed of aqueous acceptor solutions and vinyl acetate as the acetyl group donor. Mono- and diacetyl derivatives of mono and oligosaccharides were formed in relatively high yields (50–80%). With methyl or 4-nitrophenyl β-D-glucopyranoside as acceptors, the monoacetylated product was identified as the 3-*O*-Ac derivative. Position 3 is apparently the main target of the enzyme because 3-*O*-methyl-D-glucose was found not to serve as an acetyl group acceptor (Kremnický L, Biely P, Côté GL, unpublished data).

When series of different oligosaccharides were used as acetyl group acceptors, it was found that β-1,4-xylooligosaccharides and β-1,4-glucosaccharides (cellooligosaccharides) were transformed to mainly mono-*O*-acetyl derivatives, with the formation of lesser amounts of di-*O*-acetyl-derivatives. When xylobiose, xylotriose and cellobiose were replaced by β-methyl glycosides of these oligosaccharides, or

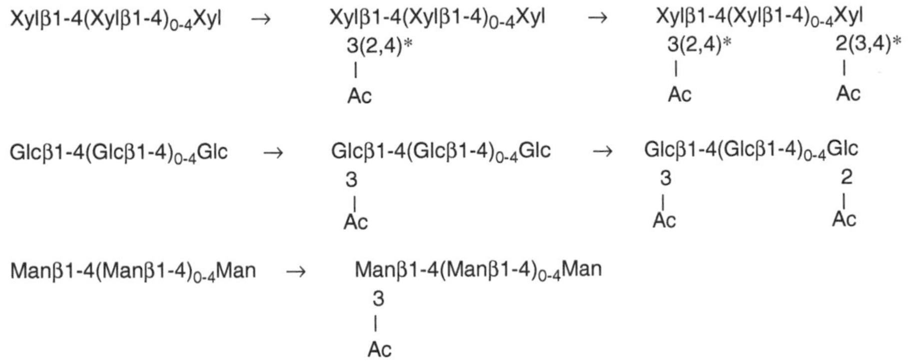

Xylβ1-4(Xylβ1-4)$_{0-4}$Xyl   →   Xylβ1-4(Xylβ1-4)$_{0-4}$Xyl   →   Xylβ1-4(Xylβ1-4)$_{0-4}$Xyl
                                                        3(2,4)*   3(2,4)*   2(3,4)*
                                                          | | |
                                                         Ac   Ac   Ac

Glcβ1-4(Glcβ1-4)$_{0-4}$Glc   →   Glcβ1-4(Glcβ1-4)$_{0-4}$Glc   →   Glcβ1-4(Glcβ1-4)$_{0-4}$Glc
                                                          3   3   2
                                                          | | |
                                                         Ac   Ac   Ac

Manβ1-4(Manβ1-4)$_{0-4}$Man   →   Manβ1-4(Manβ1-4)$_{0-4}$Man
                                                          3
                                                          |
                                                         Ac

FIGURE 21.7 Mode of esterification of oligosaccharides with acetylesterase from *T. reesei* in water/vinyl acetate two-phase system. *Other positions arise due to acetyl group migration.

when cellobiitol was used instead of cellobiose, only the formation of monoacetylated products was observed. Apparently for similar reasons, D-mannose in contrast to D-glucose afforded only two instead of three products. The formation of diacetates was never significant with β-1,4-mannooligosaccharides, indicating that the acetyl transferase does not recognize the configuration of the reducing mannose residue, which has an axial OH-group at position 2 (Kremnický L, Biely P, Côté GL, unpublished data). These results suggested that the second position acetylated in oligosaccharides was the C-1 or C-2 position of the reducing-end residue. The position of the acetyl groups in the cellobiose monoacetate and diacetate was established by NMR spectroscopy after isolation of both acetylated derivatives of cellobiose. The main acetylation takes place at position 3 of the nonreducing glucopyranosyl residue. The second acetylation of cellobiose takes place at the C-2 hydroxyl group of the reducing-end glycopyranose (Figure 21.7). This result is in a full accord with resistance of the cellobiose diacetate to periodate oxidation. This pattern of acetylation seems to occur in all xylo- and gluco-oligosaccharides β-1,4-oligosaccharides. Additional proof for the presence of the acetyl group at position 3 of the nonreducing residues of xylo- and cellooligosaccharides was obtained using β-xylosidase and β-glucosidase. None of the acetylated oligosaccharides served as substrate for the glycosidases, which act from the nonreducing end. The acetylation of the nonreducing end of the oligosaccharides prevents their degradation by the corresponding glycosidases.

The type of esterase active in the foregoing regioselective exo-type acetylation of oligosaccharides was identified by comparing the reactions catalyzed by crude preparations containing also AcXE and by AcE of *T. reesei* purified as described by Poutanen and Sundberg.[68] AcE is an esterase acting on acetylated xylooligosaccharides liberated from acetylated 4-*O*-methyl-glucuronoxylan by endo-β-1,4-xylanases.[12] It will be of interest to establish the mode of action of this enzyme on partially acetylated oligosaccharides related to physiological role of the enzyme in aqueous media. The enzyme could be an esterase acting in an "exo-fashion," i.e., deesterifying only the terminal glycopyranosyl residues in oligosaccharides. Deacetylation of 3′(2′)-*O*-acetylated xylobiose by this enzyme has already been demonstrated.[12]

21.3 Acetyl Galactoglucomannan Esterase

The major hemicellulose of softwood galactoglucomannan is partially acetylated on its D-mannopyranosyl residues (Figure 21.8). Acetylgalactoglucomannan esterases have been found mainly among extracellular proteins of *Aspergillus* strains.[6,7,43,69] An enzyme purified from *A. niger* was found to be highly specific for acetylated galactoglucomannan6 while the *A. oryzae* esterase was also active on acetylated xylan.[7,43] Both enzymes act synergistically with endo-β-1,4-mannanase with respect to the rate of cleavage of glycosidic linkages and the rate of deacetylation. Neither of these esterases has so far been classified or examined for application in hydrolytic or transesterification reactions.

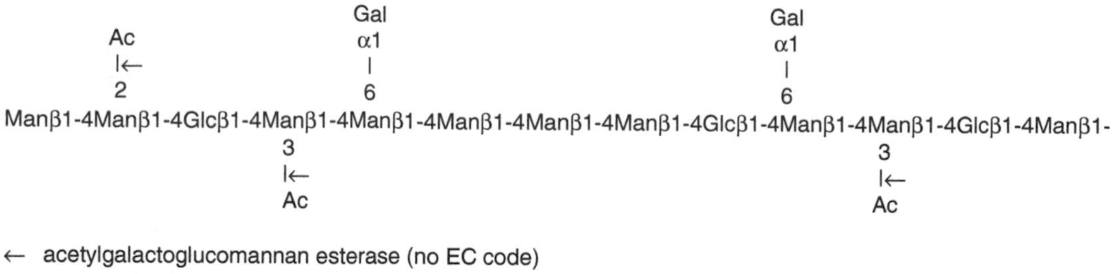

```
                    Gal                              Gal
   Ac               α1                               α1
   |←                |                                |
   2                 6                                6
Manβ1-4Manβ1-4Glcβ1-4Manβ1-4Manβ1-4Manβ1-4Manβ1-4Manβ1-4Glcβ1-4Manβ1-4Manβ1-4Glcβ1-4Manβ1-
                     3                                3
                    |←                               |←
                    Ac                               Ac
```

← acetylgalactoglucomannan esterase (no EC code)

FIGURE 21.8 Hypothetical plant acetylgalactoglucomannan and sites of the esterase action. Abbreviations: Man, D-mannopyranosyl residue; Glc, D-glucopyranosyl residue; Gal, D-galactopyranosyl residue; Ac, acetyl group.

21.4   Feruloyl Esterases (FeEs)

Hydroxycinnamic acids associated with plant hemicelluloses play an important role in cell wall integrity and protection of the plant tissues against digestion by plant-invading microorganisms.[70–72] Their protective role in plant cell walls consists of: 1) esterification of polysaccharide main chains leading to reduction of sites for attack by depolymerizing endoglycanases, and 2) cross-linking due to the formation of ferulic acid dimers connecting both hemicellulose molecules and hemicellulose and lignin molecules. The principle of the cross-linking is an *in vivo* oxidative coupling reaction of polymer-bound ferulic acid to form various types of dehydrodimers.[73–75]

The most abundant hydroxycinnamic acid is *trans*-ferulic acid, (E)-4-hydroxy-3-methoxycinnamic acid. The ferulic acid esterified with carbohydrates is found either nonsubstituted or linked to another esterified ferulic acid to form several types of di-feruloyl bridges connecting two polysaccharide chains[72,74–76] or lignin and hemicellulose.[77] Ferulic acid is usually esterified to position C-5 of α-L-arabinofuranosyl side chains in arabinoxylans, to position C-2 of α-L-arabinofuranosyl residues in arabinans, to position C-6 of β-galactopyranosyl residues in pectic substances and galactans[78] and position C-4 of α-D-xylopyranosyl residues in xyloglucans.[79] These four types of carbohydrate esters of ferulic acid are shown in Figure 21.9.

Nature has evolved enzymes that can attack ester linkages between hydroxycinnamic acids and carbohydrates in the process of biodegradation of plant cell walls. Since the first report on the occurrence of a *Streptomyces olivochromogenes* esterase releasing ferulic acid from wheat bran,[80] esterases liberating hydroxycinnamic acids from plant cell walls or cleaving ester linkages of diferuloyl bridges have been recognized as common components of hemicellulolytic enzyme systems of many microorganisms.[57,1,3–5] FeE activity has also been reported to be endogenous to plants, e.g., in germinating barley,[81] where it probably participates in the cell wall extension process.

FeEs constitute an important group of enzymes that have the potential for the use over a broad range of application in agri-food industries. Ferulic acid has long been used as a food preservative to inhibit microbial growth.[4,82,83] Phenolic acids released from plant cell walls by feruloyl esterases can be used as antimicrobial and anti-inflammatory agents,[84] antioxidants,[85] or photo-active compounds in sunscreens,[86] or can be transformed into vanillin, an essential flavor in the food industry.[83]

21.4.1   Properties as Proteins

Feruloyl esterases produced by microorganisms occur as single catalytic modules and also as a part of multimodular protein structures.[5] Some enzymes contain cellulose-binding modules,[87,88] and some are part of multimodular complex.[89] Only the separated modules of bifunctional *Clostridium* enzyme complexes have so far been crystallized and their three-dimensional structure established as an α/β barrel (Figure 21.2).[65,66] The crystal structure also confirmed the serine-type character of these feruloyl esterases. Despite wide diversity in molecular mass,[5] all feruloyl esterases thus far sequenced or partially sequenced have been classified as members of CE family 1.[25]

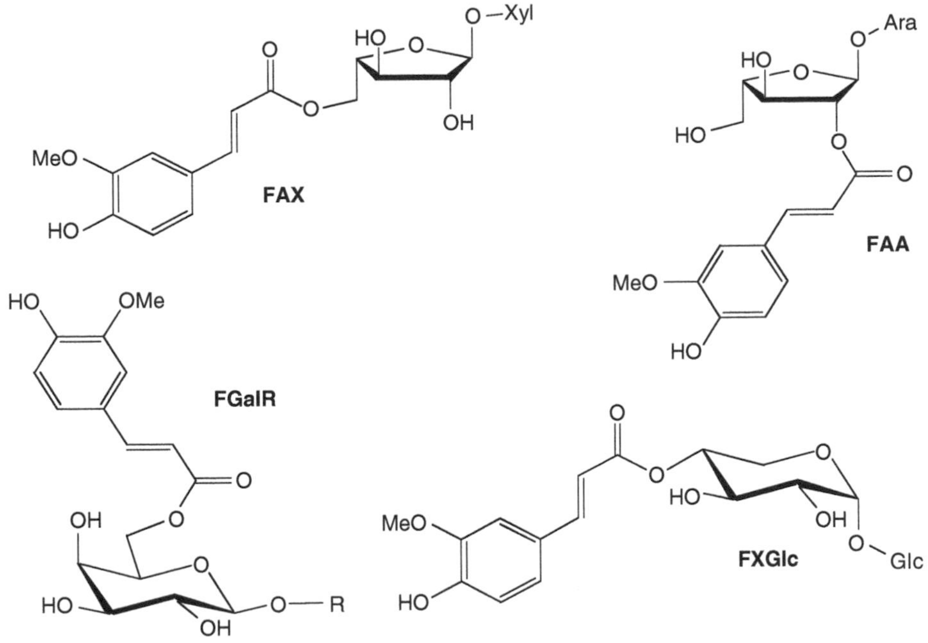

FIGURE 21.9 Formulas of four most frequent feruloyl ester linkages in plant polysaccharides: FAX, 5'-feruloyl α-1,3-L-arabinofuranosyl-D-xylopyranose occurring in plant arabinoxylans; FAA, 2'-feruloyl α-1,5-L-arabino-furanosyl-L-arabinose occurring in arabinans and pectic substances; FGalR, 6'-feruloyl-β-D-galactopyranose in β-galactans; FXGlc, 4'-feruloyl-α-1,6-D-xylopyranosyl-D-glucopyranose occurring in xyloglucans.

21.4.2 Substrate Specificity and Assay Methods

The evidence accumulated in the last 15 years suggests that microorganisms produce several types of feruloyl esterases that differ in affinity for 5-O- and 2-O-feruloylated α-L-arabinofuranosyl residues.[4,5,90–93] Naturally occurring feruloylated oligosaccharides have been obtained by controlled enzymatic digestion of wheat bran or sugar beet pulp by enzymatic systems that do not contain feruloyl esterases,[92,94–96] and by acid hydrolysis of wheat bran[97] and corn hulls.[98] These feruloylated oligosaccharides have subsequently been used to assay and differentiate the enzymes according to their affinity for the position of α-L-arabinofuranose feruloylation.

Important information on the diversity of FeEs has been obtained using methyl esters of a variety of phenolic acid as substrates. Their hydrolysis is connected with a significant change of absorbance in UV light, this being the principle of a simple spectrophotometric assay for feruloyl esterases. Individual enzymes show different affinity for esters of phenolic acids with hydroxyl or methoxy groups attached on the aromatic ring.[5]

Despite certain progress on differentiation of feruloyl esterases on a variety of synthetic and natural substrates,[4,5,99] our current knowledge about specialization of these enzymes for particular types of ester linkages within plant cell walls remains limited. The studies of substrate specificity and differentiation of feruloyl esterases is hampered by the lack of proper natural substrates, which are feruloylated oligosaccharides, plant cell wall fragments difficult to isolate. The preparation of feruloylated oligosaccharides requires proper enzymes, proper hydrolysis conditions, and subsequent purification of feruloylated polysaccharide fragments. This must be followed by determination of their structure, including the identity of the carbohydrates, the nature of the cinnamic acid, and the type of linkage.

Recently a chemoenzymic synthesis of a series of new feruloyl esterase substrates has been described[100] as an alternative to the tedious isolation of naturally occurring feruloylated oligosaccharides. The new substrates are composed of esters of ferulic acid to different positions of 4-nitrophenyl glycosides of α-L-arabinofuranose and β-xylopyranose. The newly synthesized feruloylated nitrophenyl glycosides are useful

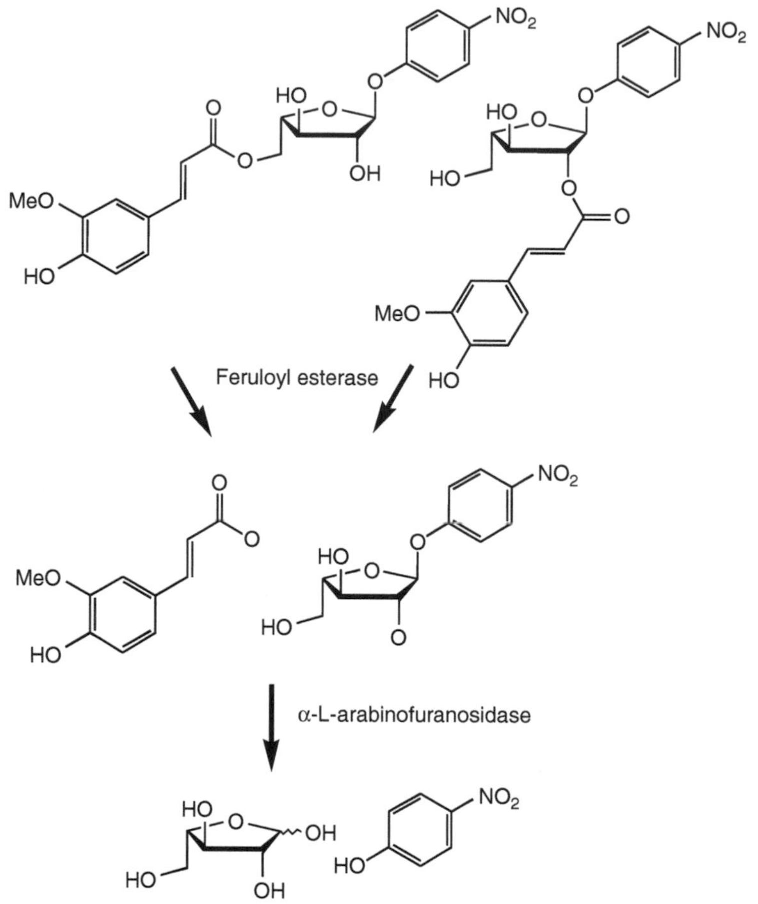

FIGURE 21.10 Principle of the α-L-arabinofuranosidase-coupled assay of FeE on two differently feruloylated 4-nitrophenyl α-L-arabinofuranosides. Reproduced from Reference 100 with permission of the publisher.

as general substrates for assays and for differentiation of carbohydrate esterases.[101,102] The 4-nitrophenyl aglycon in the compounds is advantageous in several respects. It not only mimics the second carbohydrate ring, but also makes these compounds chromogenic substrates for feruloyl esterases. The deesterification by feruloyl esterases can be coupled to a suitable α-L-arabinofuranosidase or β-xylosidase (Figure 21.10).[101] The compounds can also be used as substrates in feruloyl esterase assays that explores changes in molar absorbance at 340 nm due to ester hydrolysis.[103,104] A simpler but less specific spectrophotometric assay for feruloyl esterase employs 4-nitrophenyl ferulate.[105]

21.4.3 Current State of Classification of Feruloyl Esterases

The first attempt to classify FeEs was based on their activity toward hydroxycinnamic methyl esters.[106] An article dedicated to cinnamoyl esterases[5] presents an attempt to classify FeEs according to their action depending on two different substrate moieties. One moiety is the type of phenolic acid and the other moiety the type of the ester linkage to carbohydrate, specifically the position of esterified OH-group in L-arabinose. Based on selective hydrolytic activity of methyl esters of phenolic acids substituted with methoxy and hydroxyl groups, the following classification of FeEs has been proposed: type A: FeEs hydrolyzing methoxy-substituted methyl esters of phenolic acids, and type B: FeEs hydrolyzing hydroxyl-substituted methyl esters. Both types of FeEs hydrolyze the methyl ester of ferulic acid, which contains both methoxy and hydroxyl groups.

The above grouping of FeEs matches their affinity toward 5-feruloylated and 2-feruloylated α-L-arabinofuranosides. Type A FeEs prefer 5-feruloylated α-L-arabinofuranosides, whereas type B FeEs operate more efficiently on 2-feruloylated α-L-arabinofuranosides. Some FeEs of type A do not seem to attack 2-feruloylated substrates at all. This type of classification is quite complex because the catalytic activity of enzymes also depends on the number and nature of sugars to which the feruloylated α-L-arabinofuranoside is linked.[5] Recently Faulds[99] proposed another type of classification of FeEs. It is based on substrate specificities and amino sequence similarities, and leads to separation of the enzymes into four different groups (A, B, C, and D). FeEs capable of releasing dimers from agro-industrial materials and hydrolyzing synthetic ferulate dehydrodimers were assigned as type A. Types B and C are unable to release the dimers. The fourth group of FeEs assigned as type D was created mainly on the basis of sequence similarity.[99]

Despite certain progress in understanding the diversity of FeEs, all feruloyl esterases thus far described (with the exception of NodD-like proteins) were classified in family 1 of carbohydrate esterases as serine-type esterases. These enzymes utilize the classical mechanism of hydrolysis of the ester linkage that requires the involvement of the catalytic triad serine, histidine, and aspartic acid. All FeEs thus far described feature the peptide motif G-X-S-X-G, which occurs in all serine-type esterases.[107,108]

21.4.4 Transacetylations and Reverse Reactions Catalyzed by FeEs

The first report on reverse reactions catalyzed by a FeE and its synthetic potential was published in 2001 by Giuliani et al.[109] The authors obtained relatively high yields of pentylferulate (50–60%) synthesized from 1-pentanol and free ferulic acid via catalysis of *A. niger* FeE. The reaction was carried out in a microemulsion composed of hexane, 1-pentanol, water, and cetyltrimethyl ammonium bromide (CTAB) as surfactant. The product yields were strongly dependent on water content in the system. The reported yields of 1-pentylferulate appear good; however, it should be noted that the concentration of reactants used was at the micromolar level.

A crude FeE preparation from *Humicola insolens* (Novo Nordisk preparation Pentopan 500 BG) was examined for its ability to catalyze transesterification with substrates that bear structural similarity to the natural substrates for this enzyme.[110] The enzyme was suspended in the secondary substrate alcohol and the suspension mixed with a large excess of vinyl acetate, the most frequently used activated acyl donor.[111] The analysis of the reaction mixtures by chiral GC showed acceptable yields (7–52% conversion) and high enantioselectivity, up to 98%, with R-enantiopreference as determined by optical rotation measurements. The results encourage further interest in this biocatalyst for enantioselective synthesis of various alcohols and esters. This reaction could have interesting practical applications.

FeE II from *Fusarium oxysporum*[112] was found to catalyze esterification of phenolic acids with 1-propanol in surfactant-less microemulsions[113] composed of hexane-1-propanol-water. The conversion of acids to esters was dependent on the type of the acid. While the p-hydroxyphenylacetic and -propionic acid were converted to ester by 70–75%, ferulic and p-coumaric acid only by 15%. Since the concentration of phenolic acids in this work were about 100 times higher than in the work of Giuliani et al.,[109] the surfactant-less microemulsions offer a good option to esterify phenolic acids on a preparative scale. The enzyme shows higher hydrolytic activity toward methyl ferulate and sinapinate than to p-coumarate and caffeate.[112] The enzyme is specialized for 5-feruloyl esters of L-arabinofuranose.

The second FeE from the same fungus (FeE I) differs from FeE II in having a higher affinity toward methyl p-coumarate and caffeate.[114] The enzyme prefers the 2-feruloyl ester of L-arabinofuranose and not the 5-feruloyl ester. FeE I was tested for its ability to transesterify methyl esters of various phenolic acids with 1-butanol. The reactions were carried out in surfactant-less microemulsion systems of n-hexane-1-butanol-water. Conversion to butyl esters was higher with p-coumaric and caffeic acids than with their methoxy derivatives, ferulic, and sinapinic acid esters. This was in agreement with the hydrolytic action of the enzyme against various methyl esters.

It is surprising that so far no results have been published on enzymatic feruloylation of saccharides. The reason may be due to the different solubility of sugars and phenolic acids or their esters in organic solvent media. Such reactions could be useful for preparation of expensive feruloyl esterase substrates that would correspond to naturally occurring structures. In addition, feruloylated oligo- and polysaccharides could have wide applications as antioxidants and compounds blocking harmful UV-light.[84,86]

21.5 Conclusion

Catalytic abilities of carbohydrate esterases described in this chapter are different from those of lipases and other types of esterases. The main target of AcXEs in polysaccharides or oligosaccharides is the acetyl group in position 2 and/or 3, which are acyl esters of secondary alcohol groups. This mode of action is preserved on low molecular mass carbohydrates and thus is complementary to the pattern of deacetylation of carbohydrates by nonhemicellulolytic esterases, proteases, and lipases. Lipases have been used for a long time for deprotection and protection of carbohydrate functional groups.[10] Application of AcXE may lead to other types of carbohydrate derivatives that are difficult to prepare by the action of lipases or by classical organic chemistry. For example, numerous applications of sucrose esters exist, and enzymic methods for their synthesis have been patented.[115] The use of AcXE could complement methods already described. Cloned and overexpressed AcXE from *A. niger* has been patented for use in treatment of animal feeds, where it may serve to enhance the digestibility and fermentability of hemicellulosic substrates.[116,117] An unusual and potentially very interesting application of a glucurono-xylomannan acetyl esterase is in the treatment of cryptococcal infections.[118] Another important finding that may lead to commercial applications is that AcXEs can perform efficiently on partially acetylated cellulose.

Feruloyl esterases are enzymes structurally closely related to AcXE members of carbohydrate esterase family 1. However, they have a different function in biodegradation of plant cell walls. They hydrolyze ester linkages between ferulic acid or its dehydrodimers and hydroxyl groups of carbohydrates. The liberation of ferulic acid is important in view of a number of useful applications of the phenolic acid. Food and feed applications of *A. niger* feruloyl esterase have been patented.[119] Feruloyl esterases may find important application in cereal pulp bleaching. An enzyme produced in a recombinant *A. niger* strain was efficient in the wheat pulp bleaching when applied together with xylanase.[120] Other uses of cloned *Aspergillus sp.* feruloyl esterases have been claimed in a recent US patent, including as a feed supplement, in textiles finishing, in starch processing, and in baking.[121]

The hemicellulolytic esterases have attracted new attention due to their potential to catalyze transesterifications and reverse reactions. Neither type of these reactions has been carefully studied yet. Our current knowledge on the synthetic potential of these enzymes has emerged just from a few recent investigations. Particularly stimulating are the examples of reverse reactions and transesterifications that are carried by some of the enzyme species in aqueous media. Perhaps the enzymes operating in reverse in water utilize a different type of reaction mechanism than the serine-type ester hydrolysis. Evidence is emerging that at least members of family 4, where we find some unique AcXEs and chitin deacetylases, are not serine type esterases. In view of the immense diversity of hemicellulolytic esterases, particularly of AcXEs that belong to 7 of 13 carbohydrate esterase families,[28] the synthetic potential of these enzymes seems to be enormous and therefore deserves further attention. There could be enzymes that catalyze efficient acylation of carbohydrates and polyhydroxylated surfaces of natural fibers under environmentally friendly conditions that would lead to dramatic changes in their solubility and hydrophilic properties. Polysaccharides or natural fibers modified in such a way would be attractive as biodegradable blends of polymers and composites. A continued search for new enzymes and studies of their structure/function relationships might lead to the discovery of carbohydrate esterases that could be classified as acyl transferases.

Acknowledgment

This work was supported by a grant from the Slovak Grant Agency for Science (VEGA No. 2/3079/23). The authors thank L. Kremnický for permission to include results of unpublished work.

References

1. LP Christov, BA Prior. Esterases of xylan-degrading microorganisms: Production, properties, and significance. *Enzyme Microb Technol* 15: 460–474, 1993.
2. P Biely. Xylanolytic enzymes. In: JR Whitaker, AGJ Voragen, DWS Wong, eds. *Handbook of Food Enzymology*. New York and Basel: Marcel Dekker, 2003, pp. 879–915.
3. WS Borneman, LG Ljundahl, RD Hartley, DE Atkin. Feruloyl and p-coumaroyl esterases from the anaerobic fungus *Neocallimastix* strain MC-2: properties and functions inplant cell wall degradation. In: MP Coughland, GP Hazlewood, eds. *Hemicellulose and Hemicellulases*. London and Chapel Hill: Portland Press. 1993, pp. 85–102.
4. G Williamson, PA Kroon, CB Faulds. Hairy plant polysaccharides: a close shave with microbial esterases. *Microbiology* 144: 2011–2023, 1998.
5. CB Faulds, G Williamson. Feruloyl esterases. In: JR Whitaker, AGJ Voragen, DWS Wong, eds. *Handbook of Food Enzymology*. New York and Basel: Marcel Dekker, 2003, pp. 657–666.
6. J Puls, B Schorn, J Schuseil. Acetylmannanesterase: a new component in the arsenal of wood mannan degrading enzymes. In: M Kuwahara, M Shimada, eds. *Biotechnology in Pulp and Paper Industry*. Tokyo: Uni Publishers Co., 1992, pp. 352–363.
7. M Tenkanen, J Thornton, L Viikari. An acetylglucomannan esterase of *Aspergillus oryzae*; characterization and role in the hydrolysis of O-acetyl-galactoglucomannan. *J Biotechnol* 42: 197–206, 1995.
8. MJF Seerle-van Leeuwen, LAM van den Broek, HA Schols, G Beldman, AGJ Voragen. Rhamnogalacturonan acetylesterase: a novel enzyme from *Aspergillus aculeatus*, specific for the deacetylation of hairy (ramified) regions of pectin. *Appl Microbiol Biotechnol* 38: 347–349, 1992.
9. A Kauppinen, S Christgau, LV Kofold, T Halkier, K Dorreich, H Dalboge. Molecular cloning and characterization of a rhamnogalacturonan acetylesterase from *Aspergillus aculeatus*. Synergism between rhamnogalacturonan degrading enzymes. *J Biol Chem* 270: 27172–27178, 1995.
10. B La Ferla. Lipases as useful tools for stereo- and regioselective protection and deprotection of carbohydrates. *Monatshefte für Chemie* 133: 351–368, 2002.
11. P Biely, J Puls, H Schneider. Acetylxylan esterases in fungal cellulolytic systems. *FEBS Lett* 186: 80–84, 1985.
12. K Poutanen, M Sundberg, H Korte, J Puls. Deacetylation of xylans by acetyl esterases of *Trichoderma reesei*. *Appl Microbiol Biotechnol* 33: 506–510, 1990.
13. FJM Kormelink, B Lefebvre, F Strozyk, AGJ Voragen. Purification and characterization of an acetyl xylan esterase from *Aspergillus niger*. *J Biotechnol* 27: 267–282, 1993.
14. P Biely, CR MacKenzie, J Puls, H Schneider. Cooperativity of esterases and xylanases in the enzymatic degradation of acetyl xylan. *Bio/technology* 4: 731–733, 1986.
15. P Biely, CR MacKenzie, H Schneider. Production of acetylxylan esterase by *Trichoderma reesei* and *Schizophyllum commune*. *Can J Microbiol* 34: 767–772, 1988.
16. K Poutanen, M Rätto, J Puls, L Viikari. Evaluation of microbial xylanolytic systems. *J Biotechnol* 6: 49–60, 1987.
17. L Egaña, R Gutiérrez, V Caputo, A Peirano, J Steiner, J Eyzaguirre. Purification and characterization of two acetyl xylan esterases from *Penicillium purpurogenum*. *Biotechnol Appl Biochem* 24: 33–39, 1996.
18. F Shareck, P Biely, R Morosoli, D Kluepfel. Analysis of DNA flanking the xlnB locus of *Streptomyces lividans* reveals genes encoding acetylxylan esterase and the RNA componentof ribonuclease P. *Gene* 153: 105–109, 1995.
19. E Hägglund, B Lindberg, J McPherson. Dimethylsulfoxide, a solvent for hemicelluloses. *Acta Chem Scand* 10: 1160–1164, 1956.
20. HO Bouveng, PJ Garegg, B Lindberg. Position of O-acetyl groups in birch xylan. *Acta Chem Scand* 14: 742–748, 1960.
21. AW Khan, KA Lamb, RP Overend. Comparison of natural hemicellulose and chemically acetylated xylan as substrates for the determination of acetyl-xylan esterase activity in *Aspergilli*. *Enzyme Microb Technol.* 12: 127–131, 1990.

22. KG Johnson, JD Fontana, CR MacKenzie. Measurements of acetylxylan esterase in *Streptomyces*. *Methods Enzymol* 160: 551–560.

23. N Halgašová, E Kutejová, J Timko. Purification and some characteristics of the acetylxylan esterase from *Schizophyllum commune*. *Biochem J* 751–755, 1994.

24. G Debrassi, BC Okeke, CL Bruschi, V Venturi. Purification and characterization of an acetyl xylan esterase from *Bacillus pumillus*. *Appl Environ Microbiol* 64: 789–792, 1998.

25. P Biely, GL Côté, L Kremnický, RV Greene. Differences in catalytic properties of acetylxylan esterases and non-hemicellulolytic esterases. In: HJ Gilbert, GJ Davies, B Henrissat, B Svensson, eds. *Recent Advances in Carbohydrate Bioengineering*. Cambridge, UK: Royal Society of Chemistry, 1999, pp. 73–81.

26. H Lee, RJB To, K Latta, P Biely, H Schneider. Some properties of extracellular acetylesterase produced by the yeast *Rhodotorula mucilaginosa*. *Appl Environ Microbiol* 53: 2831–2834, 1987.

27. M Rosenberg, V Roegner, FF Becker. The quantitation of rat serum esterases by densitometry of acrylamide gels stained for enzyme activity. *Anal Biochem* 66: 206–212, 1975.

28. B Henrissat, P Coutinho. http://afmb.cnrs-mrs.fr/~pedro/CAZY/index.html.

29. BP Dalrymple, DH Cybinski, I Layton, CS McSweeney, G-P Xue, YJ Swadling, JB Lowry. Three *Neocallimastix patriciarum* esterases associated with the degradation of complex polysaccharides are members of a new family of hydrolases. *Microbiology* 143: 2605–2614, 1997.

30. W Shao, J Wiegel. Purification and characterization of two thermostable acetyl xylan esterases from *Thermoanaerobacterium* sp. strain JW/SL YS485. *Appl Environ Microbiol* 61: 729–733, 1995.

31. LMA Ferreira, TM Wood, G Williamson, CB Faulds, GP Hazlewood, HJ Gilbert. A modular esterase from *Pseudomonas fluorescens* subsp. *cellulosa* contains a non-catalytic binding domain. *Biochem J* 294: 349–355, 1993.

32. E Margolles-Cark, M Tenkanen, H Söderlund, M Penttilä. Acetyl xylan esterase from *Trichoderma reesei* contains an active site serine and a cellulose binding domain. *Eur J Biochem* 237: 553–560, 1996.

33. C Dupont, N Daignault, F Shareck, R Morosoli, D Kluepfel. Purification and characterization of an acetyl xylan esterase produced by *Streptomyces lividans*. *Biochem J* 319: 881–886, 1996.

34. JI Laurie, JH Clarke, A. Ciruela, CB Faulds, G. Williamson, HJ Gilbert, JE Rixon, J Millward-Sadler, GP Hazlewood. The NodB domain of a multifunctional xylanase from *Cellulomonas fimi* deacetylates acetylxylan. *FEMS Microbiol Lett* 148: 261–264, 1997.

35. GP Hazlewood, HJ Gilbert. Molecular biology of hemicellulases. In: MP Coughland, GP Hazlewood, eds. *Hemicellulose and Hemicellulases*. London and Chapel Hill: Portland Press. 1993, pp. 103–126.

36. AC Fernandes, CMGA Fontes, HJ Gilbert, GP Hazlewood, TH Fernandes, LMA Ferreira. Homologous xylanases from *Clostridium thermocellum*: evidence for bi-functional activity, synergism between xylanase catalytic modules and the presence of xylan-binding domains in enzyme complexes. *Biochem J* 342: 105–110, 1999.

37. A Kosugi, K Murashima, RH Doi. Xylanase and acetyl xylan esterase activities of XynA, a key subunit of the *Clostridium cellulovorans* cellulosome for xylan degradation. *Appl Environ Microbiol* 68: 6399–6402, 2002.

38. D Ghosh, M Sawicki, P Lala, M Erman, W Pangborn, J Eyzaguirre, R Gutiérrez, H Jörnvall, DJ Thiel. Multiple conformations of catalytic serine and histidine in acetylxylan esterase at 0.90 Å. *J Biol Chem* 276: 11159–11166, 2001.

39. N Hakulinen, M Tenkanen, J Rouvinen. Three-dimensional structure of the catalytic core of acetylxylan esterase from *Trichoderma reesei*: insight into the deacetylation mechanism. *J Struct Biol* 132: 180–190, 2000.

40. HM Sweers, C-H Wong. Enzyme-catalyzed regioselective deacetylation of protected sugars in carbohydrate synthesis. *J Am Chem Soc* 108: 6421–6422, 1986.

41. J-F Shaw, AM Klibanov. Preparation of various glucose esters via lipase-catalyzed hydrolysis of glucose pentaacetate. *Biotechnol Bioeng* 29: 648–651, 1987.

42. Tenkanen, M, Eyzaguirre, J, Isoniemi, R, Faulds, CB, Biely, P. Comparison of catalytic properties of acetyl xylan esterases from three carbohydrate esterase families. In SD Mansfield and JN Saddler, eds.: *Applications of Enzymes to Lignocellulosics*. American Chemical Society, Washington, DC., 2003, pp. 211–229.

43. M Tenkanen. Action of *Trichoderma reesei* and *Aspergillus oryzae* esterases in the deacetylation of hemicelluloses. *Biotechnol Appl Biochem* 27: 19–24, 1998.

44. 232. P Biely, GL Côté, L Kremnický, RV Greene, C Dupont, D Kluepfel. Substrate specificity and mode of action of acetylxylan esterase from *Streptomyces lividans*. *FEBS Lett* 396: 257–260, 1996.

45. P Biely, GL Côté, L Kremnický, D Weisleder, RV Greene. Substrate specificity of acetylxylan esterase from *Schizophyllum commune*: mode of action on acetylated carbohydrates. *Biochim Biophys Acta* 1298: 209–222, 1996.

46. P Biely, GL Côté, L Kremnický, RV Greene, M Tenkanen. Action of acetylxylan esterase from *Trichoderma reesei* on acetylated methyl glycosides. *FEBS Lett* 420: 121–124, 1997.

47. M Mastihubová, P Biely. A common access to 2- and 3-substituted methyl β-D-xylopyranosides. *Tetrahedron Letters* 42: 9065–9067, 2001.

48. P Biely, M Mastihubová, GL Côté, RV Greene. Mode of action of acetylxylan esterase from *Streptomyces lividans*: A study with deoxy and deoxy-fluoro analogues of acetylated methyl β-D-xylopyranoside. *Biochem Biophys Acta* 1622: 82–88, 2003.

49. LP Christov, P Biely, E Kalogeris, P Christakopoulos, BA Prior, MK Bhat. Effects of purified endo-β-1,4-xylanases and acetylxylan esterases on eucalyptus sulfite dissolving pulp. *J Biotechnol* 83, 231–244, 2000.

50. ET Reese. Biological degradation of cellulose derivatives. *Ind Engin Chem* 49: 89–93, 1987.

51. B Saake, S Horner, J Puls. Progress in enzymatic hydrolysis of cellulose derivatives. In: Th Heninze, WG Glasser, Eds. *Cellulose Derivatives: Modificatiuon, Characterization and Nanostructures*. Washington DC, American Chemical Society, 1998, pp. 201–216.

52. K Moriyoshi, T Ohmoto, T Ohe, K Sakai. Purification and characterization of an esterase involved in cellulose acetate degradation by *Neisseria sicca* Sb. *Biosc Biotechnol Biochem* 63: 1708–1713, 1999.

53. K Moriyoshi, T Ohmoto, T Ohe, K Sakai. Role of endo-1,4-β-glucanases from *Neisseria sicca* SB in synergistic degradation of cellulose acetate. *Biosc Biotechnol Biochem* 67: 250–257, 2003.

54. C Altaner, B. Saake, J. Puls. Mode of action of acetylesterases associated with endoglucanases towards water-soluble and -insoluble cellulose derivatives. *Cellulose* 8: 159–265, 2002.

55. K Kowasaka, K Okajima, K Kamide. Determination of the distribution of the substituent groupos in cellulose acetate by full assignment of all carbonyl carbon peaks of ^{13}C $\{^1H\}$ NMR spectra. *Polymer J* 20: 827–836, 1988.

56. CM Buchanan, KJ Edgar, JA Hyatt, AK Wilson. Preparation of cellulose 1-[^{13}C]acetates and determination of monomer composition by NMR spectroscopy. *Macromolecules* 24: 3055–3059, 1991.

57. M Tenkanen, J Schuseil, J Puls, K Poutanen. Production, purification and characterization of an esterase liberating phenolic acids from lignocellulosics. *J Biotechnol* 18: 69–84, 1991.

58. C Altaner, B Saake, J Puls. Specificity of an *Aspergillus niger* esterase deacetylating cellulose acetate. *Cellulose* 10: 85–95, 2003.

59. C Altaner, B Saake, M Tenkanen, J Eyzaguirre, CB Faulds, P Biely, L Viikari, M Siika-aho, J Puls. Regioselective deacetylation of cellulose acetates by acetyl xylan esterases. *J Biotechnol* 105: 95–104, 2003.

60. W Cui, WT Winter, SW Tatenbaum, JP Nakas. Purification and characterization of an intracellular carboxylesterase from *Arthrobacter viscosus* NRRL B-1973. *Enzyme Microb Technol* 24: 200–208, 1999.

61. JKM Sanders, BK Hunter. *Modern NMR Spectroscopy: A Guide for Chemists*, 2nd Ed., Oxford University Press, Oxford, 1993.

62. K Tokuyasu, H Ono, K Hayashi, Y Mori. Reverse hydrolysis reaction of chitin deacetylase and enzymatic synthesis of beta-D-GlcNAc-(1 → 4)-GlcN from chitobiose *Carbohydr Res* 322: 26–31, 1999.

63. K Tokuyasu, H Ono, M Mitsutomi, K Hayashi, Y Mori. Synthesis of a chitosan tetramer derivative, beta-D-GlcNAc-(1 → 4)-beta-D-GlcNAc-(1 → 4)-beta-D-GlcNAc-(1 → 4)-D-GlcNAc through a partial N-acetylation reaction by chitin deacetylase. *Carbohydr Res* 325: 211–215, 2000.

64. P Biely, KKY Wong, ID Suckling, Špániková. Transacetylations to carbohydrates catalyzed by acetylxylan esterase in the presence of organic solvent. *Biochem Biophys Acta* 1623: 62–71, 2003.

65. FD Schubot, IA Kataeva, DL Blum, AK Shah, LG Ljundahl, JP Rose, B-C Wang. Structural basis for the substrate specificity of the feruloyl esterase domain of the cellulosomal xylanase Z from *Clostridium thermocellum*. *Biochemistry* 40: 12524–12532, 2001.

66. JAM Prates, N Turbauriech, SJ Charnock, CMGA Fontes, MA Ferreira, GJ Davies. The structure of the feruloyl esterase module of xylanase 10B from *Clostridium thermocellum* provides insights into substrate recognition. *Structure* 9: 1183–1190, 2001.

67. L Kremnický, V Mastihuba, GL Côté. *T. reesei* acetylxylanesterase catalyzed transesterification in water. *J Molec Catal*. B: Enzym. 30:229–239, 2004.

68. K Poutanen, M Sundberg. An acetylesterase of *Trichoderma reesei* and its role in the hydrolysis of acetyl xylan. *Appl Microbiol Biotechnol* 28: 419–424, 1988.

69. M Tenkanen, J. Puls, M Rätto, L Viikari. Enzymatic deacetylation of galactoglucomannans. *Appl Microbiol Biotechnol* 39: 159–165, 1993.

70. T Jeffries. Biodegradation of lignin-carbohydrate complexes. *Biodegradation* 1: 163–176, 1990.

71. A Cornu, JM Besle, P Mosoni, E Grenet. Lignin-carbohydrate complexes in forages: structure and consequences in the ruminal degradation of cell-wall carbohydrates. *Reprod Nutr Dev* 34: 385–398, 1994.

72. T Ishi. Structure and functions of feruloylated polysaccharides. *Plant Sci* 127: 111–127, 1997.

73. RD Hartley, CW Ford. Phenolic constituents of plant cell walls and wall biodegradability. In NG Lewis, MG Paice, eds. *Plant Cell Wall Polymers, Biogenesis and Biodegradation ACS Symp Ser* 399: 137–149, 1989.

74. SC Fry. Cross-linking of matrix polymers in the growing cell walls of angiosperms. *Ann Rev Plant Physiol* 37: 165–186, 1986.

75. J Ralph, RF Helm. Lignin/hydroxycinnamic acid polysaccharide complexes: synthetic models for regiochemical characterization. In HG Jun., DR Buxton, RD Hatfield, J Ralph, Eds. *Forage Cell Wall Structure and Digestibility*. American Society for Agronomy, Crop Science Society of America, Soil Science Society of America, Madison, 1992, pp. 119–127.

76. K Iiyama, TBT Lam, BA Stone. Phenolic bridges between polysaccharides and lignin in wheat internodes. *Phytochemistry* 29: 733–737, 1990.

77. TBT Lam, K Kadoya, K Iiyama. Bonding of hydroxycinnamic acids to lignin: ferulic and p-coumaric acids are predominantly linked at the benzyl position of lignin, not the position, in grass cell walls. *Phytochemistry* 57: 987–992, 2001.

78. L Salnier, J-F Thibault. Ferulic acid and diferulic acid as components of sugar-beet pectins and maize bran heteroxylans. *J Sci Food Agric* 79: 396–402, 1999.

79. T Ishii, T Hiroi, JR Thomas. Feruloylated xyloglucan and p-coumaroyl arbinoxylan oligosaccharides from bamboo shoot cell-walls. *Phytochemistry* 29: 1999–2003, 1990.

80. CR MacKenzie, D Bilous, H Schneider, KG Johnson. Induction of cellulolytic and xylanolytic enzyme systems in *Streptomyces* spp. *Appl Environ Microbiol* 53: 2835–2839, 1987.

81. AI Sancho, CB Faulds, B Bartolome, G Williamson. Characterization of feruloyl esterase activity in barley. *J Sci Food Agric* 79: 447–449, 1999.

82. PA Kroon, G Williamson. Hydroxycinnamates in plants and food: current and future perspectives. *J Sci Food Aric* 79: 355–361, 1999.

83. C Faulds, B Clarke, G Williamson. Ferulic acid unearthed. *Chemistry in Britain* 36: 48–50, 2000.

84. E Graff. Antioxidant potential of ferulic acid. *Free Radic Biol Med* 13: 435–448, 1992.

85. Garcia-Conesa, M.T., Wilson, P.D., Plumb, G.W., Ralph, J., Williamson, G. (1999) Antioxidant properties of 4,4′-dihydroxy-3,3′-dimethox-β,β′-bicinnamic acid (8-8-diferulic acid, non-cyclic form). *J Sci Food Agric* 79: 379–384.

86. A Saija, A Tomaino, R Lo Cascio, D Trombetta, A Proteggente, A De Pasquale, N Ucella, F Bonina. Ferulic and caffeic acids as potential protective agents against photooxidative skin damage. *J Sci Food Agric* 79: 476–480, 1999.

87. IJ Fillingham, PA Kroon, G Williamson, HJ Gilbert, GP Hazlewood. A modular cinnamoyl ester hydrolases from the anaerobic fungus *Piromyces equi* acts synergistically with xylanase and is a part of a multiprotein cellulose-binding cellulase-hemicellulase complex. *Biochem J* 343: 215–224, 1999.

88. PA Kroon, G Williamson, NM Fish, DA Ancher, NJ Belshaw. A modular esterase from *Penicillium funiculosum* which releases ferulic acid from plant cell walls and binds crystalline cellulose contains a carbohydrate binding module. *Eur J Biochem* 267: 6740–6752, 2000.

89. DL Blum, IA Kataeva, X-L Li, LG Ljundahl. Feruloyl esterase activity of the *Clostridium thermocellum* cellulosome can be attributed to previously unknown domains of XynY and XynZ. *J Bacteriol* 182: 1346–1352, 2000.

90. G Williamson, CB Faulds, PA Kroon. Specificity of ferulic acid (feruloyl) esterases. *Biochem Soc Trans* 26: 205–209, 1998.

91. M-C Ralet, CB Faulds, G Williamson, J-F Thibault. Degradation of feruloylated oligosaccharides from sugar-beet pulp and wheat bran by ferulic acid esterases from *Aspergillus niger*. *Carbohydr Res* 263: 257–269, 1994.

92. WS Borneman, RD Hartley, DS Himmelsbach, LG Ljungdahl. Assay for *trans*-coumaroyl esterase using a specific substrate from plant cell walls. *Anal Biochem* 190: 129–133, 1990.

93. WS Borneman, LG Ljungdahl, RD Hartley, DE Akin. Purification and partial characterization of two feruloyl esterases from the anaerobic fungus *Neocallimastix* strain MC-2. *Appl Environ Microbiol* 58: 3762–3766, 1990.

94. MC Ralet, J-F Thibault, CB Faulds, G Williamson. Isolation and purification of feruloylated oligosaccharides from cell walls of sugar-beet pulp. *Carbohydr Res* 263: 227–241, 1994.

95. I Colquhoun, MC Ralet, J-F Thibault, CB Faulds, G Williamson. Structure identification of feruloylated oligosaccharides from sugar-beet pulp by NMR-spectroscopy. *Carbohydr Res* 263: 243–256, 1994.

96. PA Kroon, MT Garcia-Conesa, IJ Colhoun, G Williamson. Process for isolation of preparative quantities of [2-*O*-(trans-feruloyl)-α-L-arabinofuranosyl]-(1→5)-L-arabinofuranose from sugar-beet. *Carbohydr Res* 300: 351–354, 1997.

97. JA McCallum, IEP Taylor, GHN Towers. Spectrophotometric assay and electrophoretic detection of *trans*-feruloyl esterase activity. *Anal Biochem* 196: 360–366, 1991.

98. M Hosny, JPN Rosazza. Structures of ferulic acid glycoside esters in corn hulls. *J Nat Prod* 60: 219–222, 1997.

99. CB Faulds. Feruloyl esterases for studying plant cellwall structure and polymer interactions. In: CM Courtin, WS Veraverbeke, JA Delcour, eds. *Recent Advances in Enzymes in Grain Processing*. Leuven, Belgium: Katolieke Universiteit Leuven, 2003, pp. 83–88.

100. M Mastihubová, J Szemesová, P Biely. Two efficient ways to 2- and 5-feruloylated 4-nitrophenyl α-L-arabinofuranosides as substrates for differentialtion of feruloyl esterases. *Tetrahedron Letters* 44: 1671–1673, 2003.

101. P Biely, M Mastihubová, WH van Zyl, BA Prior. Differentiation of feruloyl esterases on synthetic substrates in α-L-arabinofuranosidase-coupled and ultraviolet-spectrophotometric assays. *Anal Biochem* 311: 68–75, 2002.

102. K Rumbold, P Biely, M Mastihubová, M Gudelj, G Gubitz, K-H Robra, BA Prior. Purification and properties of a feruloyl esterase involved in lignocellulose degradation by *Aureobasidium pulllans*. *Appl Environ Microbiol* 69:5622–5626, 2003.

103. JA McCallum, IEP Taylor, GHN Towers. Spectrophotometric assay and electrophoretic detection of *trans*-feruloyl esterase activity. *Anal Biochem* 196: 360–366, 1991.

104. RD Hatfield, RF Helm, J Ralph. Synthesis of methyl 5-*O*-feruloyl-α-L-arabinofuranoside and its use as a substrate to assess feruloyl esterase activity. *Anal Biochem* 194: 25–33, 1991.

105. V Mastihuba, L Kremnický, M Mastihubová, JL Willett, GL Côté. A spectrophotometric assay for feruloyl esterases. *Anal Biochem* 286: 96–101, 2002.

106. PA Kroon, MT Garcia-Conesa, IJ Fillingham, GP Hazlewood, G Williamson. Release of ferulic acid dehydrodimers from plant cell walls by feruloyl esterases. *J Sci Food Agric* 79: 428–434, 1999.

107. RP de Vries, J Visser. *Aspergillus* enzymes involved in degradation of plant cell walls polysaccharides. *Microbiol Mol Biol Rev* 65: 497–522, 2001.

108. VF Crepin, CB Faulds, IF Connerton. Production and characterization of the *Talaromyces stipitatus* feruloyl esterase FAEC in *Pichia pastoris:* identificationof the nucleophilic serine. *Prot Expr Purif* 29: 176–184, 2003.

109. S Giuliani, C Piana, L Setti, A Hochkoeppler, PG Pifferi, G Williamson, CB Faulds. Synthesis of pentylferulate by a feruloyl esterase from *Aspergillus niger* using water-in-oil microeulsions. *Biotechnol Lett* 23: 325–330, 2001.

110. NS Hatzakis, D Daphnomili, I Smonou. Ferulic acid esterase from *Humicola insolens* catalyzes enantioselective transesterification of secondary alcohols. *J Mol Cat B: Enz* 21: 309–311, 2003.

111. Y-F Wang, JJ Lalonde, M Momongan, DE Bergbreiter, C-H Wong. Lipase-catalyzed irreversible transesterifications using enol esters as acylating reagents: Preparative enantio- and regioselective syntheses of alcohols, glycerol derivatives, sugars, and organometallics. *J Am Chem Soc* 110: 7200–7205, 1988.

112. E Topakas, H Stamatis, P Biely, D Kekos, BJ Macris, P Christakopoulos. Purification and characterization of a feruloyl esterase from *Fusarium oxysporum* catalyzing esterification of phenolic acids in ternary water-organic solvent mixtures. *J Biotechnol* 102: 33–44, 2003.

113. YL Khmelnitsky, R Hilhorst, C Veeger. Detergentless microemulsions as media for enzymatic reactions. *Eur J Biochem* 176: 265–271, 1988.

114. E Topakas, H Stamatis, M Mastihubová, P Biely, D Kekos, BJ Macris, P Christakopoulos. Purification and characterization of a feruloyl esterase (FAE-I) from *Fusarium oxysporum* catalyzing transesterification of phenolic acids esters. *Enzyme Microb Technol* 33: 729–737, 2003.

115. DC Palmer, F Terradas. Regioselective enzymatic deacylation of sucrose esters in anhydrous organic media. United States Patent 5445951 (1995).

116. LH DeGraaff, J Visser, HC VanDenBroeck, F Strozyk, FJM Kormelink, JCP Boonman. Cloning and expression of acetyl xylan esterases from fungal origin. United States Patent 5,426,043 (1995).

117. LH DeGraaff, J Visser, HC VanDenBroeck, F Strozyk, FJM Kormelink, JCP Boonman. Method to alter the properties of acetylated xylan. United States Patent 6,010,892 (2000).

118. AC Savoy, SL Bloomer, TR Kozel. Glucuronoxylomannan (GXM)-*O*-acetylhydrolase of *Cryptococcus neoformans* and uses thereof. United States Patent 6,284,508 (2001).

119. B Michelsen, RP De Vries, J Visser, JB Soe, CH Poulsen, MR Zargahi. Enzyme system comprising ferulic acid esterase from *Aspergillus*. United States Patent 6,143,543 (2000).

120. E Record, M Asther, C Sigoillot, S Pages, PJ Punt, M Delattre, M Haon, CA Van Den Hondel, JC Sigoillot, L Lesage-Meessen, M Asther. Overproduction of the *Aspergillus niger* feruloyl esterase for pulp bleaching application. *Appl Microbiol Biotechnol* 62: 349–355, 2003.

121. WS Borneman, BS Bower. Esterases, DNA encoding therefore and vectors and host incorporating same. United States Patent 6,368,833 (2002).

22

Application of Cyclodextrin Glucanotransferase to the Synthesis of Useful Oligosaccharides and Glycosides

Hirofumi Nakano

Sumio Kitahata

22.1 Introduction

Cyclodextrin glucanotransferase (1,4-α-D-glucan 4-α-D-(1,4-α-D-glucano)-transferase, [EC 2.4.1.19], CGTase) is an enzyme that catalyzes the conversion of α-glucans such as amylose and starch to cyclo-dextrins (CDs).[1,2] CDs are the cyclic oligosaccharides composed of α-1,4-linked 6, 7, and 8 glucose residues, which are generally referred as α-, β-, and γ-CDs, respectively.[3,4] CGTase produces CDs by intramolecular transglycosylation (cyclization reaction), where the enzyme cleaves the α-1,4-linkage of the substrates in an endo manner and transfers the newly produced reducing end to the nonreducing end of its own molecule (Figure 22.1). CGTase also catalyzes two intermolecular transglycosylation reactions called *coupling* and *disproportionation*. In coupling, the enzyme utilizes CDs as glycosyl donors, where the CD ring is opened and the resulting reducing end is transferred to a linear maltooligosaccharide or some other acceptor molecule to form a linear product. Disproportionation is a transfer reaction between two maltooligosaccharide molecules to yield linear oligosaccharides with various degrees of polymerization. In the presence of suitable acceptors such as glucose and sucrose, the enzyme also transfers the glycosyl residues produced from glucans to the acceptors. In addition to these transglycosylations,

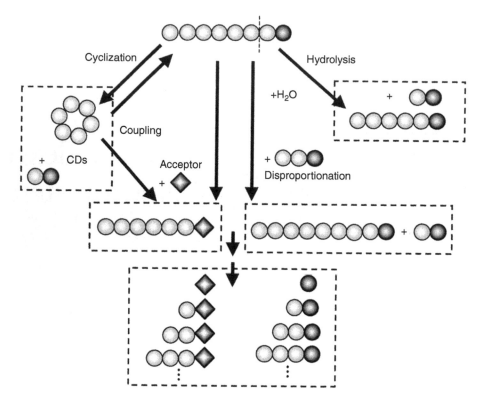

FIGURE 22.1 Schematic diagram of actions of CGTase: The white circles represent glucose residues; the shaded circles, the reducing-end glucose; the diamonds, acceptors other than glucose and maltooligosaccharides.

CGTase exhibits a weak hydrolysis activity on starch and CDs, which is possibly understood as the transfer of the glycosyl residue to a water molecule. CGTase possesses extreme industrial usefulness because of the various products, CDs, other linear oligosaccharides and glycosides, which are diversely and increasingly applied to food, pharmaceutical, and cosmetic industries.

CGTase belongs to a large and well-known group of starch degrading enzymes called α-amylase family or glycoside hydrolase family 13, which is classified according to similarity in protein structure such as primary amino acid sequence, and in the catalytic mechanism.[5-8] Extensive molecular biological studies, mutagenesis and crystallographic ones, are revealing the catalytic mechanism and reaction specificities of the enzyme in detail, and a number of CGTase mutants have been constructed to alter the enzyme characteristics; for example, reaction specificity, product specificity, and stability. Recent aspects of such subjects can be found in a number of references.[5-8] This chapter focuses on the actions of CGTase and practical usefulness of the various products.

22.2   Some Properties and Actions of CGTases

22.2.1   Some Properties of CGTase

Since Tilden and Hudson had first obtained a cell-free enzyme preparation from a culture of *Bacillus macerans* that converted starch to CDs in 1939,[9] more than 30 bacteria have been identified as sources of CGTases, which have different product specificity, acceptor preference, or other characteristics such as optimum reaction conditions and stability. No enzyme from fungal and yeast origins has been reported so far.

Most CGTases are active at a pH range from slightly acidic to neutral. Although several alkalophilic *Bacillus* strains have been isolated as sources of CGTases, their enzymes do not necessarily show alkaline pH optima: *Bacillus* No. 38-2, which grows at pH 10 using starch as a carbon source, secretes two enzymes that are active at pH 4.5–4.7 and pH 7.0.[10] Strains of *Bacillus* sp. A2-5a[11] and *Brevibacterium* sp. No.9605[12] cultivated in alkaline media produce CGTases that have optimum pH 5.5 with a broad shoulder in an alkaline side and a broad optimum pH (6–10), respectively. The reaction of CGTase at alkaline pH is preferable for the glycosylation of vitamin-like flavonoids such as hesperidin and rutin, which become more soluble at an alkaline range, as described in 22.4.1.[11,13]

The CD production by most CGTases proceeds efficiently at 40–70°C. The *B. stearothermophilus* enzyme is more thermophilic with an optimum temperature of 75°C.[14] Thermophilic anaerobic bacteria, *Thermoanaerobacter* sp. and *T. thermosulfuringenes*, produce more stable CGTases with optimum temperatures at 90–95°C and 80–90°C, respectively.[15] A hyperthermophilic archaeon, *Thermococcus* sp., produces an extremely thermostable enzyme that exhibits optima for starch-degrading activity and CD synthesis at 110°C and 90–100°C, respectively.[16] Thermostable enzymes enable CD production from raw starch without liquefaction with α-amylase. Some CGTases are stabilized against heat treatment by calcium ion. Moreover, the total yields of CDs and the ratio of γ-CD produced by the *Brevibacterium* CGTase are strongly enhanced by the addition of approximately 5 mM $CaCl_2$.[17]

22.2.2 Intramolecular Transglycosylation

22.2.2.1 Production of CDs by Various CGTases

Most CGTases produce a mixture of CDs, although the production preference for CDs greatly differs among the individual enzymes. CGTases may be grouped into three types depending on abundant CDs produced in an initial reaction stage (Table 22.1)[1,2,18] : The enzymes of *B. macerans* type, *B. megaterium* type, and *Bacillus* sp. AL 6 type mainly produce α-CD, β-CD and γ-CD, respectively. The production ratio of CDs, however, changes with the progress of the reaction: As shown in Figure 22.2, the CGTases from *B. megaterium* and *B. circulans* mainly produce β-CD in an initial stage, although the ratio of β-CD decreases afterwards.[18] The α-CD/β-CD ratios also decrease in the prolonged reactions of the

TABLE 22.1 Ratio of CDs Produced by Various CGTases

Enzyme source	α-CD	:	β-CD	:	γ-CD
α-CD producing					
B. macerans	5.7	:	1.0	:	0.4
B. stearothermophilus	1.7	:	1.0	:	0.3
Klebsiella oxytoca			α-CD > β-CD, γ-CD		
Thermococcus sp.	23	:	1.0	:	1.0
β-CD producing					
B. megaterium	1.0	:	6.3	:	1.3
B. circulans	1.0	:	6.4	:	1.4
Bacillus sp. (alkalophilic)					
acid			β-CD > α-CD, γ-CD		
neutral			β-CD > α-CD, γ-CD		
B. obensis	0	:	5.0	:	1.0
Thermoanaerobacter sp.			β-CD > α-CD ≧ γ-CD		
B. autolyticus	2.0	:	6.0	:	1.0
γ-CD producing					
Bacillus sp. AL 6	0	:	1.0	:	2.7
B. subtilis No. 313			γ-CD only		
B. firms 290-3			γ-CD ≧ β-CD		
Brevibacterium sp.	0	:	6.5	:	16
B. coagulans	1.0	:	0.9	:	1.3

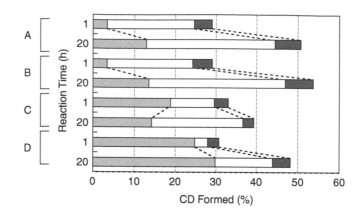

FIGURE 22.2 Change in the ratios of CDs during the reactions of several CGTases: A, B, C, and D; the enzymes from *B. megaterium*, *B. circulans*, *B. stearothermophilus*, and *B. macerans*, respectively. The gray bar represents α-CD; the white bar, β-CD; the black bar, γ-CD.

B. stearothermophilus and *B. macerans* enzymes. These results suggest the mutual conversion of CDs during the reaction. The enzyme from *Thermococcus* sp. accumulates α-CD predominantly even in a later reaction stage (79% of total CDs after 24 h-incubation) as well as in an initial stage (92% after 1 h).[16] The CGTase from *B. subtilis* No. 313 is an exceptional enzyme that produces γ-CD as the only cyclization product in addition to a large amount of maltooligosaccharides.[19]

22.2.2.2 Effects of Surfactants and Other Additives

Yields and ratio of CDs are affected by the addition of surfactants that change the helical conformation of starch, or by the addition of agents that selectively form inclusion complexes with CDs, to the reaction mixtures of CGTase. Surfactants with straight hydrocarbon chains significantly increase the yields of α-CD[20]; for instance, sodium dodecyl sulfate (SDS) enhances the formation of α-CD 1.6 times from starch. In contrast, surfactants with bulkier aromatic hydrophobic moieties such as isooctyl phenyl polyoxyethylene (Triton) enhance the formation of β-CD.

The formation of γ-CD by *Klebsiella pneumoniae* is enhanced by the addition of 3% bromobenzene as a complexing agent and 200 mM sodium acetate, both of which modify starch to a certain favorable conformation for the production of γ-CD.[21] The addition of glycyrrhizic acid to the reaction of the *B. ohbensis* enzyme boosts the yield of γ-CD 3.6-fold.[22] In the presence of macrocyclic complexing agents such as cyclohexadec-8-en-1-one, γ-CD is selectively produced by the *B. macerans* enzyme in a yield of 46% from 10% starch, as shown in Table 22.2.[4] This method is employed for the industrial production

TABLE 22.2 Effect of Addition of Selective Macrocyclic Complexing Agents on Production of CDs by the *B. macerans* CGTase

Coupling Agent	Yield of CDs (%)	Ratio (%)	
		β-CD	γ-CD
Cyclodecanone	58	98.5	1.5
Cyclotridecanone	43	2	98
Cyclotetradec-7-en-1-one	45	1	99
Cyclohexadec-8-en-1-one	46	1	99
Cycolhexadeca-1,9-dione	43	1	99
2,8-Dioxa-1-oxo-cycloheptadecane	40	1	99
2,5-Dioxa-1,6-dioxo-cyclohexadecane	38	1	99

Concentrations of starch and complexing agents are 10% and 1%, respectively.

of γ-CD. The yield of γ-CD by the *Brevibacterium* enzyme from 5% starch increases more than twofold by the addition of 20% ethanol.[17] The CGTase from *Bacillus* sp. AL 6 produces γ-CD in a yield of 35% from 2.5% potato starch in the presence of 100 mM potassium phosphate buffer (pH 8.0) and 35% ethanol.[23] Ethanol may depress the degradation of γ-CD that is due to be consumed as a donor for the intermolecular transglycosylation (coupling reaction) to form linear dextrins.

22.2.2.3 Production of Large Cyclic Glucans

The CGTases from *B. macerans* and alka-ophilic *Bacillus* sp. produce considerable amounts of large cyclic α-1,4-linked glucans with degrees of polymerization (DP) from 6 to more than 60 as initial products from synthetic amylose.[24,25] Such large cyclic glucans are termed as cycloamyloses (CAs), which are first recognized in the reaction of the disproportionating enzyme (D-enzyme, [EC 2.4.1.25]) from potato tuber. CGTase converts amylose into such CAs and finally into conventional CDs by repeating cyclization and linearization reactions, as shown in Figure 22.3. Several CGTases exhibit different productivity of CAs.[25] The CGTase from *B. macerans*, for example, converts larger CAs into smaller CAs and finally into its major product, α-CD. The CGTase from alkalophilic *Bacillus* sp. produces CAs only at an initial reaction stage and most of the CAs rapidly decrease with an increase of β-CD. On the other hand, the *B. stearothermophilus* CGTase forms only a trace amount of CAs in addition to conventional CDs even in an initial stage.

X-ray crystallography shows that CA molecules are comprised of two antiparallel single helices of 12 glucose units each, connected by "band-flip" motifs at each end. The helices, which have 6 glucose per turn, contain central cavities with a similar diameter to that of α-CD, but are much longer. CAs are highly soluble in water, have a low propensity for retrogradation, and form inclusion complexes with some guest molecules. Moreover CAs exhibit an artificial chaperon-like property to promote the refolding of several denatured enzyme proteins.[26] CAs produced by D-enzyme from recombinant *E coli* are utilized as a reagent of commercialized kit for protein refolding.

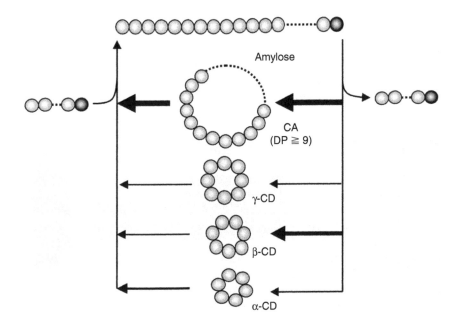

FIGURE 22.3 Schematic diagram of cyclization reaction of the CGTase from alkalophilic *Bacillus* sp. strain A2-5a on amylose. The right and left arrows indicate cyclization and coupling reactions, respectively, and the relative width of each arrow represents the relative rate of the reaction. The white circles represent glucose residues and the shaded circles indicate the reducing-end glucose.

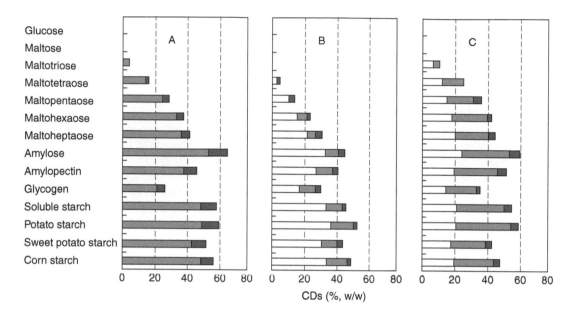

FIGURE 22.4 Yields of CDs from various α-glucans by several CGTases. A, B, and C represent the enzymes from *B. ohbensis*, *B. macerans*, and *B. circulans*; the white bar, α-CD; the gray bar, β-CD; the black bar; γ-CD. Incubation conditions; substrate concentration 1.0% (w/v), pH 6.0, 50°C, 24 h.

22.2.2.4 Production of CDs from Various Glucans

α-Glucans such as starch, amylose, amylopectin, and glycogen are favorable for the production of CDs by CGTase (Figure 22.4).[1,2,27] A high yield from amylose as well as lower yields from glycogen or β-limit dextrin[1] indicates that a high content of 1,6-branching in the substrates is undesirable for the production of CDs. CGTase is able to produce α-, β-, and γ-CD directly from maltoheptaose, maltooctaose, and maltononaose, respectively, which have one more glucose residue than the respective CDs synthesized. Furthermore, the *B. megaterium* enzyme forms CDs from smaller oligosaccharides such as maltose and maltotriose.[1] The enzyme initially catalyzes intermolecular transglycosylation (disproportionation reaction) of these substrates to produce linear oligosaccharides larger than maltohexaose and then produce CDs from the elongated oligosaccharides.

22.2.3 Intermolecular Transglycosylation

22.2.3.1 Acceptor Specificity

Preference of CGTase for an acceptor in intermolecular transglycosylation can be evaluated by the stimulation of starch degradation (dextrogenic activity[1]).[1,2,28,29] Figure 22.5 compares the dextrogenic activity of the CGTases from *B. macerans*, *B. megaterium*, and *B. stearothermophilus* in the presence of various monosaccharides.[28] Considerable stimulation with D-glucose suggests high effectiveness as acceptors. L-Sorbose is also such an extremely preferred acceptor.[1,18] D-Xylose and 6-deoxy-D-glucose are a group of effective acceptors. In the presence of these effective acceptors, the enzymes merely catalyze intermolecular tranglycosylation and produce no CDs from starch. Effective sugar acceptors are suggested to have the pyranose structure with free equatorial OH groups at C2, C3, and C4 on $^4C_1$ or $^1C_4$ conformation.[29]

D-Galactose, D-ribose, D-mannose, D-arabinose, D-fructose, and L-fucose are poor acceptors that give rise to a faint stimulation of the dextrogenic activities (Figure 22.5).[28] On actual chromatographic analysis, however, transglycosylation products are detected from the poor acceptors. These results may be explained as follows: In an initial reaction stage, CGTase mainly produces CDs. Over a long period of incubation, however, CGTase is obliged to catalyze the intermolecular tranglycosylation of such poor

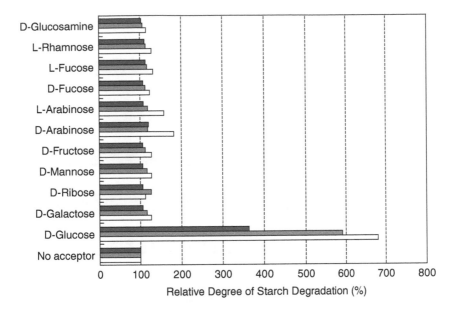

FIGURE 22.5 Degree of starch degradation by several CGTases in the presence of various monosaccharides. The relative degree of starch degradation in the presence of acceptors to that in its absence is shown. The black, gray, and white bars represent the enzymes from *B. macerans*, *B. circulans*, and *B. stearothermophilus*, respectively.

acceptors using CDs as donor substrates to accumulate the products gradually. In the presence of ineffective acceptors such as L-rhamnose, CGTase produces exclusively CDs in addition to trace amounts of glucose and maltooligosaccharides. The CGTases from *Brevibacterium* and *B. stearothermophilus* have a relatively broad acceptor specificity not only on saccharides but also on phenolic and polyhydroxyl compounds as described in Section 22.4.3.[28,30]

22.2.3.2 Structural Profiles of Products

CGTases transfer glycosyl residues specifically to the C4-OH group of good acceptors such as D-glucose, D-xylose, and 6-deoxy-D-glucose. L-Sorbose, a good acceptor, is exceptionally glycosylated at the C3-OH group. D-Mannose, a poor acceptor, is also glycosylated at the C4-OH group. D-Galactose, a poor acceptor, is glycosylated at the C1-, the C3-, and the C2- (or C4-) OH groups in a ratio of 26: 10: 1, respectively.[31] The transglycosylation of α-linked glucobioses, which have two free C4-OH groups, occurs at the C4-OH group of the nonreducing end glucose residue exclusively in 1,2-linked kojibiose or mainly in 1,3-linked nigerose and 1,6-linked isomaltose.[2]

22.2.3.3 Disproportionation

The *B. megaterium* CGTase disproportionates oligosaccharides larger than maltose to produce a series of maltooligosaccharides and CDs, whereas the enzyme from *B. macerans* and *B. ohbensis* act on maltooligosaccharides larger than matlotriose to give the products similar to those of the *B. megaterium* enzyme.[18,27] Figure 22.6 shows the distribution of disproportionation products by several CGTases on maltotriose as a starting substrate.[18] The disproportionation activity of the *B. macerans* enzyme is lower than those of the enzymes from *B. megaterium* and *B. stearothermophilus*.

22.2.4 Hydrolysis

CGTase is classified as a transferase, nevertheless, the enzyme shows hydrolytic action to a certain limited degree to produce linear dextrins from starch and CDs. This can be understood as the transfer of glycosyl residue to a water molecule. The ratios of hydrolysis frequency to a total catalysis (sum of cyclization and

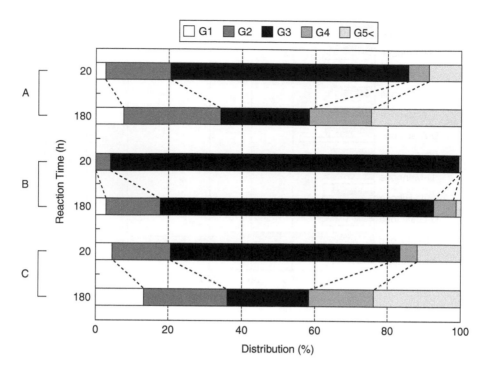

FIGURE 22.6 Distribution of disproportionation products by several CGTases on maltotriose. Initial substrate, 1% maltotriose. A, B, and C represent the enzymes from *B. megaterium*, *B. macerans*, and *B. stearothermophilus*; G1, glucose; G2, maltose; G3, maltotriose; G4, maltotetraose; G5<, maltooligosaccharides larger than maltotetraose.

hydrolysis) are 1.9%, 2.0%, 8.3%, and 2.0% for the CGTases from *B. megaterium*, *B. circulans*, *B. stearo-thermophilus*, and *B. macerans*, respectively (Figure 22.7).[18] The Km values for hydrolysis of α-CD by the above CGTases are 7.0 mM, 6.7 mM, 3.1 mM, and 2.5 mM, respectively. The *B. stearothermophilus* enzyme hydrolyzes CDs with the considerable rates,[8,18,25] which are almost comparable to that of hydrolysis of soluble starch: The relative initial velocities of hydrolysis of α-, β- and γ-CDs are estimated to be 100, 1.4, and 1.0

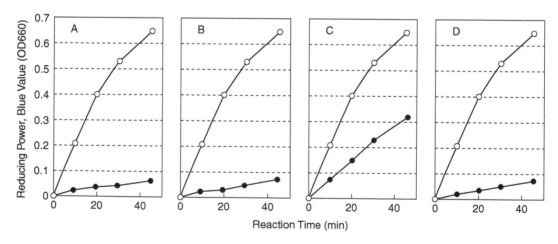

FIGURE 22.7 Hydrolysis of soluble starch by several CGTases. A, B, C, and D represent the CGTases from *B. megaterium*, *B. circulans*, *B. stearothermophilus*, and *B. macerans*, respectively; closed circle, reducing sugar determined by Somogyi-Nelson that represents the hydrolysis activity; open circle, blue value (OD at 660 nm) that represents total catalysis.

for the *B. megaterium* enzyme, and 92, 1.7, and 0.8 for the *B. circulans* enzyme, and 90, 9.1, and 15 for the *B stearothermophilus* enzyme, and 28, 0.07, and 0.03 for the *B. macerans* enzyme, respectively.[18]

22.3 Production and Application of Oligosaccharides

22.3.1 CDs

22.3.1.1 Industrial Production

α- and β-CDs or mixtures mainly containing one of these CDs are produced industrially from starch by the CGTases from *B. macerans* and alkalophilic *Bacillus* sp., respectively. γ-CD is produced in the presence of the macrocyclic complexing agents,[4] as described in Section 22.2.2.2, and is applied to industrial areas other than food manufacturing.

The first step in the production of CDs is the liquefaction of starch, where a starch slurry is treated with a bacterial thermostable α-amylase in a jetcooking process at 105–110°C and pH 6.0–6.5. The yields of CDs decrease linearly with an increase in the starch concentration and the degree of liquefaction (DE), because these factors simultaneously elevate the concentrations of noncyclic dextrins, which enhance the intermolecular transglycosylation and depress the cyclization.[1,2] The starch concentration also influences the production ratio of CDs; for example, in the reaction of the *B. macerance* enzyme, which is used for the production of α-CD, the ratio of β-CD to α-CD increases with starch concentrations. Therefore, starch concentration and DE are adjusted to 10–15% and below 1.0, respectively, in usual industrial processes. The yields of CDs by batch system are as high as 40% of the starch used. However the yields are increased by 30% in a process involving ultrafiltration:[3] CDs and small dextrins are permeated through the UF membrane and further refined to obtain a final product, whereas larger dextrins and the enzyme are retained over the membrane, circulated back to the reactor tank, and utilized again as raw materials.

The content of CDs is increased by the treatment of the reaction mixtures with glucoamylase for hydrolyzing noncyclic dextrins to glucose, which is then removed by UF membrane filtration. From the mixture of CDs, β-CD is first separated as crystals by making use of the lowest solubility. α-CD is recovered from the remaining solution after the digestion with *A. oryzae* α-amylase and glucoamylase, where only γ-CD is hydrolyzed.

22.3.1.2 Properties and Application

CDs have a hydrophilic outside and a hydrophobic cavity in the center of the molecule. CDs form inclusion complexes with various hydrophobic substances (guests) into the cavity, and this unique function enables CDs and their derivatives for diverse application in such areas as food, cosmetic, and pharmaceutical industries.[3,4,32,33] Some properties of CDs are summarized in Table 22.3. The cavity sizes of α- β-, and γ-CDs are 5.7, 7.8, and 9.5Å, respectively, and accordingly, the inclusion ability differs. The industrial benefits of complex formation by CDs are as follows:

1. Stabilization of volatile materials: a) conversion of liquid materials to dry form, b) stabilization of flavors and spices, c) deodorization of foods and drugs

TABLE 22.3 Some Physicochemical Properties of CDs

Property	α-CD	β-CD	γ-CD
Number of glucose residue	6	7	8
Molecular weight	973	1135	1297
Diameter of cavity (Å)	5 ~ 6	7 ~ 8	9 ~ 10
Depth of cavity (Å)	7 ~ 8	7 ~ 8	7 ~ 8
$[\alpha]^{25}_D$ (°, in H_2O)	+150.5	+162.5	+177.4
Solubility (g/100ml, at 25°C)	14.5	1.85	23.2
Color of iodine complex	blue	yellow	purplish brown
Shape of crystal	needle	plate	plate

2. Protection against oxidation and UV-degradation during storage or processing
3. Modification of physical and chemical properties: a) enhancement of the solubility of water-insoluble materials, b) masking of bitterness in foods and drugs, c) stabilization of deliquescent chemicals, d) catalysis of chemical reaction
4. Emulsification of hydrocarbons, steroids, fats and fatty acids

Table 22.4 summarizes the industrial application of CDs. Nowadays more than 90% of CDs produced is applied to food industries.

22.3.1.3 Production and Properties of Branched CDs

Glucosyl or maltosyl side chains are introduced to CD rings (Figure 22.8) to improve the low aqueous solubility of CDs, especially that of β-CD (1.8%).[34,35] Maltosyl CDs are manufactured from maltose and parent CDs by the reverse hydrolysis action of the thermostable *Bacillus* sp. pullulanase: The concentrated substrates at 70–80%, in which the ratio of maltose and CDs is 5:1 by weight, are incubated with the pullulanase at 60–70°C. Glucosyl CDs are produced by the removal of one glucose residue from maltosyl CDs with glucoamylase. Glucosyl and maltosyl β-CDs, generally known as "branched CDs," have about

TABLE 22.4 Industrial Application of CDs

Uses	Guests
Foods	
1) Emulsification	Oils, fats
2) Stabilization	Flavors, spices, colors, pigments
3) Masking of taste and order	
4) Improvement of quality	
5) Reduce volatility	Ethanol
6) Others	
Cosmetics and toiletries	
1) Emulsification	Oils, fats
2) Stabilization	Flavors and fragrances
Agrochemicals	
1) Stabilization	Pyrolnitrin, pyrethroids
2) Reduce volatility	Organic phosphates, thiocarbamic acid
3) Reduce toxicity	2-Amino 4-methyl-phosphynobutyric acid
Clothing	
1) Moisture retainment	Squalene
2) Antibacterial	Hinokitiol
Pharmaceuticals	
1) Solubility, dissolution rate	Prostaglandines, steroids, cardiac glycosides, nonsteroidal anti-inflammatory agents, barbiturates phenytoin, sulfonamides, sulfonylurea, benzodiazepines
2) Chemical stability	
A) Hydrolysis	Prostacyclin, cardiac glycosides, aspirin, atropine, procaine
B) Oxidation	Aldehydes, epinephrine, phenothiazines
C) Photolysis	Phenothiazines, ubiquinones, vitamins
D) Dehydration	Prostaglandine E_1, ONO-802
3) Bioavailability	Aspirin, phenytoin, digoxine, acetohexamide, barbiturates, non-steroidal antiinflammatories
4) Powdering	ONO-802, Clofibrate, benzaldehyde, nitroglycerin, vitamin K1, K2, mehylsalicylate
5) Volatility	Iodine, naphthalene, *d*-camphor, *l*-menthol, methylcinnamate
6) Taste, smell	Prostaglandines, alkylparabens
7) Irritation to stomach	Nonsteroidal antiinflammatory agents
8) Hemolysis	Phenothiazines, flufenamic acid, benzylalcohol, antibiotics

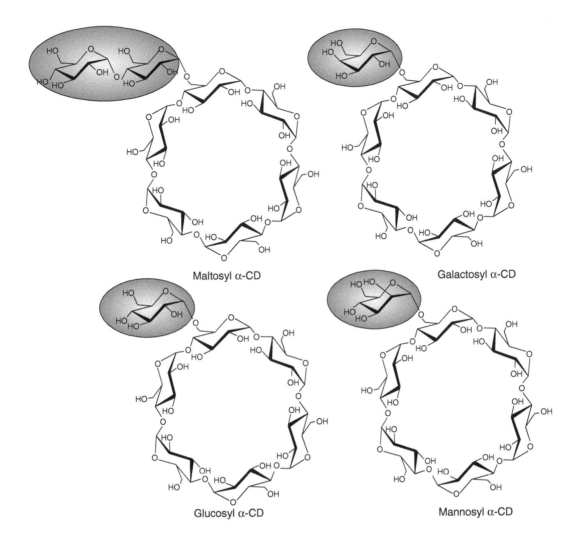

FIGURE 22.8 The structure of several branched CDs.

40 and 100 times higher aqueous solubility than β-CD, respectively. The branched CDs have almost the same inclusion ability as parent CDs, possess highly improved solubility not only in water but also in organic solvents, and exhibit lower hemolytic activity against human erythrocytes compared to the parent CDs.

D-Galactose, D-mannose, N-acetyl D-glucosamine, and N-acetyl D-galactosamin residues are also introduced to the CD rings or to the side chains of the conventional branched CDs by transfer or reverse hydrolysis actions of various exo-glycosidases such as α-galatosidase, β-galactosidase, α-mannosidase, and N-acetyl β-hexosaminidase.[34] Several plant enzymes glycosylate the CD rings more effectively than microbial enzymes. For example, microbial α-galactosidases from *Mortierella*, *Candida*, and *Absidia* transgalactosylate maltosyl CDs, but not CDs or glucosyl CDs. On the other hand, the enzyme from coffee bean exhibits a stronger transfer activity and a wide acceptor specificity to produce galactosyl derivatives of not only the branched CDs but also CDs. The products termed "heterogeneous branched CDs" (Figure 22.8) have a potential application to drug delivery carriers with affinity to a certain targeted organ tissues depending on the glycosyl residues introduced. Glucuronyl-glucosyl β-CD is derived from maltosyl β-CD by the oxidation with the microbial cells of *Pseudogluconobacter saccharroketogenes*, a new isolate that oxidizes the hydroxymethyl and hemiacetal OH groups of various aldoses, aldose oligomers,

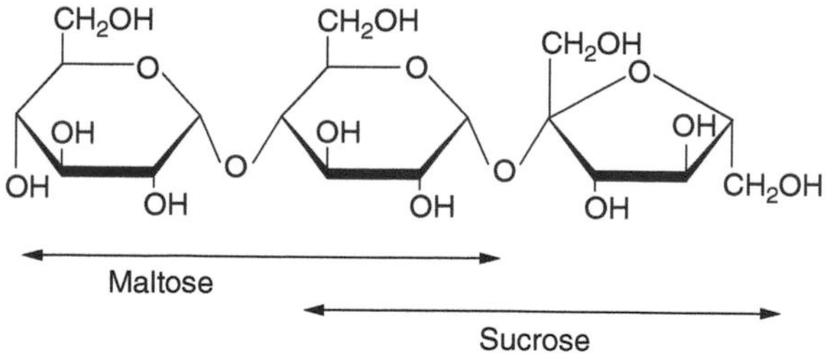

FIGURE 22.9 The structure of glucosyl sucrose.

and their derivatives. This anionic heterogeneous branched CD exhibits high aqueous solubility, resistance to amylases, less hemolytic activity, and greater affinity for basic drugs.[36]

22.3.2 Maltooligosyl Sucrose

Maltooligosyl sucrose is manufactured from starch and sucrose by the intermolecular transglycosylation of CGTase and is brought to market by the commercial name of "coupling sugar."[37,38] Glucosyl sucrose (Figure 22.9) and maltosyl sucrose are naturally occurring saccharides found in honey. Corn starch is suspended in two parts of water by weight and liquefied by a thermostable α-amylase. After inactivating the amylase, 0.5 parts of sucrose are added and the *B. stearothermophilus* CGTase and the incubation at 65°C is continued. The resulting mixture is refined with active carbon and ion-exchange resin, and concentrated. A typical sugar composition of coupling sugar is shown in Table 22.5.

Coupling sugar has about half sweetness of sucrose, induces less Maillard reaction with proteins and amino acids, prevents starch ingredients from retrogradation, and retains moisture of foods. Coupling sugar is digested by human α-amylase and α-glucosidase, and absorbed. Caloric value is 4 kcal/g. Coupling sugar is less prone to fermentation and production of acid by *Streptococcus mutans* than sucrose. It does not act as a substrate for the synthesis of insoluble glucan, and furthermore inhibits the synthesis of the insoluble glucan from sucrose by *S. mutans*. Accordingly, coupling sugar is used as an anticariogenic sweetener that does not induce dental caries, and also inhibits the cariogenicity of sucrose.

22.3.3 Other Oligosaccharides

Neotrehalose (O-α-D-glucosyl-(1→1)-β-D-glucoside), a nonreducing disaccharide, is produced with a yield of 20–30% in a reaction system containing the *B. macerans* CGTase and 20–30% maltose.[2] The *B. stearothermophilus* enzyme also catalyzes the formation of nonreducing glucosyl lactoside (α-D-glucosyl O-β-D-galactosyl-(1→4)-β-D-glucoside) from lactose and starch as a major trisaccharide.[28] For glycosylation of such poor acceptors, reaction systems containing a considerably high enzyme activity and long reaction time are generally employed. The *B. stearothermophilus* CGTase synthesizes insoluble amylose (DP 37–43) with a yield of 82% after a long-term incubation with α-CD at a low temperature.[2]

TABLE 22.5 Sugar Composition of Coupling Sugars

Fructose	0.5 ~ 1.5%
Glucose	5 ~ 7%
Sucrose	11 ~ 15%
Maltose	10 ~ 12%
Glucosyl sucrose	11 ~ 15%
Maltotriose	7 ~ 9%
Maltosylsucrose	7 ~ 11%
Higher oligosaccharides	35 ~ 41%

22.4 Production and Application of Useful Glycosides

22.4.1 Glycosides of L-Ascorbic Acid and Vitamin-Like Flavonoids

L-Ascorbic acid (AsA), well known as "Vitamin C," is being utilized as preservative and antioxidant additives in foods, medicine, cosmetics, and feeds. To overcome the liability of AsA to oxidation, 2-*O*-α-D-glucosyl AsA (AsA-2G, Figure 22.10) is produced by transglycosylation with CGTase on a commercial scale.[39] A mixture of AsA and α-CD (1:2, w/w) at 40% concentration is incubated with the enzyme from *B. stearothermophilus* at 60°C for 24 h. After removing the CGTase with ultrafiltration, the product is digested with glucoamylase and the resulting monoglucoside, AsA-2G, is purified by ion-exchange chromatography and crystallization. AsA-2G has neither reducing power nor antiscorbutic activity, and is very stable even in aqueous solution. AsA-2G functions as active AsA after hydrolysis *in vivo*. In the cosmetics industry, AsA-2G is utilized as an antioxidant, a suppressor for melanogensis of skin, or an enhancer of collagen synthesis in human fibroblasts. For the stabilization of AsA against oxidation, either the C2- or C3-OH group should be glycosylated, whereas exo-glycosidases such as α-galactosidase usually transglycosylate C6-OH[40] to give unstable derivatives.

Rutin is a quersetin glycoside spreading widely in plants and is known as "Vitamin P." Hesperidin obtained from mandarin orange peel also shows vitamin-like activity that decreases capillary permeability and fragility.[11,13] These flavonoids are transglycosylated successfully by the CGTases from *B. stearothermophilus*[41] and alkalophilic *Bacillus* sp.,[11,13,42] respectively. α-1,4-Glycosylation specifically occurs to the glucose residue in rutinose (6-*O*-α-L-rhamnosyl-D-glucose) moiety of the flavonoids. The aqueous solubility of rutin and hesperidin increased 4000 and 300 times, respectively, and the products show stabilizing activity for natural pigments against oxidation or ultraviolet radiation.

22.4.2 Glycosylation of Natural Glycosides with High Sweetness

Highly sweet glycosides such as stevioside, rubusoside, glycyrrhitin, and mogroside V that are extracted from Chinese medical plants are utilized as low-calorific sweeteners for foods and beverages (Figure 22.11). In spite of their intensive high sweetness, the quality of taste is not necessarily satisfactory. CGTase-catalyzed transglycosylation of these natural glycosides, therefore, are conducted to improve unfavorable taste.

Stevioside from the leaves of *Stevia rebaudiana* Bertoni (Compositae) exhibits 140-fold sweetness of sucrose. To improve its bitter taste and aftertaste, glycosyl stevioside is manufactured on a commercial scale.[43,44] A mixture of stevioside and liquefied starch (1:10, w/w) at about 37% (w/w) concentration is incubated with the *B. stearothermophilus* CGTase at 60°C for 40 h. 1,4-α-Glycosylation occurs both at the 13-*O*-sophorosyl and 19-*O*-β-glucosyl moieties. The intensity of sweetness is further improved by the treatment of the above mixture with β-amylase. The derivatives of stevioside either glucosylated or maltosylated at the sophorose moiety exhibit a favorable quality of taste.

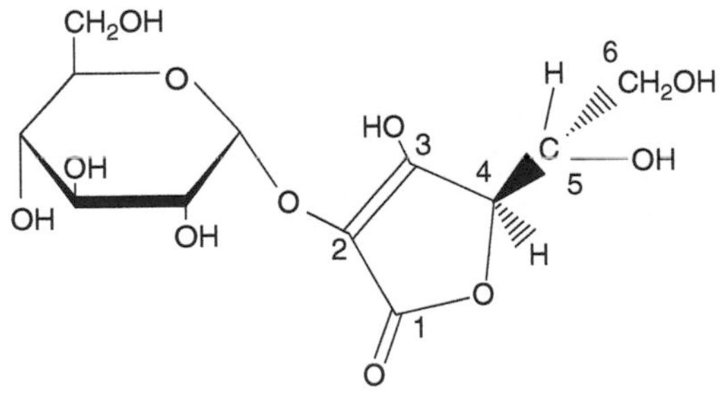

FIGURE 22.10 The structure of 2-*O*-α-glucosyl L-ascorbic acid.

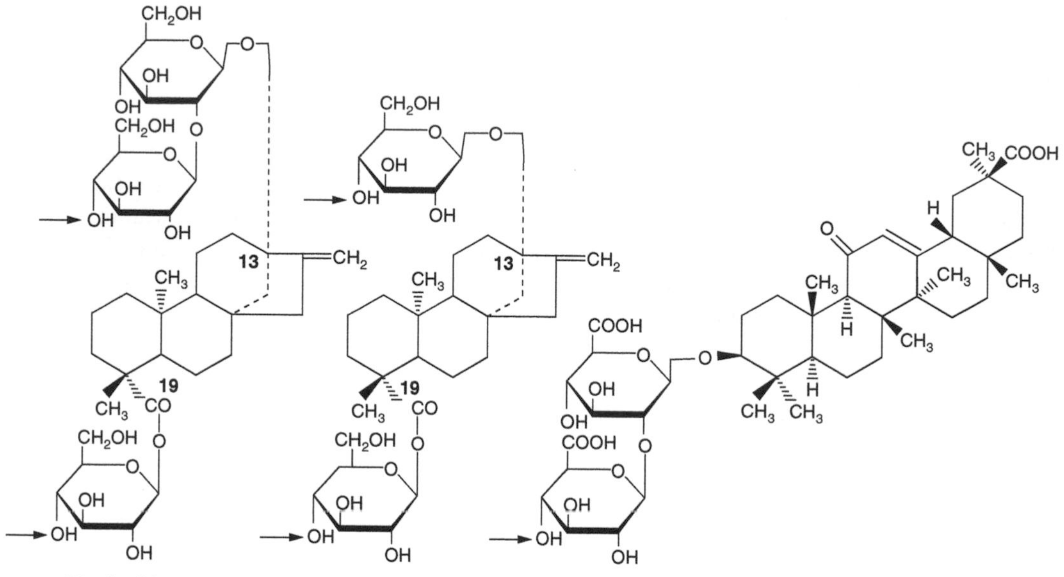

FIGURE 22.11 The structure of several natural glycosides with high sweetness. Arrows show OH groups that mainly glycosylated with CGTases.

Glycosylation of stevioside and rebaudioside A (steviol-tetraglucoside) by a CGTase gives the products with the almost same intensity of sweetness as the original glycosides. The sweetness of rubusoside (steviol-bisglucoside) increases 1.5 times with the introduction of glucose residue to the 13-*O*-glucose moiety.[45] CGTase-catalyzed transglycosylation of glycyrrhitin, a saponin of licorice, and of mogroside V,[46] a triterpene glycoside in sweet extract of a Chinese cucurbitaceous medical plant, diminishes the intensity of sweetness in spite of an improvement in the quality of taste.

Several glucosides such as salicin (*p*-hydroxymethyl phenyl β-glucoside), rubusoside, stevioside, and rutin glucoside act as effective acceptors for the *B. macerans* CGTase to give products with long glucose chains under controlled reaction conditions. The products, polyglucosylated glucosides, formed precipitates by keeping the reaction mixture at low temperatures and thus effectively synthesized.[42]

22.4.3 Polyphenols and Other Polyhydroxyl Compounds

Some glycosides of polyphenols are potentially used as antiseptics, inhibitors of melanogensis, and antioxidants, for ingredients of cosmetics. The CGTase from *Brevibacterium* sp. effectively transglycosylates various phenolic compounds.[30] For example, the enzyme forms tranglycosylation product of (+)-catechin, a main polyphenolic compound in tea leaves, with a yield of 42%, while the yields with the enzymes from *B. macerans* and *B. stearothermophilus* are 25% and 5%, respectively. The major product is identified as (+)-catechin-3'-*O*-α-D-glucoside, which inhibits tyrosinase, a key enzyme of melanogensis of human skin.[47] A glucoside of kojic acid, which is also well known as skin-whitening compounds, is produced by the *Brevibacterium* enzyme, in which glycosylation occurs not at the phenolic OH group (C-3) but at the hydroxymethyl group (C-7).[30]

CGTases transglycosylate polyhydroxyl compounds such as myo-inositol, trimethylolpropane, and glycerol.[48,49] Glucosyl glycerol and glucosyl trimethylolpropane act as effective substrates for lipase-catalyzed esterification to yield glycolipids.[48] Glucosyl glycerol, in which the glucosyl residue binds to one of the primary OH groups of glycerol, inhibits the action of rat intestinal disaccharidases. Maltosyl and maltotriosyl glycerols inhibit the activity of porcine pancreas α-amylase *in vitro*.[49]

22.5   Concluding Remarks

CGTase is one of the most important industrial biocatalysts, particularly because of its unique intramolecular transglycosylation action to produce CDs that form the inclusion complexes with various hydrophobic guest molecules. CGTase is also regarded as an excellent tool for the glycosylation owing to the intermolecular transglycosylation activity: The enzyme shows an extremely high glycosylation efficiency, exhibits a broad tolerance for acceptors that enables the glycosylation of a number of saccharides and other related compounds, and has high stereo- and regiospecificities to provide merely α-1,4-glycosylation products from most sugar acceptors. In addition, the enzyme utilizes starch, a cheap and reproductive resource, as a preferable substrate. Varieties of microbial enzymes are commercially available. Furthermore, CGTase is a well-studied member of α-amylase family enzymes, and its detailed structure and the reaction mechanism have been elucidated at molecular level. It is hoped, therefore, that CGTase will be applied more extensively and increasingly to industrial fields.

References

1. S Kitahata. Cyclomaltodextrin glucanotransferase. In: The Amylase Research Society of Japan, ed. *Handbook of Amylases and Related Enzymes*. Tokyo: Pergamon Press, 1988, pp. 154–164.

2. S Kitahata. Cyclomaltodextrin glucanotransferase. In: The Amylase Research Society of Japan, ed. *Enzyme Chemistry and Molecular Biology of Amylases and Related Enzymes*. Tokyo: CRC Press, 1995, pp. 6–17.

3. H Hashimoto. Production and application of cyclomaltodextrins. In: The Amylase Research Society of Japan, ed. *Handbook of Amylases and Related Enzymes*. Tokyo: Pergamon Press, 1988, pp. 233–238.

4. G Schmidt. Preparation and application of γ-cyclodextrin. In: D Duchene, ed. *New Trends in Cyclodextrins and Derivatives*. Paris: Editions de Sante, 1990, pp. 27–54.

5. T Kuriki, T Imanaka. The concept of the α-amylase family: structure and common catalytic mechanism. *J Biosci Bioeng* 87: 557–565, 1999.

6. A Nakamura, K Yamane. Cyclodextrin glucanotransferase. In: The Amylase Research Society of Japan, ed. *Enzyme Chemistry and Molecular Biology of Amylases and Related Enzymes*. Tokyo: CRC Press, 1995, pp. 100–107.

7. Y Matsuura, M Kubota. Crystal structure of cyclodextrin glucanotransferase from Bacillus stearothermophilus and its carbohydrate binding sites. In: The Amylase Research Society of Japan, ed. *Enzyme Chemistry and Molecular Biology of Amylases and Related Enzymes*. Tokyo: CRC Press, 1995, pp. 153–162.

8. H Leemhuis, L Dijkhuizen. Hydrolysis and transglycosylation reaction specificity of cyclodextrin glucosyltransferases, *J Appl Glycosci* 50: 263–271, 2003.

9. EB Tilden, CS Hudson. Conversion of starch to crystalline dextrins by the action of a new type of amylase separated from culture of Areobacillus macerans. *J Am Chem Soc* 63: 2900–2902, 1939.

10. N Nakamura, K Horikoshi. Purification and properties of cyclodextrin glycosyltransferase of alkalophilic Bacillus sp. *Agric Biol Chem* 40: 935–941,1976.

11. T Kometani, Y Terada, T Nishimura, H Takii, S Okada. Purification and characterization of cyclodextrin glucanotransferase from an alkalophilic Bacillus species and transglycosylation at alkaline pHs. *Biosci Biotechnol Biochem* 58: 517–520, 1994.

12. S Mori, S Hirose, T Oya, S Kitahata. Purification and properties of cyclodextrin glucosyltransferase from Brevibacterium sp. No. 9605. *Biosci Biotech Biochem* 58: 1968–1972, 1994.

13. T Kometani, Y Terada, T Nishimura, H Takii, S Okada. Transglycosylation to hesperidin by cyclodextrin glucanotransferase from an alkalophilic Bacillus species in alkaline pH and properties of hesperidin glycosides. *Biosci Biotechnol Biochem* 58: 1990–1994, 1994.

14. S Kitahata, S Okada. Purification and some properties of cyclodextrin glucanotransferase from Bacillus stearothermophilus TC-60. *J Jpn Sco Starch Sci* 29: 7–12, 1982.

15. BE Norman, ST Joergensen. Thermoanaerobacter sp. CGTase: its properties and application. *Denpun Kagaku* 39: 101–108, 1992.

16. Y Tachibana, A Kuramura, N Shirasaka, Y Suzuki, T Yamamoto, S Fujiwara, M Takagi, T Imanaka. Purification and characterization of an extremely thermostable cyclomaltodextrin glucanotransferase from a newly isolated hyperthermophilic archaeon, a Thermococcus sp. *Appl Environ Microbiol* 65: 1991–1997, 1999.

17. S.Mori, M Goto, T Mase, A Matsuura, T Oya, S Kitahata. Reaction condition for the production of γ-cyclodextrin by cyclodextrin glucanotransferase from Brevibacterium sp. No. 9605. *Biosci Biotech Biochem* 59: 1012–1015, 1995.

18. S Kitahata, S Okada. Comparison of action of cyclodextrin glucanotransferase from Bacillus megaterium, B. circulans, B. stearothermophilus, and B. macerans. *J Jpn Soc Starch Sci* 29: 13–18, 1982.

19. T Kato, K Horikoshi. A new γ-cyclodextrin forming enzyme produced by Bacillus subtilis No. 313. *J Jpn Soc Starch Sci* 33: 137–143, 1986.

20. S Kobayashi, K Kainuma, D French. Effect of surfactants on cyclization of Bacillus macerans cyclodextrin glucanotransferase. *J Jpn Soc Starch Sci* 30: 62–68,1983.

21. H Bender. An improved method for the preparation of cyclooctaamylose, using starches and the cyclodextrin glycosyltransferase of Klebsiella pneumoniae M5al. *Carbohydr Res* 124: 225–233, 1983.

22. M Sato, Y Yagi. Properties of CGTases from three types of Bacillus and production of cyclodextrins by the enzymes. *ACS Symp Ser* 458: 125–137, 1991.

23. Y Fujita, H Tsubouchi, Y Inagi, K Tomita, A Ozaki, K Nakanishi. Purification and properties of cyclodextrin glycosyltransferase from Bacillus sp. AL-6. *J Ferment Bioeng* 70: 150–154, 1990.

24. Y Terada, M Yanase, H Takata, T Takaha, S Okada. Cyclodextrins are not the major cyclic α-1,4-glucans produced by the initial action of cyclodextrin glucanotrasnferase on amylose. *J Biol Chem* 272: 15729–15733, 1997.

25. Y Terada, H Sanbe, T Takaha, S Kitahata, K Koizumi, S Okada. Comparative study of the cyclization reactions of three bacterial cyclomaltodextrin glucanotransferases. *Appl Environ Microbiol* 67: 1453–1460, 2001.

26. S Machida, S Ogawa, S Xiaohua, T Takaha, K Fujii, K Hayashi. Cycloamylose as an efficient artificial chaperone for protein refolding. *FEBS Let* 486: 131–135, 2000.

27. Y Yagi, M Sato, T Ishikura. Comparative study of CGTase from Bacillus ohbensis, Bacillus macerans, Bacillus circulans and production of cyclodextrins using those CGTases. *J Jpn Stach Sci*: 33, 144–151, 1986.

28. S Kitahata, K Hara, K Fujita, H Nakano, N Kuwahara, K Koizumi. Acceptor specificity of cyclodextrin glucosyltransferase from Bacillus stearothermophilus and synthesis of α-D-glucosyl O-β-D-galactosyl- (1→4)-β-D-glucoside. *Biosci Biotechnol Biochem* 56: 1386–1391,1992.

29. S Kitahata, S Okada, T Fukui. Acceptor specificity of transglycosylation catalyzed by cyclodextrin glucosyltransferase. *Agric Biol Chem* 42: 2369–2374, 1978.

30. S Mori, M Goto, H Tsuji, T Mase, A Matsuura, T Oya, S Kitahata. Acceptor specificity of cyclodextrin glucanotransferase from Brevibacterium sp. No. 9605 to monosaccharides and phenolic compounds. *J Appl Glycosci* 44: 23–32, 1997.

31. S Kitahata, S Okada, A Misaki. Intermolecular transglycosylation of cyclodextrin glucosyltransferase to D-galactose. *Agric Biol Chem* 43: 151–154, 1979.

32. W Saenger. Cyclodextrin inclusion compounds in research and industry. *Angew Chem Int Ed* 19: 344–362, 1980.

33. RL Starnes. Industrial potential of cyclodextrin glucosyl transferases. *Cereal Food World* 35: 1094–1099, 1990.

34. K Koizumi, S Kitahata, H Hashimoto. Preparation isolation and analysis of heterogeneous branched cyclodextrins. In: YCD Lee, RTD Lee, ed. *Methods in Enzymology 247*. New York: Academic Press, 1994, pp. 64–87.

35. S Kitahata, H Hashimoto, K Koizumi. Synthesis of various branched cyclodextrins by transglycosylation. *Oyo Toshitsu Kagaku* 41: 451–456, 1994.

36. S Tavornvipas, F Hirayama, H Arima, K Uekama, T Ishiguro, M Oka, K Hamayasu, H Hashimoto. 6-O-α-(4-O-α-D-glucuronyl)-D-glucosyl-β-cyclodextrin: solubilizing ability and some cellular effects. *Int J Pharm* 249: 199–209, 2002.

37. S Okada, S Kitahata. Maltooligosylsucrose. In: T Nakakuki ed. *Oligosaccharides: Production, Properties, and Applications*. Switzerland: Gordon and Breach Science Publishers, 1993, pp. 118–129.

38. S Okada. Malto-oligosylsucrose. In: The Amylase Research Society of Japan, ed. *Handbook of Amylases and Related Enzymes*. Tokyo: Pergamon Press, 1988, pp. 226–228.

39. H Aga, M Yoneyama, S Sakai, I Yamamoto. Synthesis of 2-O-α-D-glucopyranosyl L-ascorbic acid by cyclodextrin glucanotransferase from Bacillus stearothermophilus. *Agric Biol Chem* 55: 1751–1756, 1991.

40. S Kitahata, S Kawanaka, M Dombou, C Katayama, M Goto, H Hashimoto. Synthesis of 6-O-α-D-galactosyl L-ascrobic acid by Candida guilliermondii H-404 α-galactosidase. *J Appl Glycosci* 43: 173–177, 1996.

41. Y Suzuki, K Suzuki. Enzymatic formation of 4^G-α-D-glucopyranosyl-rutin. *Agric Biol Chem* 55: 181–187, 1991.

42. T Kometani, Y Terada, T Nishimura, H Takii, S Okada. A new method for precipitation of various glucosides with cyclodextrin glucanotransferase. *Biosci Biotechnol Biochem* 57: 1185–1187, 1993.

43. M Darise, K Mizutani, R Kasai, O Tanaka, S Kitahata, S Okada, S Ogawa, F Murakami, FH Chen. Enzymic transglucosylation of rubusoside and the structure-sweetness relationship of steviol-bisglycosides. *Agric Biol Chem* 48: 2483–2488, 1984.

44. Y Fukunaga, T Miyata, N Nakayasu, K Mizutani, R Kasai, O Tanaka. Enzymic transglycosylation products of stevioside: Separation and sweetness-evaluation. *Agric Biol Chem* 53: 1603–1607, 1989.

45. K Ohtani, Y Aikawa, H Ishikawa, R Kasai, S Kitahata, K Mizutani, S Doi, M Nakaura, O Tanaka. Further study on the 1,4-α-transglucosylation of rubusoside, a sweet steviol-bisglucoside from Rubus suavissimus. *Agric Bio Chem* 55: 449–453, 1991.

46. R Kasai, RL Nie, K Nashi, K Ohtani, J Zhou, GD Tao, O Tanaka. Sweet cucurbitane glycosides from fruits of Siraitia siamensis (chi-zi luo-han-guo), a Chinese folk medicine. *Agric Biol Chem* 53: 3347–3349, 1989.

47. M Funayama, T Nishino, A Hirota, S Murao, S Takenishi, H Nakano. Enzymatic synthesis of (+)catechin-α-glucoside and its effect on tyrosinase activity. *Biosci Biotechnol Biochem* 57: 1666–1669, 1993.

48. H Nakano, S Kitahata, Y Shimada, M Nakamura, Y Tominaga. Esterification of glycosides by a mono- and diacylglycerol lipase from Penicillium camembertii and comparison of the products with Candida cylindracea lipase. *J Ferment Bioeng* 80: 24–29, 1995.

49. H Nakano, T Kiso, K Okamoto, T Tomita, MBA Manan, S Kitahata. Synthesis of glycosyl glycerol by cyclodextrin glucanotransferases. *J Biosci Bioeng* 95: 583–588, 2003.

23

Converting Herbaceous Energy Crops to Bioethanol: A Review with Emphasis on Pretreatment Processes

Bruce S. Dien

Loren B. Iten

Christopher D. Skory

23.1 Introduction

There has been increasing interest in using bioethanol as an automotive fuel because of national security, environmental, and economic concerns associated with using petroleum. Forage or herbaceous energy crops represents a promising source of carbohydrates for producing bioethanol. Their merits include lower production costs compared with grain crops, lower energy inputs (e.g., require fewer fertilizer and pesticide applications), and the ability to grow on marginal crop lands, and because they are perennial, they are sustainable and will improve soil quality over time.

Both the U.S. and the European Union have had ongoing research programs directed at the production of herbaceous perennials for use in energy generation since the mid-1980s. In the U.S., 35 herbaceous crops were evaluated and switchgrass (*Pancium virgaum*) was selected for further development based on its high yields, broad geographical growing range, and large genetic pool for breeding.[1] Research associated with the European Union screened 20 perennial grasses and selected four for further study: miscanthus (*Miscanthus* spp.), reed canarygrass (*Phalaris arundinacea*), giant reed (*Arundo donax*), and switchgrass.* Similar to the U.S. study the primary criterion was biomass yield, which are miscanthus (5-44 t DM ha-1 a-); reed canarygrass (7–23); giant reed (3–37); and switchgrass (5–23 in Europe and

---

*Names are necessary to report factually on available data; however, the USDA neither guarantees nor warrants the standard of the product, and the use of the name by USDA implies no approval of the product to the exclusion of others that may also be suitable.

0.9–34.6 in the U.S.).[1,2] Miscanthus and switchgrass have the highest potential yields of this group because they are C4 grasses. While miscanthus tends to have higher production yields than switchgrass,[3] the latter has other advantages, including its status as a native grass, better drought resistance, stands can be started from seed, and its status as a large genetic pool.[1] As C3 grasses, reed canary grass and giant reed are better adapted to colder climates and wintering. In a recent study, production costs were compared for four energy crops growing on marginal crop land in Iowa.[2] The authors concluded that switchgrass was the least expensive to produce $38.90 (switchgrass), followed by $48.06 (canary reed grass), $55.67 (alfalfa), and $55.74 (big bluestem). Intercropping alfalfa (a legume) with either canary reed or big bluestem was also found to reduce production costs because less fertilizer was needed for high yields.[19]

While much effort has been expended on developing improved cultivars and agronomic practices for energy crops, much less attention has been paid to developing technology for converting these biomasses to fuel ethanol. Though much of the needed technology is not specific to herbaceous crops, this is not true of the pretreatment step. Pretreatment is the first step in the process that allows for subsequent enzymatic saccharification of the cellulose fraction. Most of the literature on pretreatment has concentrated on using woody biomass. In fact, early U.S. research on biomass conversion was led by the Forest Products Laboratory (Madison, WI). This review will address pretreatment options for converting herbaceous crops to fuel ethanol.

23.2 Pretreatment Methods for Energy Crops

While both hardwood and herbaceous crops contain lignocellulose, important differences remain in their properties as related to biomass conversion. In general, alkali pretreatments are much more effective for herbaceous derived feedstocks than woody biomass. This difference is evidenced by the wide spread use of alkali treatments for increasing *in vitro* digestion in forage crops (e.g., ammoniating).[4] More care also needs to be paid to preserving xylan derived sugars because they represent one half of the available carbohydrates (Table 23.1). The severity of conditions commonly used for pretreating woody biomass (e.g., steam explosion) results in substantial degradation of arabinose and xylose. Finally, while those working in pretreating woody biomass have been able to turn to research conducted by the pulp and paper field for inspiration, those working with herbaceous feedstocks have been influenced by forage research. Almost without exception, every pretreatment method tested for biomass conversion of herbaceous biomass to ethanol has also been researched for its effect on increasing forage nutrition.

23.2.1 Cell Wall Structure and Cellulose Digestibility

Pretreating biomass before adding cellulase is required to open up the cell wall structure and allow the enzyme access to its target (cellulose). Plant cell walls are composed of three fractions: cellulose, hemicellulose, and lignin. Penner et al. (1996)[5] compares the native structure of these three components to reinforced concrete with the cellulose serving as the iron reinforcing rods, hemicellulose as the concrete filler, and lignin as a binder. The β-D-glucopyranose units that comprise the cellulose polymer

TABLE 23.1 Percent Dry Weight Composition for Selected Herbaceous Crops

Feedstock	Glucan	Xylan	Arabinan	Galactan	Lignin	Ref
Canary Reed Grass	29.1	9.3	2.6	0.9	5.6	44
Macanthus	45.9	18.1	1.3	0.5	nr[a]	45
Sericea lespedeza	31.5	14.5	1.6	0.9	31.6	46
Switchgrass	31.0	20.4	2.8	0.9	17.6	46
Weeping lovegrass	36.7	17.6	2.6	1.7	21.2	46

[a] not reported, usually appox. 17%

form up to six hydrogen bonds, two of which are with other cellulose polymers. This intermolecular bonding allows for cellulose to associate into tightly organized, and crystalline, microfibers. Six factors have been identified as affecting access of cellulase to its target. These are the presence of hemicellulose and lignin, accessible area or available pore volume, cellulose crystallinity, available cellulose chain ends, and hemicellulose acetylation (see References 6, 7, and 8 for excellent discussions). Removing the hemicellulose and dislodging or partially removing the lignin appears to strongly promote cellulose digestion. Lignin also inhibits cellulase by binding the protein. Likewise, pretreatments also tend to swell the cellulose fiber, thereby presumably increasing surface area. While cellulose crystallinity also probably inhibits cellulase activity,[9] perhaps by steric hindrance, the evidence of it in inhibiting cellulase is mixed. Neither dilute acid or alkaline peroxide reduce cellulose crystallinity,[8,10] yet both are very effective at promoting digestibility. Finally it has been suggested that as hydrolysis progresses, the formation of chain ends might become rate limiting and might be one reason digestion slows dramatically as the reaction proceeds.

23.2.2 Selected Pretreatments Processes

Research on pretreating biomass dates to before the beginning of the last century and, in fact, woody biomass was being converted to ethanol in this country as early as 1910; the last production plant closed down after World War I when wood supplied became more limited. Therefore, for the sake of brevity, this review will be confined to discussing the most relevant pretreatments to converting forage type material to ethanol. Table 23.2 lists the types of pretreatments that have been performed on energy crops. This section will focus on the most well-developed pretreatments, which are dilute acid, buffered hot water, alkaline peroxide, ammonia fiber explosion, and lime. Where possible, results are given for forage crops and, where not, examples are given for the most relevant tested feedstocks.

23.2.3 Dilute Acid Pretreatment

Dilute acid is the only pretreatment listed here that was once used for commercial production of fuel ethanol and is by far the most extensively researched. From a practical perspective, dilute acid can be very effective at producing highly digestible cellulose from a wide range of feedstocks and is the only pretreatment discussed in this review that completely hydrolyzes the xylan sugars. Disadvantages of pretreating with dilute acid includes formation of more inhibitory side products compared to other pretreatments, a requirement for special (expensive) reactors capable of withstanding low pHs at high temperatures, and production of chemical waste in the form of gypsum.

Dilute sulfuric acid has been evaluated for the following energy crops: switchgrass, weeping lovegrass (*Eragrotis curvula*), and Sericea lespedeza (*Lespedeza cuneata*), which is a legume.[11] The dilute acid

TABLE 23.2 Summary of Pretreatments Tested on Selected Energy Crops

Feedstock	Pretreatment
Alfalfa hay	AFEX,[34] Dilute Acid and Autohydrolysis[47]
Canary Reed Grass	AP[28]
Macanthus	Alkali[48]
Sericea lespedeza	Dilute acid[11]
Switchgrass	AFEX,[33,39] AP,[28] Dilute-Acid,[11] Hot-Ammonia,[49,50] Hot-Water,[23] Lime[42,43]
Weeping lovegrass	Dilute-Acid[11]
Others	AP[28] and AFEX[39,40]

treatments were optimized to remove the maximum amount of hemicellulose while avoiding complete chemical hydrolysis of cellulose. It was determined that at pH 1.35–1.45 (0.45%v H_2SO_4), only 5 min at 160°C or 30 min at 140°C was sufficient to obtain the maximum cellulose digestibility for the grasses. At these conditions, 90–100% of the hemicellulose and 3–18% of the lignin were removed, and cellulose digestibility was increased to 85–90%. Unfortunately, the legume feedstock proved to be more resilient and had to be heated at 180°C for 20 min to obtain 70% cellulose digestibility. The legume also had greater buffering capacity, and 0.50%v H_2SO_4 was needed to reach the target pH. Interestingly, there was only a slight increase in solids removed for the legume between 160°C and 180°C and no difference in hemicellulose removal, indicating other factors than hemicellulose and lignin removal played a role in increasing digestibility. Still, the major mechanisms for dilute-acid increasing cellulose digestibility are removing hemicellulose, dislodging lignin, and increasing accessible surface area.

In a subsequent study, the dilute acid pretreated biomasses were converted to ethanol by simultaneous saccharification and fermentation (SSF) with *Saccharomyces*.[12] For this study, the biomasses were all treated at 140°C for 1 hr at the acid loadings cited above and fermented at 37°C for 7 days. At cellulase loadings of 19 and 26 FPU/g biomass,* the final ethanol yields for weeping lovegrass and switchgrass were 81–89% of theoretical based upon cellulose content and only 44–52% for the legume Sericea lespedeza. (Note: Usual cellulase loading is only 15 FPU/g cellulose; ca. 2.5–3.5x less activity assuming 50%wt cellulose.)

Under sufficiently harsh conditions, dilute acid pretreatment will directly hydrolyze cellulose to glucose; however, milder conditions are usually chosen to avoid glucose formation. The reason is that once glucose forms under these conditions, it quickly dehydrates and forms the furan 5-hydroxylmethlyfurfural (HMF). Not only does this lower the eventual yield of glucose, but HMF is a potent inhibitor of yeast fermentations. Sugars released from hydrolysis of the xylan also can undergo further destructive reactions. Xylose is largely converted to the furan furfural,** which is also a microbial inhibitor. (Note: Other sugars released from the hemicellulose either form furfural [pentoses] or HMF [hexoses]). As release of the hemicellulose is necessary for effective pretreatment, the formation of free xylose is unavoidable. However, by studying the kinetics of xylose formation and disappearance into furfural, a combination of temperature and time can be chosen to maximize xylose's yield.

Xylan hydrolysis is biphasic.[13,14] In the case of switchgrass xylan, 76.8% is readily hydrolyzed and the remainder hydrolyzes at a much slower rate.[14] Therefore, xylan hydrolysis to xylose can be modeled as two parallel first-order rate reactions:

$$\begin{array}{c} \text{Xylan}_f \searrow^{K_f} \\ \qquad\qquad \text{Xylose} \xrightarrow{K_d} \begin{array}{l} \text{Degradation} \\ \text{Products} \end{array} \\ \text{Xylan}_s \nearrow_{K_s} \end{array}$$

Degradation of xylose to furfural can be described by a first-order reaction. When Esteghlaian et al. (1997) measured the system's kinetic parameters, they determined that the energy of activation for the hydrolysis reaction was higher than that for xylose degradation (169 and 211 vs. 99.5 kJ/mol). This suggests higher reaction temperatures (170–180°C) and shorter reaction times (0.5–1.0 min) are needed to optimize xylose yield. The authors also examined corn stover and predicted similar conditions for its optimal xylose yield, which was validated in a subsequent study.[15]

An important aspect of any pretreatment is whether the favorable conditions modeled and confirmed in laboratory sized reactor can be replicated in units processing tones of biomass per day. While energy crops have not been processed at larger scales, the National Renewable Energy Laboratory (NREL) in Golden, Colorado has in the last few years examined dilute acid pretreatment of corn stover and their findings give cause for optimism. In one study, corn stover was treated in a 4 L steam explosion reactor

---

*Usual cellulase loading is only 15 FPU/g cellulose; ca. 2.5–3.5x less activity assuming 50% wt cellulose.

**Other sugars released from the hemicellulose either form furfural (pentoses) or HMF (hexoses).

(e.g., acid catalyzed steam explosion). The biomass was treated at an incredible solids loading of 37–47% wt and the optimal conditions were found to be 190°C for 90–130 sec. The very high solids loading allowed for only 11–16 g H_2SO_4 to be used per kg biomass, * db. The high solids also mean better energy efficiency because less energy is devoted to heating excess water. The final results were xylose yields >90% and ethanol yields for SSF (15 FPU/g cellulose) of 85%.[15] In a second study, corn stover was pretreated in a 1 t/day continuous vertical pulp digester. The xylose yield was 77% and 76% of the cellulose was digestible.[16]

23.2.4 Buffered Hot Water

As the name implies, for the buffered hot water pretreatment, the biomass is just cooked in water buffered at a neutral pH. It is an attractive alternative to dilute acid because it avoids many of the latter's problems, including extensive inhibitor formation, excess chemical usage, gypsum waste (from neutralization with $Ca(OH)_2$), and the need for constructing the pretreatment reactor out of expensive metal alloys capable of withstanding the harsh operating conditions. A major disadvantage of the buffered hot water method versus dilute acid is that xylan is not completely hydrolyzed and, therefore, would require integrating a postenzymatic or dilute acid hydrolysis step prior to fermenting them. This is a new pretreatment strategy and has only been tested using corn fiber and yellow popular wood with varying success.[17–19]

The recommended operating conditions are 160–240°C and pHs 5–7. Cooking times are 20–60 min (including the time to heat the batch reactor) depending upon the substrate.[17,19] The pH can be controlled either by automatic base addition or using a buffer. For automatic pH control, the pH is measured externally using an automatic sampler, which allows the sample to be cooled prior to measuring the pH. Cooking yellow popular slurry (6.6% wt) to 240°C dissolved 44.5% of the solids and upon scarification gave 79% conversion to glucose.[17] An 84% glucose yield was realized from corn fiber (corn hulls recovered from corn wet milling) cooked to 220°C and digested with cellulase.

Recently, a 40 gal/min continuous pilot scaled process was tested for pretreating corn fiber.[20] The corn fiber was heated to 160°C by direct steam and pumped through a 15–20 min holding tube. Light corn steep was used to buffer the reaction at pH > 4. Glucans were 100% digestible when treated with a combination of cellulase and glucoamylase, the latter added to hydrolyze residual starch extracted from the corn fiber.**

The pretreated fiber was also fermented rapidly by *Saccharomyces* in the presence of cellulase to a final yield of 11 g/l ethanol (unpublished data).

It is theorized that hot water disrupts the hydrogen bonds binding the cellulose polymers together into microfibers, thereby, swelling the cellulose structure.[21] Microcrystalline cellulose treated with water heated to 190°C swells to 4x its initial size.[17] The hot water also dissolves the hemicellulose and loosens the lignin as the temperatures used exceed its glass transition temperature. Buffering at a higher pH prevents the xylan from being completely hydrolyzed. When biomass is treated with unbuffered hot water (e.g., 160–220°C), the water disassociates to form a weak acid; at 220°C, water has a pH of 5.5.[22] The pH is further decreased once xylan is deactylelated. Eventually the xylan is completely hydrolyzed and furfural begins to form. This process is referred to as autohydrolysis. Buffering retards autohydrolysis and prevents complete hydrolysis of the xylan sugars. Actually, buffering minimizes, but does not eliminate formation of inhibitors. The yellow popular slurry cooked to 240°C produced 1 g/l HMF and 2.5 g/l furfural.[17] Furthermore, enough inhibitors were produced to lower the ethanol yield when fermented to 55% of theoretical.[17] However, fewer inhibitors would have been formed if a reactor design had been employed that did not take an hour to warm up.

Hydrothermoylsis, which uses unbuffered water, has been applied to a wide range of substrates with very good yields of sugars.[23] In this case, 10 biomass samples, including four herbaceous species, were

---

*The switchgrass and weeping lovegrass discussed earlier were treated at a 10% wt solids loading and an acid loadings of 81 g/Kg biomass, db.

**Corn hulls recovered from corn wet milling.

treated with water heated to 200–230°C for 0–15 min. All of the hemicellulose was removed as well as 39–46% of the solids. What was surprising, however, was that 76–100% of the hemicellulose sugars were subsequently recovered in the supernatant. While in this case the water was not buffered, it appears that the short time of the reaction halted complete xylan hydrolysis and subsequent dehydration to furfural. An ethanol yield from cellulose of 90% of theoretical was obtained when sugar cane bagasse was treated with hot water (220°C, 2 min, 10% solids loading), with a cellulase loading of 13 FPU/g. The final ethanol concentration was only 14.6 g/l because of the low starting cellulose concentration.[24] However, this result still demonstrates that little in the way of inhibitors are generated by this process. In two more recent studies, using compressed hot water and steam were compared for treating sugar cane bagasse and corn fiber. Steam allows for running at high solid loadings (5–8%wt vs. 50–70%) because of its high enthalpy. However, replacing liquid water with steam resulted in poor pentosan yields (e.g., <50%) and either partially or completely inhibited fermentations.[25,26] The high yields of xylan and digestible cellulose make compressed hot water a promising technology. The eventual utility of either of these approaches will depend on the limits in solids loading, concentration of recovered pentosan sugars (diluted for hydro-thermoylsis), and effective reactor design so as to maintain the highly favorable kinetics.

23.2.5 Alkaline Peroxide

Alkaline peroxide (AP) is one of the most effective pretreatments for herbaceous biomass. Cellulose from AP treated biomass, such as wheat straw, can be converted quantitatively to ethanol by SSF. The AP pretreatment consists of treating biomass in a NaOH solution at pH 11.5–11.6 in the presence of 1% v/v H_2O_2. The alkali pH removes the hemicellulose and the peroxide removes the lignin. Deliginification occurs in a two-step process: H_2O_2 disassociates to H+ and HOO– (pKa = 11.5–11.6) and the hydroperoxy anion (HOO–) further reacts with H_2O_2 to form hydroxyl radicals (•OH) and superoxide (O_2^-). Therefore, the specific alkali pH is set to favor the H_2O_2 reaction as opposed to being just sufficiently alkali to extract hemicellulose. Treatment conditions for AP are mild compared to acid catalyzed pretreatments. Under laboratory suitable conditions, 10% w/w of ground sample is treated for 24 hr in NaOH/1% v/v peroxide solution at ambient temperature while stirring.[27] Increasing the temperature to 60°C did not influence saccharification efficiency.[27]

Alkaline peroxide, as is generally true of alkali treatments, works best on herbaceous biomasses. The pretreatment has been evaluated on a wide variety of herbaceous type plant matter including: corn stalks, wheat straw, alfalfa hay, and switchgrass.[27,28] For wheat straw, corn stalks, and alfalfa hay, the efficiencies of scarification (e.g., cellulase enzyme assay) were 93.0–100% efficient. When AP was tested on a wide variety of plant material, glucose yields (g glucose/g biomass, db) following treatment with cellulase were typically double for the monocots vs. dicots. AP is also inefficient at treating woody biomass, i.e., 52.5% efficient. AP pretreated biomass has also been evaluated for conversion to ethanol. SSF with *Saccharomyces* gave ethanol yields between 80% and nearly 100% of theoretical (based on glucan contents) for corn cobs, husks, or stalks, wheat straw, and kenaf. For oak shavings the ethanol yield was still over 60%.[29]

AP treated material is unusual in one respect from biomass treated by other methods. Quantitative saccharification to glucose or fermentation to ethanol in SSF was possible after the treated material had been completely dried.[29] Normally, biomass following pretreatment is not dried because it lowers the cellulose digestibility, presumably by collapsing pores needed by cellulase enzymes to access the cellulose fibers. The unusually high digestibility of the biomass can be attributed to a number factors, including: hemicellulose removal, delignification, and swelling of the cellulose fibers. Alakali extraction is the accepted method for isolating hemicellulose from biomass and it has been long known that treating forage type material with NaOH, ammonia, or lime increases *in vitro* digestibility.[4] However, AP biomass yields twice the ethanol yield as alkali treated biomass in SSF experiments.[29] The increased efficiency can be ascribed to peroxide delignification. AP treatment of wheat straw removed up to 60% of the lignin compared to 10% with only alkali (pH 11.5) treatment.[30] Saccharification efficiency with AP treatment increases corresponding with greater delignification up to 40% lignin removal.[27] AP also swells the cellulose structure and increases available surface area. Cell wall fibers observed prior and after AP treatment, changed from an ordered array to an amorphous structure.[30] Interestingly, the one factor associated with increased digestibility that was not affected was cellulose crystallinity. Cellulose crystallinity was unchanged for AP treated versus intact wheat straw.[10]

A practical shortcoming of the AP method is the large amount of alkali and peroxide needed for pretreating biomass. An effective method for lowering reagent usage is to increase solid loading. In laboratory experiments, the solid loading was limited to 10% w/w because of mixing limitations. However, by using a modified single screw extruder, the loading was increased to 40% w/w.[31] Alkali was added through the downstream port. Peroxide was determined to be more effective when added in the downstream port — less than 15 sec prior to discharge. Following the discharge the material was incubated for 24 hr to allow for delignification. The effectiveness of the treatment was proven by increased rumen digestibility. Unfortunately, the text is ambiguous as to whether the peroxide loading had been reduced compared to the laboratory protocol.[31] Also introduction of the peroxide was complicated because the mild steel used to manufacture the extruder was found to be incompatible with H_2O_2; it appeared to absorb H_2O_2 from the slurry. Therefore, even though solids could be greatly increased by using an extruder for mixing, chemical usage appeared to remain high and the addition of peroxide will necessitate using an extruder manufactured from better quality steel.

23.2.6 Ammonia Fiber/Freeze Explosion

Ammonia Fiber Explosion (AFEX) produces some of the most digestible herbaceous cellulose fractions of any of the pretreatments. Ammonization has long been considered an effective method for increasing forage quality.[4] In the AFEX process, pressurized liquid ammonia is mixed with the biomass and after a brief holding time (5–30 min) explosively released by venting. Most remaining ammonia is evaporated away, leaving only 1–1.5% wt ammonium, which can serve as a source of nitrogen for fermentation or rumen feed. AFEX has been evaluated for pretreatment of a wide variety of herbaceous feedstocks including: alfalfa, barley, and wheat straws, corn stover, elephant grass, Bermuda grass, rice hulls, and switchgrass. AFEX has also been used for pretreating woody biomass, but has been found to be relatively ineffective.[32]

AFEX conditions need to be optimized separately for each feedstock. For switchgrass the optimized conditions were found to be an ammonium solid loading of 15% moisture, chemical loading of 1:1 g NH_3 per g biomass, db, temperature of 90°C, and a holding time of 30 min.[33] Two lots of switchgrass were evaluated in this study, one harvested in the spring (early) and the other in the fall (late). Significant differences were discovered for how the two lots reacted to AFEX treatment. The late harvested switchgrass produced 25% less reducing sugars upon saccharification compared to the other lot. The different sugar yields was attributed to the greater lignin content of the fall harvesting because both lots had similar amounts of total carbohydrates. AFEX is also effective for pretreating more recalcitrant biomasses. Saccharification of AFEX pretreated alfalfa yielded 68% of the glucans and 85% of the hemicellulose as monomers.[34] For AFEX treated rice straw, 61% of the total carbohydrates were enzymatically converted to free sugars.[35] Compared to switchgrass, AFEX conditions were modified for these harder to digest substrates by increasing ammonium loading (1.5–2 g/g). For alfalfa it was also found that soluble sugars present in the biomass (especially galactose) were partially degraded by AFEX; interestingly fructose and glucose also appeared to condense into sucrose.

Pretreated AFEX biomass has been determined to be compatible for ethanol fermentations by yeast and bacteria.[32,36–38] *Saccharomyces* readily fermented AFEX pretreated alfalfa cake and wheat straw with good ethanol yields. However, the sugar concentrations used were dilute and the final ethanol concentrations were 10 g/l or less.[32] AFEX pretreated costal bermudagrass and switchgrass were converted to ethanol using recombinant *Klebiella ocytoca* strain P2. *K. oxytoca* can ferment hexose and pentose sugars and strain P2 has been engineered to selectively produce ethanol. The strain was able to ferment both AFEX treated biomasses. To increase the final ethanol concentration for switchgrass, the sugars were concentrated by reverse osmosis to 80 g/l. In the more concentrated culture, strain P2 had a 24 hr lag phase and did not ferment the xylose, but it did achieve a final ethanol concentration of approximately 20 g/l.[39] Therefore, fermentability of AFEX treated material at higher solids may be a consideration.

AFEX increases cellulose digestibility through multiple mechanisms. First, ammonium combines with the water to form NH_4OH, which removes the hemicellulose; for AFEX to be effective, biomass needs to contain at least 15% wt moisture.[32] While hemicellulose removal is not 100% complete, maximum fractionation of it has been associated with maximum cellulose digestibility for elephant grass, rice straw, and alfalfa.[34,40] Second, the explosive force of ammonium's release at the end of the pretreatment literally blows apart the fibers, greatly increasing the surface area available for enzymatic attack.[*] The fractured fibers can be clearly observed by SEM[36] and are also evidenced by the halving in bulk density following AFEX treatment.[36] Further evidence comes from a control experiment in which ammonium was slowly released and digestibility was negatively impacted.[8] Third, the ammonium is believed to intercalate between the cellulose polymers and disrupt hydrogen bonding, which leads to reduced crystallinity.[8] Evidence that disrupting H bonds is important comes from an early control experiment that replaced pressurized NH_3 with CO_2. Even though the explosive force was similar with CO_2, the digestibility was reduced compared to NH_3.[8] While destruction of xylan sugars does not appear to be a concern because free xylose is not formed, AFEX conditions do need to be adjusted to ensure minimal sugar destruction of soluble sugars already present, especially galactose.[34]

Though AFEX was invented in the mid-1980s, it has only recently been scaled and tested with a continuous reactor to generate AFEX[**] treated rice straw for a dairy feeding trial.[41] Several tons of rice straw were treated in a pilot scale Sunds hydrolyzer. The ammonia loading was 0.3 g per g dry straw and the reaction temperature 110–132°C. The final product was dried to 85% DM. When test for *in vitro* digestion, AFEX pretreatment improved digestibility compared to untreated rice straw. Incorporating AFEX rice straw into dairy feed increased milk production by 1.3 kg/d compared and lowered milk fat (not good) compared to the control group, which was fed feed containing no rice straw. From our perspective, the primary result of this work is that the AFEX process is scalable, though unfortunately ammonium recycle did not appear to be tested in this demonstration.

23.2.7 Lime Pretreatment

Alkali chemicals have long been used to increase the digestibility of forage crops, and lime is no exception.[4] Lime has generally been considered less effective than NaOH or ammonium. However, recently lime has been studied for pretreatment of switchgrass with good results.[42] The switchgrass (10% solids) was treated at 120°C for 2 hr with 0.1g $Ca(OH)_2$/g dried biomass. The treatment removed 26% of the xylan and 29% of the lignin. The efficiency of cellulose digestion (5 FPU/g dry biomass) was 85.6% for total sugars. When converted to ethanol by SSF, lime treated switchgrass gave of yield of 72% of theoretical (cellulose concentration = 30 g/l; cellulase loading = 25 FPU/g cellulose).[43] Though the final ethanol yield is lower than observed for some of the other pretreatments, no inhibition of fermentation was observed.

While the operating conditions are very simple and readily scalable, being able to recycle the lime is essential at a 10% w/w lime loading. One suggested method given is to precipitate the Ca with CO_2 and recycle it using a lime kiln. Adding large amounts of lime in a process can also lead to scaling of process equipment surfaces and further deposits during the fermentation from calcium combining with organic acid metabolites produced by *Saccharomyces*.

The principle behind the effectiveness of the lime pretreatment is simple. Lime raises the pH to 11.5,[42] which dissolves the hemicellulose and partially removes the lignin. The significance of these recent studies is determining that with the proper loading and temperature lime is as effective as other bases (e.g., NaOH) for treating herbaceous biomasses.

---

[*]When biomass is treated with steam explosion, the explosive action is not thought to increase cellulose digestibility.
[**]Referred to as FIBEX in this study.

23.3 Summary

Herbaceous energy crops represent an immense potential feedstock for fuel ethanol production. These crops can be grown in soil not favorable for grain crops and would serve as a sustainable supply of biomass. Several favorable pretreatments can be used to either partially or completely hydrolyze biomass for fermentation. Treating with dilute acid can produce a concentrated solution of pentosan sugars and readily saccharified cellulose. Hot water pretreatments, while still in the early stages of research, are advantageous in retarding destruction of pentosans and not generating waste chemicals (e.g., gympsum). Herbaceous feedstocks also respond favorably to alkali treatments. AP is extremely effective and allows for quantitative conversion of cellulose to ethanol. AFEX is also very effective at pretreating cellulose and would result in lower chemical usage compared to AP. Finally, pretreating with lime is also capable of digesting biomass. Grasses responded well to each of these pretreatments. Legumes, which are more resilient, responded better to alkali pretreatments or a two-stage dilute acid pretreatment, where the first stage is optimized for nondestructive xylan removal. Except for dilute acid, none of these pretreatments produce free pentosan sugars. Therefore, each of these pretreatments would need to be combined with either dilute-acid or enzymatic treatment to complete hydrolysis of the pentosan oligomers. An enzymatic treatment would be more specific and environmentally benign; however, commercial hemicellulase is not currently aggressive enough to give good yields.

References

1. Lewandowski I, Scurlock JMO, Lindvall E, Christou M: The development and current status of perennial rhizomatous grasses as energy crops in the US and Europe. *Biomass & Bioenergy* 2003, 25: 335–361.
2. Hallam A, Anderson IC, Buxton DR: Comparative economic analysis of perennial, annual, and intercrops for biomass production. *Biomass & Bioenergy* 2001, 21: 407–424.
3. Heaton E, Voigt T, Long SP: A quantitative review comparing the yields of two candidate C4 perennial biomass crops in relation to nitrogen, temperature and water. *Biomass & Bioenergy* 2004, in press.
4. Han YW, Catalano EA, Ciegler A: Treatments to improve the digestibility of crop residues. Wood and agricultural residues: Research on use for feed, fuels, and chemicals: Proceedings Conference, Kansas City, Missouri, Sept 1982, 1983: 217–238.
5. Penner MH, Hashimoto A, Esteghlalian A, Fenske JJ: Acid catalyzed hydrolysis of lignocellulosic materials. In *Agricultural Materials as Renewable Resources*. Edited by Fuller G, McKeon TA, Bills DD: ACS; 1996: 13–29. Vol ACS Symposium Series 647.
6. Converse AO: Substrate factors limiting enzymatic hydrolysis. In *Bioconversion of forest and agricultural plant residues*. Edited by Saddler JN: CAB International; 1993: 93–106.
7. Hsu T-A: Pretreatment of Biomass. In *Handbook on Bioethanol: Reduction and Utilization*. Edited by C.E. Wyman: Taylor and Francis; Applied Energy Technology Series, 1996: 179–212.
8. Joseph Weil PW, Karen Kohlmann and Michael R. Ladisch: Cellulose pretreatments of lignocellulosic substrates. 1994, 16: 1002–1004.
9. Chang VS, Holtzapple MT: Fundamental factors affecting biomass enzymatic reactivity. *Applied Biochemistry and Biotechnology* 2000, 84–6: 5–37.
10. Martel P, Gould JM: Cellulose stability and delignification after alkaline hydrogen peroxide treatment of straw. *Journal of Applied Polymer Science* 1990, 39: 707–714.
11. Torget R, Werdene P, Himmel M, Grohmann K: Dilute acid pretreatment of short rotation woody and herbaceous crops, Eleventh Symposium on Biotechnology for Fuels and Chemicals. *Applied Biochemistry and Biotechnology* 1990, 24–25: 115–126.
12. Wyman Charles E: Simultaneous saccharification and fermentation of several lignocellulosic feedstocks to fuel ethanol. *Biomass and Bioenergy* 1992, 3: 301–307.
13. Kim SB, Lee YY: Kinetics in acid-catalyzed hydrolysis of hardwood hemicellulose. Biotechnology and Bioengineering Symposium 1987, 17: 71–84.

14. Esteghlalian A, Hashimoto AG, Fenske JJ, Penner MH: Modeling and optimization of the dilute-sulfuric-acid pretreatment of corn stover, poplar and switchgrass. *Bioresource Technology* 1997, 59: 129–136.

15. Tucker MP, Kim KH, Newman MM, Nguyen QA: Effects of temperature and moisture on dilute-acid steam explosion pretreatment of corn stover and cellulase enzyme digestibility. *Applied Biochemistry and Biotechnology* 2003, 105: 165–177.

16. Schell DJ, Farmer J, Newman M, McMillan JD: Dilute-sulfuric acid pretreatment of corn stover in pilot-scale reactor — Investigation of yields, kinetics, and enzymatic digestibilities of solids. *Applied Biochemistry and Biotechnology* 2003, 105: 69–85.

17. Weil J, Brewer M, Hendrickson R, Sarikaya A, Ladisch MR: Continuous pH monitoring during pretreatment of yellow poplar wood sawdust by pressure cooking in water. *Applied Biochemistry and Biotechnology* 1998, 70–2: 99–111.

18. Weil JR, Dien B, Bothast R, Hendrickson R, Mosier NS, Ladisch MR: Removal of fermentation inhibitors formed during pretreatment of biomass by polymeric adsorbents. *Industrial & Engineering Chemistry Research* 2002, 41: 6132–6138.

19. Weil JR, Sarikaya A, Rau SL, Goetz J, Ladisch CM, Brewer M, Hendrickson R, Ladisch MR: Pretreatment of corn fiber by pressure cooking in water. *Applied Biochemistry and Biotechnology* 1998, 73: 1–17.

20. Mosier NS, Hendrickson R, Welch G, Dreschel R, Dien BS, Ladisch MR: Corn fiber pretreatment scale-up and evaluation in an industrial corn to ethanol facility. In *Symposium on Biotechnology for Fuels and Chemicals*; Breckenridge: 2003: 19.

21. Weil J, Westgate P, Kohlmann K, Ladisch MR: Cellulose pretreatments of lignocellulosic substrates. *Enzyme and Microbial Technology* 1994, 16: 1002–1004.

22. Allen SG, Kam LC, Zemann AJ, Antal MJ, Jr.: Fractionation of sugar cane with hot, compressed, liquid water. *Industrial & Engineering Chemistry Research* 1996, 35: 2709–2715.

23. Mok WSL, Antal MJ: Uncatalyzed solvolysis of whole biomass hemicellulose by hot compressed liquid water. *Industrial & Engineering Chemistry Research* 1992, 31: 1157–1161.

24. Van Walsum GP, Allen SG, Spencer MJ, Laser MS, Antal MJ, Jr., Lynd LR: Conversion of lignocellulosics pretreated with liquid hot water to ethanol. *Applied Biochemistry and Biotechnology*. Spring 1996: 157–170.

25. Allen SG, Schulman D, Lichwa J, Antal MJ, Laser M, Lynd LR: A comparison between hot liquid water and steam fractionation of corn fiber. *Industrial & Engineering Chemistry Research* 2001, 40: 2934–2941.

26. Laser M, Schulman D, Allen SG, Lichwa J, Antal MJ, Lynd LR: A comparison of liquid hot water and steam pretreatments of sugar cane bagasse for bioconversion to ethanol. *Bioresource Technology* 2002, 81: 33–44.

27. Gould JM: Alkaline peroxide delignification of agricultural residues to enhance enzymatic saccharification. *Journal of Biotechnology and Bioengineering* 1984, 26: 46–52.

28. Gould JM: Enhanced polysaccharide recovery from agricultural residues and perennial grasses treated with alkaline hydrogen peroxide. *Journal of Biotechnology and Bioengineering* 1985, 27: 893–896.

29. Gould JM, Freer SN: High-efficiency ethanol production from lignocellulosic residues pretreated with alkaline peroxide. *Journal of Biotechnology and Bioengineering* 1984, 26: 628–631.

30. Gould JM: Studies on the mechanism of alkaline peroxide delignification of agricultural residues. *Journal of Biotechnology and Bioengineering* 1985, 27: 225–231.

31. Gould JM, Jasberg BK, Fahey GC, Jr., Berger LL: Treatment of wheat straw with alkaline hydrogen peroxide in a modified extruder. *Journal of Biotechnology and Bioengineering* 1989, 33: 233–236.

32. Dale BE, Henk LL, Shiang M: Fermentation of lignocellulosic materials treated by ammonia freeze-explosion. *Journal of Biotechnology and Bioengineering* 1985, 26: 223–233.

33. Dale BE, Leong CK, Pham TK, Esquivel VM, Rios I, Latimer VM: Hydrolysis of lignocellulosics at low enzyme levels: application of the AFEX process. *Bioresource Technology* 1996, 56: 111–116.

34. Ferrer A, Byers FM, Sulbaran-de-Ferrer B, Dale BE, Aiello C: Optimizing ammonia processing conditions to enhance susceptibility of legumes to fiber hydrolysis — Alfalfa. *Applied Biochemistry and Biotechnology* 2002, 98: 123–134.

35. Ferrer SF, Aristiguieta M, Dale BE, Ferrer A, Rodriguez GO: Enzymatic hydrolysis of ammonia-treated rice straw. *Applied Biochemistry and Biotechnology* 2003, 105–108.

36. Dale BE: A freeze-explosion technique for increasing cellulose hydrolysis Biomass conversion. *3rd annual Solar and Biomass Workshop,* April 1983: 188–191.

37. Moniruzzaman M, Dale BE, Hespell RB, Bothast RJ: Enzymatic hydrolysis of high-moisture corn fiber pretreated by AFEX and recovery and recycling of the enzyme complex. *Applied Biochemistry and Biotechnology.* 1997, 67: 113–126.

38. Moniruzzaman M, Dien BS, Ferrer B, Hespell RB, Dale BE, Ingram LO, Bothast RJ: Ethanol production from AFEX pretreated corn fiber by recombinant bacteria. *Biotechnology Letters* 1996, 18: 985–990.

39. Reshamwala S, Shawky BT, Dale BE: Ethanol production from enzymatic hydrolysates of AFEX-treated coastal bermudagrass and switchgrass. *Applied Biochemistry and Biotechnology.* Spring 1995: 43–55.

40. Ferrer A, Byers FM, Sulbaran-de-Ferrer B, Dale BE, Aiello C: Optimizing ammonia pressurization/ depressurization processing conditions to enhance enzymatic susceptibility of dwarf elephant grass. *Applied Biochemistry and Biotechnology* 2000, 84–6: 163–179.

41. Weimer PJ, Mertens DR, Ponnampalam E, Severin BF, Dale BE: FIBEX-treated rice straw as a feed ingredient for lactating dairy cows. *Animal Feed Science and Technology* 2003, 103: 41–50.

42. Chang VS, Burr B, Holtzapple MT: Lime pretreatment of switchgrass. *Applied Biochemistry and Biotechnology.* Spring 1997: 3–19.

43. Chang VS, Kaar WE, Burr B, Holtzapple MT: Simultaneous saccharification and fermentation of lime-treated biomass. *Biotechnology Letters* 2001, 23: 1327–1333.

44. Collings GF, Yokoyama MT: Analysis of fiber components in feeds and forges using gas-liquid chromatograhy. *Journal of Agricultural and Food Chemistry* 1979, 27: 373–377.

45. Faix O, Meier D, Beinhoff O: Analysis of lignocelluloses and lignins from *Arundo donax L.* and *Miscanthus sinensis* Anderss., and Hydroliquefaction of *Miscanthus. Biomass* 1989, 18: 109–126.

46. Wiselogel A, Tyson S, Johnson D: Biomass feedstock resources and composition. In *Bioethanol Handbook.* Edited by Wyman CE: Taylor and Francis; 1996: 105–118.

47. Sreenath HK, Koegel RG, Moldes AB, Jefferies TW, Straub RJ: Enzymatic saccharification of alfalfa fibre after liquid hot water pretreatment. *Process Biochemistry* 1999, 35: 33–41.

48. de Vrije T, de Haas GG, Tan GB, Keijsers ERP, Claassen PAM: Pretreatment of *Miscanthus* for hydrogen production by Thermotoga elfii. *International Journal of Hydrogen Energy* 2002, 27: 1381–1390.

49. Kurakake M, Kisaka W, Ouchi K, Komaki T: Pretreatment with ammonia water for enzymatic hydrolysis of cornhusk, bagasse, and switchgrass. *Applied Biochemistry and Biotechnology* 2001, 90: 251–259.

50. Iyer PV, Wu Z, Kim SB, and Lee YY: Ammonia recycled percolation process for pretreatment of herbaceous biomass. *Applied Biochemistry and Biotechnology* 1996, 57/58: 121–132.

24

Enzymes as Biocatalysts for Conversion of Lignocellulosic Biomass to Fermentable Sugars

Badal C. Saha

24.1 Introduction

In 2003, about 2.81 billion gallons of ethanol were produced in the U.S., with approximately 95% derived from fermentation of cornstarch. With increased attention to clean air and oxygenates for fuels, opportunities exist for rapid expansion of the fuel ethanol industry. Various lignocellulosic biomass such as agricultural residues, wood, municipal solid wastes, and wastes from pulp and paper industry can serve as low-cost and abundant feedstocks for production of fuel ethanol or value-added chemicals. It is estimated that approximately 50 billion gallons of ethanol could be produced from current biomass wastes with the potential to produce up to 350 billion gallons from dedicated energy farms in the U.S.[1] At present, the degradation of lignocellulosic biomass to fermentable sugars represents significant technical and economic challenges, and its success depends largely on the development of highly efficient and cost-effective enzymes for conversion of pretreated lignocellulosic substrates to fermentable sugars. This chapter reviews the current knowledge on the use of enzymes as biocatalysts for the production of fermentable sugars from lignocellulosic biomass.

24.2 Composition of Lignocellulosic Biomass

Lignocellulosic biomass includes various agricultural residues (straws, hulls, stems, stalks), deciduous and coniferous woods, municipal solid wastes (MSW, paper, cardboard, yard trash, wood products), waste from pulp and paper industry, and herbaceous energy crops (switchgrass, Bermudagrass). The compositions of these materials vary. The major component is cellulose (35–50%), followed by hemicellulose (20–35%), and lignin (10–25%). Proteins, oils, and ash make up the remaining fraction of lignocellulosic biomass.[1] The structures of these materials are complex with recalcitrant and heterogeneous characteristics, and native lignocellulose is resistant to an enzymatic hydrolysis. In the current model of the structure of lignocellulose, cellulose fibers are embedded in a lignin-polysaccharide matrix. Xylan may play a significant role in the structural integrity of cell walls by both covalent and noncovalent associations.[2]

Cellulose is a linear polymer of D-glucose units linked by 1,4-ß-D-glucosidic bonds. Hemicelluloses are heterogeneous polymers of pentoses (xylose, arabinose), hexoses (mannose, glucose, galactose), and sugar acids. Unlike cellulose, hemicelluloses are not chemically homogeneous. Hardwood hemicelluloses contain mostly xylans, whereas softwood hemicelluloses contain mostly glucomannans.[3] Xylans of many plant materials are heteropolysaccharides with homopolymeric backbone chains of 1,4-linked β-D-xylopyranose units. Besides xylose, xylans may contain arabinose, glucuronic acid or its 4-O-methyl ether, and acetic, ferulic, and p-coumaric acids. The frequency and composition of branches are dependent on the source of xylan.[4] The backbone consists of O-acetyl, α-L-arabinofuranosyl, α-1,2-linked glucuronic, or 4-O-methylglucuronic acid substituents. However, unsubstituted linear xylans have also been isolated from guar seed husk, esparto grass, and tobacco stalks.[5] Xylans can thus be categorized as linear homoxylan, arabinoxylan, glucuronoxylan, and glucuronoarabinoxylan.

Xylans from different sources, such as grasses, cereals, softwood, and hardwood, differ in composition. Birchwood (Roth) xylan contains 89.3 % xylose, 1% arabinose, 1.4% glucose, and 8.3% anhydrouronic acid.[6] Rice bran neutral xylan contains 46% xylose, 44.9% arabinose, 6.1% galactose, 1.9% glucose, and 1.1% anhydrouronic acid.[7] Wheat arabinoxylan contains 65.8% xylose, 33.5% arabinose, 0.1% mannose, 0.1% galactose, and 0.3% glucose.[8] Corn fiber xylan is one of the complex heteroxylans containing b-(1,4)-linked xylose residues.[9] It contains 48–54% xylose, 33–35% arabinose, 5–11% galactose, and 3–6% glucuronic acid.[10] About 80% of the xylan backbone is highly substituted with monomeric side chains of arabinose or glucuronic acid linked to O-2 and/or O-3 of xylose residues and also by oligomeric side chains containing arabinose, xylose, and sometimes galactose residues.[11] The heteroxylans, which are highly cross-linked by diferulic bridges, constitute a network in which the cellulose microfibrils may be imbedded.[12] Structural wall proteins might be cross-linked together by isodityrosine bridges and with feruloylated heteroxylans, thus forming an insoluble network.[13] Ferulic acid is covalently cross-linked to polysaccharides by ester bonds and to components of lignin mainly by ether bonds.[14] In softwood heteroxylans, arabinofuranosyl residues are esterified with p-coumaric acids and ferulic acids.[15] In hardwood xylans, 60–70% of the xylose residues are acetylated.[16] The degree of polymerization of hardwood xylans (150-200) is higher than that of softwoods (70–130).

24.3   Pretreatment of Lignocellulosic Biomass

The pretreatment of any lignocellulosic biomass is crucial before enzymatic hydrolysis. The objective of pretreatment is to decrease the crystallinity of cellulose, which enhances the hydrolysis of cellulose by cellulases.[17] Various pretreatment options are available to fractionate, solubilize, hydrolyze, and separate cellulose, hemicellulose, and lignin components.[1,18–20] These include concentrated acid,[21] dilute acid,[22] SO_2,[23] alkali,[24,25] hydrogen peroxide,[26] wet-oxidation,[27] steam explosion (autohydrolysis),[28] ammonia fiber explosion (AFEX),[29] CO_2 explosion,[30] liquid hot water,[31] and organic solvent treatments.[32] In each option, the biomass is reduced in size and its physical structure is opened.

The effectiveness of dilute acids to catalyze the hydrolysis of hemicellulose to its sugar components is well known. Two categories of dilute acid pretreatment are used: high temperature (>160°C) continuous-flow for low solids loading (5–10%, w/w) and low temperature (<160°C) batch process for high solids loading (10–40%, w/w).[33] Dilute acid pretreatment at high temperature usually hydrolyzes hemicellulose to its sugars (xylose, arabinose, and other sugars) that are water soluble.[18] The residue contains cellulose and often much of the lignin. The lignin can be extracted with solvents such as ethanol, butanol, or formic acid. Alternatively, hydrolysis of cellulose with lignin present produces water-soluble sugars and the insoluble residues that are lignin plus unreacted materials. Torget et al.[34] achieved both high xylan recovery and high simultaneous saccahrification and fermentation (SSF) conversion while applying extremely dilute H_2SO_4 (0.07 wt%) in a countercurrent flow through configuration. A major problem associated with the dilute acid hydrolysis of lignocellulosic biomass is the poor fermentability of the hydrolyzates. A drawback of the concentrated acid process is the costly recovery of the acid.

Steam explosion provides effective fractionation of lignocellulosic components at relatively low costs.[35] Optimal solubilization and degradation of hemicellulose are generally achieved by either high temperature and short residence time (270°C, 1 min) or lower temperature and longer residence time (190°C, 10 min) steam explosion.[36] The use of SO_2 as a catalyst during steam pretreatment results in the enzymatic accessibility of cellulose and enhanced recovery of the hemicellulose derived sugars.[37] Steam pretreatment at 200–210°C with the addition of 1% SO_2 (w/w) was superior to other forms of pretreatment of willow.[38] A glucose yield of 95%, based on the glycan available in the raw material, was achieved. Steam explosion can induce hemicellulose degradation to furfural and its derivatives and modification of the lignin-related chemicals under high severity treatment (>200°C, 3–5 min, 2–3% SO_2).[39] Boussaid et al.[40] recovered about 87% of the original hemicellulose component in the water-soluble stream by steam explosion of Douglas fir softwood under low severity conditions (175°C, 7.5 min, 4.5% SO_2). More than 80% of the recovered hemicellulose was in monomeric form. Enzymatic digestibility of the steam-exploded Douglas-fir wood chips (105°C, 4.5 min, 4.5% SO_2) was significantly improved using an optimized alkaline peroxide treatment (1% H_2O_2, pH 11.5 and 80°C, 45 min).[41] About 90% of the lignin in the original wood was solubilized by this procedure, leaving a cellulose-rich residue that was completely hydrolyzed within 48 h, using an enzyme (cellulase) loading of 10FPU/g cellulose. Saccharification of 100 g sugarcane bagasse with enzymes after steam explosion with 1% H_2SO_4 at 220°C for 30 sec at water to solid ratio of 2:1 yielded 65.1g sugar.[42]

A pretreatment method involves steeping of the lignocellulosic biomass (using corncob as a model feedstock) in dilute NH_4OH at ambient temperature to remove lignin, acetate, and extractives.[43] This is followed by dilute acid treatment that readily hydrolyzes the hemicellulose fraction to simple sugars, primarily xylose. The residual cellulose fraction of biomass can then be enzymatically hydrolyzed to glucose. Sugarcane bagasse, corn husk, and switchgrass were pretreated with ammonia water to enhance enzymatic hydrolysis.[44] Garrote et al.[45] treated *Eucalyptus* wood substrates with water under selected operational conditions (autohydrolysis reaction) to obtain a liquid phase containing hemicellulose decomposition products (mainly acetylated xylooligosaccharides, xylose, and acetic acid). In a further acid catalyzed step (posthydrolysis reaction), xylooligosaccharides were converted into xylose. Wet oxidation method can be used for fractionation of lignocellulosics into solubilized hemicellulose fraction and a solid cellulose fraction susceptible to enzymatic saccharification. Bjerre et al.[46] found that a combination of alkali and wet oxidation did not generate furfural and 5-hydroxymethyl furfural (HMF). Klinke et al.[47] characterized the degradation products from alkaline wet oxidation (water, sodium carbonate, oxygen, high temperature, and pressure) of wheat straw. Apart from CO_2 and water, carboxylic acids were the main degradation products from hemicellulose and lignin. Aromatic aldehyde formation was minimized by the addition of alkali and temperature control. Oxygen delignification of kraft pulp removed up to 67% of the lignin from softwood pulp and improved the rate and yield from enzymatic hydrolysis by up to 111% and 174%, respectively.[48] Palm and Zacchi[49] extracted 12.5 g of hemicellulose oligosaccharides from 100 g of dry spruce using a microwave oven at 200°C for 5 min.

Supercritical CO_2 explosion was found to be effective for pretreatment of cellulosic materials before enzymatic hydrolysis.[50,51] Zheng et al.[52] compared CO_2 explosion with steam and ammonia explosion for pretreatment of sugarcane bagasse and found that CO_2 explosion was more cost-effective than ammonia explosion and did not cause the formation of inhibitory compounds that could occur in steam explosion.

24.4 Detoxification of Lignocellulosic Hydrolyzates

Phenolic compounds from lignin degradation, furan derivatives (furfural and HMF) from sugar degradation, and aliphalic acids (acetic acid, formic acid, and levulinic acid) are considered to be fermentation inhibitors generated from pretreated lignocellulosic biomass.[53] The formation of these inhibitors depends on the process conditions and the lignocellulosic feedstocks.[54] Various methods for detoxification of the hydrolyzates have been developed.[55] These include treatment with ion-exchange resins, charcoal or ligninolytic enzyme laccase, pre-fermentation with the filamentous fungus *Trichoderma reesei*, removal of nonvolatile compounds, extraction with ether or ethyl acetate, and treatment with alkali (lime) or sulfite. Treatment with alkali (overliming) has been widely used for detoxification of lignocellulosic hydrolyzates prior to alcohol fermentation. However, overliming is a costly method that also produces low-value byproducts such as gypsum.[56] Softwood hydrolyzate, when overlimed with wood ash, improved its fermentability to ethanol, which is due to the reduction of the inhibitors such as furan and phenolic compounds and to nutrient effects of some inorganic components from the wood ash on the fermentation.[57] Persson et al.[58] employed countercurrent flow supercritical fluid extraction to detoxify a dilute acid hydrolyzate of spruce prior to ethanol fermentation with baker's yeast. Weil et al.[59] developed a method for the removal of furfural from biomass hydrolyzate by using a polymeric adsorbent, XAD-4, and desorption of the furfural to regenerate the adsorbent using ethanol. Bjorklund et al.[60] explored the possibility of using lignin residue left after acid hydrolysis of lignocellulosic material for detoxification of spruce dilute acid hydrolyzates prior to fermentation with *Saccharomyces cerevisiae*. Treatment with the lignin residue removed up to 53% of the phenolic compounds and up to 68% of the furan aldehydes in a spruce dilute acid hydrolyzate. Up to 84% of the lignin-derived compounds can be extracted with organic solvents (ethyl acetate and diethyl ether) from *Eucalyptus* wood acid hydrolyzate.[61] The phenolic compounds extracted by solvents showed antioxidant activity.

Each pretreatment method offers distinct advantages and disadvantages. The pretreatment of lignocellulosic biomass is an expensive procedure with respect to cost and energy.

24.5   Enzymes for Cellulose Biodegradation

Effective hydrolysis of cellulose to glucose requires the cooperative action of three enzymes: endo-1, 4-ß-glucanase (EC 3.2.1.4), exo-1, 4-ß-glucanase (EC 3.2.1.91), and ß-glucosidase (EC 3.2.1.21). Cellulolytic enzymes with ß-glucosidase act sequentially and cooperatively to degrade crystalline cellulose to glucose. Endoglucanase acts in a random fashion on the regions of low crystallinity of the cellulosic fiber, whereas exoglucanase removes cellobiose (ß-1,4 glucose dimer) units from the nonreducing ends of cellulose chains. Synergism between these two enzymes is attributed to the endo-exo form of cooperativity, and has been studied extensively between cellulases in *T. reesei* in the degradation of cellulose.[62] Besides synergism, the adsorption of the cellulases on the insoluble substrates is a necessary step prior to hydrolysis. Cellobiohydrolase appears to be the key enzyme for the degradation of native cellulose.[63] The catalytic site of the enzyme is covered by long loops, resulting in tunnel morphology.[64] The loops can undergo large movements, leading to the opening or closing of the tunnel roof.[65] An endo type attack of the polymeric substrates becomes possible when the roof is open, and once entrapped inside the catalytic tunnel, a cellulose chain is threaded through the tunnel and sequentially hydrolyzes one cellobiosyl unit at a time. Kleywegt et al.[66] revealed the presence of shorter loops that create a groove rather than a tunnel in the structure of the enzyme EGI from *T. reesei*. In most organisms, cellulases are modular enzymes that consist of a catalytic core connected to a cellulose-binding domain (CBD) through a flexible and heavily glycosylated linker region.[67] The CBD is responsible for bringing the catalytic domain in an appropriate position for the breakdown of cellulose. Binding of cellulases and the formation of cellulose-cellulase complexes are considered critical steps in the hydrolysis of insoluble cellulose.[68] ß-Glucosidase hydrolyzes cellobiose and in some cases cellooligosaccharides to glucose. The enzyme is generally responsible for the regulation of the whole cellulolytic process and is a rate-limiting factor during enzymatic hydrolysis of cellulose as both endoglucanase and cellobiohydrolase activities are often inhibited by cellobiose.[69–71] Thus, β-glucosidase not only produces glucose from cellobiose but also reduces cellobiose inhibition, allowing the cellulolytic enzymes to function more efficiently. However, like ß-glucanases, most ß-glucosidases are subject to end-product (glucose) inhibition.[72] Saha and Bothast[73] reported that *Candida peltata* produces a highly glucose tolerant β-glucosidase with a K_i value of 1.4 M (252 mg/ml) for glucose. The kinetics of the enzymatic hydrolysis of cellulose, including adsorption, inactivation, and inhibition of enzymes, have been studied extensively.[74] For a complete hydrolysis of cellulose to glucose, the enzyme system must contain the three enzymes in right proportions. *T. reesei* (initially called *T. viride*) produces at least five endoglucanases (EGI,

EGII, EGIII, EGIV, and EGV), two exoglucanases (CBHI and CBHII), and two β-glucosidases (BGLI and BGLII).[75] An exo-exo synergism between the two cellobiohydrolases was also observed.[76] The fungus produces up to 0.33 g protein per g of utilizable carbohydrate.[77]

Product inhibition, thermal inactivation, substrate inhibition, low product yield, and high cost of cellulase are some barriers to commercial development of the enzymatic hydrolysis of cellulose. Many microorganisms are cellulolytic. However, only two microorganisms (*Trichoderma* and *Aspergillus*) have been studied extensively for cellulase. A newly isolated *Mucor circinelloides* strain produces a complete cellulase enzyme system.[78] The endoglucanase from this strain was found to have a wide pH stability and activity. There is an increasing demand for the development of thermostable, environmentally compatible product, and substrate tolerant cellulases with increased specificity and activity for application in the conversion of cellulose to glucose in the fuel ethanol industry. Thermostable cellulases offer certain advantages, such as higher reaction rate, increased product formation, less microbial contamination, longer shelf life, easier purification, and better yield.

The cellulose hydrolysis step is a significant component of the total production cost of ethanol from wood.[79] Achieving a high glucose yield is necessary (>85% theoretical) at high substrate loading (>10% w/v) over short residence times (<4 days). Simultaneous saccharification (hydrolysis) of cellulose to glucose and fermentation of glucose to ethanol (SSF) improve the kinetics and economics of biomass conversion by reducing accumulation of hydrolysis products that are inhibitory to cellulase and β-glucosidase, reducing the contamination risk because of the presence of ethanol, and reducing the capital equipment requirements.[80] An important drawback of SSF is that the reaction has to operate at a compromised temperature of around 30°C instead of enzyme optimum temperature of 45–50°C. Enzyme recycling, by ultrafiltration of the hydrolyzate, can reduce the net enzyme requirement and thus lower costs.[81] A preliminary estimate of the cost of ethanol production for SSF technology based on wood-to-ethanol process is $1.22/gal of which the wood cost is $0.46/gal.[82] A separate fungal enzyme hydrolysis and a fermentation process for converting lignocellulose to ethanol were also evaluated.[83] The cellulase enzyme was produced by the fungal mutant *Trichoderma* Rut C-30 (the first mutant with greatly increased β-glucosidase activity) in a feed-batch production system that is the single most expensive operation in the process. The conversion of lignocellulosic biomass to fermentable sugars requires the addition of complex enzyme mixtures tailored for the process and parallel reuse, and recycle the enzymes until the cost of enzymes comes down. Enzyme recycling may increase the rates and yields of hydrolysis, reduce the net enzyme requirements, and thus lower costs.[84] As mentioned earlier, the first step in cellulose hydrolysis is considered to be the adsorption of cellulase onto cellulosic substrate. As the cellulose hydrolysis proceeds, the adsorbed enzymes (endo- and exo-glucanase components) are gradually released in the reaction mixture. The β-glucosidase does not adsorb onto the substrate. These enzymes can be recovered and reused by contacting the hydrolyzate with the fresh substrate. However, the amount of enzyme recovered is limited because some enzymes remain attached to the residual substrate, and some enzymes are thermally inactivated during hydrolysis. Poor recovery of cellulase was achieved in the case of substrates containing a high proportion of lignin.[85] Addition of surfactant to enzymatic hydrolysis of lignocellulose increases the conversion of cellulose to soluble sugars. Castanon and Wilke[86] reported a 14% increase in glucose yield and more than twice as much recovered enzyme from newspaper saccharification when Tween 80 was added. Karr and Holtzapple[87] studied the effect of Tween on the enzymatic hydrolysis of lime pretreated corn stover and concluded that Tween improves corn stover hydrolysis through three effects: enzyme stabilizer, lignocellulose disrupter, and enzyme effector. The enhancement is due to reduction of the unproductive enzyme adsorption to the lignin part of the substrate as a result of hydrophobic interaction of surfactant with lignin on the lignocellulose surface, which releases nonspecifically bound enzyme.[88]

Cellolignin is an industrial residue obtained during the production of furfural from wood and corncobs when pretreated by dilute H_2SO_4 at elevated temperature. It was completely converted to glucose by cellulase from *T. viride* and *A. foetidus*.[89] The concentration of glucose in the hydrolyzate reached 4–5.5% with about 80% cellulose conversion. Kinetic analysis of cellolignin hydrolysis, using a mathematical model of the process, has shown that, with product inhibition, nonspecific adsorption of cellulase onto lignin and substrate-induced inactivation seem to have a negative effect on the hydrolysis efficiency. Borchert and Buchholz[90] investigated the enzymatic hydrolysis of different cellulosic materials (straw, potato pulp, sugar beet pulp) with respect to reactor design. The kinetics were studied including enzyme adsorption, inhibition, and inactivation. The results suggest the use of reactors with plug flow characteristics to achieve high substrate and product concentrations and to avoid back-mixing to limit the effect of product inhibition. For efficient use of cellulases, a reactor with semipermeable hollow fiber or an ultrafilter membrane was used, and this allowed cellulases to escape end-product inhibition.[91–94]

24.6   Enzymes for Hemicellulose Biodegradation

Hemicellulases are either glycosyl hydrolases or carbohydrate esterases. The total biodegradation of xylan requires endo-β-1,4-xylanase (EC 3.2.1.8), β-xylosidase (EC 3.2.1.37), and several accessory enzymes, such as α-L-arabinofuranosidase (EC 3.2.1.55), α-glucuronidase (EC 3.2.1.131), acetylxylan esterase (EC 3.1.1.72), ferulic acid esterase (EC 3.1.1.73), and p-coumaric acid esterase, which are necessary for hydrolyzing various substituted xylans.[95] The endo-xylanase attacks the main chains of xylans and β-xylosidase hydrolyzes xylooligosaccharides to xylose. The α-arabinofuranosidase and α-glucuronidase remove the arabinose and 4-O-methyl glucuronic acid substituents, respectively, from the xylan backbone. The esterases hydrolyze the ester linkages between xylose units of the xylan and acetic acid (acetylxylan esterase) or between arabinose side chain residues and phenolic acids, such as ferulic acid (ferulic acid esterase) and p-coumaric acid (p-coumaric acid esterase). It is stated that hindrance of lignocellulose biodegradation is associated with phenolic compounds.[96] The phenolic acids are produced via the phenylpropanoid biosynthetic pathway.[97] They act as a cross-linking agent between lignin and carbohydrates or between carbohydrates. β-Mannanase (EC 3.2.1.78) hydrolyzes mannan-based hemicelluloses and liberate β-1,4-manno-oligomers, which can be further degraded to mannose by β-mannosidase (EC 3.2.1.25).

Many microorganisms, such as *Penicillium capsulatum* and *Talaromyces emersonii*, possess complete xylan degrading enzyme systems.[98] Significant synergistic interactions were observed among endo-xylanase, β-xylosidase, α-arabinofuranosidase, and acetylxylan esterase of the thermophilic actinomycete *Thermomonospora fusca*.[99] Synergistic action between depolymerizing and side-group cleaving enzymes has been verified using acetylated xylan as a substrate.[100] Many xylanases do not cleave glycosidic bonds between xylose units, which are substituted. The side chains must be cleaved before the xylan backbone can be completely hydrolyzed.[101] On the other hand, several accessory enzymes only remove side chains from xylooligosaccharides. These enzymes require a partial hydrolysis of xylan before the side chains can be cleaved.[102] Although the structure of xylan is more complex than cellulose and requires several different enzymes with different specificities for complete hydrolysis, the polysaccharide does not form tightly packed crystalline structures like cellulose and is thus more accessible to enzymatic hydrolysis.[103]

Corn fiber, a byproduct of corn wet milling facility, contains about 20% starch, in addition to 15% cellulose and 35% hemicellulose.[104] The xylan from corn fiber is highly resistant to enzymatic degradation by commercially available hemicellulases.[22] Dilute acid (1% H_2SO_4 v/v, 15% solids) pretreatment at a relatively low temperature (120°C, 1 h) to minimize the formation of inhibitory compounds, followed by enzymatic saccharification of the cellulosic portion, is an excellent workable process for generating fermentable sugars (85–100% yield) from corn fiber.[22] A partial saccharification of corn fiber was achieved using a crude enzyme preparation from *Aureobasidium* sp.[105] Christov et al.[106] showed that crude enzyme preparation from *A. pullulans* was only partially effective in the removal of xylan from dissolving pulp. Two newly isolated fungal cultures (*Fusarium proliferatum* NRRL 26517, *F. verticillioides* NRRL Y-26518) have the capability to utilize corn fiber xylan as growth substrate.[107–110] The crude enzyme preparations from these fungi were able to degrade corn fiber xylan well, but the purified endo-xylanases could not degrade corn fiber xylan. The purified β-xylosidase released xylose from xylobiose and other short-chain xylooligosaccharides. For effective hydrolysis of xylan substrates, a proper mix of endo-xylanase with several accessory enzymes is essential. *A. pullulans* produces a highly thermostable novel extracellular α-L-arabinofuranosidase that has the ability to rapidly hydrolyze arabinan, debranched arabinan, and release arabinose from various arabinoxylans.[111] Arabinose-rich lignocellulosic hydrolyzates can be used for production of the enzyme.[112] Spagnuolo et al.[113] reported that incubation of beet pulp with α-L-arabinofuranosidase and end-arabinase produced a hydrolyzate consisting mainly of arabinose.

TABLE 24.1   Enzymes Involved in Cellulose and Lignocellulose Degradation to Fermentable Sugars

Enzyme	Systematic Name	EC Number	Mode of Action
Cellulose:			
Endo-1,4-β-glucanase	1,4-β-D-Glucan-4-glucanohydrolase	3.2.1.4	Endo-hydrolysis of 1,4-β-D-glucosidic linkages
Exo-1,4-β-glucanase	1,4-β-D-Glucan cellobiohydrolase	3.2.1.91	Hydrolysis of 1,4-β-D-glucosidic linkages releasing cellobiose
β-Glucosidase	β-D-Glucoside glucohydrolase	3.2.1.21	Hydrolyzes cellobiose and short chain cello-oligosaccharides to glucose
Hemicellulose:			
Endo-1,4-β- xylanase	1,4-β-D-Xylan xylanohydrolase	3.2.1.8	Hydrolyzes mainly interior β-1,4-xylose linkages of the xylan backbone
α-L-Arabinofuranosidase	α-L-Arabinofuranoside arabinofurano-hydrolase	3.2.1.55	Hydrolyzes terminal nonreducing α-arabinofuranose from arabinoxylans
α-Glucuronidase	α-Glucuronoside glucanohydrolase	3.2.1.31	Releases glucuronic acid from glucuronoxylans
Acetylxylan esterase	Acetyl-ester acetylhydrolase	3.1.1.6	Hydrolyzes acetylester bonds in acetyl xylans
Ferulic acid esterase	Carboxylic ester hydrolase	3.1.1.1	Hydrolyzes feruloylester bonds in xylans

Ferulic acid esterase breaks the ester linkage between ferulic acid and the attached sugar and release ferulic acid from complex cell walls such as wheat bran, sugar beet pulp, barley spent grain, and oat hull.[114–117] Table 24.1 lists some of the enzymes involved in lignocellulose degradation to fermentable sugars.

24.7   Concluding Remarks

Lignocellulose biodegradation and its conversion to a wide variety of commodity chemicals holds enormous potential. At present, the conversion of lignocellulosic biomass to fermentable sugars is not cost-effective. Some of the emerging pretreatment methods such as alkaline peroxide and AFEX generate solubilized and partially degraded hemicellulosic biomass that need to be treated further with enzymes or other means to produce fermentable sugars from them. With the development of a suitable pretreatment method minimizing the formation of inhibitory compounds for fermentative organisms and use of proper mixture of cellulases and hemicellulases (enzyme cocktail) tailored for each biomass conversion, this vast renewable resource can be utilized for production of fuels and chemicals by fermentation or enzymatic means. Research should be focused on developing efficient and cost-effective pretreatment method, enzymes for use in cellulose and hemicellulose conversion on an industrial scale, robust efficient microorganisms to ferment lignocellulosic hydrolyzates in a cost-competitive way, and methods for cost-effective recovery of fermentation products. Finally, integration of various process steps such as biomass pretreatment, enzymatic saccharification, detoxification, fermentation of the hydrolyzates, and recovery of products will greatly aid in reducing the overall cost of using lignocellulose for practical purposes.

References

1. CE Wyman. Ethanol from lignocellulosic biomass: technology, economics, and opportunities. *Bioresour Technol* 50: 3–16, 1994.
2. JA Thomson. Molecular biology of xylan degradation. *FEMS Microbiol Rev* 104: 65–82, 1993.
3. JD McMillan. Pretreatment of lignocellulosic biomass. In: ME Himmel, JO Baker, RP Overend, Eds. *Enzymatic Conversion of Biomass for Fuel Production.* Am Chem Soc, Washington, DC, 1993, pp. 292–323.

4. GO Aspinall. Chemistry of cell wall polysaccharides. In: J Preiss, ed. *The Biochemistry of Plants (A Comprehensive Treatise)*, Vol. 3. Carbohydrates: Structure and Function. Academic Press, New York, 1980, pp. 473–500.

5. S Eda, A Ohnishi, K Kato. Xylan isolated from the stalk of *Nicatiana tobacum*. *Agric Biol Chem* 40: 359–364, 1976.

6. FJM Kormelink, AGJ Voragen. Degradation of different [(glucurono)arabino] xylans by a combination of purified xylan-degrading enzymes. *Appl Microbiol Biotechnol* 38: 688–695, 1993.

7. N Shibuya, T Iwasaki. Structural features of rice bran hemicellulose. *Phytochem* 24: 285–289, 1985.

8. H Gruppen, RJ Hamer, AGJ Voragen. Water-unextractable cell wall material from wheat flour. 2. Fractionation of alkali-extracted polymers and comparison with water-extractable arabinoxylans. *Cereal Sci* 16: 53–67, 1992.

9. BC Saha α-L-Arabinofuranosidase, biochemistry, molecular biology, and application in biotechnology. *Biotechnol Adv* 2000, pp. 403–423.

10. LW Doner, KB Hicks. Isolation of hemicellulose from corn fiber by alkaline hydrogen peroxide extraction. *Cereal Chem* 74: 176–181, 1997.

11. L Saulnier, C Marot, E Chanliaud, JF Thibault. Cell wall polysaccharide interactions in maize bran. *Carbohydr Polymers* 26: 279–287, 1995.

12. L Saulnier, JF Thibault. Ferulic acid and diferulic acids as components of sugar-beet pectins and maize bran heteroxylans. *J Sci Food Agric* 79: 396–402, 1999.

13. EE Hood, KR Hood, SE Fritz. Hydroxyproline-rich glycoproteins in cell walls of pericarp from maize. *Plant Sci* 79: 13–22, 1991.

14. A Scalbert, B Monties, JY Lallemand, E Guittet, C Rolando. Ether linkages between phenolic acids and lignin fractions from wheat straw. *Phytochem* 24: 1359–1362, 1985.

15. I Mueller-Hartley, RD Hartley, PJ Harris, EH Curzon. Linkage of *p*-coumaroyl and feruloyl groups to cell-wall polysaccharides of barley straw. *Carbohydr Res* 148: 71–85, 1986.

16. TE Timell. Recent progress in the chemistry of wood hemicelluloses. *Wood Sci Technol* 1: 45–70, 1967.

17. B Focher, A Marzetti, M Cattaneo, PI Beltrame, PJ Carniti. Effects of structural features of cotton cellulose on enzymatic hydrolysis. *Appl Polym Sci* 26: 1989–1999, 1981.

18. H Bungay. Product opportunities for biomass refining. *Enzyme Microb Technol* 14: 501–507, 1992.

19. BE Dale, MJ Moreira. A freeze-explosion technique for increasing cellulose hydrolysis. *Biotechnol Bioeng Symp* 12: 31–43, 1982.

20. J Weil, P Westgate, K Kohlmann, MR Ladisch. Cellulose pretreatments of lignocellulosic substrates. *Enzyme Microb Technol* 16: 1002–1004, 1994.

21. IS Goldstein, JM Easter. An improved process for converting cellulose to ethanol. *Tappi J* 75: 135–140, 1992.

22. BC Saha, RJ Bothast. Pretreatment and enzymatic saccharification of corn fiber. *Appl Biochem Biotechnol* 76: 65–77, 1999.

23. DP Clark, KLJ Mackie. Steam explosion of the softwood Pinus radiata with sulphur dioxide addition. 1. Process optimization. *Wood Chem Technol* 7: 373–403, 1987.

24. DP Koullas, PE Christakopoulos, D Kekos, EG Koukios, BJ Macris. Effect of alkali delignification on wheat straw saccharification by *Fusarium oxysporum* cellulases. *Biomass Bioenergy* 4: 9–13, 1993.

25. WE Kaar, MT Holtzaple. Using lime pretreatment to facilitate the enzymatic hydrolysis of corn stover. *Biomass Bioenergy* 18: 189–199, 2002.

26. JM Gould. Alkaline peroxide delignification of agricultural residues to enhance enzymatic saccharification. *Biotechnol Bioeng* 26: 46–52, 1984.

27. AS Schmidt, AB Thomsen. Optimization of wet oxidation pretreatment of wheat straw. *Bioresour Technol* 64: 139–151, 1998.

28. J Fernandez-Bolanos, B Felizon, A Heredia, R Rodriguez, R Guillen, A Jimenez. Steam-explosion of olive stones: hemicellulose solubilization and enhancement of enzymatic hydrolysis of cellulose. *Bioresour Technol* 79: 53–61, 2001.

29. BE Dale, CK Leong, PK Pham, VM Esquivel, L Rios, VM Latimer. Hydrolysis at low enzyme levels: application of the AFEX process. *Bioresour Technol* 56: 111–116, 1996.

30. BE Dale, MJ Moreira. A freeze-explosion technique for increasing cellulose hydrolysis. *Biotechnol Bioeng Symp* 12: 31–43, 1982.

31. M Laser, D Schulman, AG Allen, J Lichwa, MA Antal Jr., LR Lynd. A comparison of liquid hot water and steam pretreatments of sugar cane bagasse for bioconversion to ethanol. *Bioresour Technol* 81: 33–44, 2002.

32. HL Chum, DK Johnsoon, S Black. Organosolv pretreatment for enzymatic hydrolysis of poplars: 1, enzyme hydrolysis of cellulosic residues. *Biotechnol Bioeng* 31: 643–649, 1988.

33. Y Sun, J Cheng. Hydrolysis of lignocellulosic materials for ethanol production: a review. *Bioresour Technol* 83: 1–11, 2002.

34. RW Torget, KL Kadam, T-A Hsu, GP Philippidis, CE Wyman. Prehydrolysis of lignocellulose. US Patent 5,705,369, 1998.

35. QA Nguyen, JN Saddler. An integrated model for the technical and economic evaluation of an enzymatic biomass conversion process. *Bioresour Technol* 35: 275–282, 1991.

36. SJB Duff, WD Murray. Bioconversion of forest products industry waste cellulosics to fuel ethanol: a review. *Bioresour Technol* 55: 1–33, 1996.

37. HH Brownell, JN Saddler. Steam explosion pretreatment for enzymatic hydrolysis. *Biotechnol Bioeng Symp* 14: 55–68, 1984.

38. R Eklund, G Zacchi. Simultaneous saccharification and fermentation of steam-pretreated willow. *Enzyme Microb Technol* 17: 255–259, 1995.

39. S Ando, I Arai, K Kiyoto, S Hanai. Identification of aromatic monomers in steam-exploded poplar and their influence on ethanol production by *Saccharomyces cerevisiae*. *J Ferment Technol* 64: 567–570, 1986.

40. A Bosssaid, J Robinson, Y-J Cal, DJ Gregg, JN Saddler. Fermentability of the hemicellulose-derived sugars from steam-exploded softwood (Douglas-fir). *Biotechnol Bioeng* 64: 284–289, 1999.

41. B Yang, A Boussaid, SD Mansfield, DJ Gregg, JN Saddler. Fast and efficient alkaline peroxide treatment to enhance the enzymatic digestibility of steam-exploded softwood substrates. *Biotechnol Bioeng* 77: 678–684, 2002.

42. PJ Morjanoff, PP Gray. Optimization of steam explosion as method for increasing susceptibility of sugarcane bagasse to enzymatic saccharification. *Biotechnol Bioeng* 29: 733–741, 1987.

43. NJ Cao, MS Krishnan, JX Du, CS Gong, NWY Ho, ZD Chen, GT Tsao. Ethanol production from corn cob pretreated by the ammonia steeping process using genetically engineered yeast. *Biotechnol Lett* 18: 1013–1018, 1996.

44. M Kurakake, W Kisaka, K Ouchi, T Komaki. Pretreatment with ammonia water for enzymatic hydrolysis of corn husk, bagasse, and switchgrass. *Appl Biochem Biotechnol* 90: 251–259, 2001.

45. G Garrote, H Dominguez, JC Parajo. Generation of xylose solutions from *Eucalyptus globulus* wood by autohydrolysis-posthydrolysis processes: posthydrolysis kinetics. *Bioresour Technol* 79: 155–164, 2001.

46. AB Bjerre, AB Olesen, T Fernqvist, A Ploger, AS Schmidt. Pretreatment of wheat straw using combined wet oxidation and alkaline hydrolysis resulting in convertible cellulose and hemicellulose. *Bioresour Technol* 49: 568–577, 1996.

47. HB Klinke, BK Ahring, AS Schmidt, AB Thomson. Characterization of degradation products from alkaline wet oxidation of wheat straw. *Bioresour Technol* 82: 15–26, 2002.

48. K Draude, CB Kurniawan, SJB Duff. Effect of oxygen delignification on the rate and extent of enzymatic hydrolysis of lignocellulosic material. *Bioresour Technol* 79: 113–120, 2001.

49. M Palm, G Zacchi. Extraction of hemicellulosic oligosaccharides from spruce using microwave oven or steam treatment. *Biomacromolecules* 4: 617–623, 2003.

50. KH Kim, J Hong. Supercritical CO_2 pretreatment of lignocellulose enhances enzymatic cellulose hydrolysis. *Bioresour Technol* 77: 139–144, 2001.

51. Y Zheng, HM Lin, J Wen, N Cao, X Yu, GT Tsao. Supercritical carbon dioxide explosion as a pretreatment for cellulose hydrolysis. *Biotechnol Lett* 17: 845–850, 1995.

52. YZ Zheng, HM Lin, GT Tsao. Pretreatment of cellulose hydrolysis by carbon dioxide explosion. *Biotechnol Prog* 14: 890–896, 1998.

53. E Palmqvist, B Hahn-Hagerdal. Fermentation of lignocellulosic hydrolyzates. II. Inhibitors and mechanism of inhibition. *Bioresour Technol* 74: 25–33, 2000.

54. MJ Taherzadeh, R Eklund, L Gustafsson, C Niklasson, G Linden. Characterization and fermentation of dilute bacid hydrolyzates froim wood. *Ind Eng Chem Res* 36: 4659–4665, 1997.

55. E Palmqvist, B Hahn-Hagerdal. Fermentation of lignocellulosic hydrolyzates. I. Inhibition and detoxification. *Bioresour Technol* 74: 117–124, 2000.

56. M Von Silver, G Zucchi, I Olsson, B Hahn-Hagerdal. Cost analysis of ethanol production from willow using recombinant *Escherichia coli*. *Biotechnol Prog* 10: 555–560, 1994.

57. H Miyafuji, H Danner, M Neureiter, C Thomasser, R Braun. Effect of wood ash treatment on improving the fermentability of wood hydrolyzate. *Biotechnol Bioeng* 84: 390–393, 2003.

58. P Persson, S Larsson, LJ Jonsson, NO Nilvebrant, B Sivik, F Munteanu, L Thorneby, L Gorton. Supercritical fluid extraction of a lignocellulosic hydrolyzate of spruce for detoxification and to facilitate analysis of inhibitors. *Biotechnol Bioeng* 79: 694–700, 2002.

59. JR Weil, B Dien, R Bothast, R Hendrickson, NS Mosier, MR Ladisch. Removal of fermentation inhibitors formed during pretreatment of biomass by polymeric adsorbents. *Ind Eng Chem Res* 41: 6132–6138, 2002.

60. L Bjorklund, S Larsson, LJ Jonsson, A Reimann, N-O Nilvebrant. Treatment with lignin residue: A model method for detoxification of lignocellulose hydrolyzates. *Appl Biochem Biotechnol* 98-10: 563–575, 2002.

61. JM Cruz, JM Dominguez, H Dominguez, JC Parajo. Xylitol production from barley bran hydrolyzates by continuous fermentation with *Debaryomyces hansenii*. *Food Chem* 67: 147–153, 1999.

62. B Henrissat, H Driguez, C Viet, M Schulein. Synergism of cellulases from *Trichoderma reesei* in the degradation of cellulose. *Bio/Technol* 3: 722–726, 1985.

63. C Divne, J Stahlberg, L Ruohonen, G Pettersson, JKC Knowles, TT Teeri, TA Jones. The three-dimensional crystal structure of the catalytic core of cellobiohydrolase I from *Trichoderma reesei*. *Science* 265: 524–528, 1994.

64. C Divne, J Stahlberg, TT Teeri, T Jones. High-resolution crystal structure reveal how a cellulose chain is bound in the 50A long tunnel of cellobiohydrolase I from *Trichoderma reesei*. *J Mol Biol* 275: 309–325, 1998.

65. J-Y Zou, GJ Kleywege, J Stalberrg, H Driguez, W Nerinckx, M Clayssens, A Koivula, T Teeri, TA Jones. Crystalllographic evidence for ring distortion and protein conformational changes during catalysis in Cellobiohydrolase Cel6A from *Trichoderma reesei*. *Structure* 7: 1035–1042.

66. GJ Kleywegt, JY Zou, C Divne, GJ Davies, I Sinning, J Stahlberg, T Reinikainen, M Sridodsuk, TT Teeri, TA Jones. The crystal structure of the catalytic core domain of endoglucanase I from *Trichoderma reesei* at 3.6 angstroms resolution, and a comparison with related enzymes. *J Mol Biol* 272: 383–397, 1997.

67. NR Gilkes, B Henrissat, DC Kilburn, RC Miller Jr., RA Warren. Domains in microbial β-1,4-glycanases: sequence, conservation, function, and enzyme families. *Microbiol Rev* 55: 303–315, 1991.

68. HE Grethlein. The effect of pore size distribution on the rate of enzymatic hydrolysis of cellulosic substrates. *Bio Technol* 3: 155–160, 1985.

69. J Woodward, A Wiseman. Fungal and other β-D-glucosidases—their properties and applications. *Enzyme Microb Technol* 4: 73–79, 1982.

70. MP Coughlan. β-1,4-xylan-degrading enzyme systems: biochemistry, molecular biology, and applications. *Biotechnol Genet Eng Rev* 3: 39–109, 1985.

71. SK Kadam, AL Demain. Softwood forest thinnings as a biomass source for ethanol production: A feasibility study for California. *Biochem Biophys Res Commun* 161: 706–711, 1989.

72. BC Saha, SN Freer, RJ Bothast. In: JN Saddler, MH Penner, Eds. *Enzymatic Degradation of Insoluble Carbohydrates*, Am Chem Soc, Washington, DC, 1995, pp. 197–207.

73. BC Saha, RJ Bothast. Production, purification, and characterization of a highly glucose-tolerant novel β-glucosidase from *Candida peltata. Appl Environ Microbiol* 62: 3165–3170, 1996.

74. MR Ladisch, KW Lin, M Voloch, GT Tsao. Process considerations in the enzymatic hydrolysis of biomass. *Enzyme Microb Technol* 5: 82–102, 1983.

75. CP Kubicek, ME Penttila. In: CP Kubicek, G Harman, K Ondik, eds. *Trichoderma and Gliocladium*, Vol. 2, Taylor & Francis, Ltd, London, UK, 1998, pp. 49–72.

76. LG Fagerstam, G Pettersson. The 1,4-β-glucan cellobiohydrolases of *Trichoderma reesei* QM9414. *FEBS Lett* 119: 97–101, 1980.

77. H Esterbauer, W Steiner, I Laudova, A Hermann, M Hayn. Production of *Trichoderma* cellulase in laboratory and pilot scale. *Bioresour Technol* 36: 51–65, 1991.

78. BC Saha. Production, purification, and properties of endoglucanase from a newly isolated strain of *Mucor circinelloides. Process Biochem* 39: 1871–1876, 2004.

79. QA Nguyen, JN Saddler. An integrated model for the technical and economic evaluation of an enzymatic biomass conversion process. *Bioresour Technol* 35: 275–282, 1991.

80. GP Philippidis, TK Smith, CE Wyman. Study of the enzymatic hydrolysis of cellulose for production of fuel ethanol by the simultaneous saccharification and fermentation process. *Biotechnol Bioeng* 41: 846–853, 1993.

81. LUL Tan, EKC Yu, P Mayers, JN Saddler. Column cellulose hydrolysis reactor: the effect of retention time, temperature, cellulase concentration, and exogeneously added cellobiose on the overall process. *Appl Microbiol Biotechnol* 26: 21–27, 1987.

82. ND Hinman, DJ Schell, CJ Rieley, PW Bergeron, PJ Walter. Preliminary estimate of the cost of ethanol production for SSF technology. *Appl Biochem Biotechnol* 34/35: 639–649, 1992.

83. JD Wright, AJ Power, LJ Douglas. Design and parametric evaluation of an enzymatic hydrolysis process (separate hydrolysis and fermentation). *Biotechnol Bioeng Symp* 17: 285–302, 1986.

84. D Lee, AHC Yu, JN Saddler. Evaluation of cellulase recycling strategies for the hydrolysis of lignocellulosic substrates. *Biotechnol Bioeng* 45: 328–336, 1995.

85. M Tanaka, M Fukui, R Matsuno. Effect of pore size in substrate and diffusion of enzyme on hydrolysis of cellulosic materials with cellulases. *Biotechnol Bioeng* 32: 897–902, 1988.

86. M Castanon, C Wilke. Effects of the surfactant Tween 80 on enzymatic hydrolysis of newspaper. *Biotechnol Bioeng* 23: 1365–1372, 1981.

87. WE Kaar, MT Holtzapple. Benefits from Tween during enzymatic hydrolysis of cellulose by surfactant. *Biotechnol Bioeng* 59: 419–426, 1998.

88. T Eriksson, J Borjesson, F Tjerneld. Mechanism of surfactant effect in enzymatic hydrolysis of lignocellulose. *Enzyme Microb Technol* 31: 353–364, 2002.

89. AV Gusakov, AP Sinitsyn, JA Manenkova, OV Protas. Enzymatic saccharification of industrial and agricultural lignocellulosic wastes. Main features of the process. *Appl Biochem Biotechnol* 34/35: 625–637, 1992.

90. A Borchert, K Buchholz. Enzymatic hydrolysis of cellulosic materials. *Process Biochem* 22: 173–180, 1987.

91. RG Henley, RYK Young, PF Greenfield. Enzymatic saccharification of cellulose in membrane reactors. *Enzyme Microb Technol* 2: 206–208, 1980.

92. HE Klei, DW Sundstrom, KW Coughlin, K Ziolkowski. Hollow-fiber enzyme reactors in cellulose hydrolysis biomass energy process, *Aspergillus phoenicis. Biotechnol Bioeng Symp* 11: 593–601, 1981.

93. I Ohlson, G Trayardh, B Hahn-Hagerdal. Enzymatic hydrolysis of sodium-hydroxide-pretreated sallow in an ultrafiltration membrane reactor. *Biotechnol Bioeng* 26: 647–653, 1984.

94. S Kinoshita, JW Chua, N Kato, T Yoshida, T Taguchi. Hydrolysis of cellulose by cellulases of *Sporotrichum cellulophilumm* in an ultrafilter membrane reactor. *Enzyme Microb Technol* 8: 691–695, 1986.

95. BC Saha, RJ Bothast. Enzymology of xylan degradation. In: SH Imam, RV Greene, BR Zaidi, eds. *Biopolymers: Utilizing Nature's Advanced Materials,* Am Chem Soc, Washington, DC, 1999, pp. 167–194.

96. RD Hartley, CW Ford. Phenolic constituents of plant cell walls and wall biodegradability. In: LG Lewis, MG Paice, eds. *Plant Cell Wall Polymers, Biogenesis, and Biodegradation*, Am Chem Soc, Washington, DC, 1989, pp. 135–145.

97. CB Faulds, G Williamson. Novel biotransformations of agro-industrial cereal waste by ferulic acid esterases. *J Gen Microbiol* 137: 2339–2345, 1991.

98. EXF Filho, MG Touhy, J Puls, MP Coughlan. The xylan-degrading enzyme systems of *Penicillium capsulatum* and *Talaromyces emersonii*. *Biochem Soc Trans* 19: 25S, 1991.

99. SL Bachmann, AJ McCarthy. Purification and cooperative activity of enzymes constituting the xylan-degrading system of *Thermomonospora fusca*. *Appl Environ Microbiol* 57: 2121–2130, 1991.

100. K Poutanen, J Puls. The xylanolytic enzyme system of *Trichoderma reesei*. In: G Lewis, M Paice, Eds. *Biogenesis and Biodegradation of Plant Cell Wall Polymers*. Am Chem Soc, Washington, DC, 1989, pp. 630–640.

101. SF Lee, CW Forsberg. Purification and characterization of an α-L-arabinofuranosidase from *Clostridium acetobutylicum* ATCC 824. *Can J Microbiol* 33: 1011–016, 1987.

102. K Poutanen, M Tenkanen, H Korte, J Puls. Accessory enzymes involved in the hydrolysis of xylans. In: GF Leatham, M Himmel, Eds. *Enzymes in Biomass Conversion*, Am Chem Soc, Washington, DC, 1991, pp. 426–436.

103. HJ Gilbert, GPJ Hazlewood. Bacterial cellulases and xylanases. *Gen Microbiol* 139: 187–194, 1993.

104. BC Saha, BS Dien, RJ Bothast. Fuel ethanol production from corn fiber: Current status and technical prospects. *Appl Biochem Biotechnol* 70-72: 115–125, 1998.

105. TD Leathers, SC Gupta. Saccharification of corn fiber using enzymes from *Aureobasidium* sp. strain NRRL Y-2311-1. *Appl Microbiol Biotechnol* 59: 337–347, 1997.

106. LP Christov, J Myburgh, A van Tonder, BA Prior. Hydrolysis of extracted and fiber-bound xylan with *Aureobasidium pullulans* enzymes. *J Biotechnol* 55: 21–29, 1997.

107. BC Saha. Xylanase from a newly isolated *Fusarium verticillioides* capable of utilizing corn fiber xylan. *Appl Microbiol Biotechnol* 56: 762–766, 2001.

108. BC Saha. Production, purification, and properties of xylanase from a newly isolated *Fusarium proliferatum*. *Process Biochem* 37: 1279–1284, 2002.

109. BC Saha. Purification and characterization of an extracellular β-xylosidase from a newly isolated *Fusarium verticillioides*. *J Ind Microbiol Biotechnol* 27: 241–145, 2001.

110. BC Saha. Purification and properties of an extracellular β-xylosidfase from a newly isolated *Fusarium proliferatum*. *Bioresour Technol* 90: 33–38, 2003.

111. BC Saha, RJ Bothast. Purification and characterization of a novel thermostable α-L-arabinofuranosidase from a color-variant strain of *Aureobasidium pullulans*. *Appl Environ Microbiol* 64: 216–220, 1998.

112. BC Saha, RJ Bothast. Effect of carbon source on production of α-L-arabinofuranosidase by *Aureobasidium pullulans*. *Curr Microbiol* 37: 337–340, 1998.

113. M Spagnuolo, C Crecchio, MDR Pizzigallo, P Ruggiero. Fractionation of sugar beet pulp into pectin, cellulose, and arabinose by arabinases combined with ultrafiltration. *Biotechnol Bioeng* 64: 685–691, 1999.

114. CB Faulds, G Williamson. Release of ferulic acid from wheat bran by a ferulic acid esterase (FAEIII) from *Aspergillus niger*. *Appl Microbiol Biotechnol* 43: 1082–1087, 1995.

115. PA Kroon, CB Faulds, G Williamson. Release of ferulic acid from sugar-beet pulp by using arabinanase, arabinofuranosidase, and an esterase from *Aspergillus niger*. *Biotechnol Appl Biochem* 23: 255–262, 1996.

116. B Bartolome, CB Faulds, GJ Williamson. Enzymic release of ferulic acid from barley spent grains. *Cereal Sci* 25: 285–288, 1997.

117. P Yu, DD Maenz, JJ McKinnon, VJ Racz, DA Christensen. Release of ferulic acid from oat hulls by *Aspergillus* ferulic acid esterase and *Trichoderma xylanase*. *J Agric Food Chem* 50: 1625–1630, 2002.

25

Bioelectrocatalysis: Electroactive Microbial and Enzyme Technologies for Detection and Synthesis of Chemicals, Fuels, and Drugs

J. Gregory Zeikus

25.1 Introduction

By and large, most microbiologists, biochemists, and biotechnologists today don't consider microbes and their oxidoreductases as electronic devices. That is, they sustain function (life activities) by producing and consuming electricity. The goal here is to address this paradigm shift in thinking so as to show that microbes and oxidoreductases do produce or consume electricity, and to demonstrate the impact of bioelectrocatalysis on developing biotechnology.

The past decade of basic bioelectronic structure-function research supports this opinion. The bacterial flagellar motor is driven by electricity.[1] Microbial oxidoreductases are electronic machines that tunnel electrons through biomolecular wires.[6–8]

Figure 25.1 provides a simple bioelectronic model for cell function. Namely, a microbe operates like a battery. Here, the main cathodic reaction is glucose dehydrogenation by glyceraldehyde 3-phospate dehydrogenase, which is coupled to the main anodic reaction cytochrome C3 oxidase, which oxidizes pyridine nucleotides by reducing O_2 to H_2O. Associated with this process, a proton motive force is generated (interior alkaline) that enables electricity generation and synthesis of ATP. Notably, the cell is just like a battery since it cannot flow current unless it can maintain charge, and vice versa. Microbiologists should recognize that when they transfer a culture into fresh medium, they are in essence recharging the cellular battery. Of historical note here, Volta in 1800 discovered the voltatic pile (the first battery) and described it as an

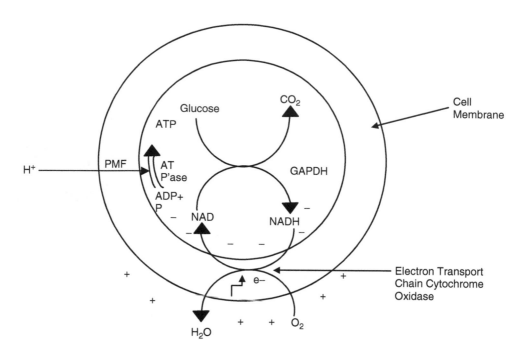

FIGURE 25.1 Simple bioelectronic model for cell life (i.e., current and charge are interlinked as in a battery).

artificial electric organ, which he compared to the electric organ of a torpedo fish.[12] Biologists have ignored this fact for too long.

Bioelectrocatalysts are electroactive (i.e., electrode reacting-oxidizing or -reducing) microbes and enzymes that have been designed to produce or consume electricity in electrode systems engineered to detect or produce chemicals, fuels, and drugs. Most previous studies have used electroactive enzymes and microbial systems to detect chemicals.[2,3,4,5,9,10,11] Recently, interest has also focused on the bioelectrosynthesis of chemicals, fuels, and drugs. By and large, most bioelectrochemical research has been in the realm of electrochemistry and physics. This brief review will cover bioelectrocatalysis from a microbiology, biochemistry, and biotechnology perspective, and it focuses largely on the relation of recent bioelectrosynthesis and bioelectrodetection studies in my lab[13-18] to the published literature primarily after 1993.

25.2 Bioelectrochemical Detection

Bioelectrochemical sensors for detection of chemicals (e.g., blood glucose or lactic acid) have been commercialized. The $1 billion market for these biosensors is rapidly growing.[19]

A bioelectrochemical detector is a type of biosensor that uses cells or enzymes as substrate recognition catalysts with a chemical transducer that couples electron flow between the biocatalyst and electrode. The chemical transducer or electron mediator converts the amount of chemical detected by the enzyme into a current that can be measured amperometrically. Thus, bioelectrochemical detectors are amperometric electrodes containing immobilized enzymes or cells combined with cofactors and electron mediators.

There are several biosensor design approaches for using electron mediators to couple enzymes to electrodes, including: diffusion of electron mediators, relaying electron mediators, and direct electrical wiring of enzymes.[24] A variety of soluble electron mediators have been shown effective for oxidoreductases, such as monomeric ferrocene, quinones, thiazine dyes, and osmium salts.[24] When using diffusionally mediated amperometric biosensors, the rate of electron transfer from the enzyme to the electrode by the mediator has to be greater than the transfer from the substrate to the electrode. The need for diffusional mediators is avoided if insultated redox centers of the enzyme are directly connected to the electrode. In biosensors utilizing relaying electron mediators, the electron mediator and enzyme are in essence immobilized or attached to the electrode in a film. Several examples are: 1) a glucose sensor using glucose oxidase covalently attached to polytheneglycol (on the electrode),[31] or to electropolymerized methylene blue (PMB) on the electrode with enzyme attachment on the PMB layer,[32] or to polyaminoanaline with copolymerized NADP on the electrode;[33] 2) an ethanol sensor using PQQ dependent alcohol dehydrogenase entrapped in electron conductive polymer polypyrrole and connected to the electrode[23] and NAD(P)H sensors using yeast alcohol dehydrogenase immobilized on a nylon mesh and attached to an electrode modified with electropolymerized a [Fe(phen-dione)$_3$]$^{2+}$ complex;[34] or glutathione reductase immobilized into a redox gel of copolymerized vinyl ferrocene with acrylamide and N, N′ methylene bioacrylamide. This immobilized enzyme-gel was connected to the carbon paste electrode by a dialysis membrane.[35]

The direct wiring of enzyme biosensors was pioneered by Heller.[24] To eliminate the membrane used in relay electron mediator–based biosensors, redox centers are uniquely oriented. Here a "wire" or redox macromolecule is designed to (a) complex the enzyme protein, (b) physically attach the enzyme to the electrode surface, and (c) electrically connect the redox center of the enzyme to the electrode. An example of a direct wired glucose sensor[24] employs a complex between glucose oxidase and 10^2KDa[Os(bpy)$_2$Cl]$^{+/2+}$ containing redox polyamine and cross-linking the complex with diepoxide to form a hydrophilic substrate and product permeable (1μM thick) redox epoxy film.

The diversity of chemicals detected by electrochemical enzyme sensors is shown in Table 25.1. By and large, most of the above amperometric enzyme sensors were developed for medical chemical detection, and they use sophisticated techniques often employing expensive gold or glassy carbon electrode materials. Very few studies have been reported on using enzyme amperometric detectors for sensing biotechnology products (e.g., fermentation or biotransformation products) online. Here more robust and less expensive systems are required. Yamazaki et al.[36] developed an amperiometric enzyme sensor for pyruvate and applied it to understanding glycolytic oscillations in yeast fermentation. They used a carbon paste electrode containing 2-methyl-1, 4-napthoquinone on the surface. Pyruvate dehydrogenase was layered onto the surface and covered with a dialysis membrane.

TABLE 25.1 Some Representative Examples of Enzyme-Based Electrochemical Sensors

Chemical Detected (Sensitivity Range)	Enzyme	Electron Mediator	Reference
ethanol(0.33-9.5mM) or NADH (0.5μM-3mM)	yeast alcohol dehydrogenase	tetrarhodoanato diammine chromates (Reineckates)	20
ethanol(1μM-1mM)	gluconobacter alcohol dehydrogenase (PQQ)	polypyrrole	23
lactic acid (0.04 mM^{-1} – 5 mM)	lactate dehyrogenase	pyrroloquinone	21
glycerol (4 × 10^{-7}M – 1 × 10^{-4}M)	glycerol dehydrogenase	polylophenylenediamine	22
glucose (1μM-15mM)	glucose oxidase	poly(vinyl pyridine) [os(bpy)(2)(L)$^{+/2+}$	24
glucose (1μM-7 mM)	Aspergillus glucose oxidase	ferrocene redox polymer	25
NAD (5-200 μM) or NADP (5-200 μM)	Clostridium pyridinine nucleotide oxidoreductase	polymerized pyrrole and viologen	26
malic acid (0.1-3.5 mM)	bovine heart malic dehydrogenase	polytyramine	27
phenylalanine (20-150 μM)	L phenylalanine dehydrogenase	O-quinone	28
aromatic aldehydes (5 μM-0.6 mM)	yeast aldehyde dehydrogenase	polymerized 3-4 dihydroxy-benzaldehyde	29
galactose (5 μM-40 mM)	galactose oxidase	polyvinyl ferrocene	30
succinic acid (5 μM-300 mM)	*Actinobacillus* fumarate reductase	neutral red	14

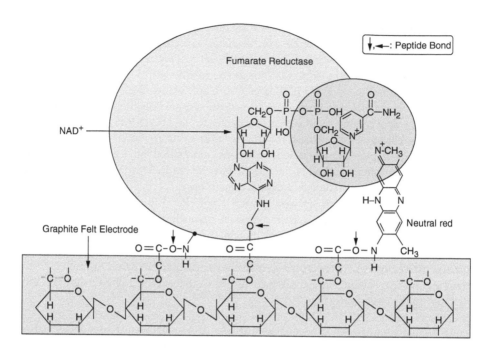

FIGURE 25.2 Diagram of the graphite felt electrode modified with the CMC-NR-NAD$^+$-fumarate reductase complex.

We have been developing an industrial succinate fermentation from glucose plus CO_2, and required an online detector for succinic acid.[75] Our approach[13] to developing a robust fermentation sensor was based on utilization of neutral red as the electron mediator and covalently linking it with the electrode NAD, and fumarate reductase (i.e., a "wired" enzyme electrode system). We had previously shown that neutral red (NR) replaced menaquinone in fumarate reductase and chemically reduced this enzyme or NAD.[14] Figure 25.2 shows a schematic of how we sequentially used immide reagent to link NR to the graphite felt electrode followed by a carboxymethyl cellulose coating and then covalently linking by peptide bonds via immide reagent more NR, NAD, and the fumarate reductase. A membrane-purified preparation of fumarate reductase from *Actinobacillus succinogenes* was employed. Figure 25.3 shows how this bioelectrocatalyst was used as a succinate sensor. The bioelectrocatalyst was put into a flow cell (see Figure 25.4) that continuously received various succinate concentrations. Figure 25.5 demonstrates that the linearity of electrical response to succinate concentration was linear between 5 μM and 400 mM. The enzyme was still stable after 3 weeks of storage at 4°C. Our current studies are aimed at developing this technology into a generic, stable, bioelectrocatalytic system for using various oxidoreductases in detection of ethanol, glutamate, lactate, lysine, citrate, and other fermentation and biotransformation products. We are also developing this direct electrochemical detection system using microbial cells as the biocatalyst. Interestingly, microbial sensors have used bioluminesce[37] or O_2 detection systems[38] but not direct electrical detection.

25.3   Bioelectrosynthesis

25.3.1   Electricity from Fuel Cells

Bioelectrosynthesis of electricity by microbial and enzyme systems has been studied much longer and in more depth than the bioelectrosynthesis of chemicals, fuels, and drugs. Cohen (1931)[37] is generally given credit for demonstrating the first direct biofuel cell and describing a bacterium as a half cell. Concerted research efforts on microbial and enzymatic fuel cells were conducted in the early 1980s in the British labs of Turner and Higgins[40] using methanol and of Allen and Bennetto[41] using glucose as energy sources.

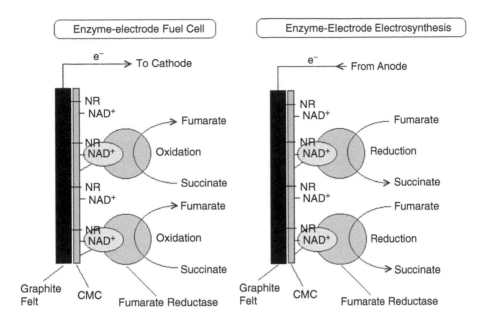

FIGURE 25.3 Mechanisms for succinate oxidation to fumarate coupled to electricity production (left) and for fumarate reduction to succinate coupled to electricity consumption (right) by the CMC-NR-NAD$^+$-fumarate reductase complex immobilized on the graphite felt electrode.

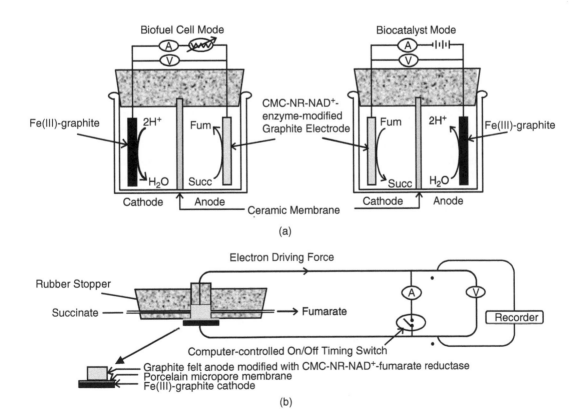

FIGURE 25.4 Diagrams of the two-compartment biocatalyst (A) and the one-compartment biosensor (B) systems.

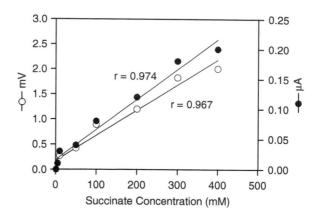

FIGURE 25.5 Amperiometric and potentiometric detection of succinate using the fumarate reductase bioelectro-catalyst system. The lowest detection limit was confirmed to be 5 μM succinate.

There have been several reviews on this topic in the 1980s and 1990s[42,43,44] and a general conceptual review in 2003.[45] The recent interest in microbial fuel cells centers around electricity recovery from biological waste treatment systems or *in situ* electrical generation in sediment (i.e., biobatteries) as a means to replace marine batteries.[55,56] Enzymatic fuel cells are thought to have uses in powering *in vivo* medical devices such as pacemakers.[5,24] The use of redox enzymes in biofuel cells requires engineering electrical contact between the redox enzyme and electrode support. Recently, new approaches have been employed to functionalize the electrode surface with layers consisting of enzymes and electrocatalysts.[46,47] These approaches include anodes based on bioelectrocatalyzed oxidation of NAD(P)H, FAD, cytC, and PQQ functionalized electrodes. Notably, the enzymatic fuel cell has been miniaturized to 7μ diameter, 2-cm long carbon fibers.[48] Previously (before 2000), microbial fuels utilized electron mediators that did not have very low redox potentials (below −100 mV) and did not generate high amount of electricity because the size of the electrochemical gradient was low and the rate of electron transfer between cells and anodes was extremely rate limiting. Recently, "mediator-less" fuel cells using metal-reducing bacteria that generate their own electron mediators, the low redox potential dye, neutral red, and novel fuel cell designs have been used in an attempt to improve the rate of electron transfer and the number of electrons transferred; and to lower the operational costs of microbial fuel cells. Allen and Bennetto[49] improved on their previous fuel cell performance using thionine (+64 mV) and *Proteus* by immobilizing cells on the graphite anode using carbodiimide and by using a stacked fuel cell system that could be coupled in series. Immobilized cells provided a 20% increase in current production from glucose when compared to using free cells.

More recently our lab[15,16] used soluble neutral red as an electron mediator (−325 mV) in microbial fuel cells and showed that this low redox electron mediator enabled a threefold increase in current over that reported by Allen and Bennetto with thionine.[49] Notably, we also showed that growing *E coli* cells produce 12-fold less electricity than stationary cells. In an attempt to make microbial fuel cells using electron mediators more practical, we improved the fuel cell design to a single compartment and incorporated the electron mediators into the graphite electrodes.[16] Figure 25.6 compares the old two- chamber with the new single-chamber fuel cell systems. In the single-chamber system, the cathode is mounted as a window with a porcelain interior septum for proton translocation. Electron transfer is enhanced in the cathode by adding iron to the graphite clay electrode, and, oxygenation is not required since O_2 oxidizes the cathode on the outer surface and H_2O is formed. The single-chamber fuel cell performed as well or better than the two-chamber system using either *E coli* or sewage sludge as biocatalyst. Table 25.2 shows that the highest current and power densities were seen with anodes containing bound electron mediators (i.e., NR or Mn^{4+}) and bound Fe^{3+} in the cathode. Sewage sludge produced more electricity than *E coli* because lactate was completely oxidized to CO_2 by sewage sludge, whereas, *E coli* forms acetate.

FIGURE 25.6 The two- (left) and one-compartment (right) fuel cells used in this study. The one-compartment system does not require a ferricyanide solution or aeration. A proton-permeable layer coats the inside of the window-mounted cathode.

Consequently, both neutral red or Mn^{4+}, which can both chemically oxidoreduce NAD(H), are suitable electron mediators to put into graphite electrons in order to enhance electron transfer between electrodes and microbial cells. More work is required to optimize the exact chemical composition and orientation of these electron mediators in graphite electrodes. There appear to be two routes of electron transfer in or out of microbes when using neutral red. Soluble neutral red (reduced) appears to be pumped-out of *E coli* using the Tol C plasmid mechanism (MacKinlay and Zeikus, unpublished results) as has been reported for the soluble electron mediator-shuttle (i.e., AQDS) produced by the iron-reducing bacterium *Shewanella*.[59] Electrode bound neutral red appears to be oxidized or reduced by a cell bound biochemical mechanism analogous to the iron bacterium *Geobacter*,[50] but not by a surface cytochrome C.[59] The exact electron system that enables electrode bound neutral red to be oxidoreduced is not known but it may involve siderophores, which have been suggested to have a role in electron transfer.[51]

Mediator-less fuel cells have drawn interest recently because they don't require the added expense of the electron mediators..[52] The first mediator-less fuel cells used *Shewanella*, an organism that can reduce either insoluble iron or a graphite electrode.[52] In other words, it uses the electrode as the electron acceptor for energy metabolism. *Geobacter* species also work in mediator-less fuel cells and appear superior to *Shewanella* species because they produce more electricity since *Geobacter* species can completely oxidize glucose or acetate.[53] Most notably the discovery of biobatteries (i.e., microbial fuel cell systems comprised of graphite anodes placed in marine sediments) can generate significant levels of electricity and may replace marine batteries as energy (power) sources in remote locations.[55,56] It also appears, however, that mediator-less fuel cells are greatly limited in the rate of electron transfer between the microbes and electrode. We demonstrated that by incorporating either NR or Mn^{4+} as electron mediators into graphite anodes enhances electricity production by *Shewanella* sixfold. (See Table 25.3.)

TABLE 25.2 Electricity Production in Biofuel Cells from a Lactate Complex Medium Using Resting Cells of Anaerobic Sewage Sludge or *E. Coli* as the Biocatalyst and Four Different Anode-cathode Combinations

Biocatalyst (bacteria)	Anode composition	Cathode composition	Current (mA)	Potential (V)	Electricity production[a]	
					Approximate current density efficiency(mA/m^2 electrode[b])	Approximate power density (mW/m^2 of anode surface)
Sewage sludge	Woven graphite	Woven graphite[c]	0.34	0.6	0.27	0.17
	Woven graphite	Fe^{3+}-graphite	1.30	0.6	1.02	0.65
	NR-woven graphite[d]	Fe^{3+}-graphite	11.0	0.58	1087.5	844.6
	Mn^{4+}-graphite	Fe^{3+}-graphite	14.0	0.45	1,750.0	787.5
E. coli	Woven graphite	Woven graphite[c]	0.6	0.6	0.47	0.30
	Woven graphite	Fe^{3+}-graphite	1.5	0.35	1.18	0.44
	NR-woven graphite[d]	Fe^{3+}-graphite	3.3	0.35	412.5	152.4
	Mn^{4+}-graphite	Fe^{3+}-graphite	2.6	0.28	325.0	91.0

TABLE 25.3 Influence of Anode Compossition and Electron Donors on Electricity Generation by *Shewanella Putrefaciens* in a Single-Compartment Fuel Cell, Using a Fe^{3+} Graphite Cathode

Electron donor	Anode material	Electricity production			
		Initial current (mA)	Final current (mA)	Coulombic value (Axs)	Power density (mW/m² anode surface)
Lactate	Woven graphite	0.02	0.18	2.16	0.02
	NR-woven graphite	0.603	1.062	17.982	9.1
	Mn^{4+} graphite	0.611	1.238	19.969	10.2

Iron-reducing bacteria are not the only bacteria that are electroactive in the absence of exogenous electron mediators. Recently, it has been shown that *Pseudomonas* species that produce pyocyanin can grow anaerobically using the electrode as the final electron acceptor.[57] Their *Pseudomonas* fuel cell system uses novel graphite electrodes, and it has generated the highest electrical current reported to date of 3.6 Wm^{-2} of electrode surface area and converted 89% of the glucose into electricity.[57]

A lot more research is required to identify the optimal microbes for electricity production in fuel cells, including genetic engineering of "electrobacters." Although metal-reducing and *Pseudomonas* species are promising, others including phototrophs[58] and especially sulfate reducers[60–63] have unique capabilities for electron transfer. Furthermore, the exact mechanism(s) of electron transfer between cells and electrodes remain to be full elucidated.

25.3.2 Chemicals, fuels, and drugs

25.3.2.1 Microbial Systems

Bioelectrosynthesis techniques have been used to increase production of a variety of reduced fermentation and biotransformation products. (See Table 25.4.) Hongo and Iwahara first demonstrated the electrical enhancement of reduced fermentation product formation.[64,65] They placed a platinum electrode with 1.5V into a *Brevibacterium* glutamate fermentation containing neutral red and increased its yield by 10%.[64,65] This electroenergized fermentation was not properly controlled since a potentiostat was not used to control the exact electrode potential and H_2 was probably generated. We showed much later that this same technique also worked to increase butanol production by 20% in *Clostridium acetobutylicum* fermentations.[60] Kim and Kim,[67] using the electroenergizing method and methyl violgen, showed that butanol production by *C. acetobutylicum* could be increased 26% while reducing acetone by 25%. Peguin et al.[58] later showed, in a system using a potentiostat to maintain the exact redox of the electrode, that they could increase butanol production by *C. acetolbutylicum* in the presence of electricity and methyl

TABLE 25.4 Representative Examples of Microbial Electrosynthesis of Chemicals, Fuels, and Drugs

Product	Organism	Reference
glutamate	Brevibacterium flavium	64–65
butanol	Clostridium acetobutylicum	66–68
proprionate	Proprionibacterium freudenreichii	69
succinate	Actinobacillus succinogenes	14
ethanol	Clostridium thermocellum Saccharomyces cerevisiae	85
methane	methanogen culture	17
6-bromo-2-tetralo	Trichosporan capitatum	39

TABLE 25.5   Comparison of Glucose Fermentation by A. Succinogenes in the
Absence and Presence of Electrically Reduced NR

Parameter	Metabolite concn (mM)	
	Minus electrical energy	Plus electrical energy
Glucose consumption	47.33	60.44
Cell mass	45.49	57.84
Succinate production	51.27	82.88
Formate production	36.95	9
Acetate production	6.40	3.75
Ethanol production	5.21	21.1
Carbon recovery	0.984	1.001
Electron recovery	0.932	1.036

viologen by 56%.[68] Emde and Schink[69] demonstrated, using a three-electrode system and anthroquinone as electron mediator, that proprionate production by *Proprionibacterium freudenreichii* was enhanced by electrical utilization.

Park and Zeikus demonstrated[14] that electricity dramatically increased the rate (by 30%) and yield (by 60%) of succinate production by *Actinobacillus succinogenes* in electrochemical reactor systems containing neutral red while drastically reducing (by 70%) waste byproducts (acetate and formate) formation. Table 25.5 illustrates the effect of electricity utilization on the succinate fermentation from glucose by *A. succinogenes.* Succinate has a variety of specialty chemical uses and can be an intermediary chemical in the synthesis of several billion pounds of nylon, engineered plastics, and polyesters.[75] Succinate is an important product for biorefineries (i.e., a corn-processing plant producing chemicals and fuels). Figure 25.7 shows that the existing ethanol fermentations waste 40% of the C in glucose as CO_2, which can be recovered when this fermentation is integrated with the succinate fermentation, which requires the use of electricity or other suitable electron donor[75] to produce a homo-succinate yield.

More recently in our lab, we have shown that by using neutral red and electrochemical reactor systems, we could significantly enhance ethanol production yield from cellulose and glucose fermentation by *C. thermocellum* and *C. ceravisae* by 61% and 12%, respectively.[85] We later showed that by using NR and an electrochemical reactor with *Trichosporon capitatum* cell suspensions we could significantly enhance the rate and yield of 6-bromo 2-tetralol production from 6-bromo-2-tetralone by *Trichosporon capitatum* cell suspensions.[39] This drug precursor is used to produce a K^+ channel blocker.

Our lab was able to show that mehanogens can use electricity as the sole electron donor for CO_2 reduction to methane by mixed methanogens.[17] This may be a potential biocatalytic method for lowering atmosphere CO_2 levels. There are very few other reports demonstrating that microbes can grow by using electricity as the sole electron donor in electrochemical reactors. This was first demonstrated by Blake et al.[70] and others[71] using *Thiobacillus ferroxidans* where the organism grew on electrically reduced iron.

$$1C_6H_{12}O_6 \longrightarrow 2CH_3CH_2OH + 2CO_2 \text{ (40\% C loss)}$$

$$1C_6H_{12}O_6 + 2CO_2 + 6XH_2 \longrightarrow 2CH_3CH_2CH_2COO^- + 6H_2O$$

$$2C_6H_{12}O_6 + 6XH_2 \longrightarrow 2CH_3CH_2CH_2COO^- + 6H_2O + 2CH_3CH_2OH$$
$$\text{(Electricity)}$$

FIGURE 25.7   Electrically enhanced chemical and fuel production yields from combined corn bioprocessing into succinate and ethanol.

We also demonstrated that succinate producers could grow in electrochemical reactors using neutral red with electricity as the sole electron donor and fumarate as the electron acceptor.[17]

25.3.2.2 Enzymatic Systems

Perhaps the biggest potential for bioelectrosynthesis is in the use of oxidoreductases in synthesis of chiral chemicals. Alcohol dehydrogenases and other oxidoreductases have many potential applications,[72,73] but their use requires cofactor—NAD(P)H—recycling and hence their commercial value has been limited to date. Nonetheless, major advances in bioelectrode systems have been made with novel electron mediators, to enable electrical recycling of cofactors. These systems have been largely applied to chemical detection in the past, but they are now being implemented for chemical synthesis. A wide variety of novel electron mediators that oxidize or reduce NAD(P)/H have been incorporated into electrode systems including: polyanaline[74] and polyanaline-viologen;[76] benzylviologen and diaphorase;[77] polyazure and other phenothiazenes;[78,79] medota blue;[80–83] polyneutral red[84] and neutral red;[14] methlviologen;[88,89] polyvinylpyridine[86] functionalized with osmium bis (bipyridine chloride); and dichlorophenolindophenol.[87] Laane et al.[87] reported an early study that suggested the importance of using electrochemical bioreactors for synthesis of biochemicals. These authors demonstrated the synthesis of gluconic acid from glucose in an enzyme electrode system using glucose oxidase and DCIP as electron mediator.[87] Yuan et al.[88] reported electrosynthesis of 2-propanol from acetone using an electrode and methyl viologen to recycle NADPH with *Thermoanaerobium brockii* 2° alcohol dehydrogenase. This system was also used to reduce acetophenone to 1-phenylethanol.[88] Sommers et al.[86] have demonstrated enantioselective oxidation of a wide variety of alcohols (i.e., 12 different substrates) using a quinohaemo protein alcohol dehydrogenase in a bioelectrode system utilizing polyvinyl pyridine complexed with osmium bis(bipyridine) as electron mediator.

Recently, Guiral-Brugna et al.[89] used an electrochemical electrode system employing *Desulfovibrio vulgaris* hydrogenase and methyl viologen to synthesize hydrogen from protons. This may provide a novel way to produce H_2, if this fuel is commercially utilized in the future. Our lab has shown that the same fumarate reductase electrode that was developed as a succinate sensor system (Figure 25.3) also works fine for a fumarate synthesis. Figure 25.8 shows the direct conversion of succinate to fumarate by this bioelectrocatalyst.

25.4 Conclusions and Directions for Future Research

The previous literature clearly demonstrates that microbes and their oxidoreductases are electrical devices that produce and consume electricity. Thus, when designed as bioelectrocatalysts they may have diverse application in chemical detection and synthesis. The future for biocatalysis certainly appears to be electrifying. The development of bioelectrocatalysts offers great promise for improving biochemical detection systems and biosynthesis systems for chemical, fuel, and drug synthesis. Nonetheless, the science and applications of bioelectrocatalysts are at their infancy. Much more science and engineering studies are needed to

1. Understand the biochemical mechanisms for electron transfer in and out of microbes and oxidoreductases.
2. Identify improved electron mediators and their "correct" location and orientation in microbial and enzyme-based electrodes.
3. Understand and improve cellular and oxidoreductase electroactivity by genetic and protein engineering techniques.
4. Develop practical (i.e., commercial) biofuel cells and bioelectrode systems.
5. Understand the impact of microbial and oxidoreductase diversity in selection of optimal bioelectrocatalysts.

Notwithstanding these knowledge gaps, it is exciting to speculate on the importance of bioelectrosynthesis in the future of industrial biocatalysis and biotechnology. Several examples are given in Table 25.6. One major limitation of using microbes in biosynthesis (e.g., recombinant protein production) is the

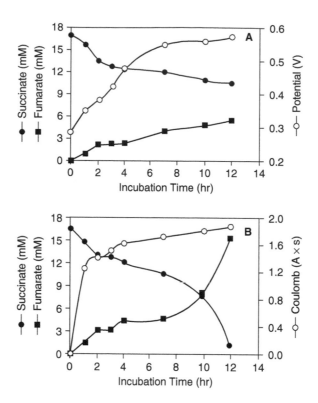

FIGURE 25.8 Succinate oxidation to fumarate by the bioelectrocatalyst coupled to electricity production in an open circuit (A) and a close circuit (B) biofuel cell systems. The anode was graphite felt modified with the CMC-NR-NAD$^+$-fumarate reductase complex, and the cathode was Fe(III)-graphite. The anode surface area was 0.14 m^2 (0.3 g electrode at 0.47 m^2/g).

mass transfer limitation of O$_2$ in fermentors. Now, one needs to evaluate what cell densities can be achieved with *Pseudomonas* and other "aerobic species" that can grow in bioelectrochemical reactors without O$_2$ using the anode as the final electron acceptor. Also, using anode redox values much more positive than oxygen should enhance cell yields.

Electrochemical recycling of NAD(P), FAD, and other enzymatic cofactors offers the potential for using a wide variety of oxidoreductases or cells to synthesize a diversity of biochemicals, including chiral

TABLE 25.6 Potential Impact of Bioelectrosynthesis on Industrial Biocatalysis and Biotechnology

1. High-density "aerobic" fermentations without O^2 (i.e., protein expression in *Pseudomonas* which makes Pyocyanin as electron mediator).
2. Commercialization of oxidoreductases for chiral alcohol synthesis (e.g., electrical recycling of NAD/H in alcohol dehydrogenases).
3. Electrically enhanced rate and yields of reduced biochemical fermentations (i.e., ethanol, succinate, glutamate, etc.).
4. Simplified processes for polyol production (e.g., xylitol and mannitol fermentations replaced by a single immobilized enzyme electrode process).
5. Generation of electricity during waste water treatment (e.g., microbial fuel cells producing less sludge).
6. Enhanced bioremediation of toxic organics (e.g. bioelectrochemical reactor systems for enhancing dechlorination).
7. New microbes growing at extreme redox potential for enhanced biotransformations (e.g., microbes able to use oxidants more positive than +800 mV or reductants more negative than –600 mV).
8. Photobiosynthesis of electricity (e.g., biocatalytic conversion of light to electricity using low redox microbial photo centers for high solar efficiency).

alcohols that are utilized in drug synthesis. Electrochemically enhanced biocatalysis may drastically lower the cost of producing specialty and commodity chemicals from biomass by fermentation and biotransformation processes. This could have a major impact on development of a corn biorefinery where both rate and yield of ethanol and succinate production can be enhanced by electroenergized fermentations. Many fermentation processes are complex and can be simplified to a single bioelectrosynthesis step; for example, the polyols xylitol and mannitol could be produced by bioelectrosynthesis of xylitol from xylose using an electron mediator and xylitol dehydrogenase, and, mannitol from fructose, using mannitol dehydrogenase and an electron mediator to recycle NAD(P).

Microbial waste treatment systems may be improved by bioelectrocatalysis. First, in aerobic waste treatment systems the O_2 transfer rate is limited and requires energy input, and 50% of the carbon and electrons end up as sludge (i.e., cells). One can envisage using a very high redox potential anode and making electricity in lieu of cellular sludge that requires disposal. Likewise, enhancement in biodegradation of toxic wastes may be enhanced by use of bioelectrocatalytic methods using either cathodic or anodic systems for electrical enhancement of biodegradation rates.

The utility of microbial cell bioelectrocatalysts for detection of chemicals needs to be examined. The direct bioelectrochemical sensing method has by and large been applied to analysis of blood chemicals using enzyme electrodes. A microbial electrocatalyst may be more practical for monitoring chemicals in industrial and environmental systems. The unknown diversity of microbes may be understood better now using bioelectrocatalysis systems to enrich for novel species that can use either the cathode as electron donor or anode as electron acceptor for cellular energy metabolism. Microbes that can grow at extreme redox potentials may provide industry with novel microbes, enzymes, and electron transfer systems for biocatalysis.

Finally, perhaps the current low efficiency of solar conversion to electricity can be improved by bioelectrosynthesis using novel microbial phototrophs or their novel light harvesting pigments with extremely low redox potentials to drive a higher electrochemical gradient than is obtained today.

Acknowledgments

This research was supported by the Office of Naval Research Grant Number (N001 14-01-1-0190).

References

1. D Fung, H Berg. Powering the flagellar motor of *Escherichia coli* with an external voltage source. *Nature* 375: 809–812, 1995.
2. M Prodromidis, MC Karayannis. Enzyme based amperiometric biosensors for food analysis. *Electroanal* 14(4): 241–261, 2001.
3. I Willner. Integration of polyaniline/poly(acrylic acid) films and redox enzymes on electrode supports. An *in situ* electrochemical/surface plasmon resonance study in the bioelectro catalyzed oxidation of glucose or lactate in the integrated bioelectrocatalytic systems. *J Am Chem Soc* 124: 6487–6496, 2002.
4. A Shipway, E Katz, I Willner. Nanoparticle arrays on surfaces for electronic, optical and sensor applications. *Chem Phschem* 1: 18–52, 2000.
5. I Willner, E. Katz. Integration of layered redox proteins and conductive supports for bioelectronic application. *Angew Chem Int Ed* 39: 1180–1218.
6. H Gray, J Winkler. Electron transfer in proteins. *Ann Rev Biochem* 65: 537–561, 1996.
7. E T Bowden. Wiring Mother Nature interfacial electrochemistry of proteins. *Interface* 6:40–44, 1997.
8. C Page, L Moser, X Chen, PL Dutton. Natural engineering principals of electron tunneling in biological oxidation-reduction. *Nature* 402: 47–52, 1999.
9. EAH Hall. Biosensors in Context, in *Biosensors*, Hall EAH (ed.), Buckingham, Open University Press, pp. 67, 1990.

10. GG Guilbault, GJ Lubrano. An Enzyme Electrode for the Amperometric Determination of Glucose, *Anal Chim Acta* , 64: 439–455, 1972.
11. A Heller A. Electrical Connection of Enzyme Redox Centers to Electrodes, *J Phys Chem*, 96: 3579–3587, 1988.
12. E Katz. History of electrochemistry. E Katz website. http://www.geocities.com/bioelectrochemistry/faq.htm (accessed August 2003).
13. DH Park, C Vieille, JG Zeikus. Bioelectrocatalysts: engineered oxidoreductase system for utilization of fumarate reductase in chemical synthesis, detection, and fuel cells. *Appl Biochem Biotech* 111: 1–13, 2003.
14. DH Park, JG Zeikus. Utilization of electrically reduced neutral red by *Actinobacillus succinogenes* : physiological function of neutral red in membrane-driven fumarate reduction and energy conservation. *J Bacteriol* 181: 2403–2410, 1999.
15. DH Park, JG Zeikus. Electricity generation in microbial fuel cells using neutral red as an electronophore. *Appl Environ Microbiol* 66: 1292–1297, 2000.
16. DH Park, JG Zeikus. Improved fuel cell and electrode designs for producing electricity from microbial degradation. *Bioeng Biotechnol* 81(3): 348–355, 2003.
17. DH Park, M Laivcnicks, MV Guettler, MK Jain, JG Zeikus. Microbial utilization of electrically reduced neutral red as the sole electron donor for growth and metabolite production. *Appl Environ Microbiol* 65: 2912–2917, 1999.
18. DH Park, JG Zeikus. Impact of electrode composition on electricity generation in a single-compartment fuel cell using *Shewanella putrefaciens*. *Appl Microbiol Biotechnol* 59: 58–61, 2002.
19. M Vreeke. Eletrochemical biosensors for affinity assays. *IVD Technology Magazine* . http://www.devicelink.com/ivdt/archive/97/07/010.html (accessed August 2003).
20. B Grundig, G Wittstock, U Rudel, B Strehlitz. Mediator-modified electrodes for electrocatalytic oxidation of NADH. *J of Electroanal Chem* 395: 143–157, 1995.
21. A Bardea, E Katz, A Buckmann, I Willner. NAD^+-Dependent enzyme electrodes: electrical contact of cofactor-dependent enzymes and electrodes. *J Am Chem Soc* 119: 9114–9119, 1997.
22. M Alvarez-Gonzalez, S Saidman, M Lobo-Castanon. Electrocatalytic detection of NADH and glycerol by NAD^+-modified carbon electrodes. *Anal Chem* 72: 520–527, 2000.
23. J Razumiene, M. Niculescu, A Ramanovicius, V. Laurinavicius, E. Csoregi. Direct bioelectrocatalysis at carbon electrodes modified with quinohemoprotein alcohol dehydrogenase from *Gluconobacter sp. 33*. *Electroanal* 14: 43, 2002.
24. A Heller. Electrical Connection of Enzyme Redox Centers to Electrodes. *J Phys Chem* 96: 3579–3587, 1992.
25. S Koide, K Yokoyama. Electrochemical characterization of an enzyme electrode based on a feerocene-containing redox polymer. *J Electronanal Chem* 468: 193–201, 1999.
26. S Cosnier, K Le Lous. Amperometric detection of pyridine nucleotides via immobilized viologen-accepting pyridine nucleotide oxidoreductase or immobilized diaphorase. *Talanta* 43: 331–337, 1996.
27. M Situmorang, J J Gooding, D B Hibbert, D Barnett. Development of potentiometric biosensors using electrodeposited polytyramine as the enzyme immobilization matrix. *Electroanal* 13: 1469–1474.
28. T Huang, A Warsinke, T Kuwana, F Scheller. Determination of L-phenylalanine based on an NADH-detecting biosensor. *Anal Chem* 70: 991–997, 1998.
29. F Pariente, E Lorenzo, F Tobalina, HD Abruna. Aldehyde biosensor based on the determination of NADH enzymatically generated by aldehyde dehydrogenase. *Anal Chem* 67: 3936–3944, 1995.
30. H. Gulce, I Atamna, A Gulce, A Yildiz. A new amperometric enzyme electrode for galactose determination. *Enzyme and Microbial Technology* 30: 41–44, 2002.
31. B Piro, LA Dang, MC Pham, S Fabiano, C Tran-Minh. A glucose biosensor based on modified-enzyme incorporated within electropolymerised poly(3,4-ethylenedioxythiophene) (PEDT) films. *J Electro Anal Chem* 512: 101–109, 2001.

32. A Silber, N Hampp. Poly(methylene blue)-modified thick-film gold electrodes for the electrocatalytic oxidation of NADH and their application in glucose biosensors. *Biosensors & Bioelectronics* 11: 215–223, 1996.

33. S Suye, Y Aramoto, M Nakamura, I Tabata, M Sakakibara. Electrochemical reduction of immobilized NADP$^+$ on a polymer modified electrode with a co-polymerized mediator. *Enzyme and Microbial Technology* 30: 139–144, 2002.

34. Q Wu, M Maskus, F Pariente, F Tobalina, VM Fernandez, E Lorenzo, HD Abruna. Electrocatalytic oxidation of NADH at glassy carbon electrodes modified with transition metal complexes containing 1,10-phenanthroline-5,6-dione ligands. *Anal Chem* 68: 3688–3696, 1996.

35. H Bu, SR Mikkelsen, AM English. NAD(P)H sensors based on enxyme entrapment in ferrocene-containing polyacrylamide-based redox gels. *Anal Chem* 70: 4320–4325, 1998.

36. S Yamazaki, K Miki, K Kano, T Ikeda. Mechanistic study on the role of the NAD$^+$ –NADH ratio in the glycolytic oscillation with a pyruvate sensor. *J of Electroanal Chem* 516: 59–65, 2001.

37. B Cohen. The bacterial cell as an electrical half cell. *J Bacteriol* 21: 18–19, 1931.

38. SF Peteu, D Emerson, RM Worden. A Clark-type oxidase enzyme-based amperometric microbiosensor for sensing glucose, galactose, or choline. *Biosensors & Bioelectronics* 11: 1059–1071, 1996.

39. JS Shin, MK Jain, M Chartrain, J.G. Zeikus. Evaluation of an electrochemical bioreactor system in the biotransformation of 6-bromo-2 tetralone to 6-bromo-2-tetralol. *Appl Microbiol Biotechnol* 57: 506–510, 2001.

40. APF Turner, WJ Aston, IJ Higgins. Applied aspects of bioelectrochemistry: fuel cells, sensors, and bioorganic synthesis. *Biotechnology & Bioengineering Symp* 12: 401–412, 1982.

41. P Bennetto, JL Stirling, K Tanaka, C Vega. Anodic reactions in microbial fuel cells. *Biotechnology & Bioengineering* 25: 559–568, 1983.

42. WJ Aston, APF Turner. Biosensors and biofuel cells. *Biotechnology and Genetic Engineering Reviews* 1: 89–120, 1984.

43. LB Wingard, Jr, CH Shaw, JF Castner. Bioelectrochemical fuel cells. *Enzyme Microb Technol* 4: 137–142, 1982.

44. GTR Palmore, GM Whitesides. Microbial and enzymatic biofuel cells. In: *Enzymatic Conversion of Biomass for Fuels Production* . Washington, DC: Amcrican Chemical Society, 1994, pp. 271–290.

45. E Katz. Biofuel cells – review. E Katz web site. http://www.geocities.com/bioelectrochemistry/faq.htm (accessed August 2003).

46. E Katz, I Willnar, A Kotlyar. A non-compartmentalized glucose| O$_2$ biofule cell by bioengineered electrode surfaces. *J Electroanal Chem* 479: 64–68, 1999.

47. I Willnar, G Arad, E Katz. A biofuel cell based on pyrroloquinoline quinone and microperoxidase-11 monolayer-functionalized electrodes. *Bioelectrochemistry and Bioenergetics* 44: 209–214, 1998.

48. T Chen, SC Barton, G Binyamin, Z Gao, Y Zhang, H Kim, A Heller. A miniature biofuel cell. *J Am Chem Soc* 123: 8630–8631, 2001.

49. R Allen, HP Bennetto. Microbial fuel-cells. *Applied Biochem and Biotech* 39/40: 27–40, 1993.

50. K Nevin, DR Lovley. Lack of production of electron-shuttling compounds or solubilization of Fe(III) during reduction of insoluble Fe(III) oxide by *Geobacter metallireducens* . *Appl and Environ Microbio* 66: 2248–2251, 2000.

51. ME Hernandez, DK Newman. Extracellular electron transfer. *Cell Mol Life Sci* 58: 1562–1571, 2001.

52. HJ Kim, HS Park, MS Hyun, IS Chang, M Kim, BH Kim. A mediator-less microbial fuel cell using a metal reducing bacterium, *Shewanella putrefaciens. Enzyme and Microbial Technol* 30: 145–152, 2002.

53. DR Bond, DR Lovley. Electricity production by *Geobacter sulfurreducens* attached to electrodes. *Appl and Environ Microbiol* 69: 1548–1555, 2003.

54. TS Magnuson, N Isoyama, AL Hodges-Myerson, G Davidson, MJ Maroney, GG Geesey, DR Lovely. Isolation, characterization and gene sequence analysis of a membrane-associated 89 kDa Fe(III) reducing cytochrome *c* from *Geobacter sulfurreducens* . *Biochem J* 359: 147–152, 2001.

55. DR Bond, DE Holmes, LM Tender, DR Lovley. Electrode-reducing microorganisms that harvest energy from marine sediments. *Science* 295: 483–485, 2002.

56. LM Tender, C Reimers, HA Stecher III, DE Holmes, DR Bond, DA Lowy, K Pilobello, SJ Fertig, DR Lovley. Harnessing microbially generated power on the seafloor. biotech.nature.com 20: 821–825, 2002.

57. K Rabaey, G Lissens, SD Siciliano, W Verstraete. A microbial fuel cell capable of converting glucose to electricity at high rate and efficiency. *Biotechnol Letters* , 2003 (in press.)

58. N Martens, EA Hall. Diaminodurene as a mediator of a photocurrent using intact cells of cyanobacteria. *Photochemistry and Photobiology* 59: 91–98, 1994.

59. J Bruce, H Shyu, DP Lies, DK Newman. Protective role of tolC in efflux of the electron shuttle anthraquinone-2,6-disulfonate. *J Bacteriol* 184: 1806–1810, 2002.

60. OV Karnachuk, SY Kurochkina, OH Tuovinen. Growth of sulfate-reducing bacteria with solid-phase electron acceptors. *Appl Microbiol Biotechnol* 58: 482–486, 2002.

61. S Tsujimura, M Fujita, H Tatsumi, K Kano, T Ikeda. Bioelectrocatalysis-based dihydrogen/dioxygen fuel cell operating at physiological pH. *Phys Chem Chem Phys* 3: 1331–1335, 2001.

62. W Habermann, EH Pommer. Biological fuel cells with sulphide storage capacity. *Appl Microbiol Biotechnol* 35: 128–133, 1991.

63. MJ Cooney, E Roschi, IW Marison, Ch Comninellis, U von Stockar. Physiologic studies with the sulfate-reducing bacterium *Desulfovibrio desulfuricans* : evaluation for use in a biofuel cell. *Enzyme and Microbial Technol* 18: 358–365, 1996.

64. M Hongo, M Iwahara. Application of electro-energizing method to L-glutamic acid fermentation. *Agric Biol Chem* 43: 2075–2081, 1979.

65. M Hongo, M Iwahara. Determination of electro-energizing conditions for L-glutamic acid fermentation. *Agric Biol Chem* 43: 2083–2086, 1979.

66. BK Ghosh, JG Zeikus. Electroenergized butanol fermentation. Am Chem Soc Meeting New Orleans, LA, September 1987.

67. T Kim, BH Kim. Electron flow shift in *Clostridiumacetobutylicum* fermentation by electrochemically introduced reducing equivalent. *Biotechnol Letters* 10: 123–128, 1988.

68. S Penguin, P Delorme, G Goma, P Soucaille. Enhanced alcohol yields in batch cultures of *Clostridium acetobutylicum* using a three-electrode potentiometric system with methyl viologen as electron carrier. *Biotechnol Letters* 16: 269–274, 1994.

69. R Emde, B Schink. Enhanced propionate formation by *Propionibacterium freudenreichii* subsp. *freudenreichii* in a three-electrode amperometric culture system. *Appl and Environ Microbio* 56: 2771–2776, 1990.

70. RC Blake II, GT Howard, S McGinness. Enhanced yields of iron-oxidizing bacteria by *in situ* electrochemical reduction of soluble iron in the growth medium. *Appl and Environ Microbiol* 60: 2704–2710, 1994.

71. N Ohmura, N Matsumoto, K Sasaki, H Saiki. Electrochemical regeneration of Fe(III) to support growth on anaerobic iron respiration. *Appl and Environ Microbiol* 68: 405–407, 2002.

72. R Devaux-Basseguy, A Bergel, M Comtat. Potential applications of NAD(P)-dependent oxidoreductases in synthesis: a survey. *Enzyme and Microbial Technol* 20: 248–258, 1997.

73. W Hummel. Lareg-scale applications of NAD(P)-dependent oxidoreductases: recent developments. *Tibtech* 17: 487–496, 1999.

74. PN Bartlett, ENK Wallace. The oxidation of -nicotinamide adenine dinucleotide (NADH) at poly(aniline)-coated electrodes Part II. Kinetics of reaction at poly(aniline)-poly(styrenesulfonate) composites. *J Electroanal Chem* 486: 23–31, 2000.

75. J G Zeikus, MK Jain, P Elankovan. Biotechnology of succinic acid production and markets for derived industrial products. *Appl Microbiol* 51: 545–552, 1999.

76. S Suye, Y Aramoto, M Nakamura, I Tabata, M Sakakibara. Electrochemical reduction of immobilized NADP+ on a polymer modified electrode with a co-polymerized mediator. *Enzyme and Microbial Technol* 30: 139–144, 2002.

77. G Tayhas R Palmore, H Bertschy, SH Bergens, G Whitesides. A methanol/dioxygen biofuel cell that uses NAD+ dependent dehydrogenases as catalysts: applicaton of an electroenzymatic method to

regenerate nicotinamide adenine dinucleotide at low overpotentials *J Electroanal Chem* 443: 155–161, 1998.

78. CX Cai, KH X. Electrocatalysis of NADH oxidation with electropolymerized films of azure I. *J Electroanal Chem* 427: 147–153, 1997.

79. M Ohtani, S Kuwabata, H Yoneyama. Electrochemical oxidation of reduced nicotinamide coenzymes at Au electrodes modified with phenothiazine derivative monolayers. *J Electroanal Chem* 422: 45–54, 1997.

80. PD Hale, HS Lee, Y Okamoto. Redox polymer-modified electrodes for the electrocatalytic regeneration of NAD⁺. *Analytical Letters* 26: 1073–1085, 1993.

81. FD Munteanu, LT Kubota, L Gorton. Effect of pH on the catalytic electrooxidation of NADH using different two-electron mediators immobilized on zirconium phosphate. *J Electroanal Chem* 509: 2–10, 2001.

82. S Sampath, O Lev. Electrochemical oxidation of NADH on sol–gel derived, surface renewable, non-modified and mediator modified composite-carbon electrodes. *J Electroanal Chem* 446: 57–65, 1998.

83. B Grundig, G Wittstock, U Rudel, B Strehlitz. Mediator-modified electrodes for electrocatalytic oxidation of NADH. *J Electroanal Chem* 395: 143–157, 1995.

84. AA Karyakin, OA Bobrova, EE Karyakina. Electroreduction of NAD⁺ to enzymatically active NADH at poly(neutral red) modified electrodes. *J Electroanal Chem* 399: 179–184, 1995.

85. HS Shin, JG Zeikus, MK Jain. Electrically enhanced ethanol fermentation by *Clostridum thermocellum* and *Saccharomyces cerevisiae*. *Appl Microbiol Biotech* 58: 476–481, 2002.

86. WAC Somers, ECA Stigter, W van Hartingsveldt, JP van der Lugt. Enantioselective oxidation of secondary alcohols at a quinohaemoprotein alcohol dehydrogenase electrode. *Appl Biochem and Biotechnol* 75: 151–162, 1998.

87. C Laane, W Pronk, M Franssen, C Veeger. Use of a bioelectrochemical cell for the synthesis of (bio)chemicals. *Enzyme Microb Technol* 6: 165–168, 1984.

88. R Yuan, S Kuwabata, H Yoneyama. Fabrication of novel electrochemical reduction systems using alcohol dehydrogenase as a bifunctional electrocatalyst. *Chemistry Letters* 2: 137–138, 1996.

89. M Guiral-Brugna, MT Giudici-Orticoni, M Bruschi, P Bianco. Electrocatalysis of the hydrogen production by [Fe] hydrogenase from *Desulfovibrio vulgaris* Hildenborough. *J Electroanal Chem* 510: 136–143, 2001.

26

Biocatalysis: Synthesis of Chiral Intermediates for Pharmaceuticals

Ramesh N. Patel

26.1   Introduction

Chirality is a key factor in the efficacy of many drug products and agrochemicals, and thus the production of single enantiomers of chiral intermediates has become increasingly important in the pharmaceutical industry.[1] Single enantiomers can be produced by chemical or chemo-enzymatic synthesis. The advantage of biocatalysis over chemical synthesis is that enzyme-catalyzed reactions are often highly enantioselective and regioselective. They can be carried out at ambient temperature and atmospheric pressure, thus avoiding the use of more extreme conditions, which could cause problems with isomerization, racemization, epimerization, and rearrangement. Microbial cells and enzymes derived therefrom can be immobilized and reused for many cycles. In addition, enzymes can be overexpressed to make biocatalytic processes economically efficient, and enzymes with modified activity can be tailor-made. The preparation of thermostable and pH stable enzymes by random and site-directed mutagenesis has led to the production of novel biocatalysts. A number of review articles[2–10] have been published on the use of enzymes in organic synthesis. This chapter provides examples of the use of enzymes for the synthesis of single enantiomers of key intermediates for drugs.

26.2   Anticancer Drugs

26.2.1   Paclitaxel

Among the antimitotic agents, paclitaxel (taxol®) **1** (Figure 26.1), a complex, polycyclic diterpene, exhibits a unique mode of action on microtubule proteins responsible for the formation of the spindle during cell division. Paclitaxel is the only compound known to inhibit the depolymerization process of microtubulin. Various types of cancers have been treated with paclitaxel, and the results in treatment of ovarian cancer and metastatic breast cancer are very promising. Paclitaxel was originally isolated from the bark of the yew, *Taxus brevifolia* , and has also been found in other *Taxus* species in relatively low yields. Taxol was initially obtained from *T. brevifolia* bark at about 0.07% yield. It required cumbersome purification of paclitaxel from the other related taxanes. It is estimated that about 20,000 pounds of yew bark (the equivalent of some 3000 trees) are needed to produce 1 Kg of purified paclitaxel.[11,12] The development of a semi-synthetic process for the production of paclitaxel from baccatin III **2** (paclitaxel without the C-13 side-chain) or 10-deacetylbaccatin III **3** (10-DAB, paclitaxel without the C-13 side-chain and the C-10 acetate) and C-13 paclitaxel side-chain **5** or **9** (Figure 26.1) was a very promising approach. Taxanes, Baccatin III, and 10-DAB can be derived from the renewable resources such as extract of needles, shoot, and young *Taxus* cultivars.[11] Thus the preparation of paclitaxel by a semi-synthetic process eliminates the cutting of yew trees.

By using selective enrichment techniques, two strains of *Nocardioides* were isolated from soil samples that contained novel enzymes C-13 taxolase and C-10 deacetylase.[13,14] The extracellular C-13 taxolase derived from filtrate of fermentation broth of *Nocardioides albus* SC 13911 catalyzed the cleavage of the C-13 side chain from paclitaxel and related taxanes such as taxol C, cephalomannine, 7-β-xylosyltaxol, 7-β-xylosyl-10-deacetyltaxol, and 10-deacetyltaxol (Figure 26.2a). The intracellular C-10 deacetylase

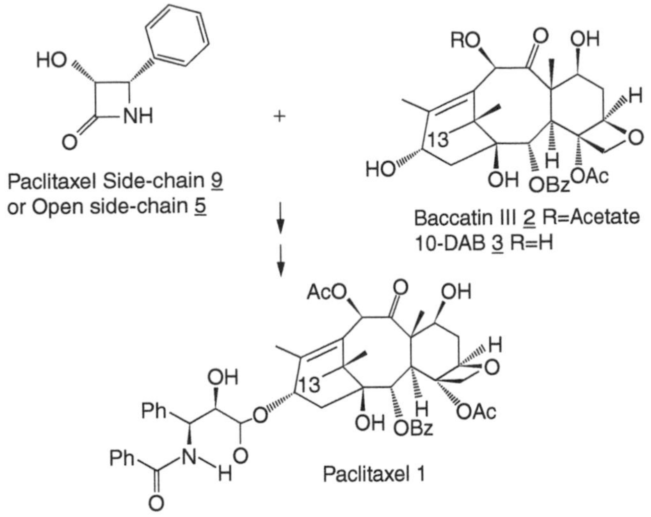

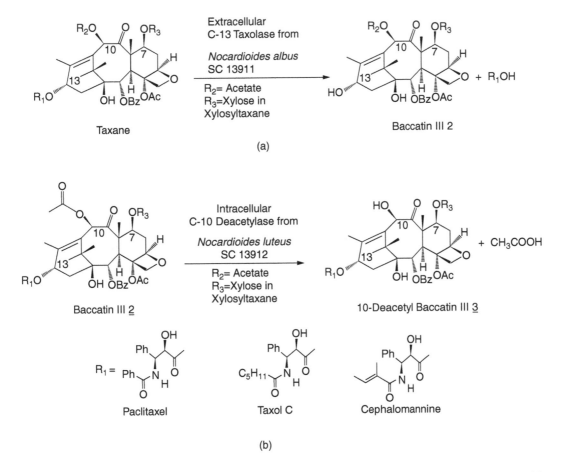

FIGURE 26.1 (A) Semisynthesis of paclitaxel **1**, an anticancer agent. Coupling of baccatin III **2** and C-13 taxol side-chain synthons **5** or **9**.

FIGURE 26.2 Enzymatic hydrolysis of C-13 side chain of taxanes by C-13 taxolase from *Nocardioides albus* SC 13911 (C). Enzymatic hydrolysis of C-10 acetate of taxanes and baccatin III **2** by C-10 deacetylase from *Nocardioides luteus* SC 13912.

derived from fermentation of *Nocardioides luteus* SC 13912 catalyzed the cleavage of C-10 acetate from paclitaxel, related taxanes. and baccatin III to yield 10-DAB (Figure 26.2b). Fermentation processes were developed for growth of *N. albus* SC 13911 and *N. luteus* SC 13912 to produce C-13 taxolase and C-10 deacetylase, respectively, in 5000 L batches, and a bioconversion process was demonstrated for the conversion of paclitaxel, related taxanes in extracts of *Taxus* cultivars to a single compound 10-DAB using both enzymes. In the bioconversion process, ethanolic extracts of the whole young plant of five different cultivars of *Taxus* were first treated with a crude preparation of the C-13 taxolase to give complete conversion of measured taxanes to baccatin III and 10-DAB in 6 hr. *Nocardioides luteus* SC 13192 whole cells were then added to the reaction mixture to give complete conversion of baccatin III to 10-DAB. The concentration of 10-DAB was increased by 5.5- to 24-fold in the extracts treated with the two enzymes. The bioconversion process was also applied to extracts of the bark of *T. bravifolia* to give a 12-fold increase in 10-DAB concentration. The enhancement of 10-DAB concentration in yew extracts was potentially useful in increasing the amount and purification of this key precursor for the paclitaxel semi-synthetic process using renewable resources.

Another key precursor for the paclitaxel semi-synthetic process is the preparation of the chiral C-13 paclitaxel side chain. Two different enantioselective enzymatic processes were developed for the preparation of the chiral C-13 pclitaxel side-chain synthon.[15,16] In one process, the enantioselective microbial reduction of 2-keto-3-(N-benzoylamino)-3-phenyl propionic acid ethyl ester **4** to yield (2R,3S)-N-benzoyl-3-phenyl isoserine ethyl ester **5** (Figure 26.3) was demonstrated.[15] Two strains of *Hansenula* were identified that catalyzed the enantioselective reduction of ketone **4** to the desired product **5** in >80% reaction yield and >94% enantiomeric excess (e.e.) Preparative-scale bioreduction of ketone **4** was demonstrated using cell suspensions of *Hansenula polymorpha* SC 13865 and *Hansenula fabianii* SC 13894 in independent experiments. In both batches, a reaction yield of >80% and e.e.'s of >94% were obtained for **5**. A 20% yield of undesired antidiastereomers content was obtained with *H. polymorpha* SC 13865 compared with a 10% yield with *H. fabianii* SC 13894. A 99% e.e. was obtained with *H. polymorpha* SC 13865 compared with a 94% e.e. with *H. fabianii* SC 13894. In a single-stage fermentation/bioreduction process (15 L scale), cells of *H. fabianii* were grown in a 15 L fermentor for 48 hours; the bioreduction process was then initiated by adding 30 g of substrate and 250 g of glucose and continued for 72 hr. A reaction yield of 88% and an e.e. of 95% were obtained for **5**.

In an alternate enzymatic resolution process for the preparation of C-13 paclitaxel side-chain, the enantioselective enzymatic hydrolysis of racemic acetate cis-3-(acetyloxy)-4-phenyl-2-azetidinone **6**

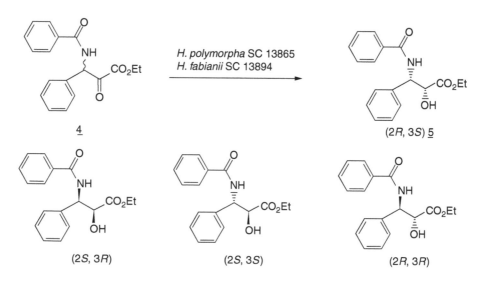

FIGURE 26.3 (A) Enzymatic synthesis of C-13 side chain **5** of paclitaxel **1**: enantioselective reduction of 2-keto-3-(N-benzoylamino)-3-phenyl propionic acid, ethyl ester **4**.

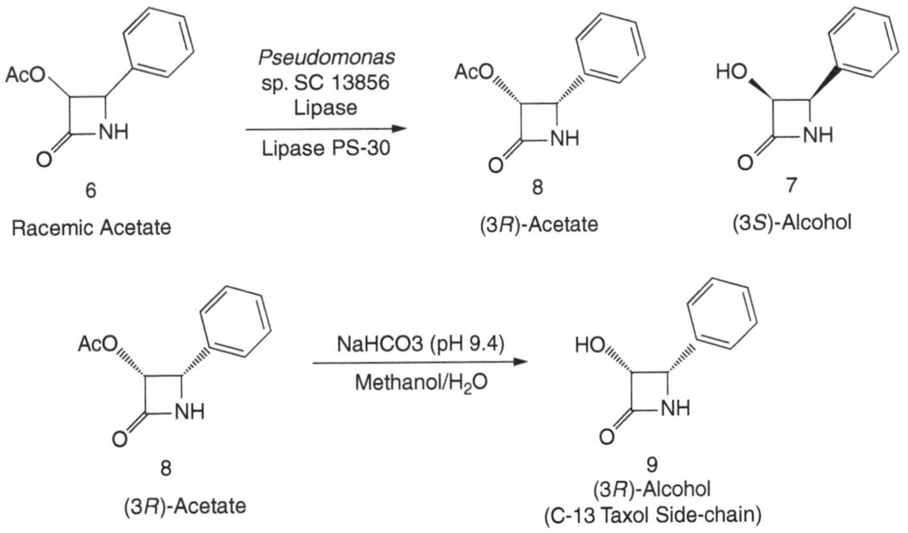

FIGURE 26.4 Enzymatic synthesis of C-13 side chain **9** of paclitaxel **1**: enantioselective hydrolysis of cis-3-acetyloxy-4-phenyl-2-azetidinone **6**.

(Figure 26.4) to the corresponding (S)-alcohol **7** and the unreacted desired (R)-acetate **8** was demonstrated[16] by lipase PS-30 from *Pseudomonas cepacia* (Amano International Enzyme Company) and BMS lipase (extracellular lipase derived from the fermentation of *Pseudomonas* sp. SC 13856). Reaction yields of >48% (theoretical maximum yield 50%) and e.e. of >99.5% were obtained for (R)-acetate. BMS lipase and lipase PS-30 were immobilized on Accurel polypropylene (PP) and immobilized lipases were reused (10 cycles) without loss of enzyme activity, productivity, or the e.e. of the product **8** in the resolution process. The enzymatic process was scaled up to 250 L (2.5 Kg substrate input) using immobilized BMS lipase and lipase PS-30, respectively. From each reaction batch, R-acetate **8** was isolated in 45 M% yield (theoretical maximum yield 50%) and 99.5% e.e. (R)-acetate was chemically converted to (R)-alcohol **9**. The C-13 paclitaxel side-chain synthon (**5** or **9**) produced either by the reductive or resolution process could be coupled to bacattin III **2** after protection and deprotection to prepare paclitaxel by the semi-synthetic process.[11]

26.2.2 Orally Active Taxane

Due to the poor solubility of paclitaxel, various groups are involved in the development of water-soluble taxane analogs.[17–21] Taxane **10** (Figure 26.5) is a water-soluble taxane derivative, which when given orally was as effective as i.v. paclitaxel in five tumor models (murine M109 lung and C3H mammary 16/C cancer, human A2780 ovarian cancer cells [grown in mice and rats], and HCT/pk colon cancer).[18] Compound **10** was also active in a human, hormone-dependent, prostate tumor model CWR-22 and just as effective as antiandrogen chemotherapy.[18]

Various approaches to the synthesis of the C-13 paclitaxel side chain from chiral synthons have been demonstrated, including the lipase catalyzed enantioselective esterification of methyl *trans*-β-phenylgly-cidate,[22] the enantioselective hydrolysis and transesterification of racemic esters and alcohols to prepare enantiomerically pure β-lactams,[23] and the enantioselective microbial reduction of 2-keto-3-N-benzoy-lamino-3-phenyl propionic acid ethyl ester to yield (2R,3S)-(-)-N-benzoyl-3-phenyl isoserine ethyl ester.[16]

An enzymatic process has been developed for the preparation of a key chiral intermediate (3R-cis)-3-acetyloxy-4-(1,1-dimethylethyl)-2-azetidinone **11** for the semi-synthesis of the new orally active taxane **10** (Figure 26.5). The enantioselective enzymatic hydrolysis of cis-3-acetyloxy-4-(1,1-dimeth-ylethyl)-2-azetidinone **12** to the corresponding undesired (S)-alcohol **13** was carried out using immo-bilized lipase PS-30 or BMS lipase to yield unreacted desired (R)-acetate **11**. Reaction yields of >48%

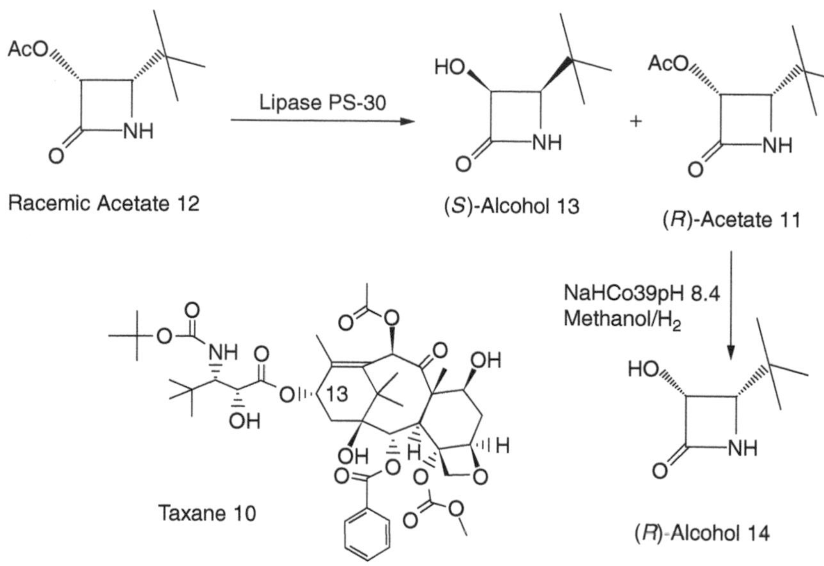

FIGURE 26.5 Enzymatic synthesis of C-13 side chain **11** of an orally active taxane **10**: enantioselective hydrolysis of cis-3-acetyloxy-4-dimethylethyl-2-azetidinone **12**.

and enantiomeric excesses of >99% were obtained for the **11**. Acetoxy β-lactam **11** was converted to hydroxy β-lactam **14** for use in the semisynthesis of oral taxane **10**. This enzymatic process was scale-up to prepare kg of hydroxy β-lactam **11**.[24]

26.2.3 Deoxyspergualin

An antitumor antibiotic spergualin was discovered in the culture filtrate of a bacterial strain and its structure was determined to be 15S-1-amino-10-guanidino-11,15-dihydroxy-4,9,12-triazanonadecane-10,13-dione.[25] The total synthesis was accomplished by the acid-catalyzed condensation of 11-amino-1,1-dihydroxy-3,8-diazaundecane-2-one with (S)-7-guanidino-3-hydroxy-heptanamide followed by the separation of the 11-epimeric mixture. Antibacterial or antitumor activity of the enantiomeric mixture of spergualin was about half of that of the natural spergualin,[26] indicating the importance of the configuration at C-11 for antitumor activity.

The lipase-catalyzed enantioselective acetylation of racemic 7-[N,N′-Bis(benzyloxycarbonyl) N-(guanidinoheptanoyl)]-α-hydroxy-glycine **15** to the corresponding (S)-acetate **16** and unreacted alcohol (R)-**17**.[27] (S)-acetate **16** (Figure 26.6) is a key intermediate for the total chemical synthesis of (-)-15-deoxyspergualin **18**, a related immunosuppressive agent and antitumor antibiotic.[26] The reaction was carried out in methyl ethyl ketone (MEK) using lipase from *Pseudomonas sp.* (lipase AK), and vinyl acetate was used as an acylating agent. A reaction yield of 48% (theoretical maximum yield 50%) and an e.e. of 98% were obtained for (S)-acetate **16**. The unreacted alcohol (R)-**17** was obtained in 41% yield and 94% e.e.

25.2.4 Antileukamic Agent

Compound **19** (Figure 26.7), a prodrug of 9-β-D-arabinofuranosyl guanine **20**, was developed for the potential treatment of leukamia. Compound **19** is poorly soluble in water and its synthesis by conventional technique is difficult. An enzymatic demethoxylation process was developed using adenosine deaminase.[28] Compound **19** was prepared enzymatically from 6-methoxyguanine **21** and ara-uracil **22** using uridine phosphorylase and purine nucleotide phosphorylase. Each protein has been cloned and overexpressed in independent *Escherichia coli* strains. Fermentation conditions were optimized for production of both enzymes, and co-immobilzed enzyme preparation was used in the biotransformation process at 200 g/L

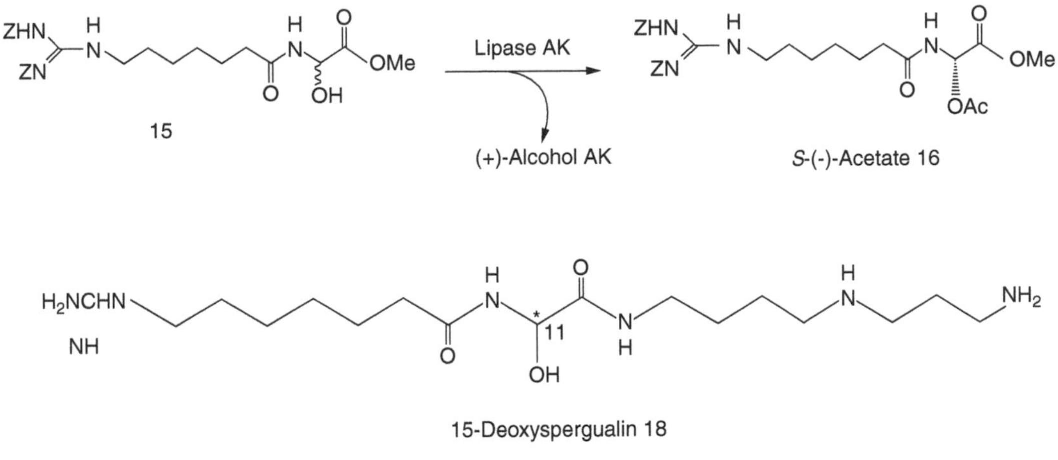

FIGURE 26.6 (A) Synthesis of chiral intermediates for antitumor antibiotic 15-deoxyspergualin **18**: enantioselective enzymatic acylation of racemic **15** to yield (*S*)-acetate **16**.

substrate input. Enzyme was recovered at the end of the reaction by filtration and reused several cycles. More water soluble 5′-acetate ester of compound **19** was subsequently prepared by the enzymatic acylation process using immobilized *Candida antarctica* lipase in 1,4-dioxane (100 g/L substrate) with vinyl acetate acyl donor.[29]

26.2.5 Farnesyl Transferase Inhibitor

An enzymatic process has been developed for the preparation of chiral intermediates (*R*)-**23** (Figure 26.8) for the synthesis of compound **24**, a selective farnesyl transferase inhibitor (involved in binding Ras, which moderates cell proliferation). Resolution of racemic substituted (6,11-dihydro-5H-benzo-[5,6]cyclohepta[1,2-b]pyridin-11-yl) piperidines **25** has been demonstrated to yield desired (*R*)-**23** and unreacted (*S*)-**26** by enzymatic acylation process using a Toyobo lipase LIP-3000 in t-butyl methyl ether with trifluoroethyl isobutyrarte.[30] A reaction yield of 46% and an e.e. of 97% were obtained for N-isobutyryl derivative (*R*)-**23**, which was isolated and subsequent hydrolyzed to yield the desired product.

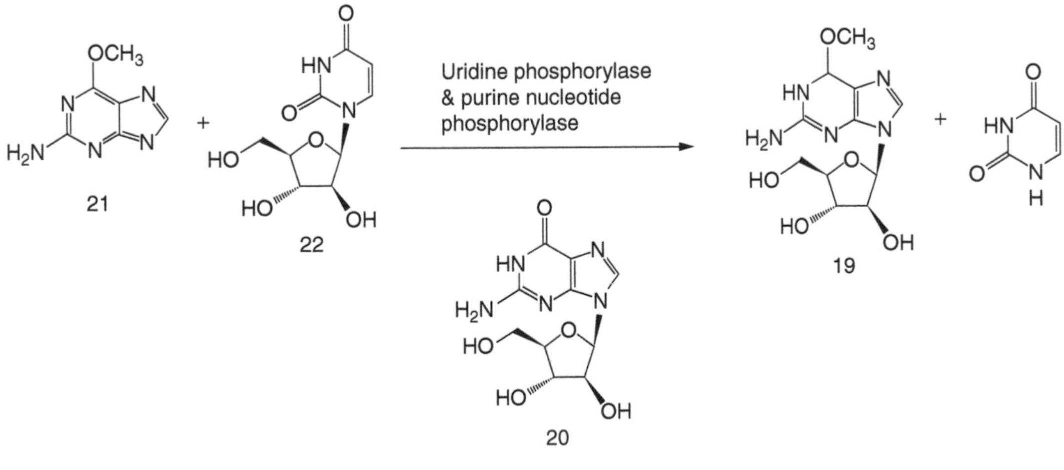

FIGURE 26.7 Enzymatic synthesis antileukemic agent **19**.

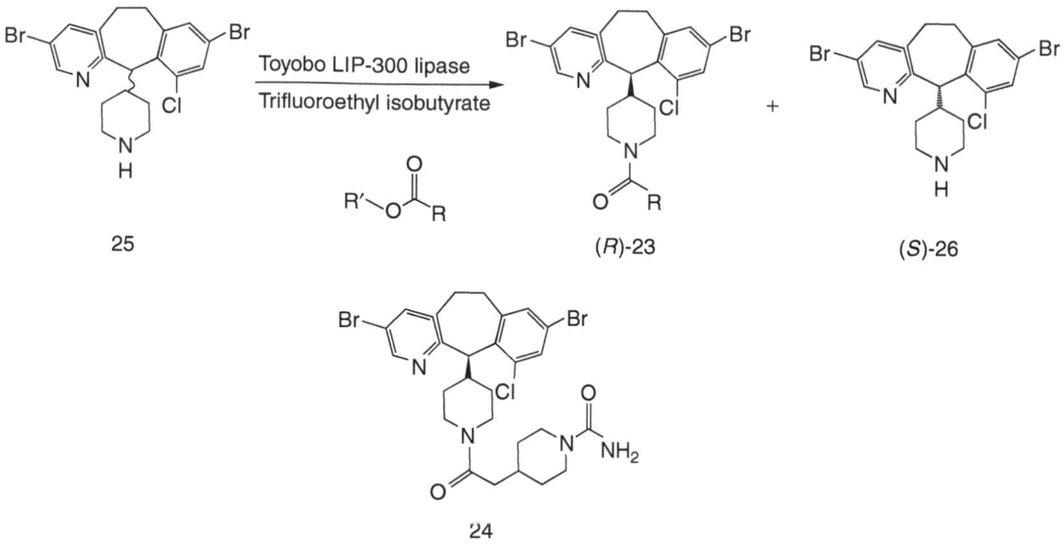

FIGURE 26.8 Synthesis of chiral intermediates for farnesyl transferase inhibitor **24**: enantioselective enzymatic acylation of racemic **25** to yield (*R*)-**23**.

26.3 Antiviral Drugs

26.3.1 BMS-186318 (HIV Protease Inhibitor)

An essential step in the life cycle of the human immunodeficiency virus (HIV-1) is the proteolytic processing of its precursor proteins by HIV-1 protease, a virally encoded enzyme. Inhibition of HIV-1 protease arrests the replication of HIV *in vitro*, and thus, HIV-1 protease is an attractive target for chemotherapeutic intervention. The discovery of a new class of selective HIV protease inhibitors that incorporate a C2 symmetric aminodiol core as its key structural feature have been reported.[31] Members of this class, particularly compound **27**, display potent anti-HIV activity in cell culture. The diastereoselective microbial reduction of (1*S*)-[3-chloro-2-oxo-1-(phenylmethyl)propyl] carbamic acid, 1,1-dimethyl-ethyl ester **28** (Figure 26.9) to (1*S*,2*S*)-[3-chloro-2-hydroxy-1-(phenylmethyl)propyl]carbamic acid, 1,1-dimethylethyl ester **29**, a key intermediate in the total chemical synthesis of compound **27** has been demonstrated.[32] Among 100 microorganisms screened for the reduction, the two best cultures, *Streptomyces nodosus* SC 13149 and *Mortierella ramanniana* SC 13850, were used to convert ketone **28** to the corresponding chiral alcohol **29**. A reaction yield of 67%, an e.e. of 99.9%, and a diastereomeric purity of >99% were obtained for **29** using cells of *Streptomyces nodosus* SC 13149. *Mortierella ramanniana* SC 13850 gave a reaction yield of 54%, an e.e. of 99.9%, and a diastereomeric purity of 92% for **29**. A single-stage fermentation-biotransformation process was developed using cells of *Streptomyces nodosus* SC 13149. A reaction yield of 80%, a diasteromeric purity of >99%, and an e.e. of 99.8% were obtained.

26.3.2 Atzanavir (HIV Protease Inhibitor)

Atazanavir **30** is an acyclic aza-peptidomimetic (Figure 26.10), a potent HIV protease inhibitor.[33–35] An enzymatic process has been developed for the preparation of (1*S*,2*R*)-[3-chloro-2-hydroxy-1-(phenylmethyl)propyl]carbamic acid, 1,1-dimethylethyl ester **31** for the total synthesis of the HIV protease inhibitor Atazanavir. The diastereoselective reduction of (1*S*)-[3-chloro-2-oxo-1-(phenylmethyl)propyl] carbamic acid, 1,1-dimethyl-ethyl ester **28** was carried out using microbial cultures among, which *Rhodococcus, Brevibacterium, and Hansenula* strains reduced **28** to **31** (Figure 26.10). Three strains of *Rhodococcus* gave >90% yield. A diastereomeric purity of >98% and enantiomeric excess of 99.4% were obtained for chiral

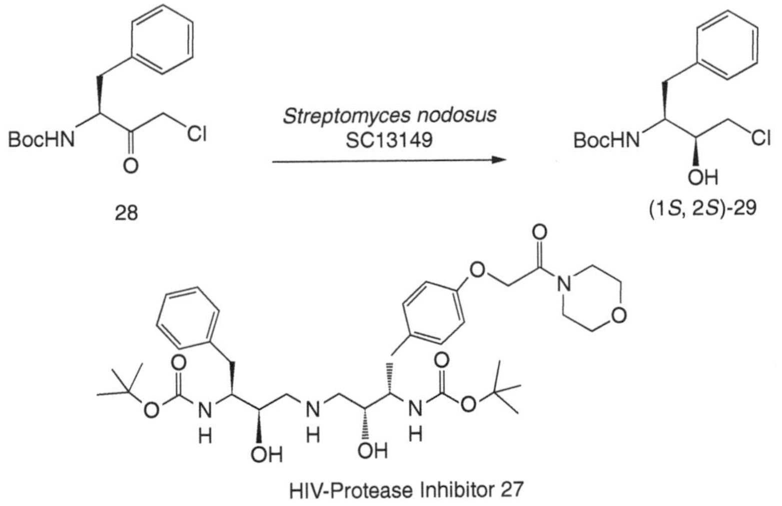

FIGURE 26.9 Synthesis of chiral intermediates for antiviral agent **27**: enantioselective enzymatic reduction of (1S)[3-chloro-2-oxo-1(phenylmethyl)propyl]carbamic acid 1,1-dimethyl-ethyl ester **28** to the corresponding chiral alcohol **29** by *Streptomyces nodosus* SC 13149.

alcohol **31**.[36] Chemical reduction of chloroketone **28** using $NaBH_4$ produces primarily the undesired chlorohydrin.[37]

An efficient single-stage fermentation-biotransformation process was developed for the reduction of ketone **28** with cells of *Rhodococcus erythropolis* SC 13845. A reaction yield of 95%, diasteromeric purity of 98.2%, and e.e. of 99.4% for alcohol **31** were obtained from a 12-L reaction mixture, 9.1 g of alcohol **31** was isolated in 76% overall yield.

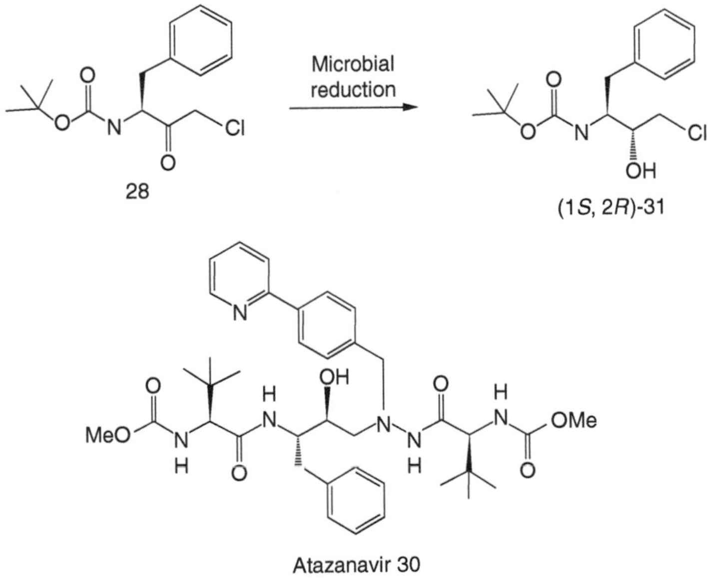

FIGURE 26.10 Synthesis of chiral intermediates for atzanavir **30**: enantioselective enzymatic reduction of (1S)[3-chloro-2-oxo-1(phenylmethyl)propyl]carbamic acid 1,1-dimethyl-ethyl ester **28** to the corresponding chiral alcohol **31** by *Rhodococcus erythropolis* SC 13845.

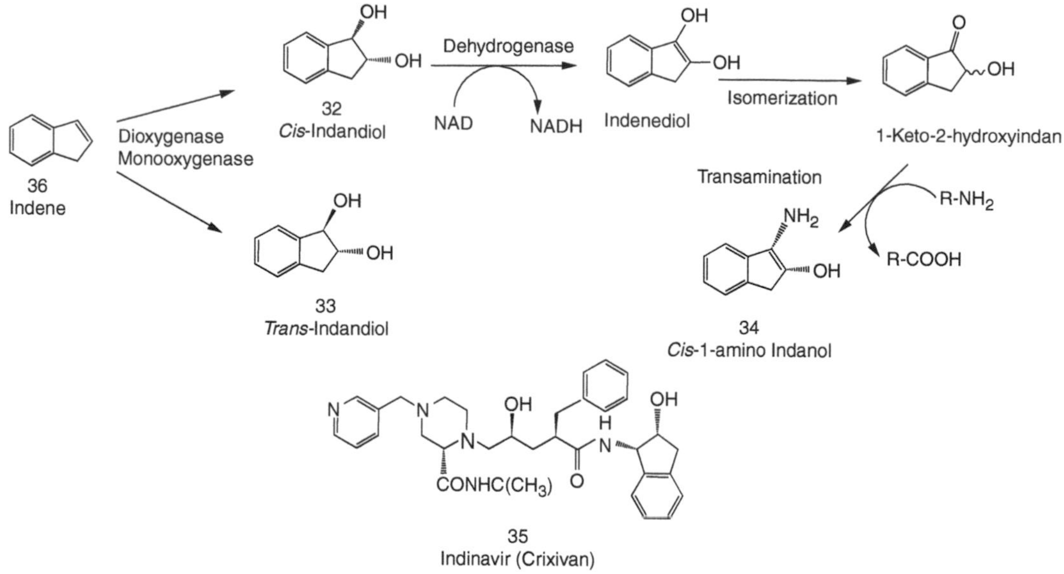

FIGURE 26.11 Synthesis of chiral intermediates for indinavir **35**: microbial oxygenation of indene **36** to *cis*-indandiol **32** and *trans*-indandiol **33**.

26.3.3 Crixivan (HIV Protease Inhibitor)

Cis-(1*S*,2*R*)-indandiol **32** or trans-(1*R*,2*R*)-indandiol **33** are both potential precursors to cis-(1*S*,2*R*)-1-aminoindan-2-ol **34**, a key chiral synthon for Crixivan (Indinavir) **35**, a leading HIV protease inhibitor. Enrichment and isolation of microbial cultures yielded two *Rhodococcus* sp. strain B 264-1 (MB 5655) and strain I-24 (MA 7205) capable of biotransforming indene **36** to cis-(1*S*,2*R*) indandiol and trans-(1*R*,2*R*) indandiol (Figure 26.11), respectively.[38] Isolate MB 5655 was found to have a toluene dioxygenase, while isolate MA 7205 was found to harbor both toluene and naphthalene dioxygenases as well as a naphthalene monooxygenase, which catalyzes above biotransformation. When scaled up in a 14-L fermentor, MB5655 produced up to 2.0 g/L of cis-(1*S*,2*R*) indandiol **32** with an e.e. >99%. *Rhodococcus* sp MA 7205 cultivated under similar conditions produced up to 1.4 g/L of trans-(1*R*,2*R*) indandiol **33** with an e.e. >98%. Process development studies yielded titers >4.0 g/L of **33**.[39] A metabolic engineering approach[40] and a directed evolution technique[41] were evaluated to avoid side reactions and blocking the degradative pathway and to enhance the key reaction to convert indene to cis-amino indanol or cis-indanediol.

26.3.4 Abacavir (Reverse Transcriptase Inhibitor)

Abacavir (Ziazen™) **37** (Figure 26.12a), a 2-aminopurine nucleoside analogue, is a selective reverse transcriptase inhibitor for the treatment of human HIV and hepatitis B virus (HBV). The γ–lactam (2,-azabicyclo[2.2.1]hept-5-en-3-one **38**, is potentially intermediate useful in the synthesis of Abacavir (Figure 26.11b). A biocatalytic process has been developed for the resolution of racemic γ-lactam **39** to yield desired **38** and amino acid **40** using γ–lactamase containing organisms *Pseudomonas solonacearum* NCIMB 40249 and *Rhodococcus* NCIMB 40213.[42] However, because of a lack of commercial availability of lactamase, an enzymatic process was developed for the enantioselective hydrolysis (in phosphate buffer pH 8.0 containing 50% tetrahydrofuran) of racemic [tert butyl-3-oxo-2-azabicyclo (2.2.1)hept-5-ene-2-carboxylate] **41**. A number of commercially available hydrolytic enzymes hydrolyzed the lactam bond of **41** to yield the corresponding N-acyl amino acid **42** and leaving unreacted desired (1*R*,4*S*)-**43** (Figure 26.12b).

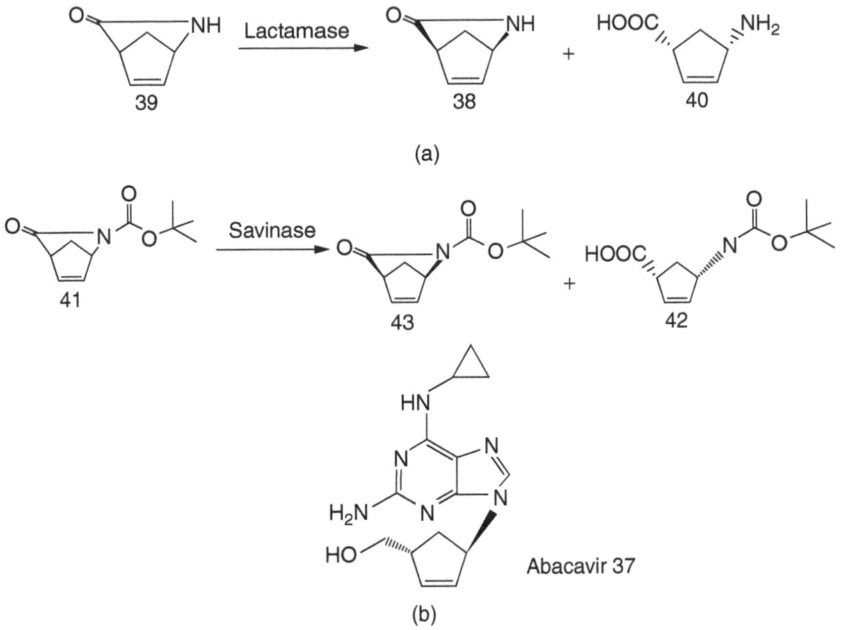

FIGURE 26.12 Synthesis of chiral intermediates for reverse transcriptase inhibitor **37**: enzymatic resolution of racemic γ-lactams **39** and **41**.

A reaction yield of 50% and an e.e. of 99% were obtained when the reaction was carried out at 100 g/L substrate input using savinase.[43]

26.3.5 Lobucavir

Lobucavir **44** (Figure 26.13) is a cyclobutyl guanine nucleoside analog recently under development as an antiviral agent for treatment of herpes virus and hepatitis B.[44] A prodrug in which one of the two hydroxyls is coupled to valine, **45** has also been considered for development. Regioselective aminoacylation is difficult to achieve by chemical procedures; however, it appeared to be suitable for an enzymatic approach.[45] Synthesis of lobucavir prodrug, L-valine,[1*S*,2*R*,3*R*)-3-(2-amino-1,6-dihydro-6-oxo-9H-purin-9-yl)-2-(hydroxymethyl) cyclobutyl]methyl ester monohydrochloride **45**, requires regioselective coupling of one of the two hydroxyl groups of lobucavir **44** with valine. Enzymatic processes were developed for aminoacylation of either hydroxyl group of lobucavir.[45] The selective hydrolysis of N,N′-bis[(phenylmethoxy)carbonyl]bis[L-valineO,O′-[(1*S*,2*R*,3*R*)-3-(2-amino-6-oxo-1H-purin-9-yl)cyclobuta-1,2-diyl]methyl ester **46** with lipase M (Figure 26.12) to yield N-[(phenyl-methoxy)carbonyl]-L-valine,[(1*R*,2*R*,4*S*)-2-amino-6-oxo-1H-purin-9-yl)-4-(hydroxymethyl) cyclobutyl]methyl ester **47** (83% yield). When bis[valine],O,O′-[(1*S*,2*R*,3*R*)-3-(2-amino-6-oxo-1H-purin-9-yl)cyclobuta-1,2-diyl]methyl ester **48**, dihydrochloride was hydrolyzed with lipase from *Candida cylindraceae*, L-valine, [(1*R*,2*R*,4*S*)-2-amino-1,6-dihydro-6-oxo-9H-purine-9-yl)-4-(hydroxymethyl)cyclobutyl] methyl ester monohydrochloride **49** (87% yield) was obtained. The final intermediates for lobucavir prodrug, N-[(phenylmethoxy)carbonyl]-L-valine, [(1*R*,2*R*,4*S*)-3-(2-amino-6-oxo-1H-purin-9-yl)-2-(hydroxymethyl) cyclobutyl]methyl ester **50** could be obtained by transesterification of lobucavir using ChiroCLEC™ BL (61% yield) or more selectively by using lipase from *Pseudomonas cepacia* (84% yield).

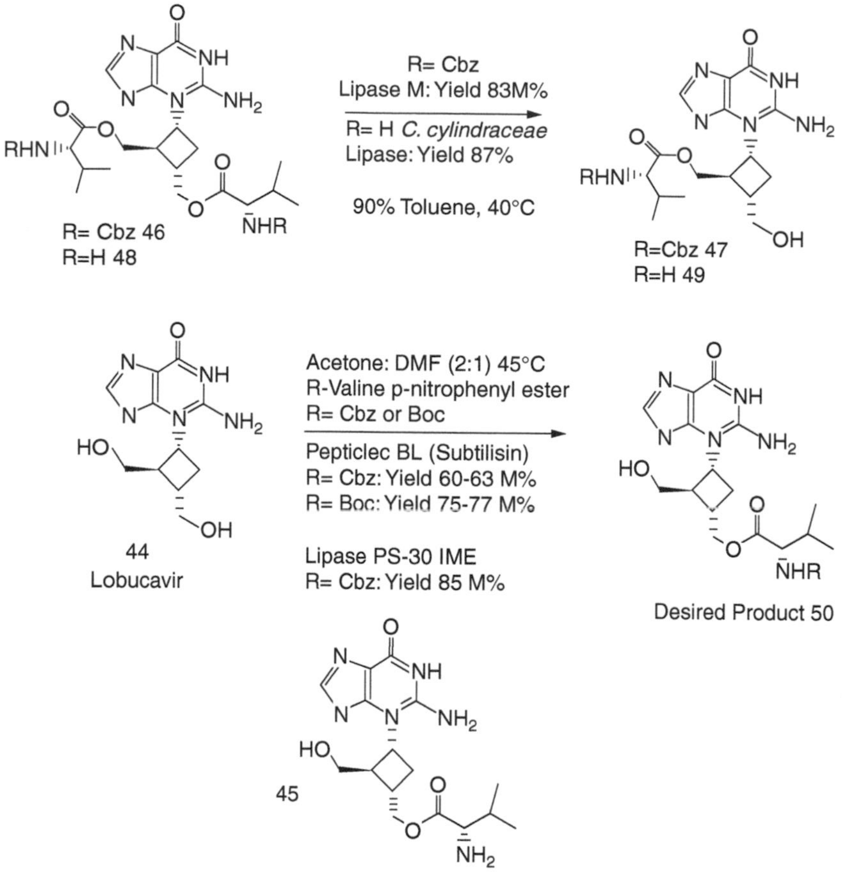

FIGURE 26.13 Synthesis of chiral intermediates for lobucavir **44**: (A) regioselective anzymatic aminoacylation of lobucavir **44** and hydrolysis of **46** and **48**.

26.4 Antihypertensive Drugs

26.4.1 Angiotensin Converting Enzyme Inhibitors

26.4.1.1 Captopril

Captopril is designated chemically as 1-[(2S)-3-mercapto-2-methylpropionyl]-L-proline **51** (Figure 26.14). It is used as an antihypertensive agent through suppression of the renin-angiotensin-aldosterone system.[46,47] Captopril prevents the conversion of angiotensin I to angiotensin II by inhibiting ACE. The potency of captopril **51** as an inhibitor of ACE depends critically on the configuration of the mercaptoalkanoyl moiety; the compound with the S-configuration is about 100 times more active than its corresponding R-enantiomer.[48] The required 3-mercapto-(2S)-methylpropionic acid moiety has been prepared from the microbially derived chiral 3-hydroxy-(2R)-methylpropionic acid, which is obtained by the hydroxylation of isobutyric acid.[49]

The synthesis of chiral side-chain of captopril by the lipase-catalyzed enantioselective hydrolysis of the thioester bond of racemic 3-acetylthio-2-methyl propanoic acid **52** to yield (*S*)-**53** has been demonstrated.[50] Among various lipases evaluated, lipase from *Rhizopus oryzae* ATCC 24563 (heat-dried cells) and lipase PS-30 from *Pseudomonas cepacia* in organic solvent system (1,1,2-trichloro-1,2,2-trifluoroethane or toluene) catalyzed the hydrolysis of thioester bond of undesired enantiomer of racemic **52** to yield desired (S)-**53**, (*R*)-3- mercapto-2-methyl propanoic acid **54** and acetic acid (Figure 14). The reaction yield of >24% (theoretical max. is 50%) and e.e. of >95% were obtained for (*S*)-**53** using each lipase.

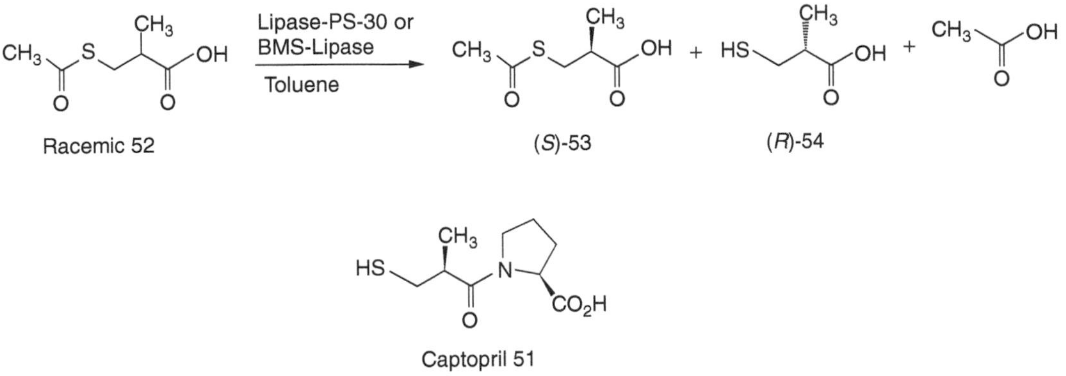

FIGURE 26.14 Synthesis of captopril **51** side-chain (*S*)-**53**: enantioselective enzymatic hydrolysis of racemic 3-acylthio-2-methyl propanoic acid **52**.

In an alternative approach to prepare the chiral side chain of captopril **51**, and zofenopril **55**, the lipase-catalyzed enantioselective esterification of racemic 3-benzoylthio-2-methylpropanoic acid **56** (Figure 26.15) in an organic solvent system was demonstrated to yield *R*-(+) methyl ester **57** and unreacted acid enriched in the desired (*S*)-**58**.[51] Using lipase PS-30 with toluene as solvent and methanol as nucleophile, the desired (*S*)-**58** was obtained in 37% reaction yield (maximum theoretical yield is 50%) and 97% e.e. Substrate was used at 22 g/L concentration. Crude lipase PS-30 was immobilized on Accurel polypropylene (PP) in absorption efficiencies of 98.5%. The immobilized lipase catalyzed an efficient esterification reaction, giving 45% reaction yield and 97.7% e.e. of (*S*)-**58**. The immobilized enzyme under identical conditions gave similar e.e. and yield of product in 23 additional reaction cycles without any loss of activity and productivity. (*S*)-**58** is a key chiral intermediate for the synthesis of captopril[52] or zofenopril.[53]

26.4.1.2 Monopril

The (*S*)-2-cyclohexyl 1,3-propanediol monoacetate **59** and the (*S*)-2-phenyl- 1,3-propanediol monoacetate **60** are key chiral intermediates for the chemo-enzymatic synthesis of Monopril **61**(Figure 26.16), a new antihypertensive drug that acts as an ACE inhibitor. The asymmetric hydrolysis of 2-cyclohexyl-1,3-propanediol diacetate **62** and 2-phenyl-1,3-propanediol diacetate **63** to the corresponding (*S*)-monoacetate

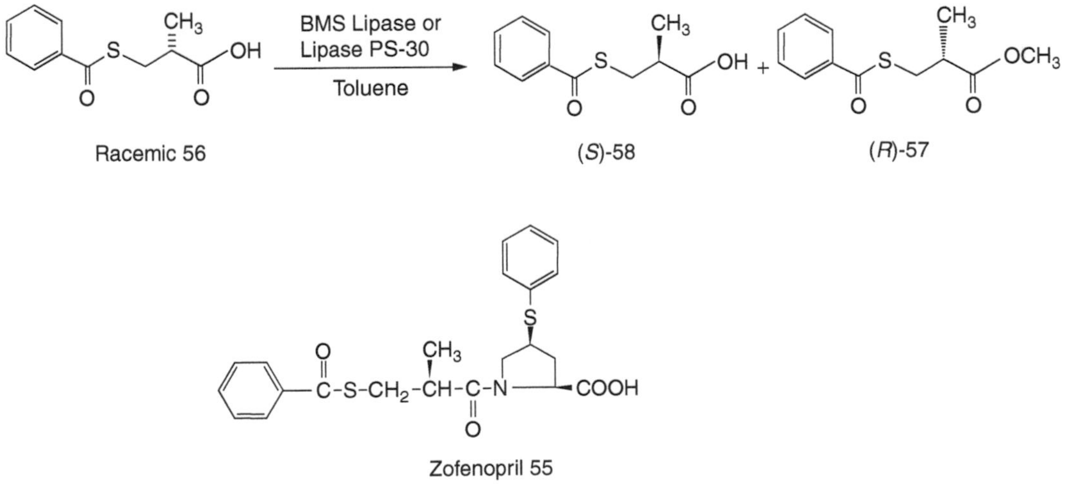

FIGURE 26.15 Synthesis of zofenopril **55** side-chain (*S*)-**58**: enantioselective enzymatic esterification of racemic 3-benzylthio-2-methyl propanoic acid **56**.

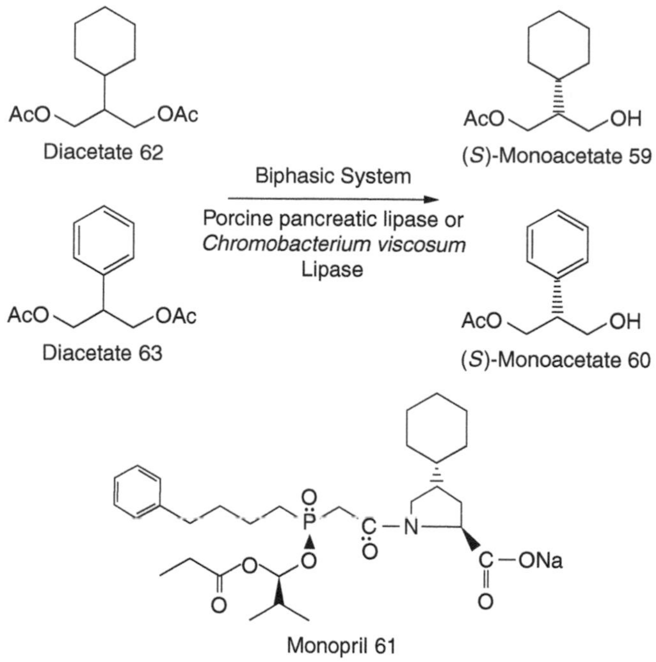

FIGURE 26.16 Preparation of chiral synthon for monopril **61**: asymmetric enzymatic hydrolysis of 2-cyclohexyl-**62** and 2-phenyl-1,3-propanediol diacetate **63** to the corresponding (S)-monoacetates **59** and **60**.

59 and (S)-monoacetate **60** by PPL and *Chromobacterium viscosum* lipase have been demonstrated by Patel et al.[54] In a biphasic system using 10% toluene, the reaction yield of >65% and e.e. of 99% were obtained for (S)-**59** using each enzyme. (S)-**60** was obtained in 90% reaction yield and 99.8% e.e. using *C. viscosum* lipase under similar conditions.

26.4.1.3 Ceranopril

Ceranopril **64** is another ACE inhibitor[55] which requires chiral intermediate carbobenzoxy(Cbz)-L-oxylysine **65** (Figure 26.17). A biotransformation process was developed by Hanson et al.[56] to prepare the Cbz-L-oxylysine. N-ε-carbobenzoxy(Cbz)-L-lysine **66** was first converted to the corresponding keto acid **67** by oxidative deamination using cells of *Providencia alcalifaciens* SC 9036, which contained L-amino acid oxidase and catalase. The keto acid **67** was subsequently converted to **65** using L-2-hydroxy-isocaproate dehydrogenase (HIC) from *Lactobacillus confusus*. The NADH required for this reaction was regenerated using formate dehydrogenase from *C. boidinii*. The reaction yield of 95% with 98.5 % e.e. was obtained in the overall process.

26.4.2   Neutral Endopeptidase Inhibitors

The (S)-α-[(acetylthio)methyl]benzenepropanoic acid **68** (Figure 26.18) is a key chiral intermediate for the neutral endopeptidase inhibitor **69**.[57] The lipase PS-30 catalyzed enantioselective hydrolysis of thioester bond of racemic α-[(acetylthio)methyl] benzenepropanoic acid **70** in organic solvent to yield (R)-α-[(mercapto)methyl] benzenepropanoic acid **71** and (S)-**68**. A 40% reaction yield (theoretical maximum is 50%) and 98% e.e. was obtained for (S)-**68**.[50]

26.4.3   ACE and NEP Inhibitors

Omapatrilat **72** (Figure 26.19) is an antihypertensive drug that acts by inhibiting angiotensin-converting enzyme (ACE) and neutral endopeptidase (NEP).[58] Effective inhibitors of ACE have been used not only

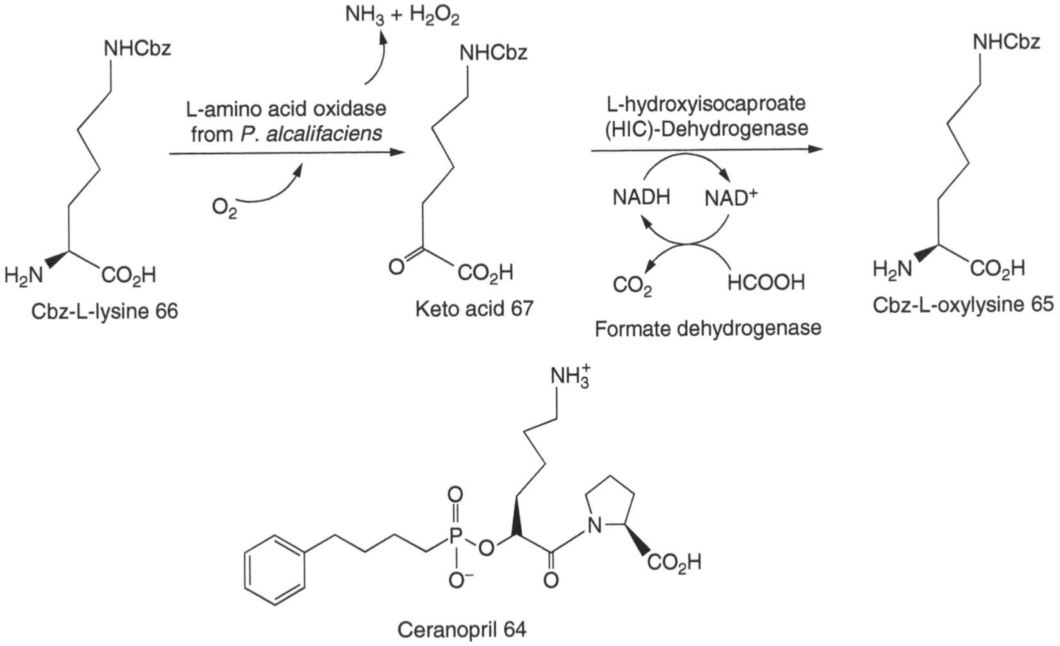

FIGURE 26.17 Synthesis of chiral synthon for ceranopril **64**: enzymatic conversion of Cbz-L-lysine **66** to L-Cbz-oxylysine **65**.

in the treatment of hypertension but in the clinical management of congestive heart failure. NEP, like ACE, is a zinc metalloprotease and is highly efficient in degrading atrial natriuretic peptide (ANP), a 28-amino acid peptide secreted by the heart in response to atrial distension. By interaction with its receptor, ANP promotes the generation of cGMP via guanylate cyclase activation, thus resulting in vasodialatation, natriuresis, diuresis, and inhibition of aldosterone. Therefore, simultaneous potentiation of ANP via NEP inhibition and attenuation of angiotensin II (AII) via ACE inhibition should lead to complementary effects in the management of hypertension and congestive heart failure.[59]

26.4.3.1 Enzymatic Synthesis of (S)-6-Hydroxynorleucine

(S)-6-Hydroxynorleucine **73** (Figure 26.19) is a key intermediate in the synthesis of Omapatrilat. Reductive amination of ketoacids using amino acid dehydrogenases has long been known to be a useful method for the synthesis of natural and unnatural amino acids.[60,61] The synthesis and complete conversion of

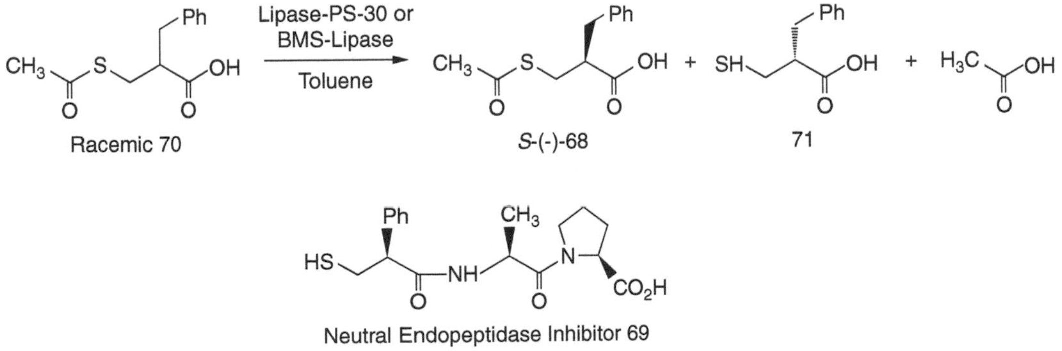

FIGURE 26.18 Preparation of chiral synthon for neutral endopeptidase inhibitor **69**: enantioselective enzymatic hydrolysis of racemic α-[(acetylthio) methyl] benzenepropanoic acid **70**.

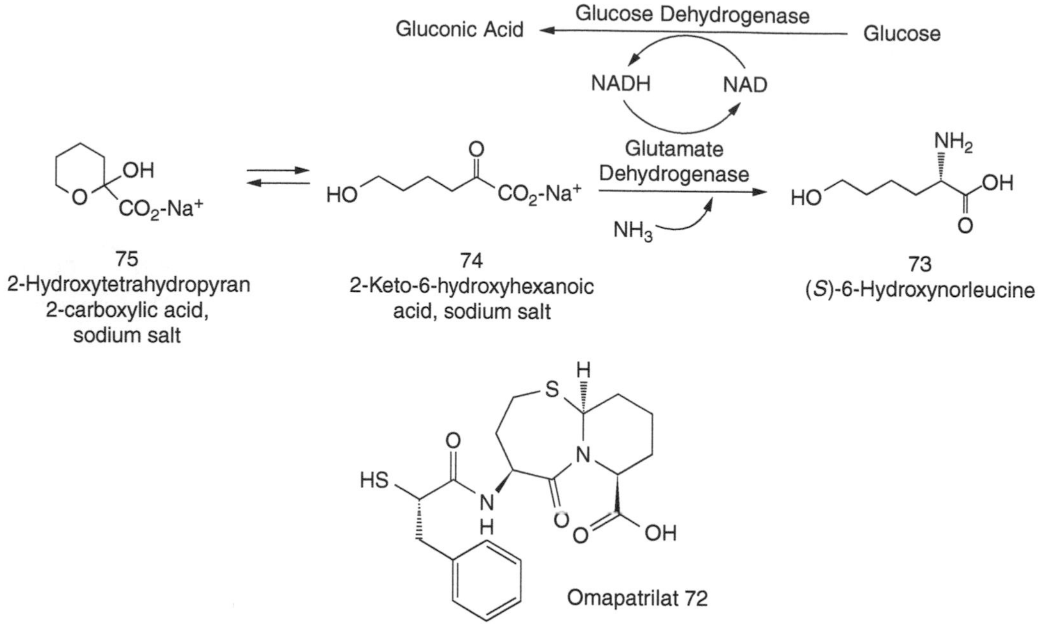

FIGURE 26.19 Enzymatic synthesis of chiral synthon for omapatrilat **72**: reductive amination of sodium 2-keto-6-hydroxyhexanoic acid **74** to (S)-6-hydroxynorleucine **73** by glutamate dehydrogenase.

2-keto-6-hydroxyhexanoic acid **74** to (S)-6-hydroxynorleucine **73** by reductive amination using phenylalanine dehydrogenase (PDH) from *Sporosarcina* sp. or beef liver glutamate dehydrogenase.[62] Beef liver glutamate dehydrogenase was used for preparative reactions at 100 g/L substrate concentration. As depicted, 2-keto-6-hydroxyhexanoic acid sodium salt **74**, in equilibrium with 2-hydroxytetrahydropyran-2-carboxylic acid sodium salt **75**, was converted to S-6-hydroxynorleucine **73**. The reaction requires ammonia and NADH. NAD produced during the reaction was recycled to NADH by the oxidation of glucose to gluconic acid using glucose dehydrogenase from *Bacillus megaterium* . The reaction was complete in about 3 hr with reaction yields of 92% and e.e. of >99% for (S)-6-hydroxynorleucine.

The synthesis and isolation of 2-keto-6-hydroxyhexanoic acid **74** required several steps. In a second, more convenient process (Figure 26.20) the ketoacid was prepared by treatment of racemic 6-hydroxynorleucine **76** (produced by hydrolysis of 5-(4-hydroxybutyl)hydantoin **77**) with D-amino acid oxidase and catalase. After the e.e. of the remaining S-6-hydroxynorleucine had risen to >99%, the reductive amination procedure was used to convert the mixture containing 2-keto-6-hydroxyhexanoic acid **74** entirely to (S)-6-hydroxynorleucine **73** in 97% yield and of 98% e.e. from racemic 6-hydroxynorleucine at 100 g/L. Porcine kidney D-amino acid oxidase and beef liver catalase or *T. variabilis* whole cells (source of both the oxidase and catalase) were used successfully for this transformation.[62] The (S)-6-hydroxynorleucine **73** prepared by the enzymatic process was converted chemically to Omapatrilat **72** as described previously.[58]

26.4.3.2 Enzymatic Synthesis of Allysine Ethylene Acetal

(S)-2-Amino-5-(1,3-dioxolan-2-yl)-pentanoic acid (S)-allysine ethylene acetal, **78** (Figure 26.21), is one of three building blocks used in an alternative synthesis of Omapatrilat **72**. It previously had been prepared via an eight-step chemical synthesis from 3,4-dihydro[2H]pyran.[63] An alternate synthesis of **78** was demonstrated by reductive amination of ketoacid acetal **79** using phenylalanine dehydrogenase (PDH) from *Thermoactinomyces intermedius*.[64] The reaction required ammonia and NADH; NAD produced during the reaction was recyled to NADH by the oxidation of formate to CO2 using formate dehydrogenase (FDH).

T. intermedius gave useful activity on a small scale (15 L) but lysed soon after the end of the growth period, making recovery of activity difficult or impossible on a large scale (4000 L). This problem was solved

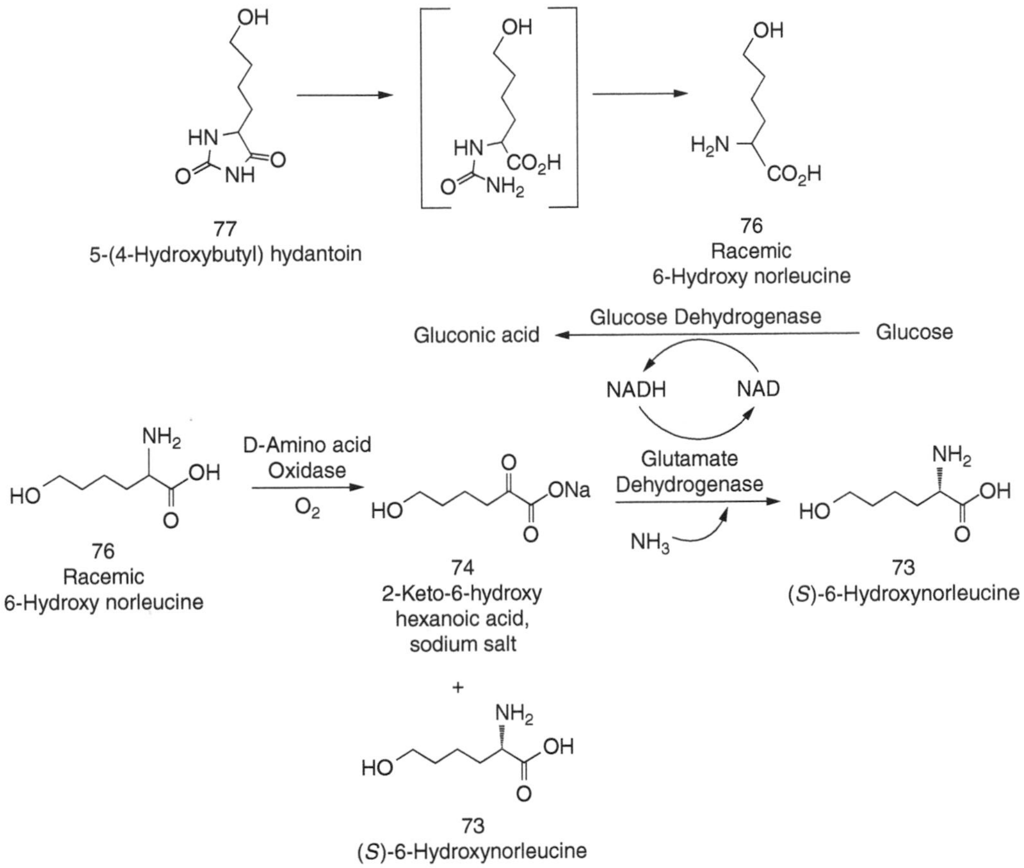

FIGURE 26.20 Conversion of racemic 6-hydroxynorleucine **76** to (*S*)-6-hydroxynorleucine **73** by D-amino acid oxidase and glutamate dehydrogenase.

by cloning and expressing the *T. intermedius* PDH in *Escherichia coli*, inducible by β–D-isopropylthio-galactoside (IPTG). Fermentation of *T. intermedius* yielded 184 units of PDH activity per liter of whole broth in 6 hr. In contrast, *E coli* BL21 (DE3) (pPDH155K) produced over 19,000 units per liter of whole broth in about 14 hr.

C. *boidinii*[65] or *P. pastoris*[66] grown on methanol are useful sources of FDH. Expression of *T. intermedius* PDH in *P. pastoris*, inducible by methanol, allowed to obtain both enzymes from a single fermentation. Formate dehydrogenase activity/g wet cells in *P. pastoris* was 2.7-fold greater than for *C. boidinii*, and fermentor productivity was increased by 8.7-fold compared to *C. boidinii*. Fermentor productivity for PDH in *P. pastoris* was about 28% of the *E coli* productivity.

Reductive amination reactions were carried out at pH 8.0. A procedure using heat-dried cells of *E coli* containing cloned PDH and heat-dried *C. boidinii* was scaled up. A total of 197 kg of **78** was produced in three 1600-L batches using a 5% concentration of substrate **79** with an average yield of 91 M% and e.e. >98%. A second-generation procedure, using dried recombinant *P. pastoris* containing *T. intermedius* PDH inducible with methanol and endogenous FDH induced when *P. pastoris* was grown in medium containing methanol, allowed both enzymes to be produced during a single fermentation. The procedure with *P. pastoris* was also scaled up to produce 15.5 kg of **78** in 97 M% yield and >98% e.e. in a 180-L batch using 10% ketoacid **79** concentration. The (*S*)-allysine ethylene acetal **78** produced by the enzymatic process was converted to Omapatrilat **72**.[67]

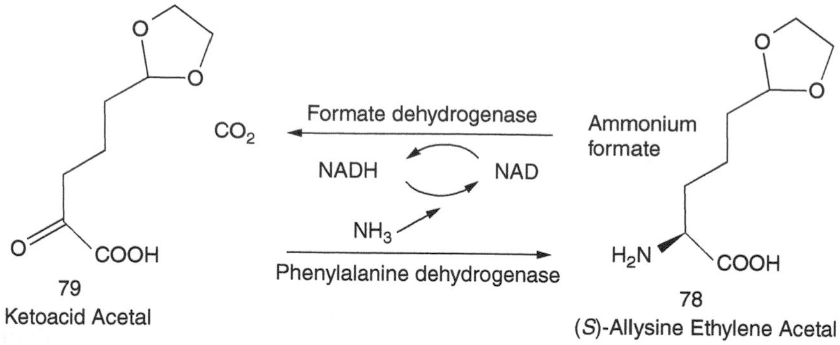

FIGURE 26.21 Enzymatic synthesis of chiral synthon for omapatrilat **72**: reductive amination of keto acid acetal **79** to amino acid acetal **78** by phenylalanine dehydrogenase. Regeneration of NADH was carried out using formate dehydrogenase.

26.4.3.3 Enzymatic Synthesis of Thiazepine

[4*S*-(4a,7a,10ab)]-1-Octahydro-5-oxo-4-[[(phenylmethoxy)carbonyl]amino]-7H-pyrido-[2,1-b][1,3]thiazepine-7-carboxylic acid **80** is a key intermediate in the synthesis of Omapatrilat **72**. An enzymatic process (Figure 26.22) was developed for the oxidation of the ε-amino group of (*S*)-lysine in the thiol **81** generated *in situ* from disulfide N²-[N[[(phenylmethoxy)carbonyl] L-homocysteinyl] L-lysine)-1,1-disulphide **82** to produce compound **80** using L-lysine ε–aminotransferase (LAT) from *S. paucimobilis* SC 16113.[68] This enzyme was overexpressed in *E coli* and a biotransformation process was developed using the recombinant enzyme. The aminotransferase reaction required α-ketoglutarate as the amine acceptor. Glutamate formed during this reaction was recycled back to α-ketoglutarate by glutamate oxidase (GOX) from *Streptomyces noursei* SC 6007.

A selective culture technique was used to isolate eight different types of microbial cultures able to utilize N-α-Cbz-S-lysine as the sole source of nitrogen. Cell extracts prepared from cell suspensions were evaluated for oxidation of the ε-amino group of (*S*)-lysine in the substrate **81** generated from compound **82** by treatment with DTT. Product **80** formation was observed with four cultures. One of the cultures, Z-2, was later identified as *S. paucimobilis* SC 16113. Due to the low activity of LAT in *S. paucimobilis* SC 16113 and to minimize **81** hydrolysis, enzyme was overexpressed in *E coli* strain GI724(pAL781-LAT).

Screening of microbial cultures led to the identification of *S. noursei* SC 6007 as a source of extracellular GOX. *S. noursei* SC 6007 was grown in 380-L fermentors. GOX activity correlated with growth of the culture in the fermentor and reached 0.75 units/mL at harvest. Starting from the extracellular filtrate, the GOX was purified 260-fold to homogeneity, determined its amino terminal and internal peptide sequences, and expressed in *Streptomyces lividans* .

Biotransformation of compound **81** to compound **80** was carried out using LAT from *Escherichia coli* GI724[pal781-LAT] in the presence of α-ketoglutarate and dithiothreitol (DTT or tributylphosphine) and glutamate oxidase. Reaction yields of 65–67 M% were obtained. To reduce the cost of producing two enzymes, the transamination reactions were carried out in the absence of GOX and with higher levels of α-ketoglutarate. The reaction yield in the absence of GOX averaged only about 33–35 M%. However, the reaction yield increased to 70 M%, by increasing the α-ketoglutarate to 40 mg/ml of (10X increase in concentration) and conducting the reaction at 40°C, equivalent to that in the presence of GOX.[68]

26.5   Anticholesterol Drugs

26.5.1   Hydroxymethyl Glutaryl CoA (HMG-CoA) Reductase Inhibitors

Pravastatin **83** and Mevastatin **84** are anticholesterol drugs that act by competitively inhibiting HMG CoA reductase.[69] Pravastatin sodium is produced by two fermentation steps. The first step is the production of

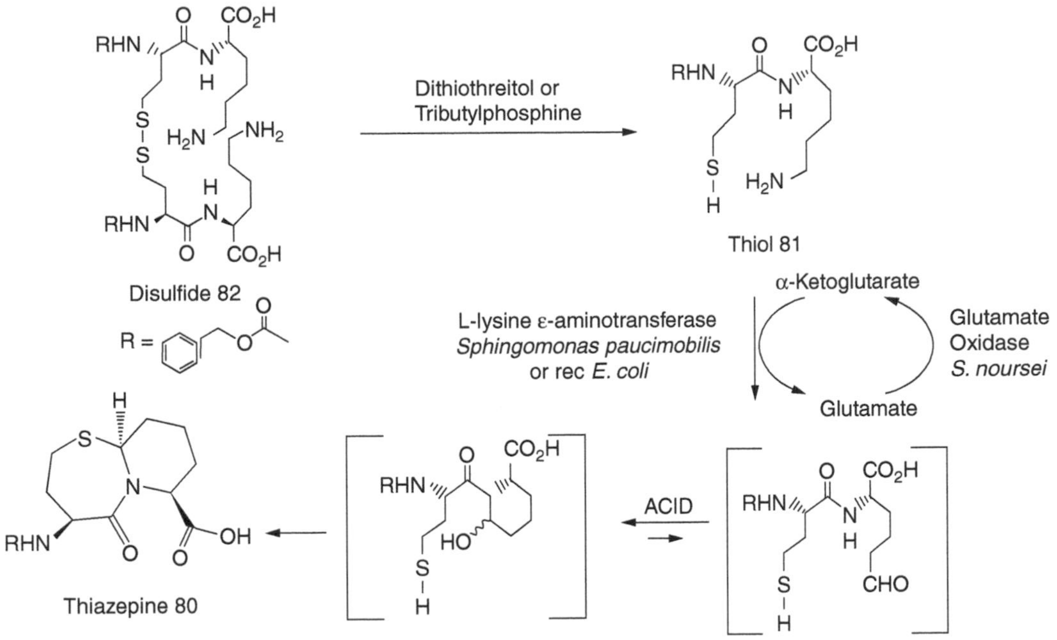

FIGURE 26.22 Enzymatic synthesis of chiral synthon for omapatrilat **72**: conversion of disulfide **82** to thiazepine **80** by L-lysine ε-aminotransferase.

compound ML-236B by *Penicillium citrinum*.[69,70] The purified compound was converted to its sodium salt **85** with sodium hydroxide and in the second step was hydroxylated to Pravastatin sodium **83** (Figure 26.23) by *Streptomyces carbophilus*.[71] A cytochrome P-450 containing enzyme system has been demonstrated from *S. carbophilus* that catalyzed the hydroxylation reaction.[72]

Chiral β-hydroxy esters are versatile synthons.[73,74] The well-known asymmetric reduction of carbonyl compounds using baker's yeast has been reviewed.[75,76] We have described the reduction of methyl 4-chloro-3-oxobutanoate **86** (Figure 26.24) to (*S*)-methyl 4-chloro-3-hydroxybutanoate **87** by cell suspensions of *Geotrichum candidum* SC 5469.[77] (*S*)-**87** is a key chiral intermediate in the total synthesis of **88** (Figure 26.24), a cholesterol antagonist that acts by inhibiting HMG CoA reductase. In the biotransformation process, a reaction yield of 95% and e.e. of 96% were obtained for (*S*)-**87** using glucose-, acetate-, or glycerol-grown cells (10% w/v) of *G. candidum* SC 5469. The e.e. of (*S*)-**87** was increased to 98% by heat treatment (55°C for 30 min) of cell suspensions prior to conducting the bioreduction. The purified oxido-reductase was immobilized on Eupergit C and used to catalyze the reduction of **86**. The cofactor NAD+ required for the reduction reaction was regenerated by glucose dehydrogenase. A 90% reaction yield and 98% e.e. were obtained for (*S*)-**87**.

Many microorganisms and enzymes derived therefrom have been used in the reduction of a single keto group of β-keto or α-keto compounds.[78,79] We have demonstrated the stereoselective reduction of the diketone[80] ethyl 3,5-dioxo-6-(benzyloxy)hexanoate **89** to the diol ethyl (3*R*,5*S*)-dihydroxy-6-(benzyloxy) hexanoate **90a** (Figure 26.25). Compound **90a** is a key intermediate in synthesis of [4-[4 a, 6ß(E)]]-6-[4, 4-Bis [4-fluorophenyl]-3-(1-methyl-1H-tetrazol-5-yl)-1,3-butadienyl]-tetrahydro-4-hydroxy-2H-pyren-2-one, **91**, a potential new anticholesterol drug that acts by inhibition of HMG CoA reductase.[81] Among various microbial cultures evaluated for the stereoselective reduction of diketone **89**, glycerol-grown cell suspensions of *Acinetobacter calcoaceticus* SC 13876 were shown to give a reaction yield of 85% and e.e. of 97%. The substrate and cells were used at 2 g/L and 20% (w/v, wet cells) concentration, respectively.

Cell extracts of *A. calcoaceticus* SC 13876 in the presence of NAD+, glucose, and glucose dehydrogenase reduced **89** to the corresponding isomeric monohydroxy compounds **92** and **93**, which were further

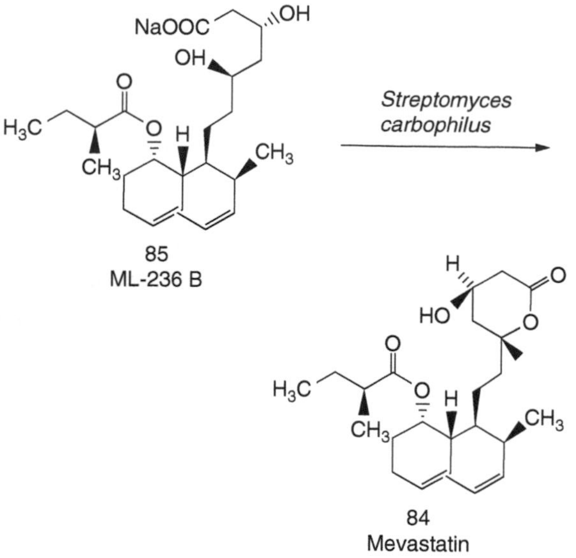

FIGURE 26.23 (A) Structure of anticholesterol drugs pravastatin **83** and mevatsatin **84**. (B) Microbial hydroxylation of ML236B.

reduced to (3R,5S)-dihydroxy compound **90a**. A reaction yield of 92% and an e.e. of 98% were obtained at 10g/L in a 1-L batch. The product was isolated from the reaction mixture in 72% overall yield with HPLC purity of 99% and e.e. of 98.5%. The reductase (mol. Wt. 33,000 daltons) from cell extracts of *A. calcoaceticus* SC 13876 was purified to homogeneity.

Using an enzymatic diastereoselective acetylation process, (5R,3R)-alcohol **91** was prepared from **94**.[82] We evaluated various lipases, among which lipase PS-30 and BMS lipase (produced by fermentation of *Pseudomonas* strain SC 13856) efficiently catalyzed the acetylation of **94** to yield (5R,3S)-acetate **95** and unreacted desired (5R,3R)-alcohol **91** (Figure 26.26). A reaction yield of 49 M% and an e.e. of 98.5% were obtained for (5R,3R)-alcohol **91** when the reaction was conducted in toluene in the presence of isopropenyl acetate as an acyl donor.

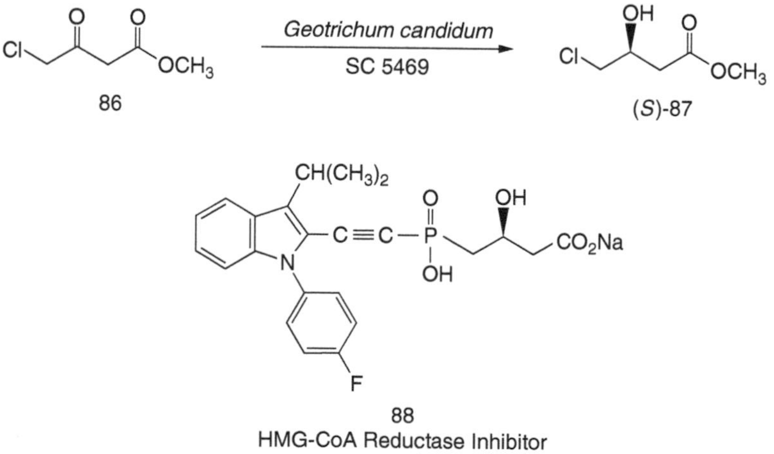

FIGURE 26.24 Synthesis of chiral synthon for anticholesterol drug **88**: enantioselective microbial reduction of 4-chloro-3-oxobutanoic acid methyl ester **86**.

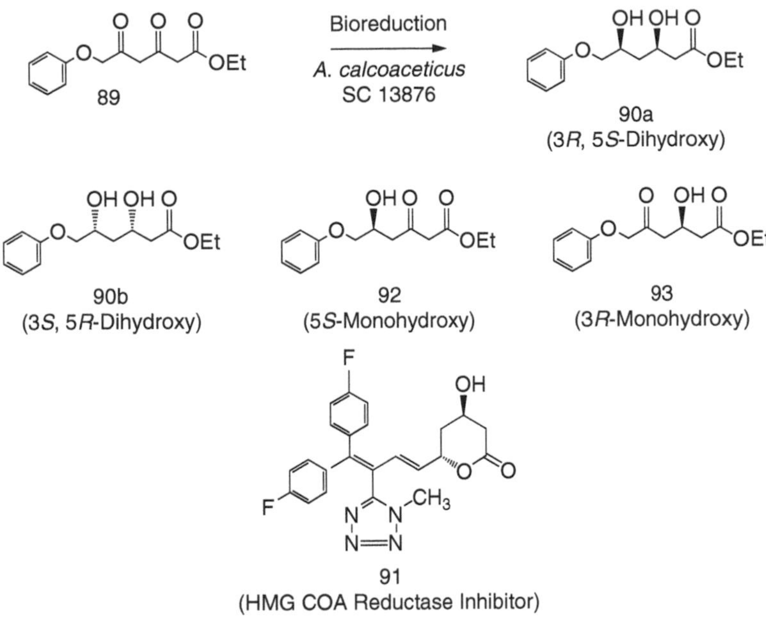

FIGURE 26.25 Synthesis of chiral synthon for anticholesterol drug (R)-**91**: enantioselective microbial reduction of 3,5-dioxo-6-(benzyloxy) hexanoic acid, ethyl ester **89**.

The substrate was used at 4 g/L concentration. In methyl ethyl ketone at 50 g/L substrate concentration, a reaction yield of 46 M% and e.e. of 96% were obtained for **91**. The enzymatic process was scaled up to a 640-L preparative batch using immobilized lipase PS-30 at 4 g/L of **94** in toluene as solvent. From the reaction mixture (5R,3R)-alcohol **91** was isolated in 35 M% overall yield (theoretical maximum

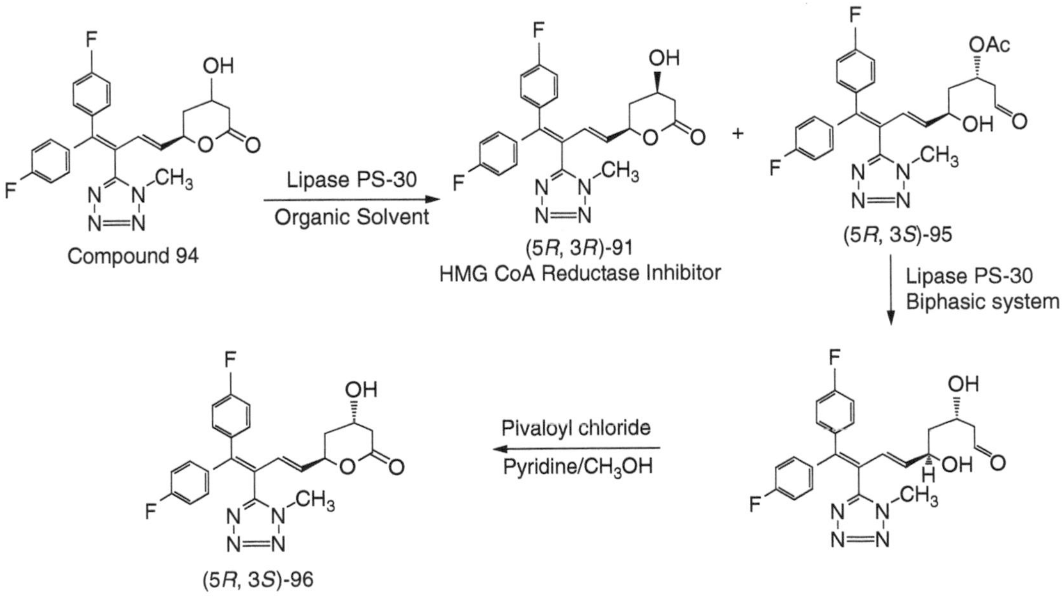

FIGURE 26.26 Synthesis of anticholesterol drug (R)-**91**: diastereoselective enzymatic acetylation of **94**.

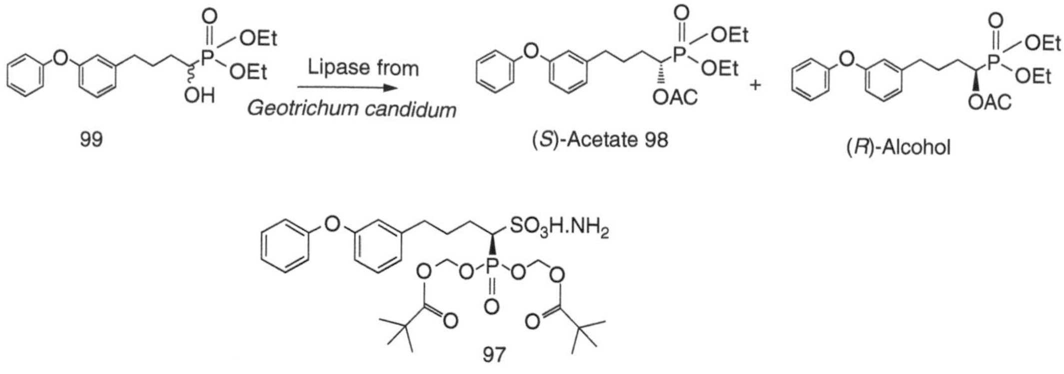

FIGURE 26.27 Enzymatic synthesis of chiral synthon for squalene synthase inhibitor **98**: enantioselective enzymatic acetylation of racemic **99**.

yield 50%) with 98.5% e.e. and 99.5% chemical purity. The (5R,3S)-acetate **95** produced by this process was enzymatically hydrolyzed by lipase PS-30 in a biphasic system to prepare the corresponding (5R,3S)-alcohol **96**.

26.5.2 Squalene Synthase Inhibitors

Squalene synthase is the first pathway-specific enzyme in the biosynthesis of cholesterol and catalyzes the head-to-head condensation of two molecules of farnesyl pyrophosphate (FPP) to form squalene. It has been implicated in the transformation of FPP into presqualene pyrophosphate. FPP analogs are a major class of inhibitors of squalene synthase.[83] However, this class of compounds lacks specificity, and they are potential inhibitors of other FPP-consuming transferases such as geranyl geranyl pyrophosphate synthase. To increase enzyme specificity, analogs of PPP and other mechanism-based enzyme inhibitors have been synthesized such as **97**.[84] S-[1-(acetoxyl)-4-(3-phenyl) butyl] phosphonic acid, diethyl ester **98** is a key chiral intermediate required for the total chemical synthesis of **97**. The enantioselective acetylation of racemic [1-(hydroxy)-4-(3-phenyl)butyl]phophonic acid, diethyl **99** (Figure 26.27) was carried out using *Geotrichum candidum* lipase in tolune as solvent and isopropenyl acetate as acyl donor.[85] A reaction yield of 38% (theoritical max. 50%) and an e.e.of 95% were obtained for chiral **98**.

26.6 Thromboxane A2 Antagonist

26.6.1 Enzymatic Synthesis of Lactol [3aS-(3aα, 4α, 7α, 7aα)]-Hexahydro-4,7-Epoxyisobenzo-Furan-1-(3H)-ol and Corresponding Lactone

Thromboxane A2 (TxA2) is an exceptionally potent vasoconstrictor substance produced by the metabolism of arachidonic acid in blood platelets and other tissues. Together with its potent anti-aggregatory and vasodilator activities, TxA2 plays an important role in the maintenance of vascular homeostasis, and contributes to the pathogenesis of a variety of vascular disorders. Approaches to limiting the effect of TxA2 have focused on either inhibiting its synthesis or blocking its action at its receptor sites by means of an antagonist.[86,87] The lactol [3aS-(3aα, 4α, 7α, 7aα)]-hexahydro-4,7-epoxyisobenzo-furan-1-(3H)-ol **100** or the corresponding chiral lactone **101** (Figure 26.28a) are key intermediates in the total synthesis of [1S-[1α,2α(Z),3α,4α[[-7-[3-[[[[1-oxoheptyl)-amine] acetyl] methyl]-7-oxabicyclo-[2.2.1] hept-2-yl]-5-heptanoic acid **102**, a new cardiovascular agent of potential use in the treatment of thrombotic disease.[88]

The enantioselective oxidation of (exo, exo)-7-oxabicyclo [2.2.1] heptane-2,3-dimethanol **103** to the corresponding chiral lactol **100** and lactone **101** by cell suspensions (10% w/v, wet cells) of *Nocardia*

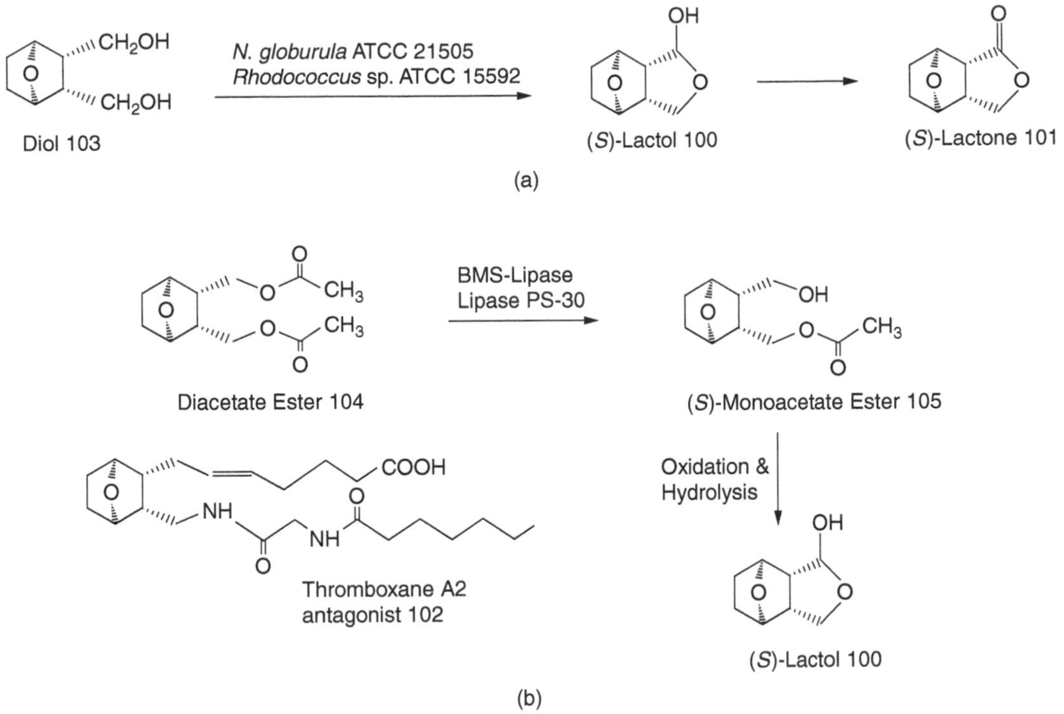

FIGURE 26.28 Synthesis of chiral synthon for thromboxane A2 antagonist **102**: (A) Stereoselective microbial oxidation of (exo, exo)-7-oxabicyclo [2.2.1] heptane-2,3-dimethanol **103** to the corresponding lactol **100** and lactone **101**. (B) Asymmetric enzymatic hydrolysis of (exo, exo)-7-oxabicyclo[2.2.1] heptane-2,3-dimethanol, diacetate ester **104** to the corresponding (S)-monoacetate ester **105**.

globerula ATCC 21505 and *Rhodococcus* sp. ATCC 15592 have been described.[89] Lactone **101** was obtained in 70 M% yield and 96% e.e. after 96 hr at 5 g/L substrate concentration using cell suspensions of *N. globerula* ATCC 21505. An overall reaction yield of 46 M% (lactol and lactone combined) and e.e. of 96.7% and 98.4% were obtained for lactol **100** and lactone **101**, respectively, using cell suspensions of *Rhodococcus* sp. ATCC 15592; substrate **103** was used at 5 g/L concentration.

The enantioselective hydrolysis of the diacetate (exo,exo)-7-oxabicyclo [2.2.1] heptane-2,3-dimethanol, **104** to the corresponding S-monoacetate ester **105** (Figure 26.28b) has been demonstrated with lipases.[90] Lipase PS-30 from *P. cepacia* was the most effective in the enantioselective hydrolysis to the desired S-monoacetate. A reaction yield of 75 M% and e.e. of >99% was obtained when the reaction was conducted in a biphasic system with 10% toluene at 5 g/L of the substrate. Lipase PS-30 was immobilized on Accurel polypropylene (PP) and the immobilized enzyme was reused (5 cycles) without loss of enzymic activity, productivity, or e.e. of product **105**. The reaction process was scaled up to 80 L (400 g of substrate) and the product **105** was isolated in 80 M% yield with 99.3% e.e. The S-monoacetate was oxidized to its corresponding aldehyde, which was hydrolyzed to the lactol **100**, which was used in the chemoenzymatic synthesis of thromboxane A2 antagonist **102**.

26.7 Calcium Channel Blockers

26.7.1 Enzymatic Synthesis of [(3R-cis)-1,3,4,5-Tetrahydro-3-Hydroxy-4-(4-Methoxyphenyl)-6-(Trifluromethyl)-2H-1-Benzazepin -2-one]

Diltiazem **106** (Figure 26.29), a benzothiazepinone calcium channel-blocking agent that inhibits influx of extracellular calcium through L-type voltage-operated calcium channels, has been widely used clinically in

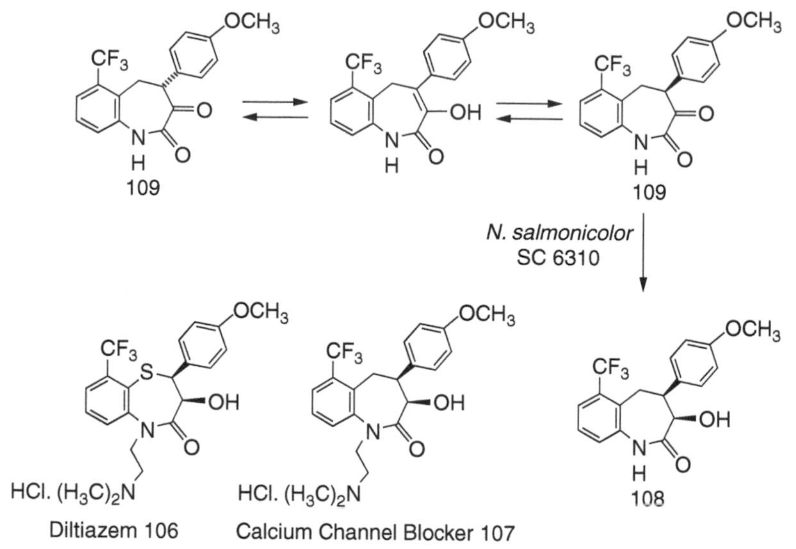

FIGURE 26.29 Synthesis of chiral synthon for calcium channel blocker **107**: mMicrobial reduction of 4,5-dihydro-4-(4-methoxyphenyl)-6-(trifluoromethyl)-1H-benzazepin-2,3-dione **109**.

the treatment of hypertension and angina.[91] Since diltiazem has a relatively short duration of action,[92] an 8-chloroderivative recently has been introduced into the clinic as a more potent analogue.[93] Lack of extended duration of action and little information on the structure-activity relationships in this class of compounds led Floyd et al.[94] to prepare isosteric 1-benzazepin-2-ones; this led to identification of [(cis)-3-(acetoxy)-1-[2-(dimethylamino)ethyl]-1,3,4,5-tetrahydro-4-(4-methoxyphenyl)-6-trifluoromethyl)-2H-1-benzazepin-2-one] **107** as a longer-lasting and more potent antihypertensive agent. A key intermediate in the synthesis of this compound was [(3R-cis)-1,3,4,5-tetrahydro-3-hydroxy-4-(4-methoxyphenyl)-6-(trifluromethyl)-2H-1-benzazepin-2-one] **108**. A enantioselective microbial process (Figure 26.29) was developed for the reduction of 4,5-dihydro-4-(4-methoxyphenyl)-6-(trifluoromethyl)-1H-1benzazepin-2,3-dione **109**, which exists predominantly in the achiral enol form in rapid equilibrium with the two enantiomeric keto forms. Reduction of **109** could give rise to formation of four possible alcohol stereoisomers. Remarkably, conditions were found under which only the single alcohol isomer **108** was obtained. Among various cultures evaluated, microorganisms from the genera *Nocardia, Rhodococcus, Corynebacterium,* and *Arthobacter* reduced compound **109** with 60–70% conversion yield at 1 g/L substrate input. The most effective culture, *Nocardia salmonicolor* SC 6310, catalyzed the bioconversion of **109** to **108** in 96% reaction yield with 99.8% e.e. at 2 g/L substrate concentration. A preparative-scale fermentation process for growth of *N. salmonicolor* and a bioreduction process using cell suspensions of the organism were demonstrated.[95]

26.8 Potassium Channel Openers

26.8.1 Microbial Oxygenation of 6-Cyano-2,2-Dimethyl-2H-1-Benzopyran to Chiral Epoxide and Trans-diol

The study of potassium (K) channel biochemistry, physiology, and medicinal chemistry has flourished; numerous papers and reviews have been published in recent years.[96,97] It has long been known that K channels play a major role in neuronal excitability and a critical role in the basic electrical and mechanical function of a wide variety of tissues, including smooth muscle and cardiac muscle.[98] A new class of highly specific compounds that either open or block K channels has been developed.[99,100] The synthesis and antihypertensive activity of K-channel openers based on monosubstituted trans-4-amino-3,4-dihydro-2,2-dimethyl-2H-1-benzopyran-3-ol **110** (Figure 26.30) have been demonstrated.[101,102] Chiral epoxide **111** and

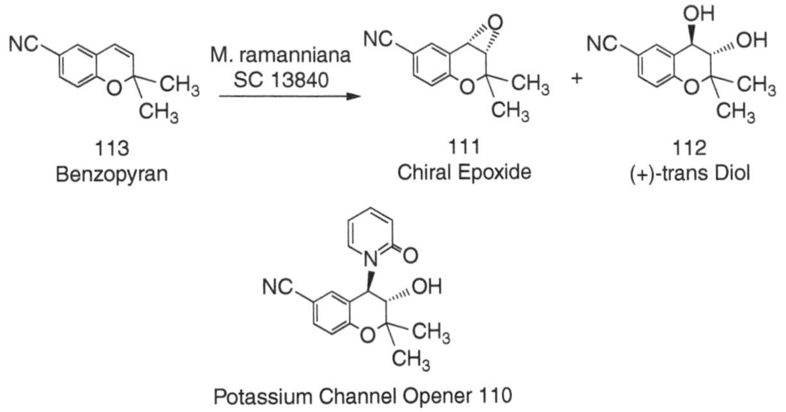

FIGURE 26.30 Preparation of chiral synthons for potassium channel openers **110**: oxygenation of 2,2-dimethyl-2H-1-benzopyran-6-carbonitrile **113** to the corresponding chiral epoxide **111** and (+)-trans diol **112** by *M. ramanniana* SC 13840.

diol **112** are potential intermediates for the synthesis of **110**. The enantioselective microbial oxygenation of 6-cyano-2,2-dimethyl-2H-1-benzopyran **113** to the corresponding chiral epoxide **111** and chiral diol **112** has been demonstrated.[103] Among microbial cultures evaluated, *Mortierella ramanniana* SC 13840 and *Corynebacterium* sp. SC 13876 gave 67.5 M% and 32 M% yield and 96% and 89% e.e.'s, respectively, for (+)-trans diol **112**. *Corynebacterium* sp. SC 13876 also gave chiral epoxide **111** in 17 M% yield and 88% e.e.

A single-stage process (fermentation/epoxidation) for the biotransformation of **113** was developed using *M. ramanniana* SC 13840. In a 25-L fermentor, (+)-trans diol **112** was obtained in a 61 M% yield and e.e. of 92.5%. In a two-stage process using a cell-suspension (10% w/v, wet cells) of *M. ramanniana* SC 13840, the (+)-trans diol **112** was obtained in 76 M% yield with an e.e. of 96%.

In an enzymatic resolution approach, (+)-trans diol **112** was prepared by the enentioselective acetylation of racemic diol with lipases from *Candida cylindraceae* and *P. cepacia*. Both enzymes catalyzed the acetylation of the undesired enantiomer of the racemic diol to yield the monoacetylated product and unreacted (+)-trans diol **112**. A reaction yield of 40% (theoretical max yield is 50%) and an e.e. of >90% were obtained using each lipase.[104]

26.9 β3-Receptor Agonists

β3-Adrenergic receptors are found on the cell surface of both white and brown adipocytes and are responsible for lipolysis, thermogenesis, and relaxation of intestinal smooth muscle.[105] Consequently, several research groups are engaged in developing selective β3 agonists for the treatment of gastrointestinal disorders, type II diabetes, and obesity.[106,107] Three different biocatalytic syntheses of chiral intermediates required for the total synthesis of β3 receptor agonists **114** (Figure 26.31a) have been investigated.[108]

26.9.1 Microbial Reduction of 4-Benzyloxy-3-Methanesulfonylamino-2'-Bromoacetophenone

The microbial reduction of 4-benzyloxy-3-methanesulfonylamino-2'-bromoacetophenone **115** (Figure 26.31b) to the corresponding (*R*)-alcohol **116** has been demonstrated using *Sphingomonas paucimobilis* SC 16113. The growth of *S. paucimobilis* SC 16113 was carried out in a 750-L fermentor and cells (60 Kg) harvested from the fermentor were used to conduct the biotransformation in 10-L and 200-L preparative batches. The cells were suspended in 80 mM potassium phosphate buffer (pH 6.0) at 20% (wt/vol, wet cells) concentration and supplemented with compound **115** (2 g/L) and glucose (25 g/L) and the reduction was carried out at 37°C. In some batches, the fermentation broth was concentrated

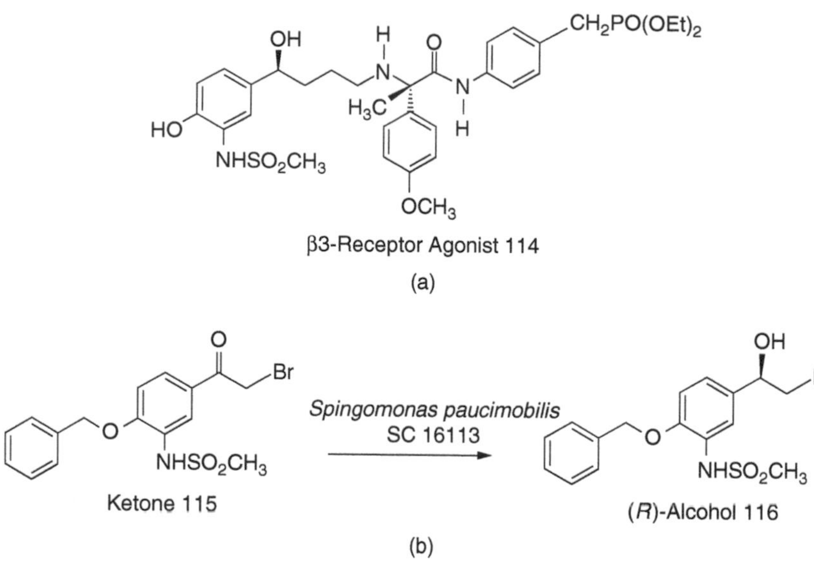

β3-Receptor Agonist 114

(a)

(b)

FIGURE 26.31 (A) Enzymatic synthesis of chiral synthon for β-3-receptor agonist **114**: enantioselective reduction of 4-benzyloxy-3-methanesulfonylamino-2'-bromoacetophenone **115** to (*R*)-alcohol **116**.

threefold by microfilteration and subsequently washed with buffer by diafiltration and were used directly in the bioreduction process. In all the batches, reaction yields of >85% and e.e.'s. of >98% were obtained. The isolation of alcohol **116** from the 200-L batch gave 320 g (80% yield) of product with an e.e. of 99.5%.

In an alternate process, frozen cells of *S. paucimobilis* SC 16113 were used with XAD-16 hydrophobic resin (50 g/L) adsorbed substrate at 10 g/L concentration. In this process, an average reaction yield of 85% and an e.e. of >99% were obtained for alcohol **116**. At the end of the biotransformation, the reaction mixture was filtered on a 100 mesh (150 μ) stainless steel screen, and the resin retained by the screen was washed with water. The product was then desorbed from the resin with acetonitrile and crystallized in an overall 75 M% yield and 99.8% e.e.

26.9.2   Enzymatic Resolution of Racemic α-Methyl Phenylalanine Amides

The chiral amino acids **117** and **118** (Figure 26.32a) are intermediates for the synthesis of β3-receptor agonists.[106,107] These are available via the enzymatic resolution of racemic α-methyl phenylalanine amide **119** and α-methyl-4-methoxyphenylalanine amide **120**, respectively, by an amidase from *Mycobacterium neoaurum* ATCC 25795.[108] Wet cells (10% wt/vol) completed reaction of amide **119** in 75 min with a yield of 48 M% (theoretical max. 50%) and an e.e. of 95% for the desired (*S*)-amino acid **117**. Alternatively, freeze-dried cells were suspended in 100 mM potassium phosphate buffer (pH 7.0) at 1% concentration to give complete reaction in 60 min with a yield of 49.5 M% (theoretical max. 50%) and an e.e. of 99% for the (*S*)-amino acid **117**.

Freeze-dried cells of *M. neoaurum* ATCC 25795 and partially purified amidase (amidase activity in cell extracts purified 5-fold by diethyl aminoethyl celluose column chromatography) were used for the biotransformation of compound **120**. A reaction yield of 49 M% and an e.e. of 78% were obtained for the desired product **118** using freeze-dried cells. The reaction was completed in 50 hr. Using partially purified amidase, a reaction yield of 49 M% and an e.e. of 94% were obtained after 70 hr.

26.9.3   Enantioselective Hydrolysis of Diethyl Methyl-(4-Methoxyphenyl-Propanedioate

The (*S*)-monoester **121** (Figure 26.32b) is a key intermediate for the syntheses of β3-receptor agonists. The enantioselective enzymatic hydrolysis of diester **122** to the desired acid ester **121** by pig liver

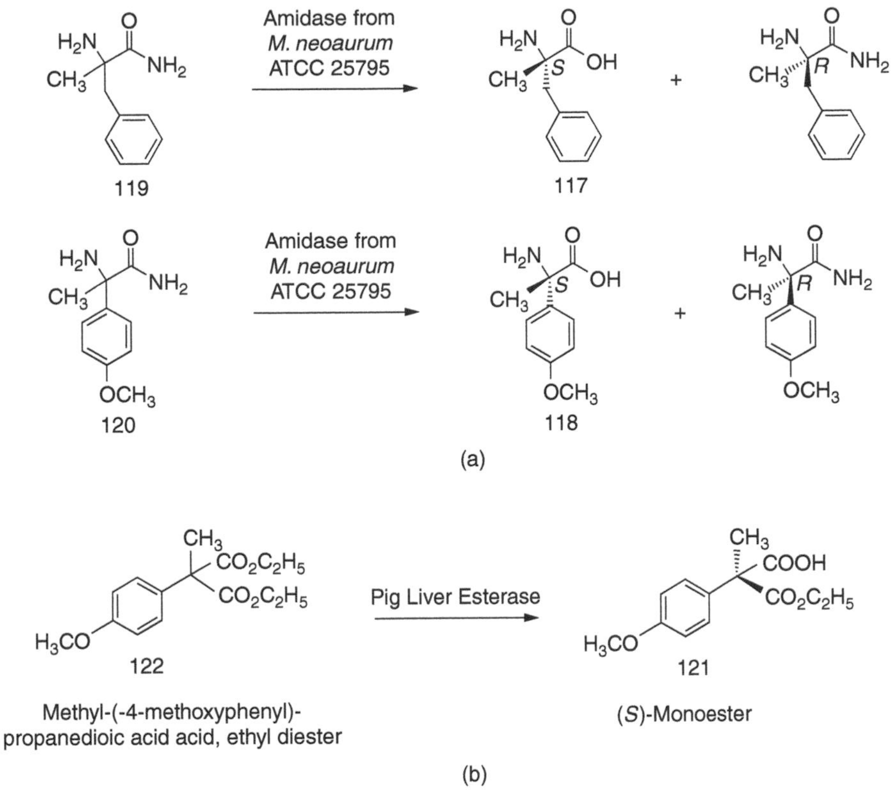

(a)

(b)

FIGURE 26.32 (A) Enzymatic synthesis of chiral synthon for β-3-receptor agonist **114**: (B) Enantioselective hydrolysis of α-methyl phenylalanine amide **119** and α-methyl-4-hydroxyphenylalanine amide **120** by amidase. (C) Enantioselective hydrolysis of methyl-(4-methoxyphenyl)-propanedioic acid ethyl diester **122** to (*S*)-monoester **121**.

esterase108 has been demonstrated. In various organic solvents the reaction yields and e.e. of monoester **121** were dependent upon the solvent used. High e.e.'s. (>91%) were obtained with methanol, ethanol, and toluene as a cosolvent. Ethanol gave the highest reaction yield (96.7%) and e.e. (96%) for the desired acid ester **121**.

It was observed that the e.e. of the (*S*)-monoester **121** was increased by decreasing the temperature from 25°C to 10°C, when biotransformation was conducted in a biphasic system using ethanol as a cosolvent. A semi-preparative 30 g scale hydrolysis was carried out using 10% ethanol as a cosolvent in a 3L reaction mixture (pH 7.2) at 10°C for 11 hr. A reaction yield of 96 M% and an e.e. of 96.9% were obtained. From the reaction mixture, 26 grams (86 M% overall yield) of (*S*)-monoester **121** of 96.9% e.e. were isolated.

26.10 Melatonin Receptor Agonists

26.10.1 Enantioselective Hydrolysis of Racemic 1-{2′,3′-Dihydro Benzo[b]Furan-4′-yl}-1,2-Oxirane

Epoxide hydrolase catalyzes the enantioselective hydrolysis of a epoxide to the corresponding enantiomerically enriched diol and unreacted epoxide.[109,110] The (*S*)-epoxide **123** is a key intermediate in the synthesis of a number of prospective drug candidates.[111] The enantiospecific hydrolysis of the racemic 1-{2′, 3′-dihydro benzo[b]furan-4′-yl}-1,2-oxirane **124** to the corresponding (*R*)-diol **125** and unreacted *S*-epoxide **123** (Figure 26.33a) was demonstrated by Goswami et al.[112] Among cultures evaluated, two *A. niger* strains

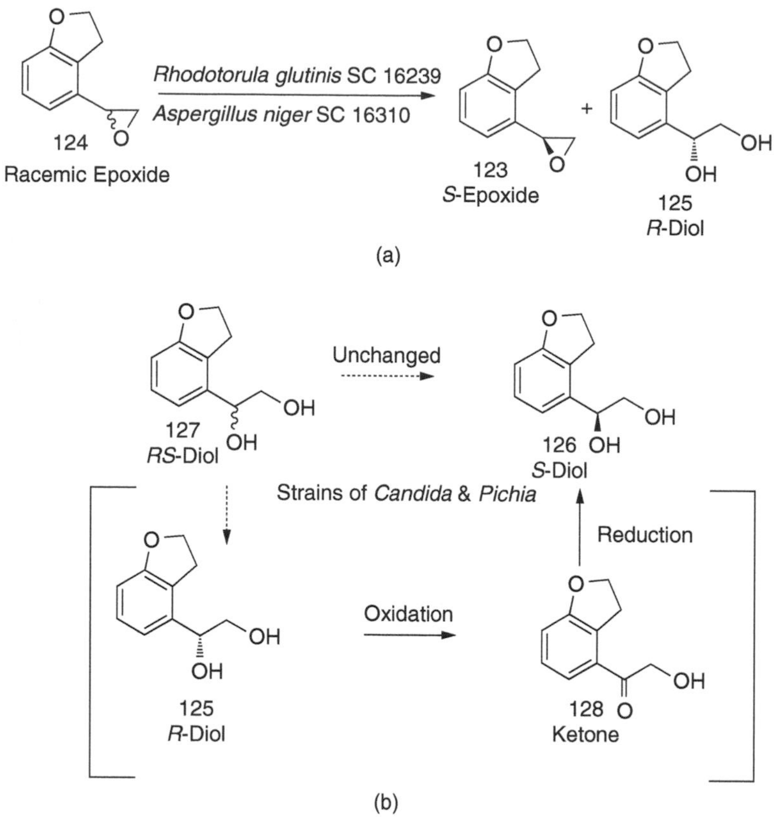

(a)

(b)

FIGURE 26.33 (A) Synthesis of chiral intermediates for melatonin receptor agonist: Enantioseective microbial hydrolysis of racemic epoxide **124** to the corresponding (R)-diol **125** and unreacted (S)-epoxide **123**. (B) Stereoinversion of racemic diol **127** to (S)-diol **126** by *Candida boidinii* and *Pichia methanolica*.

(SC 16310, SC 16311) and *Rhodotorula glutinis* SC 16293 selectively hydrolyzed the (R)-epoxide, leaving behind the (S)-epoxide **123** in >95% e.e. and 45% yield (theoretical maximum yield is 50%).

Several solvents at 10% vol/vol were evaluated in an attempt to improve the e.e. and yield. Solvents had significant effects on both the extent of hydrolysis and the e.e. of unreacted (S)-epoxide **123**. Most solvents gave a lower e.e. product and slower reaction rate than that of reactions without any solvent supplement. MTBE gave a reaction yield of 45% (theoretical maximum yield 50%) and an e.e. of 99.9% for unreacted (S)-epoxide **123**.

26.10.2 Biocatalytic Dynamic Kinetic Resolution of (R,S)-1-{2′,3′DihydroBenzo[b]Furan-4′-yl}-Ethane-1,2-Diol

Most commonly used biocatalytic kinetic resolution racemates often provide compounds with high e.e., and the maximum theoretical yield of the product is only 50%. In many cases, the reaction mixture contains a ca. 50:50 mixture of reactant and product, which possess only slight differences in physical properties (e.g., a hydrophobic alcohol and its acetate), and thus separation may be very difficult. These issues with kinetic resolutions can be addressed by employing a dynamic kinetic resolution process involving a biocatalyst or biocatalyst with metal-catalyzed *in situ* racemization.[113,114]

S-1-{2′,3′-dihydrobenzo[b]furan-4′-yl}-ethane-1,2-diol **126** (Figure 26.33b) is a potential precursor of epoxide **123**.[112] The dynamic kinetic resolution of the racemic diol **127** to the (S)-enantiomer **125** was demonstrated.[115] Seven cultures (*Candida boidinii* SC 13821, SC 13822, SC 16115, *Pichia methanolica* SC

13825, SC 13860 and *Hansenula polymorpha* SC 13895, SC 13896) were found to be promising for dynamic kinetic resolution. During biotransformation, the relative proportion of (*S*)-diol **126** increased with time and at the end of one week, the e.e. was found to be in the range of 87–100% (yield 60–75%) with these microorganisms. A new compound was formed during these biotransformations as evidenced by the appearance of a new peak in the HPLC of the reaction mixture. The identity of this compound was established as the hydroxy ketone **128** from an LC-MS peak at mass 178. The area of the HPLC peak for hydroxy ketone **128** first increased with time, reached a maximum, and then decreased, as expected for the proposed dynamic kinetic resolution pathway. *C. boidinii* SC 13822, *C. boidinii* SC 16115, and *P. methanolica* SC 13860 transformed the racemic diol **127** in 3–4 days, to S-diol **126** in yields of 62–75% and e.e.'s of 90–100%.

26.11 Anti-Alzheimer's Drugs

26.11.1 Resolution of Racemic Secondary Alcohols

(*S*)-2-Pentanol is an intermediate in the synthesis of several potential anti-Alzheimer's drugs, which inhibit β-amyloid peptide release and/or its synthesis.[116] The enzymatic resolution of racemic 2-pentanol and 2-heptanol by lipase B from *Candida antarctica* has been demonstrated.[117]

Commercially available lipases were screened for the enantioselective acetylation of racemic 2-pentanol in an organic solvent (hexane) in the presence of vinyl acetate as an acyl donor. *C. antarctica* lipase B efficiently catalyzed this reaction, giving yields of 49% (theoretical maximum yield 50%) and 99% e.e. for (*S*)-2-pentanol. Preparative scale acetylation (100 g input) was carried out. At the end of the reaction, 44.5 g of (*S*)-2-pentanol was estimated by HPLC analysis with an e.e. of 98%. Among acylating agents tested, succinic anhydride was found to be of choice due to easy recovery of the (*S*)-2-pentanol at the end of the reaction. Reactions were carried out using racemic 2-pentanol as solvent as well as substrate. Using 0.68 mole equivalent of succinic anhydride (Figure 26.34a) and 13 g of

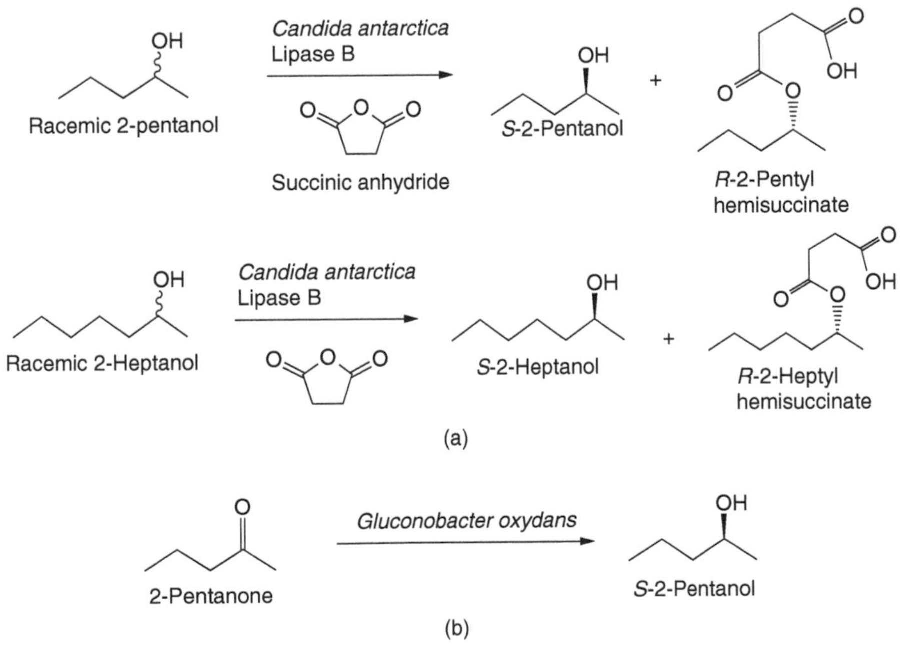

FIGURE 26.34 (A) Synthesis of chiral intermediates for anti-Alzheimer's drug: Enzymatic resolution of racemic secondary alcohols by *Candida antarctica* lipase. (B) Enantioselective microbial reduction of 2-pentanone.

lipase B per Kg of racemic 2-pentanol, a reaction yield 43 M% and ee of >98% were obtained for (S)-2-pentanol, isolated in overall 38% yield. The resolution of 2-heptanol was also carried out using lipase B under similar conditions to give a reaction yield of 44 M% and an e.e. of >99% of (S)-2-heptanol, isolated in overall 40 M% yield.

In an alternate approach, the enantioselective reduction of 2-pentanone to the corresponding (S)-2-pentanol (Figure 26.34b) has been demonstrated by *Gluconobacter oxydans*. Using triton X-100 treated cells of *G. oxydans* preparative scale reduction of 2-pentanone was carried out and 1.06 kg of (S)-2-pentanol was prepared.[118] A 2-ketoreductase (29,000 mol. wt.) has been purified to homogeneity from *G. oxydans*.

26.11.2 Enzymatic Reduction of 5-Oxohexanoate and 5-Oxohexanenitrile

(S)-5-hydroxyhexanoate **129** and (S)-5-hydroxyhexanenitrile **130** (Figure 26.35a) are key chiral intermediates in the synthesis of pharmaceuticals. Both chiral compounds have been prepared by enantioselective reduction of 5-oxohexanoate **131** and 5-oxohexanenitrile **132** by *Pichia methanolica* SC 16116. Reaction yields of 80–90% and >95% enantiomeric excess were obtained for each chiral compound. In an alternate approaches, the enzymatic resolution of racemic 5-hydroxyhexanenitrile **134** (Figure 26.35b) by enzymatic succinylation was demonstrated using immobilized lipase PS-30 to obtain (S)-5-hydroxyhexanenitrile **130** in 35% yield (maximum yield is 50%). In contast, (S)-5-acetoxy-hexanenitrile **135** was prepared by enantioselective enzymatic hydrolysis of racemic 5-acetoxyhexanenitrile **136** by *Candida antarctica* lipase. A reaction yield of 42% and an e.e. of >99% were obtained.[119]

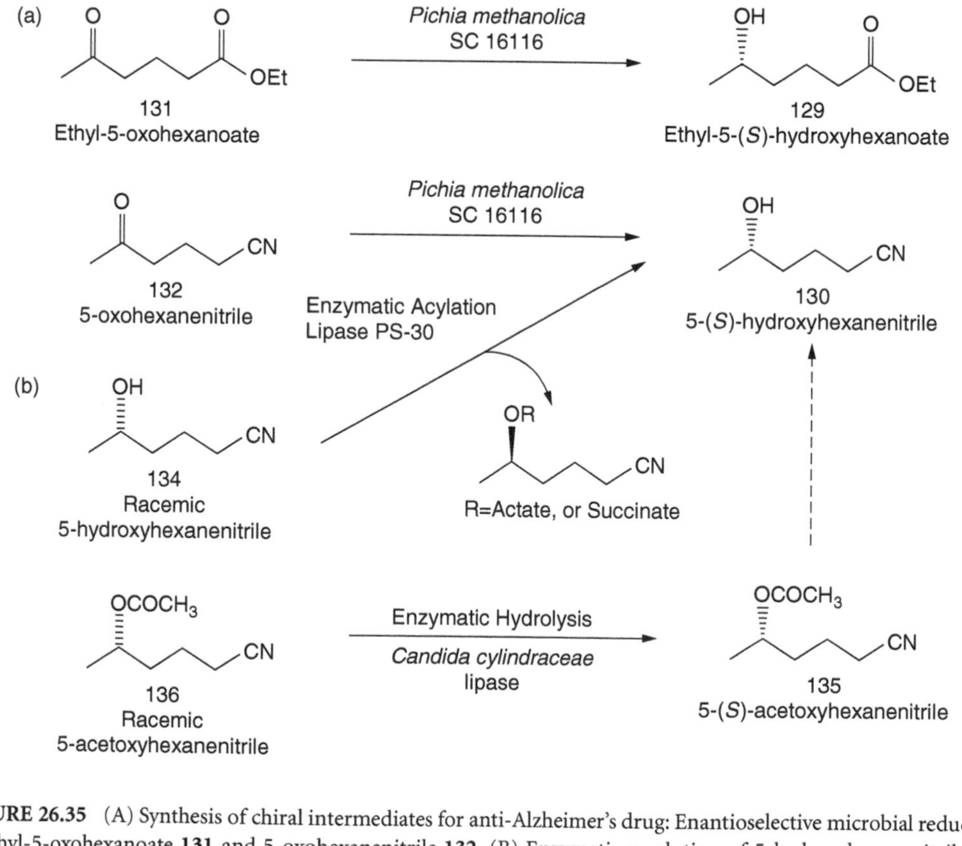

FIGURE 26.35 (A) Synthesis of chiral intermediates for anti-Alzheimer's drug: Enantioselective microbial reduction of ethyl-5-oxohexanoate **131** and 5-oxohexanenitrile **132**. (B) Enzymatic resolution of 5-hydroxyhexanenitrile **134** and 5-acetoxyhexanenitrile **136**.

26.12 Antiinfective Drugs

26.12.1 Pleuromutilin and Mutilin

Pleuromutilin **137** (Figure 26.36) is an antibiotic from *Pleurotus* or *Clitopilus* basidiomycetes strains that kills mainly gram-positive bacteria and mycoplasms. A more active semisynthetic analogue, tiamulin, has been developed for the treatment of animals and poultry infection and has been shown to bind to prokaryotic ribosomes and inhibits protein synthesis.[120] Metabolism of pleuromutilin derivatives results in hydroxylation by microsomal cytochrome P-450 at the 2- or 8-position and inactivates the antibiotics.[121] Modification of the 8-position of pleuromutilin and analogues is of interest as a means of preventing the metabolite hydroxylation. Microbial hydroxylation of pleuromutilin **137** or mutilin **138** would provide a functional group at this position to allow further modification at this site to avoid metabolic hydroxylation. The target analogues would maintain the biological activity of the parent compounds but not susceptible to metabolic inactivation.

Biotransformation of mutilin and pleuromutilin by microbial cultures was investigated to provide a source of 8-hydroxymutilin or 8-hydroxypleuromutilin.[122] LC/MS analysis of culture broths showed that several strains gave M+16 products from mutilin and one culture gave an M+16 product from pleuromutilin, suggesting addition of oxygen. Biotransformation products were extracted from culture broths with ethylacetate, dried, and purified by chromatography on silica gel. *Streptomyces griseus* strains SC 1754 and SC 13971 (ATCC 13273) converted mutilin to (8S)-, (7S)-, and (2S)-hydroxymutilin. *Cunninghamella echinulata* SC 16162 (NRRL 3655) gave (2S)-hydroxymutilin or (2R)-hydroxypleuromutilin from biotransformation of mutilin or pleuromutilin, respectively. The biotransformation of mutilin by *S. griseus* strain SC 1754 was scaled up in 15-, 60-, and 100-L fermentation to produce a total of 49 g of (8S)-hydroxymutilin (BMS-303786), 17 g of (7S)-hydroxymutilin (BMS-303789) and 13 g of (2S)-hydroxymutilin (BMS-303782) from 162 g of mutilin.[122]

A C-8 ketopleuromutilin derivative has been synthesized from the biotransformation product 8-hydroxymutilin.[123] A key step in the process was the selective oxidation at C-8 of 8-hydroxymutilin using

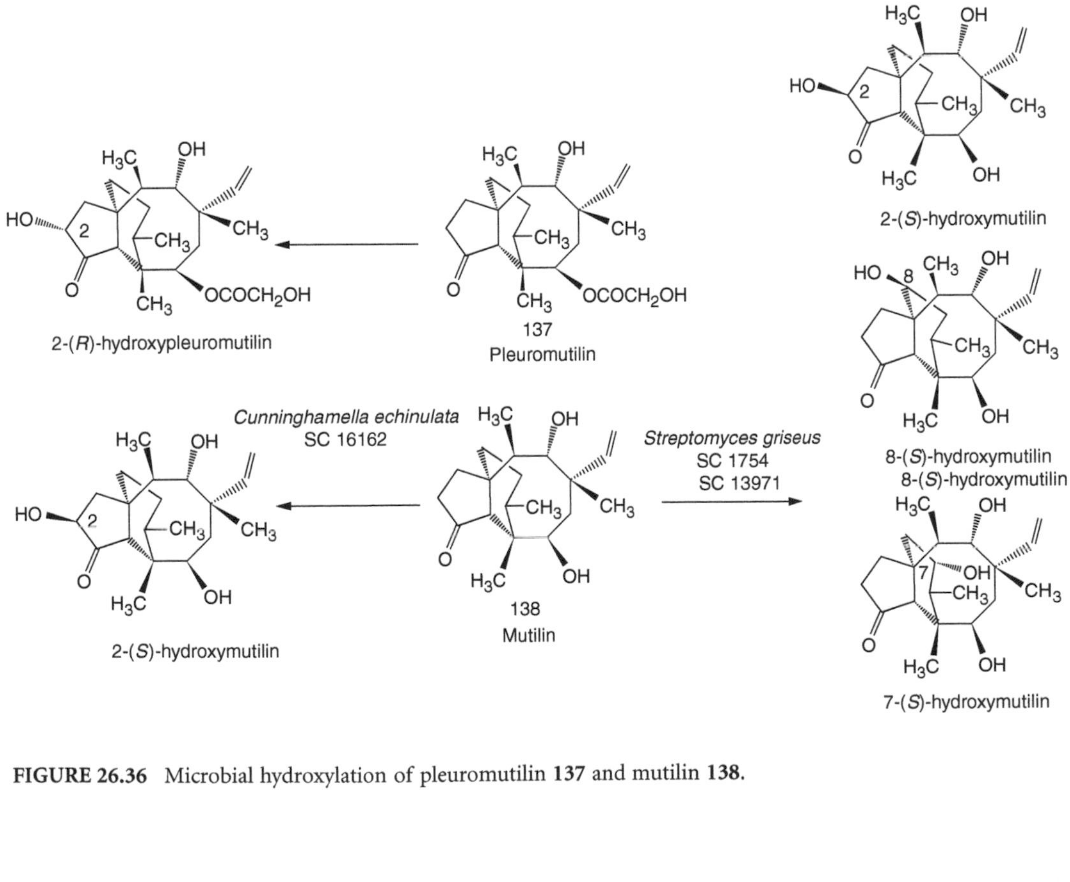

FIGURE 26.36 Microbial hydroxylation of pleuromutilin **137** and mutilin **138**.

tetrapropylammonium perruthenate. The presence of the C-8 keto group precipitated interesting intramolecular chemistry to afford a compound with a novel pleuromutilin-derived ring system.

26.12.2 Tigemonam

During the past several years synthesis of α-amino acids has been pursued intensely[124,125] because of their importance as building blocks of compounds of medicinal interest particularly antiinfective drugs. The asymmetric synthesis of β-hydroxy-α-amino acids by various methods has been demonstrated[126,127] because of their utility as starting materials for the total synthesis of monobactam antiobiotics.

L-β-hydroxyvaline **139**,[128] is a key chiral intermediate required for the total synthesis of orally active monobactam, Tigemonam **140** (Figure 26.37). The resolution of carbobenzyloxy (Cbz)-β-hydroxyvaline by chemical methods has been demonstrated.[129] Leucine dehydrogenase from strains of Bacillus[130] has been used for the synthesis of branched-chain amino acids. Hanson et al.[131] have described the synthesis of L-β-hydroxyvaline **139** from α-keto-β-hydroxy isovalerate **141** by reductive amination using leucine dehydrogenase from B. sphaericus ATCC 4525 (Figure 26.31). NADH required for this reaction was regenerated by either formate dehydrogenase from Candida boidinii or glucose dehydrogenase from B. megaterium. The immobilized cofactor such as polyethylene glycol-NADH and dextrans-NAD were effective in the biocatalytic process. The required substrate **141** was generated either from α-keto-β-bromo isovalerate or its ethyl esters by hydrolysis with sodium hydroxide *in situ*. In an alternate approach, the substrate **141** was also generated from methyl-2-chloro-3,3-dimethyloxiran carboxylate and the corresponding isopropyl and 1,1-dimethyl ethyl ester. These glycidic ester are converted to substrate **141** by treatment with sodium bicarbonate and sodium hydroxide. In this process, an overall reaction yield of 98% and an e.e. of 99.8% were obtained for the L-β-hydroxyvaline **139**.

26.13   Respiratory and Allergic Diseases

26.13.1   Resolution of *R,S*-N-(tert-butoxycarbonyl)-3-hydroxymethylpiperidine

S-N (tert-butoxycarbonyl)-3-hydroxymethylpiperidine **142** (26.38) is a key intermediate in the synthesis of a potent tryptase inhibitor, which are effective against respiratory diseases and allergic diseases.[132]

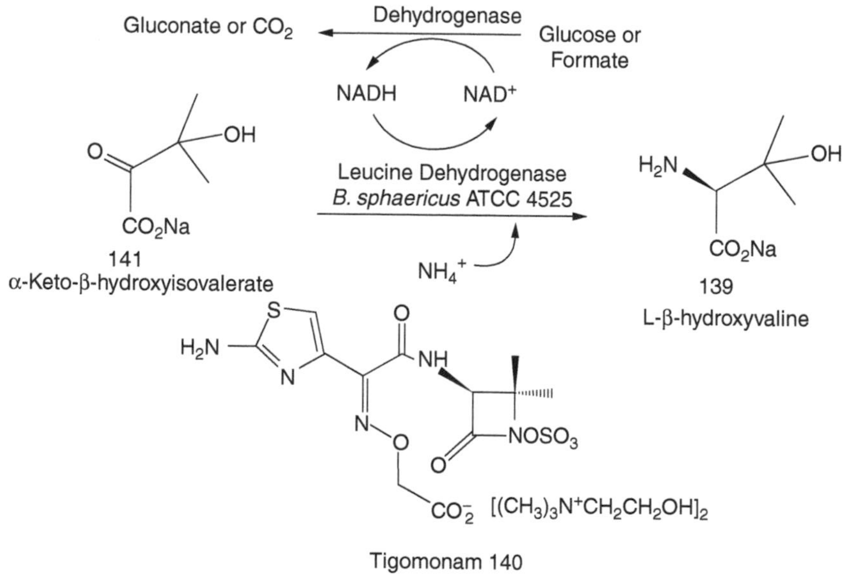

Tigomonam 140

FIGURE 26.37 Preparation of chiral synthon of tigemonam **140**: enzymatic conversion of α-keto-β-hydroxy isovalerate **141** to L-β-hydroxyvaline **139**.

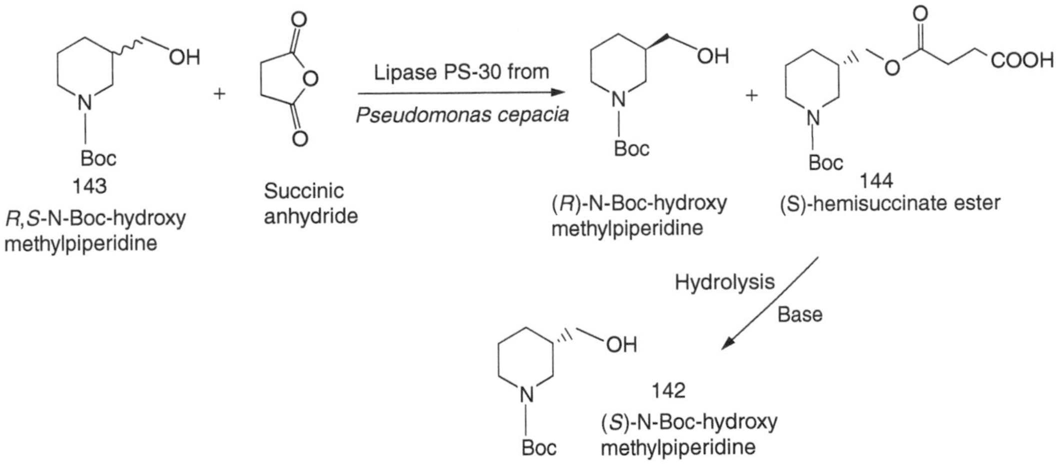

FIGURE 26.38   Preparation of chrial synthon for tryptase inhibitor. Enzymatic resolution of racemic (tert-butoxy-carbonyl)-3-hydroxymethyl-piperidine **143**.

Previous reports on the enzymatic resolution of R,S-3-(hydroxymethyl)piperidine by pig liver esterase and acylase showed only marginal enantiospecificity.[133] Enzymatic resolution of R,S-N-(tert-butoxycarbonyl)-3-hydroxymethylpiperidine **143** by lipase P from *Pseudomonas fluorescens* was reported for the preparation of R-N-(tert-butoxycarbonyl)-3-hydroxy methylpiperidine.[134] This process with low substrate concentration and chromatographic separation will have limited practical use in large-scale industrial application.

Goswami et al.[135] developed an alternate method for the preparation of S-N (tert-butoxycarbonyl)-3-hydroxymethylpiperidine **142**. Lipase PS-40 (Amano Enzymes) catalyzed esterification of the R,S-N-(tert-butoxycarbonyl)-3-hydroxymethylpiperidine **143** (Figure 26.38) with succinic anhydride provided the S-hemisuccinate ester **144**, which could be easily separated and hydrolyzed by base to the S-N-(tert-butoxycarbonyl)-3-hydroxymethylpiperidine **142**. The yield and e.e. could be improved greatly by repetition of the process. Using the repeated esterification procedure S-N-(tert-butoxycarbonyl)-3-hydroxymeth-ylpiperidine **142** was obtained in 32% yield (maximum theoretical yield 50%) and 98.9% e.e.

26.14   Acyloin Condensation

Asymmetric α-hydroxyketones (acyloin) are important classes of intermediates in organic synthesis due to their bi-functional aspect, especially having one chiral center amenable to further modification. Enzyme mediated acyloin formation could provide an advantageous, environment-friendly method to prepare optically active asymmetric acyloins.[136] Acyloin formations mediated by yeast pyruvate decarboxylase[137] and bacterial benzoylformate decarboxylase[138] have been reported. Though phenylpyruvate decarboxylase (PPD)[136] for decarboxylaton of phenylpyruvic acid was known for a long time, recently we reported the acyloin condensation catalyzed by PPD.[139] *Achromobacter eyrydice* PPD was used to catalyze the asymmetric acyloin condensation of phenylpyruvate **145** with various aldehydes **146** to produce optically active acyloins Ph CH₂COCH(OH)R **147** (Figure 26.39). The acyloin condensation yield decreased with increasing chain length for straight-chain aliphatic aldehydes from 76% for acetaldehyde to 24% for valeraldehyde. The e.e's of the acyloin products were 87–98%. Low yields of acyloin products were obtained with chloroacetaldehyde (13%) and glycoaldehyde (16%). Indole-3-pyruvate was a substrate of the enzyme and provided acyloin condensation product 3-hydroxy-1-(3-indolyl)-2-butanone **148** with acetaldehyde in 19% yield.

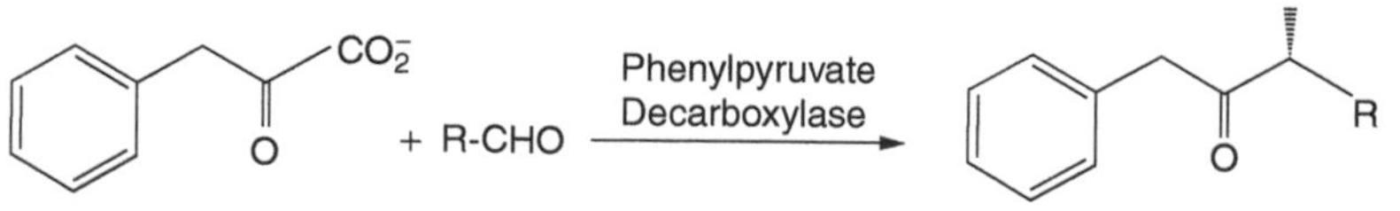

(R): a (Me), b (Et), c (n-Pr), d (n-Bu), e (PhCH$_2$), f (Ph), g (H),
h (Me(CH$_2$)$_6$), i (Me(CH$_2$)$_8$), j (Me(CH$_2$)$_{10}$), k (PhCH=CH), l (Br$_3$C), m (Me$_2$CH),
n (Me$_3$CCH$_2$), o (BrCH$_2$), p (BrCH$_2$CH$_2$), q (CH$_2$=CH), r (ClCH$_2$), s (HOCH$_2$).

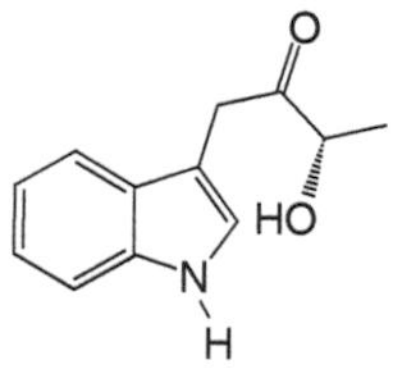

FIGURE 26.39 Enzymatic asymmetric acyloin condensation reactions catalyzed by phenylpyruvate decarboxylase.

26.15 Enantioselective Enzymatic Deprotection

Amino groups often require protection during synthetic transformations elsewhere in the molecule; at some point, the protecting group must be removed. Enzymatic protection and deprotection under mild conditions have been demonstrated previously. Penicillin G amidase and phathalyl amidase have been used for the enzymatic deprotection of the phenylacetyl and phthaloyl groups from the corresponding amido or imido compounds.[140–144] Acylases have been used widely in the enantioselective deprotection of N-acetyl-DL-amino acids.[145] Enzymatic deprotection of N-carbamoyl L-amino acids and N-carbamoyl D-amino acids has been demonstrated by microbial L-carbamoylases and D-carbamoylases, respectively.[146,147]

The cabobenzyloxy (Cbz) group is commonly used to protect amino and hydroxyl groups during organic synthesis. Chemical deprotection is usually achieved by hydrogenation with a palladium catalyst.[148–150] However, during chemical deprotection some groups are reactive under (e.g., carbon-carbon double bonds) or may interfere with (e.g., thiols or sulfides) to the hydrogenolysis conditions. In addition, the process is not enantioselective. Recently we have developed enantioselective enzymatic deprotection process that can be performed under mild conditions without damaging any otherwise susceptible groups in the molecule. A selective culture technique was used to isolate and identified *Sphingomonas paucimobilis* strain SC 16113 catalyzed the enantioselective cleavage of cabobenzyloxy groups from various Cbz-protected amino acids. Only Cbz-L-amino acids were deprotected, giving complete conversion to the corresponding L-amino acid. Cbz-D-amino acids gave <2% reaction yield.

Racemic Cbz-amino acids were also evaluated as substrates for hydrolysis by cell extracts of *S. paucimobilic* SC 16113. As anticipated only the L-enantiomer was hydrolyzed, giving the L-amino acids in >48% yield and >99% e.e. The unreacted Cbz-D-amino acids (Table 26.1) were recovered in >48% yield and >98% e.e.[151]

TABLE 26.1 Enzymatic Resolution of Cbz-Amino acids

Substrate	Product	HPLC Yield (%)	L-Amino Acids e.e. (%)	Cbz-D-Amino Acids	
				HPLC Yield (%)	e.e. (%)
N-α-Cbz-DL-tyrosine	L-tyrosine	48	99.4	49.5	99.4
N-α-Cbz-DL-Proline	L-Proline	49.8	99.8	49	98
N-α-Cbz-DL-phenylalanine	L-phenylalanine	48	99.4	49.5	99.3
N-α-Cbz-DL-Lysine	L-Lysine	49	99.9	48.8	99.4

26.16 Conclusion

The production of single enantiomers of drug intermediates is increasingly important in the pharmaceutical industry. Organic synthesis is one approach to the synthesis of single enantiomers, and biocatalysis provides an added dimension and an enormous opportunity to prepare pharmaceutically useful chiral compounds. The advantages of biocatalysis over chemical catalysis are that enzyme-catalyzed reactions are stereoselecive and regioselective and can be carried out at ambient temperature and atmospheric pressure. The use of different classes of enzymes for the catalysis of many different types of chemical reactions is capable of generating a wide variety of chiral compounds. This includes the use of hydrolytic enzymes such as lipases, esterases, proteases, dehalogenases, acylases, amidases, nitrilases, lyases, epoxide hydrolases, decarboxylases, and hydantoinases in the resolution of racemic compounds and in the asymmetric synthesis of enantiomerically enriched chiral compounds. Oxido-reductases and aminotransferases have been used in the synthesis of chiral alcohols, aminoalcohols, amino acids, and amines. Aldolases and decarboxylases have been effectively used in asymmetric synthesis by aldol condensation and acyloin condensation reactions. Monoxygenases have been used in enantioselective and regioselective hydroxylation and epoxidation reactions and dioxygenases in the chemo-enzymatic synthesis of chiral diols. In the course of the last decade, progress in biochemistry, protein chemistry, molecular cloning, random and site-directed mutagenesis, directed evolution of biocatalysts, and fermentation technology has opened up unlimited access to a variety of enzymes and microbial cultures as tools in organic synthesis.

References

1. Food & Drug Administration: FDA's statement for the development of new stereoisomeric drugs. *Chirality* 4: 338–340, 1992.
2. BC Buckland, DK Robinson, M Chartrain. Biocatalysis for pharmaceuticals: Status and prospects for a key technology. *Metabolic Engineering* 2: 42–48, 2000.
3. MK O'Brien, B Vanasse. Asymmetric processes in the large-scale preparation of chiral drug candidates. *Curr Opin Drug Discov Dev* 3: 793–806, 2000.
4. JA Pesti, R Dicosimo. Recent progress in enzymatic resolution and desymmetrization of pharmaceuticals and their intermediates. *Curr Opin Discov Dev* 3: 764–778, 2000.
5. RN Patel. Microbial/Enzymatic synthesis of chiral intermediates for pharmaceuticals. *Enzyme Microb Technol* 31: 804–826, 2002.
6. V Gotor. Pharmaceuticals through enzymatic transesterification and enzymatic aminolysis reactions. *Biocat Biotrans* 18: 87–103, 2000.
7. D Stewart. Dehydrogenases and transaminases in asymmetric synthesis. *Curr Opin Chem Biol* 5: 120–129, 2001.
8. A Zak. Industrial biocatalysis. *Curr Opin Chem Biol* 5: 130–136, 2001.
9. RN Patel. Enzymatic synthesis of chiral drug intermediates. *Adv Synth Cata* 343: 527–546, 2001.
10. RN Patel. Stereoselective biocatalysis for synthesis of some chiral pharmaceutical intermediates. In *Stereoselective Biocatalysis* . Edited by Patel RN. New York, Marcel Dekker, 2000; 87–130.
11. M Suffness, ME Wall. Discovery and development of taxol. In *Taxol: Science and Application* ed. Suffness M. New York, CRC Press, 1995.
12. RN Patel. Tour de paclitaxel: Biocatalysis for semisynthesis. *Ann Rev Microbiol* 98: 361–395, 1995.
13. RL Hanson, JM Wasylyk, VB Nanduri, DL Cazzulino, RN Patel, LJ Szarka. Site-specific enzymatic hydrolysis of taxanes at C-10 and C-13. *J Biol Chem* 269: 22145–22149, 1994.
14. VB Nanduri, RL Hanson, TL LaPorte, RY Ko, RN Patel, LJ Szarka. Fermentation and isolation of C-10 deacetylase for the production of 10-DAB from baccatin III. *Biotech Bioeng* 48: 547–550, 1995.
15. RN Patel, A Banerjee, JM Howell, CG McNamee, DB Brzozowski, VB Nanduri, JK Thottathil, LJ Szarka. Stereoselective microbial reduction of 2-keto-3-(N-benzoylamino)-3-phenyl propionic acid ethyl ester. Synthesis of taxol side-chain synthon. *Tetrahedron*: Asymmetry 4: 2069–2084, 1993.

16. RN Patel, A Banerjee, RY Ko, JM Howell, WS Li, FT Comezoglu, RA Partyka, LJ Szarka. Enzymic preparation of (3R-cis)-3-(acetyloxy)-4-phenyl-2-azetidinone: a taxol side-chain synthon. *Biotech and Appl Biochem* 20: 23–33, 1994.

17. V Guillemard, H Saragovi. Taxane-antibody conjugates afford potent cytotoxicity, enhanced solubility, and tumor target selectivity. *Cancer Research* 61: 694–699, 2001.

18. WC Rose, BH Long, CR Fairchild, FY Lee, JF Kadow, Preclinical pharmacology of BMS-275183, an orally active taxane. *Clin. Cancer Res.* 7: 2016–2021, 2001.

19. ML Rothenberg, Taxol, Taxotere and other new taxanes. *Current Opinion in Investigational Drugs* . 2: 1269–1277, 1993.

20. RA Holton, H Nadizadeh, RJ Beidiger, Preparation of furyl- and thienyl-substituted taxanes as antitumor agents. Eur. Pat. Appl. 1993, 17 pp. EP 534708 A1 19930331 CAN 119: 49693 AN 1993:449693.

21. AE Mathew, MR Mejillano, JP Nath, RH Himes, VJ Stella, Synthesis and evaluation of some water-soluble prodrugs and derivatives of taxol with antitumor activity. *J Med Chem* 1992, 35(1), 145–151.

22. D-M Gou, Y-C Liu, C-S Chen. *J Org Chem* 58: 1287–1289, 1993.

23. R Brieva, JZ Crich, C Sih, Chemoenzymic synthesis of the C-13 side chain of taxol: optically active 3-hydroxy-4-phenyl β-lactam derivatives. *J Org Chem* 58: 1068–1072, 1993.

24. RN Patel, JM Howell, R Chidambaram, S Benoit, J Kant. Enzymatic preparation of (3R-*cis*)-3-Acetyloxy-4-(1,1-dimethylethyl)-2-azetidinone: A side-chain synthon for an orally active taxane. *Tetrahedron : Asymmetry* (in press).

25. H Umezawa, S Kondo, H Iinuma, Y Kunimoto, H Iwasawa, D Ikeda, T Takeuchi. Structure of an antitumor antibiotic, spergualin. *J Antibiotics* 34: 1622–1624, 1981.

26. Y Umeda, M Moriguchi, I Katsushige, H Kuroda, T Nakamura, A Fujii, T Takeuchi, H Umezawa. Synthesis and antitumor activity of spergualin analogues. III. Novel method for synthesis of optically active 15-deoxyspergualin and 15-deoxy-11-O-methylspergualin. *J Antibiotics* 40: 1316–1324, 1987.

27. RN Patel, A Banerjee, LJ Szarka. Stereoselective acetylation of racemic 7-[N,N'-bis-benzyloxycarbonyl)-N-(guanidinoheptanoyl)]-α-hydroxyglycine. *Tetrahedron: Asymmetry* 8: 1767–1771, 1997.

28. TA Krenitsky, DR Averett, JD Wilson, AR Moorman, GW Koszalka, SD Chamberlain, DW Porter. Preparation of 2-amino-6-alkoxy-9-(β-D-arabinofuranosyl)-9H-purines and esters as antitumor agents. 1992, *PCT Int Appl* WO 9201456, CAN 116: 236100.

29. M Mahmoudian, J Eaddy, M Dawson. Enzymic acylation of 506U78 (2-amino-9-D-arabinofuranosyl-6-methoxy-9H-purine), a powerful new anti-leukemic agent. *Biotechnol Appl Biochem* 29: 229–233, 1999.

30. B Morgan, A Zaks, DR Dodds, J Liu, R Jain, S Megati, FG Njoroge, VM Girijavallabhan. Enzymatic kinetic resolution of piperidine atropisomers: synthesis of a key intermediate of the farnesyl protein transferase inhibitor, SCH66336. *J Org Chem* 2000, 65: 5451–5459.

31. JC Barrish, E Gordon, M Alam, PF Lin, GS Bisacchi, PT Cheng, AW Fritz, JA Greytok, MA Hermsmeier, WG Humphreys, KA Lis, MA Marella, Z Merchant., T Mitt, RA Morrison, MT Obermeier, J Pluscec, M Skoog, WA Slusarchyk, S Spergel, JM Stevenson, CQ Sun, JE Sundeen, P Taunk, JA Tino, BM Warrack, R Colono, R Zahler. Aminodiol HIV protease inhibitors. 1. Design, synthesis, and preliminary SAR. *J Med Chem* 37: 1758–1771, 1994.

32. RN Patel, A Banerjee, C McNamee, D Brzozowski, LJ Szarka. Preparation of chiral synthon for HIV protease inhibitor: stereoselective microbial reduction of N-protected α-aminochloroketone. *Tetrhedron: Asymmetry* . 1997; 8: 2547–2552.

33. A Fassler, G Bold, HG Capraro, R Cozens, J Mestan, B Poncioni, J Rosel, M Tintelnot-Blomley, M Lang. Aza-Peptide Analogs as Potent human immunodeficiency virus type-1 protease inhibitors with oral bioavailability. *J Med Chem* 39: 3203–3216, 1996.

34. BS Robinson, KA Riccardi, Y-F Gong, Q Guo, DA Stock, WS Blair, BJ Terry, CA Deminie, F Djang, RJ Colonno, P-F Lin. BMS-232632, a highly potent human immunodeficiency virus protease inhibitor that can be used in combination with other available antiretroviral agents. *Antimicrobial Agents and Chemotherapy* 44: 2093–2099, 2000.

35. Y-F Gong, BS Robinson, RE Rose, C Deminie, TP Spicer, D Stock, RJ Colonno, P-F Lin, *In vitro* resistance profile of the human immunodeficiency virus type 1 protease inhibitor BMS-232632. *Antimicrobial Agents and Chemotherapy* 44: 2319–2326, 2000.

36. RN Patel, L Chu, R Mueller. Diastereoselective Microbial Reduction of (*S*)-[3-Chloro-2-oxo-1-(phenylmethyl)propyl]carbamic acid, 1,1-dimethylethyl ester. *Tetrahedron: Asymmetry* (in press).

37. X Rabasseda, J Silvestre, J Castaner. BMS-232632: anti-HIV HIV-1 protease inhibitor. *J Drugs of the Future* 24: 375–380, 1999.

38. M Chartrain, B Jackey, C Taylor, V Sandfor, K Gbewonyo, L Lister, L Dimichele, C Hirsch, B Heimbuch, C Maxwell, D Pascoe, B Buckland, R Greasham. Bioconversion of indene to cis-(1S,2R)-Indandiol and trans- (1R,2R)-indandiol by Rhodococcus sp. *J Ferment Technol* 86: 550-558, 1998.

39. BC Buckland, DK Robinson, M Chartrain. Biocatalysis for pharmaceuticals — status and prospects for a key technology. *Metabolic Engineering* 2: 42–48, 2000.

40. M Chartrain, PM Salmon, DK Robinson, BC Buckland. Metabolic engineering and directed evolution for the production of pharmaceuticals. *Curr Opin Biotechnol* 11: 209–214, 2000.

41. N Zhang, BG Stewart, JC Moore, RL Greasham, DK Robinson, B Buckland, C Lee. Directed evolution of toluene dioxygense from *Pseudomonas putida* for improved selectivity toward cis-(1S,2R)-indanediol during indane bioconversion. *Metab Eng* 2: 339–348, 2000.

42. CT Evans, SM Roberts, KA Shoberu, AG Sutherland. Potential use of carbocyclic nucleosides for the treatment of AIDS, chemoenzymatic synthesis of the enantiomers of carbovir. *J Chem Soc Perkin Trans* 1: 589–592, 1992.

43. M Mahmoudian, A Lowdon, M Jones, M Dawson, C Wallis. A practical enzymic procedure for resolution of N-substituted 2-azabicyclo[2.2.1]hept-5-en-3-ones. *Tetrahedron: Asymmetry* 10: 1201–1206, 1999.

44. C Ireland, PA Leeson, J Castaner. Lobucavir: *Antiviral Drugs Future* 22: 359–370, 1997.

45. RL Hanson, Z Shi, DB Brzozowski, A Banerjee, TP Kissick, J Singh, AJ Pullockaran, JT North, J Fan, J Howell, SC Durand, MA Montana, DR Kronenthal, RH Mueller, RN Patel. Regioselective enzymatic aminoacylation of lobucavir to give an intermediate for lobucavir prodrug. *Bioorg Med Chem* 8: 2681–2687, 2000.

46. MA Ondetti, DW Cushman. Inhibition of renin-angiotensin system: a new approach to the theory of hypertension. *J Med Chem* 24: 355–361, 1981.

47. MA Ondetti, B Rubin, DW Cushman. Design of specific inhibitors of angiotensin-converting enzyme: new class of orally active antihypertensive agents. *Science* 196: 441–444,1977.

48. DW Cushman, MA Ondetti. Inhibitors of angiotensin-converting enzyme for treatment of hypertension. *Biochem Pharm* 29: 1871–1875, 1980.

49. CT Goodhue, JR Schaeffer. Preparation of L-(+)-β-hydroxyisobutyric acid by bacetrial oxidation of isobutyric acid. *Biotechnol Bioeng* 13: 203–214, 1971.

50. RN Patel, JM Howell, CG McNamee, KF Fortney, LJ Szarka. Stereoselective enzymatic hydrolysis of α-[acetylthio)methyl]benzenepropanoic acid and 3-acetylthio-2-methylpropanoic acid. *Biotechnol Appl Biochem* 1992; 16: 34–47.

51. RN Patel, JM Howell, A Banerjee, KF Fortney, LJ Szarka. Stereoselective enzymatic esterification of 3-benzoylthio-2-methylpropanoic acid. *Appl Microbiol Biotechnol* 36: 29–34, 1991.

52. JL Moniot. Preparation of N-[2-(mercaptomethyl)propionyl]-L-prolines. U.S. Patent Application; 1988 CN 88-100862.

53. MA Ondetti, A Miguel, J Krapcho. Mercaptoacyl derivatives of substituted prolines. U.S. Patent 4316906, 1982.

54. RN Patel, RS Robison, LJ Szarka. Stereoselective enzymic hydrolysis of 2-cyclohexyl- and 2-phenyl-1,3-propanediol diacetate in biphasic systems. *Appl Microbiol Biotechnol* 34: 10–14, 1990.

55. DS Karenewsky, MC Badia, DW Cushman, JM DeForrest, T Dejneka, MJ Loots, MG Perri, EW Petrillo, JR Powell. (Phosphinyloxy)acyl amino acid inhibitors of angiotensin converting enzyme (ACE). 1. Discovery of (*S*)-1-[6-amino-2-[[hydroxy(4-phenylbutyl)-phosphinyl]oxy]-1-oxohexyl]-L-proline, a novel orally active inhibitor of ACE. *J Med Chem* 31: 204–212, 1988.

56. RL Hanson, KS Bembenek, RN Patel, LJ Szarka. Transformation of N-ε-CBZ-L-lysine to CBZ-L-oxylysine using L-amino acid oxidase from *Providencia alcalifaciens* and L-2-hydroxy-isocaproate dehydrogenase from *Lactobacillus confusus. Appl Microbiol Biotechnol* 37: 599–603, 1992.

57. NG Delaney, EN Gordon, JM DeForrest, DW Cushman. Amino acid and peptide derivatives as inhibitors of neutral endopeptidase and their use as antihypertensives and diuretics. European Patent EP361365, 1988.

58. JA Robl, C Sun, J. Stevenson, DE Ryono, LM Simpkins, MAP Cimarusti, T Dejneka, WA Slusarchyk, S Chao, L Stratton, RN Misra, MS Bednarz, MM Asaad, HS Cheung, BE Aboa-Offei, PL Smith, PD Mathers, M Fox, TR Schaeffer, AA Seymour, NC Trippodo. Dual metalloprotease inhibitors: mercaptoacetyl-based fused heterocyclic dipeptide mimetics as inhibitors of angiotensin-converting enzyme and neutral endopeptidase. *J Med Chem* 40: 1570–1577, 1997.

59. AA Seymour, JN Swerdel, BE Abboa-Offei. Antihypertensive activity during inhibition of neutral endopeptidase and angiotensin converting enzyme. *J Cardiovasc Pharmacol* 17: 456–465, 1991.

60. R WichmanR, C Wandrey, AF Bueckmann, M-R Kula. Continuous enzymic transformation in an enzyme membrane reactor with simultaneous NAD(H) regeneration. *Biotechnol Bioeng* 23: 2789–2802, 1981.

61. U Kragl, D Vasic-Racki, C Wandrey. Continuous processes with soluble enzymes. *Chem Ing Tech* 64: 499–509, 1992.

62. RL Hanson, MD Schwinden, A Banerjee, DB Brzozowski, B-C Chen, BP Patel, C McNamee, G Kodersha, D Kronenthal, RN Patel, LJ Szarka. Enzymatic synthesis of L-6-hydroxy-norleucine. *Bioorg Med Chem* 7: 2247–2252, 1999.

63. A Rumbero, JC Martin, MA Lumbreras, P Liras, C Esmahan. Chemical synthesis of allysine ethylene acetal and conversion *in situ* into 1-piperideine-6-carboxylic acid: key intermediate of the α-aminoadipic acid for β-lactam antibiotics biosynthesis. *Bioorg Med. Chem* 3: 1237–1240, 1995.

64. RL Hanson, JM Howell, TL LaPorte, MJ Donovan, DL Cazzulino, V Zannella, MA Montana, Vb Nanduri, SR Schwarz, RF Eiring, SC Durand, JM Wasylyk, WL Parker, LJ Szarka, RN Patel. Synthesis of allysine ethylene acetal using phenylalanine dehydrogenase from Thermoactinomyces intermedius. *Enzyme Microbial Technol* 26: 348–358, 2000.

65. H Schütte, J Flossdorf, H Sahm, M-R Kula. Purification and properties of formaldehydedehydrogenase and formate dehydrogenase from *Candida boidinii. Eur J Biochem* 62: 151–160, 1976.

66. CT Hou, RN Patel, AI Laskin, N Barnabe. NAD-linked formate dehydrogenase from methanol-grown Pichia pastoris NRRL-Y-7556. *Arch Biochem Biophys* 216: 296–305, 1982.

67. RN Patel. Enzymatic synthesis of chiral intermediates for Omapatrilat, an antihypertensive drug. *Biomolecular Engineering* 17: 167–182, 2001.

68. RN Patel, A Banerjee, V Nanduri, S Goldberg, R Johnston, R Hanson, C McNamee, D Brzozowski, T Tully, R Ko, T LaPorte, D Cazzulino, S Swaminathan, L Parker, J Venit. Biocatalytic preparation of a chiral synthon for a vasopeptidase inhibitor: Enzymatic conversion of N2-{N-[(phenylmethoxy)carbonyl]L-homocysteinyl]-L-lysine (1>1')-disulfide to [4S-(4a,7a,10ab)]1-octahydro-5-oxo-4-[(phenylmethoxy) carbonyl]amino]-7H-pyrido-[2,1-b][1,3]thiazepin-7-carboxylic acid methyl ester by a novel L-lysine ε-aminotransferase. *Enzyme Microb Technol* 27: 376–389, 2000.

69. A Endo, M Kuroda, Y Tsujita. ML-236A, ML-236B, and ML-236C, new inhibitors of cholesterogenesis produced by Penicillum citrinium. *J Antibiot* 29: 1346–1348, 1976.

70. A Endo, M Kuroda, K Tanzawa. Competitive inhibition of 3-hydroxy-3-methylglutaryl coenzyme A reductase by ML-236A and ML-236B fungal metabolites, having hypocholesterolemic activity. *FEBS Lett* 72: 323–326, 1976.

71. M Hosobuchi, K Kurosawa, H Yoshikawa. Application of computer on monitoring and control of fermentation process: microbial conversion of ML-236B sodium to pravastatin. *Biotechnol and Bioeng* 42: 815–820, 1993.

72. N Serizawa, S Serizawa, K Nakagawa, K Furuya, T Okazaki, A Tarahara. Microbial hydroxylation of ML-236B (compactin). Studies on microorganisms capable of 3β-hydroxylation of Ml-236B *J Antibiot* 36: 887–891, 1983.

73. K Mori, H. Mori, H. Sugai. Biochemical preparation of both the enantiomers of methyl-3- hydroxypentanoate and their conversion to the enantiomers of 4-hexanolide, the pheromone of *Trogoderma glabrum* .*Tetrahedron* 41: 919–925, 1985.

74. AS Gopalan, CJ Sih. Bifunctional chiral synthons via biochemical methods. 5. Preparation of (S)-ethyl hydrogen-3-hydroxyglutarate, key intermediate to (R)-4-amino-3- hydroxybutyric acid and L-carnitine. *Tetrahedron Letters* 25: 5235–5238, 1984.

75. CJ Sih, BN Zhan, AS Gropalan, CS Chen, G Girdaukais, F vanMiddlesworth. Enantioselective reduction of β-ketoesters by Baker's yeast. *Ann N Y Acad Sci* 434: 186–193, 1984.

76. OP Ward, CS Young. Reductive biotransformations of organic compounds by cells or enzymes of yeast. *Enzyme Microb Technol* 12: 482–493, 1990.

77. RN Patel, CG McNamee, A Banerjee, JM Howell, RS Robison, LJ Szarka. Stereoselective reduction of β-ketoesters by *Geotrichum candidum* . *Enzyme Microb Technol* 14: 731–738, 1992.

78. E Keinan, EK Hafeli, KK Seth, RR Lamed. Thermostable enzymes in organic synthesis. 2. Asymmetric reduction of ketones with alcohol dehydrogenase from *Thermoanaerobium brockii*. *J Am Chem Soc* 108: 162–168, 1986.

79. RN Patel, CT Hou, AI Laskin, P Derelanko. Microbial production of methylketones: properties of purified yeast secondary alcohol dehydrogenase. *J Appl Biochem* 3: 218–232, 1981.

80. RN Patel, A Banerjee, CG Mc Namee, D Brzozowski, RL Hanson, LJ Szarka. Enantioselective microbial reduction of 3,5-dioxo-6-(benzyloxy) hexanoic acid, ethyl ester. *Enzyme Microb Technol* 15: 1014–1021, 1993.

81. SY Sit, RA Parker, I Motoe, HW Balsubramanian, CD Cott, PJ Brown, WE Harte, MD Thompson, J Wright. Synthesis, biological profile, and quantitative structure-activity relationship of a series of novel 3-hydroxy-3-methylglutaryl coenzyme A reductase inhibitors. *J Med Chem* 33: 2982–2999, 1990.

82. RN Patel, CG McNamee, LJ Szarka. Enantioselective enzymic acetylation of racemic [4-[4α,6β (E)]]-6-[4,4-bis(4-fluorophenyl)-3-(1-methyl-1H-tetrazol-5-yl)-1,3-butadienyl] tetrahydro-4-hydroxy-2H-pyran-2-one. *Appl Microbiol Biotechnol* 38: 56–60, 1992.

83. A Steiger, HJ Pyun, RM Coates. Synthesis and characterization of aza analog inhibitors of squalene and geranylgeranyl diphosphate synthases. *J Org Chem* 57: 3444–3450, 1992.

84. M Lawrence, SA Biller, OM Fryszman. Preparation of α-phosphonosulfinic squalene synthetase inhibitors. 1995, U.S. Patent 5447922.

85. RN Patel, A Banerjee, LJ Szarka. Stereoselective acetylation of 1-(hydroxy)-4-(3-phenoxyphenyl)butyl]phosphonic acid diethyl ester. *Tetrahedron Asymmetry* 8: 1055–1059, 1997.

86. M Nakane. Preparation and formulation of 7-oxabicycloheptane substituted sulfonamide prostaglandin analogs useful in treatment of thrombolic disease. 1987, U.S. Patent 4,663, 336.

87. AW Ford-Hutchinson. Innovations in drug research: inhibitors of thromboxane and leukotrienes. *Clin and Exper Allergy*. 21: 272–276, 1991.

88. J Das, MF Haslanger, JZ Gougoutas, MF Malley. Synthesis of optically 7-oxabicyclo[2.2.1]heptanes and assignment of absolute configuration. *Synthesis*. 12: 1100–1112, 1987.

89. RN Patel, M Liu, A Banerjee, JK Thottathil, J Kloss, LJ Szarka. Stereoselective microbial/enzymatic oxidation of (exo,exo)-7-oxabicyclo[2.2.1]heptane-2,3-dimethanol to the corresponding chiral lactol and lactone. *Enzyme Microb Technol* 14: 778–784, 1992.

90. RN Patel, M Liu, A Banerjee, LJ Szarka. Stereoselective enzymic hydrolysis of (exo,exo)-7-oxabicyclo[2.2.1]heptane-2,3-dimethanol diacetate ester in a biphasic system. *Appl Microbiol Biotechnol* 37: 180–183, 1992.

91. M Chaffman, RN Brogden. Diltiazem. A review of its pharmacological properties and therapeutic efficacy. *Drugs* 1985 29: 387–390.

92. C Kawai C, T Konishi T, E Matsuyama E, H Okazaki H. Comparative effects of three calcium antagonists, diatiazem, verapamil and nifedipine, on the sinoatrial and atrioventricular nodes. Experimental and clinical studies. *Circulation* 63: 1035–1038, 1981.

93. T Isshiki, B Pegram, E Frohlich. Immediate and prolonged hemodynamic effects of TA-3090 on spontaneously hypertensive (SHR) and normal Wistar-Kyoto (WKY) rats. *Cardiovasc Drug Ther* 2: 539–544, 1988.

94. DM Floyd, RY Moquin, KS Atwal, SZ Ahmed, SH Spergel, JZ Gougoutas, MF Malley. Synthesis of benzazepinone and 3-methylbenzothiazepinone analogs of diltiazem. *J Org Chem* 55: 5572–5575, 1990.

95. RN Patel, RS Robison, LJ Szarka, J Kloss, JK Thottathil, RH Mueller. Stereospecific microbial reduction of 4,5-dihydro-4-(4-methoxyphenyl)-6-(trifluoromethyl-1H-1)-benzazepin-2-one. *Enzyme Microb Technol* 13: 906–912, 1991.

96. G Edwards, AH Weston. Structure-activity relationships of K+ channel openers. TIPS 11: 417–422, 1990.

97. DW Robertson, MI Steinberg. Potassium channel openers: new biological probes. *Ann Med Chem* 24: 91–100, 1989.

98. TC Hamilton, AH Weston. Cromakalim, nicorandil and pinacidil: novel drugs which open potassium channels in smooth muscle. *Gen Pharmac* 20: 1–9, 1989.

99. VA Ashwood, RE Buckingham, F Cassidy, JM Evans, TC Hamilton, DJ Nash, G Stempo, KJ Willcocks. Synthesis and antihypertensive activity of 4-(cyclic amido)-2H-1- benzopyrans. *J Med Chem* 29: 2194–2201, 1986.

100. R Bergmann, V Eiermann, RJ Gericke. 4-Heterocyclyloxy-2H-1-benzopyran potassium channel activators. *J Med Chem* 33: 2759–2761, 1990.

101. K Atwal, GJ Grover, KS Kim. Preparation of benzopyranylcyanoguanidine derivatives. 1991, U.S. Patent 5140031.

102. JM Evans, CS Fake, TC Hamilton, RH Poyser, EA Watts. Synthesis and antihypertensive activity of substituted trans-4-amino-3,4-dihydro-2,2-dimethyl-2H-1-benzopyran-3-ols. *J Med Chem* 26: 1582–1589, 1983.

103. RN Patel, A Banerjee, B Davis, JM Howell, CG McNamee, D Brzozowski, J North, D Kronenthal, LJ Szarka. Stereoselective epoxidation of 2,2-dimethyl-2H-1. benzopyran-6-carbonitrile. *Bioorg Med Chem* 2: 535–542, 1994.

104. RN Patel, A Banerjee, CG McNamee, LJ Szarka. Stereoselective acetylation of 3,4-dihydro-3,4-dihydroxy-2,2-dimethyl-2H-1-benzopyran-6-carbonitrile. *Tetrahedron: Asymmetry* 6: 123–130, 1995.

105. JRS Arch. β3-adrenoceptors and other putative atypical β-adrenoceptors. *Pharmacol Rev Commun* 9: 141–148, 1997.

106. JD Bloom, MD Datta, BD Johnson, A Wissner, MG Bruns, EE Largis, JA Dolan, TH Claus. Disodium (R,R)5-[2-(3-chlorophenyl)-2-hydroxyethyl]amino]propyl]-1,3-benzodioxole-2,2-dicarboxylate. A potent β-adrenergic agonist virtually specific for β3 receptors. *J Med Chem* 35: 3081–3084, 1989.

107. LG Fisher, PM Sher, S Skwish, IM Michael, S Seiler, KEJ Dickinson. BMS-187257, a potent, selective, and novel heterocyclic β-3 adrenergic receptor agonist. *Bioorg Med Chem Lett* 6: 2253–2258, 1994.

108. RN Patel, A Banerjee, L Chu, D Brzozowski, V Nanduri, LJ Szarka. Microbial synthesis of chiral intermediates for β-3-receptor agonists. *J Am Oil Chem Soc* 75: 1473–1482, 1998.

109. A Archelas, R Furstoss. Biocatalytic approaches for the synthesis of enantiopure epoxides. *Top Curr Chem* 200: 159–191, 1999.

110. M Mischitz, W Kroutil, U Wandel, K Faber. Asymmetric microbial hydrolysis of epoxides. *Tetrahedron: Asymmetry* 6: 1261–1272, 1995.

111. JD Catt, G Johnson, DJ Keavy, RJ Mattson, MF Parker, KS Takaki, JP Yevich. Preparation of benzofuran and dihydrobenzofuran melatonergic agents. 1999, U.S. 5856529, CAN 130:110151.

112. A Goswami A, MJ Totleben MJ, AK Singh AK, RN Patel RN. Stereospecific enzymatic hydrolysis of racemic epoxide: a process for making chiral epoxide. *Tetrahedron: Asymmetry* 10: 3167–3175, 1999.

113. H Stecher, K Faber. Biocatalytic deracemization techniques. Dynamic resolutions and stereoinversions. *Synthesis* 1: 1–16, 1997.

114. G Fantin, M Fogagnolo, A Medici, A Pedrini, S Fontana. Kinetic resolution of racemic secondary alcohols via oxidation with Yarrowia lipolytica strains. *Tetrahedron: Asymmetry* 11: 2367–2373, 2000.

115. A Goswami, KD Mirfakhrae, RN Patel. Deracemization of racemic 1,2-diol by biocatalytic stereoinversion. *Tetrahedron: Asymmetry* 10: 4239–4244, 1999.

116. JE Audia, TC Britton, J Droste, BK Folmer, GW Huffman, VL John, H Lee, TE Mabry, JS Nissen. Preparation of N-(phenylacetyl)di- and tripeptide derivatives for inhibiting β-amyloid peptide release. *PCT Int Appl* 1998; WO 9822494 A2 19980528 CAN 129: 41414.

117. RN Patel, A Banerjee, V Nanduri, A Goswami, FT Comezoglu. Enzymatic resolution of racemic secondary alcohols by lipase B from *Candida antarctica* . *J Am Oil Chem Soc* 77: 1015–1019, 2000.

118. VB Nanduri, R Johnston, SL Goldberg, PM Cino, RN Patel. Sequences of a *Gluconobacter oxydans* 2-ketoreductase. Eur. Pat. Appl., 2003, 28 pp. EP 1321517, A2 20030625 AN 2003: 488643.

119. VB Nanduri, RL Hanson, A Goswami, JM Wasylyk, TL LaPorte, K Katipally, H-J Chung, RN Patel. Biochemical approaches to the synthesis of ethyl 5-(S)-hydroxyhexanoate and 5-(S)-hydroxyhexanenitrile. *Enzyme and Microbial Technology* 28: 632–636, 2001.

120. G Hoegenauer. Mechanism of action of antibacterial agents. Tiamulin and pleuromutilin. *Antibiotics* 5: 344–360, 1979.

121. H Berner, H Vyplel, G Schulz, P Stuchlik. Chemistry of pleuromutilin. IV. Synthesis of 14-O-acetyl-8α-hydroxymutilin. *Tetrahedron* 39: 1317–1321, 1983.

122. RL Hanson, JA Matson, DB Brzozowski, TL LaPorte, DM Springer, RN Patel. Hydroxylation of mutilin by *Streptomyces griseus* and *Cunninghamella echinulata*. *Organic Process Research & Development* 6: 482–487, 2002

123. DM Springer, ME Sorenson, S Huang, TP Connolly, JJ Bronson, JA Matson, RL Hanson, DB Brzozowski, TL LaPorte, RN Patel. Synthesis and activity of a C-8 keto pleuromutilin derivative. *Bioorg & Med Chem Letters* 13: 1751–1753, 2003.

124. RM Williams. In JE Baldwin, PD Magnus, ed. *Synthesis of Optically Active α-Amino Acids* , Pergamon, Oxford, New York. 7: 130–150, 1989.

125. MJ O'Donnell, DW Bennett, S Wu. The stereoselective synthesis of α-amino acids by phase-transfer catalysis. *J Am Chem Soc* 111: 2353–2355, 1989.

126. G Bold, RO Duthaler, M Riediker. Enantioselective synthesis with titanium carbohydrate complexes. Part 3. Enantioselective synthesis of (D)-threo-β-hydroxy-α-amino acids with titanium carbohydrate complexes. *Angew Chem Int Ed Engl* 28: 497–498, 1989.

127. EM Gordon, MA Ondetti, J Pluscec, CM Cimarusti, DP Bonner, RB Sykes, O-Sulfated β-lactam hydroxamic acids (monosulfactams). Novel monocyclic β-lactam antibiotics of synthetic origin. *J Amer Chem Soc* 104: 6053–6060, 1982.

128. RB Sykes, CM Cimarusti, DP Bonner, K Bush, DM Floyd, NH Georgopadakou, WH Koster, WC Liu, WL Parker, PA Principle, ML Rathnim, WA Slusarchyk, WH Trejo, JS Wells. Monocyclic β-lactam antibiotics produced by bacteria. *Nature* 291: 489–491, 1981.

129. JD Godfrey, RH Mueller, DJ Von Langen. β-Lactam synthesis: cyclization versus 1,2-acyl migration-cyclization. The mechanism of the 1,2-acyl migration-cyclization. *Tetrahedron Lett* 27: 2793–2796, 1986.

130. H Schuette, W Hummel, H Tsai, M-R Kula. L-2-Hydroxyisocaproate dehydrogenase — a new enzyme from Lactobacillus confusus for the stereospecific reduction of 2-ketocarboxylic acids. *Appl Microbiol Biotechnol* 22: 306–317, 1985.

131. RL Hanson, J Singh, TP Kissick, RN Patel, LJSzarka, RH Mueller. Synthesis of L-β-hydroxyvaline from α-keto-β-hydroxyisovalerate using leucine dehydrogenase from *Bacillus* species. *Bioorg Chem* 18: 116–130, 1990.

132. G Bisacchi, WA Slusarchyk, U Treuner, JC Sutton, R Zahler, S Seiler, DR Kronenhal, ME Rondazzo, Z Xu, Z Shi, MD Schwinden. Preparation of amidino and guanidino azetidinone compounds as tryptase inhibitors. WO 9967215 A1 19991229 CAN 132: 64103 AN 1999: 819347, 1999.

133. B Herradon, S Valverde. Biocatalysis in organic synthesis. 6. First use of an acylase as catalyst in the irreversible transacylation of alcohols and amines: application to selective transformation. *Synlett* 6: 599–602, 1995.

134. B Wirz, W Walther. Enzymic preparation of chiral 3-(hydroxymethyl)piperidine derivatives. *Tetrahedron: Asymmetry* 3: 1049–1054, 1992.

135. A Goswami, JM Howell, EY Hua, KD Mirfakhrae, MC Soumeillant, S Swaminathan, X Qian, FA Quiroz, TC Vu, X Wang, B Zheng, DR Kronenthal, RN Patel. Chemical and Enzymatic Resolution of (R,S)-N-(tert-Butoxycarbonyl)-3-hydroxymethylpiperidine. *Organic Process Research & Development*. 5: 415–420, 2001.

136. OP Ward, MV Baev. Decarboxylases in stereoselective catalysis. In *Stereoselective Biocatalysis*. Ed RN Patel. New York, Marcel Dekker, 267–287, 2000.

137. H Iding, P Siegert, K Mesch, M Pohl. Application of α-keto acid decarboxylases in biotransformations. *Biochim Biophys Acta* 1385: 307–322, 1998.

138. H Iding, T Dunnwald, L Greiner, A Liese, M Muller, P Siegert, J Grotzinger, AS Demir, M Pohl. Benzoylformate decarboxylase from *Pseudomonas putida* as stable catalyst for the synthesis of chiral 2-hydroxy ketones. *Chem Eur J* 6: 1483–1495, 2000.

139. Z Guo, A Goswami, VB Nanduri, RN Patel. Asymmetric acyloin condensation catalyzed by phenylpyruvate decarboxylase. *Tetrahedron: Asymmetry*. 12: 571–578, 2001.

140. D Sebastian, A Heuser, S Schulze, H Waldmann. Selective enzymic removal of protecting groups from phosphopeptides. Chemoenzymic synthesis of a characteristic phosphopeptide fragment of the Raf-1 kinase. *Synthesis* 9: 1098–1108, 1997.

141. H Waldmann, A Heuser, S Schulze. Selective enzymic removal of protecting groups: the phenylacetamide as amino protecting group in phosphopeptide synthesis. *Tetrahedron Letters* 37: 8725–8728, 1996.

142. H Waldmann, A Reidel. The phenylacetyl group—the first amino protecting group that can be removed enzymically from oligonucleotides in solution and on a solid support. *Angewandte Chemie*, International Edition in English 36: 647–649, 1997.

143. J Barankiewicz, MM Jezewska. Purine-nucleoside phosphorylase and adenosine aminohydrolase activities in fibroblasts with the Lesch-Nyhan mutation. *Adv Exp Med Biol* 76A, 391–397, 1997.

144. CA Costello, AJ Kreuzman, MJ Zmijewski. Selective deprotection of phthalyl protected amines. *Tetrahedron Letters* 37: 7469–7472, 1996.

145. AS Bommarius, K Drauz, K Gunther, G Knaup, M Schwarm. L-methionine related L-amino acids by acylase cleavage of their corresponding N-acetyl-DL-derivatives. *Tetrahedron: Asymmetry* 8: 3197–3200, 1997.

146. J Ogawa, S Shimizu, H Yamada. N-carbamoyl-D-amino acid amidohydrolase from Comamonas sp. E222c purification and characterization. *Eur J Biochemistry* 212: 685–691, 1993.

147. J Ogawa, H Miyake, S. Shimizu. Purification and characterization of N-carbamoyl-L-amino acid amidohydrolase with broad substrate specificity from Alcaligenes xylosoxidans. *Appl Microbiol Biotechnol* 43: 1039–1043, 1995.

148. H Sajiki, K Hattori, K Hirota. The Formation of a Novel Pd/C-Ethylenediamine Complex Catalyst: Chemoselective Hydrogenation without Deprotection of the O-Benzyl and N-Cbz Groups. *J of Organic Chemistry* 63: 7990–7992, 1998.

149. H Yamada, T Tagawa, S Goto. Hydrogenolysis for deprotection of amino acid in a stirred tank reactor containing gas-liquid-liquid-solid four phases. *J of Chem Eng of Japan* 29: 373–376, 1996.

150. GP Royer, W Chow, KS Hatton. Palladium/polyethylenimine catalysts. *J. Mol. Catal.* (1985), 31: 1–13, 1985.

151. RN Patel, VB Nanduri, D Brzozowski, C McNamee, A Banerjee. Enantioselective enzymatic cleavage of N-benzyloxycarbonyl groups. *Advanced Synthesis & Catalysis* 345: 830–834, 2003.

27

Nutrition Delivery System: A Novel Concept of Nutrient Fortification

T.P. Rao

N. Sakaguchi

L.R. Juneja

27.1 Introduction

Nutrient deficiency commonly referred to as malnutrition is widespread in most of the developing countries. In the developed world, the malnutrition particularly for micronutrients is observed in children, adolescents, and pregnant and lactating mothers (Pilch and Senti 1984; Dallman et al. 1984). Iron deficiency anemia (IDA), vitamin A deficiency, and iodine deficiency disorders are the most common disorders found in malnutrition. Thirty percent of the world's population is affected by iron deficiency (WHO 1992a). Nearly 4–5 billion people are suffering from either clinical or subclinical forms of iron deficiencies.

In Asia, 60% of women during pregnancy and 50% of children are suffering from IDA (DeMaeyer et al. 1985; Joseph 2000). Anemia in children is associated with retardation of physical growth and intellectual and psychomotor development (Dallman et al. 1980; Levin 1986). Anemia during pregnancy may be fatal due to loss of blood in childbirth. This causes more than half the maternal deaths in the world (DeMaeyer et al. 1989). Maternal anemia may also lead to fetal growth retardation, low-birth-weight infants, and increased rates of early neonatal mortality. Iron deficiency has a profound affect on productivity and therefore has economic implications for countries in which it is a significant public health problem (Scholz et al. 1997).

A lack of dietary iron is the main cause for anemia (MacPhail and Bothwell, 1992). Increasing the content and bioavailability of iron in the diet can prevent iron deficiency. Dietary supplements with minerals and vitamins are often recommended to overcome the malnutrition, but its practical implementation in daily life is seldom observed. Fortified food is a wise alternative to deliver the nutrients. Recently, demand for nutritionally fortified foods (iron, calcium, vitamins, carotenoids, etc.) has increased as more consumers look for fortified foods to avoid taking dietary supplements or altering their eating habits (Allen 1998; Hallberg et al. 1966; Hallberg 1995; Hurrell and Cook 1990; Hurtado et al. 1999).

Many countries have launched national projects to enrich iron in daily foods. In response to public health goals of reducing deficiency, the Food and Drug Administration (FDA) in the U.S. established standards of identity for enriched staple foods (e.g., flour, bread, rice, cornmeal) and specified levels of iron in addition to other vitamins to be added (Crane et al. 1995). Chile in the 1950s was the first Latin American country to fortify the diet with iron, although not explicitly for anemia control. In 1993 Venezuela legislated fortification of wheat flour with iron (Layrisse et al. 1996). At least 18 governments in the region are now implementing fortification to provide iron compounds to their people through fortified wheat flour. The Philippines is trying to launch iron-enriched rice to be supplied. Besides staple foods, fortified dairy and liquid products are also recommended for effective delivery of nutrients in daily life (Zoller et al. 1980).

Though the concept of fortified foods is gaining popularity for the delivery of nutrients, the difficulties in fortification of nutrients in various food products is hampering its effective use. For example, iron has more potential for use in many food vehicles than iodine or vitamin A, but fortification with iron is technically more difficult than other nutrients as it reacts with several food ingredients. The biggest challenge with iron is to identify a form that is adequately bioavailable and yet does not alter the appearance or taste of the food vehicle. The buff-colored, insoluble, iron phosphate compounds are stable under a variety of storage conditions, but its bioavailability is poor in normal conditions. Soluble iron salts like ferrous sulfate are well absorbed but easily discolored with other food components like tannins, etc. Therefore, for effective fortification the nutrient must not only have the merit of good bioavailability but also it should be pleasant with taste and flavor, and stable and safe to apply in various applications. Also, it should show adequate dispersible property to apply in various liquid or dairy products.

The concept of the Nutrient Delivery System (NDS) developed at Taiyo Kagaku Co. Ltd. Japan, is to develop nutrients suitable for fortification by overcoming the problems in their taste, flavor, stability, solubility and safety, and bioavailability. To achieve this goal we have adopted the blend of nanotechnology and encapsulation to develop nutrients for NDS. Using NDS, it has become possible to deliver minerals, vitamins, and other nutrients in a certain place under a certain condition. In Taiyo Kagaku Co. Ltd. Japan, we have developed various nutrients under the brand name SunActive for NDS.

27.2 SunActive Fe

The soluble iron compounds like ferrous sulfate exhibit high bioavailability but adversely affect food quality by accelerating lipid oxidation or by producing an unfavorable color or flavor (Bothwell et al. 1992; Disler et al. 1975; Douglas et al. 1981; Edmonson et al. 1971; Hurrell et al. 1985, 1992; Viteri et al. 1995). In various food products, compatible and nonreactive iron compounds are attractive for fortification because they have less of an iron taste compared to soluble iron. Ferric pyrophosphate is considered one such compound because of its nonreactivity. However, due to its low bioavailability and insolubility, precipitation of the iron prevents fortification. For NDS, we have developed SunActive Fe from ferric pyrophosphate with the merits of high bioavailability, dispensability, and stability to use in various food fortifications.

27.2.1 Preparation and Physical Properties of SunActive Fe

SunActive Fe is a super dispersible ferric pyrophosphate preparation. Ferric pyrophosphate was prepared by adding sodium pyrophosphate solution to ferric chloride solution. The ferric pyrophosphate suspension was dispersed with lecithin and other emulsifiers. The conventional insoluble ferric pyrophosphate

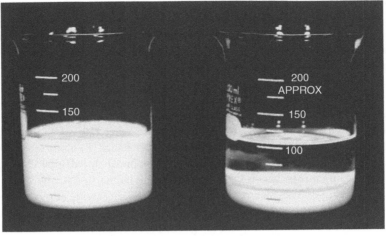

SDFe Solution
(Fe 12 mg/g)

Ferric Pyrophosphate Solution
(Fe 12 mg/g)

FIGURE 27.1 Solubility of SunActive Fe (left) and ferric pyrophosphate (right) in water.

was made soluble in water by this technique (Figure 27.1). SunActive Fe is available in both liquid (12 mg Fe/g) and powder (80 mg Fe/g) forms.

The particle size distribution of SunActive Fe showed a sharp particle size distribution between 0.1 to 2.6 μm, with an average particle size of 0.5 μm, which is several times smaller than that of commercial ferric pyrophosphate, which has an average particle size of 5.2 μm (Figure 27.2). Unlike commercial ferric pyrophosphate, SunActive Fe did not aggregate and stablize in the fine particle state.

27.2.2 Stability of SunActive Fe

SunActive Fe features the following:

Storage stability: Unlike commercial ferric pyrophosphate, SunActive Fe was not precipitated in liquid solutions. SunActive Fe was completely dispersed in liquid solution and the dispensability and iron content (12 mg/g) was maintained for 6 months (Figure 27.1).

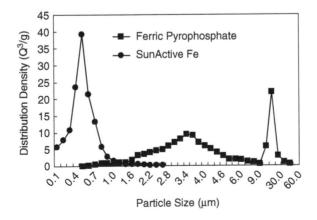

FIGURE 27.2 Particle size distribution of SunActive Fe and ferric pyrophosphate.

Color stability: Various iron compounds such as ferric pyrophosphate, ferrous sulfate, and sodium ferrous citrate and SunActive Fe were dispersed in distilled water at the rate of 5 mg iron per 100 g of distilled water, heated at 70°C for 10 min and then kept at 40°C. SunActive Fe solution was transparent and no precipitation observed even after 3 months of storage. In contrast, commercial ferric pyrophosphate sedimented immediately; in ferrous sulfate, a brownish precipitate was formed; and in sodium ferrous citrate, the solution turned to brownish color after 2 days.

In another study, 100 ml of liquid samples containing 2.6 g of fat, 3.7 g of protein, 17.2 g of carbohydrate, 2.5 mg of iron (sodium ferrous citrate or SunActive Fe) and a small amount of vitamins and minerals were prepared. The samples were pasteurized at 120°C for 20 min and stored in a dark place at 50°C for 4 months. After storage, the color change (E) of the samples was measured. SunActive Fe was stable and color change was very low under stored at 50°C in a dark place.

Stability of food components: Iron is highly reactive with vitamin C, salt, and sugars causing unfavorable taste and color. The stability of vitamin C, salt, and sugar with SunActive Fe was examined. The test was performed with SunActive Fe, ferric pyrophosphate, sodium ferrous citrate, and ferrous sulfate as 1 g Fe and mixed with 100 g ascorbic acid, respectively. The samples were stored in the dark at room temperature for 4 weeks. After storage, highest stability of vitamin C and lowest residual vitamin C was observed with SunActive Fe (Figure 27.3).

In another study, the stability of salt and sugar with SunActive Fe and other iron sources was examined. SunActive Fe, ferric pyrophosphate, ferrous sulfate, or sodium ferrous citrate as 6 mg and 12 mg iron were mixed with 10 g of salt and 10 g of sugar, respectively. After storage at room temperature, the color was observed. The color of salt and sugar was stable with SunActive Fe and did not show any reactivity as compared to ferrous sulfate and ferrous citrate.

Stability of flavor: Sensory evaluation tests for unpleasant flavor of food components with different iron sources were conducted with 10 panelists. A solution containing 5% fructose-glucose-liquid sugar was added with SunActive Fe, ferric pyrophosphate, ferrous sulfate, or sodium ferrous citrate, respectively. SunActive Fe did not cause any unpleasant flavor as compared to other iron sources (Figure 27.4). Similarly, the changes in taste of yogurt drink with SunActive Fe were examined with 10 panelists. SunActive Fe and vitamin C were mixed in yogurt drink. Final concentrations of iron and vitamin C were ranged from 1.0 to 10 mg and from 5.0 to 100 mg in yogurt drink, respectively. After mixing, the tasting test was done. The results indicated no unpleasant taste of yogurt drink with SunActive Fe containing 10 mg of iron.

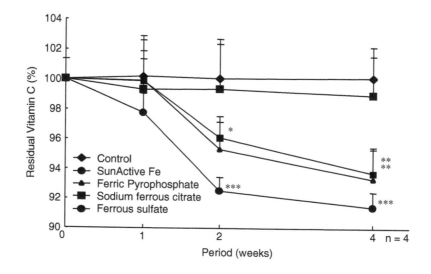

FIGURE 27.3 Stability of SunActive Fe and other iron sources with vitamin C.

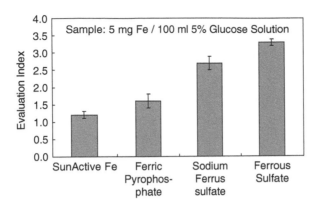

FIGURE 27.4 Sensory test of different iron solution. Evaluation index: 0: no odd flavor or taste; 1: No iron flavor or taste; 2: iron flavor and taste; 3: strong iron flavor and taste and 4: extremely strong iron flavor and taste.

Stability against heat: SunActive Fe (1.6 mg Fe) was mixed with 100 g of concentrated prune juice and heated at 121°C for 30 min. After cooling, the sample was centrifuged at 8000 g for 30 min. Free iron concentration in the supernatant was determined by atomic absorption spectrometer. There was almost no release of free iron from SunActive Fe.

SunActive Fe was heat stable and maintained its emulsifying and flavor masking properties during baking and retort processes. SunActive Fe was found to have no effects on the color or flavor of the final food products.

27.2.3 Bioavalability of SunActive Fe

Absorption and bioavailability of SunActive Fe (Sakaguchi et al. 2004) was investigated by the serum iron method (Pagella et al. 1984), hemoglobin regeneration efficiency (HRE) method (Forbes et al. 1989), and modified AOAC method in rats.

Serum iron curve method: This method was based on serum iron concentration (SIC) in normal rats (Ekenved et al. 1976). Serum iron concentration was measured after oral administration of 2 mg Fe/kg body weight as SunActive Fe, ferric pyrophosphate, sodium ferrous citrate, ferrous sulfate, and commercial heme iron. After 0.5, 1, 2, 4, and 12 hr of the oral administration, blood samples were taken from the carotid section for determination of SIC. Iron absorption determined by SIC curve showed a mean value of 113.4 μg/dl of SIC in the control, which was unchanged during the study period. In iron fortification, SIC rapidly increased and decreased after iron administration of all test Fe compounds. The peak of SIC was 340.0 μg/dl in ferric pyrophosphate and 441.2 μg/dl in sodium ferrous citrate after 30 min of oral administration, and 444.4 μg/dl in ferrous sulfate after 60 min of oral administration, and then SIC rapidly decreased. In the case of SunActive Fe and commercial heme iron, the Fe absorption peak of SIC was delayed as compared to other iron sources, reaching 388.8 μg/dl and 471.6 μg/dl after 2 h of oral administration, respectively. High serum iron concentrations were observed even after 8 h of oral administration of SunActive Fe (Figure 27.5). The average values of AUC as shown in Figure 27.6 were 2839, 1573, 2108, 2001, 2294 and 909 μg/dl for SunActive Fe, ferric pyrophosphate, sodium ferrous citrate, ferrous sulfate, commercial heme iron, and control, respectively. SIC after oral iron administration of SunActive Fe showed a lag in peak time and sustained release of iron in the serum. This could be due to the fact that SunActive Fe preparation is nano-sized encapsulated iron formulation. AUC values confirmed the sustained release of iron from SunActive Fe.

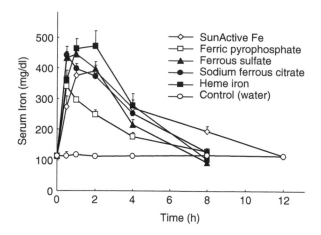

FIGURE 27.5 Serum iron level in normal rats after oral administration of SunActive Fe and other iron compounds (2 mg Fe/kg body weight).

HRE method: Iron bioavailability from SunActive Fe and other test iron compounds was determined by the hemoglobin regeneration efficiency (HRE) based on hemoglobin repletion assay in anemic rats (Zhang et al. 1989). Experimental diets were prepared by adding 3.5 mg Fe/ kg to iron deficient diets with SunActive Fe, ferric pyrophosphate, sodium ferrous citrate or ferrous sulfate. Experimental diets were fed ad libitum for 28 days. Control rats were fed a standard diet. The HRE value was calculated as previously reported (Forbes et al. 1989; Zhang et al. 1989). The relative biological value (RBV) was calculated by HRE of SunActive Fe, ferric pyrophosphate, or sodium ferrous citrate relative to the mean HRE of ferrous sulfate. After 2 weeks of iron fortification, the hemo- globin regeneration efficiencies were 55.36, 41.11, 52.97, and 52.81 for SunActive Fe, ferric pyro- phosphate, sodium ferrous citrate, and ferrous sulfate, respectively. The relative biological values (RBV) of iron sources per ferrous sulfate were 1.05, 0.78, and 1.00 for SunActive Fe, ferric pyrophosphate, and sodium ferrous citrate, respectively (Figure 27.7). SunActive Fe showed the highest value of HRE as well as RBV among the iron compounds tested.

AOAC method: Iron bioavailability from SunActive Fe was determined by the AOAC method (Forbes, 1989). Experimental diets were prepared by adding 0, 6, 12, 18, or 24 mg Fe/kg to iron deficient diets with SunActive Fe, ferric pyrophosphate, or ferrous sulfate. Experimental diets were fed ad libitum for 2 weeks. The bioavailability of each test iron source relative to ferrous sulfate was

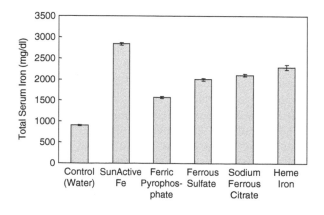

FIGURE 27.6 Incremental area under the curves following intake of iron (2 mg Fe / kg body weight).

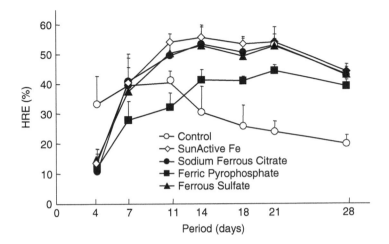

FIGURE 27.7 Hemoglobin regeneration efficiency (HRE) value in iron deficient anemic and normal rats (dose: 3.5 mg Fe/100 g).

calculated by comparing gain in hemoglobin content with the iron level in the diet by the slope ratio procedure (relative biological value, RBV). The slope values of each test iron were 0.270, 0.145, and 0.259 for SunActive Fe, ferric pyrophosphate, and ferrous sulfate respectively, and RBVs were 1.04 and 0.56 for SunActive Fe and ferric pyrophosphate, respectively (Figure 27.8).

SunActive Fe showed higher iron bioavailability as compared to other iron sources by the SIC method, HRE method, and Modified AOAC method.

SunActive Fe absorption from milk: Hallberg et al. (1992) reported that iron absorption from the milk was restrained due to the presence of high calcium content. The effect of calcium on iron absorption from SunActive Fe was examined in collaboration with Nutrition College of Japan. Thirteen young female students having less than 12 g/ dl of hemoglobin level were served with 200 ml of milk containing SunActive Fe (5 mg as Fe) for 72 days (net 50 days). After the experiment, the hemoglobin and hematocrit level were significantly higher than initial values (Figure 27.9). These results suggest that the bioavailability of SunActive Fe from milk is also high despite high concentrations of calcium in milk.

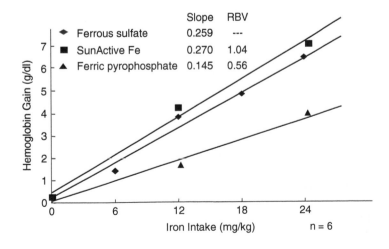

FIGURE 27.8 Relative biological value of SunActive Fe and other test iron sources (Modified AOAC method) in iron-deficient anemic rats.

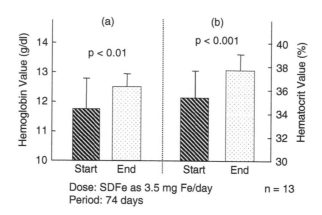

FIGURE 27.9 Hemoglobin and Hematocrit values after the intake of SunActive Fe (3.5 mg Fe/day) for 74 days in 13 young females.

Absorption mechanism: The mechanism of iron absorption was different for heme iron and non-heme iron compounds. It is well known that iron is absorbed in the proximal small intestine. However, Terato et al. (1973) reported that iron could be absorbed in the distal small intestine too, but the mechanism was not clearly established. It seems the dietary inorganic iron such as ferric pyrophosphate is solubilized in acidic gastric juices of the stomach where it chelates mucins and certain dietary constituents to keep them soluble and available for absorption at alkaline duodenum (Conrad et al. 1993). Gupta (1999) demonstrated that ferric pyrophosphate is transported from the dialysate to the blood compartment during simulated, *in vitro* hemodialysis.

Dietary constituents that solubilize iron may enhance absorption, whereas compounds that cause precipitation or molecular aggregation of iron decrease absorption (Benjamin et al. 1967; Conrad et al. 1993). The reason for the higher iron absorption and bioavailability of SunActive Fe as compared to ferric pyrophosphate may be due to high solubility of micronized particles and their sustained release from the encapsulation.

27.2.4 Safety Aspects of SunActive Fe

Normally, when iron is continuously consumed or when high doses of iron are taken, several side effects such as nausea, emesis, anorectic, abdominal pain, diarrhea, and constipation were reported (Hallberg et al. 1966; WHO 1992b). We examined the safety of SunActive Fe by evaluating its gastric tolerance and acute toxicity in rats and by Ames mutagenesity test.

Gastric tolerance study: In this experiment, 8-week-old rats were fed a standard diet for 5 days. All rats were fasted for 48 hr prior to the examination and were randomized into 4 groups of 10 rats each so that mean body weight was approximately equal among the groups. Thereafter, SunActive Fe, ferric pyrophosphate, ferrous sulfate, and sodium ferrous citrate dissolved in distilled water (30 mg Fe/kg body weight) was orally administered to rats with a conductor 3 times in 24 hr. Approximately 5 hr after the final administration of iron, the stomach of the rat was enucleated and gastric tolerance showing the extent of gastric ulcers induced by the iron compounds was evaluated by the method of Adami et al. (1964). Hemorrhagic suffusion or ulcers were observed in sodium ferrous citrate and ferrous sulfate groups, but the SunActive Fe group did not show any lesions or toxicity. The results showed that SunActive Fe was mild on the stomach and did not have any harmful effects on the gastrointestinal system as compared with other iron compounds.

Acute toxicity test: In this study, rats of both sexes were treated orally by giving as much as 635 mg Fe/kg as SunActive Fe. The results showed no significant differences between the control group and SunActive Fe group for the general behavior, mortality, body weight increase, and food and water intake for 14 days. The LD_{50} of SunActive Fe was >635 mg Fe/kg body weight. The acceptable daily intake level of iron phosphate was 70 mg/kg in humans, so the LD_{50} of SunActive Fe is quite high (635 mg Fe/kg body weight), suggesting the high safety levels of SunActive Fe.

Ames mutagenicity test: SunActive Fe was examined for mutagenic activity in two independent Ames tests using the histidine-requiring *Salmonella typhmurium* strains TA1535, TA1537, TA98, and TA100, the tryptophan-requiring *Escherichia coli* strain WP2 *uvrA*, and a liver fraction of Aroclor-induced rats for metabolic activation (S9-mix). SunActive Fe was not toxic to any of the strains, as was evidenced by the absence of a drastic decrease in the mean number of revertants. SunActive Fe was not found to be mutagenic under the conditions employed in this study.

27.3 SunActive Mg

Magnesium is another essential micronutrient known to be required for many functions in the body. The role of magnesium in the transmission of hormones (such as DHEA, estrogen, insulin, testosterone, thyroid, etc.) and neurotransmitters (such as dopamine, catecholamines, GABA, serotonin, etc.) was well established. Many common diseases such as migraines, asthma, allergy, attention deficient disorder, fibromyalgia, and mitral valve prolapse were linked to the Mg deficiency.

Similar to iron, fortification of Mg also has many difficulties, like taste and solubility. By using NDS, we have developed SunActive Mg to overcome the above difficulties for effective fortification in many food applications.

27.3.1 Preparation and Physical Properties of SunActive Mg

SunActive Mg is an encapsulated micronised tri-magnesium phosphate preparation. The particle size distribution of SunActive Mg is very sharp (between 0.15 and 1.0 μm) with an average particle size of 0.35 μm, whereas tri-magnesium phosphate has very wide particle size distribution (between 0.35 and 28.0 μm) with an average particle size of 3.2 μm. Micronized particle size of SunActive Mg enabled its effective dispersion in liquid solutions.

27.3.2 Bioavailability of SunActive Mg

The bioavailability of SunActive Mg was compared with magnesium phosphate and magnesium oxide in rats. The above compounds were orally administrated at the rate of 20 mg Mg/kg and the plasma Mg levels were monitored for 24 hr. The Mg content reached as high as 2.2 mg/dl within 4 hr as compared to 1.8 mg/dl for magnesium oxide and magnesium phosphate. The plasma Mg levels were sustained for a longer period (more than 12 h) in the SunActive Mg group than the other two sources. The area under curve (AUC) was more than two times for SunActive Mg than for the other two Mg sources (Figure 27.10). The encapsulated, micronized particle size of SunActive Mg ensured its active absorption and sustained release.

In another experiment, the absorption of Mg and Ca was measured after 3 weeks of feeding of SunActive Mg, tri-magnesium phosphate, magnesium oxide, and magnesium hydrochloride to 3-week-old male Fischer rats. After the experimental period, the absorption of Mg as well as Ca was significantly high in SunActive Mg group (Table 27.1) .

27.3.3 Stability of SunActive Mg

SunActive Mg is completely dispersed in liquid solutions and the dispersion was found stable for more than 6 months. SunActive Mg does not contain the typical taste of Mg. The advantages such as greater bioavailability, clear solubility, and no flavor activity of SunActive Mg allow its effective use in NDS.

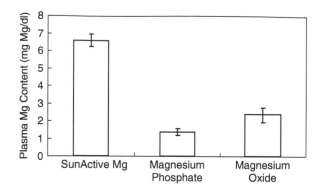

FIGURE 27.10 Plasma Mg content after 24 of oral administration of SunActive Mg and other iron sources. Values measured from AUC of time course plasma Mg content.

27.4 SunActive DHA

The omega-3 long-chain polyunsaturated fatty acid docosahexaenoic acid (DHA) is the primary building block and most abundant fat in the brain and retina of the eye (Martinez 1992). DHA therefore is a vital nutrient for optimal eye and brain function. Recent studies also relate the deficiency of this nutrient with Attention Deficit Hyperactivity Disorder (ADHD) in children. Breast milk and fish are the major sources for DHA. Women should include sufficient DHA in their daily diets to ensure sufficient levels of DHA in their breast milk (Jensen et al. 1992). DHA levels in U.S. women are among the lowest in the world. The Food and Agriculture Organization/World Health Organization (FAO/WHO) expert committee has recommended the minimum levels, which should at least equal to the levels of breast milk in infant formulas (FAO/WHO Expert Committee 1994).

Given that fish oil is the major source for DHA, fortification of foods with DHA often results in such problems as fishy odor, easy oxidation, and solubility in water. Using NDS we have overcome these problems and prepared SunActive DHA to fortify various food products.

27.4.1 Preparation and Physical Properties of SunActive DHA

We have prepared micronized encapsulated DHA, with an average particle size of 0.29 μm, which is several times smaller than commercially available emulsified DHA, whose average particle size is 1.86 μm. SunActive DHA is easily soluble in water. SunActive DHA is also devoid of fish odor and stable in many storage conditions. SunActive DHA is available in both powder and liquid formulations containing 5% and 6% of DHA, respectively.

SunActive DHA did not show any oxidation even up to 30 days of storage in a dark place (Figure 27.11).

TABLE 27.1 Apparent Magnesium and Calcium Absorption after the Intake of SunActive Mg and Other Mg Sources

Source	Magnesium absorption (%)	Calcium absorption (%)
SunActive Mg	89.2 ± 2.6	70.6 ± 0.8
Magnesium phosphate	75.6 ± 1.4	58.6 ± 1.5
Magnesium oxide	80.4 ± 2.4	52.7 ± 1.2
Magnesium hydrochloride	78.5 ± 1.9	61.8 ± 1.4

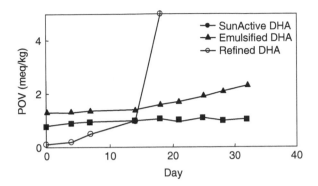

FIGURE 27.11 Changes of peroxide values (POV) with SunActive DHA and other DHA products stored at 40°C in a dark place.

27.4.2 Bioavailability of SunActive DHA

The bioavailability of SunActive DHA was examined in a human study. The powder and liquid forms of SunActive DHA and refined DHA were fed as 1g DHA in milk shakes. The plasma DHA levels were monitored at regular intervals for 24 hr. The plasma DHA reached a peak level within in 2 hr in powder from SunActive group, similar to the refined DHA group; however, in liquid form for the SunActive DHA group the plasma DHA levels increased gradually and remained at high levels for more than 12 hr. The AUC values suggest that there is no significant difference in total absorption among the treatments (unpublished data).

The two forms of SunActive DHA offer a good source of fish odor–free DHA with additional benefits such as solubility and longer hours of bioavailability for fortification.

27.5 Conclusion

Minerals, vitamins, and several nutrients play a crucial role in the growth, development, and proper function of the human body. The lack of these essential nutrients invariably causes dysfunction of the body and may lead to diseases. Daily supplementation of these essential nutrients are necessary to maintain their proper levels in the body and to ensure a healthy life. Daily intake of these nutrients in the form of tablets or capsules is often ignored and impractical in most cases. The use of nutrient fortified foods is practically viable to deliver nutrients effectively to the body.

Fortification of nutrients in various food applications is a challenging task because of the problems in taste, flavor, stability safety, and bioavailability of the nutrients. For example, a nutrient like iron is highly reactive with food substances and therefore the fortification of iron causes unpleasant taste, color, and flavor. So preparation of nutrients is a necessary step for effective fortification and proper delivery of the nutrients. We at Taiyo developed a unique concept, the Nutrition Delivery System (NDS), by combining the nano-technology and encapsulation to prepare the nutrients for fortification. Using NDS, it has become possible to fortify foods with nutrients having the following merits:

Stability: Stable against heat and oxidation in food processing
Taste: Masking of unpleasant taste and flavor
Safety: Mild on the stomach because of low reactivity and sustained release
Bioavailability: Sustained release and high absorption

The products developed for NDS under the brand name SunActive, like SunActive Fe, SunActive Mg, and SunActive DHA for Fe, Mg, and DHA, respectively, allow their effective fortification in any given conditions. Thus, NDS is a breakthrough concept in the fortification and delivery of the nutrients at the right place in the right condition.

References

Adami, E., Marazzi-Uberti, E., and Turba, C. 1964. Pharmacological Research on Gefanate. A new synthetic isopenoid with an anti-ulcer action, *Arch. International Pharmacodyn*, 147: 113–145.

Allen, L.H. 1998. Pregnancy and iron deficiency, *Nutrition Reviews*, 55(4): 91–101.

Benjamin, B.I., Cortel, S., and Conrad, M.E. 1967. Bicarbonate induced iron complexes and iron absorption: One effect of pancreatic secretions, *Gastroenterology*, 53(3): 389–396.

Bothwell, T.H., and McPhil, P. 1992. Prevention of iron deficiency by food fortification, in *Nutritional Anemias*, Fomon S and Zlotokin S, New York: Raven Press Ltd., Nestle Nutrition Workshop Series 30, New York: pp. 183–192.

Conrad, M.E., Umbreit, J.N., and Moore, E.G. 1993. Regulation of iron absorption: proteins involved in duodenal mucosal uptake and transport, *J. American College of Nutrition*, 12: 720–728.

Crane, N.T., Wilson, D.B., Cook, A., Levis, C.J., Yetley, E.A., and Radar, J.L. 1995. Evaluation of food fortification options; general principles revisited with folic acid, *American J. Public Health*, 85: 660–666.

Dallman, P.R., Siimes, M.A., and Stekel, A. 1980. Iron deficiency in infancy and childhood, *American. J. Clinical. Nutrition,* 33: 86–118.

Dallman, P.R., Tip, R., and Johnson C. 1984. Prevalence and causes of anemia in the United States, *American J. Clinical. Nutrition*, 39: 437–445.

DeMaeyer, E.M., Adiels-Tegman, M., and Rayston, E. 1985. The prevalence of anemia in the world, World Health Stat Q, 38: 302–316.

DeMaeyer, E.M., Dallman, P., Gumey, J.M., Hallberg, L., Sood, S.K., and Srikantia, S.G. 1989. Preventing and controlling iron deficiency anemia through primary health care: A guide for health administrators and program managers, World Health Organization, Geneva, Switzerland, pp. 5–58.

Disler, P.B., Lynch, S.R., Charton, R.W., Bothwell, T.H., Walker, R.B., and Mayet, F. 1975. Studies on the fortification of cane sugar with iron and ascorbic acid, *British J. Nutrition*, 34: 141–152.

Douglas, F.W., Rainey, N.H., and Wong, N.P. 1981. Color, flavor, and iron bioavailability in iron-fortified chocolate milk, *J. Dairy Science* 64: 1785.

Edmonson, L.F., Douglas, F.W., and Avants, J.K. 1971. Enrichment of pasteurized whole milk with iron, *J. Dairy Science* 54(10): 1422–1426.

Ekenved, G., Norrby, A., and Sollvell, L. 1976. Serum iron increase as a measure of iron absorption-studies on the correlation with total absorption, *Scandinavia J. Haematol*, 28 (suppl): 31–49.

FAO/WHO Expert Committee 1994. Fats and oils in human nutrition, Food and nutrition paper, FAO, Rome, Italy.

Forbes, A.L., Adams, C.E., Arnaud, M.J., Chichester Co, Cook, J.D., Harrison, B.N., Hurrell, R.F., Kahn, S.G., Morris, E.R., Tanner, J.T., and Whittaker, P. 1989. Comparison of *in vitro*, animal, and clinical determinations of iron bioavailability International Nutritional Anemia Consultative Group Task Force report on iron bioavailability, *American J. Clinical Nutrition*, 49: 225–238.

Gupta, A., Amin, N.B., Besarab, A., Vogel, S.E., Divine, G.W., Yee, J., and Anandan, J.V. 1999. Dialysate iron therapy: Infusion of soluble ferric pyrophosphate via the dialysate during hemodialysis, *Kidney International*, 55: 1891–1898.

Hallberg, L., Hogdahl, A.M., Nilsson, L., and Rybo, G. 1966. Menstrual blood loss and iron deficiency, *Acta Medica, Scandinavica*, 180(5): 639–650.

Hallberg, L., Rassander-Hulten, L., Brune, M., and Gleerup, A. 1992. Bioavailability in man of iron in human milk and cow's milk in relation to their calcium content, *Pediatric Research*, 31, 524–527.

Hallberg, L. 1995. Results of surveys to Assess iron status in Europe, *Nutrition Reviews*, 53(11): 314–322.

Hurrel, R.F. 1985. Types of iron fortifications: nonelemental sources, *In Iron Fortification of Foods*, Clydesdale, F.M., and Wiemer, K.L., Orlando, FL: Academic Press, pp. 39–53.

Hurrell, R.F., and Cook, J.D. 1990. Strategies for iron fortification of foods, *Trends in Food Science & Technology*, 9: 56–61.

Hurrel, R.F. 1992 Prospects of improving the iron fortification of foods. In *Nutritional Anemias*, Fomon S & Zlotokin S, New York: Raven Press Ltd, Nestle Nutrition Workshop Series 30, New York, pp. 193.

Hurtado, E.K., Claussen, A.H., and Scott, K.G. 1999. Early childhood anemia and mild or moderate mental retardation, *American J. Clinical Nutrition* 69: 115.

Jensen, R.G., Lammi-Keefe, C.J., Henderson, R.A., Bush, V.J., and Ferris, A.M. 1992. Effect of dietary intake of n-6 and n-3 fatty acids on the fatty acid composition of human milk in North America. *J. Pediatr.* 120: 87–92.

Joseph, M.H. 2000 Why countries and companies should invest to eliminate micronutrient malnutrition, Manila Forum 2000: Strategies to Fortify Essential Foods in Asia and the Pacific, 32–41.

Layrisse, M., Chavez, F., Mendez–Castellano, H., Bosch, V., Tropper, E., Bastardo, B., and Gonzalez, E. 1996. Early response to the effect of iron fortification in the Venezuela population, *American J. Clinical Nutrition*, 64: 903–907.

Levin, H.M. 1986. A benefit-cost analysis of nutritional programmers of anemia reduction, *World Bank Res. Observer*, 1(2): 219–245.

MacPhail, A.P., and Bothwell, T.H. 1992. The prevalence and cause of nutritional iron deficiency anemia, in Fomon SJ, Zlotkin S (eds), *Nestle Nutrition Workshop Series*, vol. 30. Raven Press, New York, 1–12.

Martinez, M.J. 1992. Tissue levels of polyunsaturated fatty acids during early human development. *J. Pediatr.* 120: 129–138.

Pagella, P.G., Bellavitte, O., Agozzino, S., and Dona, G.C. 1984. Pharmacological and toxicological studies on an iron succinyl–protein complex (ITF 282) for oral treatment of iron deficiency anemia, *Arezneim Forsch/Drug Research*, 34(9): 952–958.

Pilch, S.M., and Senti, F.R. 1984. Assessment of the iron nutritional status of the U.S. population based on the data collected in the second National Health and Nutrition Examination Survey 1976–1980, *Federation of American Societies for Experimental Biology*, Bethesda, MD, p. 65.

Sakaguchi, N., Rao, T.P., Nakata, K., Nanbu, H., and Juneja, L.R. Iron absorption and bioavailability in rats of micronised dispersible ferric pyrophosphate. *Int. J. Vit. Nutr. Res* 74(1): 3–9.

Scholz, B.D., Gross, R., Schultink, W., and Sastroamidjojo, S. 1997. Anemia is associated with reduced productivity of women workers even in less-physically-strenuous tasks, *British J. Nutrition* 77: 47–57.

Terato, K., Hiramatsu, Y., and Yoshino, Y. 1973. Studies on iron absorption, *Digestive Disease*, 18(2): 129–134.

Viteri, F.E., Alvarez, E., Batres, R., Torun, B., Pineda, O., Mejia, L.A., and Sylvi, J. 1995 Fortification of sugar with iron sodium ethlenediaminotetraacetate (NaFeEDTA) improves iron status in semi rural Guatemalan population, *American J. Clinical Nutrition* 61: 1153–1163.

WHO. 1992a. The prevalence of anemia in women, 2nd ed. WHO/NCH/MSM/92.2. WHO, Geneva.

WHO. 1992b National strategies for overcomeing micronutrient malnutrition, EB 89/27.45th World Health Assembly Provisional Agenda Item 21, WHO/A45/3. WHO, Geneva.

Zhang, D., Hendricks, D.G., and Mahoney, A.W. 1989. Bioavailability of total iron from meat, spinach (*Spinacea oleracea* L.) and meat-spinach mixture by anemic and non-anemic rats, *British J. Nutrition*, 61: 331–343.

Zoller, J.M., Wolinsky, I., Paden, C.A., et al. 1980. Fortification of non–staple food items with iron, *Food Technology*, 38–47.

28

Renewable Resources for Production of Aromatic Chemicals

Sima Sariaslani

Tina Van Dyk

Lisa Huang

Anthony Gatenby

Arie Ben-Bassat

28.1 Introduction

The latter part of the 20th century and the beginning of the 21st have been marked by the increasing drive from both the chemical industry and government agencies to move away from petroleum-based nonrenewable feedstocks for production of commodity and specialty chemicals to the application of renewable resources such as carbohydrates, oils, and fats. The use of renewable resources for production of industrial chemicals are often referred to as "green chemistry" or "green processes."[1,2] Additional attractive features of chemistry based on using renewable resources include development of environment-friendly processes that will not require organic solvents, consuming a lot of energy, generating undesirable byproducts and/or waste.[1]

Among various potential biological systems, the use of microorganisms as production platforms for chemical synthesis has been pursued most intensely. Industrial application of microorganisms for production of industrial chemicals, often called "metabolic engineering," requires design and generation of optimal biocatalyst(s) for yield and productivity. Useful tools such as metabolic flux analysis assist understanding of metabolic pathways and provide insight for the rational design of the optimal biocatalyst(s). These tools will allow identification of, within each pathway, critical genes/enzymes that need to be manipulated for maximum productivity and yield. Successful integration of new analytical methodologies, metabolic engineering, and cutting-edge molecular biology technologies such as DNA microarrays and directed evolution have dramatically altered approaches used for strain improvement and biocatalyst design resulting in the development of strains with very specific characteristics.[3] Further development of bioprocesses is envisioned to allow the most efficient vehicle for harnessing the agricultural raw materials as rich resources of several renewable materials for production of many chemicals. Examples of compounds of interest that have been produced via bioengineering include: lactate,[4] lysine,[5] 1,2-propanediol,[6] and polyhydroxyalkanoate biopolymer.[7] One of the most recent success stories of bioengineering of microorganism for production of an industrial chemical is production of 1,3- propanediol, which is used as an intermediate in the synthesis of fibers such as polyesters and polyurethanes .[8] In this example, genes from various microorganisms have been transferred to *Escherichia coli* to allow conversion of glucose to 1,3-propanediol. The ability to transfer genes for exotic reactions from uncommon microorganisms to *E. coli* or other hosts most commonly used for production of chemicals has opened up new possibilities for more extensive application of green chemistry.

This chapter describes studies to use renewable resources for production of aromatic chemicals. As a model system we have focused on conversion of aromatic amino acids phenylalanine and tyrosine to the bifunctional aromatic compound *trans- para*-hydroxycinnamic (pHCA) acid.

The enzyme phenylalanine ammonia lyase (PAL) converts phenylalanine to cinnamic acid (CA).[9] Certain PAL enzymes, such as the PAL from *Rhodotorula glutinis (Rhodosporidium toruloides)*, can use tyrosine as substrate in addition to phenylalanine. Conversion of tyrosine via tyrosine ammonia lyase activity (TAL) results in the production of pHCA in one step.

28.2 Pathway Engineering

pHCA production in *E. coli* can be effectively built onto the aromatic amino acid synthesis pathway with minor additions of heterologous genetic material (Figure 28.1 and Figure 28.2), with the additions comprising a TAL enzyme if tyrosine is used as the precursor molecule, or a PAL enzyme and cytochrome P-450/P-450 reductase if phenylalanine is the precursor. Clearly, pHCA from tyrosine is the favored route. The aromatic amino acid synthesis pathway is well understood in *E. coli* and in a number of other important microorganisms. Furthermore, the commercial interest in tryptophan as a feed supplement and phenylalanine as an ingredient in the sweetener aspartame has resulted in intensive studies to maximize the efficient production of these amino acids from glucose.[10,11,12] Other aromatic derivatives such as indigo are also made in *E. coli*.[13]

A number of inherent control points are present in the aromatic amino acid biosynthesis pathway to ensure the regulated synthesis of the required levels of tryptophan, tyrosine, and phenylalanine for maximum cell growth. The various control mechanisms include those that govern transcription, and also feedback inhibition of several enzymes by aromatic amino acid end products. To manipulate this

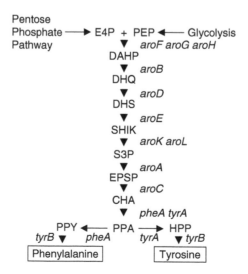

FIGURE 28.1 Biosynthesis of phenylalanine and tyrosine.

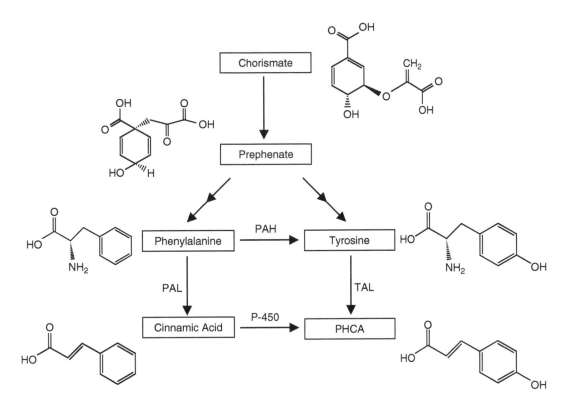

FIGURE 28.2 Bioroutes for production of *p*-hydroxycinnamic acid.

biosynthetic pathway, therefore, requires a detailed understanding of the mechanisms of regulation so that appropriate genetic changes can be made that will result in high yields of the desired product. These changes primarily focus on committing carbon to the aromatic amino acid pathway, removing transcriptional and allosteric control, and preventing rate-limiting pathway steps. In addition to manipulation of the aromatic amino acid pathway itself, there are numerous other alterations to cellular metabolism that have important roles in improving the energetics and efficiency of the bioprocess. Amongst these can be included glucose transport and metabolism, and synthesis of cofactors and precursors.

The modifications to central metabolism, which enhance carbon flow to aromatic compounds, have been reviewed by Berry.[11] In essence, the focus is on increasing the commitment of the two aromatic precursors erythrose-4-phosphate (E4P) and phosphoenolpyruvate (PEP) by increasing their concentration in cells. Enhancing expression of the *tktA* gene, encoding transketolase, or *talB*, encoding transaldolase, increases E4P concentration, which are components of the pentose phosphate pathway. Normally, only 3.3% of the cellular PEP is diverted to aromatic compound synthesis.[14] PEP availability is improved by several genetic alterations. These include inactivation of the phosphotransferase transport system (PTS), which converts PEP to pyruvate;[15] inactivation of the pyruvate kinase genes *pykA* and *pykF*, which also convert PEP to pyruvate;[16] and increased expression of PEP synthase (*pps*), which converts pyruvate to PEP.[17] These modifications have a synergistic effect when combined.[16] Inactivation of PTS enables significant conservation of PEP since *E. coli* consumes about 50% of the cellular PEP when growing on glucose as a sole carbon source,[14] and cells devoid of PTS can redirect 50% more PEP into the aromatic pathway.[15]

A key, and the first enzymatic step, in directing carbon flow from central metabolism to the aromatic amino acid biosynthesis pathway is the condensation of E4P and PEP by the enzyme 3-deoxy-D-arabinoheptulosonate 7-phosphate (DAHP) synthase. In *E. coli* there are three DAHP synthase isozymes encoded by *aroF*, *aroG*, and *aroH*, and these are regulated at both the transcriptional level and by allosteric control, by the amino acids tyrosine, phenylalanine and tryptophan, respectively. As an initial practical exercise, transcriptional regulation by any of the amino acids can be abolished by the simple expedient of replacing the natural promoters with foreign promoters, usually on a plasmid construct, which incidentally also enhances gene copy number and expression level of the cloned DAHP synthase gene. In minimal media the *aroG* gene in *E. coli* is responsible for about 80% of cellular DAHP synthase, and this gene is most often used to drive flux through the first part of the aromatic amino acid synthesis pathway. Allosteric control of *aroG*-encoded DAHP synthase can be removed by changing certain amino acids in the active site of the enzyme.[18] We carried out site-specific mutagenesis to change several different amino acids in the active site, and found that changing aspartic acid to asparagine at position 146 resulted in a DAHP synthase insensitive to allosteric control.

Six enzymatic steps are required to take DAHP to chorismate, which is a branch-point compound from which separate pathways are used for tyrosine and phenylalanine via the common intermediate prephenate (PPA). The six enzymes used for chorismate synthesis, with genes in parenthasis, are 3-dehydroquinate synthase (*aroB*), 3-dehydroquinate dehydratase (*aroD*), shikimate dehydrogenase (*aroE*), shikimate kinase (*aroK, aroL*), 5-enolpyruvylshikimate 3-phosphate synthase (*aroA*), and chorismate synthase (*aroC*). These six steps between DAHP and chorismate, with the exception of the *aroL*-encoded shikimate kinase, are not regulated. However, when present in the genome as single copies, the above genes are insufficient to allow high flux through this part of the pathway when increased carbon flow is directed into the aromatic pathway by enhancing PEP and E4P availability,[19] but it has been demonstrated that cloning of the *aroBELAC* genes on a low copy number plasmid, and under the control of a *lac* promoter avoids rate-limiting steps in chorismate synthesis.[11]

The chorismate to tyrosine steps make use of a *tyrA*-encoded bifunctional chorismate mutase and prephenate dehydrogenase to take chorismate via prephenate to 4-hydroxyphenylpyruvate. The *tyrB*-encoded tyrosine aminotransferase then converts 4-hydroxyphenylpyruvate to tyrosine. The *aspC*-encoded aspartate aminotransferase also converts 4-hydroxyphenylpyruvate to tyrosine when the intracellular pool of 4-hydroxy-phenylpyruvate is elevated. Inactivation of the *tyrR* regulatory gene results in deregulated expression of *tyrA* and *tyrB*, as well as *aroF*, *aroG*, and *aroL*. Further elevation of *tyrA* and *tyrB* expression can be achieved by multicopy expression or expression from heterologous promoters. Inactivation of *pheA* will direct chorismate flux to tyrosine. Finally, the tyrosine may be used as a substrate by a cloned TAL to form pHCA.

28.3 Biochemical Studies of Phenylalanine and Tyrosine Ammonia Lyases

As discussed earlier, the last fermentation step in conversion of glucose to pHCA is catalyzed by the PAL and/or TAL enzymes. While some TALs specifically catalyze conversion of tyrosine to pHCA, certain PAL enzymes catalyze both conversion of tyrosine to pHCA as well as formation of CA from phenylalanine. In contrast, some other PALs have very low TAL activity.

While PAL (EC 4.3.1.5) is ubiquitous in higher plants,[9] it is also found in fungi,[20] yeast,[21] and *Streptomyces*,[22,23] but not in *E. coli* or mammalian cells.[24] In plants, PAL catalyzes formation of cinnamic acid from L-phenylalanine, the first committed step toward the phenylpropanoid pathway. In the presence of a P-450 enzyme system, CA can then be converted to pHCA, which serves as the common intermediate for production of various secondary metabolites such as stilbenes, flavones, and isoflavonoids.[25] The plant PAL activities are elevated at certain developmental stages such as flowering and fruit ripening, or induced by a variety of environmental stresses such as pathogen attack, wounding, or UV irradiation.

In microbes, CA and not pHCA acts as the precursor for secondary metabolite formation. In *Streptomyces maritimus*, the PAL enzyme (encoded by *encP* gene) is involved in the biosynthesis of enterocin polyketide.[23] In the red yeast, *Rhodotorula glutinis (Rhodosporidium toruloides)*, PAL activity is induced by L-phenylalanine or L-tyrosine, while growth with glucose or ammonium salts represses

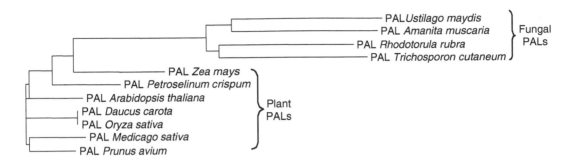

FIGURE 28.3   PAL/TAL evolution tree homology.

enzyme synthesis. The function of PAL in this organism is to degrade phenylalanine as a catabolic function and the CA formed by the action of this enzyme is converted to benzoate and other cellular materials.[26]

The gene sequences of PAL from various sources, including *R. glutinis*, have been determined and published.[27–31] Figure 28.3 depicts a sequence comparison of some of the currently known PAL enzymes. There is significant sequence diversity among the plant PALs, e.g., the sequence identity between maize PAL and PALs from rice, tomato, parsley, and *Arabidopsis thaliana* is 89%, 69%, 62%, and 68%, respectively. A larger diversity exists in the fungal PALs, with sequence homology ranging from 54% to 86% when compared to *R. glutinis* PAL (Table 28.1). The PAL genes from various sources have been expressed heterologously as active PAL enzyme in yeast, *E. coli* and insect cell culture.[32–35] Among the plant enzymes, catalytically active parsley PAL has been expressed in *E. coli*. The expression level is low due to the presence of rare *E. coli* codons in the parsley PAL gene. A further improvement of the PAL expression level was

TABLE 28.1   PAL Homologies

Organism	Length (a.a)	Homology	References
Rhodotorula glutinis	716	100%	Anson, J. G. et al., Gene 58 (2–3), 189–199 (1987) Complete nucleotide sequence of the *Rhodosporidium toruloides* gene coding for phenylalanine ammonia lyase
Ustilago maydis	724	54%	Kim, S. H., Gene bank direct deposit
Amanita muscaria	740	56%	Nehis, U., Gene bank direct deposit
Rhodotorula rubra	713	86%	Vaslet, C. A. et al., Nucleic Acids Res 16(23). 11982 (1968) cDNA and genomic cloning of yeast phenylalanine ammonia-lyase genes reveal genomic intron deletions
Trichosporon cutaneum	689	60%	Sariaslani et al., US Patent Application Publication No. 2004-0023357A1

TABLE 28.2　PAL Kinetics Parameters

Enzyme	Organism	Substrate	K_m(NM)	K_{cat}(S^{-1})	K_{cat}/K_m($M^{-1}S^{-1}$)	TAL/PAL ratio
PAL	R. glutinis	Tyrosine	110	0.46	4.14×10^3	0.5
		Phenylalanine	250	2.09	8.36×10^3	
mPAL (1540T)	R. glutinis	Tyrosine	50	0.63	1.26×10^4	1.5
		Phenylalanine	330	2.45	7.36×10^3	
PAL	Zea may	Tyrosine	19	0.92	4.84×10^4	1.1
		Phenylalanine	270	10.6	3.93×10^4	
PAL	P. crispum	Tyrosine	2500	0.3	120	9.4×10^{-5}
		Phenylalanine	17.2	22	1.27×10^6	

accomplished by redesigning the whole gene based on *E. coli* codon usage, as well as coexpressing with the chaperone HSP60 protein.[36]

It is known that PAL from some plants and microorganisms, in addition to its ability to convert phenylalanine to CA, can accept tyrosine as substrate. However, all natural PAL enzymes prefer phenylalanine rather than tyrosine as their substrate. The level of TAL activity is always lower than PAL, but the magnitude of this difference varies over a wide range. For example, the parsley enzyme has a K_m for phenylalanine of 15–25 μM and for tyrosine 2.0–8.0 mM with turnover numbers 22/sec and 0.3/sec respectively.[37] This indicates that most likely phenylalanine is the only substrate for this enzyme under physiological conditions. In contrast, the maize enzyme has a K_m for phenylalanine only 15 times higher than for tyrosine, and turnover numbers about tenfold higher.[38] The K_m of 19 μM for tyrosine suggests that it probably acts as a natural substrate for maize PAL. *R. glutinis* PAL is another enzyme that has relatively good TAL activity in addition to its PAL activity, with a ratio of TAL/PAL activity of approximately 0.5[39] (Table 28.2).

A recent publication by Kyndt et al.[40] described a "true" TAL enzyme isolated from *Rhodobacter capsulatus*, a photosynthetic bacterium. This enzyme is reported to be highly specific toward tyrosine as substrate with negligible activity toward phenylalanine. A similar enzyme has also been identified by us at DuPont's Central Research and Development from *Rhodobacter sphaeroides*.[41]

Both PAL and TAL belong to the same ammonia lyase family, which also includes the histidine ammonia lyases (HAL), which catalyzes the deamination of histidine to urocanic acid.[42] Although PAL and TAL enzymes share very similar substrates, the sequences of the TAL enzymes are more closely related to the HAL compared to the PALs. However, no shared activity has been found between the HAL and TAL enzymes. All enzymes in this family contain a conserved *Ala-Ser-Gly* amino acid motif that undergoes autocatalytic cyclization to generate a 3,5-dihydro-5-methylidene-4H-imidazol-4-one (MIO) group. This MIO group acts as the catalytic electrophile that carries out the concerted elimination of ammonia and a non-acidic-proton from the amino acid substrate. This proposed mechanism is supported by the X-ray crystal structure of the *Pseudomonas putida* HAL[43] and the structure of the *R. glutinis* PAL enzyme that has recently been elucidated.[44]

The biotechnological application of the PAL enzyme was first attempted in the enzymatic production of L-phenylalanine from cinnamic acid in a reverse reaction catalyzed by the *R. glutinis* PAL enzyme.[45] Recently, we have been studying the potential of using PAL/TAL enzyme activities for bioproduction of pHCA. We first demonstrated the ability to clone the *R. glutinis* PAL enzyme and the plant (*Heliantus tuberosus*) P-450 and P-450 reductase in *Saccharomyces cerevisiae* for conversion of glucose to pHCA. To bypass the need for using the P-450 enzyme system, we determined to exploit the TAL activity of the *R. glutinis* PAL enzyme that would give us a direct route from tyrosine to pHCA. Using protein engineering approaches, Gatenby et al. have generated a mutant *R. glutinis* PAL with improved specificity toward tyrosine.[46] It is noteworthy that most PAL and TAL enzymes are strongly inhibited by their products, pHCA, and cinnamic acid.

28.4 Efflux Transport of pHCA

Metabolic engineering of microbial strains to produce industrial chemicals should include attention to transport of the final product out of the cell. Optimizing the expression of efflux transport in biocatalyst cells can lead to increased end product yield, improved end product tolerance of production organism, and maximized product recovery from the extracellular medium. In recent years, consideration of active efflux transport in bioprocess development has taken an increasing role in the development of production organisms for small molecules.[47] For example, genetic engineering of a specific efflux transporter for lysine in *Corynebacterium glutamicum* has been a key to increasing lysine production in industrial fermentations.[48] Likewise, increased expression of an efflux system for cysteine metabolites in an industrial *E. coli* cysteine-producing strain led to increased cysteine yield,[49] and increased expression of threonine exporters enhanced threonine production in an *E. coli* threonine-producing strain.[50] Furthermore, efflux systems have been found to play a role in removal of *p*-hydroxybenzoate from cell membranes in *Pseudomonas putida* DOT-T1E[51] and are thus important for tolerance of this organism used for *p*-hydroxybenzoate production.[52] Accordingly, it was desirable to discover mechanisms of pHCA efflux with the hope that improvement of efflux transport would aid development of production organisms.

Evidence that pHCA is actively effluxed from *E. coli* was initially obtained by testing the effect of a *tolC* mutation. The TolC protein is a member of the outermembrane factor (OMF) family of proteins, which are outermembrane accessory proteins used in efflux transport in Gram negative bacteria.[53] In *E. coli*, TolC functions as an outermembrane channel[54] and is used by several different efflux systems.[55] Mutation in *tolC* renders *E. coli* hypersensitive to inhibition by a wide variety of compounds.[56] Thus, hypersensitivity of a *tolC* mutant to a specific chemical suggests that the chemical is a substrate for an efflux pump that uses TolC as the outermembrane channel. An otherwise isogenic pair of strains, differing only in the *tolC* allele was tested for sensitivity to pHCA by a disk diffusion assay on plates made with Vogel-Bonner minimal medium with glucose as a carbon source. The results are shown in Table 28.3. The dramatic increase in the size of the zone of pHCA inhibition for the *tolC* mutant indicates that it is actively effluxed from *E. coli*.

The major multidrug efflux system in *E. coli* is AcrAB.[57,58] AcrB is an inner membrane efflux pump in the Resistance/Nodulation/Cell Division Superfamily (RND) of proteins;[59] it catalyzes substrate efflux by an H⁺ antiport mechanism. AcrB mediates efflux transport of a broad range of compounds that include cationic, neutral, and anionic small molecules.[60] AcrA is a periplasmic protein in the membrane fusion protein family (MFP); it functions as an accessory molecule in efflux.[61] The AcrA and AcrB proteins require the TolC channel for function[62] and thus together form a tripartite efflux system (Figure 28.4). To test if pHCA is a substrate for the AcrB efflux pump, the tolerance of *E. coli* strains with and without a deletion of *acrAB* were compared. Also tested were an otherwise isogenic *tolC* mutant and a *tolC ΔacrAB* double mutant. The minimum inhibitory concentration (MIC) for growth in LB medium was determined (Table 28.4). The increased sensitivity of the *ΔacrAB* mutant strain indicates that pHCA is a substrate for the AcrB efflux pump. Interestingly, the sensitivity of the *tolC* mutant is greater than that of the *acrB* mutant, thus suggesting that *E. coli* has at least one additional efflux pump for pHCA that uses the TolC channel. As expected, the sensitivity of the *tolC ΔacrAB* double mutant is not greater than that of the *tolC* mutant, which is consistent with function of AcrA, AcrB, and TolC as a complex.

TABLE 28.3 Hypersensitivity of *E. coli tolC* Mutants to pHCA

E. coli strain	*tolC* allele	Zone of growth inhibition, mm diameter, pHCA (3.3 mg/disk)
RFM443	*tolC⁺*	9.0 clear
DE112	*tolC⁻*	20.0 clear, 24 turbid

TABLE 28.4 Hypersensitivity of *E. coli* Δ*acrAB*
and *tolC* Mutants to pHCA

E. coli strain	genotype	pHCA MIC, mM
K12	+	100
KZM120	Δ*acrAB*	25
DPD2032	*tolC⁻*	6
DPD2034	*tolC⁻* Δ*acrAB*	6

To find candidate genes encoding other pHCA efflux systems in *E. coli*, the transcriptional alterations induced by sublethal pHCA stress were characterized by DNA microarray analysis.[63] The underlying assumption of this approach is that bacterial cells often respond to stress by inducing expression of defense proteins that combat the stress condition. In some cases, the defense mechanisms include the increased expression of efflux pumps. Thus, treatment with a chemical may trigger expression of efflux pumps for which that chemical is a substrate.

A culture of *E. coli* strain MG1655 grown in Vogel-Bonner minimal medium with glucose as a carbon source to an OD600 of 0.2 was split and one culture was treated with 60 mM pHCA. The pHCA treatment resulted in 55% growth inhibition. Samples from the treated and control cultures were harvested at 30 and 60 min after pHCA addition. Following RNA isolation, labeling, hybridization, and data acquisition,[64] the expression of all known or putative efflux genes was examined. Table 28.5 lists all such genes for which the expression was elevated by twofold or greater at 30 min. In some instances, the expression was no longer induced after 60 min of pHCA treatment while in other cases, the expression remained elevated, but to a lesser extent than at 30 min. Accordingly, the genes listed in Table 28.5 are candidates for systems that efflux pHCA. Evidence for efflux function can be obtained by examining the consequence of loss of function or gain of function genetic alterations on pHCA tolerance. The dramatic upregulation of *aaeB* and *aaeA* at both 30 and 60 min of pHCA treatment suggested that these genes were excellent candidates for further testing.

The *E. coli* gene *aaeB* encodes a member of the putative efflux protein (PET) family.[66] Efflux function for members of the PET family has been suggested because of the likely inner membrane location and because they are often cotranscribed with members of the MFP family that function as accessory proteins for efflux. The product of *aaeA* is a member of the MFP family. The *E. coli aaeA* and *aaeB* genes are predicted to be cotranscribed with an adjacent gene, *aaeX*, encoding a protein for which no prediction of function has been made. The order of genes in this putative operon is *aaeXAB*. Expression of the *E. coli aaeXAB* operon under control of the *trc* promoter in a multicopy plasmid in *E. coli* MG1655 resulted in a twofold increase in tolerance to pHCA. This result indicates that pHCA is a substrate for the AaeB efflux pump and furthermore demonstrates that increased expression of *aaeXAB* is an effective way to improve tolerance of *E. coli* to pHCA.

Figure 28.4 summarizes the current state of knowledge about pHCA efflux systems of *E. coli*. AcrA/AcrB works with the TolC channel. AaeA/AaeB does not require the TolC channel for function.[67] Thus, there must be a third system for pHCA efflux that uses the TolC channel, but currently the inner membrane and periplasmic components of this system are not defined. In contrast, overexpression of AcrA/AcrB, which

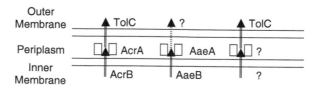

FIGURE 28.4 Summary of pHCA efflux transport systems in *E. coli*.

TABLE 28.5 Known or Putative Efflux Genes in *E. coli* Upregulated by pHCA Treatment

b#	Gene Name	Protein Family	Fold Induced, 30 *min*	Fold Induced, 60 *min*
b0127	yadG	ABC	4	1.3
b0128	yadH	ABC	4	2
b0449	mdlB	ABC	4	1.5
b0478	ybaL	CPA2	3	3
b0479	fsr	MFS	5	4
b0543	emrE	DMT/SMR	3	0.8
b0572	cusC	OMF	3	0.6
b0879	ybjZ	ABC	2	1.6
b1053	yceE	MFS	3	3
b1469	narU	MFS	2	2
b1496	yddA	ABC	3	3
b1533	eamA	DMT/DME	3	1.1
b1644	ydjH	MFP	3	3
b1690	ydiM	MFS	3	1
b1959	yedA	DMT/DME	2	1.5
b2074	mdtA	MFP	2	3
b2138	yohG	OMF	2	1.9
b2506	yfgI	MFP	3	2
b2685	emrA	MFP	2	1
b2686	emrB	MFS	2	1.4
b3051	yqiK	MFP	3	ND
b3240	aaeB	PET	179	19
b3241	aaeA	MFP	85	8
b3265	acrE	MFP	3	1.5
b3266	acrF	RND	2	1.6
b3473	yhhS	MFS	3	1
b3486	rbbA	ABC	3	0.9
b3487	yhiI	MFP	4	0.8
b3597	yibH	MFP	3	1.3
b3754	yieO	MFS	3	0.9
b3827	yigM	DMT/DME	3	1.9
b4044	dinF	MATE	2	1
b4080	yjcP	OMF	3	1

has broad substrate specificity, may result in efflux of necessary intracellular metabolites,[68] which could have deleterious consequences for the production organism.

28.5 Fermentation and pHCA Tolerance Studies

As described in previous sections, genetic modification of the aromatic amino acid biosynthesis pathway in recombinant strains of *E. coli* allows fermentation of glucose to tyrosine and pHCA. Byproducts such as phenylalanine and CA may also be produced. Compounds such as pHCA and CA are toxic to cell growth and metabolism and the severity of the effect is very much influenced by the strain used, e.g., wild type versus recombinant production strain, the fermentation conditions and the media used. We used growth rate to estimate toxicity and cellular tolerance of these products. The fermentation studies were conducted in 250 ml baffled flasks and 10L fermentors with mineral medium in which glucose and $(NH_4)_2SO_4$ were the C and N source, respectively.

The strain used for tyrosine production was a recombinant tyrosine overproducing *E. coli* strain. The 10 L fermentation was used in a fed-batch mode where glucose was first supplied in batch and later with a specific feed program, the NH_4 was first supplied in batch and later as part of the pH control.

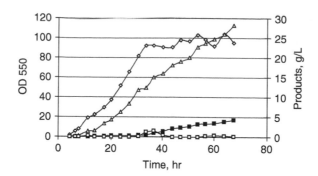

FIGURE 28.5 Kinetics of growth and tyrosine production. 10 liter fermentation in minimal medium, IPTG was added at 8 hr. Diamonds = OD at 550 nm; black squares = phenylalanine; triangles = tyrosine and white squares = acetic acid.

Figure 28.5 shows the kinetics of growth and tyrosine production: Tyrosine accumulates to 28.2 g/L and phenylalanine to 4.3 g/L in 67.6 h. Although the strain used is a phenylalanine auxotroph (*pheA⁻*) it is known that prephenate, a common intermediate in both tyrosine and phenylalanine synthesis can be nonenzymatically converted to phenylpyruvate,[69] which can subsequently be converted by a transaminase enzyme to phenylalanine. In these studies glucose is usually maintained at low concentrations (<0.5 g/L), because high concentrations of glucose enhance acetic acid formation, resulting in inhibition of metabolism and product formation.

An *E. coli* tyrosine producing strain expressing TAL/PAL activity was used to test pHCA tolerance and production. Due to pHCA toxicity, growth inhibition was observed from 1.0 g/L (Figure 28.6) and became severe at 8 g/L. It was observed that addition of complex nutrients enhanced growth and tolerance of the *E. coli* strains to pHCA. Fermentation of glucose to pHCA with this strain is associated with significant production of tyrosine, phenylalanine, and CA. Accumulation of tyrosine, and phenylalanine may be due to insufficient TAL/PAL activity and/or its inhibition by pHCA or CA. Production of CA is due to the PAL activity of the enzyme that converts phenylalanine to CA. The kinetics of growth and pHCA production in 10L fermentor are depicted in Figure 28.7. The final concentrations of tyrosine, phenylalanine, pHCA and CA after 72 hr were 5.6, 1.9, 1.9, and 1.2 g/L, respectively.

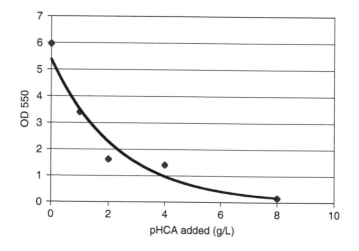

FIGURE 28.6 Growth inhibition by pHCA.

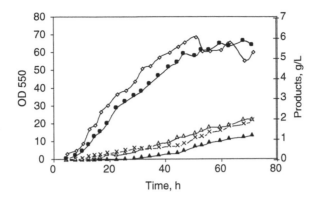

FIGURE 28.7 Kinetics of growth and pHCA production. 10L fermentation in minimal medium; IPTG was added at 8 hr. White diamonds = OD at 550 nm; Black circles = Tyrosine; white triangles = Phenylalanine; X = pHCA and Black triangles = CA.

28.6 Conclusion

The studies outlined in this chapter describe an example of the potential application of genetic engineering for use of microbes in conversion of renewable resources to materials of industrial interest. While this embryonic research program is far from a successful biotechnological process, some lessons can be learned from it for application of microbial systems for production of industrial chemicals. A) One needs to identify the properties of an ideal cell that converts a substrate to a valuable product. For example, such a cell would grow rapidly to high density, can be sustained at high viability under no or minimal growth, utilizes an inexpensive substrate and secretes the product of interest at high rates and maximum yield while making little or no side product(s). B) Attempts to develop such bioprocesses need to elucidate the properties of the metabolic reaction networks and rigorously evaluate the physiological state of the resulting recombinant cells (i.e., fluxes, small metabolites, mRNA transcripts, and proteins). C) It is possible to use a combination of enzymatic conversion and traditional organic chemistry to produce the desired materials.[70]

In the not-too-far future, novel biosynthetic routes through engineering microorganisms having genes of enzymes from several sources will become available for production of industrial chemicals. While some of the products could be used as produced by microorganisms, others might have to be interfaced with conventional chemical reactions for application.

References

1. K. Urata, N. Takaishi. A perspective on the contribution of surfactants and lipids toward "Green Chemistry": present status and future potential," *J. Surfactants and Detergents*. 4: 191–200, 2001.
2. A. Aristodou, M. Penttila, Metabolic engineering applications to renewable resource utilization, *Curr. Topics Biotechnol.* 11: 187–198, 2000.
3. D.E. Stafford, G. Stephanopoulos. Metabolic engineering as an integrating platform for strain development. *Current Topics Microbiol.* 4: 336–340, 2001.
4. C. Dong-Eun, J. Heung-Chae, R. Joon-Shick, P. Jae-Gu. Homofermentative production of D- or L-lactate in metabolically engineered *Escherichia coli* RR1. *Appl. Environ. Microbiol.* 65: 1384–1389, 1999.
5. J.J. Vallino, G. Stephanopoulos. Metabolic flux distributions in *Corynebacterium glutamicum* during growth and lysine overproduction. *Biotechnol. Bioeng.* 41: 633–646, 1993.
6. N.E. Altaras, D.C. Cameron. Metabolic engineering of a 1,2-propanediol pathway in *Escherichia coli. Appl. Environ. Microbiol.* 65: 1180–1185, 1999.

7. H. Mee-Jung, Y.S. Sun, L.S. Yup. Proteome analysis of metabolically engineered *Escherichia coli* producing poly(3-hydroxybutyrate). *J. Bacteriol.* 183: 301–308, 2001.

8. C.E. Nakamura, G.M. Whited. Metabolic engineering for the microbial production of 1,3-propanediol, *Curr. Op. Biotechnol.* 14: 454–549, 2003.

9. J. Koukol, E.E. Conn. The metabolism of aromatic compounds in higher plants. IV. Purification and properties of the phenylalanine deaminase of *Hordeum vulgare. J. Biol. Chem.* 236: 2692–2698, 1961.

10. J.W. Frost, K.M. Draths. Biocatalytic syntheses of aromatics from D-glucose: renewable microbial sources of aromatic compounds. *Annu. Rev. Microbiol.*, 49: 557–579, 1995.

11. A. Berry. Improving production of aromatic compounds in *Escherichia coli* by metabolic engineering. *Trends Biotech.* 14: 250–256, 1996.

12. J. Bongaerts, J.M. Kramer, M.U. Muller, U.L. Raeven, L.M. Wubbolts Metabolic engineering for microbial production of aromatic amino acids and derived compounds. *Metabol. Eng.*, 3: 289–300, 2001.

13. G. Chotani, T. Dodge, A. Hsu, M. Kumar, R. LaDuca, D. Trimbur, W. Weyler, K. Sanford, K. The commercial production of chemicals using pathway engineering. *Biochim. Biophys. Acta.* 1543: 434–455, 2000.

14. F. Valle, E. Munoz, E. Ponce, N. Flores, F. Bolivar. Basic and applied aspects of metabolic diversity: the phosphoenolpyruvate node. *J. Ind. Microbiol.* 17: 458–462, 1996.

15. N. Flores, J. Xiao, A. Berry, F. Bolivar, F. Valle. Pathway engineering for the production of aromatic compounds in *Escherichia coli. Nat. Biotechnol.* 14: 620–623, 1996.

16. G. Gosset, J. Yong-Xiao, A. Berry. A direct comparison of approaches for increasing carbon flow to aromatic biosynthesis in *Escherichia coli. J. Ind. Microbiol.* 17: 47–52, 1996.

17. R. Patnaik, W.D. Roof, R.F. Young, J.C. Liao. Stimulation of glucose catabolism in *Escherichia coli* by a potential futile cycle. *J. Bacteriol.* 174: 7527–7532, 1992.

18. Y. Kikuchi, K. Tsujimoto, O. Kurahashi. Mutational analysis of the feedback sites of phenylalanine-sensitive 3-deoxy-D-arabino-heptulosonate-7-phosphate synthase of *Escherichia coli. Appl. Environ. Microbiol.* 63: 761–762, 1997.

19. K.A. Dell, J.W. Frost. Identification and removal of impediments to biocatalytic synthesis of aromatics from D-glucose: Rate-limiting enzymes in the common pathway of aromatic amino acid biosynthesis. *J. Am. Chem. Soc.* 115: 11581–11589, 1993.

20. K.K. Kalghatgi, P.V. Subba Rao. Microbial L-phenylalanine ammonia-lyase. Purification, subunit structure and kinetic properties of the enzyme from *Rhizoctonia solani. Arch. Biochem. Biophys.* 149: 65–75, 1975.

21. J.R. Parkhurst, D.S. Hodgins. Yeast phenylalanine ammonia-lyase. Properties of the enzyme from *Sporobolomyces pararoseus* and its catalytic site. *Arch. Biochem. Biophys.* 152: 597–605, 1972.

22. G.S. Bezanson, D. Desaty, A.V. Emes, L.C. Vining. Biosynthesis of cinnamamide and detection of phenylalanine ammonia-lyase *in Streptomyces verticillatus. Can. J. Microbiol.* 16: 147–151, 1970.

23. L. Xiang, B.S. Moore. Inactivation, complementation, and heterologous expression of *encP*, a novel bacterial phenylalanine ammonia-lyase gene. *J. Biol. Chem.* 277: 32505–32509, 2002.

24. K.R. Hanson and E.A. Havir. The enzymatic elimination of ammonia. In *The Enzymes*, 3rd ed.; Boyer, P., Ed.; Academic: New York, 7: pp. 75–167, 1972.

25. K. Hahlbrock, D. Scheel. Physiology and Molecular Biology of Phenylpropanoid Metabolism. *Annu. Rev. Plant Phys. Plant Mol. Biol.* 40: 347–369, 1989.

26. J.F. Kane, M.J. Fiske. Regulation of phenylalanine ammonia lyase in *Rhodotorula glutinis. J. Bacteriol.* 161: 963–966, 1985.

27. G.P. Bolwell, J.N. Bell, C.L. Cramer, W. Schuch, C.J. Lamb, R.A. Dixon. L-Phenylalanine ammonia-lyase from *Phaseolus vulgaris*. Characterisation and differential induction of multiple forms from elicitor-treated cell suspension cultures. *Eur. J. Biochem.* 149: 411–419, 1985.

28. R. Lois, A. Dietrich, K. Hahlbrock, W. Schulz. A phenylalanine ammonia-lyase gene from parsley: structure, regulation and identification of elicitor and light responsive *cis*-acting elements. *EMBO J.* 8: 1641–1648, 1989.

29. E. Minami, Y. Ozeki, M. Matsuoka, N. Koizuka, Y. Tanaka. Structure and some characterization of the gene for phenylalanine ammonia-lyase from rice plants. *Eur. J. Biochem.* 185: 19–25, 1989.

30. J.G. Anson, H.J. Gilbert, J.D. Oram, N.P. Minton. Complete nucleotide sequence of the *Rhodosporidium toruloides* gene coding for phenylalanine ammonia-lyase. *Gene* 58: 189–199, 1987.

31. O.F. Rasmussen, H. Oerum. Analysis of the gene for phenylalanine ammonia-lyase from *Rhodosporidium toruloides*. *DNA Sequence.* 1: 207–211, 1991.

32. J.D. Faulkner, J.G. Anson, M.F. Tuite, N.P. Minton. High-level expression of the phenylalanine ammonia lyase-encoding gene from *Rhodosporidium toruloides* in *Saccharomyces cerevisiae* and *Escherichia coli* using a bifunctional expression system. *Gene* 143: 13–20, 1994.

33. B. Langer, D. Rother, J. Retey. Identification of essential amino acids in phenylalanine ammonia-lyase by site-directed mutagenesis. *Biochemistry* 36: 10867–10871, 1997.

34. G.R. McKegney, S.L. Butland, D. Theilmann, B.E. Ellis. Expression of poplar phenylalanine ammonia-lyase in insect cell cultures. *Phytochemistry* 41: 1259–1263, 1996.

35. W. Schulz, H.G. Eiben, K. Hahlbrock. Expression in *Escherichia coli* of catalytically active phenylalanine ammonia-lyase from parsley. *FEBS Letters* 258: 335–338, 1989.

36. M. Baedeker, G.E. Schulz. Overexpression of a designed 2.2 kb gene of eukaryotic phenylalanine ammonia-lyase in *Escherichia coli*. *FEBS Letters* 457: 57–60, 1999.

37. C. Appert, E. Logemann, K. Hahlbrock, J. Schmid, N. Amrhein. Structural and catalytic properties of the four phenylalanine ammonia lyase isoenzymes from parsley (*Petroselinum crispum Nym.*). *Eur. J. Biochem.* 225: 491–499, 1994.

38. J. Rösler, F. Krekel, N. Amrhein, J. Schmid. Maize phenylalanine ammonia-lyase has tyrosine ammonia-lyase activity. *Plant Physiol.* 113: 175–179, 1997.

39. K.R. Hanson and E.A. Havir. Phenylalanine ammonia lyase, In *The Biochemistry of Plants*; Academic: New York, 7: pp. 577–625, 1981.

40. J.A. Kyndt, T.E. Meyer, M.A. Cusanovich, J.J. Van Beeumen Characterization of a bacterial lyrosine ammonia lyase expressed, a biosynthetic enzyme for the photoactive yellow protein. *FEBS Letters* 512: 240–244, 2002.

41. US Patent Publication No. 2004-0023357A1.

42. D. Hernandez, A.T. Phillips. Purification and characterization of *Pseudomonas putida* histidine ammonia-lyase expressed in *Escherichia coli*. *Protein Expr Purif.* 4: 473–478, 1993.

43. T.F. Schwede, J. Retey, G.E. Schulz. Crystal structure of histidine ammonia-lyase revealing a novel polypeptide modification as the catalytic electrophile. *Biochemistry* 38: 5355–5361, 1999.

44. J.C. Calabrese, D.B. Jordan, A. Boodhoo, S. Sariaslani, T. Vannelli. Crystal structure of phenylalanine ammonia lyase: Multiple helix dipole implicated in catalysis. *Biochemistry* 43: 11403–11416, 2004.

45. A.I. El-Batal. Continuous production of L-phenylalanine by *Rhodotorula glutinis* immobilized cells using a column reactor. *Acta Microbiol. Pol.* 51: 139–152, 2002.

46. A.A. Gortenby, S. Sariaslani, X-S Tang, W.W. Qi, T. Vannelli. US Patent No. 6,368,837.

47. A. Burkovski, R. Kramer. Bacterial amino acid transport proteins: occurrence, functions, and significance for biotechnological applications. *Appl. Microbiol. Biotechnol.* 58: 265–274, 2002.

48. W. Pfefferle, B. Mockel, B. Bathe, A. Marx. Biotechnological manufacture of lysine. *Adv. Biochem. Eng. Biotechnol.* 79: 59–112, 2003.

49. T. Dassler, T. Maier, C. Winterhalter, A. Brock. (2000) Identification of a major facilitator protein from *Escherichia coli* involved in efflux of metabolites of the cysteine pathway. *Mol. Microbiol.* 36: 1101–1112, 2000.

50. D. Kruse, R. Kramer, L. Eggeling, M. Rieping, W. Pfefferle, J.H. Tchieu, Y.J. Chung, Jr., M.H. Saier Jr., A. Burkovski. Influence of threonine exporters on threonine production in *Escherichia coli*. *Appl Microbiol Biotechnol* 59: 205–210, 2002.

51. M.I. Ramos-Gonzalez, P. Godoy, M. Alaminos, A. Ben-Bassat, J.L. Ramos. Physiological characterization of *Pseudomonas putida* DOT-T1E tolerance to p-hydroxybenzoate. *Appl. Environ. Microbiol.* 67: 4338–4341, 2001.

52. M.I. Ramos-Gonzalez, A. Ben-Bassat, M.J. Campos, J.L. Ramos. Genetic engineering of a highly solvent-tolerant *Pseudomonas putida* strain for biotransformation of toluene to p-hydroxybenzoate. *Appl. Environ. Microbiol.* 69, 5120–5127, 2003.

53. I. Paulsen, J. Park, P. Choi, M. H. Saier Jr. A family of G-negative bacterial outer membrane factors that function in the export of proteins, carbohydrates, drugs and heavy metals from gram-negative bacteria. *FEMS Microbiol. Lett.* 156: 1–8, 1997.

54. R. Benz, E. Maier, I. Gentschev. *TolC* of *Escherichia coli* functions as an outer membrane channel. *Int. J. Med. Microbiol. Virol. Parasitol. Inf. Dis.* 278: 187–196, 1993.

55. A. Sharff, C. Fanutti, J. Shi, C. Calladine, B. Luisi. The role of the TolC family in protein transport and multidrug efflux. From stereochemical certainty to mechanistic hypothesis. *Eur. J. Biochem.* 268: 5011–5026, 2001.

56. C. Schnaitman. Improved strains for target-based chemical screening. *ASM News* 57: 612, 1991.

57. K. Nishino, H. Yamaguchi. Analysis of a complete library of putative drug transporter genes in *Escherichia coli. J. Bacteriol.* 183: 5803–5812, 2001.

58. M.C. Sulavik, C. Houseweart, C. Cramer, N. Jiwani, N. Murgolo, J. Greene, B. DiDomenico, K. J. Shaw, G. H. Miller, R. Hare. G. Shimer. Antibiotic susceptibility profiles of *Escherichia coli* strains lacking multidrug efflux pump genes. *Antimicrob. Agents Chemother.* 45: 1126–1136, 2001.

59. T.T. Tseng, K.S. Gratwick, J. Kollman, D. Park, D.H. Nies, A. Goffeau, M.H. Saier Jr. The RND permease superfamily: an ancient, ubiquitous and diverse family that includes human disease and development proteins. *J. Mol. Microbiol. Biotechnol.* 1: 107–125, 1999.

60. H. Nikaido, H. I. Zgurskaya. AcrAB and related multidrug efflux pumps of *Escherichia coli. J. Mol. Microbiol. Biotechnol.* 3: 215–218, 2001.

61. T. Dinh, I. T. Paulsen, M.H. Saier, Jr. A family of extracytoplasmic proteins that allow transport of large molecules across the outer members of Gram-negative bacteria. *J. Bacteriol.* 176: 3825–3831, 1994.

62. J.A. Fralick. Evidence that TolC is required for functioning of the Mar/AcrAB efflux pump of *Escherichia coli. J. Bacteriol.* 178: 5803–5805, 1996.

63. S.K. Picataggio, L.J., Templeton, D.R. Smulski, R.A. LaRossa. Transcript profiling of *Escherichia coli* using high-density DNA microarrays. *Methods Enzymol.* 358:177–188, 2002.

64. D.R. Smulski, L.L. Huang, M.P. McCluskey, M.J. Reeve, A.C. Vollmer, T.K. Van Dyk, R.A. LaRossa. Combined, functional genomic-biochemical approach to intermediary metabolism: interaction of acivicin, a glutamine amidotransferase inhibitor, with *Escherichia coli* K-12. *J. Bacteriol.* 183: 3353–3364, 2001.

65. Y. Wei, J. –M. Lee, C. Richmond, F. Blattner, J. A. Rafalaski, R. A. LaRossa. High-density microarray mediated gene expression profiling of *Escherichia coli. J. Bacteriol.* 183: 545–556, 2001.

66. K.T. Harley, M.H. Saier, Jr. A novel ubiquitous family of putative efflux transporters. *J. Mol. Microbiol. Biotechnol.* 2: 195–198, 2000.

67. T.K. Van Dyk, L.J. Templeton, K.A Cantera, P.L. Sharpe, F.S. Sariaslani. Characterization of the *Escherichia col: AaeAB* efflux pump: A metabolic relief valve? *J. Bacteriol.* 186: 7196–7204, 2004.

68. R.B. Helling, B.K. Janes, H. Kimball, T. Tran, M. Bundesmann, P. Check, D. Phelan, C. Miller. Toxic waste disposal in *Escherichia coli. J. Bacteriol.* 184: 3699–3703, 2002.

69. L. Zamir, R. Tiberio, R. Jensen. Differential acid-catalyzed aromatization of prephenate, arogenate and spiro arogenate, *Tetrahedron Lett.* 24: 2815–2818, 1983.

70. K.M. Draths, S. Kambourakis, K. Li, J.W. Frost, 3-Dehydroshikimic acid: A building block for chemical synthesis from renewable feedstocks, *in* J. J. Bozell Ed., *Chemicals and Materials from Renewable Resources*, ACS Press, 133–146, 2001.

29

Extremophiles from the Origin of Life to Biotechnological Applications

Chiara Schiraldi

Mario De Rosa

29.1 Introduction

29.1.1 Microorganisms from Extreme Environments

The discovery of microbial life in seemingly prohibitive environments continues to challenge conventional concepts of the growth-limiting conditions of cellular organisms. In the last few decades, it has become clear that microbial communities can be found in the most diverse conditions, including extremes of temperature, pressure, salinity, and pH. The diversity of extremophiles has barely been tapped, and estimates generally agree that <1% of the microorganisms in the environment have been cultivated in pure cultures to date. Many parts of the world are defined as extreme, such as geothermal environments, polar regions, acid and alkaline springs, and the cold, pressurized depths of the ocean. As conditions become more demanding, the microbial population is constituted exclusively of microorganisms belonging to the bacterial and archaeal domain.

The discovery of novel microorganisms with peculiar features and characteristics has rearranged the phylogenetic tree. The latter, which is 16S rDNA-based, consists of a tripartite division of the living organisms into Bacteria, Eucaria, and Archaea.[1] The Archaea consist of two major kingdoms: the Crenarcheota (*Sulfolobus, Pyrodictium, Pyrolobus, Thermoproteus*, etc.) and the Euryarchaeota, which include hyperthemophiles

TABLE 29.1 Major Examples of Enzymes Isolated from Extremophilic Microorganisms and Their Foreseen Applications in Industrial Processes or Commercial Products

Microorganisms	Enzymes	Applications	References
Thermophiles (50–110°C)	a) Amylase Glycosidase, pullulanase, glucoamylases, cellulases, xylanases	a) starch, cellulose, chitin, pectin processing, oligosaccharides synthesis textiles	[2–8]
	b) Chitinases	b) chitin modification for food and health products	
	c) Xylanases	c) paper bleaching	
	d) Lipases, esterases	d) waste water treatments, detergents, stereospecific reactions (trans-esterifications, organic biosynthesis)	
	e) Proteases	e) detergents, hydrolysis in food and feed, brewing and baking	
	f) DNA-polymerases	f) Molecular biology	
	g) dehydrogenases	g) Oxidation reactions	
Psychrophiles (0-20°C)	a) amylase	a) detergents and bakery	[3,9]
	b) cellulases	b) detergents feed and textiles	
	c) dehydrogenases	c) biosensors	
	d) lipases	d) detergents food and cosmetics	
	e) proteases	e) detergents, dairy industry	
Alkaliphiles pH > 9	a) amylase, cellulases, lipases	a) Polymers degrading agents in detergents	[10–12]
	b) cyclodextrins synthase	b) Food additives	
Acidophiles	a) amylase, glucoamylases, cellulases	a) Starch processing	[12,13]
pH < 3.5	b) proteases	b) Feed components	
	c) oxidases	c) Desulfurization of coal	
Halophiles (2–5 M NaCl)	a) Proteases	a) Peptide synthesis	[12,14]
	b) dehydrogenases	b) Biocatalysis in organic media	

(e.g., *pyrococcus*), methanogens (*methanococcus*), and holophiles (*Halobacterium*). The realization that extreme environments harbor a different kind of prokaryote lineage has resulted in a complete reassessment of our concept of microbial evolution and has given considerable impetus to extremophilic research.

In fact, it is worth mentioning that modern biotechnology, which provides a whole new repertoire of methods and products, still tries to mimic nature, thus demanding continuous effort in the isolation and characterization of novel microorganisms.

The peculiar features of extremophiles and their enzymatic patrimony prove highly interesting for the development of novel biotechnological applications (Table 29.1).[2–14]

29.1.2 Psychrophiles

Microorganisms that colonize cold environments, such as polar regions or oceanic depths, are known as Psychrophiles. They are able to grow at temperatures close to 0°C, and their enzymes allow them to compensate for the negative effects of low temperature. Most psychrophiles have been isolated from Arctic or Antarctic seawaters where, despite the harsh conditions, the density of bacterial cells is as high as that reported for temperate waters.

Cold-adapted microorganisms can be either psychrotolerant, if their optimal growth temperature is around 20°C, or psychrophilic if they can grow at 0°C, and the optimum is at temperatures lower than 15°C.

The first psychrophilic bacterium was isolated by Foster (1887) from preserved fish. A systematic investigation was recently carried out in order to understand the rules governing their molecular adaptation to low temperatures. Unraveling these key features will certainly help in the biotechnological exploitation of these microorganisms and their cell components, such as polysaccharides, membranes and enzymes (Table 29.1).[2-14]

29.1.3 Thermophiles

Microorganisms that have adapted to grow at high temperatures (60–110°C) have been isolated from volcanic areas, geothermal heated hydrothermal vents, solfataric fields, hot springs, and submarine hot vents.

Because of their ability to convert volcanic gases and sulphur compounds at high temperatures, hyperthermophilic communities living in such hydrothermal vents are expected to play an important role in marine ecological, geochemical, and volcanic processes.[15]

Moderate thermophiles are able to grow at temperatures of 50–60°C, and most of these microorganisms belong to many different taxonomic groups of eu- and prokaryotic microorganisms, such as protozoa, fungi, algae, and cyanobacteria. They appear to be closely related phylogenetically to mesophiles, and adapted to life in hot environments.

Extreme thermophiles grow optimally at 60–80°C, and hyperthermophiles at 80–108°C. The majority of hyperthermophiles isolated to date belong to the archaeal domain.

29.1.4 Acidophiles and Alkaliphiles

The preferred biotopes for microorganisms that are able to thrive in thermophilic and acidic conditions are Solfataric fields. Because of the geophysical composition of the soils, it is possible to isolate different kinds of microorganisms from these environments. In the upper layer of the soil various thermoacidophilic aerobes were isolated (optimum growth conditions pH between 0.7 and 5, and T 60–90°C) belonging to the genera Sulfolobus, Acidianus, Thermoplasma, and Pichrophilus. Moderate acidophilic or even neutrophilic anaerobs are mostly isolated in a second deeper soil layer and mainly belong to the genera Thermoproteus and Methanothermus.

Sulfolobales are strict aerobes growing either autotrophically, heterotrophically, or facultative heterotrophically. They oxidize S^0, S^{2-}, and H_2 to sulphuric acid or water during autotrophic growth.

Microorganisms that grow at high pH, the so-called alkaliphiles, are widely distributed on earth, especially in carbonate rich springs and alkaline soils, whose pH can be higher than 10. Several species of cyanobacteria and bacillus are normally abundant in these biotopes and provide organic matter for various groups of heterotrophs. Alkaliphiles need high pH and sodium ions not only for growth but also for sporulation and germination. They generally require different types of nutrients, although a few bacillus strains have been demonstrated to duplicate in minimal media containing glycerol, glutammic acid, and citric acid. In general, the cultivation temperature ranges of 20–55°C.

Many haloalkaliphiles have been isolated from alkaline saline lakes that can grow at high pH in the presence of 20% NaCl. Recently, thermoalkaliphilic strains have also been isolated in Lake Bogoria (Kenya) and the Yellowstone Park (U.S.), and these represent a new line within the *Clostridium/Bacillus* phylum.[16]

The two archaeal thermoalkaliphiles isolated, which are *Thermococcus alcaliphilus* and *T.acidoaminivorans*, are growing at 85°C pH 9.

Alkaliphilic enzymes are mainly employed in the detergent industry and can also be exploited in the de-hairing process when a pH higher than 8 is required (Table 29.1).

29.1.5 Halophiles

Halophilic microorganisms grow optimally at NaCl concentrations higher than 0.6 M and generally belong either to the bacterial or to the archaeal domain. They are identified for the range of salinity they require for growth into moderate halophiles (salt concentration of 0.4–3.5 M) and extreme halophiles (NaCl concentration higher than 2 M).[17]

Most of these microorganisms have been isolated from saline lakes, evaporitic lagoons, and coastal salterns with NaCl concentrations ranging between 1 and 2.6 M. Saline soils have been less explored as they are less stable biotopes given the fact they are subject to periodical dilutions due to the rainy periods.

Although most hypersaline environments have been demonstrated to contain significant populations of adapted microorganisms, the nutrient source is still unclear. A carbon concentration of about 1/g· l⁻¹ was frequently found, even though organic compounds may also derive from primary producers such as cyanobacteria or algae.

Despite Hypersaline lakes presenting a high surface/volume ratio, dissolved oxygen was found to be as low as 2 mg· l⁻¹, indicating that the environmental conditions are almost anaerobic.[17]

It has been shown that aerobic heterotrophs living in marine salterns are bacterial halophiles if the salt concentration is lower than 2 M, whereas at higher concentrations the population is almost completely composed by archaea. The variety of heterotrophic halophilic bacteria gives rise to the assumption that all kinds of metabolic features may be found in high-salinity environments. These microorganisms do not belong to an homogeneous group but to different bacterial taxa, and their ability to grow in the presence of salt is a secondary adaptation.

The archaeal family of the "halobacteria" comprises the red pigmented extremely halophilic microorganisms. They have membranes with ether-linked lipids which is a specific characteristic of archaeal domain microorganisms. The orange color of halobacteria is due to the presence of carotenoids; some species are colorless while others have gas vescicles that result in opaque pink colonies. When salinity approaches saturation, halobacteria remain the only life form. They also influence evaporation given that carotenoid pigments of halobacteria trap solar radiation, thus increasing the environment temperature.

It has been demonstrated that halophiles are able to counterbalance osmotic stress by accumulating intracellularly compatible solutes. These molecules have also been tested as stress protectants outside the mother cells and gave interesting results for a wide range of biotechnological applications.

It was argued that the peculiar physiology of halophiles, which have to handle concentrations as high as 4 M ions, could be responsible for the evolution of enzymes that can work in low water activity environments (e.g., organic solvents). However, to date, halophiles have not had the impact expected on the biotechnological world. It is worth mentioning that 20% of patent applications for extremophiles concerns halophiles, thus confirming the potentiality of these microorganisms and their biomolecules for industrial applications (Table 29.1).

29.1.6 Archaeal Lipids

Archaea, one of the three domains of life, comprises a variety of extremophiles classified in two kingdoms: Euryarcheota and Crenaecheota.

The lipids from archaea can be used taxonomically to distinguish between the three phenotypic subgroups of archaea (halophiles, methanogens, hyperthermophiles) and to distinguish them clearly from other organisms.[18,19] In fact, archaea contain very different membrane lipid structures to their bacterial and eukaryotic counterparts. Archaeal membrane lipids are mainly composed of saturated isoprenoic chains of different length, in ether linkage to glycerol carbons with sn-2,3 configuration, and not to fatty acyl chains, which are often unsaturated and are esterified to glycerol at carbons sn-1,2.[18,20] Archaeal membrane lipids are derivatives of the C_{20}-C_{20} isopranyl glycerol diether, archaeol, ubiquitously found in varying amounts in all archaeal lipids, and its dimer, the dibiphytanyldiglycerol tetraether, caldarchaeol.[21] Lipids derived from archaeol and caldarchaeol or their variants have not been found in bacteria or eukarya. Variations in the polar head groups occur frequently and may often provide molecular taxonomic fingerprints identifying archaea both to define large groups and to distinguish subgroups within species.[20,22]

The uniqueness of archaeal lipids is reflected in their topology and function within membranes and evolutionary relationships within the archaea and between archaea, bacteria and eukarya.[23] This uniqueness has boosted interest in the biotechnological applications of such compounds. They can provide an excellent material to form liposomes with remarkable thermostability and tightness against solute leakage.[24] In addition, the archaeal lipids and in particular bipolar tetraethers, offer novel structural opportunities in terms of protein lipid interactions, which are of interest for the assembly of electronic devices based on redox proteins or enzymes.[25]

29.2 Cultivation of Extremophilic Microorganisms

The cultivation of extremophiles require the identification of suitable media, especially in relation to oligoelements, as well as the development of unconventional equipment to sustain the long-term fermentation in environmental conditions comprising very high temperatures, or acidic/alkaline pHs, high pressure, and high salinity. In addition, other issues such as gas-liquid mass transfer and material corrosion have to be tackled to safely complete any fermentation experiment. In fact, it can be argued that despite the outstanding features of extremophiles, the industrial application of their enzymes and metabolites is hampered by the low productivity typical of the fermentation processes of these microorganisms. This is the result of both slow growth rate and low biomass yield. In an attempt to tackle these issues, several research groups have focused their attention on the physiology of extremophilic microorganisms of biotechnological interest, as well as on designing bioreactors and bioprocesses that were able to positively influence the productivity once the growth conditions were improved. The most interesting approaches suggested in recent literature are reported here.

29.2.1 Special Equipment

As mentioned earlier, one of the main challenges to scientists and engineers in developing novel fermentation equipment is the necessity to re-create the "extreme" environments from which extremophiles were collected in the lab and at a later stage in a production site (plant). This means that materials are required to withstand high temperatures and low pHs in the cultivation of thermoacidophiles such as *Sulfolobus solfataricus*, but also high salinity or high alkaline solutions for halophiles and alkaliphiles. Probably the most demanding microorganisms come from the deep sea, where high pressure and volcanic areas are colonized. To our knowledge only Professor Clark's group at University of California at Berkeley (UCB) is working to develop a pilot-scale bioreactor, able to continuously operate under extreme temperature and pressure, to overcome the limitation of cultivating these extremophilic microorganisms on a large scale. This system will allow the continuous input of growth medium and removal of biomass at pressures exceeding 600 bar and temperatures up to 200°C. The bioreactor, which has a capacity of 30 L, will make a significant contribution toward evaluating the culture conditions for deep-sea extremophiles, thus permitting research under the optimal growth conditions.[26]

A novel corrosion-resistant bioreactor made of highly resistant high-tech materials was specifically constructed to allow the repeated cultivation of two newly isolated extreme halophiles, producing poly-γ-glutamic acid (PGA) that can be used as a biodegradable thickener or humectant and poly-β-hydroxybutyric acid (PHB), respectively.[27] Batch fermentations of the two halophiles on n-butyric acid as carbon source gave a cell density of 2.3 g cell dry matter per l, with an accumulation of PHB acid up to 53% of dry weight.

In collaboration with Schmid et al. (1998),[28] Bioengineering AG (Wald, Switzerland) developed a high-pressure explosion-proof bioreactor that proved useful in two-liquid phase bioconversions at pilot scale in the presence of a bulk amount of organic solvents.

29.2.2 Continuous Culture

Continuous cultures have often been reported as effective tools in physiology studies. In fact, the unconventional environmental conditions minimize contamination problems and long-term experiments can be successfully carried out. Continuous culture experiments have provided insights into the significance of specific enzymes in the metabolism of particular substrates, as well as giving a better understanding of the stress response and unusual physiological characteristics of hyperthermophilic and extremely thermoacidophilic microorganisms. One of the most difficult microorganisms to grow is *Pyrococcus furiosus*, an hyperthermophilic anaerobe. An increase in cell density up to 10^8 cells·ml^{-1} was achieved in continuous cultures on minimal media containing starch.[29,30]

Metallosphera sedula (T-opt 74°C, pH-opt 2.0), *Pyrococcus furiosus* (T-opt 98°C), and *Thermococcus litoralis* (T-opt 88°C) have been cultivated in a chemostat[31] and the production of exopolysaccharides by these hyperthermophilic heterotrophs was also evaluated in relation to biofilm formation. An extremely thermophilic methanogen, *Methanococcus jannashii*, was grown in continuous culture to evaluate the hydrogenase activity in relation to growth rate and pressure.[32] Another example of continuous culture of methanogens was reported by Schills et al. (1996);[33] in particular, *Methanobacterium thermoautotropicum* was cultivated at different gassing rates using online measurements of growth limiting factor H_2. The results made it possible to develop a simple unstructured mathematical model that satisfactorily fitted the experimental growth curve.

The optimization of growth media is also of key importance in the production of extremophilic biomasses for subsequent industrial exploitation. Particular interest has been given to the cultivation of anaerobic archaea because their sensitivity to oxygen is a major obstacle to their cultivation in lab scale and also for the development of archaeal genetic exchange systems. Rothe and Thomm (2000)[34] reported the replacement of Na_2S with Na_2SO_3 for the protection of cultures from oxygen even outside the anaerobic chamber, thus simplifying the cultivation method. A gas-lift bioreactor was exploited in a continuous culture of *Pyrococcus abyssi* ST549 and the experiments investigated the growth on carbohydrates, proteinaceous substrates, and amino acids.[35] The results suggested that the disaccharides maltose and cellobiose did not support growth, whereas proteinaceus material, such as peptone or amino acid in the presence of sulphur, proved good substrates. More recently, Biller et al.[36] demonstrated that *Pyrococcus furiosus* grew to a higher density on soluble starch rather than on maltose. This result supports the possibility of using renewable resources for extremophilic biomass production and is useful when making an economic assessment of the commercial exploitation of the related enzymes. In this respect *Sulfolobus solfataricus* MT4 was also demonstrated to grow on whey supplied directly from a dairy farmer (87°C) in a semi-pilot experiment (100 L fermenter),[37] giving a yield coefficient of 80 g of dry biomass per g-mol of lactose.

Furthermore, the ability of certain extremophiles to grow on pollutants is a very interesting feature. Recently, a thermophilic strain, *Bacillus thermoleovorans* sp. A2, was investigated for its capacity to degrade phenol. The authors demonstrated that growth rates were fourfold higher than those reported for mesophilic microorganisms. The specific growth rates and the yield coefficients were also evaluated in this study.[38] Another thermophilie, *Bacillus* (sp. IHI-91), was cultivated in batch and continuous mode on olive oil for the production of a thermostable lipase. Productivity proved 50% superior in the chemostat experiments with respect to the batch cultures.[39]

29.2.3 Fed-Batch Cultures

In order to improve extremophilic biomass and related enzymes or metabolites productivity, scientists focused their attention on designing novel bioprocesses, such as fed-batch cultures, and on developing special bioreactors that allow easy removal of toxic metabolites and the replacement of nutrients. To date, both of these ways have been evaluated and successfully applied in the cultivation of thermophilic and halophilic microorganisms.

Efficient biomass production was achieved through a fed-batch technique in the cultivation of *Sulfolobus solfataricus* DSM 1617.[40,41] These studies reported the effects of yeast extract and other complex components on metabolite formation (e.g., ammonia) and cell yield. During fed-batch processes, a maximum cell yield of 23 g of dry biomass per l was achieved when constant volume was maintained through the continuous replacement of evaporated water. The kinetic parameters and the mathematical model of the fermentation process were evaluated.

Although the application of fed-batch techniques coupled with media optimization permitted physiological studies and the production of a sufficient amount of biomass for the isolation and characterization of numerous enzymes of potential biotechnological interest, the levels of productivity achieved are not compatible with large-scale applications of these biocatalysts. In fact, it often happens that the accumulation of low molecular weight metabolites is responsible for growth inhibition in a similar fashion as for mesophile cultures (e.g., acetate accumulation in *E coli* fermentation, lactate for lactic acid bacteria). For this reason, fed-batch techniques as well as medium optimization will always result in a *plateau* in the growth curve. However, the improvement of biomass yield remains of great interest for the economically feasible exploitation of extremozymes and novel organic solutes or membrane components from these peculiar microbes. To address this issue, several groups have been working on designing bioreactors to couple fermentation and separation processes, thus overcoming problems stemming from the toxicity of accumulated metabolites.

29.2.4 Novel Bioreactors and Bioprocesses

29.2.4.1 Dialysis Bioreactor

A novel dialysis bioreactor was designed and developed at the Technical University of Harburg-Hamburg.[42] This technique is centered on the principle that if two solutions with different concentrations of a certain compound are kept in contact through a specific membrane (cutoff 10/30 KDa), there will be a spontaneous mass transport from the highly concentrated to the lower concentration solution that is proportional to the concentration difference. However, it also depends on the diffusivity of the molecule of interest (i.e., inhibitors, given that the membrane is permeable to the compound). It can be easily understood that the efficiency of the process is strictly dependent on the concentration drop at the membrane sides. This means that either a large volume of dialysate or a specially designed external dialysis module[43] should be used for an efficient removal of the molecules of interest. An extensive study on a laboratory scale, exploiting a dialysis process, was reported by Krahe et al. (1996),[42] in which high cell densities were achieved in the cultivations of three extremophilic strains. In the study by Krahe, several issues typical of extremophile cultures are addressed. Special attention is focused on chemical reactions likely to occur in the medium and gas-liquid mass transfer at high temperature. In the dialysis experiments, a final cell increase of 30- to 40-fold compared to the batch process was achieved in 350 h of fermentation for *Sulfolobus shibatae*. In the cultivation of *Pyrococcus furiosus*, a 20-fold increase with respect to the batch process was reported, corresponding to a maximal density of $3 \cdot 10^{10}$ cells·ml^{-1}. Lastly, an halophilic eubacterium, *Marinococcus* M52, producing compatible solutes was cultivated in a dialysis bioprocess achieving over 100 g·l^{-1} of dry biomass. The paper substantially demonstrated the applicability of the novel bioreactor in the cultivation of different types of microorganisms isolated from extreme environments, in aerobic and anaerobic conditions and also at acidic pHs. To our knowledge, the bioprocess described earlier has so far not been developed on a large scale, at least for extremophile cultivation.

29.2.4.2 External Filtration

A novel process has been developed by Bitop (Bonn) for the production of compatible solutes from halophilic bacteria. The process consists of three successive steps: (1) the high density cultivation of selected species through the use of external filtration modules and cell recycling; (2) the medium exchange with osmotic downshock to obtain the release of compatible solutes from the cells; (3) a second medium exchange to newly introduce high salinity growth medium. The production process was called "bacterial milking" and permitted the large scale production of these peculiar biomolecules. Compatible solutes have proven to be excellent stabilizers of biological materials, and they are progressing in the cosmetic and pharmaceutical fields.[44] From this innovative idea translated into large-scale production, we learn that there are niches in which the industrial exploitation of extremophiles is valuable. However, once the biotechnological potential of new biomolecules of extremophilic origin is demonstrated, their actual delivery on the market is strictly dependent on the design of custom-tailored processes aiming at economically feasible scale-up of their production.

29.2.4.3 Microfiltration Bioreactor

In our group, a novel microfiltration (MF) technique was exploited for the high-density cultivation of thermoacidophiles and halophiles. Microfiltration is based on the principle that by applying a difference of pressure across the membrane, there would be a flux that is dependent on the permeability and the

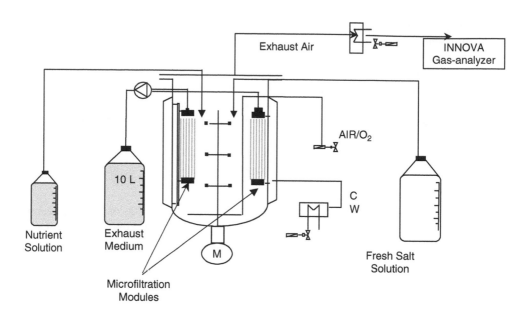

FIGURE 29.1 Schematic view of the MF bioreactor modified from Schiraldi et al. (2000).

resistance characteristics of the membrane, and this will remove the whole exhausted medium while retaining the cells that are larger than the average pore size (0.20–0.25 μm). The membrane bioreactor developed consisted of a traditional fermentation vessel modified next to the baffles by the insertion of one/four microfiltration (MF) modules made of polypropylene capillaries (Figure 29.1). The final assembly of the modules was carried out after testing several glue-components to ensure its stability at repeated sterilization cycles and to "extreme" growth parameters (30°C < T < 90°C, 2 < pH < 7, ionic strength > 1.5 M) and the compatibility with the growth of the microorganisms. The modules were placed along the baffles to ensure high turbulence. In addition, considering the mixing profiles within a stirred tank reactor (STR) equipped with rushton turbines, a cross-flow filtration is obtained using this configuration consequently limiting the fouling processes. The fermentation experiment consisted of three consecutive phases, i.e., batch, fed-batch, and microfiltration, prolonging the microorganisms' growth phase. The bioreactor was exploited in more than 120 experiments and was able to keep the filtration potentiality for about 2 years. In particular, we obtained high biomass concentration of *Sulfolobus solfataricus* Gθ and *Sulfolobus solfataricus* MT4, showing that the MF bioreactor could improve yield 10- to 15-fold (maximal cell density 38 g of cell dry weight·L^{-1} and 37 g·L^{-1}, respectively) with respect to the batch experiments. Fewer nutrients were wasted, demonstrating the effectiveness of this experimental setup with respect to repeated batch experiments. During MF experiments, *Sulfolobus* maintained an active synthesis of reporter enzymes of great industrial interest, such as α-glucosidase, ß-glycosidase, and alcohol dehydrogenase.[45,46] In spite of the outstanding results, the productivity of the enzymes was still low owing to both the long doubling time of *Sulfolobus* (e.g., 7–15 h) and the poor expression level of these biocatalysts. These issues have been frequently encountered in the production of extremophilic biomasses and their enzymes. Therefore the application of genetic engineering to improve productivity would be strategic, and the recent completion and publication of many archaeal genomes will simplify this task.

29.3 Enzymes of Biotechnological Interest from Extremophilic Sources

It is worth mentioning that the difficulties encountered in the cultivation of extremophiles suggest that the real breakthrough for biotechnology will lie in the production, even by means of molecular biology techniques of enzymes and/or special biomolecules (e.g., lipids, compatible solutes, etc.) rather than the application of these unique microbes to obtain common primary metabolites (e.g., organic acids, ethanol, etc.). In fact, it is well known that biocatalysis present a far better level of chemical precision compared to organic synthesis, providing the opportunity to achieve a stereospecific production of peculiar molecules with high yield and purity while minimizing the occurrence of side reactions, thus resulting in a lower environmental burden.[47] Despite the fact that to date more than 3000 different enzymes have been identified and that many of these have been applied in industrial processes, the current enzyme toolbox is still not sufficient to meet all demands. The main reason for this is that many available enzymes do not withstand industrial reaction conditions. As a result, the characterization of new extremophilic microorganisms and their enzymatic patrimony may certainly contribute to an increase in the use of biocatalysis in commercial production processes.

29.3.1 Strategies for the Discovery of Novel Enzymes

Traditionally, many extremozymes have been detected through the cultivation of newly isolated strains and classical protein extraction and purification. However, this method requires a sufficiently large amount of biomass (pure culture) and entails a downstream processes that may prove long and cumbersome. It is also possible to screen certain activities, for instance the hydrolytic activities, by phenotypic detection using dyed substrates such as red-amylopectine, red-pullulan, and azo-casein. Strains producing the desired enzymes will be distinguished by the formation of clearing zones around colonies. This simple method was successfully applied to screen a recombinant pullulanase from *Fervodobacterium pennivorans* Ven5.[48] with the added advantage that a large number of strains can be screened in a reasonable time, unlike the typical enzymatic assays on crude extracts. The novel screening techniques at a genetic level are even more promising. In fact, genes encoding a particular enzyme can be detected and isolated by radioactive/non-radioactive probes. This technique was exploited to clone, express, and characterize a DNA-polymerase from *Pyrobaculum islandicum*.[49] Genetic libraries may offer another opportunity to detect biotechnologically relevant genes and their corresponding enzymes. However, a cDNA cloning experiment has more chances to be successful if it is known that a particular gene is transcribed at reasonably high levels. The major drawback is that a cDNA clone will never give details of important signals structures such as enhancers, promoters, transcription factors, binding sites, etc.

Frequently, biotech companies prefer to follow another approach. They collect environmental samples containing heterogeneous populations of uncultured microbes from diverse ecosystems and extract genetic material without needing to grow and maintain the microorganisms. The validity of the method is also due to the fact that even small samples yield sufficient DNA, and therefore the sensitivity to the environment is minimized. Gene expression libraries created from the microbial DNA make it possible to produce the biomolecules of interest at a later stage. One of the companies most involved in this kind of screening is Diversa (San Diego, CA,), which currently has gene libraries with the complete genomes of over 1 million unique microorganisms. This enormous resource of genetic material can be screened for valuable commercial products (Eric Mathur, Personal communication, 2002). However, it is impossible to trace the nucleotide sequence back to the original microorganisms, and therefore the source of the biocatalyst of interest remains unknown.

29.4 Case Study: Production and Applications of Thermophilic Enzymes in Carbohydrate Processing

Over the last decade, our group has been working on the isolation, characterization, production, and application of four enzymes from Sulfolobales of specific interest in the biotranformation of carbohydrates, namely, β-glycosidase, α-glucosidase, trehalosyl dextrin forming enzymes, and trehalose-forming enzyme. The following case studies describe the protocols developed to obtain these enzymes from the wild-type and recombinant organisms. In addition, the high cell density cultivation processes based on the microfiltration technique will be discussed and the results obtained in immobilized cells biotranformation presented.

TABLE 29.2 Characteristics of the Enzymes Isolated from *S. solfataricus* (Ss) and *S. shibatae* (Ssh)

ENZYME	Native MW (Subunits)	T_{opt} (°C)	pH_{opt}	pI	No of Subunits	K_m (mM)	k_{cat} (s^{-1})	Half-life at 75°C (h)	Residual Activity (40°C)
β-glycosidase (βgly)	Ss 240 (60)	80	6.5	4.5	4	0.5	542	>24	5
α-glucosidase (αgly)	Ssh 313 (80)	85	5.5	4.7	4	8	45.5	156	5
	Ss 320 (80)	100	5		4	n.d.	n.d.	n.d.	5
Trehalosyl-dextrin forming enzyme (TDFE)	Ssh 80	70	4.5	5	1	2	33	96	60
	Ss 80	75	5.5		1	n.d.	n.d.	n.d.	n.d.
Trehalose forming enzyme (TFE)	Ssh 65	85	4.5	5	1	1	83	96	30
	Ss 76	85	5.5	6.1	1	n.d.	n.d.	n.d.	n.d.

29.4.1 Wild-Type Enzymes

29.4.1.1 β-glycosidase

β-glycosidases (EC 3.2.1.x) are enzymes with the greatest potential in biotechnology. This group of enzymes, which catalyzes the hydrolysis of glycosidic bonds among a large variety of saccharides, includes β-gluco, β-galacto, and phospo-β-galactosidases. Such ample substrate specificity allows their utilization in several fields, and in particular, thermophilic β-glycosidases can prove an ideal tool for peculiar application in chemistry, biochemistry, and molecular biology. The saccarification of cellulosic materials, the hydrolysis of lactose at high temperature in dairy products, the use of gene reporter in molecular biology, and the synthesis of drugs and chemicals in the pharmaceutical industry by reversed-catalyzed reactions are some of the applicative examples under evaluation (Table 29.1).

Nucci et al.[50] purified a β-glycosidase from the extreme thermoacidophilic archaeon *Sulfolobus solfataricus*, strain MT-4 (optimal growth conditions: T = 87°C pH = 3). The enzyme is a tetramer of four identical subunits with a total native molecular weight of 240,000 Da. The enzyme is able to catalyze the hydrolysis of a large number of disaccharides derived from the β-linking of the glucose with aryl, alkyl, or 6-carbon-atom sugars. Furthermore, in the presence of β-linked glucose oligomers from two to five glucose units, the enzyme produces glucose from the nonreducing end of the oligosaccharide at high temperature. Consequently, it was classified as a β-glycosidase with remarkable exo-glucosidase activity and was named Ssβgly. The enzyme has been fully characterized for its structural and kinetic features (Table 29.2). The ample substrate specificity of Ssβgly includes the ability of the enzyme to synthesize different glycosides by trans-glycosidase activity, using several carbohydrates and alcohols as donors and acceptors respectively. Primary, secondary, but not tertiary alcohols are good acceptors especially if used in molar excess.[51] Furthermore, noticeable enantioselectivity was found using phenyl-β–galactosides as donors and various 1,2 diols as acceptors. These results confirmed the great versatility of Ssβgly and the possibility of using it not only in the synthesis of different glycosides but also in the study of the enzyme's stereochemistry. In order to understand the structural basis of the extreme thermostability and thermophilicity of the enzyme, the protein purified from *S. solfataricus* was crystallized and the 3-D structure of Ssβgly was resolved by x-ray diffraction of crystals with low solvent content.

29.4.1.2 α-glucosidase

The starch industry, the second-biggest consumer of enzyme sales, counting for 10–15% of the total world enzyme market, is one production process that could benefit from biocatalysis at high temperatures since the production of glucose from starch is a multistage process involving different microbial enzymes in successive steps. In the first step, the starch is liquefied at 105°C for 5 min and later at 95°C for 90 min at pH 5.5–6 by a thermostable α-amylase. Then, during the second step for dextrin saccharification, a α-glucosidase from *Aspergillus niger* is used, working optimally at pH and temperature of 4.5 and 60°C, respectively. The current process could be improved by finding efficient and thermostable α-glucosidases acting at the pH and temperature of the liquefaction step. The α-glucosidase activity (EC 3.2.1.20) found in the thermoacidophilic microorganisms *Sulfolobus shibatae*[2] and *Sulfolobus solfataricus* MT-4 (Ssαgly) (data not published) have a high thermal stability and an optimal pH at 5.5. Both characteristics make these enzymes good candidates for industrial applications in starch processing.

Following cell disruption and debris separation via centrifugation, the purification of Ssαgly involved precipitation with ammonium sulfate and the application of several chromatographic steps. In particular, the chromatographic media used were: CM 50, DEAE, phenyl sepharose, and Sephacryl S-200. The enzyme final yield was 21% with a specific activity of 2.8 U·mg^{-1} of proteins.

Ssαgly was demonstrated to be active on starch, amylose, and a series of maltodextrins with a degree of polymerization from 2 to 7. In contrast, the enzyme isolated from *S. solfataricus* was unable to degrade starch. The main characteristics of the enzyme are reported in Table 29.2.

29.4.1.3 Trehalosyl Dextrin-Forming Enzymes and Trehalose-Forming Enzyme

Trehalose, a nonreducing disaccharide widely found in nature, has multiple roles because it acts as an energy source in the blood of insects and protects some plants and organisms from the damage caused by freezing and desiccation. Trehalose is a highly stable and nonhygroscopic disaccharide that does not caramelize and does not undergo the Maillard reactions. It is used to stabilize enzymes, antibodies, vaccines, and hormones, etc., as well as in the production of novel types of food in which it maintains the properties and aroma of the fresh products. Because various applications are foreseen for trehalose and trehalosyl derivatives and given that the traditional extraction from baker's yeast is too expensive, interest in its industrial production has generated a search for organisms having biosynthetic pathways for its synthesis, or expressing enzymes that could be used for its production from suitable substrates.[8]

Lama et al.[52] first reported the ability of the archaea *Sulfolobus solfataricus* to produce trehalose from starch. A Japanese research group later isolated and characterized two enzymatic activities from different hyperthermophilic sources involved in the production of trehalose.[53,54]

Two enzymes that convert starch and dextrins into α,α-trehalose were found in the cell homogenates of the hyperthermophilic acidophilic archaeon *Sulfolobus shibatae* DMS 5389.[2] The first, trehalosyl dextrin-forming enzyme (SsTDFE), transformed maltodextrines to the corresponding trehalosyl derivatives with an intramolecular transglycosylation process that converts the glucosidic linkage at the reducing end from α-1,4 to α-1,1. The second, the trehalose forming enzyme (SsTFE), hydrolyses the α-1,4 linkage adjacent to the α-1,1bond of trehalosyl dextrins forming trehalose and lower molecular weight dextrins (Figure 29.2). *S. shibatae* DSM 5389 was grown in batch cultures in a 100 l fermenter at 87°C with an air flowrate of 20 l·min^{-1} on a semidefined culture medium containing yeast extract and casaminoacids (1 g·l^{-1}). The cells harvested in the stationary growth phase (100 g wet weight) were mechanically disrupted, the cell debris was removed by centrifugation, and the supernatant was extensively dialysed. The downstream procedure then consisted of the application of different chromatographic steps using CM50 in batch and DEAE fast flow chromatography for purification of SsTFE, while phenyl sepharose and Sephacryl S-200 column chromatography, followed by MonoQ FPLC, were used for SsTDFE purification. The yield was about 20%, with a purification factor of 110, the specific activity for the coupled enzymes was 550 U·mg^{-1} of proteins. The enzymes were extensively characterized[2] and proved active on oligosaccharides with 7-3 glucose residues, with the best conversion kinetic found on maltohexaose. The structural and kinetic characteristics of the two enzymes are shown in Table 29.2.

29.4.2 Heterologous Expression

As for many other extremophiles, the optimal growth conditions of *Sulfolobus solfataricus* and its low biomass yield present serious drawbacks to obtaining a quantitative amount of enzymes not only for basic research studies but also for industrial applications. The development of novel bioprocesses, such

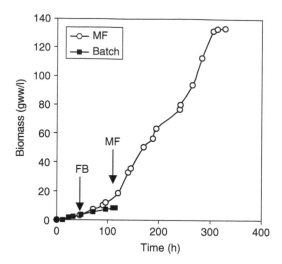

FIGURE 29.2 Growth curve of *S. solfataricus* MT4 in batch and MF experiments (Biostat CT 2L working volume equipped with two MF modules), the maximal biomass yield was 2.3 g dry weight and 37.4 g cdw per litre, respectively.

as dialysis and microfiltration, contributed to improve throughput, but productivities are still not compatible with the large-scale production due to the low growth rate of this microorganism. Cloning and overexpression of thermophilic genes into mesophilic hosts is now a common tool to overcome these issues.

29.4.2.1 β-glycosidase

The gene encoding for Ssβgly (namely, lacS) failed to show significant homology with the *Escherichia coli* lacZ genes. The overexpression of the lacS gene in *E coli* was obtained by cloning it in a plasmid under the control of the b10 gene promoter of the T7 bacteriophage. When this recombinant vector is introduced in the appropriate *E coli* strain, the gene is transcribed after induction with isopropyl-β-D-thiogalactopyranoside (IPTG) by the T7 RNA polymerase present in the host genome. Consequently the enzyme was overexpressed in its functional form and the purification procedure proved much simpler and faster. In fact, crude extracts were simply heated at 80°C for 20 min, the denatured host proteins were eliminated by centrifugation, and all of the Ssβgly was recovered in the supernatant. This treatment also allowed the complete elimination of the thermally unstable endogenous β-galactosidase activity. Subsequent conventional chromatography gave a purified protein with the same specific activity as the enzyme from *S. solfataricus* but with better final purification fold and yield. A typical purification resulted in more than 1.5 mg·l⁻¹ of homogeneous enzyme representing about 150 U·g⁻¹ biomass in the thermoprecipitated fraction. Generally, *S. solfataricus* crude extracts yield 15–20 U per g of biomass and the heterologous expression is tenfold that in the wild-type. This thermoprecipitation procedure allows the use of the Ssβgly for all the applications where the protein in homogeneous form is not required. In fact, after only one simple and fast purification step, the enzyme already shows high specific activity (34 U·mg⁻¹ at 75°C) and all the other contaminating activities result were inactivated because of heat shock.

The Ssβgly expressed in *E coli* has been extensively characterized structurally and kinetically, demonstrating that it maintains the features of its native form.[55]

In order to evaluate the possibility of expressing this enzyme in a generally recognized as safe (GRAS) system in view of its application in the food industry, the Ssβgly was cloned and expressed in the yeast *Saccharomyces cerevisiae*.

The functional expression of the enzyme was obtained using a plasmid (pYE87) in which the lacS gene was under the control of a yeast galactose-inducible promoter (UAS$_{GAL}$).[56] Furthermore, in order to evaluate the use of heat-inducible promoters for the expression of biotechnological enzymes, we tested whether a sequence called poly-HSE (Heat Shock Element) could be an appropriate expression tool for our thermostable β-glycosidase. This sequence is able to mimic a heat-inducible Upstream Activating Sequence (UAS) in *S. cerevisiae*. We obtained the expression of the gene from a plasmid in which UAS_{GAL} was substituted with the poly-HSE. Consequently, Ssβ-gly was expressed in yeast from both these promoters, the classical galactose inducible promoters and a new heterologous heat-inducible promoter.

The protein expressed was purified and characterized. This was the first protein from an extreme thermophile expressed in a functional form in yeast, which also showed retention of the thermostability and thermophilicity of the wild-type. However, the amount of protein produced in both cases accounts for 1% of the total yeast proteins and is unfortunately comparable to the level of expression in *S. solfataricus* (15–20 $U \cdot g^{-1}$ of biomass). This low yield has to be attributed to some post-transcriptional event that prevents higher expression of the enzyme.

Of the two strategies, poly-HSE proved more convenient. In fact, induction is accomplished by a simple temperature shift and does not require any peculiar medium whereas induction by GAL promoter occurs in expensive media containing galactose as the sole carbon source. As this is not one of the nutrients preferred by yeast it causes a very slow growth rate (e.g., doubling time 45 min vs. 24 h). However, the poly-HSE did not permit a sound control of expression; in fact, the basal enzyme level in experiments at room temperature was relatively high. This aspect may prove a problem when expressing proteins toxic for the cells; fortunately this is not the case for thermophilic proteins that are inactive at low temperatures.

Further fundamental research was focused on the study of the secretion mechanism in *S. cerevisiae*. In fact, protein secretion allows the recovery of the product directly in the culture medium, greatly improving the purification procedure. One further advantage offered by the secretion of the Ssβgly in the medium is the possibility of directly applying a heat treatment to inactivate any other protein in the medium thus simplifying the enzyme recovery. A secretion-expression vector pYSL11 derived from the Yepsec vector was used to clone lacS gene in frame with the yeast *Kluyveromyces lactis* killer toxin and under the control of the UASGAL promoter. This vector was used to transform the *S.cerevisiae* strain KY117 and the presence of Ssβgly in intracellular, membrane, and extracellular fractions was monitored. Thermophilic and thermostable enzymatic activity was detected in all fraction after 36 h growth in equivalent amounts, whereas the intracellular activity for the pYE87 proved about 95% of the total Ssβgly production. An increase in the experimental time (e.g., 144 h vs. 36 h) led to an accumulation of the enzyme in the extracellular fraction with respect to the periplasmic and intracellular one. It was also demonstrated that the enzyme was stable against attack by the yeast proteases.

29.4.2.2 α-glucosidase

The α-glucosidase open reading frame was isolated by PCR amplification using *S. solfataricus* MT4 genomic DNA. The PCR product of approximately 2100 bp, was ligated to the *NdeI* e *BamHI* cloning sites of the expression vector pT7-SCII obtaining the recombinant vector pT7αgly. In this construct, the α-glucosidase open reading frame is under the control of an IPTG inducible promoter. The pT7αgly plasmid was transformed in the *E coli* strain BL21(DE3). Recombinant cells were grown in a Luria-Bertani medium supplemented with ampicillin as a selection marker.

The recombinant enzyme was produced in batch experiments in a 2L fermenter and the induction was obtained during the exponential phase by adding 1 mM IPTG. The induction period lasted about 18 h, giving a biomass yield of 25–30 $g \cdot l^{-1}$ (wet weight) corresponding to an enzymatic activity of 85–90 $U \cdot g^{-1}$ of wet cells. The cells were collected by centrifugation and lysed by two cycles at high pressure (2000 psi) in the French press. Purification to homogeneity was achieved by simple heat denaturation steps, i.e., five cycles of 30 min were completed at increasing temperature from 50°C to 80°C. The crude extracts were centrifuged after each step to separate the mesophilic host proteins that precipitated after denaturation. The last thermoprecipitation step consisted of incubating the enzymatic solution 24 h at 85°C, which gave a purification factor of 14.5 with a specific activity of the enzyme of interest of 18.8 $U \cdot mg^{-1}$.[57]

The expression of Ssαgly was about 80-fold the one obtained from the wild-type microorganisms. The recombinant enzyme was extensively characterized (its optimal pH was 5.0 and the optimal temperature 100°C). Furthermore, the enzyme showed very high stability at high temperatures, and in particular the purified enzyme had a half-life of 120 h when incubated at 85°C at pH 5.5.

The optimal pH feature of Ssαgly makes it suitable for industrial applications. In fact, in the current biotransformation process, salts that are generally added to the starch mixture to adjust pH have to be removed later by exchange resins. These procedures make the production process more complex, increasing duration, costs, and environmental pollution.

Recently, the research work was focused on the evaluation of novel GRAS expression systems to produce the Ssαgly. In particular, *Lactococcus lactis* was used as a host for protein production. The strain was kindly donated by NIZO Food research, Ede, Netherlands, within the framework of a scientific collaboration. The Ssαgly gene was amplified by PCR from the *E coli* expression vector pT7-SCII to insert an *NcoI* and *SpeI* sites at the 5′ and 3′ ends, respectively. The PCR product was cloned into the *L.lactis* expression vector pNZ8148 under the control of a nisin inducible promoter. *L.lactis* cells were transformed by electroporation. Induction trials were performed with increasing nisin concentrations (1–4 ng/mL⁻¹) to find the optimal conditions and enzyme production was assayed either on the cell extracts or on the permeabilized cells on a synthetic substrate, p-nitrophenyl-α-D-glucopiranoside (pNPG).

The recombinant strain was transferred into the fermenter for production experiments. Compared with the wild-type production, the results achieved were quite impressive, leading to an approximately 1000-fold increase.[58]

29.4.2.3 Trehalosyl Dextrin-Forming Enzyme and Trehalose-Forming Enzyme

A gene bank for the MT4 strain of *S.solfataricus* was constructed and *E coli* BO3310-competent cells were transformed with the ligation mixture and grown in a Luria-Bertani medium containing ampicillin for 4 h for propagation and amplification of the gene bank.

The suitably modified pTrc99A expression vector was used to clone the entire TDFE sequence. Details on the procedure are reported by de Pascale et al. (2001).[59]

The complete coding sequence of the TFE gene was amplified using genomic DNA from *S. solfataricus* MT4. After digestion with *RcaI* and *BamHI*, it was cloned in the similarly digested pTrc99A vector. The recombinant expression vector was designed pTrcTFE.[59]

E coli Rb791 competent cells were transformed with pTrcTDFE and pTrcTFE expression vectors and grown at 37°C in shake flasks. Induction was performed by adding IPTG to a final concentration of 1 mM. The induction time varied from 8 to 18 h.

Crude extracts were submitted to progressive and selective thermoprecipitations in order to separate the recombinant enzymes from *E coli* proteins. The heating cycles lasted 10 min, and were carried out at 65°C and 70°C. After each cycle, the suspension was centrifuged to separate precipitated host proteins. The supernatant was then ultrafiltered for concentration and one step on a Sephacryl S-200 was sufficient to achieve purification to homogeneity.

The plasmid pTrcTFE showed a high-level expression of the protein in *E coli* after induction with IPTG, indicating that they were not sequences interfering with the codon usage.

The yield of recombinant TFE was 3600 U·l⁻¹ about 1000-fold the yield obtained in the *S. solfataricus* MT4. The recombinant enzyme represented more than 50% of the soluble proteins after the first heat treatment of the cell-free extract and the removal of the denatured *E coli* proteins.

IPTG induction of the *E coli* RB791 transformend with pTrcTFE yielded more than a twofold increase in enzyme content.

However, TDFE protein was not expressed at a high level and the IPTG induction on the pTrcTDFE gene failed to increase the synthesis of the recombinant enzymes. A number of other different expression systems were tested but did not deliver better results. The highest level of protein expression was obtained with a pTrc99A vector in the absence of IPTG induction. The corresponding enzyme yield was 180 U·l⁻¹ i.e., about 100-fold with respect to the wild-type strain.[59] The recombinant enzymes were characterized with respect to the stability at pH and temperature and optimal biocatalytic parameters in order to be compared to the wild-type counterpart.

More recently, the construction of a fusion protein merging TFE and TDFE coding sequences was performed. In fact, trehalose is formed by the coupled reaction of the two thermophilic enzymes; therefore the design of a fusion protein containing both enzymatic activities is very valuable.

It was demonstrated that the bifunctional fusion enzyme is able to produce trehalose starting from maltooligosaccharides at 75°C. The recombinant fusion protein was partially purified from crude extract and insoluble fractions, as it was found that the expression product aggregated in inclusion bodies[60](de Pascale et al., 2002). It was exploited for the conversion of maltohexaose into trehalose and showed a faster kinetic with respect to the equimolar mixture of the two separate enzymes. In agreement with this observation, the intermediate product formed by TDFE activity alone (trehalosyl-maltotetraose), which is easily detectable in the sequential bioconversion, was absent in the reaction mixture containing the fusion protein.

29.4.3 High-Yield Production

29.4.3.1 β-glycosidase

To improve the Ssβgly expression yields in yeast, we performed fermenter experiments on a rich medium up to the pilot scale of 100 l and carried out a rapid purification, giving an enzyme solution suitable for most industrial applications. The yeast KY117 transformed with the plasmid pEY87 was grown at 30°C in a complex medium containing peptone (2% w/v), yeast extract (1% w/v) and glucose (1% w/v) with galactose as the inductor. During fermentation, growth was monitored by absorbance measurements (540 nm) and Ssβgly activity assays were performed on whole cells to evaluate the maximum yield and therefore the optimal duration of the experiment. It was found that the enzyme is accumulated until the late exponential phase. Fermentation was stopped after 50 h with a biomass yield of 13 g cww/l^{-1}, and in the crude extract (mechanical cell disruption) Ssβgly activity of 42 U·g^{-1} was found corresponding to a final specific activity of 1.35 U·mg^{-1} of proteins.[61] However, a purification protocol based on autolysis was developed in order to simplify the downstream of large-scale production. In particular, when the yeast culture was incubated at 37°C, the enzyme was slowly released into the medium by autolysis of the cells, reaching a maximum yield after 72 h with a specific activity that was 15-fold that found in the crude extracts. The addition of a detergent such as triton X-100 resulted in a faster release of the enzyme, but its specific activity was 5- to 10-fold lower compared with the previous protocol.

Although the enzyme activity for gram of cells is not very different, the enzymatic productivity in recombinant yeast is 35-fold higher than that obtained by *S. solfataricus*.

The strategy described here could be adopted for the recovery of other thermophilic proteins expressed in yeast, since it can easily be scaled up and can deliver enzyme solutions generally with a higher degree of purity with respect to the conventional crude extracts.[61]

29.4.3.2 α-Glucosidase, TDFE, and TFE Production Exploiting a MF Bioreactor

On the one hand, the efficient production of proteins is closely linked to their expression level in the microorganism, while on the other to the maximum biomass concentration that can be achieved during cultivation. Genetic engineering is certainly the most powerful tool when the target is to improve the protein expression level by using a host organism, while bioreactor design plays a key role in improving biomass yield. For this reason, the recombinant microorganisms expressing either α-glucosidase, or TDFE or TFE, were grown in the microfiltration bioreactor developed in our laboratory.[46,62] The improvement in biomass yield is caused by the removal of acetic acid from the fermentation broth during cultivation; in fact it was demonstrated that an acetate concentration higher than 5 g·l^{-1} is responsible for a severe growth inhibition in *E coli*. The growth conditions and the optimal induction time and duration were identified separately for each recombinant strain. Similar strategies delivered different throughputs for the three enzymes probably due to the different residual activities of the enzymes at the growth temperature (37°C). In fact, the best improvement in biomass and enzyme yield was obtained with Ssαgly, which shows less than 10% of the activity at low temperature, while the lowest impact was found on

TABLE 29.3 Comparison of the Enzymatic Activities in the Cell Free Extracts with Respect to the One Recovered with the Heat Treatment of Recombinant Whole Cells

	Wild-Type			Recombinant *E Coli*		
Enzyme	Specific activity $U \cdot (g_{cdw})^{-1}$	Throughput $U \cdot l^{-1}$	Productivity $U \cdot (l \cdot h)^{-1}$	Specific activity $U \cdot (g_{cdw})^{-1}$	Throughput $U \cdot l^{-1}$	Productivity $U \cdot (l \cdot h)^{-1}$
Ssαgly	2.0	70.0	0.2	250	12500	420
SsTDFE	3.3	115.5	0.4	44	792	20
SsTFE	11.6	406.0	1.4	667	25300	640

Enzyme	Ssαgly	SsTDFE	SsTFE
	$U \cdot g_{ww}^{-1}$ (%)	$U \cdot g_{ww}^{-1}$ (%)	$U \cdot g_{ww}^{-1}$ (%)
Cell free extract	87.5 (60)	36 (100)	132 (100)
(cell membranes)	(58.3) (40)		
Cellular suspension	116.6 (80)	7 (20)	33 (25)
Permeabilized cell suspension	163.2 (110)[a]	30 (83)[a]	58 (44)[a]
Permeabilized cell suspension in presence of triton-X100	n.d.	30 (83)	99 (75)

[a]The incubation time for *E coli* expressing αgly was 6 h, while for TDFE and TFE recombinant cells were 1 h and 2 h, respectively.

TDFE, which preserved more than 60% of its activity at the same temperature.[46] In particular, the MF process made it possible to achieve about 50 $g \cdot l^{-1}$ of dry biomass in the cultivation of recombinant *E coli* producing Ssαgly, 18 $g \cdot l^{-1}$ cdw for *E coli* expressing TDFE, and 38-41 $g \cdot l^{-1}$ of the recombinant strain containing TFE.

Table 29.3 compares the enzyme productivities in the wild-type and recombinant microorganisms, also by exploiting the microfiltration process. The results clearly highlight that the combination of genetic engineering and innovative bioprocess design substantially improves enzyme production, as we obtained a 7-fold increase in SsTDFE, a 62-fold increase in SsTFE, while Ssαgly production was about 180-fold with respect to the wild-type.

Ssαgly was successfully cloned and expressed in *Lactococcus lactis* using a GRAS strain as host and nisin, a natural peptide, as inducer. This induction system makes the enzyme produced safer and consumer friendly, especially in view of the current application of this biocatalyst in the carbohydrate industry.

The production of Ssαgly in *L.lactis* was investigated in batch, fed-batch, and preliminary MF experiments.

Batch experiments were performed using the complex medium M17, modifying the induction time and duration and a biomass yield as high as 15–20 g wet weight per l was obtained. However, inhibition due to the accumulation of lactic acid (LA) (over 35 $g \cdot L^{-1}$), was easily recognizable due to the lowering in the growth rate of the microorganisms. During the experiments, the maximal enzymatic activity achieved was about 610 $U \cdot L^{-1}$. As fed-batch experiments did not substantially enhance the enzyme production, we tried to exploit the MF technique to limit LA accumulation. In fact, the productivities were increased and, even if the process is still not optimized with respect to the medium and the feeding solution, a significant improvement with respect to the wild-type productivities was obtained (Table 29.3).[58]

Because of the encouraging results obtained during our research, we further characterized the enzymes of interest by developing bioconversion processes based on immobilized recombinant *E coli* cells.

29.4.4 Biotransformation Exploiting Immobilized Cells

Immobilized cell reactors have been extensively used due to their suitability for confining a high cell density, thus making it possible to improve productivities and the superior stability of the immobilized biocatalysts. Generally, the use of a packed-bed reactor is recommended for continuous production.[63,64]

In order to evaluate the bioconversion performances of the enzymatic activities of interest, we investigated the possibility of exploiting immobilized cell biocatalysts for the manufacturing of nonreducing saccharides, glucose, and trehalose at high temperatures. In particular, several well-known immobilization methods were compared. The biocatalyst performances were evaluated in relation to the simplicity of the protocols, availability of materials, and more importantly, the stability of matrices at temperatures higher than 70°C.

After selecting the most appropriate carrier (Ca-alginate), we proceeded to design and develop an optimal reactor configuration for the bioprocess of interest. Theoretical considerations on the processes suggested a tubular reactor as the optimal configuration. In fact, simple mass balance analyses were sufficient to demonstrate that, for enzymes following the typical Michaelis and Menten kinetic, an ideal plug flow tubular reactor (PFR) performs a greater conversion than a continuous stirred tank reactor (CSTR) of the same volume.[65]

29.4.4.1 Immobilization Protocols

The activity of recombinant enzymes was determined both in cell-free extract and in whole-cells of *E coli*. As shown in Table 29.4, the activity detected in whole-cells corresponded to only 20–40% of the cell-free extract activity for TDFE and TFE and 50–60% for Ssαgly. In our opinion, this depended on the transport resistance across the cell membranes. To overcome this limitation, the cellular suspensions were permeabilized by thermal treatment at 75°C (1 h for cells expressing TDFE and TFE, and 6 h for Ssαgly). The latter greatly increased the specific activity, reaching a yield of 80–100% (Table 29.4). It was possible to operate this permeabilization method given the enzymes thermostability; this also determined the selective denaturation of mesophilic proteins. A further advantage lies in the fact that the organic solvent treatments normally used for breaking down the selective barrier of the cell membrane are eliminated.[66,67] The high recovery of enzymatic activities permitted immobilization of whole cells instead of free enzyme, thus avoiding both cell disruption and enzyme purification.

In particular, the permeabilized whole-cells suspended in sodium/calcium acetate buffer pH 5.5 were entrapped using different supports (hen egg white, chitosan, etc.). However the best matrix was found to be alginate[68] because it is food grade and presents good stability at high temperatures. Ca-alginate beads (0.3 ± 0.1 cm diameter, 2% [w/v] gel final concentration) were prepared using the Smidsrod and Skjak-Braek method.[69]

29.4.4.2 Bioreactor Configuration

The process designed was based on the application of recombinant permeabilized *E coli* cells; therefore, adding oxygen or nutrients was not necessary. The pH remained unchanged during the reaction and the reactor configuration was very simple. Hence, different jacketed glass columns (with two different volumes: V = 5 ml; V = 125 ml; internal diameter 0.7 cm and 2 cm, respectively) fixed horizontally and connected to a thermostatic bath were packed with the biocatalyst beads. The packing was improved by

TABLE 29.4 Comparison of the Characteristics of the Immobilized Recombinant Enzymes

Enzyme	T_{opt} (°C)	pH_{opt}	Half-life at 75°C(days)	Operational stability (75°C) (days)	Storage stability (4°C) (months)	K_m (mM)	Productivity* mM $(min \cdot g_{biocatalyst})^{-1}$
Ssαgly	100	5.5	12	7	12	n.d.	2.9
SsTDFE	75	5.5	60	20	12	5.4	9.0
SsTFE	85	5.5	50	10	12	8.7	2.7

*These values were calculated from bioconversion experiments carried out using 30 mM of substrate (maltohexaose or maltrehalosyltetraose) and 4–5 g of biocatalyst beads.

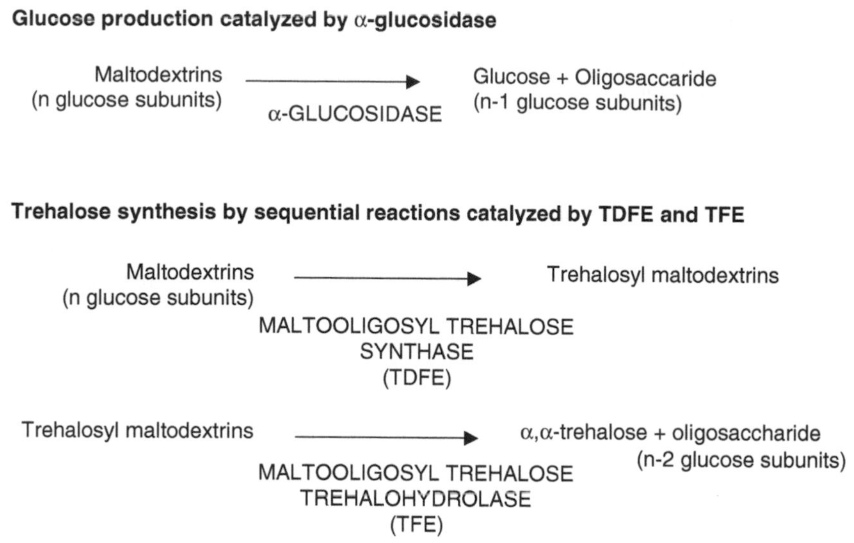

Glucose production catalyzed by α-glucosidase

Maltodextrins ──────────────▶ Glucose + Oligosaccharide
(n glucose subunits)   (n-1 glucose subunits)
                   α-GLUCOSIDASE

Trehalose synthesis by sequential reactions catalyzed by TDFE and TFE

Maltodextrins ──────────────▶ Trehalosyl maltodextrins
(n glucose subunits)
              MALTOOLIGOSYL TREHALOSE
                      SYNTHASE
                       (TDFE)

Trehalosyl maltodextrins ──────────────▶ α,α-trehalose + oligosaccharide
                                                 (n-2 glucose subunits)
              MALTOOLIGOSYL TREHALOSE
                  TREHALOHYDROLASE
                       (TFE)

FIGURE 29.3 Reaction substrates, products, and intermediates in the biosynthesis of glucose, nonreducing saccharides, and trehalose.

using glass beads of a similar diameter at the inlet and outlet of the column that was not surrounded by the jacket, in order to prevent the enzymes from working at different temperatures. The substrate solutions containing either maltohexaose (M6) or dextrin mixtures, or starch dissolved in a calcium acetate buffer at pH 5.5, was fed in the bioreactor using a peristaltic pump, which regulated the flow rate and therefore the linear velocity. Once the solution passed through the catalytic bed, it was collected and eventually recycled until maximal conversion was achieved.

29.4.4.3   Bioprocesses Performances

The practical use of the biocatalyst was tested in preliminary experiments in a 5 ml tubular reactor. In order to equilibrate the matrix with the substrate solution, two subsequent adsorption cycles (2 mL) were performed. The operational parameters were 75°C and pH 5.5. Bioconversion experiments were carried out using 10 mM and 30 mM maltohexaose (M6) inlet stream (4 mL), which was recycled to the reactor in order to obtain an overall residence time of 15–20 min. The small bioreactor was exploited for apparent kinetic parameters evaluation and to define the optimal operational conditions with regard to enzyme/substrate ratio and linear velocity.

In addition, in order to evaluate the bioreactor performances with regard to industrial applications, several experiments were carried out on the 125 ml bioreactor, using commercial dextrin mixtures at high concentration (up to 30% w/v) as substrate. The bioreactor was operated in total recycle mode and samples were withdrawn at predefined intervals once an integral number of cycles was completed (e.g., 1, 5, 10, 20, etc.).

29.4.4.4   Glucose Production

The results of the experiments carried out on maltohexaose showed that the immobilized *E coli* cells expressing Ssαgly were able to completely convert the substrate into glucose (Figure 29.3). A few runs were completed using increasing linear velocities (same reactor but higher flow rate) to establish whether the external mass transfer or the intrinsic kinetic were limiting the biotransformation rate. In particular, conversion curves were completed at increasing fluxes from 0.5 ml·min^{-1} to 1.6 ml·min^{-1} showing the latter as the optimal flow rate in relation to the biotransformation for all substrate concentrations (10–30 mM).[70] These results suggested that a flow rate of 1.6 ml·min^{-1} was sufficient to minimize external mass transfer resistance.

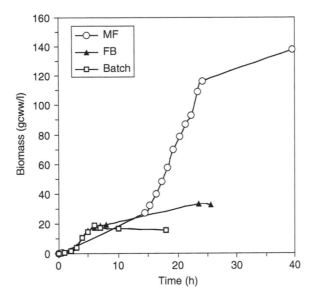

FIGURE 29.4 Comparison between growth curves (cell wet weight) in batch, fed-batch, and MF experiments of recombinant *E coli* Rb 791 expressing SsTDFE. (Modified from Schiraldi et al., 2001.)

After a set of experiments (10 days) the beads retained their initial shape. This finding was satisfactory and allowed us to deduce that the compressibility of the matrix was very low at the operated fluxes.

Further experiments were carried out to better characterize the bioreactor potential. In particular, the bioconversion of a highly concentrated dextrin mixture (30% w/v) in 5 ml reactor was performed at the optimal flow rate.

The conversion (%) was calculated as the ratio of final glucose versus the maximal achievable concentration. Commercial dextrin mixture includes oligosaccharides with a different number of glucose residues (i.e., 2–7 about 70%). The remaining part (30%) was made up of residues at a higher number

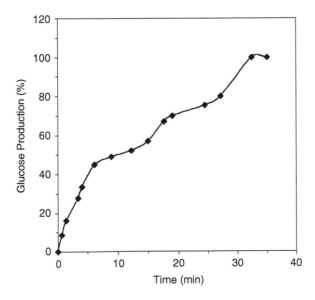

FIGURE 29.5 Conversion curve for the biotransformation of maltodextrines (30% w/v) into glucose at 75°C using the 125 ml bioreactor packed with Ca-alginate beads entrapping *E coli* expressing Ssαgly.

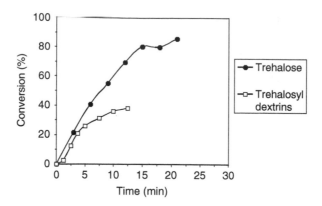

FIGURE 29.6 Trehalosyl dextrins and trehalose production from maltodextrines (30% w/v) at 75°C using the 125 ml bioreactor packed with immobilized *E coli* expressing TDFE and co-immobilized biocatalysts (TDFE + TFE). The percentage is referred to the maximal theoretical yield considering that maltotriose and maltose are not substrates for the enzymes.

of glucose units. Bearing this in mind, stoichiometry must be considered when calculating the maximal theoretical yield. In particular, two glucose molecules can be obtained from maltose, while six from M6. This experiment gave a 98% conversion and a corresponding glucose productivity of $1 \text{ g} \cdot \text{h}^{-1} \cdot \text{g}^{-1}_{\text{biocatalyst}}$. To verify the applicability of the process on a larger scale, the hydrolysis of dextrin at industrial concentrations was carried out in a 125 ml tubular reactor. Biotransformation was completed using the same linear velocity as the lower scale experiments and a 30% w/v dextrin solution. In about 9 min residence time, 98–100% conversion was achieved corresponding to a glucose productivity of $1.75 \text{ g} \cdot \text{h}^{-1} \cdot \text{h}^{-1}_{\text{biocatalyst}}$ (Figure 29.5). At the end of the reaction, a small amount of maltodextrin (1.5% w/v) was recovered probably due to the slow desorption process.

The operational stability of the biocatalyst was also studied. The bioreactor completed several biotransformations maintaining the maximal activity with an half-life of more than 7 days. The production obtained in the 125 ml bioreactor packed with 150 g of biocatalyst (corresponding to about 0.2 g of enzyme) was $290\text{--}300 \text{ g} \cdot \text{l}^{-1}$ (98–100%) in 32 min of residence time (Figure 29.6). This value corresponded to a productivity of $550\text{--}570 \text{ g} \cdot \text{l}^{-1} \cdot \text{h}^{-1}$ and proved higher than that obtained using the glucoamylase from *Aspergillus niger*, which in the bioconversion of maltodextrins ($100 \text{ g} \cdot \text{l}^{-1}$, 0.03 g of enzyme), showed a productivity of $1.8 \text{ g} \cdot \text{l}^{-1} \cdot \text{h}^{-1}$ at 70°C in batch.[71] It resulted in almost 50-fold lower than the value obtained in our bioprocess, when normalizing data with respect to the enzyme amount.

The biocatalyst beads retained their initial activity after storage at 4°C for 1 year.

29.4.4.5 Trehalosyl Dextrins Production

Trehalosyl-dextrins are linear dextrins containing trehalose as end units. Due to the absence of reducing activity, these sugars do not undergo the Maillard reaction in the presence of proteins and peptides. Hence, these compounds are interesting alternatives to the conventional carbohydrates generally utilized in food preparations, pharmaceutics, and cosmetics. A mesophilic TDFE was firstly isolated by Nakada et al.[72] from *Arthrobacter* sp. Q36 and more recently, the corresponding thermophilic proteins were isolated from Sulfolobales.[2,53] The main advantages of using thermophilic enzymes in the carbohydrate processing are related to the drop in viscosity of the substrate solutions and the minor contamination risks. However, to date, there are no industrial processes based on the thermophilic enzymes. The aim of our research was to develop a complete procedure from the production of the enzymes to their application, thus demonstrating the feasibility of the high temperature process while also containing the production costs. Therefore after producing the enzyme at a high yield by exploiting genetic engineering and the MF bioreactor, we developed an immobilized cell biocatalyst for the manufacturing of nonreducing saccharides at high temperatures, and used the latter to pack a tubular reactor.

A few studies were completed to compare the stability at 75°C of the free enzyme, whole-cell suspension, and Ca-alginate entrapped cells. The data showed that Ca-alginate gel beads retained more than 60% of their initial activity after 2 months of incubation, while free enzyme and whole-cell suspension showed a half-life of 4 days and 1 month, respectively (Table 29.4).[70]

Before defining the setup of experiments, the intrinsic kinetic parameters were measured by exploiting a differential PFR and assuring a conversion lower than 10% for each cycle. This permitted the assumption of having a constant concentration along the biocatalyst bed. The v_{max} and K_m for the immobilized enzyme were 7.5 mM·$(min·g_{biocatalyst})^{-1}$ and 5.4 mM, respectively.

Because the biocatalyst beads were substantially identical to those obtained for Ssαgly, the operational flow rate was maintained at 1.6 ml/min^{-1}. Another set of experiments was carried out in order to explore the influence of substrate concentration on the reaction rate, while keeping the catalytic bed length and the flow rate constant. A higher concentration of M6 led to a higher initial rate of reaction. In particular, the trehalosylmaltotetraose (TM4) formation rate was 2.5 mM·$(min·g_{biocatalyst})^{-1}$ using an inlet stream at 10 mM M6, while this increased to about 7 mM·$(min·g_{biocatalyst})^{-1}$ when a 30 mM solution of M6 was fed.

During all the experiments, the conversion never reached 100% and data analysis confirmed that independently of the flow rate, the enzyme amount and the initial substrate concentration, the production of TM4 always reached a *plateau*. The equilibrium constant was evaluated by averaging all the data generated during optimization experiments and a value of 2.6 was obtained. In addition, an experiment was performed by adding substrate to the reaction mixture after the plateau was reached. This moved the reaction, as expected, but the equilibrium was still restored after a certain period of time.

In the same way as the research experiments on immobilized Ssαgly, for all the experiments aiming to define optimal operating conditions, we used a pure substrate (M6) normally containing only maltopentaose (M5) as an impurity and therefore extremely suitable for precise rate definition. The packed-bed reactor was then exploited for the production of nonreducing saccharides from a commercially available dextrin mixture. In this case, because all the maltodextrins having 4 to 10 glucose residues were substrates for the enzyme, we observed a conversion rate that was an averaged value of the kinetics of all these reactions. Finally, we obtained a productivity of 106.5 mg/$(ml·h·g_{biocatalyst})^{-1}$ using a 30% w/v solution. The half-life of the immobilized biocatalyst was found to be 20 days under the operational conditions.

29.4.4.6 Trehalose Production

Immobilized *E coli* cells expressing TDFE were coupled to those producing TFE to obtain trehalose from maltodextrins (Figure 29.3). As mentioned before, trehalose functions as a carbohydrate reserve in many organisms and also performs stabilizing effects against environmental stresses such as desiccation. Several applications have been proposed and even patented for this sugar in the food, pharmaceutical, and cosmetics industries. In fact, trehalose can be used as a sweetener, a stabilizer for dried and frozen foods, a moisture retainer, and a drug preservative.[8,73] The development of novel biotechnological processes for trehalose production is very challenging given its widening spectrum of applications. Therefore, trehalose production at high temperature was investigated using the bioreactor and the operational parameters similar to those developed for the other two processes previously described. In the first place, the performances of a bioreactor containing alternate layers of the two enzymes were evaluated. The bioconversion curves followed the typical behavior of series reactions, first showing an accumulation of the intermediate products (e.g., trehalosylmaltotetraose) and later the conversion to the final product. The conversion rate was definitely low; hence, the two recombinant strains, containing TDFE and TFE activities, respectively, were co-immobilized considering the different expression levels of the two enzymes. Results showed that the higher proximity of the two enzymes in the co-immobilized form improved both the trehalose yield and the productivity. The experiments carried out using commercial maltodextrins mixtures yielded an 80% conversion (considering that maltotriose is not a substrate for TDFE) (Figure 29.6) corresponding to a trehalose productivity of 127 mg·$(h·g_{biocatalyst})^{-1}$ The entrapped biocatalysts showed a very high stability at 75°C (1–2 months) when incubated in a buffer at pH 5.5. The half-life of the co-entrapped biocatalyst in operating conditions mimicking industrial processes was found to be 10 days. The biocatalyst retained its initial activity after storage at 4°C for 1 year.[68]

References

1. CR Woese, GE Fox. Phylogenetic structure of the prokaryotic domain: the primary kingdoms. *Proc Natl Acad Sci* USA 74:5088–5090, 1977.

2. I Di Lernia, A Morana, A Ottombrino, S Fusco, M Rossi, M De Rosa. Enzymes from *Sulfolobus shibatae* for the production of trehalose and glucose from starch. *Extremophiles* 2:409–416, 1998.

3. DW Hough, MJ Danson. Extremozymes. *Curr Opin Mol Biol* 3:39–46, 1999.

4. F Duffner, C Bertoldo, JT Andersen, K Wagner G Antranikian. A new thermoactive pullulanase from Desulfurococcus mucosus: cloning sequencing, purification and characterization of the recombinant enzymes after expression in *Bacillus subtilis*. *J Bacteriol* 182:6331–6338, 2000.

5. M Moracci, A Trincone, B Cobucci-Ponzano, G Perugino, M Ciaramella, M Rossi. Enzymatic synthesis of oligosaccharides by two glycosyl hydrolases of Sulfolobus solfataricus. *Extremophiles* 5(3):145–52, 2001.

6. K Schumacher, E Heine, H Hocker. Extremozymes for improving wool properties. *J Biotechnol* 89:281–288, 2001.

7. C Schiraldi, M De Rosa. The production of biocatalysts and biomolecules from extremophiles. *TIBTECH* 20(12):515–521, 2002.

8. C Schiraldi, I Di Lernia, M De Rosa. Trehalose production exploiting novel approaches. *TIBTECH* 20(10):420–425, 2002.

9. NJ Russell. Toward a molecular understanding of cold activity of enzymes from psychrophiles. *Extremophiles* 4:83–90, 2000.

10. S Ito. Alkaline cellulases from alkaliphilic Bacillus: enzymatic properties, genetics and application to detergents. *Extremophiles* 1(2):61–66, 1997.

11. K Horikoshi. Alkaliphiles: some applications of their products for biotechnology. *Microbiol Mol Biol Rev* 63(4):735–750, 1999.

12. B Van den Burg. Extremophiles as a source for novel enzymes. *Curr Opin Microbiol* 6:213–218, 2003.

13. F Niehaus, C Bertoldo, M Kahler, G Antranikian. Extremophiles as a source of novel enzymes for industrial application. *Appl Microbiol Biotechnol* 51:711–729, 1999.

14. R Margesin, F Schinner. Potential of halotolerant and halophilic microorganisms for biotechnology. *Extremophiles* 5(2):73–83, 2001.

15. R Huber, P Stoffers, JL Cheminee, HH Richnow, KO Stetter. Hyperthermophilic archaebacteria within the crater and open-sea plume of erupting MacDonald Seamount. *Nature* 345:179–182, 1990.

16. SG Prowe, G Antranikian. Anaerobranca gottshallkii sp.nov., a novel thermoalkaliphilic bacterium that grows anaerobically at high pH and Temperature, 3rd Int Congr On Extremophiles 2000, Hamburg Germany, Book of abstracts, p 107, 2000.

17. WD Grant, RT Gemmell, TJ McGenity. Halophiles, in: *Halophiles* (K Horikoshi, WD Grant, eds) 93–133. New York: Wiley-Liss, 1998.

18. M De Rosa, A Gambacorta. The lipids from archaebacteria. *Progress in Lipid Research* 27:153–175, 1988.

19. H Lechevalier, MP Lechevalier. Chemotaxonomic use of lipids — an overview. In: *Microbial Lipids*, 1, eds C Ratledge SG Wilkinson. 869–902. New York: Academic press, 1988.

20. GD Sprott. Structure of archebacterial membrane lipids. *J Bioenerg Biomemb* 24:555–566, 1992.

21. Y Koga, M Akagawa-Mattsushita, Ohga M Nishihara. Taxonomic significance of the distributions of components parts of polar ether lipids in methanogens. *Syst Appl Microbiol* 16:342–351, 1993.

22. A Gambacorta, A Gliozzi, M De Rosa. Archaeal lipids and their biotechnological applications. *World J Microbiol Biotechnol* 11:115–132, 1995.

23. M De Rosa, A Morana, A Riccio, A Gambacorta, A Trincone, O Incani. Lipids of the Archaea: a new tool for bioelectronics. *Biosensors and Bioelectronics* 9:669–675, 1994.

24. PI Lelkes, D Goldenberg, A Gliozzi, M De Rosa, A Gambacorta, IR Miller. Vesicles from mixtures of bipolar archaebacterial lipids with egg phosphatidylcholine. *Biochim Biophys Acta* 732:714–718, 1983.

25. M De Rosa, A Trincone, B Nicolaus, and A Gambacorta. Archaebacteria: lipids, membrane structures and adaptation to environmental stresses. In *Life under Extreme Conditions*, ed G Di Prisco 61–87. Heidelberg: Springer-Verlag, 1991.

26. CB Park, DS Clark. Rupture of the cell envelope by decompression of the deep–sea methanogen *Methanococcus jannaschii*. *Appl Environ Microbiol* 68:1458–1463, 2002.

27. FF Hezayen, BH Rehm, R Eberhardt, A Steinbuchel. Polymer production by two newly isolated extremely halophilic Archaea: application of a novel corrosion-resistant bioreactor. *Appl Microbiol Biotechnol* 54:319–325, 2000.

28. A Schmid, A Kollemer, RG Mathys, B Witholt. Developments towards large-scale bacterial bioprocesses in the presence of bulk amounts of organic solvents. *Extremophiles* 2:249–256, 1998.

29. N Raven, N Ladwa, D Cossar, R Sharp. Continuous culture of the hyperthermophilic archaeum *Pyrococcus furiosus*. *Appl Microbiol Biotechnol* 38:263–267, 1992.

30. N Raven, RJ Sharp. Development of defined and minimal media for the growth of the hyperthermophilic archaeon *Pyrococcus furiosus* Vc.1. *FEMS Microbiol Lett* 146:135–141, 1997.

31. KD Rinker, CJ Han, RM Kelly. Continuous culture as a tool for investigating the growth physiology of heterothrophic hyperthermophiles and extreme thermoacidophiles. *J Appl Microbiol* 85:118S–127S, 1999.

32. JH Tsao, MK Kaneshiro, S Yu, DS Clark. Continuous culture of *Methanococcus jannaschii*, an extremely thermophilic methanogen. *Biotechnol Bioeng* 43:258–261, 1994.

33. N Schill, WM van Gulik, R Voisard, U von Stickar. Continuous cultures limited by gaseous substrate: development of a simple, unstructured mathematical model and experimental verification with *Methanobacterium thermoautotrophicum*. *Biotechnol Bioeng* 51:645–658, 1996.

34. O Rothe, M A Thomm. Simplified method for the cultivation of extreme anaerobic Archaea based on the use of sodium sulfite as reducing agent. *Extremophiles* 4(4):247–252, 2000.

35. A Godfroy, ND Raven, RJ Sharp. Physiology and continuous culture of the hyperthermophilic deep-sea vent archaeon *Pyrococcus abyssi* ST549. *FEMS Microbiol Lett* 186(1):127–132, 2000.

36. KF Biller, I Kato, H Markl. Effect of glucose, maltose, soluble starch, and CO2 on the growth of hyperthermophilic archaeon Pyrococcus furiosus. Extremophiles 6:161–166, 2002.

37. I Romano, V Calandrelli, E Pagnotta, R Di Maso. Whey as medium for biomass production of *Sulfolobus solfataricus*. *Biotechnol Tech* 6(5):391–392, 1992.

38. H Feitkenhauer, S Schnicke, R Muller, H Markl. Determination of the kinetic parameters of the phenol degrading thermophilic *Bacillus thermoleovorans* sp. A2. *Appl Microbiol Biotechnol* 57:744–750, 2001.

39. P Becker, I Abu–Reesh, S Markossian, G Antranikian, H Markl. Determination of the kinetic parameters during continuous cultivation of the lipase producing thermophile *Bacillus* sp. IHI-91 on olive oil. *Appl Microbiol Biotechnol* 48:184–190, 1997.

40. CB Park, SB Lee. Constant volume fed-batch operation for high density cultivation of hyperthermophilic aerobes. *Biotechnol Tech* 11(5):277–281, 1997.

41. CB Park, SB Lee. Inhibitory effect of mineral ion accumulation on high density growth of the hyperthermophilic Archaeon *Sulfolobus solfataricus*. *Biosci Bioeng* 87(3):315–319, 1999.

42. M Krahe, G Antyranikian, H Märkl. Fermentation of extremophilic microorganisms. *FEMS Microbiol Rev* 18:271–285, 1996.

43. C Fuchs, D Koster, S Wiebusch, K Mahr, G Eisbrenner, H Markl. Scale up of dyalisis fermentation for high cell density cultivation of Escherichia coli. *J Biotechnol* 93:243–251, 2002.

44. T Sauer, EA Galinski. Bacterial milking: a novel bioprocess for production of compatible solutes. *Biotechnol Bioeng* 57:306–313, 1998.

45. C Schiraldi, F Marulli, I Di Lernia, A Martino, M De Rosa. A microfiltration bioreactor to achieve high cell density in *Sulfolobus solfataricus* fermentation. *Extremophiles* 3:199–204, 1999.

46. C Schiraldi, M Acone, M Giuliano, I Di Lernia, C Maresca, M Cartenì, M De Rosa. Innovative fermentation strategies for the production of extremophilic enzymes. *Extremophiles* 5:193–8, 2001.
47. JD Rozzell. Commercial scale biocatalysis: myths and realities. *Bioorg Med Chem* 7:2253–2261, 1999.
48. C Bertoldo, F Duffner, PL Jorgensen, G Antranikian. Pullulanase type I from Fervidobacterium pennavorans Ven5: cloning, sequencing, and expression of the gene and biochemical characterization of the recombinant enzyme. *Appl Env Microbiol* 65:2084–2091, 1999.
49. M Kähler, G Antranikian. Cloning and characterization of a family B DNA polymerase from the hyperthermophilic crenarchen Pyrobaculum islandicum. *J Bacteriol* 182:655–663, 2000.
50. R Nucci, M Moracci, C Vaccaro, N Vespa, M Rossi. Exo-glucosidase activity and substrate specificity of the α-glycosidases isolated from the extreme thermophile *Sulfolobus solfataricus*. *Biotechnol Appl Biochem* 17:239–250, 1993.
51. M Moracci, M Ciaramella, R Nucci, LH Pearl, I Sanderson, A Trincone, M Rossi. Thermostable α-glycosidases from *Sulfolobus solfataricus*. *Biocatalysis* 11:89–103, 1994.
52. L Lama, B Nicolaus, A Trincone, P Morbillo, M De Rosa, A Gambacorta. Starch conversion with Immobilized Thermophilic Archaebacterium *Sulfolobus solfataricus*. *Biotechnol Lett* 12(6):431–432, 1990.
53. T Nakada, S Ikegami, H Chaen, M Kubota, S Fukuda, T Sugimoto, M Kurimoto, Y Tsujisaka. Purification and characterization of thermostable maltooligosyl trehalose trehalohydrolase from the thermoacidophilic archaebacterium Sulfolobus acidocaldarius. *Biosci Biotechnol Biochem* 60:267–270, 1996.
54. M Kato, Y Miura, M Kettoku, K Scindo, A Iwamatsu, K Kobayashi. Reaction mechanism of a new glycosyl-trehalose-producing enzyme isolated from the hyperthermophilic archaeum *Sulfolobus solfataricus* KM1. *Biosci Biotechnol Biochem* 60:921–924, 1996.
55. M Moracci, R Nucci, F Febbraio, C Vaccaio, N Vespa, F La Cara, M Rossi. Expression and extensive characterization of a α-glycosidase from the extreme thermoacidophilic archaeon *Sulfolobus solfataricus* in *Escherichia coli*: authenticity of the recombinant enzyme. *Enz Microb Tech* 17:992–997, 1995.
56. M Moracci, A La Volpe, JF Pulitzer, M Rossi, M Ciaramella. Expression of the thermostable α-galactosidase gene from the archaeobacterium *Sulfolobus solfataricus* in *Saccharomyces cerevisiae* and characterization of a new inducible promoter for heterologous expression. *J Bacteriol* 174:873–882, 1992.
57. A Martino, C Schiraldi, S Fusco, I Di Lernia, T Costabile, T Pellicano, M Marotta, M Generoso, J van der Oost, C W Sensen, R L Charlebois, M Moracci, M Rossi, M De Rosa. Properties of the recombinant α-glucosidase from *Sulfolobus solfataricus* in relation to starch processing. *J Molec Catal B: Enzym* 11:787–794, 2001.
58. M Giuliano, C Schiraldi, M Marotta, J Hugenholtz, M De Rosa. Expression of *Sulfolobus solfataricus* α-glucosidase in *Lactococcus lactis Appl Microbiol Biotechnol* 64(6): 829–832, 2004.
59. D de Pascale, MP Sasso, I Di Lernia, A Di Lazzaro, A Furia, M Cartenì Farina, M Rossi, M De Rosa. Recombinant thermophilic enzymes for trehalose and trehalosyl dextrins production *J Mol Cat B: enzym*, 11:777–786, 2001.
60. D de Pascale, I Di Lernia, MP Sasso, A Furia, M De Rosa, M Rossi. A novel thermophilic fusion enzyme for trehalose production. *Extremophiles* 6:463–468, 2002.
61. A Morana, M Moracci, A Ottobrino, M Ciaramella, M Rossi, M De Rosa. Industrial-scale production and rapid purification of an archaeal β-glycosidase expressed in *Saccharomyces cerevisiae*. *Biotechnol Appl Biochem* 22:261–268, 1995.
62. C Schiraldi, A Martino, M Acone, I Di Lernia, A Di Lazzaro, F Marulli, M Generoso, M Cartenì, M De Rosa. Effective Production of a Thermostable α-Glucosidase from *Sulfolobus solfataricus* in *Escherichia coli* Exploiting a Microfiltration Bioreactor. *Biotechnol Bioeng* 70 (6):670–676, 2000.
63. HY Cho, AE Yousef, ST Yang. Continuous production of pediocin by immobilized *Pediococcus acidilactici* PO2 in a packed-bed bioreactor. *Appl Microbiol Biotechnol* 45:589–594, 1996.

64. JN Nigam. Continuous ethanol production from pineapple cannery waste using immobilized yeast cells. *J Biotechnol* 80:189–193, 2000.

65. JE Bailey, DF Ollis. *Biochemical Engineering Fundamentals.* 2nd ed McGraw-Hill, New York, 1986.

66. L De Riso, E de Alteriis, F La Cara, A Sada, P Parascandola. Immobilization of *Bacillus acidocaldarius* whole-cell rhodanase in polysaccharide and insolubilized gelatin gels. *Biotechnol Appl Biochem* 23:127–131, 1996.

67. W Ryan, SJ Parulekar. Immobilization of *E. Coli* JM1038pUC8 in-carrageenan coupled with recombinant protein release by *in situ* cell membrane permeabilization. *Biotechol Prog* 7:99–110, 1991.

68. I Di Lernia, C Schiraldi, M Generoso, M De Rosa. Trehalose production at high temperature exploiting an immobilized cell reactor. *Extremophiles* 6(4):341–347, 2002.

69. O Smidsrød, G Skjåk-Bræk. Alginate as immobilisation matrix for cells. *Tibtech* 8:71–78, 1990.

70. C Schiraldi, I Di Lernia, M Giuliano, M Generoso, A D'Agostino, M De Rosa. Evaluation of a high temperature immobilised enzyme reactor for production of non-reducing oligosaccharides. *J Ind Microbiol Biotechnol* 30:302–307, 2003.

71. E Chepeda, M Hermosa, A Ballestros. Optimization of Maltodextrins hydrolysis by glucoamylase in a batch reactor. *Biotechnol Bioeng* 76(1):70–76, 2001.

72. T Nakada, K Maruta, K Tsusaki, M Kubota, H Chaen, T Sugimoto, M Kurimoto, Y Tsujisaka. Purification and properties of a novel enzyme, maltooligosyl trehalose synthase from *Arthrobacter* sp. Q36. *Biosci Biotechnol Biochem* 59:2210–2214, 1995.

73. C Paiva, A Panek. Biotechnological applications of the disaccharide trehalose. *Biotechnol Annu Rev* 2:293–314, 1996.

Index

9 780367 392673